THE CORRESPONDENCE OF
CHARLES DARWIN

Editors

FREDERICK BURKHARDT   DUNCAN M. PORTER

SHEILA ANN DEAN   SAMANTHA EVANS

SHELLEY INNES

ALISON PEARN   ANDREW SCLATER

PAUL WHITE

Associate Editors

ANNE SCHLABACH BURKHARDT

ROSEMARY CLARKSON

RUTH GOLDSTONE

MURIEL PALMER

ELLIS WEINBERGER

This edition of the Correspondence of Charles Darwin is sponsored by the American Council of Learned Societies. Its preparation is made possible by the co-operation of Cambridge University Library and the American Philosophical Society.

Advisory Committees for the edition, appointed by the Council, have the following members:

Support for editing has been received from the Alfred P. Sloan Foundation, the Andrew W. Mellon Foundation, the National Endowment for the Humanities, the National Science Foundation, the Pew Charitable Trusts, the British Academy, the British Ecological Society, the Isaac Newton Trust, the Natural Environment Research Council, the Royal Society of London, the Stifterverband für die Deutsche Wissenschaft, and the Wellcome Trust. The National Endowment for the Humanities funding of the work was under grants nos. RE-23166-75-513, RE-27067-77-1359, RE-00082-80-1628, RE-20166-82, RE-20480-85, RE-20764-89, RE-20913-91, RE-21097-93, RE-21282-95, RZ-20018-97, RZ-20393-99, and RZ-20849-02; the National Science Foundation funding of the work was under grants Nos. SOC-75-15840, SOC-76-82775, SES-7912492, SES-8517189, SBR-9020874, SBR-9616619, and SES-0135528. Any opinions, findings, conclusions, or recommendations expressed in this publication are those of the editors and do not necessarily reflect the views of the grantors.

Hypothetical sphinx moth pollinating *Angraecum sesquipedale*
in the forests of Madagascar (see pp. 384 and 385 n.5)
Quarterly Journal of Science 4 (1867): facing p. 471
Illustration by T. W. Wood
By permission of the Syndics of Cambridge University Library

THE CORRESPONDENCE OF
CHARLES DARWIN

VOLUME 15   1867

CAMBRIDGE
UNIVERSITY PRESS

CAMBRIDGE
UNIVERSITY PRESS

University Printing House, Cambridge CB2 8BS, United Kingdom

One Liberty Plaza, 20th Floor, New York, NY 10006, USA

477 Williamstown Road, Port Melbourne, VIC 3207, Australia

314-321, 3rd Floor, Plot 3, Splendor Forum, Jasola District Centre, New Delhi - 110025, India

79 Anson Road, #06-04/06, Singapore 079906

Cambridge University Press is part of the University of Cambridge.

It furthers the University's mission by disseminating knowledge in the pursuit of education, learning and research at the highest international levels of excellence.

www.cambridge.org
Information on this title: www.cambridge.org/9780521859318

© Cambridge University Press 2005

First published 2005

Citation:
Burkhardt, Frederick, *et al.*, eds. 2005. *The correspondence of Charles Darwin.* Vol. 15. Cambridge: Cambridge University Press.

A catalogue record for this publication is available from the British Library

ISBN 978-0-521-85931-8 Hardback

In Memoriam
ERNST MAYR
1904–2005

CONTENTS

ILLUSTRATIONS

CALENDAR LIST OF LETTERS

The following list is in the order of the entries in the *Calendar of the correspondence of Charles Darwin*. It includes all those letters that are listed in the *Calendar* for the year 1867, and those that have been redated into 1867. Alongside the *Calendar* numbers are the corrected dates of each letter. A date or comment printed in italic type indicates that the letter has been omitted from this volume.

Letters acquired after the publication of the first edition of the *Calendar*, in 1985, have been given numbers corresponding to the chronological ordering of the original *Calendar* listing with the addition of an alphabetical marker. Many of these letters are summarised in a 'Supplement' to a new edition of the *Calendar* (Cambridge University Press, 1994). The marker 'f' denotes letters acquired after the second edition of the *Calendar* went to press in 1994.

1394. 5 Mar [1867?]
2670. *31 Jan [1868]*
3587. 1 June [1867]
4969. 4 Jan [1867]
5205. 1 Sept [1867]
5332. *[8 May 1866 – 31 Aug 1871]*
5333. *[19 May 1864]. To be published in next supplement.*
5334. [before 19 Nov 1867]
5335. *[before 3 Feb 1868?]*
5336. *[c. August 1867?]*
5337. *[2–30 Mar 1867]*
5338. *[before 25 Jan 1872]*
5339. *[Apr–June 1865]*
5340. *[c. Feb 1868?]*
5341. 20 [1867?]
5342. 22 [Mar 1867]
5343. [before 7 Jan 1867]
5343f. 1 Jan 1867
5344 *Cancel: duplicates part of 5376.*
5344a. 1 Jan 1867
5345. 2 Jan [1867]
5346. 3 Jan [1867]
5347. 6 Jan [1867]
5348. 7 Jan [1867]
5349. 8 Jan 1867
5350. 8 Jan [1867]
5351. *8 Jan [1868]*
5352. *[after 8 Jan 1868]*
5353. 9 Jan [1867]
5354. 9 Jan [1867]
5355. 9 Jan 1867

5356. 10 Jan [1867]
5357. 11 Jan 1867
5358. [12 Jan 1867]
5359. 12 Jan 1867
5360. 14 Jan 1867
5361. 15 Jan [1867]
5362. 15 Jan [1867]
5363. 15 Jan [1867]
5364. 15 Jan [1867]
5365. 15 Jan [1867]
5366. 16 Jan 1867
5367. 17 Jan [1867]
5368. *Enclosure to 5378.*
5369. 17 [July 1867]
5370. 18 Jan 1867
5371. 19 Jan [1867]
5372. 20 Jan 1867
5373. 21 Jan [1867]
5374. 21 Jan 1867
5375. 22 Jan [1867]
5376. 22 Jan 1867
5376a. 23 Jan [1867]
5377. 24 Jan 1867
5378. 26 Jan 1867
5379. 27 Jan [1867]
5380. 28 Jan [1867]
5381. 29 Jan [1867]
5382. 29 Jan [1867]
5383. 30 Jan [1867]
5384. 31 Jan [1867]
5385. [April 1867?]

INTRODUCTION

Charles Darwin's major achievement in 1867 was the completion of his large work, *The variation of animals and plants under domestication* (*Variation*). After years of delays, caused by illness and various interruptions, the manuscript, except for the last chapter, had been delivered to the publisher in the final week of 1866. It would take all of 1867 to correct proofs, and just when completion seemed imminent, a further couple of months were needed to index the work, a task that Darwin handed over to someone else for the first time.

The completion of one book marked the beginning of two others, as Darwin decided to exclude the 'chapter on man' from the already oversized two-volume *Variation* and instead write a short (as he then expected) 'Essay on Man'. The focus of the essay was to be the role of sexual selection in forming human races, and there was also to be a chapter on the meaning and cause of the expression of emotions. The 'essay' grew into another two-volume work, *The descent of man and selection in relation to sex* (*Descent*), published in 1871, and the chapter on expression into a book, *The expression of the emotions in man and animals* (*Expression*), published in 1872. Although Darwin had been collecting material and making observations in these areas for decades, it was only now that he began to work with a view to publishing his observations.

The importance of Darwin's network of correspondents becomes vividly apparent in his work on expression in 1867, as he continued to circulate a list of questions on human expression that he may have drawn up in late 1866. His correspondents were asked to copy the list and forward it to those who might best answer the questions, with the result that Darwin began to receive replies from different corners of the world.

Darwin's work was now guaranteed to arouse interest both at home and further afield, and, with *Variation* nearing completion, he received enthusiastic offers of immediate translation, not only into German and French, but also into Russian. Whereas the earlier Russian translation of *Origin* had been based not on the original, but on the German first edition, the new Russian translation of *Variation* would be based on proof-sheets received as Darwin corrected them. Closer to home, two important works, a book by the duke of Argyll, and an anonymous review by an engineer, Henry Charles Fleeming Jenkin, challenged different aspects of Darwin's theory of transmutation as elucidated in *On the origin of species by means of natural selection* (*Origin*) and in *On the various contrivances by which British and foreign orchids are fertilised by insects* (*Orchids*). While Darwin privately gave detailed opinions of these critiques, he was more than happy to leave the public defence of the theory in the capable hands of Alfred Russel Wallace. At the same time, Darwin was persuaded by some German supporters to write to his most vociferous German champion, Ernst Haeckel, to encourage him to tone down his attacks on critics, which, they felt, were becoming counterproductive.

Throughout the year, Darwin continued to discuss now familiar topics such as dimorphism and trimorphism, self-sterility, pollination, and seed dispersal with a growing network of correspondents who worked on similar topics and were happy to supply him with information and discuss the implications of their findings in light of transmutation theory. Three important new correspondents in 1867 were Hermann Müller and Anton Dohrn in Germany, and Federico Delpino in Italy, who provided Darwin with the collegial support and rigorously scientific atmosphere that he so much needed in what was becoming a highly combative and emotional arena.

Thomas Henry Huxley sent Darwin the New Year's greeting, 'may you be eupeptic through 1867 & your friends & the world in general will profit'. Darwin's health, which had improved considerably the previous year, continued to be good, allowing him to pursue several projects at once. First and foremost was the completion of part of his long-delayed 'big book', started in January 1860, and advertised in the press since 1865 with the unwieldy title, 'Domesticated Animals and Cultivated Plants, or the Principles of Variation, Inheritance, Reversion, Crossing, Interbreeding, and Selection under Domestication'. Having just received the printer's estimate of the size of the two-volume work from his publisher, John Murray, he wrote to Murray on 3 January 1867, 'I cannot tell you how sorry I am to hear of the enormous size of my Book.' He told Murray he would not hold him to his agreement to publish and also informed him that he had finished the concluding chapter and remained doubtful whether or not to include a chapter 'on Man'. After a few days, he wrote back to Murray proposing that some of the more technical passages in the book be printed in smaller type, a plan to which Murray readily agreed. More letters were exchanged, clarifying financial arrangements and discussing the number of copies to be printed, and by the end of the month Darwin promised to send the revised manuscript to the printer as soon as he had marked out passages to be set in smaller type. He added, 'I feel a full conviction that my Chapter on man will excite attention & plenty of abuse & I suppose abuse is as good as praise for selling a Book' (letter to John Murray, 31 January [1867]).

A week later, Darwin had sent the manuscript to the printer, but without the additional chapter. In a letter written on 8 February [1867] to his close friend, Joseph Dalton Hooker, he explained, 'I began a chapter on Man, for which I have long collected materials, but it has grown too long, & I think I shall publish separately a very small volume, "an essay on the origin of mankind"'. Eventually, the chapter would expand into not one but two substantial books, *Descent* and *Expression*. In the same letter, Darwin revealed the conclusion to his newly completed book: a paragraph throwing doubt on Asa Gray's doctrine that each variation had been 'specially ordered or led along a beneficial line'. He added, 'It is foolish to touch such subjects, but there have been so many allusions to what I think about the part which God has played in the formation of organic beings, that I thought it shabby to evade the question.'

As the year progressed, the book continued to consume Darwin's time. The first proof-sheets arrived on 1 March 1867 and the tedious work of correction began. Darwin wrote to Murray on 18 March to say that he preferred the title 'invented' by the compositors, and so the book became *The variation of animals and plants under domestication*. In a letter to his son William dated 27 [March 1867], he admitted, 'I fear the book is by no means worth the confounded labour.' However, offers to translate the work came quickly, and by May, Russian, German, and French translations had been arranged. Darwin had now found sympathetic translators whom he could trust to convey his work without interjecting comments or additions of their own, as the earlier German and French translators had done. The French translator, Jean Jacques Moulinié, had been personally recommended by Carl Vogt and had translated Vogt's own *Vorlesungen über den Menschen* (Lectures on man; Vogt 1863) from German into French. With a background in natural history, native fluency in English, and a desire to devote himself 'body and soul' to the translation, Moulinié was an excellent choice. The offer to undertake a Russian translation was made by a young naturalist equally devoted to Darwin's work, Vladimir Onufrievich Kovalevsky. Kovalevsky included several of his brother's embryological papers with his first letter to Darwin of 15 March 1867, although he described some of Alexander Kovalevsky's ideas as 'a little wild'. Though primarily concerned with finding a good translator for his book, Darwin was always on the lookout for evidence to support his theory of transmutation. Darwin heavily annotated his copy of Alexander's paper on the embryology of ascidians (A. O. Kovalevsky 1866b), revealing his keen interest in one of the first studies to suggest an ancestral link between invertebrates and vertebrates.

Finally, Darwin had made sure that Julius Victor Carus, who had completely revised the German translation of *Origin* in 1866, would be called upon to translate *Variation*. Indeed, he told his publisher, John Murray, in a letter of 4 April [1867], not to send stereotypes of the illustrations to the German publisher until he was certain that Carus would undertake the translation. Darwin had received other offers, notably one from Vogt in April 1867, to translate the new work. Carus had already agreed in principle to translate the work but told Darwin, 'I am so very much occupied just now and within the next twelve months, that I should feel exceedingly obliged if you would kindly tell me, at what rate your work will be published' (letter from J. V. Carus, 5 April 1867). This hint of uncertainty caused Darwin to respond to Vogt somewhat ambiguously, as he wrote, 'Prof. Carus, though he has undertaken the translation informs me that he has much work on hand, & it is possible (though not probable) that when he hears (& I wrote to him on the subject yesterday) of the size of the book, & that several sheets will be printed immediately & sent to him, he may wish to give up the task' (letter to Carl Vogt, 12 April [1867]). Darwin need not have worried. Carus soon wrote back, assuring Darwin that he could translate the first volume by November 1867, and, apparently alerted to Vogt's approach, warning him that Vogt was not 'the fit person' to introduce the work to the German public (letter from J. V. Carus, 15 April 1867). Darwin may not have fully appreciated how politically radical Vogt was (it was his political activism that led to his forced departure from his native country), but he was quick to reassure Carus, 'The wish never for a moment crossed my mind that Vogt should translate my book in preference to you' (letter to J. V. Carus, 18 April [1867]). Darwin was not disappointed in his choice, for Carus did more than provide an accurate translation. Thanks to his knowledgeable and careful reading of the original, he alerted Darwin to errors in the text and suggested corrections, many of which Darwin was able to incorporate in the English edition.

Although Darwin longed to have the book off his hands, he continued to make additions, especially when new material came to light that seemed to support his beloved theory of heredity, the 'provisional hypothesis of pangenesis'. Such was the case, reported by Charles Victor Naudin, of a fan palm, pollinated by a date palm, whose fruit appeared to show the direct action of the male parent on the female, in that the fruit looked more like that of the date palm. Naudin had sent specimens to Hooker, who reported the 'wonderful discovery' to Darwin on 14 March 1867. Then, in April, Robert Trail wrote from Scotland about a potato he had produced by joining two different varieties at the eye, which resulted in a mottled hybrid (letter from Robert Trail, 5 April 1867). Darwin told his American friend Asa Gray, 'I am repeating this experiment on a large scale, for it seems to me, if true, a wonderful physiological fact' (letter to Asa Gray, 15 April [1867]). Although he did not succeed in duplicating Trail's results, Darwin claimed in *Variation* 1: 396, that the case apparently afforded 'clear evidence of the intimate commingling of the two varieties'. Such a case, if proved, would be evidence of 'the essential identity of sexual and asexual reproduction', Darwin wrote in his chapter on pangenesis (*Variation* 2: 365).

Darwin had developed his provisional hypothesis of pangenesis over many years apparently without discussing it or showing it to anyone until 1865, when he sent a version of it to Huxley, asking whether it should be published. Huxley had pointed out similarities to earlier theories, and, while not discouraging Darwin from publishing altogether, had advised that his views should be presented not as 'formed conclusions' but as 'hypothetical developments' (see *Correspondence* vol. 13, pp. xix–xx). Not surprisingly then, when circulating proofs of the book, Darwin had been especially anxious about the reception of pangenesis. He was happy that Charles Lyell had a positive response, commenting, 'I have been particularly pleased that you have noticed Pangenesis. I do not know whether you ever had the feeling of having thought so much over a subject that you had lost all power of judging it. This is my case with Pangen: (which is 26 or 27 years old!) but I am inclined to think that if it be admitted as a probable hypothesis, it will be a somewhat important step in Biology' (letter to Charles Lyell, 22 August [1867]). Darwin's insecurity persisted, however, and in November he told Hooker, 'I shall not be at all surprised if you attack it & me with unparalleled ferocity' (letter to J. D. Hooker, 17 November [1867]).

Even when the corrections were complete, in mid-November, the book was further delayed by the time it took William Sweetland Dallas to prepare the index. John Murray had engaged Dallas and Darwin approved the choice but asked to see a specimen of his work. Although he had initially wanted Dallas to index every name that appeared in the text, he eventually conceded, 'On reflection I fear you will find it endless labour to give all author's names in notes. So use your own discretion; anyhow most ought to be introduced' (letter to W. S. Dallas, 8 November [1867]). Dallas resisted the temptation to cut corners, and told Darwin, 'I have adopted your plan of giving every author's name', but added, 'It makes the labour very great, however, & I cannot get on so quickly as I could wish' (letter from W. S. Dallas, 20 November 1867). Dallas, like Carus, alerted Darwin to errors in the text, although some of these corrections were too late to make it into the first printing of the first edition. The relationship between the two men seems to have been uneasy, at least until Dallas finished the index. Although many of Darwin's letters to Dallas have not been found, it is clear from the defensive tone of Dallas's responses that he was under a great deal of pressure to complete his work and was torn between the desire to finish the index quickly and the desire to impress Darwin with the quality of his work.

As the 'horrid tedious dull work' of correcting *Variation* went on, Darwin was at the same time seeking information for his next project, the 'essay on man'. One of the first areas he focused on was expression. In fact, Darwin had been interested in the physical nature of the expression of the emotions in humans for a long time. From around 1838, he had begun making observations on expression, noting the difference or similarity between people and particular animals. He also recorded the expressions of some of his children from infancy, and read books on the anatomy of expression by medical experts such as Charles Bell and Guillaume Benjamin Amand Duchenne. Now Darwin was able to mobilise an ever-widening network of correspondents in an attempt to establish through observation the universality of human expressions. As early as January 1860, he had sent a list of specific queries regarding the expression of emotions in the indigenous people of Tierra del Fuego, whom he had first encountered in 1832 while on the *Beagle* voyage (see *Journal of researches*, pp. 228–9). After hearing from his former *Beagle* shipmate Bartholomew James Sulivan at Christmas 1866, Darwin had written at the end of the year asking again for information on Fuegian expressions. On 11 January 1867, Sulivan replied, enclosing belated answers from Thomas Bridges to the queries Darwin had sent in 1860 and relaying a promise from a missionary friend who was returning to South America to 'look out again' for answers.

In a letter of 22 February [1867] to Fritz Müller in Brazil, in which he asked for information on two subjects – 'sexual selection' and 'expression of countenance' – Darwin explained, 'I am thinking of writing a little essay on the origin of Mankind, as I have been taunted with concealing my opinions; & I sh^d do this immediately after the completion of my present book. In this case I sh^d add a chapter on the cause or meaning of Expression.' With this letter Darwin enclosed a list of questions, handwritten by an amanuensis, headed 'Queries about Expression'. In a postscript to the letter he added, 'But you must not plague yourself on a subject which will appear trifling to you, but has, I am sure, some considerable interest.' Darwin also introduced the subject to Alfred Russel Wallace, who suggested in his response of 11 March [1867] that Darwin send his queries to foreign newspapers. The letter also reveals that he did not share Darwin's interest in studying human expression. He briefly answered some of the queries from memory, and then added, 'But do you think these things are of much importance?' Further, he wrote, 'I would rather see your second volume on "The Struggle for Existence &c." for I doubt if we have a sufficiency of fair & accurate facts to do any thing with Man.' Darwin replied, not altogether ingenuously, 'I fully agree with you that the subject is in no way an important one: it is simply a "hobby-horse" with me about 27 years old; & *after* thinking that I would write an essay on man, it flashed on me that I could work in some "supplemental remarks on expression"' (letter to A. R. Wallace, [12–17] March [1867]). Darwin's doggedness in pursuing answers to his queries reveals a different picture about the importance of the subject to him.

Copies of the queries were sent across the globe to North and South America, Australia, New Zealand, South Africa, India, and elsewhere. Darwin requested that his correspondents pass on the queries to acquaintances in remote areas. On 26 March, Asa Gray wrote, 'You see I have *printed* your queries—privately—50 copies—as the best way of *putting* them where useful answers may be expected', and asked whether he should send more printed copies. Darwin replied, 'I wish I had thought earlier of having them printed, for in that case I might have sent a dozen to each of my few correspondents, as it is I can think of no one to send them to, so do not want any more' (letter to Asa Gray, 15 April [1867]). Nevertheless, at some point during the year he did, in fact, have the queries printed on his own account. Copies of Darwin's own printed questionnaire survive. One of these has been transcribed in Appendix IV. Robert Swinhoe, the British consul in Amoy, had a handwritten version of Darwin's queries published in *Notes and Queries on China and Japan*, 31 August 1867. Another version, possibly derived from Asa Gray's printed queries, was published in 1868 in the *Annual Report of the Board of Regents of the Smithsonian Institution for the year 1867*.

In his 15 April [1867] letter to Gray, Darwin commented, 'I have been lately getting up & looking over my old notes on Expression, & fear that I shall not make so much of my hobby-horse, as I thought I could: nevertheless it seems to me a curious subject, which has been strangely neglected.' CD's doubts did not persist for long, especially as replies to the questionnaire continued to arrive.

'Sexual selection' was the other 'new' subject Darwin devoted his energies to in 1867. Darwin had, in fact, introduced the concept of sexual selection in *Origin*; he proposed that secondary sexual characters had been accumulated by sexual selection, which, he argued, was 'less rigid' in its action than natural selection since it did not 'entail death', but only resulted in fewer offspring for less favoured males (see *Origin*, pp. 156–7). In *Variation*, Darwin had discussed changes in secondary sexual characters under domestication, noting that such modifications could be explained through sexual selection, but had not elaborated further (*Variation* 2: 75). In notes for his reply to a letter from Edward Blyth dated 19 February 1867, Darwin had written, '*Sexual Selection.*— too many questions to ask'. In the reply itself, written 23 February [1867], Darwin introduced the topic with a compliment to Blyth: 'I have picked up more facts on sexual characters ... from your writings than from those of any one else.' He then proceeded to ask for still more information on sexual differences in mammals and birds. In his letter to Fritz Müller of 22 February [1867], Darwin was more specific about what he wanted, asking for examples of sexual differences that did not relate to 'different habits of life' in males and females. In his reply of 1 April 1867, Müller supplied Darwin with information about sexual differences in crustaceans, spiders, and annelid worms that might reflect sexual selection.

Although Darwin wrote to several people in his search for material on sexual selection, the individual with whom he discussed and debated the topic on a theoretical level was Alfred Russel Wallace. In a letter to Wallace written on 23 February 1867, Darwin asked whether Wallace could suggest a solution to the puzzling problem of bright colours in caterpillars as well as in butterflies. Wallace was sure that the colours were protective and suggested that John Jenner Weir might conduct experiments in his aviary to see whether this was the case (letter from A. R. Wallace, 24 February [1867]). He also suggested a simple experiment to determine whether female butterflies preferred more colourful males. When Darwin had asked the question about bright caterpillars, the idea that bright colours in male butterflies resulted from sexual selection was implicit. Wallace's response contained much more than a possible explanation about caterpillars; it called into question the whole notion of sexual selection as an explanation of colour in adults of different sexes, and ultimately, the notion of sexual selection as an agent of change.

Darwin was obviously dismayed that his theory of sexual selection was being challenged at a fundamental level. In his response to Wallace (letter to A. R. Wallace, 26 February [1867]), Darwin defended his position about colour in adult insects but turned the discussion to the role of sexual selection in humans, remarking, 'I still strongly think ... that sexual selection has been the main agent in forming the races of Man.' The two debated the matter over the course of several months. In the 1867 correspondence, Wallace steered clear of the issue of formation of human races, but continued to build his argument about the protective function of colour in both insects and birds. Darwin conceded that Wallace had made a convincing argument concerning protective coloration, but continued to emphasise the importance of sexual selection in humans.

Through their use of polite language and willingness to give ground, Darwin and Wallace were able to sustain a dialogue in an area where they disagreed on many points. On the practical side, Darwin was happy to take advantage of Wallace's influence in the entomological community in order to gather more information on insects. Moreover, he was still able to engage in fruitful theoretical discussions by allowing a place in his own theory for Wallace's hypothesis about protection. Similarly, Wallace conceded that sexual selection might come into play in some circumstances. In a letter of 5 May [1867], Darwin admitted, referring to Wallace's argument that female birds that used open nesting sites retained dull coloured plumage because of its protective effects, 'your explanation with respect to the females had not occurred to me. I am surprized at my own stupidity, but I have long recognized how much clearer & deeper your insight into matters is than mine.' Darwin had already told Wallace that he would not have much time to devote to this new research until he had finished correcting proofs of *Variation* and added, 'when I return to the work I shall find it much better done by you than I c^d have succeeded in doing' (letter to A. R. Wallace, 29 April [1867]). Thus Darwin was able to portray his research as collaborative, with Wallace and him pursuing the same, albeit broadly defined, goal.

Almost a year would pass before Darwin was again ready to take up the subject of sexual selection in earnest. In the meantime, his work on several botanical projects continued and he gained a valuable new correspondent in this area, Hermann Müller. Darwin had already benefited enormously from his correspondence with Hermann's brother, Fritz, who, living in Brazil, could supply Darwin with much information about the rich flora and fauna of the country. Fritz had been writing to his younger brother about his correspondence with Darwin and had even suggested research projects that Hermann, a secondary-school teacher in Westphalia, could pursue with a view to supporting Darwin's theory (letter from Fritz to Hermann Müller, 11 February 1867, in Möller ed. 1915–21, 2: 111–16). Hermann sent Darwin some papers he had published on Westphalian mosses, and Darwin, in a reply that has not been found, evidently wrote encouragingly to him. In his first extant letter to Darwin, that of 23 March 1867, Hermann told Darwin of his research plans. He wrote, 'I think I could not easily find a higher enjoyment and at the same time a better preparation for the researches intended than in repeating your charming observations on the fertilisation of Orchids by insects, as far as the Westfalian Flora offers any opportunities to it and in devoting my attention in general to the fertilisation of flowers by insects.' By the summer, Hermann was already making observations on the pollination of orchids by insects that would confirm points that Darwin had only conjectured in his 1862 study, *On the various contrivances by which British and foreign orchids are fertilised by insects* (*Orchids*).

In October, Müller wrote to thank Darwin for his present of the new German edition of *Origin* and mentioned an observation that he thought might be of some interest. Müller had observed the special adaptation in the mouthparts of various species of hoverflies (Syrphidae) allowing them to eat pollen-grains as well as suck nectar from flowers they visited (letter from Hermann Müller, 23 October 1867). The letter contains two illustrations of Syrphidae mouthparts, the first known depiction of this singular adaptation. Müller's work not only confirmed many of Darwin's earlier observations on floral mechanisms for ensuring cross-pollination, it complemented Darwin's work on flower structure by focusing on the details of insect morphology such as specialised mouthparts.

Darwin continued to receive a wealth of information from Fritz Müller in Brazil. As well as providing material on sexual selection and answers to Darwin's queries about expression, Fritz continued to add observations on dimorphism and trimorphism and reported on a series of crossing experiments with orchids. Darwin commented, 'You have communicated to me many more cases than any two or three botanists put together' (letter to Fritz Müller, 7 February [1867]). Müller had written the previous year concerning the self-sterility of some orchids, and in his first letter of 1867 he reported on his experiments with an orchid whose pollen had a poisonous effect when applied to its own stigma (letter from Fritz Müller, 1 January 1867). Darwin replied, 'The fact about the own-pollen being poisonous is quite extraordinary', and wondered whether the cause of the decay might be 'parasitic cryptogams' (letter to Fritz Müller, 22 February [1867]). Müller was able to assure him that he had only once seen a parasitic cryptogam on the plant (letter from Fritz Müller, 1 April 1867). Darwin was so interested in Müller's observations on the poisonous effect of a plant's own pollen that he decided, even at such a late stage, to add an abstract of the material to the proof-sheets of *Variation* (letter to Fritz Müller, 31 July [1867]).

Darwin was also interested in experiments crossing different species of orchids that Müller had reported on, but remarked, 'I fear their interest will be greatly lessened by the crossed seeds not germinating', noting, 'One single man in Europe has found out how to make these seeds germinate, & he keeps it a secret in his trade of nurseryman' (letter to Fritz Müller, 26 May [1867]). Darwin was no doubt alluding to the success of the famous nursery of James Veitch & Son in maintaining what was, in 1867, a more than ten-year monopoly in the production of orchid hybrids (Shephard 2003). Darwin frequently received seeds and specimens as well as information from Müller, and, in a letter of 2 November 1867, was able to report on the progress of plants he had grown from some of Müller's seeds. Darwin would continue with these long-term experiments, growing several generations of plants from a wide variety of self-pollinated and cross-pollinated specimens, observing growth rates, vigour, seed production, and other features, and comparing his own results with those of trusted correspondents like Müller and Friedrich Hildebrand.

During the year, two significant critical works appeared, both of which offered very different challenges to the theory of the transmutation of species through natural selection. In January 1867, the duke of Argyll, George Douglas Campbell, published *The reign of law* (G. D. Campbell 1867), a book based on a series of articles that had appeared in 1865. In it he challenged aspects of Darwin's theory, especially the notion that beauty had an adaptive function, arguing instead that beauty in nature was designed by the creator for the aesthetic education of humans. Several of Darwin's friends commented on it. Huxley, who had not read it, referred to it dismissively as 'the Dukelets book', and noted, 'I hear he is down on both of us' (letter from T. H. Huxley, [before 7 January 1867]). In February, Hooker asked whether Darwin had read it and whether it was worth reading (letter from J. D. Hooker, 4 February 1867). In a letter to his son William, Darwin confided, 'Mamma has several times declared that the Duke did not understand the Origin, but I pooh-poohed her, & as it seems very unjustly' (letter to W. E. Darwin, 27 [March 1867]). Unfortunately, he did not elaborate further on Emma Darwin's view, so we are left to wonder about the nature of her criticisms. By June, Darwin was reading the book, which he thought interesting and 'clever', but with certain weak parts (letter to Charles Lyell, 1 June [1867]). Charles Kingsley found the book 'fair' but told Darwin, 'he [Campbell] writhes about under you as one who feels himself likely to be beat' (letter from Charles Kingsley, 6 June 1867). Darwin was inspired by Kingsley's remarks to give his own critique of the book, which he had just finished reading. He was struck by Campbell's arrogance, pointed out inconsistencies in the remarks on orchids and hummingbirds, and argued, 'With respect to the Deity having created objects beautiful for his own pleasure, I have not a word to say against it but such a view c^d hardly come into a scientific book' (letter to Charles Kingsley, 10 June [1867]).

After dealing with Campbell's book, Darwin turned to the other work that had posed serious challenges to his theory, an anonymous critique in the *North British Review*, which he described as 'one of the most telling Reviews of the hostile kind' (letter to Charles Kingsley, 10 June [1867]). Kingsley himself had remarked, 'It is a pity the man who wrote it had not studied a little zoology & botany, before writing about them' (letter from Charles Kingsley, 6 June 1867). The review had, in fact, been written by an engineer, Henry Charles Fleeming Jenkin, who had recently collaborated with William Thomson on experiments on electric cable insulation. One of his criticisms in the article, based on Thomson's calculation of the age of the earth, was that there was not enough time for the changes Darwin described to have occurred. This argument had been raised before and Darwin seemed confident in dealing with it, pointing to the uncertainty and disagreement within the mathematical community itself on the question of the age of the earth. Far more thought-provoking, however, was the argument that a variation, no matter how favourable, would tend to be swamped over time, given a theory of heredity that relied on blending inheritance, the generally accepted view of heredity, which Darwin shared. Darwin conceded that he would need to change some of the wording in the relevant section of *Origin*, commenting, 'Instead of saying, as I have sometimes incautiously done a bird suddenly appeared with a beak *[particularly]* longer than that of his fellows, I would now say that of all the birds annually born, some will have a beak a shade longer, & some a shade shorter, & that under conditions or habits of life favouring longer beak, all the individuals, with beaks a little longer would be more apt to survive than those with beaks shorter than average' (letter to Charles Kingsley, 10 June [1867]).

Typically, Darwin chose to steer clear of responding publicly to criticism, so he was pleased when Wallace published a long article, 'Creation by law' (A. R. Wallace 1867c), which responded to Jenkin's and Campbell's critiques. Darwin commended Wallace, 'You have just touched on the points which I particularly wished to see noticed. I am glad you had the courage to take up Angræcum after the Duke's attack; for I believe the principle in this case may be widely applied. I like the Figure but I wish the artist had drawn a better sphynx' (letter to A. R. Wallace, 12 and 13 October [1867]). Darwin referred to the full-page illustration at the beginning of the article, an artist's impression of the orchid *Angraecum sesquipedale* with its pollinator, a hypothetical sphinx moth (which Darwin had predicted must exist) with a proboscis long enough to reach the base of the nectary. (The predicted moth, *Xanthopan morgani praedicta*, was eventually discovered in 1903.) The illustration is reproduced as the frontispiece to this volume. In the same letter, Darwin reported on a favourable review from an unlikely source. George Warington, a writer who usually wrote on religious subjects, had written an abstract of *Origin* for the *Journal of the Transactions of the Victoria Institute or Philosophical Society of Great Britain*. The Victoria Institute had been founded a couple of years earlier with the aim of defending 'the great truths revealed in Holy Scripture' against 'the oppositions of Science, falsely so called'. Not surprisingly, Warington's article caused something of a furore within the Victoria Institute itself, which Darwin, having read the proceedings, described as 'very rich from the nonsense talked' (letter to A. R. Wallace, 12 and 13 October [1867]).

Although Darwin himself almost always shied away from public debate, many of his supporters did not. Darwin was called upon to encourage one of his most devoted supporters, the German zoologist Ernst Haeckel, to moderate his public attacks on the opposition. Darwin's German translator, Julius Victor Carus, told Darwin that while Haeckel's recent large work, *Generelle Morphologie der Organismen* (Haeckel 1866), contained much interesting material, it also contained 'personal and quite unnecessary remarks', and pleaded with Darwin to intervene, claiming, 'There is only one man, to whose judgement he would subdue; that is yours' (letter from J. V. Carus, 5 April 1867). Darwin complied, and his letter to Haeckel gives an insight into his own views on criticism in general. He wrote, 'I have long observed that much severity leads the reader to take the side of the attacked person', and concluded, 'I know that it is easy to preach & if I had the power of writing with severity I dare say I sh^d triumph in turning poor devils inside out & exposing all their imbecility. Nevertheless I am convinced that this power does no good, only causes pain. I may add that as we daily see men arriving at opposite conclusions from the same premises it seems to me doubtful policy to speak too positively on any complex subject however much a man may feel convinced of the truth of his own conclusions' (letter to Ernst Haeckel, 12 April [1867]).

On a personal level, the year was relatively peaceful. Darwin's own health continued to be good for the most part, although in September he suffered a bout of eczema, evidently accompanied by some temporary memory loss, which caused

Emma to consult his physician, Henry Bence Jones. Jones wrote reassuringly, conjecturing the cause might be 'increased mental work', and suggesting, 'the best course would be to put mental work aside *altogether* for this month; except on wet days' (letter from H. B. Jones, 1 October [1867]). There is no evidence that Darwin took Jones's advice; in fact, he had long maintained that bouts of eczema actually alleviated his other symptoms, and made him feel more alert (see *Correspondence* vol. 13, letter to J. D. Hooker, 9 February [1865] and n. 4). Darwin's wife and children also prospered; after the deaths of two of his sisters the previous year, it must have been a relief to have no serious illness in the family. However, his youngest son, Horace, had a bout of fever, which Darwin mentioned to his close friend Hooker, whose own newborn baby had been worryingly ill with convulsions. At one point Hooker confided, 'We have no hopes for our pretty little baby whose fits increase in number & duration' (letter from J. D. Hooker, 31 March 1867). Within days, however, Frances Harriet Hooker was able to reassure the Darwins that the baby was recovering quickly.

   Darwin's family continued to assist him in his work, his wife and daughters reading to him and acting as amanuenses. Henrietta Emma Darwin read and corrected proof-sheets of *Variation*, and while she was away in Cornwall in the summer took some of the work with her. Darwin wrote to commend her work, 'All your remarks, criticisms doubts & corrections are excellent, excellent, excellent' (letter to H. E. Darwin, 26 July [1867]). The year ended as it had begun, with letters about *Variation*. Darwin wrote to Carus on 10 December, informing him of errors discovered by Dallas and asking him to make the changes if not too late in the first volume, then adding, 'The book is delayed by the index-maker'. The letter Darwin sent to Dallas himself has not been found, but must have indicated his impatience, for Dallas replied on 26 December, 'I am vexed to the heart that you should have occasion to write to me again about this Index.' Dallas noted, in his own defence, 'the real cause of delay lies in the nature of the work itself.' *Variation* was published on 30 January 1868.

ACKNOWLEDGMENTS

The editors are grateful to the late George Pember Darwin and to William Huxley Darwin for permission to publish the Darwin letters and manuscripts. They also thank the Syndics of Cambridge University Library and other owners of manuscript letters who have generously made them available.

The work for this edition has been supported by grants from the National Endowment for the Humanities (NEH), the National Science Foundation (NSF), and the Wellcome Trust. The Alfred P. Sloan Foundation, the Pew Charitable Trusts, and the Andrew W. Mellon Foundation provided grants to match NEH funding, and the Mellon Foundation awarded grants to Cambridge University that made it possible to put the entire Darwin correspondence into machine-readable form. Research and editorial work have also been supported by grants from the Royal Society of London, the British Academy, the British Ecological Society, the Isaac Newton Trust, the Jephcott Charitable Trust, the Natural Environment Research Council, and the Wilkinson Charitable Foundation. The Stifterverband für die Deutsche Wissenschaft provided funds to translate and edit Darwin's correspondence with German naturalists.

Cambridge University Library, the American Philosophical Society (APS), and Cornell University have generously made working space and many services available to the editors.

Since the project began in 1975, the editors have been fortunate in benefiting from the interest, experience, and practical help of many people, and hope that they have adequately expressed their thanks to them individually as the work proceeded.

Without the expert help of John L. Dawson of the Literary and Linguistic Computing Centre of Cambridge University, the initial computerisation of the correspondence would not have been possible. Iain Burke, Peter Dunn, Ray Horne, Patricia Killiard, Adrian Miller, Chris Sendall, Merina Tuladhar, and Tomasz Waldoch have helped maintain and extend our access to the Library's electronic resources, and provided essential technical support. Simon Buck provided invaluable assistance in setting up automatic typesetting procedures, and we are grateful to Robin Fairbairns for expert help in their maintenance and development. Gratitude is expressed to the late C. A. Tripp for providing the editors with an invaluable tool for searching texts.

English Heritage has responded most generously to requests for information and for material from the collections at Down House, Downe. We are particularly grateful to Tori Reeve, curator of Down House. Richard Darwin Keynes kindly made available Darwin family material in his possession. Ursula Mommens has also provided letters and other materials that belonged to her grandfather, Francis Darwin.

Institutions and individuals all over the world have given indispensable help by making available photocopies of Darwin correspondence and other manuscripts in their collections. Those who furnished copies of letters for this volume can be found in the List of provenances. The editors are indebted to them, and to the many people who have provided information about the locations of particular letters.

The editors make daily use of the incomparable facilities of Cambridge University Library and have benefited greatly from its services and from the help and expertise of its staff, particularly the staff of the Manuscripts Department. We are especially grateful to the former University Librarian, Frederick W. Ratcliffe, to his successor, Peter K. Fox, and to the Keeper of Manuscripts and Archives, Patrick Zutshi, for their generous support. Other members of the library's staff who frequently respond to the editors' requests are: Marjolein Allen, Wendy Aylett, Gerry Bye, Kathleen Cann, Colin Clarkson, Barry Eaden, Les Goodey, David J. Hall, John Hall, Tony Harper, Brian Jenkins, Morag Law, Elisabeth Leedham-Green, David Lowe, Peter Meadows, Anne Murray, William Noblett, Adam Perkins, John Reynolds, Jayne Ringrose, Mark Scudder, Clive Simmonds, Nicholas Smith, Anne Taylor, Godfrey Waller, and Cynthia Webster. The fetchers in the Rare Books reading room have also patiently dealt with the editors' often complex requirements, as have the staff of the Map Room.

At the American Philosophical Society Library, a splendid collection of Darwiniana and works in the history of science has been available to the editors since the inception of the project. Whitfield J. Bell Jr, secretary of the society until 1983, serves on the United States Advisory Committee for the project and has done his utmost to further its work. The editors have also benefited from the co-operation of the late Edward Carter II, Robert S. Cox, Roy C. Goodman, and Martin L. Leavitt, all of the APS Library.

Thanks are due to the faculty and staff of the Department of Science and Technology Studies, Cornell University, for accomodating one of the Project's editors, and to staff at Cornell University Libraries for help with research enquiries.

The editors would like to acknowledge the assistance of Rodney Dennis, Jennie Rathbun, and Susan Halpert of the Houghton Library, Constance Carter of the Science Division of the Library of Congress, and Judith Warnement and Jean Cargill of the Gray Herbarium of Harvard University, who have all been exceptionally helpful in providing material from the collections in their charge.

In Britain, the editors regularly receive assistance from Gina Douglas, librarian of the Linnean Society of London, and Michele Losse of the Royal Botanic Gardens, Kew. We would also like to thank Anne Barrett, college archivist at the Imperial College of Science, Technology and Medicine; and Douglas Ferguson and Candace Guite of Christ's College, Cambridge.

Among the others who advise and assist the editors in their work are Nick Gill, Randal Keynes, Gene Kritsky, Gren Lucas, Carl F. Miller, Jim Moore, Jim Secord, Garry J. Tee, Béatrice Willis, and Leonard Wilson. The editors are also pleased to acknowledge the invaluable support of the members of the British and United States Advisory Committees.

For help with particular enquiries in volume 15, the editors would like to thank, besides those already mentioned, Phil Ackery, Frank T. Krell, and Robert Prys-Jones of the Natural History Museum; Shane T. Ahyong of the Australian Museum; Charles Aylmer; Michael F. Braby of the Australian National University; Lynda Brooks of the Linnean Society; Rosalind Caird, archivist at the Cathedral Library, Hereford; Robert Dressler of the University of Florida; Francis Gilbert of the University of Nottingham; Christiane Groeben of the Stazione Zoologica 'Anton Dohrn'; Donald Harris; David Kupitz; Virginia Murray of the John Murray Archive; Jamie Owen of the Natural History Museum Picture Library; and Anne Stow.

Thanks are also due to all former staff of the Darwin Correspondence Project, including: Sarah Benton, Charlotte Bowman, Heidi Bradshaw (whom we were glad to see return to the project for one year, part time, while volume 15 was under way), Janet Browne, P. Thomas Carroll, Mario Di Gregorio, Rhonda Edwards, Kate Fletcher, Hedy Franks, Joy Harvey, Arne Hessenbruch, Thomas Junker, David Kohn, Jyothi Krishnan-Unni, Kathleen Lane, Sarah Lavelle, Anna K. Mayer, William Montgomery, Perry O'Donovan, Stephen V. Pocock, Marsha L. Richmond, the late Peter Saunders, Anne Secord, Tracey Slotta, the late Sydney Smith, Nora Carroll Stevenson, Jonathan R. Topham, Tyler Veak, and Sarah Wilmot.

The editors also extend thanks to Jean Macqueen for indexing the volume.

During 2005, we said farewell to Andrew Sclater. We wish him well for the future.

During the preparation of this volume we were greatly saddened to hear of the death of Deirdre Haxton. Deirdre had corresponded with many of the editors on questions to do with Shrewsbury local history and the Darwin family. She was untiring in her interest and attention to detail, and the project is indebted to her.

Since the beginning of the Darwin Project in 1975 we have been fortunate to have the advice and guidance of Ernst Mayr, eminent systematist, Darwin scholar, and philosopher of evolutionary biology. As a member of the US Advisory Committee, he has given generously of his time and expertise. In memory of his irreplaceable service, we dedicate this volume of the *Correspondence* to him.

LIST OF PROVENANCES

The following list gives the locations of the original versions of the letters printed in this volume. The editors are grateful to all the institutions and individuals listed for allowing access to the letters in their care. Access to material in DAR 261 and DAR 263, formerly at Down House, Downe, Kent, England, is courtesy of English Heritage.

American Philosophical Society, Philadelphia, Pennsylvania, USA
Archives, The New York Botanical Garden, Bronx, New York, USA
Athenæum (publication)
Bayerische Staatsbibliothek, München, Germany
Bibliothèque Publique et Universitaire de Genève, Geneva, Switzerland
British Library, London, England
Cambridge University Library, Cambridge, England
The Charles Darwin Trust, London, England
Christ's College, Cambridge, England
Cleveland Health Sciences Library, Cleveland, Ohio, USA
DAR *see* Cambridge University Library
Edinburgh University Library, Edinburgh, Scotland
Field Museum of Natural History, Chicago, Illinois, USA
Jean-Louis Fischer (private collection)
Fitzwilliam Museum, Cambridge, England
Paul V. Galvin Library, Illinois Institute of Technology, Chicago, Illinois, USA
Gardeners' Chronicle and Agricultural Gazette (publication)
J. L. Gray ed. 1893 (publication)
Gray Herbarium of Harvard University, Cambridge, Massachusetts, USA
B. C. Guild (private collection)
Ernst-Haeckel-Haus, Friedrich-Schiller-Universität Jena, Germany
Hardwicke's Science-Gossip (publication)
S. J. Hessel (private collection)
Eilo Hildebrand (private collection)
Houghton Library, Harvard University, Cambridge, Massachusetts, USA
Imperial College of Science, Technology, and Medicine, London, England
Institut Mittag-Leffler, Djursholm, Sweden
King's College, University of London, Strand, London
Linnean Society of London, Piccadilly, London, England
K. M. Lyell ed. 1881 (publication)

McGill University Libraries, Rare Books and Special Collections Division, Montréal, Québec, Canada

Marchant ed. 1916 (publication)

ML (publication)

Möller ed. 1915–21 (publication)

John Murray Archive, London, England

Museo Civico di Storia Naturale, Milan, Italy

Muséum National d'Histoire Naturelle (Département Systématique & Evolution; Cryptogamie), Paris, France

The Natural History Museum, Cromwell Road, London, England

Notes and Queries on China and Japan (publication)

Quaritch (dealer), London, England

Royal Botanic Gardens, Kew, Richmond, Surrey, England

Royal College of Physicians, Regent's Park, London, England

Royal Entomological Society, Queen's Gate, London, England

Elizabeth Rütimeyer (private collection)

Joseph Sakmyster (dealer)

Rachel Salt (private collection)

Shrewsbury School, Shrewsbury, England

Sotheby Parke Bernet (dealer), London, England

Sotheby's London (dealer), London, England

Staatsbibliothek zu Berlin, Germany

E. Sulivan (private collection)

Alexander Turnbull Library (National Library of New Zealand), Wellington, New Zealand

University of Virginia, Charlottesville, Virginia, USA

Württembergische Landesbibliothek, Stuttgart, Germany

A NOTE ON EDITORIAL POLICY

The first and chief objective of this edition is to provide complete and authoritative texts of Darwin's correspondence. For every letter to or from Darwin, the text that is available to the editors is always given in full. The editors have occasionally included letters that are not to or from Darwin if they are relevant to the published correspondence. Volumes of the *Correspondence* are published in chronological order. Occasional supplements will be published containing letters that have come to light or have been redated since the relevant volumes of the *Correspondence* appeared. Letters that can only be given a wide date range, in some instances spanning several decades, are printed in the supplement following the volume containing letters at the end of their date range. The first such supplement was in volume 7 and included letters from 1828 to 1857. The second supplement was in volume 13, and included letters from 1822 to 1864.

Dating of letters and identification of correspondents

In so far as it is possible, the letters have been dated, arranged in chronological order, and the recipients or senders identified. Darwin seldom wrote the full date on his letters and, unless the addressee was well known to him, usually wrote only 'Dear Sir' or 'Dear Madam'. After the adoption of adhesive postage stamps in the 1840s, the separate covers that came into use with them were usually not preserved, and thus the dates and the names of many recipients of Darwin's letters have had to be derived from other evidence. The notes made by Francis Darwin on letters sent to him for his editions of his father's correspondence have been helpful, as have matching letters in the correspondence, but many dates and recipients have had to be deduced from the subject-matter or references in the letters themselves.

Transcription policy

Whenever possible, transcriptions have been made from manuscripts. If the manuscript was inaccessible but a photocopy or other facsimile version was available, that version has been used as the source. In many cases, the editors have had recourse to Francis Darwin's large collection of copies of letters, compiled in the 1880s. Other copies, published letters, or drafts have been transcribed when they provided texts that were otherwise unavailable.

The method of transcription employed in this edition is adapted from that described by Fredson Bowers in 'Transcription of manuscripts: the record of variants',

Studies in Bibliography 29 (1976): 212–64. This system is based on accepted principles of modern textual editing and has been widely adopted in literary editions.

The case for using the principles and techniques of this form of textual editing for historical and non-literary documents, both in manuscript and print, has been forcefully argued by G. Thomas Tanselle in 'The editing of historical documents', *Studies in Bibliography* 31 (1978): 1–56. The editors of the *Correspondence* followed Dr Tanselle in his conclusion that a 'scholarly edition of letters or journals should not contain a text which has editorially been corrected, made consistent, or otherwise smoothed out' (p. 48), but they have not wholly subscribed to the statement made earlier in the article that: 'In the case of notebooks, diaries, letters and the like, whatever state they are in constitutes their finished form, and the question of whether the writer "intended" something else is irrelevant' (p. 47). The editors have preserved the spelling, punctuation, and grammar of the original, but they have found it impossible to set aside entirely the question of authorial intent. One obvious reason is that in reading Darwin's writing, there must necessarily be reliance upon both context and intent. Even when Darwin's general intent is clear, there are cases in which alternative readings are, or may be, possible, and therefore the transcription decided upon must to some extent be conjectural. Where the editors are uncertain of their transcription, the doubtful text has been enclosed in italic square brackets.

A major editorial decision was to adopt the so-called 'clear-text' method of transcription, which so far as possible keeps the text free of brackets recording deletions, insertions, and other alterations in the places at which they occur. Darwin's changes are, however, recorded in the back matter of the volume, under 'Manuscript alterations and comments', in notes keyed to the printed text by paragraph and line number. All lines above the first paragraph of the letter (that is, date, address, or salutation) are referred to as paragraph 'o'. Separate paragraph numbers are used for subscriptions and postscripts. This practice enables the reader who wishes to do so to reconstruct the manuscript versions of Darwin's autograph letters, while furnishing printed versions that are uninterrupted by editorial interpolations. The Manuscript alterations and comments record all alterations made by Darwin in his letters and any editorial amendments made in transcription, and also where part of a letter has been written by an amanuensis; they do not record alterations made by amanuenses. No attempt has been made to record systematically all alterations to the text of copies of Darwin letters included in the correspondence, but ambiguous passages in copies are noted. The editors believe it would be impracticable to attempt to go further without reliable information about the texts of the original versions of the letters concerned. Letters to Darwin have been transcribed without recording any of the writers' alterations unless they reflect significant changes in substance or impede the sense; in such cases footnotes bring them to the reader's attention.

Misspellings have been preserved, even when it is clear that they were unintentional: for instance, 'lawer' for 'lawyer'. Such errors often indicate excitement or haste and may exhibit, over a series of letters, a habit of carelessness in writing to a particular correspondent or about a particular subject.

Capital letters have also been transcribed as they occur except in certain cases, such as 'm', 'k', and 'c', which are frequently written somewhat larger than others as initial letters of words. In these cases an attempt has been made to follow the normal practice of the writers.

In some instances that are not misspellings in a strict sense, editorial corrections have been made. In his early manuscripts and letters Darwin consistently wrote 'bl' so that it looks like 'lb' as in 'albe' for 'able', 'talbe' for 'table'. Because the form of the letters is so consistent in different words, the editors consider that this is most unlikely to be a misspelling but must be explained simply as a peculiarity of Darwin's handwriting. Consequently, the affected words have been transcribed as normally spelled and no record of any alteration is given in the textual apparatus. Elsewhere, though, there are misformed letters that the editors have recorded because they do, or could, affect the meaning of the word in which they appear. The main example is the occasional inadvertent crossing of 'l'. When the editors are satisfied that the intended letter was 'l' and not 't', as, for example, in 'stippers' or 'istand', then 'l' has been transcribed, but the actual form of the word in the manuscript has been given in the Manuscript alterations and comments.

If the only source for a letter is a copy, the editors have frequently retained corrections made to the text when it is clear that they were based upon comparison with the original. Francis Darwin's corrections of misreadings by copyists have usually been followed; corrections to the text that appear to be editorial alterations have not been retained.

Editorial interpolations in the text are in square brackets. Italic square brackets enclose conjectured readings and descriptions of illegible passages. To avoid confusion, in the few instances in which Darwin himself used square brackets, they have been altered by the editors to parentheses with the change recorded in the Manuscript alterations and comments. In letters to Darwin, square brackets have been changed to parentheses silently.

Material that is irrecoverable because the manuscript has been torn or damaged is indicated by angle brackets; any text supplied within them is obviously the responsibility of the editors. Occasionally, the editors are able to supply missing sections of text by using ultraviolet light (where text has been lost owing to damp) or by reference to transcripts or photocopies of manuscript material made before the damage occurred.

Words and passages that have been underlined for emphasis are printed in italics in accordance with conventional practice. Where the author of a letter has indicated greater emphasis by underlining a word or passage two or more times, the text is printed in bold type.

Paragraphs are often not clearly indicated in the letters. Darwin and others sometimes marked a change of subject by leaving a somewhat larger space than usual between sentences; sometimes Darwin employed a longer dash. In these cases, and when the subject is clearly changed in very long stretches of text, a new paragraph has been started by the editors without comment. The beginnings of letters, valedictions, and postscripts are also treated as new paragraphs regardless of whether they appear as new paragraphs in the manuscript. Special manuscript devices delimiting sections or paragraphs, for example, blank spaces left between sections of text and lines drawn across the page, are treated as normal paragraph indicators and are not specially marked or recorded unless their omission leaves the text unclear.

Occasionally punctuation marking the end of a clause or sentence is not present in the manuscript, but the author has made his or her intention clear by allowing, for example, extra space or a line break to function as punctuation. In such cases, the editors have inserted an extra space following the sentence or clause to set it off from the following text.

Additions to a letter that run over into the margins, or are continued at its head or foot, are transcribed at the point in the text at which the editors believe they were intended to be read. The placement of such an addition is only recorded in a footnote if it seems to the editors to have some significance or if the position at which it should be transcribed is unclear. Enclosures are transcribed following the letter.

The hand-drawn illustrations and diagrams that occur in some letters are reproduced as faithfully as possible and are usually positioned as they were in the original text. In some cases, however, it has been necessary to reduce the size of a diagram or enhance an outline for clarity; any such alterations are recorded in footnotes. The location of diagrams within a letter is sometimes changed for typesetting reasons. Tables have been reproduced as close to the original format as possible, given typesetting constraints.

Some Darwin letters and a few letters to Darwin are known only from entries in the catalogues of book and manuscript dealers or mentions in other published sources. Whatever information these sources provide about the content of such letters has been reproduced without substantial change. Any errors detected are included in footnotes.

Format of published letters

The format in which the transcriptions are printed in the *Correspondence* is as follows:

1. *Order of letters*. The letters are arranged in chronological sequence. A letter that can be dated only approximately is placed at the earliest date on which the editors believe it could have been written. The basis of a date supplied by the editors is given in a footnote unless it is derived from a postmark, watermark, or endorsement that is recorded in the physical description of the letter (see section 4, below). Letters with the same date, or with a range of dates commencing with that date, are printed in the alphabetical order of their senders or recipients unless their contents dictate a clear alternative order. Letters dated only to a year or a range of years precede letters that are dated to a particular month or range of months, and these, in turn, precede those that are dated to a particular day or range of dates commencing with a particular day.

2. *Headline*. This gives the name of the sender or recipient of the letter and its date. The date is given in a standard form, but those elements not taken directly from the letter text are supplied in square brackets. The name of the sender or recipient is enclosed in square brackets only where the editors regard the attribution as doubtful.

3. *The letter text*. The transcribed text follows as closely as possible the layout of the source, although no attempt is made to produce a type-facsimile of the manuscript: word-spacing and line-division in the running text are not adhered to. Similarly, the typography of printed sources is not replicated. Dates and addresses given by authors are transcribed as they appear, except that if both the date and the address are at the head of the letter they are always printed on separate lines with the address first, regardless of the manuscript order. If no address is given on a letter by Darwin, the editors have supplied one, when able to do so, in square brackets at the head of the letter. Similarly, if Darwin was writing from an address different from the one given on the letter, his actual location is given in square brackets. Addresses on printed stationery are transcribed in italics. Addresses, dates, and valedictions have been run into single lines to save space, but the positions of line-breaks in the original are marked by vertical bars.

4. *Physical description*. All letters are complete and in the hand of the sender unless otherwise indicated. If a letter was written by an amanuensis, or exists only as a draft or a copy, or is incomplete, or is in some other way unusual, then the editors provide the information needed to complete the description. Postmarks, endorsements, and watermarks are recorded only when they are evidence for the date or address of the letter.

5. *Source*. The final line provides the provenance of the text. Some sources are given in abbreviated form (for example, DAR 140: 18) but are listed in full in the List of provenances unless the source is a published work. Letters in private collections are also indicated. References to published works are given in author–date or short-title form, with full titles and publication details supplied in the Bibliography at the end of the volume.

6. *Darwin's annotations*. Darwin frequently made notes in the margins of the letters he received, scored significant passages, and crossed through details that were of no further interest to him. These annotations are transcribed or described following the letter text. They are keyed to the letter text by paragraph and line numbers. Most notes are short, but occasionally they run from a paragraph to several pages, and sometimes they are written on separate sheets appended to the letter. Extended notes relating to a letter are transcribed whenever practicable following the annotations as 'CD notes'.

Quotations from Darwin manuscripts in footnotes and elsewhere, and the text of his annotations and notes on letters, are transcribed in 'descriptive' style. In this method the alterations in the text are recorded in brackets at the places where they occur. For example:

'See Daubeny ['vol. 1' *del*] for *descriptions of volcanoes in [*interl*] S.A.' *ink* means that Darwin originally wrote in ink 'See Daubeny vol. 1 for S.A.' and then deleted 'vol. 1' and inserted 'descriptions of volcanoes in' after 'for'. The asterisk before 'descriptions' marks the beginning of the interlined phrase, which ends at the bracket. The asterisk is used when the alteration applies to more than the immediately preceding word. The final text can be read simply by skipping the material in brackets. Descriptive style is also used in the Manuscript alterations and comments.

Editorial matter

Each volume is self-contained, having its own index, bibliography, and biographical register. A chronology of Darwin's activities covering the period of each volume and translations of foreign-language letters are supplied, and additional appendixes give supplementary material where appropriate to assist the understanding of the correspondence. A cumulative index is planned once the edition is complete. References are supplied for all persons, publications, and subjects mentioned, even though some repetition of material in earlier volumes is involved.

If the name of a person mentioned in a letter is incomplete or incorrectly spelled, the full, correct form is given in a footnote. Brief biographies of persons mentioned in the letters, and dates of each correspondent's letters to and from Darwin in the current volume, are given in the Biographical register and index to correspondents. Where a personal name serves as a company name, it is listed according to the family name but retains its original order: for example, 'E. Schweizerbart'sche Verlagsbuchhandlung' is listed under 'S', not 'E'.

Short titles are used for references to Darwin's books and articles and to collections of his letters (e.g., *Descent*, 'Parallel roads of Glen Roy', *LL*). They are also used for some standard reference works and for works with no identifiable author (e.g., *Alum. Cantab.*, *Wellesley index*, *DNB*). For all other works, author–date references are used. References to the Bible are to the authorised King James version unless otherwise stated. Words not in *Chambers dictionary* are usually defined in the footnotes with a source supplied. The full titles and publication details of all books and papers referred to are given in the Bibliography. References to archival material, for instance that in the Darwin Archive at Cambridge University Library, are not necessarily exhaustive.

Darwin and his correspondents writing in English consistently used the term 'fertilisation' for the processes that are now distinguished as fertilisation (the fusion of female and male gametes) and pollination (the transfer of pollen from anther to stigma); the first usage known to the editors of a distinct term for pollination in English was in 1873 (letter from A. W. Bennett, 12 July 1873 (*Calendar* no. 8976)). 'Fertilisation' in Darwin's letters and publications often, but not always, can be regarded as referring to what is now termed pollination. In the footnotes, the editors, where possible, have used the modern terms where these can assist in explaining the details of experimental work. When Darwin or his correspondents are quoted directly, their original usage is never altered.

The editors use the abbreviation 'CD' for Charles Darwin throughout the footnotes. A list of all abbreviations used by the editors in this volume is given on p. xlii.

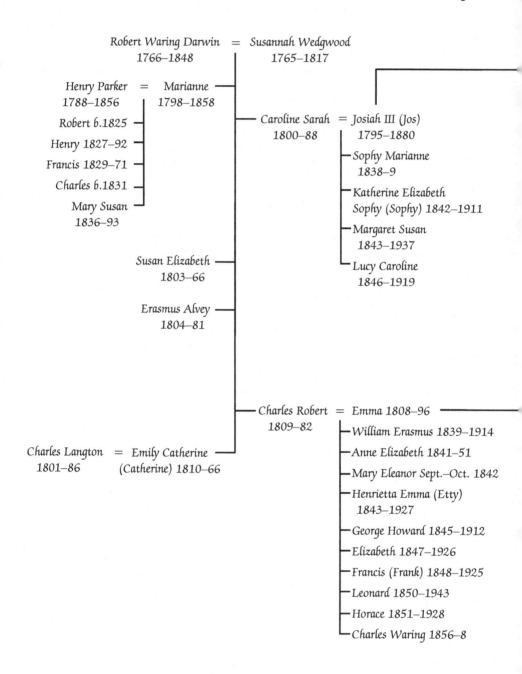

Robert Waring Darwin = Susannah Wedgwood
1766–1848   1765–1817

Henry Parker = Marianne
1788–1856   1798–1858

Robert b.1825

Henry 1827–92

Francis 1829–71

Charles b.1831

Mary Susan
1836–93

Caroline Sarah = Josiah III (Jos)
1800–88   1795–1880

Sophy Marianne
1838–9

Katherine Elizabeth
Sophy (Sophy) 1842–1911

Margaret Susan
1843–1937

Lucy Caroline
1846–1919

Susan Elizabeth
1803–66

Erasmus Alvey
1804–81

Charles Robert = Emma 1808–96
1809–82

Charles Langton = Emily Catherine
1801–86   (Catherine) 1810–66

William Erasmus 1839–1914

Anne Elizabeth 1841–51

Mary Eleanor Sept.–Oct. 1842

Henrietta Emma (Etty)
1843–1927

George Howard 1845–1912

Elizabeth 1847–1926

Francis (Frank) 1848–1925

Leonard 1850–1943

Horace 1851–1928

Charles Waring 1856–8

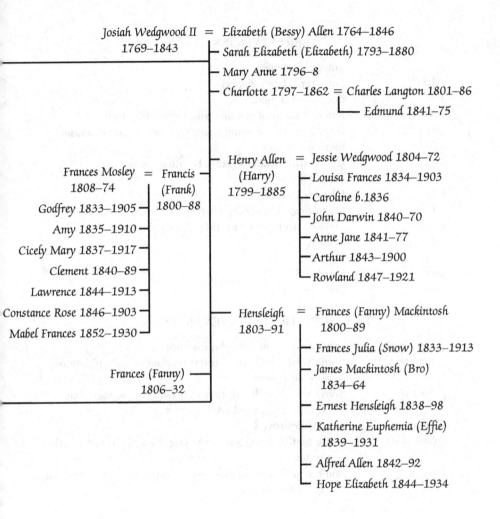

Josiah Wedgwood II = Elizabeth (Bessy) Allen 1764–1846
1769–1843
├─ Sarah Elizabeth (Elizabeth) 1793–1880
├─ Mary Anne 1796–8
├─ Charlotte 1797–1862 = Charles Langton 1801–86
│   └─ Edmund 1841–75

Frances Mosley = Francis ─┬ Henry Allen = Jessie Wedgwood 1804–72
1808–74   (Frank)   (Harry)   ├─ Louisa Frances 1834–1903
Godfrey 1833–1905 ┤ 1800–88 1799–1885   ├─ Caroline b.1836
Amy 1835–1910 ┤   ├─ John Darwin 1840–70
Cicely Mary 1837–1917 ┤   ├─ Anne Jane 1841–77
Clement 1840–89 ┤   ├─ Arthur 1843–1900
Lawrence 1844–1913 ┤   └─ Rowland 1847–1921
Constance Rose 1846–1903 ┤
Mabel Frances 1852–1930 ┤   Hensleigh = Frances (Fanny) Mackintosh
                            1803–91   1800–89
                                         ├─ Frances Julia (Snow) 1833–1913
Frances (Fanny)   ├─ James Mackintosh (Bro)
1806–32   1834–64
                            ├─ Ernest Hensleigh 1838–98
                            ├─ Katherine Euphemia (Effie)
                                1839–1931
                            ├─ Alfred Allen 1842–92
                            └─ Hope Elizabeth 1844–1934

ABBREVIATIONS

AL	autograph letter
ALS	autograph letter signed
DS	document signed
LS	letter in hand of amanuensis, signed by sender
LS(A)	letter in hand of amanuensis with additions by sender
Mem	memorandum
(S)	signed with sender's name by amanuensis
CD	Charles Darwin
CUL	Cambridge University Library
DAR	Darwin Archive, Cambridge University Library
del	deleted
illeg	illegible
interl	interlined
underl	underlined

TRANSCRIPTION CONVENTIONS

[some text]	'some text' is an editorial insertion
⌈some text⌋	'some text' is the conjectured reading of an ambiguous word or passage
[some text]	'some text' is a description of a word or passage that cannot be transcribed, e.g., '*3 words illeg*'
⟨ ⟩	word(s) destroyed
⟨some text⟩	'some text' is a suggested reading for a destroyed word or passage
⟨*some text*⟩	'some text' is a description of a destroyed word or passage, e.g., '*3 lines excised*'

THE CORRESPONDENCE OF CHARLES DARWIN
1867

To *Athenæum* 1 January 1867

> Down, Bromley, Kent
> Jan. 1, 1867.

I was glad to see in your paper of the 15th ult. that you have allowed "A Great Reader" to protest against books being sold uncut.[1] He is obliged to own that many persons like to read and cut the pages at the same time; but, on the other hand, many more like to turn rapidly over the pages of a new book so as to get some notion of its contents and see its illustrations, if thus ornamented. But "A Great Reader" does not notice three valid objections against uncut books. In the first place they sometimes get torn or badly cut, as may be seen with many books in Mudie's Library;[2] and I know a lady who is habitually guilty of cutting books with her thumb. Secondly, and which is much more important, dust accumulates on the rough edges, and gradually works in between the leaves, as the books vibrate on their shelves. Thirdly, and most important of all, for those who not merely read but have to study books, is the slowness in finding by the aid of the index any lost passage, especially in works of reference.[3] Who could tolerate a dictionary with rough edges? I have had Loudon's 'Encyclopædia of Plants' and Lindley's 'Vegetable Kingdom' in constant use during many years,[4] and the cloth binding is still so good that it would have been a useless expense to have had them bound in leather; nor did I forsee that I should have consulted them so often, otherwise the saving of time in finding passages would have amply repaid the cost of binding. The North Americans have set us the example of cutting and often gilding the edges.[5] What can be the reason that the same plan is not followed here? Is it mere Toryism?[6] Every new proposal is sure to be met by many silly objections. Let it be remembered that a deputation of paper-manufacturers waited on Sir R. Peel, when he proposed to establish the penny postage, urging that they would suffer great loss, as all persons would write on notepaper instead of on letter sheets![7] It is always easy to suggest fanciful difficulties. An eminent publisher remarked to me that booksellers would object to receiving books cut, as customers would come into their shops and read them over the counter; but surely a book worth reading could not be devoured in this hasty manner.[8] The sellers of old books seem never to object to any one studying the books on their stalls as long as he pleases. "A Discursive" remarks in your paper that booksellers would object to books being supplied to them with their edges cut, as they would thus "relinquish an obvious advantage in palpable evidence of newness."[9] But why should this objection be more valid here than in America? Publishers might soon ascertain the wishes of the public if they would supply to the same shop cut and uncut copies, or if they would advertise that copies in either state might be procured, for booksellers would immediately observe which were taken in preference from their counters. I hope that you will support this movement, and earn the gratitude of all those who hate the trouble and loss of time in cutting their books, who lose their paper-cutters, who like to take a hasty glance through a new volume, who dislike to see the edges of the pages deeply stained with dust, and who have the labour of searching for lost passages. You will not only earn the gratitude of many readers, but in not a few cases that of their children, who have to cut through dry and pictureless books for the benefit of their elders.

Charles Darwin.

Athenæum, 5 January 1867, pp. 18–19

[1] Most books and periodicals in Britain were sold with uncut pages. A letter to the *Athenæum*, 15 December 1866, p. 803, signed by 'A Great Reader', asked for the 'opinion of the literary world' as to whether a 'period of civilization' had not arrived 'when the readers of books and periodicals might reasonably ask that they should be delivered from the publishers ready cut'. CD told Joseph Dalton Hooker that he had then, 'like an ass', sent a long letter to the *Athenæum* urging publishers to cut the pages (see *Correspondence* vol. 14, letter to J. D. Hooker, 28 [December 1866] and n. 6). For Hooker's response and his criticism of British publishers as 'Penny-wise Pound foolish, Penurious, Pragmatical Prigs', see the letter from J. D. Hooker, [29 December 1866].

[2] CD refers to Mudie's Select Library of New Oxford Street, London, a subscription lending library. For the Darwin family's use of Mudie's as a source of books, see Browne 2002.

[3] CD noted this last problem when urging his publisher, John Murray, to have the pages of the fourth edition of *Origin* cut (see *Correspondence* vol. 14, letter to John Murray, 15 July [1866]).

[4] There is an annotated copy of John Claudius Loudon's *Encyclopædia of plants* (Loudon 1841) in the Darwin Library–CUL (*Marginalia* 1: 504–6). A copy of John Lindley's *Vegetable kingdom* (Lindley 1853) is listed in CD's Library catalogue (DAR 240), but it has not been found in the Darwin Library–Down or the Darwin Library–CUL.

[5] Beginning in the 1850s, North American publications had their pages trimmed after binding (see Tebbel 1972, pp. 260–1). CD had earlier written to Thomas Henry Huxley concerning the *Natural History Review*, 'Do inaugurate a great improvement, & have pages cut, like the Yankees do' (*Correspondence* vol. 11, letter to T. H. Huxley, 10 [January 1863]). See also *Correspondence* vol. 13, letter to Charles Lyell, 21 February [1865].

[6] CD's reference is to the association of the Conservative, or Tory, party with opposition to change or reform.

[7] Stationers objected to the establishment of the penny postage in 1839, allegedly because they feared that government issue of franked envelopes would affect their sales; however, this deputation protested to the Whig administration preceding Robert Peel's Tory administration (see Hill 1880, 1: 348). Peel was not prime minister until 1841, but when various other protests were raised against the penny postage in 1842, he refused to abandon it (*ibid.*, p. 449). See also Fryer and Akerman eds. 2000, 2: 715–16, 759–60.

[8] CD refers to a comment made by his publisher, Murray, in his letter of 18 July [1866] (*Correspondence* vol. 14).

[9] Another reply to the letter in the *Athenæum* of 15 December (see n. 1, above) was signed 'A Discursive'; the author suggested that only the lateral edges be cut, leaving the top edges uncut (see *Athenæum*, 22 December 1866, p. 848).

From Fritz Müller 1 January 1867

Desterro, Brazil,
January 1th 1867.

My dear Sir

In my last letter (Decbr. 1th) I told you that Oncidium flexuosum is sterile with own pollen; more than 80 flowers of 8 plants, which were fertilized with own pollen (taken either from the same flower of from a distinct flower of the same panicle or from a distinct panicle of the same plant) yielded not a single seed-capsule; the flowers fell off about a week after fertilization.— But what is still more curious, pollen and stigma of the same plant are not only entirely useless to, but even act as a poison to each other![1] Thus, four or five days after fertilization a brownish colour appears on the adjoining surface of the pollen and stigma and soon afterwards the whole pollinium is rendered dark-brown.[2]

This is not the case when you bring instead of own pollen, the pollen of widely different species on the stigma of Oncidium flexuosum. Among others I tried the pollinia of Epidendrum Zebra[3] (nearly allied to, or perhaps not specifically distinct from Ep. variegatum).[4] Of course no seed-capsules were produced; 8–9 days (in one out of about 20 flowers 12 days) after fertilization the germs began to shrink, but even then the pollen and its tubes which sometimes had penetrated in the upper part of the germ, had a perfectly fresh appearance, rarely showing a very faint scarcely perceptible brownish colour.— The pollinia of Ep. fragrans also I found to be perfectly fresh, as well as their tubes after 5 days stay in the stigmatic chamber of Oncidium flexuosum.

The poisonous action of own pollen becomes still more evident, on placing on the same stigma two different pollen-masses. In a flower of Oncidium flexuosum, on the stigma of which I had placed one own pollen-mass and one of a distinct plant of the species, I found five days after the former brown, the latter fresh; in some other flowers 4 or 5 days after both the pollen-masses were brown, and I think, although my experiments are not yet quite decisive, that own pollen will always kill the pollen of another plant when placed on the same stigma.— Now compare this destructive action of own pollen with that of Epidendrum (species allied to variegatum).

Debr. 15 I placed on the stigmas of some flowers of Onc. flexuosum one pollen-mass from a distinct plant of that species and one of Epidendrum.— Debr. 21th both the pollen-masses fresh melting with numerous tubes.— Dec. 26 both the pollen-masses dissolved into single pollen-grains, most of which have long tubes; numerous tubes of either pollen descend half way down the germen; the pollen-mass of Epidendrum is to be reconnoitred only by the unaltered caudicula. Dec. 30 the germs of the two resting flowers (all the others having been dissected) are slightly curved to one side; this side, probably that of the Ep.-pollen swelling to a lesser degree than the other.[5]

I suspect that the sterility with the same plants pollen will be very common among Vandeae and one of the principal causes of them seeding so badly; for the several specimens of most of these plants grow scattered in the forests, at great distance from one another and thus the chance of pollinia being brought from a distinct plant is not very great.[6]

I already observed a second instance of this sterility, and of the mutual poisonous action of the same plants pollen and stigma. I found a large raceme of a Notylia with more than sixty aromatic flowers. The slit lending to the stigmatic chamber is less narrow in this second species than in that mentioned in one of my former letters and a single pollen-mass might be introduced rather easily.[7] I fertilized (Dec. 12th 13th 14th) almost all the flowers with pollen from the same raceme. Two days after fertilization the flowers withered and I found that the pollen-masses were dark brown and had not emitted a single tube. You see the poisonous action of own pollen is here much more rapid, than in Oncid. flexuosum. There remained eight flowers, which had not been fertilized, and these I fertilized (Decbr. 19th 20th) with pollen-masses from a small raceme of a different plant of the species. Two of them I afterwards dissected and found the pollen fresh and having emitted numerous tubes. The other six have now fine swelling pods.[8]

Very different from the innocent pollen of Ep. Zebra, that of Notylia is as deletery to Oncidium flexuosum as are this latter plants own pollinia. Dec. 14 I placed on the same stigma of Oncidium flexuosum one pollen-mass from a distinct plant of that species and one of Notylia. Decbr. 21: the latter was brown as well as the neighbouring part of the stigma; the Oncidium pollen-mass was nearly fresh; only on the side towards the Notylia-pollen a brownish stripe began to make its appearance between pollen-mass and stigma.[9]

Strange as the destructive action of own pollen may appear, it may be easily shown to be of real use to the plant. If flowers are sterile with own pollen and if the introduction of own pollen-masses into the stigmatic chamber prevents, as it does in Oncidium and Notylia the subsequent fertilization by other pollinia, it must be injurious to the plant to waste anything in the nutrition of flowers rendered useless by the introduction of own pollinia, and useful to become rid of them as soon as possible.[10] This view is confirmed by a comparison of Oncidium and Notylia. Decbr. 21th I fertilized on a panicle of Oncidium flexuosum 36 flowers (12 with own pollen, 24 with pollen from a distinct plant). Decbr. 24th before any difference had appeared between the two kinds of pollen the peduncles and germs of 55 not fertilized flowers of this panicle were withering and discoloured yellowish, while all the fertilized flowers had green swelling germens. The panicle had about 160 flowers.

In Notylia on the contrary, when about $\frac{5}{6}$ of the flowers of a raceme were fertilized with own pollen, they all fell off in a few days without injuring even one of the not fertilized flowers. In Notylia fertilization is much easily effected a week or so after the expansion of the flowers, the entrance of the stigmatic cavity being open.

As to Notylia I may add that nectar is secreted at the base of the bracteae and also at the base of the upper sepalon. I found nectar at the base of the bracteae in a small species of Oncidium also.[11]

At last I have gratified my wish of examining myself the wonderful genus Catasetum.[12] I had three fine racemes of the male Catasetum mentosum and one raceme with only three flowers of Monachanthus (probably of the same species).[13] In this Catasetum a membrane connects the antennae with the interior margin of the stigmatic chamber. The ovula are scarcely more rudimentary than in Monachanthus, and not so much so, as in many other Vandeae. The stigmatic surface is not viscid at all; but notwithstanding pollen-masses (from the same as well as from a distinct plant and also from Cattleya Leopoldi), when introduced, began to dissolve into groups of pollen grains and to emit tubes, some of which were 2^{mm} long, ... the flowers withered.—[14]

The female flowers, of a uniform green colour, are much like those of Monachanthus viridis, but the anther is much smaller. There is a *pedicellus* and *disk*; the disk is brown, and *quite dry*; the pedicellus white, elastic, *not connected with the pollen-masses!* On touching it, the pedicellus is ejected at some distance assuming the form of a hemicylinder.[15] The anthers do not open (at least they had not done so some days after the expansion of the flowers, long after the pedicelli having been ejected). The pollen-masses consequently remain enclosed; although being much smaller, they ressembled in shape those of Catasetum and had a small caudiculus. I brought three of these pollen-masses into the stigmatic chamber of Catasetum, where they emitted numerous pollen-tubes. Infortunately I had cut off the raceme of Catasetum, in order to preserve it from insects, and thus I am unable to say whether the pollen of Monachanthus may as yet be able to fertilize the ovules of Catasetum.— Certainly insects can never effect this fertilization. At all events this seems to me to be one of the most interesting cases of rudimentary organs. We have on the one hand in Monachanthus a disk, a well developed elastic pedicellus, caudiculi and apparently good pollen, we have on the other hand in Catasetum a stigmatic surface able to cause this pollen to emit its tubes, and apparently good ovules and in spite of all this—from the dryness of the stigma and disk and from the pedicelles not connected with the enclosed pollen-masses an utter impossibility of fertilization.[16]

When the pollen-masses of Catasetum are introduced into the entrance of the stigmatic cavity of Monachanthus, they peep out at half their length; but in the course of the first days they are allowed, as it were, entirely, and the stigma is shut. This swallowing of the pollen-masses is also to be observed in Cirrhaea and here it is easy to see how it is effected. The stigmatic cavity has a very narrow transversal slit into which only the very tip of the long pollen-masses may be introduced. Under the slit the cavity widens gradually and continues into a large canal occupying the center of the columna; this canal is empty, while the upper part of the stigmatic

cavity is filled with loose viscid cells. Now the tip of the pollen-masses in contact
with the humid stigma swells and thus is forced down into the wider inferior part
of the stigmatic cavity and at last into the canal of the columna. Of course, what
at first sight appears contradictory, the thickest pollen-masses must be swallowed
first. Thus Decbr. 25th at 7[h] in the morning I fertilized two flowers of Cirrhaea
with dry pollen-masses of another plant of the species (collected Decbr. 3[d]), four
flowers of the same raceme with fresh own pollen-masses and one flower with a
much larger pollen-mass of Gongora (bufonia?).[17] This latter had disappeared at 3[h]
in the afternoon, when the others peeped out half their length; at 7[h] in the evening
all had disappeared, with exception of one of the old pollen-masses of which a
small part as yet peeped out.

   I enclose some seeds of our two species of *Gesneria*; they are, as you see, very
small and may probably be blown at a great distance by the wind. Now there is in
the seed-capsules a very fine contrivance preventing the seeds from falling to the
ground without the action of the wind.[18] The two valves remain united at the tip,
and the pod only opens by two longitudinal slits, on its upper and under surfaces.

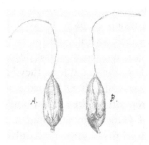

The slit on the under side (A) is shut by two rows of hairs inserted on the margins
of the valves. So you may conserve the open pods for a long time without a single
grain falling out, whereas by blowing you will drive them out in a moment. In
some other cases, in which hairs on the valves, or hair-like processes on the orifice
of the capsule are combined with exceedingly small seeds (as in a great number
of Orchids, in most Hepaticae, in the peristome of mosses) their use seems to be
different from what it is in Gesneria.[19]

   I am to start to morrow for a botanical excursion on the continent, where I
intent to spend a couple of weeks and whence I hope I shall not return without
some interesting news.

   With every good wish and profound respect believe me, dear Sir, very sincerely
yours | Fritz Müller.

ALS incomplete and draft[20]
Möller ed. 1915–21, 2: 104–9; DAR 157a: 104

CD ANNOTATION
Below signature: 'Desterro Jan 1, 1867'

[1] In his letter of 1 December 1866 (*Correspondence* vol. 14), Müller described his observations of the stigmatic chamber, pollinia, and ovaries in *Oncidium flexuosum* flowers pollinated by the same flower or by a flower from a distinct plant. CD included the information on the poisonous action of own-pollen, and noted that *O. flexuosum* was only fertile with pollen of flowers from a different plant, in *Variation* 2: 134–5. CD added a brief reference to Müller's findings in *Origin* 5th ed., p. 304. See also CD's remarks on self-sterile plants in *Cross and self fertilisation*, pp. 340–8.

[2] Müller had earlier described this occurrence in self-pollinated *Oncidium flexuosum* flowers (*Correspondence* vol. 14, letter from Fritz Müller, 1 December 1866).

[3] Alfred Möller added a drawing of *Epidendrum* made by Müller to Möller ed. 1915–21, 2: 105 (see n. 20, below).

[4] Müller never published the name *Epidendrum zebra*. The drawing of the orchid published in Möller ed. 1915–21, 2: 105 (see n. 3, above), has been identified as *Prosthechea vespa*, which has the synonym *Epidendrum variegatum* (Robert Dressler, personal communication; see also Higgins 1997, p. 381).

[5] CD reported Müller's experiment of placing the two different pollen masses on one *Oncidium flexuosum* stigma in *Variation* 2: 134–5; he noted that after eleven days the *Epidendrum* pollen was indistinguishable from the other, except for the caudicles, or attached stalks. He did not mention any later developments. See also letter from Fritz Müller, 2 February 1867.

[6] CD noted how often 'various Orchideous tribes' failed to have their flowers fertilised, and noted this observation of Müller's in regard to the Epidendreae and to *Vanilla* in Brazilian forests, in *Orchids* 2d ed., pp. 280–1.

[7] Müller wrote of the difficulty he had in pollinating another species of *Notylia* in his letter of 2 August 1866 (*Correspondence* vol. 14); he sent a drawing of that flower with specimens attached.

[8] CD included a description of the pollination of Müller's second *Notylia* by pollen from flowers from the same raceme and by pollen from flowers from a different plant in his discussions of plants that were poisoned by their own pollen (see *Variation* 2: 134–5). In *Orchids* 2d ed., p. 172, CD mentioned the poisonous effect of same-plant pollen on Müller's first *Notylia* species.

[9] See *Variation* 2: 134–5.

[10] CD reported this view of Müller's in *Variation* 2: 135.

[11] CD added Müller's observation of nectar secretion in *Notylia* and *Oncidium* to *Orchids* 2d ed., p. 266.

[12] Möller added a figure of parts of a *Catasetum* flower to Möller ed. 1915–21, 2: 107 (see n. 20, below).

[13] In his letter to Müller of 20 September [1865] (*Correspondence* vol. 13), CD had suggested Müller read his observations on *Catasetum* in *Orchids*, pp. 211–48, 'to shew how perfect the contrivances are'; CD reminded him of this in his letter of 17 October [1865], adding that Hermann Crüger of Trinidad had confirmed all that he had written (see *Correspondence* vol. 12, letter from Hermann Crüger, 21 January 1864, and letter to Daniel Oliver, 17 February [1864]; see also Crüger 1864). CD had argued in *Orchids*, pp. 236–46, and in 'Three sexual forms of *Catasetum tridentatum*', that *C. tridentatum* (now *C. macrocarpum*) was the male form of a plant that also had a female form (*Monachanthus viridis*; now *C. macrocarpum*), and a hermaphrodite form (*Myanthus barbatus*; now *C. barbatum*).

[14] CD included Müller's account of the *Monachanthus* and *Catasetum mentosum* in Brazil, as well as the failure to fertilise the *C. mentosum* with its own pollen or that of another plant, in *Orchids* 2d ed., p. 206. See also 'Fertilization of orchids', p. 154 (*Collected papers* 2: 151).

[15] Müller refers to the *Monachanthus* of the previous paragraph, which he thought was the female form of *Catasetum mentosum*. CD had found no viscid disc or pedicel in *Monachanthus viridis*, and surmised that they fell off with the rudimentary pollen-masses (*Orchids*, pp. 239–44). CD had described how, in *Catasetum*, the entire pollinium, with the viscid disc attached by the highly elastic pedicel (now known as the stipe) to the pollen-masses, was ejected with great force when the antenna was touched (*Orchids*, pp. 222–7). For diagrams of CD's 'Catasetidæ', see *Orchids*, pp. 216–17, 232, 239; for explanations of his orchid terminology, see *Orchids*, chapter 1, and pp. 220–1.

[16] CD illustrated the rudimentary nature of the pollen-masses of the female form in particular by comparing those of *Catasetum* and *Monachanthus* (see *Orchids* 2d ed., pp. 201–3). He concluded: 'At a period not far distant, naturalists will hear with surprise, perhaps with derision, that grave and learned

men formerly maintained that such useless organs were not remnants retained by inheritance, but were specially created and arranged in their proper places like dishes on a table . . . by an Omnipotent hand "to complete the scheme of nature"' (*ibid.*, p. 203). Müller later published his observations on *C. mentosum* (now *C. atratum*) (F. Müller 1868); CD cited this work in 'Fertilization of orchids', p. 154 (*Collected papers* 2: 151); see also *Orchids* 2d ed., pp. 206–7.

[17] CD cited Müller's observations in his description of 'deglutition', when pollen-masses were slowly sucked into the narrow slit of the stigma, in *Monachanthus, Cirrhaea*, and *Notylia* (see *Orchids* 2d ed., pp. 171–2, 206). See also 'Fertilization of orchids', p. 153 (*Collected papers* 2: 150).

[18] Müller first wrote to CD about *Gesneria*, evidently in regard to its possible dimorphism, in a missing section of his letter of [2 November 1866] (*Correspondence* vol. 14); see *ibid.*, letter to Fritz Müller, [late December 1866 and] 1 January 1867.

[19] In *Gesneria*, the hairs on the margin of the valve retain the seeds in the pod until they are blown out by the wind (ensuring they will be carried a distance from the original plant). However, in some orchids and the Hepaticae the movement of hygroscopic hairs ensures dispersal, as the teeth of the peristome in a spore capsule release spores with slight changes in humidity.

[20] Only the last half page of the original letter has been found (beginning 'I enclose some seeds . . .'). The remaining portion is from a draft found by Alfred Möller and published in *Fritz Müller: Werke, Briefe und Leben* (Möller ed. 1915–21). According to Möller, all of Fritz Müller's letters to CD were written in English (see Möller ed. 1915–21, 2: 72 n.); most of them have not been found. Many of the letters were later sent by Francis Darwin to Möller, who translated them into German for his book. Möller also found drafts of some letters (*ibid.*) and published these in their original English version. The present letter (except for the last half page) is one of these drafts.

From John Murray 2 January [1867][1]

Albemarle St

Jany 2

My Dear Sir

Mr Clowes' estimate of your MS.[2] shows that it will make

$$\left.\begin{array}{l} \text{Vol I. 648 pages} \\ \text{"   II 624 pages} \end{array}\right\} \text{post } 8^{vo}$$

same size as the origin of Species—independant of contents Preface Index & any addit^l Chapter—[3]

This is clearly too much for volumes of that size—& I propose therefore—with your consent to substitute the larger (Demy) 8vo size in wch your work w^d form 2 Volumes like the last Ed^n of Lyells Principles[4]

On hearing from you the Printer will be set to work

I am My Dear Sir | faithfully yours | John Murray

Ch. Darwin Esq

DAR 171: 342

[1] The year is established by the relationship between this letter and the letter to John Murray, 3 January [1867].

[2] CD sent the manuscript of *Variation* to John Murray on 22 December 1866 (see *Correspondence* vol. 14, letter to John Murray, 21 and 22 December [1866], and Appendix II). Murray also refers to William Clowes of the printing firm William Clowes & Sons.

[3] Murray refers to *Origin*. In his letter to Murray of 21 and 22 December [1866] (*Correspondence* vol. 14), CD said that he was sending all of the manuscript of *Variation* apart from the last chapter and a possible chapter on humans (see also *ibid.*, letter to J. D. Hooker, 24 December [1866]). '8vo': octavo.

[4] Murray refers to the tenth edition of Charles Lyell's *Principles of geology* (C. Lyell 1867–8).

To John Murray 3 January [1867]

Down. | *Bromley.* | *Kent. S.E.*
Jan. 3^d

My dear Sir

I cannot tell you how sorry I am to hear of the enormous size of my Book.[1] I fear it can never pay. But I cannot shorten it now; nor indeed, if I had foreseen its length, do I see which parts ought to have been omitted.

If you are afraid to publish it, say so at once I beg you, & I will consider your note as cancelled. If you think fit get anyone whose judgment you rely on, to look over some of the more legible chapters; viz the Introduction & on Dogs & on Plants; the latter chapters being in my opinion the dullest in the book.[2] There is a Hypothetical & curious Chapter called Pangenesis which is legible, & about which I have no idea what the instructed public will think; but to my own mind it has been a considerable advance in knowledge—[3] The list of Chapters, & the inspection of a few, here & there, w^d give a good judge a fair idea of the whole Book. Pray do not publish blindly, as it would vex me all my life if I led you to heavy loss. I am extremely much vexed at the size; but I believe the work has some value, though of course I am no fair judge.—

You must settle all about type & size according to your own judgment; but I will only say that I think, & *hear on all sides* incessant complaint of the fashion which is growing of publishing intolerably heavy volumes:—[4]

I have written my concluding Chapter; whether that on Man, shall appear, shall depend on size of book, on time & on my own strength.[5]

My dear Sir | Yours very sincerely | Ch. Darwin

Endorsement: '1867. Jany 3^d'
John Murray Archive

[1] CD refers to *Variation* (see letter from John Murray, 2 January [1867]).

[2] CD refers to the introduction, chapter 1 ('Domestic dogs and cats'), and chapters 9 to 11 of *Variation*.

[3] Chapter 27 of *Variation*, 'Provisional hypothesis of pangenesis', outlined CD's ideas regarding heredity; CD suggested that minute particles (gemmules) circulated in the bodily fluids and were capable of generating new cells, remaining dormant until required. He thought his hypothesis could explain both sexual and asexual reproduction, as well as reversion and the regrowth of body parts. For CD's discussion of pangenesis with correspondents, see, for example, *Correspondence* vol. 13, letter to T. H. Huxley, 27 May [1865], and *Correspondence* vol. 14, letter to J. D. Hooker, 4 April [1866].

[4] For Murray's suggestion about the page size of *Variation*, see the letter from John Murray, 2 January [1867]. For CD's own complaints about the weight of the sixth edition of Charles Lyell's *Elements of geology* (C. Lyell 1865), see *Correspondence* vol. 13, letter to Charles Lyell, 21 February [1865].

[5] For CD's possible additional chapter on humans, see *Correspondence* vol. 14, letter to John Murray, 21 and 22 December [1866]. See also *ibid*, letter to J. D. Hooker, 24 December [1866]. Ultimately, CD did not add this chapter to *Variation*, but used the material he had begun collecting for it in *Descent*.

From John William Salter 4 January [1867][1]

8, B. Road | St John's Wood

Jan. 4/66

Dear Mr Darwin

The very kind way in which you have done this takes off the feeling with which one receives aid from many.[2]

Believe me, it has been 3 years long hard struggle before I thought of asking any help but what my hands & brain could give me— The latter has given way a little I fear, but I am better now.[3]

I feel sure I shall be able in some way to return y^r. kindness

The accomp^g pamphlets you will not care for perhaps—though one refers to the new formation.[4]

It is a source of great pleasure to me to find that all the improvements in classification made in England are adopted abroad— They dont lead us as a rule, but we them.

I have a letter from old Sedgwick still as lively as ever—and I shall have a little work to do for him in the arrang^t. of his Museum soon.[5]

Could I have got my own English Botany finished, I believe I should have been tolerably independent of accidents. But the same cause that has made it necessary to write to you *prevented* me from completing & making it valuable—[6]

Should you have any neighbours who possess English Botany—or yourself care about it, I enclose a prospectus & also a jeu d'esprit of my sister's on my hard fate, as she calls it—[7] It is more amusing than my papers—

I must tell you fairly, that the further I examine with the aid of your new theory, the more facts appear to me to agree with it— There are still some very important exceptions that make me think there is another law beside it not recognized— I do not think that breaks in the geol. succession are sufficient to account for the sudden leaps in life among the old strata. e.g. from Cephalopoda to Fish.[8]

But I follow y^r. direction & make notes occasionally of these, or rather I will, for three years have passed away since I could get time for this.

Of course everybody agrees about species thats settled—but why do Entomostraca univalve Mollusca, & amphibia begin with such high forms? I would add Fish, but I might run the risk of saying something outrè & you have always Huxley[9] at command.

J W Salter

In Plants you have it all your own way. Cryptogamia—Coniferæ—Phanero-gamia, Diœcia[10]

DAR 177: 11

[1] The year is established by the relationship between this letter and the letter from J. W. Salter, 31 December 1866 (*Correspondence* vol. 14). Salter wrote '1866' in error.

[2] No letter from CD has been found; however, Salter requested financial aid from CD in his letter of 31 December 1866 (*Correspondence* vol. 14). CD sent Salter £10 on 1 January 1867, listing the amount under 'Charities' in his account book (CD's Classed account books (Down House MS)).

[3] Salter had resigned his post as palaeontologist to the Geological Survey of Great Britain in 1863, and had since been employed as an independent palaeontologist at various local museums (see *DNB*, and n. 5, below). For Salter's struggles with mental illness, see Secord 1985.

[4] Salter may have enclosed his papers on Palaeozoic fossils from Pembrokeshire (Salter 1864, and Salter and Hicks 1865); Salter collected a distinct group of Palaeozoic fossils that led him to propose the new 'Mænevian group' as the lowest stratum of the Lingula flags (Salter and Hicks 1865, p. 477 n). The Menevian formation is now considered to be late middle Cambrian. No articles of Salter's dated after 1861 have been found in the Darwin Pamphlet Collection–CUL.

[5] Salter had assisted Adam Sedgwick, Woodwardian Professor of geology at the University of Cambridge, in the 1840s; after Salter retired from the Geological Survey of Great Britain in 1863, Sedgwick employed him at times (see *DNB* and Secord 1985, pp. 62–3).

[6] Early in his career, Salter was apprenticed to the naturalist and illustrator James de Carle Sowerby; Salter then assisted him with the early volumes of the five-volume *Supplement to the English botany of the late Sir J. E. Smith and Mr. Sowerby* (W. J. Hooker, Sowerby, [*et al.*] 1831–63; see Secord 1985). Salter was the proprietor of and contributed illustrations to volumes 4 and 5; the publication dates given on the title pages of volumes 4 and 5 are 1849 and 1863 respectively, but the parts of volume 5 were published from August 1863 to June 1865. A notice in the last part of volume 5 complains that hardly any of the original subscribers could be found, and that the sale of the current numbers did not even pay for paper or print, and speculates that this might be due to the success of the third edition of *English botany* (Syme ed. 1863–92), which contained the same engravings.

[7] The enclosures have not been found in the Darwin Archive–CUL. For the present locations of Salter's papers, see Secord 1985. Salter's sister has not been identified.

[8] For CD's discussion of these topics, see his chapters 'On the imperfection of the geological record' and 'On the geological succession of organic beings' in *Origin*, pp. 279–345. For Salter's efforts at reconciling CD's theory with his religious beliefs, see Secord 1985, p. 68.

[9] Thomas Henry Huxley.

[10] Salter names some parts of the plant kingdom, ending with Dioecia, a class in the Linnaean system.

To William Bernhard Tegetmeier   6 January [1867][1]

Down. | Bromley. | Kent. S.E.
Jan. 6th

My dear Sir

I return by this Post in tin cannister & therefore I trust safely the two skulls of the Horned Fowl & White Crested Polish, & most sincerely do I thank you for their loan & for your great kindness in having allowed me to keep them so long & have them engraved.—[2]

My M.S. is all in Printers' Hands, but I do not suppose my Book will be out till next November, as Murray does not like publishing except in Spring or Autumn[3]

I shall never have finished correcting before the end of the Spring.—
My dear Sir | Yours very sincerely | Ch. Darwin
I have changed all my numerous References to your new Poultry Book—[4]

Archives of The New York Botanical Garden (Charles Finney Cox collection) (Tegetmeier 101)

[1] The year is established by the reference to the manuscript of *Variation* having been sent to the printers (see n. 3, below).

[2] Tegetmeier sent skulls of wild and domesticated fowls for CD to examine in April 1861 (see *Correspondence* vol. 9). CD returned some of the specimens in 1864, but retained some to be engraved for *Variation* (see *Correspondence* vols. 12 and 13). CD sent the skulls of the white-crested Polish and the horned fowl to George Brettingham Sowerby Jr for engraving in 1866 (see *Correspondence* vol. 14, letter to W. B. Tegetmeier, 18 [December 1866] and n. 4). In *Variation* 1: 260, CD acknowledged 'the kindness of Mr. Tegetmeier' in lending nearly half of the fifty-three skulls examined. Engravings of the skulls of a white-crested Polish cock and of a horned fowl appear in *Variation* 1: 262, 263, and 265.

[3] CD refers to the manuscript of *Variation*, and to the printing firm William Clowes & Sons (see letter from John Murray, 2 January [1867]). Murray had postponed publication of the fourth edition of *Origin* from July to early November 1866 (see *Correspondence* vol. 14, letter from John Murray, 18 July [1866]; see also letter to J. D. Hooker, 3 and 4 August [1866]). *Variation* was published in January 1868 (Freeman 1977).

[4] Tegetmeier edited a revision of the first edition of *The poultry book* (Wingfield and Johnson 1853) in 1856 and 1857; however, the publishers went out of business before the edition was complete (see *Correspondence* vol. 14, letter from W. B. Tegetmeier, 22 January [1866] and n. 10). Wingfield and Johnson 1856–7 was superseded by Tegetmeier 1867; both are in the Darwin Library–CUL, and are annotated by CD (see *Marginalia* 1: 798–803). Tegetmeier 1867 was originally published in fifteen instalments in 1866 and 1867. In 1866, Tegetmeier asked CD twice to cite Tegetmeier 1867, rather than Wingfield and Johnson 1856–7 (see *Correspondence* vol. 14). Tegetmeier 1867 is cited frequently in the chapter on fowls in *Variation* (1: 225–75); Wingfield and Johnson 1856–7 is not cited.

From Thomas Henry Huxley [before 7 January 1867][1]

Geological Survey of Great Britain

My dear Darwin

A happy new year to you— may you be eupeptic through 1867 & your friends & the world in general will profit—

I have been making holiday & though I took your letter away with me purposing to answer it—I need hardly say I did not[2]

I have read a good deal of Haeckels book—not thoroughly but as I best could— and my present judgment very much coincides with yours—[3] It is a very good methodical *ab ovo*, statement of the case—excellent for the Germanic mind— But I fear it would never do for these latitudes The terminology would frighten every Englishman who should look at it into fits—[4]

Like you, I find but little new in either fact or speculation except the attempt to reduce animal forms to geometric plans[5]—and the applications of the developmental view to the details of Botany & Zoology The former I have not read with care; indeed if I did I doubt if I have enough geometry to understand them

As regards the latter they are undoubtedly marked by very great & accurate knowledge—& are full of interesting suggestion—[6] But I have tried so many 'trees' of my own & found it is hopeless to apply any criterion by which one 'tree' could be shewn to be better than another, that I entertain a certain shyness of these speculations—[7]

I got up a very fair genealogy of the Mammalia in my last Hunterian course; but I have never mustered up courage to publish it elsewhere—[8]

I have not written to Haeckel yet as I promised to do but I think I shall tell him that it is useless to attempt an English translation— I don't believe he would sell 100 copies & the expense would be great[9]

I am glad to get a pat on the back for 'Man's Place'[10] as Giebel has been making an awful onslaught on it and on me![11] But he really says nothing which is of the least consequence or has not been said already and I really believe that the main argument is quite impregnable— I will get out a second Edition some of these days[12]

The Physiology book is purely elementary & hardly worth your reading[13]

I will read the Hybridism chapter again with all care—[14] Depend upon it the gates are wide open to any one who will storm the castle— I should be too happy to see the argument in favour of your views logically complete— But until you can show that B & C have been selectively produced from A—& that B & C are infertile in the first or second degree there must be a hole in the ballad[15] Some may think it big & some may think it little but there it is—

I take the theological line & jump over the hole, by an Act of Faith—but I cant forget the hole & I wish it were not as big as even a pins point—

Have you read the Dukelets book? I hear he is down on both of us—[16] But you know what Lord Derby said to him[17]

Ever yours faithfully | T H Huxley

Endorsement: 'Jan 67'
DAR 102: 134a–d

---

[1] The date is established by the relationship between this letter and the letter to T. H. Huxley, 7 January [1867], and by the reference to '1867' in the New Year greeting.

[2] Huxley refers to CD's letter to him of 22 December [1866] (*Correspondence* vol. 14).

[3] CD had expressed his opinion of Ernst Haeckel's *Generelle Morphologie der Organismen* (Haeckel 1866) in his letter to Huxley of 22 December [1866] (*Correspondence* vol. 14); see also letter to Ernst Haeckel, 18 August [1866] (*ibid.*). For Haeckel's discussions with CD about his work on the book, see *Correspondence* vols. 12–14. There is an annotated copy of Haeckel 1866 in the Darwin Library–CUL (see *Marginalia* 1: 355–7).

[4] In his letter to Huxley, CD commented on Haeckel 1866, and wrote: 'The number of new words . . . is something dreadful. He seems to have a passion for defining, I daresay very well, & for coining new words' (see *Correspondence* vol. 14, letter to T. H. Huxley, 22 December [1866] and n. 3).

[5] Huxley refers to what Haeckel described as 'promorphology', the description and explanation of the external form of an organism and all its constituent parts, and the reduction of this form to basic geometric forms (see Haeckel 1866, 1: 46–9).

[6] Huxley refers to Haeckel's discussion of the development of the plant and animal kingdoms; Haeckel represented the various branches of the kingdoms in eight genealogical trees inside the back cover of the second volume of Haeckel 1866; these represented the possible relationships of all living organisms. For Haeckel's theoretical discussion of his phylogenetic categories, see Haeckel 1866, 2: XVII–XX, XXXI–XXXII, XLVIII–L, 406–17. Haeckel's genealogical trees established a standard iconography for phylogeny (see, for example, S. J. Gould 1990, pp. 263–7). CD had used a tree-like diagram to illustrate the divergence of offspring from parental types (*Origin*, facing p. 117). See *Correspondence* vol. 14, letter from Ernst Haeckel, 11 January 1866 and n. 8.

[7] For Huxley's early resistance to constructing genealogical classifications in the form of a tree, see A. Desmond 1994–7, 1: 262, 354–6.

[8] As Hunterian Professor at the Royal College of Surgeons, Huxley was required to deliver an annual course of twenty-four lectures (see L. Huxley ed. 1900, 1: 254). Manuscript versions of the 1863 lectures are in the Huxley papers 39: 1 (Imperial College of Science, Technology, and Medicine Archives).

[9] CD had expressed scepticism about the possibility of translating Haeckel 1866 in the letter to Ernst Haeckel, 20 January [1866], the letter to Fritz Müller, [before 10 December 1866], and the letter to T. H. Huxley, 22 December [1866] (*Correspondence* vol. 14). Haeckel 1866 has never been translated into English.

[10] CD had just read *Man's place in nature* (T. H. Huxley 1863a) for the second time (see *Correspondence* vol. 14, letter to T. H. Huxley, 22 December [1866]). CD's annotated copy of T. H. Huxley 1863a is in the Darwin Library–CUL (see *Marginalia* 1: 424). For CD's earlier praise for the book, see *Correspondence* vol. 11.

[11] Huxley refers to Christoph Gottfried Andreas Giebel and his recent attack on Huxley's comparison between human and ape anatomy in T. H. Huxley 1863a (see Giebel 1866). He accused Huxley of prejudging the familial relationship between humans and apes and of not giving the reader sufficient facts upon which to make an independent assessment (Giebel 1866, p. 401). After comparing human and orang-utan skulls in detail, Giebel concluded with an extensive critique of Huxley's views, without, however, referring specifically to *Man's place in nature* or to its German translation (Carus trans. 1863; see Giebel 1866, pp. 415–19).

[12] A second edition of T. H. Huxley 1863a never appeared.

[13] Huxley had sent *Lessons in experimental physiology* (T. H. Huxley 1866) for Henrietta Emma Darwin to read (see *Correspondence* vol. 14, letter from T. H. Huxley, 11 November 1866); CD commented in his letter of 22 December [1866] (*ibid.*) that he had read only one chapter. For Henrietta's interest in Huxley's publications, see *Correspondence* vol. 11 and *Correspondence* vol. 13, letter from T. H. Huxley, 1 May 1865 and n. 2.

[14] CD had urged Huxley to read the revised chapter on hybridity in the fourth edition of *Origin* (*Origin* 4th ed., pp. 292–338) in his letter of 22 December [1866] (*Correspondence* vol. 14); see also n. 15, below. CD made extensive additions to the chapter in the fourth edition, especially to the section headed (in the fourth edition) 'Origin and causes of the sterility of first crosses and of hybrids'; he also added a new section, 'Reciprocal dimorphism and trimorphism' (see *Origin* 4th ed., pp. 310–26, and Peckham ed. 1959, pp. 443–59).

[15] In his review of *Origin* ([T. H. Huxley] 1860), in subsequent publications, and privately, Huxley asserted that the principle of natural selection could not be regarded as proved until varieties of a species that were infertile with one another had been produced by selective breeding, thereby creating new 'physiological' species; Huxley's critique served as an important impetus to CD's investigations into cross- and self-pollination and hybrid sterility (see *Correspondence* vols. 8–14, especially *Correspondence* vol. 10, Appendix VI). In *Origin* 4th ed., p. 323, CD observed: 'the physiological test of lessened fertility, both in first crosses and in hybrids, is no safe criterion of specific distinction'.

[16] Huxley refers to George Douglas Campbell, eighth duke of Argyll, and to *The reign of law* (G. D. Campbell 1867). For Campbell's recent criticism of CD's theory in articles that contributed to G. D. Campbell 1867, see *Correspondence* vols. 13 and 14.

[17] Huxley probably refers to Edward George Geoffrey Smith Stanley, fourteenth earl of Derby, who became prime minister for the third time in 1866. Campbell, a Liberal, and Stanley, a Conservative, sometimes opposed each other in the House of Lords (see I. E. Campbell ed. 1906 and *DNB*). Stanley's comment has not been identified.

To T. H. Huxley 7 January [1867][1]

Down. | Bromley. | Kent. S.E.
Jan 7th

My dear Huxley

Very many thanks for your letter which has told me exactly what I wanted to know.— I shall give up all thoughts of trying to get the book translated, for I am well convinced that it would be hopeless without too great an outlay.—[2] I much regret this, as I sh^d. think the work w^d be useful & I am sure it would be to me, as I shall never be able to wade through more than here & there a page of the original. To all people I cannot but think that the number of new terms would be a great evil.[3] I must write to him. I suppose you know his address but in case you do not, it is "to care of

Signor. Nocolaus Krohn
Madeira."[4]

I have sent the M.S of my Big book, & horridly disgustingly big it will be, to the Printers, but I do not suppose it will be published, owing to Murray's idea on seasons, till next November.—[5]

I am thinking of a Chapter on man, as there has lately been so much said on nat. selection in relation to man.—[6] I have not seen the Duke's (or Dukelets?, how can you speak so of a living real Duke) book, but must get it from Mudie, as you say he attacks us.—[7]

Ever yours my dear Huxley | C. Darwin

Nature never made species mutually sterile *by selection*; nor will man.—[8]

Imperial College of Science, Technology, and Medicine Archives (Huxley 5: 233)

[1] The year is established by the relationship between this letter and the letter from T. H. Huxley, [before 7 January 1867].

[2] In his letter to Huxley of 22 December [1866] (*Correspondence* vol. 14), CD had asked for his impression of Haeckel 1866, noting that he feared an English translation was out of the question. For Huxley's reply, see his letter of [before 7 January 1867].

[3] See letter from T. H. Huxley, [before 7 January 1867] and nn. 3 and 4.

[4] Ernst Haeckel told CD he would be spending the winter on zoological research on Madeira and Tenerife in his letter of 19 October 1866 (*Correspondence* vol. 14). Haeckel probably left Nicholas Krohn's address with CD when he visited him on 21 October 1866 (see *ibid.*, letter to Ernst Haeckel, [20 October 1866]). Haeckel travelled to Madeira and Tenerife in November 1866, and arrived in December at Lanzarote, where he spent three months (Krauße 1987, pp. 76–7); see also letter from Ernst Haeckel, 12 May 1867. Krohn Brothers & Co., wine merchants, was founded in Madeira in 1858 by John and Nicholas Krohn, British brothers of Russian descent.

[5] CD refers to the manuscript of *Variation* and to the printing firm William Clowes & Sons. For CD's concern about the publication date, see also the letter to W. B. Tegetmeier, 6 January [1867] and n. 3.

[6] See letter to John Murray, 3 January [1867] and n. 5. For recent discussions of human origins and natural selection, see especially A. R. Wallace 1864 and Lubbock 1865; see also *Correspondence* vols. 11–13.

[7] Huxley had mentioned George Douglas Campbell, the duke of Argyll's *Reign of law* (G. D. Campbell 1867) in his letter of [before 7 January 1867]). CD also refers to Mudie's Select Library, a subscription lending library (see letter to *Athenæum*, 1 January 1867 and n. 2).

[8] CD is responding to Huxley's argument regarding natural selection (see letter from T. H. Huxley, [before 7 January 1867] and n. 15).

To Ernst Haeckel 8 January 1867

Down. | *Bromley.* | *Kent. S.E.*

Jan 8th. 1867

My dear Prof. Haeckel

I received some weeks ago your great work.[1] I have read several parts, but I am too poor a German scholar and the book is too large for me to read it all; I cannot tell you how much I regret this, for I am sure that nearly the whole would interest me greatly, and I have already found several parts very useful, such as the discussion on cells, and on the different forms of reproduction.[2] I feel sure after considering the subject deliberately, and after consulting with Huxley, that it would be hopeless to endeavour to get a publisher to print an English translation; the work is too profound and too long for our English country-men.[3]

The number of new terms would also I am sure tell much against its sale; and indeed I wish for my own sake that you had printed a glossary of all the new terms which you use.[4] I fully expect that your book will be highly successful in Germany; and the manner in which you often refer to me in your text, and your dedication and the title I shall always look at as one of the greatest honours conferred on me during my life.[5]

I sincerely hope that you have had a prosperous expedition and have met with many new and interesting animals[6] If you have spare time, I should much like to hear what you have been doing and observing. As for myself I have sent the M.S. of my book on "Domestic Animals &c" to the printers; it turned out to be much too large; it will not be published I suppose until next November.[7] I find that we have discussed several of the same subjects, and I think we agree on most points fairly well. I have lately heard several times from Fritz Müller, but he seems now cheifly to be working on plants.[8] I often think of your visit to this house, which I enjoyed extremely, and it will ever be to me a real pleasure to remember our acquaintance.[9] From what I heard in London, I think you made many friends there,[10] Shall you return through England; if so, and you can spare the time, we shall all be *delighted* to see you here again.

With cordial good wishes for your success in every way, believe | Dear Haeckel | Yours very sincerely | Charles Darwin

LS(A)
Ernst-Haeckel-Haus (Bestand A-Abt. 1-52/12)

[1] CD refers to Haeckel's *Generelle Morphologie* (Haeckel 1866). He had earlier received a proof-sheet of the book (see *Correspondence* vol. 14, letter to Ernst Haeckel, 18 August [1866]). There is an annotated copy of Haeckel 1866 in the Darwin Library–CUL (see *Marginalia* 1: 355–7).

[2] CD also mentioned the difficulty he was having in reading Haeckel 1866 in the letter to Fritz Müller, [before 10 December 1866], the letter to T. H. Huxley, 22 December [1866] (*Correspondence* vol. 14), and the letter to T. H. Huxley, 7 January [1867] (this volume). For Haeckel's discussion of cells, see Haeckel 1866, 1: 269–89; for his discussion of different forms of reproduction see *ibid.*, 2: 32–109. These sections are lightly annotated in CD's copy (see *Marginalia* 1: 355–7).

[3] CD consulted Thomas Henry Huxley in his letter of 22 December [1866] (*Correspondence* vol. 14). For Huxley's reply, see his letter of [before 7 January 1867]; see also letter to T. H. Huxley, 7 January [1867], and L. Huxley ed. 1900, 1: 288–9.

[4] See also letter to T. H. Huxley, 7 January [1867] and n. 3.

[5] The subtitle to Haeckel 1866 reads: 'Allgemeine Grundzüge der organischen Formen-Wissenschaft, mechanisch begründet durch die von Charles Darwin reformirte Descendenz-Theorie' (General outline of the science of organic form, mechanically established through the theory of descent, reformed by Charles Darwin). The second volume of Haeckel 1866 has the following dedication: 'Den begründern der Descendenz-Theorie, den denkenden Naturforschern, Charles Darwin, Wolfgang Goethe, Jean Lamarck, widmet diese Grundzüge der allgemeinen Entwickelungsgeschichte in vorzüglicher Verehrung' (To the founders of descent-theory, the thinking naturalists, Charles Darwin, Wolfgang Goethe, Jean Lamarck, this outline of general developmental history is dedicated with greatest respect). See also *ibid.*, 2: 166–70, for Haeckel's discussion of Darwinism.

[6] In his letter to Fritz Müller of [before 10 December 1866] (*Correspondence* vol. 14) CD mentioned that Haeckel would be in Madeira over the winter working largely on the Medusae; see also this volume, letter to T. H. Huxley, 7 January [1867] and n. 4. In the event, Haeckel and two students spent most of the winter at Lanzarote, where they studied hydrozoans and siphonophores (Krauße 1987, pp. 76–7).

[7] CD refers to *Variation* (see letter to John Murray, 3 January [1867], and letter to T. H. Huxley, 7 January [1867] and n. 5).

[8] CD's most recent letters from Fritz Müller were those of 1 and 3 October 1866, [2 November 1866], and 1 December 1866 (*Correspondence* vol. 14).

[9] Haeckel visited Down House on 21 October 1866 (see *Correspondence* vol. 14).

[10] Haeckel met Huxley and Charles Lyell while in London in 1866 (see Krauße 1987, pp. 76–7).

To John Murray 8 January [1867]

Down. | Bromley. | Kent. S.E.
Jan 8th

My dear Sir

I am sorry to trouble you. I continue excessively annoyed at the absurd size of my book, more especially as I feel nearly sure that I can make a striking chapter on man & this will make it so much bigger.—[1]

A plan, which is often followed on the continent, has occurred to me, & which has some decided advantages besides reduction of bulk, namely to give details in smaller type; so that the general reader may at once pass over such details. Now in *several* chapters large portions, for instance all osteological details might be thus given.— I shᵈ not however like very small type being used. If, we will say, a $\frac{1}{3}$ of Book were printed in moderately smaller type, would this save much? In some chapters there could be no small type introduced.— What do you think of this plan?

If you approve I will send for M.S. & mark with a red line all that may be printed in smaller type; but I shᵈ like to see the two types which could be used.

In the Introduction, I have given a sketch of Natural Selection in 20 or 30 pages, for the sake of those who have not read the Origin; I could strike the whole of this out.—² If you approve, I must know whether to send to your house or to Stamford St. for the M.S.³

I could mark all the passages in a week or so.—

Of course, if you think fit, you can decline publishing altogether, & I must undergo the trouble & disappointment of looking out for some other channel of publication.

My dear Sir | Yours very sincerely | Ch Darwin

Endorsement: '1867. 8. Jany'
John Murray Archive

¹ After CD sent the manuscript of *Variation*, Murray informed him of the projected size of the book (see letter from John Murray, 2 January [1867]). For CD's anxiety regarding the size, and for his chapter on humans, see the letter to John Murray, 3 January [1867] and n. 5, and the letter to T. H. Huxley, 7 January [1867].
² CD retained a discussion of natural selection in the published introduction of *Variation* (see *Variation* 1: 1–14).
³ The address of Murray's publishing house was 50A Albemarle Street, London (*Post Office London directory*). The manuscript of *Variation* had already been examined by the printer William Clowes & Sons at Stamford Street, London (see letter from John Murray, 2 January [1867]).

To Joseph Dalton Hooker　9 January [1867]

Down.
Jan 9ᵗʰ

My dear Hooker

I like the first part of your paper in Gard. Chronicle to an *extraordinary* degree;¹ you never, in my opinion, wrote anything better. You ask for all, even minute, criticisms.—² In 1ˢᵗ column, you speak of no *Alpine* plants, & no replacement by zones, which will strike everyone with astonishment who has read Humboldt & Webb on zones on Teneriffe: : do you not mean *boreal* or *Arctic* plants?³ In 3ᵈ column you speak as if savages had generally viewed the endemic plants of the

Atlantic Islands, now, as you well know, the Canaries alone of all the Arch! were inhabited.[4] In 3ᵈ column have you really materials to speak of confirming the *proportion* of winged & wingless insects on Islands?—[5]

Your comparison of plants of Madeira with islets of Grt. Britain is admirable.—[6]

I must just allude to one of your last notes with very curious case of proportion of annuals in N. Zealand.[7] Are annuals adapted for short seasons, as in Arctic regions, or Tropical countries with dry season, or for periodically disturbed & cultivated ground? You speak of Evergreen vegetation as leading to few or confined conditions; but is not evergreen vegetation connected with *humid* & equable climate? Does not a very humid climate almost imply (Tyndall) an equable one?[8]

I have never printed a word that I can remember about orchids, & papiliona-ceous plants being few in islands on account of rarity of insects:[9] & I remember you screamed at me when I suggested this apropos to Papilionaceæ in N. Zealand, & to the statement about clover not seeding there till the Hive Bee was introduced, as I stated in my paper in Gard. Chronicle.—[10]

I have been these last few days vexed & annoyed to a foolish degree by hearing that my M.S. "on Dom. An. & Cult. Plants", will make 2 vols, both bigger than the "Origin". The volumes will have to be full-sized Octavo, & I have written to Murray to suggest details to be printed in small type.[11] But I feel that the size is quite ludicrous in relation to the subject. I am ready to swear at myself & at every fool who writes a book.—

Yours affect | C. Darwin

Seed of any Plumbago.[12]

Endorsement: '/67'
DAR 94: 3–4

[1] CD refers to the first of four parts of Hooker's article on insular floras (J. D. Hooker 1866a); the first part was published in the *Gardeners' Chronicle*, 5 January 1867, pp. 6–7. Hooker wrote at the start of the article that it was 'the substance of a Lecture' that he delivered to the British Association for the Advancement of Science, at Nottingham, on 27 August 1866. For Hooker's and CD's extensive discussions during Hooker's preparation of the lecture, see *Correspondence* vol. 14. CD's annotated copies of the *Gardeners' Chronicle* are in the Cory Library, Cambridge Botanic Garden; all four instalments of J. D. Hooker 1866a are annotated. An abstract (J. D. Hooker 1866b), and a French translation (J. D. Hooker 1866c) were also published. The lecture also appeared in condensed form in the *Journal of Botany* (J. D. Hooker 1867).

[2] In his letter of 14 December 1866 (*Correspondence* vol. 14), Hooker informed CD that his lecture was to be published in the *Gardeners' Chronicle* (J. D. Hooker 1866a), and wrote: 'I *intreat* you to overhaul it—'. Some of CD's criticisms were incorporated when the *Gardeners' Chronicle* version of the lecture was reprinted as a pamphlet; see Williamson 1984 for a reprint of the pamphlet. There are two annotated copies of the pamphlet in the Darwin Pamphlet Collection–CUL.

[3] Hooker wrote that the flora on mountains of small oceanic islands presented 'few alpine or sub-alpine species'; in the next column he wrote that in ascending the mountains of Madeira, there was found 'little or none of that replacement of species of a lower level by those of a higher northern latitude', with which observers were familiar in ascending continental mountains of equal or less height (J. D. Hooker 1866a, pp. 6 and 7). CD refers to discussions on zones of vegetation changing with altitude

in Alexander von Humboldt's *Personal narrative* (Humboldt 1814–29, 1: 263–76), and in Philip Barker Webb and Sabin Berthelot's *Histoire naturelle des Iles Canaries* (Webb and Berthelot 1836–50, 3 (part 1): 168–73). CD read Humboldt's *Personal narrative* carefully and took it on the *Beagle* voyage; in his *Autobiography*, he wrote that he copied out long passages about Tenerife and read them aloud to friends (see also *Correspondence* vol. 1, letter to W. D. Fox, [7 April 1831] and n. 2, and *Correspondence* vol. 4, Appendix IV; see Humboldt 1814–29, 1: 111–293, for his discussions relating to Tenerife). A copy of Humboldt's *Personal narrative*, made up of three different editions, is in the Darwin Library–CUL and is annotated (*Marginalia* 1: 415–20). For CD's reading of Webb and Berthelot 1836–50, see *Correspondence* vols. 2–4.

4 When considering rare and local plants on the Madeiran group of islands, Hooker wrote: 'Considerations . . . warrant our belief that such plants on oceanic islands are, like the savages which in many cases have been so long the sole witnesses of their existence, the last representatives of their several races' (J. D. Hooker 1866a, p. 7). Hooker changed the wording slightly in his 1867 pamphlet (see n. 2, above, and Williamson 1984, p. 63). 'Arch^s.': archipelagos.

5 Hooker surmised that as oceanic islands subsided and lost plant species, pollinating insects also diminished, particularly the winged species that were known to be the most active pollinators (J. D. Hooker 1866a, p. 7), citing Thomas Vernon Wollaston's observation that winged insects existed in smaller proportions than wingless on Madeira and the Canaries. CD had mentioned Wollaston's observation to Hooker, and hypothesised that winged insects were more easily blown off islands (see *Correspondence* vol. 5, letter to J. D. Hooker, 7 March [1855]). CD had discussed Wollaston's work on the proportion of wingless and winged beetles of Madeira in the manuscript of his 'big book' on species (*Natural selection*, pp. 291–3), and *Origin*, pp. 135–6. See also *Correspondence* vol. 5, letter from T. V. Wollaston, 2 March [1855] and nn. 10 and 11, and letter from J. D. Hooker, [before 17 March 1855], *Correspondence* vol. 14, letter from J. D. Hooker, [24 July 1866], and Wollaston 1854 and 1856.

6 In his article, Hooker pointed out that the flora of the Madeiran group was not dominated by species from the nearest continent, and compared the effect to that of finding unusual species isolated on single British islands or mountains (J. D. Hooker 1866a, p. 7).

7 See *Correspondence* vol. 14, letter from J. D. Hooker, 25 December 1866; in working on the second volume of his *Handbook of the New Zealand flora* (J. D. Hooker 1864–7), Hooker had found that only a small proportion of indigenous dicotyledons in New Zealand were annuals.

8 In his letter of 25 December 1866 (*Correspondence* vol. 14), Hooker commented: 'I suppose there can be no doubt but that a deciduous leaved vegetation affords more conditions for vegetable life than an evergreen one', and noted that more uniform climates tended to be characterised by more evergreen vegetation. CD refers to John Tyndall and his work on aqueous vapour and radiant heat (Tyndall 1864a and 1864b).

9 In his letter of 25 December 1866 (*Correspondence* vol. 14), Hooker wrote: 'Orchids & Leguminosæ are scarce in Islets because the necessary fertilizing insects have not migrated with the plants. Perhaps you have published this'.

10 In 1858, CD learned that clover did not seed in New Zealand until hive-bees were introduced; he wondered whether the absence of small Leguminosae had been due to the absence of small bees, and suggested this to Hooker in his letter of 12 January [1858] (*Correspondence* vol. 7). For Hooker's response, see *ibid.*, letter from J. D. Hooker, 15 January 1858. CD sought more information from various correspondents in 1858 regarding the pollination of Leguminosae in New Zealand, and also continued conducting his own experiments on cross-fertilisation in the family (see DAR 157a); he published a letter on the subject in the *Gardeners' Chronicle* in November (see *Correspondence* vol. 7, letter to the *Gardeners' Chronicle*, [before 13 November 1858] (*Collected papers* 2: 19–25)). The letter was also published as an article in *Annals and Magazine of Natural History* 3d ser. 2 (1858): 459–65. See also *Origin*, pp. 94–5, and *Correspondence* vol. 13, letter to J. D. Hooker, 22 and 28 [October 1865] and n. 14, and letter from George Henslow, 1 November 1865 and nn. 8 and 9.

11 CD refers to *Variation* (see letter to John Murray, 8 January [1867]). *Variation* was originally advertised under the title: 'Domesticated Animals and Cultivated Plants, or the Principles of Variation, Inher-

itance, Reversion, Crossing, Interbreeding, and Selection under Domestication' (*Publishers' Circular*, 1865, p. 386).

[12] After Fritz Müller wrote of finding a dimorphic *Plumbago* in Brazil, CD had asked Hooker whether he could send seed of any *Plumbago* species (see *Correspondence* vol. 14, letter from Fritz Müller, [2 November 1866], and letter to J. D. Hooker, 24 December [1866]).

From John Murray 9 January [1867][1]

Albemarle St
Jany 9

My Dear Sir

I think you take too much to heart the amount of the bulk of your MS— I am by no means disheartened & do not contemplate at all throwing up the publication[2]

Your suggestion of printing technical details in smaller type is a very good one & will relieve us of part of the dilemma[3]

But before I return your MS to be marked with red I beg you to consent to the carrying out of an Experiment suggested by yourself viz—testing the substance of the MS by its effects on a man of Letters & good information—not a man of science & awaiting the impression made on him by it—

I send the MS. yesterday to such a person a friend of mine, in about a week I shall have his answer & will give you the result[4]

I remain | My Dear Sir | Yours very sincerely | John Murray

The printer will send you rough proofs of the cuts, wch you may insert with *their titles* in the places of the MS where they come in[5]—in case this is not already done.

Ch Darwin Esq

DAR 171: 343

[1] The year is established by the relationship between this letter and the letter to John Murray, 8 January [1867].
[2] The reference is to *Variation*. See letter to John Murray, 8 January [1867] and n. 1. CD had expressed his annoyance about the length of his book to several correspondents; for his most recent comment, see the letter to J. D. Hooker, 9 January [1867].
[3] See letter to John Murray, 8 January [1867].
[4] Murray's reader was John Milton, an accountant (A. Fraser 1996, pp. 16, 37–8).
[5] William Clowes & Sons was the firm printing *Variation* (see letter from John Murray, 2 January [1867]).

From James Philip Mansel Weale 9 January 1867

Bedford, Algoa Bay | Cape Colony
9th. January 1867

My dear Sir

I send you by this mail a paper I have drawn up on a species of Bonatea I discovered here last spring, & which I believe to be new.[1] My friend M[r]. M[c]Owan of Grahams town informs me that he has looked through all the available literature he has, & that he cannot find any description corresponding with the specimen I sent him.[2] I sent also some notes on the subject to my friend, M[r] Trimen of Cape Town & he thought it a pity that I should not publish in some English Journal an account of the same.[3]

I had originally intended merely drawing up an account for the Port Elizabeth Natural History Society & botanists &c in the Colony.[4]

I send you a slightly modified paper, which if you think it worthy of further publication I will leave in your hands for the Linnean Society.

I have taken the liberty of proposing the name "Darwinii" if you will permit me to offer this humble return for the stimulus which your works have given me in the Study of Natural History.[5]

I am now busily engaged in examining the Asclepiids & have already made some drawings of dissections I have a vast collection of insects belonging to various orders, but principally Hymenoters covered with the pollinia of different species. So far as my observations go I believe the impregnation of this order to be very simple, & in no way are the contrivances so wonderful as in the Orchids.[6]

I have now a very extensive collection of Insects, most however unnamed. Two most singular moths of which the larva cases resemble most perfectly the thorns of the "Acacia horrida", are, I believe, discoveries of my own. These cases are not, as at first sight they appear, empty thorns, but most beautiful fabrications of the insects themselves, & are so deceptively like the real thorns that they would have entirely escaped my notice had I not seen them move.[7]

I believe that I have discovered four more species or varieties of Rhopalocera most from the Karoo, but have not yet been able to describe them to M[r]. Trimen.[8]

M[r]. T. has asked me to accompany him on a short tour to Natal & I write by next post to him to enquire when he starts,[9] as I am just about making a month excursion to more thoroughly examine the Bushman caves mentioned by Barrow & Burchell in the Tarka Mountains, as also some recent beds near the G[t] Fish River.[10]

I hope to be able to obtain before many months the head of a Bushman murderer, but it is difficult to convince the authorities of the interests of Science.[11] I have long been on the look-out.

Should I accompany M[r]. Trimen, to Natal I shall probably return home with him in May to England, as my friends have long been pressing me[12]

I enjoy this country so much that I do not like to leave it for ever, but again home ties, after over 4 years, influence one's feelings much

I shall endeavour to send you a copy of the "G[t] Eastern" with a letter by me signed 'Gogaje Man', a name by which I am well known out here,[13] & I do so because I think D[r] Brown's parting letter to the colonists a gross insult.

On his journey to Colesberg, he himself informed me he had collected no plants, & when I shewed him a species of "*Disperis*" of which I wanted to know the specific name, he did not even recognise it as an orchid.[14]

So far as D[r]. Brown's Blue Books are concerned they are simply compilations from other people's works, & I do not know of a single original observation in any of his Colonial works.[15]

As D[r]. Brown has received a salary equal to that of a Civil Commissioner in this country I think

AL incomplete
DAR 82: A113–14

CD ANNOTATIONS
1.1 I send ... Society. 3.2] *crossed ink*
4.1 I have ... Orchids. 5.5] *crossed blue crayon*
5.2 I have a vast ... Orchids. 5.5] *scored pencil*
6.1 I have ... move. 6.6] *enclosed in square brackets blue crayon*
7.1 I believe ... I think 15.2] *crossed blue crayon*
Top of letter: '*M*[r]. J. Mansel Weale' *ink*

[1] Weale refers to the manuscript of Weale 1867, which included a description of '*Bonatea Darwinii*'. *Bonatea darwinii* is now *B. cassidea*; for more on the taxonomy of this group of orchids, see, for example, Pridgeon *et al.* eds. 1999–2003, 2: 263–5. The manuscript is in the Linnean Society archives (SP1249). There is an annotated copy of Weale 1867 in the Darwin Pamphlet Collection–CUL.

[2] Weale's friend was the botanist Peter MacOwan, principal of Shaw College in Grahamstown (R. Desmond 1994). Grahamstown lies approximately fifty miles south-east of Bedford (Weale's place of residence) in Cape Colony (now the Eastern Cape province of the Republic of South Africa).

[3] CD had communicated two papers of Roland Trimen's on two other orchids from South Africa (Trimen 1863 and 1864) to the Linnean Society of London in 1864 (see *Correspondence* vols. 11–13).

[4] A natural history society was formed in Port Elizabeth in 1866; meetings were held in a building that also housed the Town Hall, Library, and Athenaeum Society (personal communication, Margaret Harradine, Port Elizabeth Municipal Libraries). Bedford lies inland, almost directly north of Port Elizabeth, which is on Algoa Bay, on the south-eastern coast of Cape Colony. The area near Algoa Bay was settled by the British in 1820 (*EB*).

[5] In the manuscript of Weale 1867 (see n. 1, above), Weale proposed the specific name 'Darwinii' for the *Bonatea* orchid, in honour of the 'great naturalist' from whose works he had 'derived so much enlightenment & incitement to prosecute the study of living beings'.

[6] In a letter that has not been found, Weale evidently wrote to CD about the family Asclepiadaceae; CD replied that there was still much to be discovered regarding its pollination, and that he once saw a hymenopter from America with its tarsi covered with pollen from an *Asclepias* (see *Correspondence* vol. 13, letter to J. P. M. Weale, 6 May [1865] and nn. 5 and 6). Weale later published a paper on pollination in the Asclepiadaceae (Weale 1870).

[7] The moth that Weale refers to is an unidentified species of the family Psychidae, whose members are now commonly known as bagworms. CD referred to Weale and his observation of the caterpillars' cases 'on the mimosas in South Africa' in *Descent* 1: 416. Weale later published a paper on the thorn and foliage imitators found on *Acacia horrida* (Weale 1878); he mentioned the caterpillar he had described to CD, without giving its scientific name, on page 185.

[8] Trimen had recently completed publication of his three-volume work on South African butterflies, *Rhopalocera Africæ Australis* (Trimen 1862–6). Weale later published on the Rhopalocera; see Weale 1877.

[9] In March and April 1867, Trimen visited Natal, collecting insects along the South Coast and as far inland as Pietermaritzburg and Noodsberg (Gunn and Codd 1981, p. 351).

[10] Weale refers to the accounts by John Barrow and William John Burchell of their respective travels in South Africa (J. Barrow 1801–4 and Burchell 1822–4). Barrow mentioned 'Bushman' caves and

rock drawings on the east side of the Tarka Mountains, north of Algoa Bay (J. Barrow 1801–4, 1: 239–41, 307–8). For eighteenth and nineteenth-century European perceptions of indigenous South Africans, and contemporary meanings of the term 'Bushman', see Dubow 1995, pp. 20–32; the peoples known to many nineteenth-century Europeans as 'Bushmen' are now considered to be members of the Khoisan peoples. Barrow also described the Great Fish River and some adjacent hot springs (J. Barrow 1801–4, 1: 184–91, 307–8); the Great Fish River drains into the Indian Ocean north-east of Port Elizabeth. In an 1865 letter written to the *Natural History Review* (Weale 1865), Weale also mentioned his intention of visiting some caves (see also *Correspondence* vol. 13, letter to J. P. M. Weale, 6 May [1865] and n. 7).

[11] In Weale 1865, Weale had mentioned that he was seeking a 'Bushman's' cranium for Thomas Henry Huxley; Huxley had been embroiled in a dispute (the 'hippocampus controversy') with Richard Owen regarding structures in the brains of humans and the higher apes (see A. Desmond 1982, Rupke 1994, and *Correspondence* vols. 8–11). For European interest in South African anthropology in the nineteenth century, see Dubow 1995, pp. 32–4. For Weale's attitude towards Africans, see Shanafelt 2003. For the general attitudes of Victorian colonists towards the people inhabiting the land they settled, see, for example, Stocking 1987, pp. 144–237.

[12] Trimen did travel to England later in 1867 (see letter to Roland Trimen, 24 December [1867]), returning to South Africa the following year (see *Correspondence* vol. 16, letter from Roland Trimen, 13 April 1868). Judging from Weale's 1867 correspondence with CD, it is unlikely that Weale also travelled to England in 1867.

[13] Weale's letter of 5 January 1867 appeared in the 10 January 1867 issue of the *Great Eastern*, a newspaper published in Grahamstown. Another letter of his of 12 January 1867 appeared in the 19 January 1867 issue. See n. 14, below. It is not known whether he sent either letter to CD.

[14] Weale refers to John Croumbie Brown, who had returned to the United Kingdom late in 1866 after his appointment as colonial botanist was terminated, evidently owing to budget constraints. Brown's parting letter in the 1 January 1867 issue of the *Great Eastern* stressed his work on 'the natural history, botanical characters, and economic uses of the trees, shrubs, and economic herbs of Southern Africa'; he also noted the minimal expense to the government required for his work and associated travel. According to a testimonial and a notice in the *Great Eastern*, 8 January 1867, pp. 2 and 3, the termination of Brown's appointment was controversial. Weale's two letters to the *Great Eastern* (see n. 13, above) were intensely critical of Brown's botanical knowledge and skill. Weale also accused Brown of preaching rather than collecting specimens, and of extravagant spending.

[15] Brown promoted the study of botany while in South Africa as colonial botanist; however, his own publications during this time were limited to government reports, in which he dealt with agricultural problems, forestry, and soil erosion. He later published books and papers primarily on forestry (A. C. Brown 1977, p. 465, Gunn and Codd 1981). Blue books contain the official publications of the British Parliament and Privy Council, and the term is extended to other official publications (*OED*). Weale criticised Brown's blue books in the *Great Eastern* of 10 and 19 January 1867 (see n. 14, above).

To John Murray 10 January [1867]

Down. | *Bromley.* | *Kent. S.E.*

Jan. 10th

My dear Sir

Your note has been a great relief to me; & I am very glad that you agree about type.— I will insert Wood-cuts.[1]

I am rather alarmed about the verdict of your friend, as he is not a man of science.— I think if you had sent the Origin to an unscientific man, he w[d]. have utterly condemned it. I am,, however, **very glad** that you have consulted anyone, on whom you can rely.—[2]

I must add that my Journal of Researches was seen in M.S by an eminent semi-scientific man, & was pronounced unfit for publication.[3] Let me hear at once as soon as I may send to *your* house (ie Albemarle St) for the M.S.; as I much wish to begin printing, & I will return to Me^ss Clowes the few first chapters in a day or two.—[4]

Yours very sincerely | Ch. Darwin

Look at Athenæum at my letter about "Cut Books" & "Eminent publisher".—[5]

John Murray Archive
Endorsement: '1867. Jany 10'

[1] CD refers to the type sizes and woodcuts for *Variation* (see letter from John Murray, 9 January [1867]).
[2] Murray had not identified the reader of *Variation*, John Milton, to CD (see letter from John Murray, 9 January [1867] and n. 4). CD also refers to *Origin*.
[3] No reader of the manuscript of CD's *Journal and remarks*, the third volume of the record of the *Beagle* voyage (*Narrative*), has been identified; it was published in 1839. Robert FitzRoy read the proof-sheets (see Browne 1995, p. 414). For CD's writing of *Journal and remarks*, and for its separate publication as *Journal of researches*, see *Correspondence* vol. 2.
[4] CD refers to Murray's publishing house, at 50A Albemarle Street, London, and to the printer William Clowes & Sons (see letter to John Murray, 8 January [1867] and n. 5).
[5] In his letter to the *Athenæum* of 1 January 1867, CD urged that books be published with pages already cut. The 'eminent publisher' mentioned in his letter, who had told him that booksellers would object to receiving books cut, was John Murray (see *Correspondence* vol. 14, letter from John Murray, 18 July [1866]).

From Bartholomew James Sulivan 11 January 1867

Bournemouth
Jany. 11 | 67.

My dear Darwin

I went to Southampton to see M^r. Stirling off,[1] and on giving him your paper he reminded me that I gave him a somewhat similar one from you before[2]—and from his and our catechists notes he had written some answers for you, but they were so incomplete that he did not think them worth sending. On searching his desk he found the questions & answers written by M^r. Bridges which I now send you.[3] He will look out again for the points you mention & ask M^r. Bridges the catechist to do the same.

Your last question about the cattle I think I can answer.[4] I believe the calves shew their colour from the first, the white cows had white calves. The cow that got me down once after I killed her calf was white with black head—& her calf was just like her. it was about two months old. I recollect in the South—where the cattle are nearly all white—after killing a white cow, a little calf not more than three or four days old ran alongside us to the boat; it was quite white or nearly so—[5]

when at Southampton I recollected your eldest son had settled there, so I paid him a visit at his Bank.[6] I should have known him, though I had not quite recollected how many years have passed, & therefore was surprised to see him look older than I expected.

I hope when he wants a holiday he will run down here for a few days.

With kind regards | Believe me | very sin^ly yours | B. J. Sulivan

DAR 177: 288

[1] Waite Hockin Stirling was returning to the Falkland Islands, where he would resume the post of missionary to Tierra del Fuego, based at the mission station on Keppel Island (Hazlewood 2000). Sulivan became acquainted with Stirling in 1857, when Sulivan was head of the Marine Department of the Board of Trade and Stirling was secretary of the Patagonian Mission Society (Sulivan ed. 1896, p. 387).

[2] In his letter of 25 December 1866 (*Correspondence* vol. 14), Sulivan had written that he had gone to see their 'Mission schooner' preparing to go to sea, and had mentioned Stirling (see n. 1, above). CD asked Sulivan whether he would ask Stirling to observe the expressions of Fuegians, and enclosed a list of questions in his letter of 31 December [1866] (*ibid.*). The enclosure has not been found, but the questions were probably similar to the queries about expression sent to several correspondents in February 1867. For CD's earlier questions, see *Correspondence* vol. 8, letter to Thomas Bridges, 6 January 1860.

[3] The replies to the queries on human expression and the breeding of animals that CD posed in the letter to Thomas Bridges, 6 January 1860, are published in *Correspondence* vol. 8, in the letter from Thomas Bridges, [October 1860 or after]. At the end of that letter, CD wrote 'Information from Mr Bridges, Catechist to Fuegian Mission, through Mr Stirling—'. Thomas Bridges had been a missionary based on Keppel Island in the Falklands since 1856 (see E. L. Bridges 1948).

[4] In the list of questions sent in the letter to Bridges of 6 January 1860 (*Correspondence* vol. 8), CD had asked: 'What colour are the calves of the wild *White* cattle with red ears, in the Falkland Islands?'; Bridges' reply, dated [October 1860 or after], stated that he could not answer the question. Sulivan had resided on the Falkland Islands from 1848 to 1851, and had previously surveyed them (*DNB*).

[5] CD did not publish Sulivan's comment on the Falkland Islands calves. However, he did publish in *Variation* Sulivan's observations of the colour of Falklands cattle that Sulivan sent in his letter of [10 May 1843] (*Correspondence* vol. 2; see *Variation* 1: 86).

[6] William Erasmus Darwin, CD's eldest son, was a partner in the Southampton and Hampshire Bank.

From Thomas Belt 12 January 1867

1 Ridley Terrace | New Castle on Tyne
January 12. 1867
⟨*rest of first page excised*⟩

will enclose it.

With sincere respect | I am Dear Sir

⟨*rest of second page excised*⟩[1]

[Enclosure]

On Esculent Fruits.

The intimate relations existing between the fauna and flora of a country and the dependence of each upon the other are strikingly shewn in the provision of esculent fruits.

These may be divided into two classes— 1st—those in which the seed is surrounded by or attached to a fleshy or juicy pulp more or less palatable and nutritious and 2nd those in which the seed itself is eaten, as food—

The first class may be subdivided into those where the seeds are enclosed in a juicy pulp and pass uninjured through the bodies of the animals eating them and into those where the seeds are attached to or surrounded by a pulpy or fleshy fruit but are not themselves swallowed

To the first of these subdivisions belong nearly all berries. I will take my illustration of the relations subsisting between them and the animals of a country from Nova Scotia where there is a succession of small fruits throughout the year including the long and severe winter[2]

First the wild strawberry ripens in June and for about a month is abundant throughout the province As they begin to fail the ripe raspberries appear and are even more abundant. They last until the beginning of Autumn, when a number of berries resembling the English whortleberry ripen and continue until the winter frosts set in— At the end of Autumn the Tea berry (Coltheria procumbens) ripens its red berries.[3] Throughout the long severe winter they may be found by scraping off the snow: their bright red colour making them conspicuous. I have observed them far into the spring when the strawberries were ripening again.

Thus a fare of small berries is provided throughout the year for the small birds & squirrels. At first sight this appears to be entirely for the benefit of the animals but in reality the first intention is the benefit of the species providing the fruits.

In all these berries the seeds are small and hard cased and pass uninjured through the bodies of the birds and are distributed about by them— The pulpy mass surrounding the seeds is a provision to ensure their dissemination. It attains the same object as the hooks on the fruit of the "burs" or the pappus on that of the compositae[4]

Some considerations follow from this view. A slow improvement of the fruits ought to be brought about for the largest and finest fruits will be eaten in preference to inferior ones. The plant that produced the finest fruits would hold out the greatest premium for the dissemination of its seeds—[5]

The succession of berries and the ripening of some at an apparently unpropitious season of the year is easily explained— It is manifestly to the interest of an inferior fruit not to come into competition with a superior one The raspberry has a better chance of being eaten *after* the strawberry has done fruiting. The tea berry is not a palatable fruit and if it appeared in the Summer or Autumn would be neglected for the more luscious berries then ripening. But in the Winter it has no competitors and by offering itself when there is little else to get it attains the same object as the more attractive fruits in the Summer and Autumn

To the second subdivision, including these fruits whose seeds are *not* swallowed, belongs the apple, Orange Peach and most of the large fruits— A study of any of these will shew how admirably the distribution of the seed is ensured. The fruit of the Cashew in Brazil is a good illustration.[6] It is formed by an enlargement of the seed stalk and is about the size of an apple   The seed is bean shaped and is attached to the end of the fruit by fibres that run through the pulpy mass—

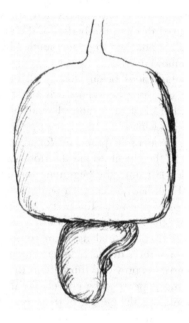

It forms a natural handle to hold the fruit by. To prevent its being eaten itself, it contains underneath the first skin a most acrid oil that ensures its being thrown away as soon as the fruit is consumed— The provision of a handle to a fruit could only be of use in a country where the animals *hired* to disseminate the seeds had hands. The monkies are I believe the great distributors of these & other Brazilian seeds. Some of them carry away with them as many of the fruits as they can manage—

Many of the Brazilian palms have their hard nuts surrounded by a fleshy pulp to entice animals to carry them away but the cocoa nut palm thriving only on the sea shore and depending for the distribution of its seeds on their being floated away has its nuts covered with a light, bulky envelope.[7]

In the second division of fruits the seeds are themselves eaten   In many cases this must be prejudicial to the species and there are many expedients to prevent it

The oak the beech and the chestnut are probably benefited by their seeds being buried by squirrels— If only one seed out of a hundred escapes being eaten, it will be amply sufficient for the welfare of the species.

Many small seeds are unintentionally swallowed by herbivorous animals and the plants producing them are thus distributed   The seeds of the common white clover are thus being rapidly spread over Australia.

AL incomplete
DAR 47: 181–9

CD ANNOTATIONS
Top of letter: 'Thomas Belt | Has visited Brazil'
Enclosure:
10.1 The succession ... ripening. 10.6] *scored pencil*
11.1 fruits ... swallowed,] *scored pencil*; 'Oranges are carried away (Rengger) & peeled by monkeys.'[8]
Sketch of cashew: 'Handles for Monkeys' *pencil*
12.1 Many of ... pulp] *scored pencil*
12.4 has its ... envelope.] *scored pencil*
15.1 Many small ... Australia. 15.3] *scored pencil*
Bottom of enclosure: 'On the distribution of Seeds by Animals, Th. Belt | (good)'

[1] The extant portion of the letter has been pasted onto the last sheet of the enclosure.
[2] Belt travelled in 1863 to Nova Scotia, where he worked as superintendent of the Nova-Scotian Gold Company's mines; he remained there for two or three years (Belt 1888, pp. xvii–xviii).
[3] Belt refers to *Gaultheria procumbens*, also known as wintergreen.
[4] Burrs are prickly seedcases or flower-heads that easily stick to passing animals. The pappus is a ring of fine hairs attached to achenes or fruits of members of the Compositae; these aid in dissemination by wind.
[5] CD discussed the dispersal of seeds by birds in *Origin*, pp. 356–65, in the context of geographical distribution; he mentioned colourful seeds in relation to the idea of beauty in *Origin* 4th ed., p. 240.
[6] In 1866, Belt spent three months in the State of Maranhão, Brazil (*Transactions of the Entomological Society of London* 5 (1865–7): p. lxxxix).
[7] See *Origin*, p. 360.
[8] CD refers to a statement in Johann Rudolph Rengger's *Naturgeschichte der Saeugethiere von Paraguay* (Rengger 1830). On page 39 of CD's heavily annotated copy in the Darwin Library–CUL, an annotation next to a discussion of monkeys reads: 'Beat the oranges to losen rind' (see *Marginalia* 1: 699). See *Descent* 1: 139–40.

From J. D. Hooker [12 January 1867][1]

Kew
Saturday.

Dear Darwin

A thousand thanks for your criticisms—[2] It is not merely that there are no Boreal or Arctic plants in the Mts of Canaries &c, but no gentians, or any of the Alpine Spanish plants as Cruciferæ, Alsineæ, Compositæ &c &c—at least in so far as I believe. It is true there are species of Spartiums &c, which do not grow low down, but they are shrubs &c & do not seem to represent an *alpine* vegetation. But I must go through Webbs book & tabulate the altitudes &c.[3]

2. I tried hard so to arrange the sentence about savages in Islands, as that it should obviously not apply to Madeira & the Azores &c. I have not a copy by me, of the G. C., & fear I have failed somehow.[4]

Pray go on with such criticisms, they will be most useful

3 Of the groups of Islands with Insects I particularly remember Kerguelen's land, with 5 insects one only winged, if I remember aright, (Dipterous), the moth was apterous and had rudimentary wings.— Ross. says only 5 insects, & 2 winged.[5]

In Auckland & Campbell's Island again, winged Insects were very rare, Excepts
>   Diptera—(blood-suckers)
>   Fuegia ditto.
>   Falkland I forget—
>   Ascension—little but crickets[6]

Yes Annuals are certainly best adapted for short seasons, & they do abound in
cultivated ground *in the* more equable climates[7]—but there are lots in the unculti-
vated districts of Australia, Asia Minor Levant—N. Africa & California, & it would
not be easy to define their season as *short* Whereas in Arctic regions, as I have
somewhere remarked, there are none or next to none.—& in alpine regions there
are very few indeed.

Yes—humid season implies equability, with which an evergreen vegetation is
closely connected.[8]

I do remember some passages between us anent Bees & clover in N. Zealand
& I *don't doubt I was quite right; in screaming at you at the time:*[9] indeed I cannot doubt
it—. I could not have done so, you see, if you had not been wrong. Owen & the
Bp of Oxford[10] have accepted this, however, & so I do now believe in Bees &
Clover, but not because you said it!— What I want to know now is, whether you
have ever suggested to me that the rarity of irregular flowered plants in general or
papilionaceæ in particular in Islands, was due to the rarity of winged insects. In
plain truth I feel that I have begged borrowed & stolen such a lot from you, that
my "meum & tuum" may well be vaguely limited

I must confess that I was (Fanny has just had a fine boy, excuse the interruption)[11]
not surprized that your new book should require more space & be much bigger
than the origin, & I think it is well that it should have a different form, too.[12] As to
the size of the book being out of proportion to the subject, I do not see how that
can well be— surely domesticated animals alone would fill a large volume under
your treatment, & plants a *larger* even. This however is no reason why you should
not swear at yourself & other book writers too— it can do no harm to yourself &
may do great good to the latter

Mrs Hooker has presented me with a fine boy since this letter was begun, & is
doing well

>   Ever yrs aff | J D Hooker
>   *Plumbago* not forgotten.[13]

DAR 102: 131–4

CD ANNOTATION[14]

End of letter: 'Orig Letter— Mʳˢ H.– Violet like Pyrenees | I did suggest [*above del* 'ask'] about *proportion
of [*interl*] irregular flowers in Isls—after ['writing that letter' *del*] giving a very short discussion on
bearing of such flowers as Lythrum— But what on earth does my asking such a question signify—
it has no relation to meum & tuum— You have comforted me a little about bigness of my Book' *ink*

---

[1] The date is established by the relationship between this letter and the letters to J. D. Hooker, 9 January
[1867] and 15 January [1867]. In 1867, the Saturday following 9 January was 12 January.

[2] Hooker refers to CD's criticisms of the first part of his article on insular floras (J. D. Hooker 1866a; see letter to J. D. Hooker, 9 January [1867]).

[3] See letter to J. D. Hooker, 9 January [1867] and n. 3. Hooker refers to Webb and Berthelot 1836–50; the third volume covered the phytogeography of the Canary Islands. For Hooker's comments on this topic in 1866, see *Correspondence* vol. 14, letters from J. D. Hooker, [24 July 1866] and 31 July 1866.

[4] See letter to J. D. Hooker, 9 January [1867] and n. 4. J. D. Hooker 1866a was published in the *Gardeners' Chronicle.*

[5] See letter to J. D. Hooker, 9 January [1867] and n. 5. Hooker is recalling his observations made on Kerguelen's Land during his 1839 to 1843 Antarctic voyage on the *Erebus* and *Terror* with James Clark Ross (see J. D. Hooker 1844–7 and R. Desmond 1999). In *A voyage of discovery and research in the southern and Antarctic regions, during the years 1839–43* (Ross 1847, 1: 90), Ross noted several insects on Kerguelen's Land including a '*curculio*', a small brownish moth, and two flies; he thought, however, there would be more insects in the summer. For earlier correspondence between CD and Hooker on apterous insects, including a moth, on Kerguelen's Land, see *Correspondence* vol. 5, letter from J. D. Hooker, [before 17 March 1855] and nn. 3 and 4, and *Correspondence* vol. 14, letter from J. D. Hooker, [24 July 1866] and n. 15; see also *Natural selection*, p. 292.

[6] Hooker is evidently recalling informal observations of insects made on the Lord Auckland Islands, Campbell's Island, Hermit Island of Tierra del Fuego, the Falkland Islands, and Ascension Island during his Antarctic voyage; for Hooker's botanical observations on these islands, see J. D. Hooker 1844–7 and R. Desmond 1999.

[7] See letter to J. D. Hooker, 9 January [1867] and n. 7.

[8] See letter to J. D. Hooker, 9 January [1867] and n. 8.

[9] See letter to J. D. Hooker, 9 January [1867] and nn. 9 and 10.

[10] Richard Owen and Samuel Wilberforce.

[11] Hooker refers to his wife, Frances Harriet Hooker, and his newborn son, Reginald Hawthorn Hooker (Allan 1967).

[12] Hooker refers to *Variation* (see letter to J. D. Hooker, 9 January [1867] and n. 11).

[13] See letter to J. D. Hooker, 9 January [1867] and n. 12.

[14] CD's annotations are notes for his letter to Hooker of 15 January [1867].

From Robert Monsey Rolfe 14 January 1867

Holwood, | *Bromley.* | *S.E.*
14 Jany 1867

My dear Mr Darwin

M^r & Mrs Charles Kingsley are coming to us tomorrow— Lady Cranworth tells me that you thought of coming over here to see him on Wednesday, but it occurs to her as well as to me that in this severe weather it may be more agreeable to you that we should bring Mr Kingsley over to be introduced to you—[1] Pray send me a line by post to say what you w^d wish—[2] He is obliged to go into town on Thursday to give a lecture at the Royal Institution & they leave us on Saturday—[3] so that Wednesday w^d probably be the best day—or, if not, then Friday

yours very truly | Cranworth

DAR 161: 235

[1] Charles Kingsley had been an occasional correspondent of CD's since 1859 (see *Correspondence* vols. 7, 10, 13, and 14), and they may have met in 1854 at a meeting of the Linnean Society (see Colloms 1975, p. 185). Rolfe also refers to Frances Eliza Kingsley. There is no extant correspondence between

CD and Lady Cranworth (Laura Rolfe); however, in a letter dated [15 September 1866], Emma Darwin wrote to Henrietta Emma Darwin that Lady Cranworth had just called and expressed how much Kingsley would like to see CD (DAR 219.5: 45). Rolfe's residence, Holwood Park, was $1\frac{1}{2}$ miles north of Down House.

[2] No further correspondence has been found regarding a possible visit.

[3] Kingsley was to give three lectures at the Royal Institution, on Tuesday 15, Thursday 17, and Saturday 19 January 1867 (*Proceedings of the Royal Institution of Great Britain* 5 (1866–9): 4).

To Thomas Belt 15 January [1867][1]

Down
Jan. 15

Dear Sir

I am extremely much obliged to you for so kindly sending me your paper on the distribution of seeds.[2] It has interested me much. I hope that you will not be prevented from publishing on subject, if so inclined, by having sent me the MS., for it will be impossible for me to use your facts for some time.[3] Some of the points had occurred to me, and I have just alluded to the subject in a new edition of my Origin of Species.[4] The successive seasons is new to me. The case of the cashew fruit is most novel and curious. By an odd coincidence I have within last few months received several letters from S. Brazil on other and rather different cases of the apparent dispersal of seeds in that country.[5] I have always imagined that wild pigs and other large mammals might disperse seeds of apples and peaches; but I know not whether apple seeds would be digested.

With very sincere thanks for your great kindness, I remain Dear Sir | Yours faithfully & obliged | Ch. Darwin.

Copy
DAR 143: 76

[1] The year is established by the relationship between this letter and the letter from Thomas Belt, 12 January 1867.

[2] Enclosure to letter from Thomas Belt, 12 January 1867.

[3] Most of Belt's articles were on geological topics (*Royal Society catalogue of scientific papers*); no paper by him on esculent fruits has been found.

[4] See *Origin* 4th ed., p. 240.

[5] CD refers to his correspondence with Fritz Müller in Destêrro (now Florianópolis, Brazil); see *Correspondence* vol. 14, letters from Fritz Müller, 2 August 1866, and 1 and 3 October 1866. See also *Correspondence* vol. 14, letter from Fritz Müller, 1 December 1866, and this volume, letter from Fritz Müller, 1 January 1867.

To J. D. Hooker 15 January [1867]

Down
Jan 15

My dear Hooker

Thanks for your jolly letter.[1] We are both heartily rejoiced, & this not in a parenthesis, that Mrs Hooker is safe through her affair.[2] I have read your second article, & like it even more than the first, & more than this I cannot say.[3] By mere chance I stumbled yesterday on a passage in Humboldt that a violet grows on Peak of Teneriffe in common with the Pyrenees.[4] If Humboldt is right that the Canary I's which lie nearest to the continent have a much stronger African character than the others ought you not just to allude to this?[5] I do not know whether you admit, & if so allude to, the view which seems to me probable that most of the genera confined to the Atlantic I.s, I do not say the species, originally existed on, & were derived from, Europe, having become extinct on this continent. I sh^d thus account for the community of peculiar genera in the several Atlantic I's.[6] About the Salvages is capital; I am glad you speak of *linking*, though this sounds a little too close, instead of being continuous.[7] All about St Helena is grand. You have no faith, but if I knew any one who lived in St Helena, I w^d supplicate him to send me home a cask or two of earth from a few inches beneath the surface from the upper parts of the I., & from any little dried up pond, & thus as sure as I am a wriggler I sh^d revive a multitude of lost plants.[8]

I did suggest to you to work out proportion of plants with irregular flowers on islands; I did this after giving a very short discussion on irregular flowers in my Lythrum paper. But what on earth has a mere suggestion like this to do with meum & tuum?[9]

You have comforted me much about the bigness of my book, which yet turns me sick when I think of it.[10]

yours affectionately | Ch. Darwin | (Signed, whilst my wife writes)

Dear Dr Hooker

We shall be anxious to hear of Mrs Hooker going on comfortably.[11] In my experience I used to be very flourishing for a few days & then not quite so well. Give her my kind love & congrats.

yours very sincerely | E. D

Endorsement: '67/'
LS(A)
DAR 94: 5-6

[1] Letter from J. D. Hooker, [12 January 1867].
[2] CD refers to Frances Harriet Hooker and the birth of her son, Reginald Hawthorn Hooker; Hooker announced the event in a parenthetical statement in his letter of [12 January 1867].
[3] The second of the four parts of Hooker's article on insular floras (J. D. Hooker 1866a) appeared in the *Gardeners' Chronicle*, 12 January 1867, p. 27. See letter to J. D. Hooker, 9 January [1867], n. 1.
[4] Alexander von Humboldt wrote that he had seen *Viola cheiranthifolia*, which resembled *V. decumbens*, growing on the Peak of Tenerife (Humboldt 1814–29, 1: 183 and 273); the violet, commonly known as the Teide violet, is found only in Tenerife at altitudes up to 3500 m. CD evidently thought that the *V. decumbens* that Humboldt referred to was a Pyrenean species. For CD's earlier reference to alpine plants on Tenerife, see the letter to J. D. Hooker, 9 January [1867].
[5] Humboldt remarked that Lanzarote and Fuerteventura, the two Canary islands lying closest to Africa, resembled Africa in climate and vegetation more than the other islands did (Humboldt 1814–29, 1: 123, 274). In his paper, Hooker wrote that though the Canary Islands were closer to Africa than Madeira was, the flora of the Canary Islands was not African, and contained 'comparatively very

few of the plants of that continent'; he also wrote that the least numerous group of plants on the Canaries was 'a sprinkling of African plants' belonging to a division he called Arabo-Saharan (J. D. Hooker 1866a, p. 27).

[6] See *Origin*, pp. 397–9, 403–6. See also *Correspondence* vol. 14, letters to J. D. Hooker, 30 July [1866] and 3 and 4 August [1866]. In the margin of his copy of J. D. Hooker 1866a, p. 27 (Cory Library, Cambridge Botanic Garden), CD wrote, 'I suppose you look at whole Atlantic ['Flora' *del*] genera as having been common to Europe', and '& most on is'd from continents'.

[7] Hooker wrote that the Salvages, a group of rocky Atlantic islets, supported an Atlantic flora intermediate between that of Madeira and the Canaries, and surmised that, before subsiding, the islands had occupied an important botanical and geographical position in the Atlantic Ocean, 'more or less closely linking the Canaries with Madeira' (J. D. Hooker 1866a, p. 27). See *Origin*, p. 410.

[8] Hooker described the destruction of the indigenous flora on St Helena, first by the introduction of goats and then by the introduction of exotic plants (J. D. Hooker 1866a, p. 27). He attempted to reconstruct the original flora by collating two herbarium collections, and cautiously surmised that it had primarily African affinities, with some Indian and American plants, and was therefore unique and interesting. Hooker noted that probably 100 plants had disappeared (*ibid.*):

> Every one of these was a link in the chain of created beings, which contained within itself evidence of the affinities of other species, both living and extinct, but which evidence is now irrecoverably lost. If such be the fate of organisms that lived in our day, what folly it must be to found theories on the assumed perfection of a geological record which has witnessed revolutions in the vegetation of the globe, to which that of the Flora of St Helena is as nothing.

CD alludes to his conviction of the long-term viability of seeds; for Hooker's initial scepticism and guarded change of mind, see *Correspondence* vol. 5, letter to *Gardeners' Chronicle*, 13 November [1855], and letters to J. D. Hooker, 14 November [1855], [23 November 1855], and 9 [December 1855] and nn. 1 and 3. For CD's belief in the viability of seeds in pond mud and seawater, see *Origin*, pp. 386–8, 358–60.

[9] See letter from J. D. Hooker, [12 January 1867]. In his letter to Hooker of 7 January [1865] (*Correspondence* vol. 13), CD suggested that Hooker compare the number of plants with irregular flowers in New Zealand with the number in England. CD briefly mentioned his view of the relationship between insect pollination and the structure of irregular flowers in 'Three forms of *Lythrum salicaria*', pp. 175–6 (*Collected papers* 2: 111–12).

[10] CD refers to *Variation* (see letter from J. D. Hooker, [12 January 1867]).

[11] Emma Darwin wrote the postscript, inquiring after Frances Harriet Hooker (see n. 2, above).

To John Murray 15 January [1867]

Down.

Jan 15th

My dear Sir

The type seems to me excellent & I will force a good deal into the smaller size.[1] The page, as I understood, will be as large as in Lyell's new Edit. of Principles.—[2]

I am rather pleased to hear from D[r] Hooker that he is not surprised at my book being big.—[3]

My dear Sir | Yours sincerely | Ch. Darwin—

John Murray Archive
Endorsement: '1867. Jany 15.'

[1] CD had evidently received from Murray, or the printer William Clowes & Sons, samples of the smaller print size for which CD would mark the more technical portions of the manuscript of *Variation* (see

letters to John Murray, 8 January [1867] and 10 January [1867], and letter from John Murray, 9 January [1867]). No letter from Murray or Clowes including a sample of the print size has been found.
² Murray had mentioned that because of the length of *Variation*, the pages would have to be the same size as in the tenth edition of Charles Lyell's *Principles of geology* (C. Lyell 1867–8); see letter from John Murray, 2 January [1867].
³ Joseph Dalton Hooker had reassured CD about the size of *Variation* in his letter of [12 January 1867].

To B. J. Sulivan 15 January [1867]¹

Down Bromley | Kent.
Jan 15

My dear Sulivan

I am extremely much obliged to you for getting me Mʳ Sterling's answers. They are of much interest to me. I had quite forgotten that I had previously sent nearly the same questions:² the answer about the breeding of the dogs comes in very useful just now that I am writing upon domestic animals.³ Many thanks also about the white cattle.⁴

I am very glad you were so kind as to call on my son William; he does not have many holidays, but if he cᵈ spare the time I am sure he wᵈ enjoy seeing you at Bournemouth.⁵

My dear Sulivan, | yours very sincerely | Ch. Darwin

LS
E. Sulivan (private collection)

¹ The year is established by the relationship between this letter and the letter from B. J. Sulivan, 11 January 1867.
² In his letter of 11 January 1867, Sulivan enclosed responses from Thomas Bridges regarding the mannerisms and facial expressions of Fuegians, conveyed by Waite Hockin Stirling; these responses are published in *Correspondence* vol. 8, letter from Thomas Bridges, [October 1860 or after]. CD cited information from Bridges at least four times in *Expression*, and in *Descent* 2: 351 n. For the queries that CD had forgotten he had sent, see *Correspondence* vol. 8, letter to Thomas Bridges, 6 January 1860; see also, this volume, letter from B. J. Sulivan, 11 January 1867 and nn. 2 and 3. The latest list of queries that CD sent to Sulivan has not been found.
³ The section on breeding dogs is in a missing section of Bridges's reply (see *Correspondence* vol. 8, letter from Thomas Bridges, [October 1860 or after]); however, see *Variation* 2: 207 for the information that Bridges sent. In his letter, Bridges also wrote that, to make their dogs lighter, the Fuegians cut off their tails while they were puppies; CD mentioned the remark in *Variation* 1: 39.
⁴ See letter from B. J. Sulivan, 11 January 1867 and nn. 4 and 5.
⁵ CD refers to William Erasmus Darwin; see letter from B. J. Sulivan, 11 January 1867 and n. 6.

To William Turner 15 January [1867]¹

Down. | Bromley. | Kent. S.E.
Jan. 15ᵗʰ

My dear Sir

As you were so kind as to say that I might ask you a few more questions, & as my wishes are now rather more definite, I do so now;² but you must not suppose that I am in any hurry for an answer.

(1) One or two good cases of any rudiment of a muscle, would suffice (I have lost one good case which I copied from some French writer): if any muscle in our arms exists in a rudimentary or nearly rudimentary condition, & which would be of service to a quadruped, going on all fours; such case would perhaps be best.[3]

(2) You reminded me that there were two sets of muscles for moving the whole ear & its parts; which of such muscles are rudimentary in human ear?[4]

(3) I have used your information (from Theile etc) about muscles to Os Coccyx:[5] if my memory does not deceive me, those 4 coccygeal bones contain spinal marrow at an early embryonic age, & afterwards it retreats. If this is so, are vestiges of membranes of spinal marrow retained?[6]

(4) I have given case of Prostate gland (to which you allude in one of your papers) as a rudiment or representation of the Uterus.[7] I am sure I have read (& hope to find the reference,) some recent observations of the existence of both testes & ovaria at early embryonic age in *both* sexes of the higher animals.[8] Do you know anything on this head?

(5) Is any other gland rudimentary in mankind, besides the mammary glands in male mammals?[9]

(6) I may add that I have alluded to traces of the supra-condyloid foramen in humerus of man;[10] & to the nictitating membrane. By the way do you chance to remember whether the nictitating membrane is well developed in Marsupials?[11]

Pray forgive me, if you can, for being so very troublesome, & believe me, my dear Sir | Yours sincerely obliged | Ch. Darwin

Edinburgh University Library, Dc.2.96/5 folio 2

[1] The year is established by the relationship between this letter and the letter to William Turner, 1 February [1867].

[2] CD met Turner at a reception at the Royal Society of London on 28 April 1866, and had since received papers from him (see *Correspondence* vol. 14). Turner and CD did correspond in 1866, but Turner's letters have not been found (*Correspondence* vol. 14).

[3] CD may be referring to a paper on rudimentary muscles in the foot and hand by Adolphe Richard (Richard 1852), which he cited in *Descent* 1: 19 n. CD had long been interested in rudimentary organs (see *Origin*, pp. 450–6). He was currently working on his 'Essay on man', which he was considering adding to *Variation*, but which ultimately became part of *Descent* and *Expression* (see letter to John Murray, 3 January [1867] and n. 5, and 'Journal', Appendix II).

[4] The letter containing Turner's statement on the muscles of the ear and Turner's reply to this letter have not been found. CD wrote in *Descent* 1: 20–1 that all the muscles moving the ear were in a rudimentary condition in humans; he did not cite Turner.

[5] In *Descent* 1: 29–30, CD wrote that he had been informed by Turner that one of the associated rudimentary muscles of the os coccyx had been described by Friedrich Wilhelm Theile 'as a rudimentary repetition of the extensor of the tail', which was 'so largely developed in many mammals'. Theile made this statement in Theile 1839. For CD's earlier queries to Turner on the coccyx, see *Correspondence* vol. 14, letter to William Turner, 14 December [1866]. CD had suspected for some time that muscles were attached to the os coccyx and that it was a rudimentary tail (see *Correspondence* vol. 7, letter to Charles Lyell, 11 October [1859]).

[6] CD mentioned the vestige of the spinal cord in the os coccyx in *Descent* 1: 30, citing Turner as the source for the information.

[7] CD referred to the male prostate gland as a rudiment or 'homologue' of the female uterus in *Descent* 1: 31, but cited Leuckart 1852, an article he had been alerted to in 1861 (see *Correspondence* vol. 9, letter from George Rolleston, 16 April 1861). CD cited Turner's article on the prostate gland (Turner 1865) in his chapter on development in *Descent* 1: 123–4. A lightly annotated copy of an offprint of Turner 1865 is in the Darwin Pamphlet Collection–CUL.

[8] CD mentioned in *Descent* 1: 207 that at an early embryonic period both sexes of vertebrates possess male and female reproductive glands, and discussed the possibility of a progenitor of the vertebrates having been hermaphrodite or androgynous; he cited Gegenbaur 1870 on this point.

[9] CD noted the rudimentary mammary glands in male mammals in *Descent* 1: 30–1 and 209, citing information from Turner on page 209.

[10] In *Descent* 1: 28, CD wrote that sometimes the trace of a passage near the lower end of the humerus (bone of upper arm) was apparent in humans; he stated that the fact that, when it was present, the 'great nerve' invariably passed through it indicated that it was 'the homologue and rudiment of the supra-condyloid foramen of the lower animals'; CD added that Turner estimated that it occurred in about one per cent of recent human skeletons.

[11] The rudiment of the nictitating membrane in humans was mentioned by CD in *Descent*, as was the fairly well-developed nictitating membrane in monotremes and marsupials (*Descent* 1: 23); however, CD did not cite Turner on these points. See also *Descent* 1: 207.

From John Lubbock 16 January 1867

High Elms
16 Jan/67

My dear Mʳ. Darwin

My brother in law Robert Birkbeck[1] is going to spend some months in Corsica, & he thinks that Murray might possibly like some notes with reference to a Handbook for the Island.[2]

He is anxious therefore for a note of introduction to him & I thought I might take the liberty of asking you to give him one.

If you would not mind adding, on my authority, that he is a man who will take pains to get accurate information, & whose statements would I think be thoroughly reliable, I should be very much obliged.

Hoping you are better again | I remain, dear Mʳ. Darwin, | Very sincerely yours | John Lubbock

C. Darwin Esq

DAR 170: 54

[1] Birkbeck was married to Lubbock's elder sister, Mary Harriet (Hutchinson 1914).
[2] John Murray was CD's publisher.

To John Lubbock 17 January [1867][1]

Down
Jan 17

My dear Lubbock

I have embodied what you say, in the enclosed note of introduction to Mr Murray & I very sincerely hope it may answer its purpose.[2]

We take this opportunity of returning with many thanks the beautiful book & the curious revolving machine which Lady Lubbock was so kind as to lend us.[3]

I was so very sorry not to see you the other day when you called & hear about some wonderful new animal. You must have a knack of creating new animals down there.[4]

yours very truly | Ch. Darwin

LS
DAR 263: 64 (English Heritage 8820 6508)

[1] The year is established by the relationship between this letter and the letter from John Lubbock, 16 January 1867.
[2] The enclosed letter to John Murray has not been found (see letter from John Lubbock, 16 January 1867).
[3] The book, and the machine lent by Ellen Frances Lubbock, have not been identified.
[4] CD may refer to the animal Lubbock reported on in his paper 'On *Pauropus*, a new type of centipede' at a meeting of the Linnean Society on 6 December 1866 (Lubbock 1866). Joseph Dalton Hooker had reported Lubbock's find in his letter of [12 December 1866] (*Correspondence* vol. 14).

From Julius Victor Carus 18 January 1867

My dear Sir,

I received your letter of Dec. 26 and altered the place accordingly.[1]

To-day I beg to bring three questions before you. P. 101 of the original you mention the rock-thrush of Guiana. Bronn had translated "Rupicola aurantia".[2] Thrush is Turdus; I find a "cock" of the rock indicated which would be for aught I know the Rupicola. Our European rock-thrush is Turdus saxatilis.[3] Is the species to which you allude a nearly allied form? At any rate is it a Turdus or a Rupicola?

P. 171. (at the bottom) you speak of the tortoise-shell-colour of cats.[4] Is this a motley fur with white, yellow and black? or else what colour?

P. 366. Mentioning the way by which the wings of the Pinguins might have been modified you give as an instance of a somewhat intermediate form the "logger-headed duck". Bronn left out the apposition and gives only ". . . . Ente", i.e. duck. I find in Linnaeus Syst. nat. ed. XIII. Anas cinerea, and here the vulgar name "logger-headed goose, Latham, syn. III. 2. p. 429."[5] Is that the species you mean? He gives the Falkland isles as habitat, so it may be. Could you give me the present name? I prefer giving the systematic term instead of the vulgar scarcely known with us and thereby leading to misapprehension.

Believe me | My dear Sir | Ever yours sincerely | J. Victor Carus

Leipzig, Jan. 18. | 1867

DAR 161: 56

[1] The letter has not been found. Carus and CD had been corresponding about Carus's revised translation of *Origin* based on the fourth English edition (Bronn and Carus 1867; see *Correspondence* vol. 14).

Two earlier German editions had been translated by Heinrich Georg Bronn (Bronn trans. 1860 and Bronn trans. 1863).

[2] See Bronn trans. 1863, p. 102. Bronn used the common name Felshahn and gave only the genus, *Rupicola*. *Rupicola aurantia* is now *R. rupicola*, the common name being Guianan cock-of-the-rock or, in German, Orange Felsenhahn or Klippenvogel.

[3] *Turdus saxatilis* is now *Monticola saxatilis*, the common name being rock thrush or, in German, Steinrötel.

[4] In *Origin* 3d ed., p. 162, CD had written, 'What can be more singular than the relation between blue eyes and deafness in cats, and the tortoise-shell colour with the female sex.' (The sentence was slightly modified in *Origin* 4th ed., p. 171.) Bronn had translated the phrase, 'and the tortoise-shell colour with the female sex' as 'oder die der Farbe des Panzers mit dem weiblichen Geschlechte der Schildkröten' (literally 'or the colour of the shell with the female sex in turtles'), having misunderstood the original (see Bronn trans. 1863, p. 171).

[5] See Bronn trans. 1863, p. 333. Carus refers to Carl von Linné, John Latham, and Linnaeus 1788–93.

To Alfred Newton 19 January [1867]

Down. | *Bromley.* | *Kent. S.E.*

Jan 19.

My dear Sir

Will you have the kindness to give me some information on one point? Not long since I was speaking to Mr Wallace about his mimetic butterflies,[1] & I told him of the case of the Rhynchœa, of which the female is more beautiful than the male, with the young resembling the latter.[2] He answered me that you at Nottingham had advanced this or some such case, & that you had simply explained it by the male being the incubator.[3] I should be extremely obliged if you w^d give me any information on this head & allow me to quote you. The subject interests me greatly, as in the 4th Edit. of the Origin I gave the obvious explanation of female birds not being gaudily coloured &c on account of their incubating;[4] I knew then of the Rhynchœa but passed over the case from not having space & from its appearing to me quite inexplicable.

I hope that you will forgive me troubling you & believe me my dear Sir | yours sincerely | Charles Darwin

P.S. As I am writing, I will ask one other question, for the chance of your being able to answer it: Does the male black Australian swan, or the black & white S. American Swans, differ from the females in *plumage*? ie in the intensity of the black, or in the amount of black in the black-necked species?[5]

LS(A)

Endorsement: 'Jan.^{y}. $\frac{19}{21}$ /67.' 'Answered Jany. 21/67.'

DAR 185: 87

[1] CD may have seen Alfred Russel Wallace when he was in London from 22 to 29 November 1866; see *Correspondence* vol. 14, Appendix II. See also *ibid.*, letter from A. R. Wallace, 19 November 1866.

[2] CD refers to a species of painted snipe, *Rhynchaea* (now *Rostratula*; see *Birds of the world*). CD discussed secondary sexual characteristics in three species of *Rhynchaea* in *Descent* 2: 202–3.

[3] Wallace and Newton attended the 1866 meeting of the British Association for the Advancement of Science in Nottingham (*Report of the thirty-sixth meeting of the British Association for the Advancement of Science*

1866). In *Descent* 2: 200, CD wrote that it was Wallace who 'first called attention to the singular relation
... between the less bright colours of the males and their performing the duties of incubation'; he
cited [Wallace] 1867a.

[4] See *Origin* 4th ed., p. 241.

[5] CD saw black-necked swans in South America while on the *Beagle* voyage (see *Journal of researches*,
pp. 133, 346).

From J. D. Hooker 20 January 1867

Kew

Jany 20/67

Dear Darwin

Prof. Miquel of Utrecht begs me to ask you for your Carte— & offers his in
return. I grieve to bother you on such a subject— I am sick & tired of this Carte
Correspondence.[1]

I cannot conceive what Humboldts Pyrenean violet is, no such is mentioned in
Webb, & no alpine one at all.[2]

I am sorry that I forgot to mention the stronger African affinity of the Eastern
Canary Islds.—[3] Thank you for mentioning it.

I cannot admit without further analysis, that most of the peculiar Atlantic Isld.
genera were derived from Europe & have since become extinct there. I have rather
thought that many are only altered forms of *Existing* European genera: but this is a
very difficult point & would require a careful study of each genus & allies with this
object in view—[4] the subject has often presented itself to me as a grand one for
analytic Botany. No doubt its establishment would account for the [community] of
the peculiar genera, on the several groups & Islets, but whilst so many species are
common we must allow for a good deal of intermigration of peculiar genera too

By Jove I will write out next mail to the Governor of St Helena for boxes of
earth; & you shall have them to grow.[5]

Thanks for telling me of having suggested to me the working out of proportions
of plants with irregular flowers in Islands;[6]—I thought it was a deuced deal too
good an idea to have arisen spontaneously in my block, though I did not recollect
your having done so no doubt your suggestion was crystallized in some corner of
my sensorium. I *should* like to work out the point.

My wife goes on well but has a horrid face-ache.— & Reginald blooms &
squeaks.[7]

This awful weather has terribly damaged us.[8]

Ever Yrs aff | J D Hooker

Have you Kerguelen land amongst your Volcanic Islds.—[9] I have a curious
book of a sealer who was wrecked on the Islands & who mentions a Volcanic Mt
& hot Springs at the S.W. end: it is called the "Wreck of the Favrite"[10]

DAR 102: 135–7

[1] Hooker refers to Friedrich Anton Wilhelm Miquel, professor of botany at Utrecht. Beginning in 1862,

it became commonplace for CD and his correspondents to request a carte-de-visite from one another (see *Correspondence* vols. 10–14).

[2] See letter to J. D. Hooker, 15 January [1867] and n. 4. Hooker refers to the *Histoire naturelle des Iles Canaries* (Webb and Berthelot 1836–50).

[3] See letter to J. D. Hooker, 15 January [1867] and n. 5.

[4] For CD's query on this point, see the letter to J. D. Hooker, 15 January [1867]. For Hooker's discussion in the third instalment of his article on insular floras, see J. D. Hooker 1866a, p. 50.

[5] See letter to J. D. Hooker, 15 January [1867] and n. 8. The governor of St Helena was Charles Elliot, who co-operated with Hooker in introducing plants to St Helena (Meliss 1875, p. 35).

[6] See letter to J. D. Hooker, 15 January [1867] and n. 9.

[7] Hooker's wife was Frances Harriet Hooker, and their newborn son was Reginald Hawthorn Hooker (see letter to J. D. Hooker, 15 January [1867]).

[8] Heavy snowfalls in January 1867 destroyed many trees at the Royal Botanic Gardens, Kew (R. Desmond 1995, p. 371).

[9] CD did not mention Kerguelen's Land in *Volcanic islands*.

[10] Hooker refers to John Nunn and to his *Narrative of the wreck of the 'Favorite' on the island of Desolation* (Nunn 1850). The description of the volcano and the hot springs is in Nunn 1850, pp. 103–4.

To Charles Henry Middleton 20 [1867?][1]

Down Bromley Kent
20th

Dear Sir

I am very sorry that I can give you no information. I cannot remember where S. Filippe is & have no good charts of coast of Chile.[2] Guano might be accumulated in extreme N. part of coast of Chile, but I never heard of any.—[3]

I confess that I cannot believe that the bones of the ox, sheep & Horse could have been deposited in Guano.—

Nevertheless the Bones would of course be worth examination.— I am sorry that I can give no information & beg leave to remain | Dear Sir | Yours very faithfully | C. Darwin

Christ's College Library, Cambridge (in Middleton's copy of *Origin* 4th ed., BB.5.6)

[1] The month and year are conjectured on the basis of the letter's being pasted into Middleton's copy of the fourth edition of *Origin*, which was published in November 1866; the year 1867 is on Middleton's book plate, which is inside the front cover.

[2] No letter from Middleton has been found. Middleton may have asked about San Felipe, an inland city north of Santiago, Chile; CD mentioned the city in *Journal of researches*, pp. 311, 316. There was also a sixteenth-century settlement on the south coast of the Strait of Magellan called San Felipe; it was later renamed Port Famine (Puerto del Hambre). CD mentioned Port Famine in *ibid.*, p. 264, and in *South America*, pp. 151–2, 156.

[3] Extensive deposits of guano in Peru, north of Chile, contributed to the mid-nineteenth-century guano trade between Peru and Britain, where the seabird excrement was used as fertiliser (see Mathew 1981). The climate of northern Chile is hot and dry, like that of Peru, and therefore conducive to guano deposition; guano was also exported from Chile (*EB*).

To J. D. Hooker 21 January [1867]

Down. | Bromley. | Kent. S.E.

Jan 21st

My dear H.

Four lines from bottom of 2^{d} column apparent bad misprint "commonest" for "rarest"[1]

Fourth column instead of "*oceanic*" fish, read "*fresh-water*" fish.—[2]

You give an *excellent* abstract of arguments in favour of occasional means of transport; even such a bigot, as I, could not possibly desire anything better, clearer or more favourable.[3]

Ever yours | C.D

Endorsement: '67'
DAR 94: 7

[1] CD refers to the third instalment of Hooker's article on insular floras (J. D. Hooker 1866a) in the *Gardeners' Chronicle*, 19 January 1867, pp. 50–1. See also the letters to J. D. Hooker, 9 January [1867] and 15 January [1867], for his comments on the first and second instalments. In summarising the peculiarities of island floras, Hooker had written: 'the plants having no affinity with those of the mother continent are often the commonest of all' (*ibid.*, p. 50). In the pamphlet Hooker published from J. D. Hooker 1866a (see letter to J. D. Hooker, 9 January [1867], n. 2), Hooker changed 'the commonest of all' to 'very common' (see Williamson 1984, p. 70).

[2] In summarising CD's arguments for the trans-oceanic migration of plants, Hooker reported that CD had established a number of supporting facts, including the following: 'That oceanic fish devour seeds, and that if these become the prey of birds, the contents of their stomachs may thus be deposited on distant islands' (J. D. Hooker 1866a, p. 51). In his 1867 pamphlet (see n. 1, above), Hooker made CD's suggested change (see Williamson 1984, p. 72). See *Origin*, p. 362.

[3] Hooker summarised the arguments for the stocking of oceanic islands with plants from a continent either by former land-bridges, or by the transport of seeds by ocean currents, wind, birds, icebergs, or similar agencies in J. D. Hooker 1866a, pp. 50–1. For CD's arguments in favour of seed and animal transport rather than land-bridges as a means of species dispersal, see *Origin*, pp. 356–65, 388–410.

From Alfred Newton 21 January 1867

10, Beaufort Gardens. | S.W.

21 Jan^{y} | 1867

My dear Sir,

If I remember right the instance of *Rhynchæa* was cited at Nottingham by M^{r}. Edgar Layard (not by me) with reference to a remark of M^{r}. Wallace's to the effect that among birds the plumage of the male was invariably more brilliant than that of the female.[1]

I therefore hazarded the suggestion that *perhaps* in *Rhynchæa* the duties of incubation were performed by the male, adding that I had had no personal acquaintance with the species of that genus—abnormal & peculiar as they are in many respects—but that I had some reason from my own observation to suspect that such was the case in *Phalaropus* (including the subgenus *Lobipes*).[2]

When in East Finmark in 1855 I ascertained for myself that the female of *Phalaropus hyperboreus* was the larger & the most highly coloured of the two sexes—[3] This was no new discovery, for the fact though subsequently contradicted had been stated before. The same summer & in the same locality I found that the male of this bird shewed much greater devotion to its young when they were in danger than the female did—[4] I do not know when the idea first crossed my mind, but it was either then or three years later in Iceland, when I had abundant opportunities of observing the same species, that it struck me that there was in all probability a connexion between the facts I have mentioned. Your beautiful theory had not then been published & was of course unknown to me, or I should have probably at once seen the desirability of making precise observations to determine the truth of my hypothesis—and I have never since had the opportunity of doing so— But I still entertain a strong belief that it will be found that in those species of *Grallæ* wherein the plumage of the female far exceeds in brilliancy that of the male, the male has the greatest share in the duties of incubation— Among these species may be specially cited *Phalaropus hyperboreus* & *P. fulicarius*, *Limosa lapponica* & *L. ægocephala*, *Tringa canutus* & *T. subarquata*, but there are probably at least as many others—[5]

Of *Limosa lapponica* I have twice had sent me from Lapland the birds which had been killed (snared I believe) *on* the nests. In each case they were in the pale dull plumage which has caused the birds wearing it to be described as a distinct species (*L. meyeri* Leisler), and though the sexes were not noted by the collector, from the small size of the specimens I have little doubt of their being males— One of them I have still at Cambridge, & can send it to you when I return thither if you like to see it.[6]

It would seem from M[r]. Swinhoe's statements that a similar state of things would obtain in the genus *Turnix*.— Cf. Ibis, 1865, pp. 542, 543, & 1866, pp. 131, 403, 405.[7]

As to your enquiry respecting the Swans I am sorry to say I have no information to give, nor do I know where you could obtain any except perhaps from M[r]. Samuel Gurney—[8]

Trusting that some of the foregoing statements may be of use to you, and assuring you that it gives me great pleasure to write to you, | I remain, | Yours very truly | Alfred Newton

C. Darwin, Esq. F.R.S.

DAR 84.1: 22–5

CD ANNOTATIONS
1.1 If ... respects— 2.4] *crossed pencil*
2.4 I had some reason ... *Lobipes* 2.5] *scored pencil*; 'had some reason' *underl pencil*
2.4 my own ... that there 3.8] *crossed ink*
3.1 the female ... two sexes— 3.2] *scored pencil*
3.4 the male ... female did— 3.6] *scored pencil*
3.8 was in ... plumage of 3.14] *crossed pencil*

3.14 the female ... others— 3.17] *crossed ink*
3.14 the male has ... incubation 3.15] *double scored pencil*
4.1 Of ... killed 4.2] *double scored pencil*
4.2 (snared ... Cambridge, 4.6] *crossed ink*
4.6 & can ... see it. 4.7] *double scored pencil*
5.1 It ... to you, 7.2] *crossed pencil*
End of letter: 'J. G.— Swans & see about *Turnix*' *pencil*

[1] For CD's query regarding the brightly coloured females of a species of *Rhynchaea* (painted snipe; now *Rostratula*), see the letter to Alfred Newton, 19 January [1867]; CD wrote that Alfred Russel Wallace had referred him to Newton on the subject. The discussion with Edgar Leopold Layard at the Nottingham meeting of the British Association was not recorded either in the *Report of the thirty-sixth meeting of the British Association for the Advancement of Science* 1866, or in the brief discussion published with the full text of Wallace's paper at the meeting, A. R. Wallace 1866a.

[2] Newton refers to what is now called the red-necked phalarope, *Phalaropus lobatus*, which is sometimes still placed in the monospecific genus *Lobipes* (*Birds of the world* 3: 532).

[3] In 1855 Newton travelled to Finnmark, in the extreme north of Norway (*DNB*). Newton refers to the bird now named *Phalaropus lobatus* (Audubon 1827–38); see n. 2, above.

[4] CD quoted this statement of Newton's in *Descent* 2: 204 n. 20.

[5] The now obsolete order *Grallae* included all wading birds. CD noted Newton's comments on *Limosa lapponica* (the bar-tailed godwit) and 'some few other Waders' in *Descent* 2: 204 n. 20. Newton also refers to what is now *Phalaropus fulicaria*, the red phalarope. *Limosa aegocephala*, the black-tailed godwit, is now *L. limosa*, while *Tringa canutus*, the knot (a sandpiper), is now *Calidris canutus* (Peters *et al.*, 1931–87). *Tringa subarquata*, the curlew sandpiper, is now *Calidris ferruginea* (Audubon 1827–38).

[6] There is no evidence that Newton sent the specimen to CD.

[7] Newton refers to publications by Robert Swinhoe in *Ibis* n.s. 1 (1865): 538–46, and 2 (1866): 129–38 and 397–406. In *Ibis* n.s. 1 (1865): 542–4, Swinhoe described his discovery of a male button quail (*Turnix*) caring for its young, and published a description of the species he called *T. rostrata* (now *T. suscitator rostrata*; see *Birds of the world*). In *Ibis* n.s. 2 (1866): 131, Swinhoe suggested that the male *T. rostrata* incubated the eggs. In *Ibis* n.s. 2 (1866): 403–4, Swinhoe further discussed the classification of the species of *Turnix* found in 1865.

[8] For CD's query on the black swan and black-necked swan, see the letter to Alfred Newton, 19 January [1867] and n. 5. CD mentioned the black swans in *Descent* 2: 226–7, and the black-necked swans in *ibid.*, p. 230, when discussing the role of sexual selection in conspicuously coloured birds; he did not cite Samuel Gurney, or any other author, on the subject. His notes on the two different swans are in DAR 84.2: 200a and 200b. CD's last annotation suggests that he thought Newton may have intended to recommend the ornithologist John Henry Gurney.

To J. V. Carus   22 January [1867] [1]

Down. | Bromley. | Kent. S.E.
Jan 22

My dear Sir

I have great pleasure in answering your questions.

The Rock-thrush is the Rupicola aurantia.[2]

p. 171. The tortoise-shell cat is piebald with white, orange & black.[3]

p. 366. The Logger headed duck is the Anas brachyptera of Linn. or the Micropterus brachypterus of Eyton.[4]

I am very much obliged for all the great trouble which you are taking in the translation & believe me

my dear Sir | yours very sincerely | Ch. Darwin

LS
Staatsbibliothek zu Berlin, Sammlung Darmstaedter (Carus 34)

[1] The year is established by the relationship between this letter and the letter from J. V. Carus, 18 January 1867.
[2] See letter from J. V. Carus, 18 January 1867 and nn. 2 and 3. In spite of CD's information, Carus mistakenly used the common name Steindrossel (now more commonly Steinrötel) and did not include a scientific name (see Bronn and Carus trans. 1867, p. 113).
[3] See letter from J. V. Carus, 18 January 1867 and n. 4. Carus described the cat as 'gelb, schwarz und weiss gefleckten' (literally 'yellow, black, and white spotted'; see Bronn and Carus trans. 1867, p. 182).
[4] See letter from J. V. Carus, 18 January 1867 and n. 5. CD refers to Carl von Linné and to Thomas Campbell Eyton, who wrote a monograph on the duck family (Eyton 1838). Carus translated the name as 'Dickkopf-Ente (Micropterus brachypterus)' (see Bronn and Carus trans. 1867, p. 369). The modern name for this bird is *Tachyeres brachypterus*, the Falkland steamer duck, or Dampfschiffente in German (*Birds of the world*).

From John Scott 22 January 1867

Royal Botanic Gardens | Calcutta
22^{d} Jany. 1867.

Dear Sir,

I cant longer defer writing a note to you—though I have yet failed to accomplish the purpose of my delay. This you can only guess at just now. I shall explain in a future letter, and in the meantime, I would have you believe me though silent, in heart ever grateful to you my chiefest benefactor.[1]

My duties as Curator of the Botanic Gardens have been and yet are engrossing my time so completely, that I have been quite unable to engage in the experimental illustrations of many subjects which lie very near my heart.[2] You may be surprised at the press of duties incumbent to my appointment considering that our Gardens have been so long established. This is natural, but when you know, that we are raising a new Garden out of the wholesale devastations of the Cyclone of 1864, and this on a scientific basis, which had never been previously attempted here, you can conceive of the labour— reclaiming much actual jungle land, planning and planting systematically a tropical garden 250 acres in extent. We have now however got the outlines of all the Natural Orders fixed, and the planting of many of them in so far as we can at present represent them, done,[3] so that my duties are turning daily lighter, and will soon permit of my directing some little portion of my time to former scientific recreations.

In talking of our Garden you will I doubt not be interested in hearing, that we have now planned out on a rather extensive scale the basis of a **temperate** *herbaceous* garden. So far as my observations go, this promises to be most pregnant with interesting results: plants from an elevation of 7 & 8000 ft in the Himalayahs are luxuriating in our hands; and *many many* common **British plants** of which I have had seeds this season from the North of Scotland are proving quite at home. This experiment is interesting me greatly; and excitingly interesting is the hope that further experience may yield us something in the way of illustrating your views on the origin of species by variation.[4] It has already however shown me that our domesticated plants, are not as some would have us believe, the alone possessors of the power of habitating most diverse climes. Again it seems to show us, that there must not necessarily have been a great lowering of the temperature of the globe to account for the occurrence of temperate species of plants on tropical mountains. In so far as temperature is concerned, from the wide flexibility of constitution, which many temperate plants possess, I believe that under existing physicial arrangements, along continuous meridional tracts of land migrations of those species into tropical climes might occasionally be effected. I say occasionally for it is evident, that plants though flourishing in diverse climes under cultivation could rarely compete successfully with the respective indigens in their exodus from the temperate to the tropic zone; while there is further the absence of any directive force occasioning such migrations, which is so well explained by your views of a simultaneous glacial epoch in both hemispheres and a concomitant cooling of the tropical zone.[5]

I have hopes after a few years careful observation of the various temperate plants I am now introducing to accumulate an interesting body facts on acclimatation. This subject seems yet to excite some little discussion at home; perhaps a brief note on a few experiments may interest you— Thus the Common Sweet Pea (Lathyrus odoratus) from English grown seeds germinates freely here, producing strong rigid plants, a reduced developement of leaf and tendril, with thick prominently winged stems, rarely ever producing a **single flower** *no* **seed**.[6] Again English seeds grown in the vicinity of Darjeeling are abundantly productive of flowers and seeds, affording material for successful cultivation in the plains of India. Seedlings raised from these acclimatized up-country seeds, are readily distinguished from the direct products of English saved seed, when grown in the plains, of their lax, and naturally scandent habit, and their flowering and seeding profusely... It is worthy of note also as indicating a stage in the process of acclimatation that a small quantity of French grown Sweet Peas which I had yielded plants, markedly distinct from the English seedlings (and naturally so) *blossomed* sparingly, but all proved infertile.[7] I am curious to see how a number of other Legumes introduced from various parts of Europe during last season will behave.... Another point worthy of notice is the tendency of many of those highly cultivated productions of the florist to vary under cultivation here. Thus the double flowering Balsam from English seed, are as *fine* for the **first** season as any I ever saw at home— in the second generation there is a decided falling off. While in the third scarcely a double flower is to be seen. This is also the case with other double flowering florists flowers, and so with many single varieties of the same. Petunias and Antirrhinums for example are splendid for the first season, while in the second and third they are [surely] wortheless for the flower garden. Indian florists are in general so disheartened with these very natural results that in place of making attempts to raise varieties from their retro-verted stock, suitable to the climate and the adornment of their garden, trust almost entirely to direct imports of English grown seed. I am experimenting a little with this view.

By the way I am growing successfully *Leersia oryzoides* from the seeds you were good enough to send me.[8] It has not however as yet protruded a single flower, all are *enclosed within* the *culms* and **quite** *fertile*. I had great hopes of cultivation here inducing variation on that point, but have been as yet disappointed. Succeeding generations, however may yet afford it[9] . . . . . Curiously enough some of our Indian Violas, like the European, produce also perfect and imperfect flowers. *V. Roxburghiana*, for example, produces *perfect flowers* **only** in the cold season— these I may remark are quite fertile—during the hot, though chiefly in the rainy season an abundance of closed-fertile flowers are produced.[10]

Excuse my desultory notes, as I am as yet only able to indicate points of promise.

Yours most respectfully | John Scott

DAR 177: 117, DAR III: A91

CD ANNOTATIONS
1.1 I cant . . . recreations. 2.13] *crossed blue crayon*
2.7 and this . . . labour— 2.8] *scored pencil*
3.1 In talking] *after opening square bracket, blue crayon*
3.4 plants . . . home. 3.6] *double scored pencil*
4.1 I have . . . infertile 4.15] *crossed red crayon*; '*Acclimatisation*' *pencil*
4.15 sparingly,] '*but all proved infertile*' *added below, ink*
4.15 but all . . . view. 4.29] *crossed blue crayon*
4.26 attempts . . . seed. 4.28] *scored pencil*
5.1 By . . . disappointed. 5.4] '*Leersia*' *in margin, pencil*
5.6 *V. Roxburghiana* . . . produced. 5.9] '*Viola*' *in margin, pencil*
Top of letter: 'J. Scott' *ink*; '*Keep*' *pencil*
Top of last page: 'Jan 1. 1867' *ink*

[1] The last extant letter from John Scott is that of 21 July 1865 (see *Correspondence* vol. 13). Scott alludes to the money that CD gave him for his passage to India, where he went to seek employment.

[2] Scott became curator at the Royal Botanic Garden, Calcutta, in late May 1865 (see letter from John Scott, 21 July 1865 (*Correspondence* vol. 13). He had conducted experimental work while foreman of the propagating department of the Royal Botanic Garden, Edinburgh, and had corresponded extensively about the work with CD since 1862 (see *Correspondence* vols. 10–13).

[3] For information on the Royal Botanic Garden, Calcutta, and the destructive cyclone of 1864, see Burkill [1965], pp. 21, 66, 103, 133; the garden was on the banks of a tidal river, the Hooghly, and suffered inundations of seawater during the cyclone. The Calcutta garden covered more than 300 acres (McCracken 1997, p. 111). The garden had earlier been planted partly along the lines of a horticultural nursery, and partly according to the Linnaean system (see Burkill [1965]); Scott was arranging the beds according to the natural orders (families) promoted by Robert Brown (1773–1858) and John Lindley.

[4] Wild plants and seeds were brought down from higher elevations by the garden's collectors, and cultivated plants and seeds from plantations in Sikkim; they were then planted at sea level in the botanic garden at Calcutta (see Burkill [1965] and McCracken 1997). CD had written on acclimatisation in *Origin*, pp. 139–43.

[5] Scott alludes to CD's discussions in *Origin*, pp. 365–82, of plant dispersal during the glacial period.

[6] CD included Scott's information on *Lathyrus odoratus* in India in a section on acclimatisation in *Variation* 2: 311.

[7] CD noted Scott's information on the French sweetpeas in *Variation* 2: 311.

[8] Scott asked CD for seed of *Leersia oryzoides* in his letter of 21 July 1865 so that he, like CD, could investigate self-fertilisation in closed flowers (see *Correspondence* vol. 13). For CD's interest in the plant and his acquisition of specimens, see *Correspondence* vol. 12. CD's observations had been published in 'Three forms of *Lythrum salicaria*', pp. 191–2 n. (*Collected papers* 2: 131).

[9] Scott evidently never succeeded in cultivating *L. oryzoides* in India with flowers that opened; see *Forms of flowers*, p. 335.

[10] CD cited Scott's observations of *Viola roxburghiana* in *Forms of flowers*, p. 320.

To Alfred Newton   23 January [1867]

Down. | Bromley. | Kent. S.E.

Jan 23

My dear Sir

I thank you very sincerely for your very full & perfectly clear statement about the male plumage.[1] I wish the evidence had been a little more distinct, though certainly from the several points which you mention, the probability of the truth of your view seems great. I will look to the papers in the Ibis to which you have been so good as to refer me.[2] Your theory is to me so captivating that I much hope you will be able some day to prove it.[3]

With my cordial thanks for your very kind letter believe me | yours very sincerely | Ch. Darwin

LS
Endorsement: 'C. Darwin, Down. Jan.ʸ $\frac{23}{24}$/67'
DAR 185: 88

[1] See letter from Alfred Newton, 21 January 1867 and nn. 2–6.

[2] See letter from Alfred Newton, 21 January 1867 and n. 7.

[3] CD discussed the instances of male birds that sat on eggs being smaller and less brightly coloured than females of the same species in *Descent* 2: 200–8.

From E. Schweizerbart'sche Verlagsbuchhandlung[1]   24 January 1867

Stuttgart

24 Januar 1867

Verehrtester Herr!

Aus Ihren werthen Zeilen vom 30 October ersah ich gerne dass Ihnen Herr Prof. Carus in Leipzig als Uebersetzer Ihres Buches angenehm ist.[2]

Derselbe hat mir bereits einen Theil des Manuscriptes gesandt und ist solches schon in die Druckerei gegeben worden; ich werde veranlassen dass der Druck möglichst beschleunigt wird und werde das Vergnügen haben Ihnen die Bogen nach und nach, wie sie fertig werden zu übersenden.

Ich habe heute noch eine Bitte an Sie, verehrtester Herr; auf eine Anfrage bei *Williams & Norgate* wegen eines etwa vorhandenen Bildes von Ihnen, sofern es noch ein besseres gebe als jenes welches ich zu der $2^{t\cdot}$ deutschen Auflage nach einer Photographie machen ließ, haben mich diese Herren noch ohne Antwort gelassen;[3] ich beabsichtigte nehmlich für die 3^{t} Auflage das Porträt in Stahlstich zu bringen, wenn ich ein gutes Original davon haben könnte; nun möchte ich Sie selbst bitten, mir zu sagen ob die Photographie der 2^{t} Aufl. gut ist, oder ob Sie eine bessere in einer Copie mir dafür geben könnten.[4]

In der Hoffnung dass der so kalte Winter dieses Jahres Ihrer Gesundheit nicht nachtheilig gewesen seie bin ich mit aller Hochachtung | Ihr ergebenster | E Schweizerbart[5]

vor einigen Tagen hatten wir 25° Kälte, heute 9° Wärme (Réaumür)[6]

DAR 177: 74

[1] For a translation of this letter, see Appendix I. It was written by Christian Friedrich Schweizerbart, head of the Stuttgart publishing firm E. Schweizerbart'sche Verlagsbuchhandlung (see n. 5, below).

[2] CD's letter to E. Schweizerbart'sche Verlagsbuchhandlung of 30 October 1866 has not been found, but see the letter from E. Schweizerbart'sche Verlagsbuchhandlung, 26 October 1866 (*Correspondence* vol. 14). Julius Victor Carus was translating the fourth English edition of *Origin* for a third German edition. Beginning with Carus's letter of 7 November 1866, he and CD exchanged four letters in November 1866 discussing the new edition; see also this volume, letter from J. V. Carus, 18 January 1867, and letter to J. V. Carus, 22 January [1867].

[3] The publishers and booksellers Williams & Norgate specialised in foreign scientific literature. The photograph of CD in Bronn trans. 1863 is the same as the photograph, taken *circa* 1857, that appears as the frontispiece to *Correspondence* vol. 8.

[4] CD's reply to this letter has not been found, but see the letter from E. Schweizerbart'sche Verlagsbuchhandlung, 22 March 1867 and n. 8.

[5] In his business communications, C. F. Schweizerbart continued to use the signature of his uncle, Wilhelm Emanuel Schweizerbart, from whom he had purchased the publishing firm in 1841 (*Jubiläums-Katalog*, pp. x–xi).

[6] Schweizerbart gives the temperatures based on the Réaumur thermometer scale, which has the freezing-point of water at 0°, and the boiling-point at 80°. The approximate equivalents in Celsius would be −31° centigrade and 11° centigrade.

From Alexander F. Boardman 26 January 1867

Brunswick Maine U.S.A.
January 26. 1867

Mr Charles Darwin
Dear Sir

Having written the accompanying letter about a week since[1] & not having your address, and never having seen your work on the origin of Species although knowing something of its drift, I sent for the book partly to obtain your address and partly to get your precise views and partly to get a little acquainted with the man I had been writing to and who had done more than any one else, I suppose, to convince the world of the truth of a theory which has always since I have been old enough to have an opinion on the subject been a favorite with me and which I regard as the corner stone of a stupendous fabric reaching through the moon & planet & sun & perhaps suns suns, from the lowest organised matter to the great central sun or suns the capital of all things and the throne of the Almighty— I suppose electricity or something of that sort to pervade the whole line so that when a system is admitted to the union (I'm afraid that our system is what in the United States of America we call a territory) the lowest matter & of course all above is represented at the center of the universe. "The hairs of our head are all counted" or numbered I believe it is.

The Caspian Sea is a material isolation. The district of Columbia in this country and Switzerland in Europe appear to be mated isolations which in connection with the thousand other reasons which easily present themselves *hint* at least very strongly at Switzerland as the future Capital of Europe & its territories.

You have noticed I see the isolation of Paraguay in South A—a in one particular.[2] Their present struggle against great odds look significant in this connection.[3]

Brazil & Siberia appear to be mates and I should expect some country in the interior of Africa where peaceful animals could roam at large unmolested by wild beasts or something similar to that. Is any thing of that kind known to exist there?

May not the destruction of the Alexandrian library in Egypt have cut us off from all written knowledge of our preadamite ancestry as the flood cut off the ancestry itself making a material and spiritual parralelism?[4]

The preponderance of land and of mind and of quality of mind is largely in favor of the northern hemisphere and the compass points to the north.

Other influences of course wholly or in part produced this but it is a parralellism and I am inclined to think that magnetism pitched the tune. In the accompanying letter I supposed that the impregnation of the Black Sea by the gulf stream took place previous to the flood. I don't know that it is impossible that the flood might have been the effect of this rush of waters through the Mediteranean, in which case the Caspian Sea was reached of course.

I prefer however to suppose that the flood took place before this.

I dont know but your divine who "has gradually learnt to see that it is just as noble a conception &c" is fully up to and perhaps in advance of the times, which is encouraging,[5] but to my mind, the comparison between the two would be something like this. If the old system is a stone thrown from a sling the new is a howitzer shell. If the old is a pebble the new is an acorn. If the old is the moon the new is the sun that gives it light—

I hope, Sir, that you have had patience to read this far and that I have not wearied you. I should write thus freely to but few. I was rather disappointed in finding so little reference to man in your work, but I see that you are driving the piles for a superstructure very thick and some of them very deep and I heartily bid you God Speed.

Yours | Alex F. Boardman

[Enclosure]

Brunswick Me

Jany 17/66

Mr Charles Darwin

Dear Sir

I wish to lay before you some thought on the theory (or perhaps more properly supplementary to the theory) of the origin of man which you have done so much to elucidate and which has been christened with your name.[6] My idea is that one branch of the human race originated in the vicinity of *Borneo*, migrated north & west untill improved by change of climate food crossings &c it arrived at the vicinity of the Black Sea—

It there meets the more enterprising of (or a selection from) the African branch who have found their way through the isthmus of Suez— The two branches mingle. The waters of the Nile the Danube the D(o)n[7] &c containing the fertilizing matter of their respective regions here meet & mingle, and through their evaporations affect the climate, the plants & the inhabitants, who are also crossing the breeds the while— While this is going on I suppose the straits of Gibraltar to have been closed (and I understand that their appearance warrants or gives color to this supposition)

The Caspian Sea, receiving & *holding* an (entirely different drainage, of course assisted also, and the Arabian & Baltic Seas are not too far off to have some slight influence—

I suppose all this to have resulted in a Black Man in the right state to be affected by another influence—when, the Black & Mediteranean seas being low & their waters concentrated by drought, the *boist*ereous Atlantic bursts through the pillars of Hercules & rushes clear up to the Black Sea, bearing through the Gulf stream a selection from an entire Hemisphere.[8]

This influence from the to them unknown world I suppose to have had bleaching & other qualities & to have resulted in the white man, or Adam. From Adam to Noah the breed was *fixed* as much as possible or expedient in Noah— The flood sweeps all away & Noah & his family have a clear field—[9] Some of his progeny go back to Asia & Africa & under those influences breed back to a more or less extent, others migrating to Europe & continuing to receive the influence of the gulf stream & finding other better influences & perhaps a *virgin* soil have resulted in what they are.

I suppose a similar operation to have gone on on this continent between the waters & inhabitants of the valley of the Amazon & Missisippi &c in and around the gulf of Mexico, resulting in a race of beings similar to that produced by a similar combination in Europe. I suppose also that every thing material & immaterial in nature is differently charged as positive & negative or male & female, that this Hemisphere in this connection is male (ie the male preponderates) & in the other the female preponderates.

I do not suppose this operation of the gulf stream through the Mediteranean Sea to be an *imitation* of sexual intercourse— The resemblance is very striking however— It is a means well adapted to an end. And the similarity in the means is probably paralell with the similarity in the materials used & the object to be attained, and the different wombs you will perceive correspond with the organizations resulting. I have long considered the name of Monkey (*Man*key) as significant, and in this connection, although it may seem childish, Portugal, Spain & the Fortress of Gibraltar held by England look suspicious.

Hellespont, Marmora, Bosphorous &c do not require a very vivid imagination to perceive the hint if we are looking for one.[10] Caucus is not a bad synonym for what is supposed to have transpired on & around the Caucasian mountains and there is the plain bald unquestioned fact that here, in the vicinity of the *Black* & Caspian Seas, *Man* (not what man was made of or from) but *Man* himself was made.[11]

There is man right where those tools would have made him if they did make him. Those tools could have made him I think, and in the very way in which God makes every thing which we know any thing about and there are even the signboards which (tho' not absolutely necessary perhaps but convenient) we naturally expect to find on approaching a County Seat or similar place— And then & thus or similarly, I believe God did make him.

As intimated before I suppose this principle of opposites or (more properly perhaps) apposites to pervade all Nature, Government among the rest, that Monarchical & Democratical forms of government are both faulty, & both need an opposite as much as the male or female, in fact that they are one positive & the other negative or whatever you please to call it.

I believe that the Eastern continent or Hemisphere is and should be monarchically charged and that this Hemisphere is democratically charged. Two straight lines may always approach each other & never meet & thus I imagine that Monarchy & Democracy Good & Evil or any other two opposites may always approach each other, growing more & more like but never meeting or becoming the same.

Monarchy I conceive to be Adam and Democracy Eve or Adam or monarchy conseved or taken out of his rib.

God & Devil (Good & Evil) hold no feeble light on this Subject. Amazon & Mis Sippi & Miss Houri seem to be branded.[12] In fact as you near the focus every thing seems to tell—

This play upon words may seem puerile to you (and the whole may for ought I know) but the more I meditate upon it (and its a regular *Medite*ranaen *Sea*) the more I am convinced that there is something and a good deal too in it. I should be glad to hear from you on the subject either in the way of objection or any other way.

Of course it is a theory only and I do not undertake to prove it, otherwise than as it speaks for itself or as I may elucidate it.

Of course to carry out the Governmental part of the theory, as per the chart, the Eastern Hemisphere should be gathered together under one head, the western under another head, we should have one man & one woman who ought at once to get married and behave themselves like a good husband & wife give us peace & the millenium and all the good things which must follow in its train.

Yours respecty | Alex F. Boardman | Brunswick | Maine | USA

DAR 160: 226, 226/1, 227

[1] See enclosure.

[2] Boardman may refer to CD's discussion in *Origin*, pp. 72–3, of insects checking the increase of cattle in Paraguay, but not in adjacent countries.

[3] Paraguay's War of the Triple Alliance against Argentina, Uruguay, and Brazil had started in 1864 and would last until 1870 (*EB*).

[4] The library of Alexandria was thought to have been finally destroyed in the seventh century (*EB* 8th ed.).

[5] Boardman quotes in part an addition to the second edition, and the American edition, of *Origin*: 'A celebrated author and divine has written to me that "he has gradually learnt to see that it is just as noble a conception of the Deity to believe that He created a few original forms capable of self-development into other and needful forms, as to believe that He required a fresh act of creation to supply the voids caused by the action of His laws"' (see *Origin* 2d ed., p. 481). The author CD quoted was Charles Kingsley (see *Correspondence* vol. 7, letter from Charles Kingsley, 18 November 1859). For publication of the American edition of *Origin*, see *Correspondence* vol. 8, Appendix IV.

[6] Though CD did not discuss humans directly in *Origin*, he did write: 'Light will be thrown on the origin of man and his history' (p. 488).

[7] The Danube and Don rivers drain into the Black Sea from the north-west and north.

[8] The pillars of Hercules are the rocks flanking the Strait of Gibraltar, with Calpé to the north, and Abyla (Ceuta) to the south (*OED*). For a contemporary account of ocean currents in the Mediterranean, including flow from the Gulf Stream and the Black Sea, see C. Lyell 1853, pp. 294, 333–6.

[9] For the diminishing credibility in nineteenth-century Britain of the 'Noachian Deluge', and CD's role in this, see Browne 1983.

[10] The Hellespont (the Dardanelles) is the strait connecting the Aegean Sea and the western end of the Sea of Marmara; the Bosporus is the strait connecting the eastern end of the Sea of Marmara with the Black Sea.

[11] The Caucasian Mountains lie between the Black and Caspian Seas. Charles Lyell had recently noted that at least seventy different languages had been spoken in the area, suggesting the intersection of different peoples over time (see C. Lyell 1863, pp. 460–1).

[12] Boardman refers to the Amazon river in South America and the Mississippi and Missouri rivers in North America.

To John Murray 27 January [1867]

Down. | Bromley. | Kent. S.E.

Jan 27th

My dear Sir

I hope that you will urge your friend to give his judgment **soon**. He has now had the M.S. for nearly 3 weeks,[1] & I shall soon have nothing to do for my chapter on Man is nearly finished.[2] Please remember if your friend's verdict is against me, & you decline to publish as proposed, that I shall suffer great loss of time in arranging some other method of publication. If you decline to publish, I think you will make a mistake, for though I do not believe this book will have nearly so large a sale as the "Origin", I shall be astonished if it has not a fair sale, for as yet I have found what interests me greatly, likewise interests others, & some of the chapters in my present book have much interested me.

I earnestly beg you to do your best to come to an early decision.—[3]

My dear Sir | Yours sincerely | Ch. Darwin

John Murray Archive
Endorsement: '1867. 27 Jany'

[1] Murray's friend was John Milton. In his letter of 9 January [1867], Murray wrote that he had sent the manuscript of *Variation* to a literary man who was not a man of science (see letter from John Murray, 9 January [1867] and n. 4); Murray said he expected to hear of his impression in a week. For CD's sceptical response, see his letter to Murray of 10 January [1867].

[2] CD was composing an additional chapter on humans for *Variation* (see letter to John Murray, 3 January [1867] and n. 5). CD's chapter was not published in *Variation* but the material in it was used in *Descent* and *Expression*.

[3] CD also expressed anxiety about the publication of *Variation* in his letter to Murray of 3 January [1867]. For information on the publication of both *Origin* and *Variation*, see Freeman 1977.

From John Murray 28 January [1867][1]

Albemarle St
Jan[y] 28

My Dear Sir

Pray put yourself at ease about the publication of your new book.[2] I will publish it for you coute qui coute provided you will be content that I pay you one half the profits of the edition instead of a sum down at first—[3] This I ask because—no doubt there is considerably greater risque in this than in the publication of your former works.—[4]

This work is not intended nor likely to become generally popular but I think after the sale of 6000 of your "Origins" I can *count* upon 500 purchasers of these new volumes—the "Pièces Justificatives" on wch that work is founded & I w[d] propose to print an Edition of 750 copies[5]—in the size type & page of Lyells Principles—like wch it will make 2 volumes 8vo.[6]

I have heard from my literary friend—but have not yet got back the MS.S from him— He certainly finds it difficult of digestion but he is not a man of science so his opinion is not a fair test altogether—[7] Still in the face of it, I venture to submit to you the above proposal.

I am My Dear Sir | Yours very faithfully | John Murray

I hope to return the MS. this week | JM

Charles Darwin Esq

DAR 171: 344

[1] The year is established by the relationship between this letter and the letter to John Murray, 27 January [1867].

[2] For CD's concern regarding whether Murray would publish *Variation*, see the letter to John Murray, 27 January [1867] and n. 3. Murray had also assured CD he would publish the work in his letter of 9 January [1867].

[3] When Murray published the fourth edition of *Origin* in 1866, he agreed to pay CD two-thirds of the profits when half of the copies had been sold; for earlier editions of *Origin*, CD had received that sum on publication (see *Correspondence* vol. 14, letter from John Murray, 24 February [1866]). On CD's astute management of royalty payments for his books, see Browne 2002, pp. 97, 461. Murray misspelled *coûte que coûte* ('at all costs').

[4] CD had also expressed concern that *Variation* might not appeal to the public (see *Correspondence* vol. 13, *Correspondence* vol. 14, especially letters to John Murray, 22 February [1866] and 21 and 22 December [1866], and this volume, letter to John Murray, 3 January [1867]).

[5] According to Freeman 1977, pp. 122–3, Murray published 1500 copies of *Variation* in January 1868, but these sold out in one week, most to booksellers; another 1250 copies (with some changes inserted by CD) were issued in February 1868.

[6] Murray refers to the tenth edition of Charles Lyell's *Principles of geology* (C. Lyell 1867–8; see letter from John Murray, 2 January [1867]).

[7] Murray refers to John Milton. See letter to John Murray, 27 January [1867] and n. 1.

To J. D. Hooker 29 January [1867][1]

Down. | Bromley. | Kent. S.E.
Jan 29th

My dear Hooker

Very many thanks for 2 Plumbago; but I am very sorry that I have caused you trouble in vain, as seed alone w^d be of service to me for Dimorphism.—[2]

I have read your concluding paper & it is excellent.[3] Such papers will do far more than regular Treatises on the subject to convert people to the derivation Theory.—[4] You pay me about Distribution an *enormous* compliment & really I think much too strong. It rejoices my inward heart to find we accord so very closely.[5] I think you lay too much weight on the affinity not going *strictly* with geographical distance.— The Azores, (though I know some fragments of miocene beds have been found there) struck me when there in general aspect as a far more modern group, (with still active volcano, fumarole &c &c) than Madeira; & w^d not this account to great extent for more strictly European flora:[6] I suppose you will admit that each isl^d has received many of its plants, not from other isl^d, but from continent; I remember coming from your paper on Galapagos to this conclusion.[7] At top of third column, you hardly put case about volcanic islands quite fairly; for I do not suppose anyone would object as improbable to very many large groups of oceanic isl^{ds} being volcanic; but the difficulty arises from all oceanic islands being volcanic; & volcanos, whilst active, it may be added, characterise rising, not subsiding areas.—[8] It is a splendid paper.

Yours affect | C. Darwin

DAR 94: 8–9

[1] The year is established by the references to the publication of J. D. Hooker 1866a (see n. 3, below).

[2] CD had asked for *Plumbago* seeds in December 1866 (*Correspondence* vol. 14, letter to J. D. Hooker, 24 December [1866]), and Hooker had evidently sent specimens after 12 January 1867 (see letter to J. D. Hooker, 9 January [1867], and letter from J. D. Hooker, [12 January 1867]).

[3] CD refers to the fourth instalment in *Gardeners Chronicle*, 26 January 1867, pp. 75–6, of Hooker's article on insular floras (J. D. Hooker 1866a). See the letters to J. D. Hooker, 9 January [1867] and n. 1, 15 January [1867], and 21 January [1867], for CD's comments on the first three instalments.

[4] Hooker discussed how the 'hypothesis of trans-oceanic migration' and the 'theory of the derivative origin of species' could contribute to the understanding of the distribution of plants on small oceanic islands. He added: 'if many of the phenomena of oceanic island Floras are thus well explained by the theory of the derivative origin of species, and not at all by any other theory, it surely is a strong corroboration of that theory' (J. D. Hooker 1866a, p. 75). At the close of the paper, Hooker made a humorous analogy concerning the reception of the 'Derivative doctrine of species' at the 1860 meeting of the British Association at Oxford, and at the 1866 meeting (*ibid.*, pp. 75–6; see also *Correspondence* vol. 14).

[5] In the third and fourth instalments of J. D. Hooker 1866a, pp. 50, 51, and 75, Hooker discussed at length CD's hypothesis of trans-oceanic migration as a means of plant and animal distribution, and presented it as one of two explanatory hypotheses for plant distribution on oceanic islands (see also letter to J. D. Hooker, 21 January [1867] and n. 3). Hooker discussed the 'many powerful arguments' that CD had presented in support of his hypothesis (*ibid.*, p. 51), later adding (*ibid.*, p. 75):

> It shows a power and skill of bringing facts to bear, and a fertility of invention in devising means of verifying these facts, that almost compel me to agree with him in regarding oceanic transport to be, in the present state of science, the principal and most probable means by which oceanic islands have been stocked with plants.

Towards the end of the paper, Hooker wrote (*ibid.*):

> The great objection to the continental extension hypothesis is, that it may be said to account for everything, but to explain nothing; it proves too much: whilst the hypothesis of trans-oceanic migration, though it leaves a multitude of facts unexplained, offers a rational solution of many of the most puzzling phenomena that oceanic islands present.

[6] Hooker noted that there were more European plants on the Azores, even though the islands lay more than twice as far from the continent than did Madeira (J. D. Hooker 1866a, p. 75); he included what he called an imperfect explanation that Madeira, receiving more immigrants, exhibited 'the sharper struggle'. For CD's impressions of volcanic activity on the Azores, which he visited on the *Beagle* in September 1836, see R. D. Keynes ed. 1988, pp. 437–41, and *Correspondence* vol. 2, letter to William C. Redfield, 24 February [1840]. CD had discussed European plants on the Azores and Madeira in *Origin*, pp. 314, 363, 390–1. CD and Hooker also discussed plant distribution on the Azores and Madeira in 1866 (*Correspondence* vol. 14). The Miocene origins and some of the volcanic history of the Atlantic Islands were discussed in C. Lyell 1867–8, 2: 403–11.

[7] CD refers to Hooker's paper 'On the vegetation of the Galapagos Archipelago' (J. D. Hooker 1846), based on CD's collection of plant specimens; see also J. D. Hooker 1845. CD believed that many of the animal species, including birds, on the Galápagos Islands had derived from those in South America, but that few had dispersed from one island of the archipelago to another (see *Journal of researches* 2d ed., p. 398, and *Origin*, pp. 398–403; see also *Correspondence* vol. 3).

[8] In his article, Hooker discussed the improbability of subsiding oceanic islands containing fossil mammals (J. D. Hooker 1866a, p. 75).

To John Murray 29 January [1867]

> *Down. | Bromley. | Kent. S.E.*
>
> Jan 29th

My dear Sir

I agree to your proposal about half profits. You must be the best judge, but I cannot avoid thinking that you will make a mistake if you print only 750 copies.[1] I beg you earnestly to get M.S. back immediately & let me hear immediately & I will send a servant *the same day* for it. I want to mark the passages for small type & to make some correction, before leaving home for a week's rest & I must pay my visit in about a weeks time.—[2]

In Haste | Yours very sincerely | Ch. Darwin

P.S. You will understand that I agree for half profits only for the first Edition, for with the weakness of an author I estimate my book at a higher value than you do.—

Endorsement: '1867. Jany 29'
John Murray Archive

[1] For Murray's proposals about his payment to CD for *Variation* and the numbers of copies he would print, see the letter from John Murray, 28 January [1867].
[2] In his letter of 28 January [1867], Murray said he hoped to send the manuscript to CD in the coming week. CD visited his brother, Erasmus Alvey Darwin, in London from 13 to 21 February 1866 (see 'Journal' (Appendix II)).

From John Murray 30 January [1867][1]

Albemarle St
Jan[y]. 30

My Dear Sir

I have this day got back the first portion of your MS. wch shall be delivered to your Servant whenever he may call—

The remainder will be in my hands Sat[y]. morn[g] & will also be delivered—if called for.[2]

I will not decide on the N[o]. to be printed until I can read a good portion— Our agreement will apply only to *First* Edition[3]

My Dear Sir | Yours very faithfully | John Murray

Chas Darwin Esq

DAR 171: 345

[1] The year is established by the relationship between this letter and the letter to John Murray, 29 January [1867].
[2] For CD's request for the manuscript of *Variation*, see the letter to John Murray, 29 January [1867]. In 1867, the Saturday after 30 January was 2 February.
[3] For CD's and Murray's agreement concerning the publication of *Variation*, see the letter from John Murray, 28 January [1867], and the letter to John Murray, 29 January [1867].

To John Murray 31 January [1867][1]

Down. | Bromley. | Kent. S.E.
Jan 31st

My dear Sir

Many thanks. After all I now find, on account of my servant & Horses, I cannot send till Saturday for the M.S. & then my servant can bring all & I will, as soon as it is marked, send all to Mess. Clowes in Stamford St—[2]

I feel a full conviction that my Chapter on man will excite attention & plenty of abuse & I suppose abuse is as good as praise for selling a Book.—[3]

Yours sincerely | C. Darwin

John Murray Archive

[1] The year is established by the relationship between this letter and the letter from John Murray, 30 January [1867].

[2] See letter from John Murray, 30 January [1867]. CD refers to the printer William Clowes & Sons.

[3] CD had been writing an additional chapter on humans for *Variation* that was not included in the manuscript he had sent to Murray (see letter to John Murray, 27 January [1867] and n. 2). Material in the chapter was ultimately used in *Descent* and *Expression*.

From D. Appleton & Co. to Asa Gray 1 February 1867

Statement of Sales of Darwin's Origin of Species[1]

to 1st Feby *1867*

by D. APPLETON & CO., for account of Asa Gray

On hand last account,	290	*On hand this date,*	148
Printed since,	100	*Given away,*	
		Sold to date,	242
	390		390

Sold 242 Copies. $2^{50} *Rate* 5% $ 30.25
Bal Jany 1/66.   85.61
$115.86

Document
DAR 159: A81

[1] The presence of the statement in the Darwin Archive–CUL indicates that Gray sent it to CD; Gray sent the previous year's statement as well, and may have enclosed it with his letter of 18 July 1866 (see *Correspondence* vol. 14). CD sent a letter to Gray on 28 February 1867 that has not been found (see letter from Asa Gray, 26 March 1867), and Gray may have included the statement in a letter to CD of early February that also has not been found. For more on the sale of the US edition of *Origin*, see *Correspondence* vol. 14, letter from Asa Gray, 7 May 1866 and n. 3.

From William Darwin Fox 1 February [1867][1]

Delamere Rectory | Northwich | Cheshire
Feb 1

My dear Darwin

I have so long kept my resolution of not writing to you,[2] knowing how your time is taken up, and how small your powers are—that I have now determined to reward my goodness by sending off a few lines, hoping that you will some day find time and strength to rejoice my eyes with your handwriting again, and that you will tell me as much as possible about yourself—Mrs Darwin & your family.

I cannot help hoping that I shall hear riding has done much for you.[3] If it suits you, it will be every thing to you. You get air and exercise without fatigue and must perforce, give that big brain of yours *some* rest.

How strange it seems that you and I are left alive, and poor Susan & Catherine taken away.[4] They seemed so healthy and strong, and we such runtlings in comparison.

I had such a nice cheerful letter from Susan only a few months before her sad sufferings came on that I quite hoped to have again seen her cheery face.[5]

What is become of the old house at Shrewsbury. What heaps of genuine kindness have I met within its walls.[6] I often look back to the days of joy & sorrow I passed there.

How is Caroline, and where? I should so much like to see her and her girls.[7]

I have looked anxiously for your Book on domestic animals. Is it coming out this Spring? How gets on the great Book.[8] Facts from your numerous correspondents in the World, must keep accumulating so much that it must seem almost to go back instead of forward.

I did fear you would never live to complete it, but I cannot help hoping now that you may live to a good old age, and become strong again.

We have just broken up our Christmas Party—& dispersed 4 of our children—to Oxford—Kings College—& London—while the rest are sitting down to their routine habits—and I to mine of teaching my 4th Boy Latin &c preparatory for school.[9]

Except myself we have wintered well, & the Boys have had plenty of skating. I have not had a good winter—my lungs having kept me a prisoner almost altogether. I hope however I shall now be able to get out & take air & exercise, both of which I much want. And now I will release you from reading this sad tract. Do let me have a few lines from you soon.

I hope Mrs Darwin and your children are all well. I have a very pleasing recollection of the face of your eldest Girl, whom I saw in London at Erasmus's.[10]

With our kindest regards to Mrs Darwin Believe me very dear D^n | Ever yours affect^ly & truly W D Fox

P.S. I have no less than *3* Free Martins here who have children—1 man 2 women, but one of the latter I find from her New Man had a near escape of being barren. She has however a fine Son.[11]

DAR 164: 185

[1] The year is established by the relationship between this letter and the letter to W. D. Fox, 24 August [1866] (*Correspondence* vol. 14).

[2] The last extant letter from Fox is that of 20 August [1866] (*Correspondence* vol. 14).

[3] In his letter of 24 August [1866] (*Correspondence* vol. 14), CD told Fox that he attributed the improvement in his health partly to his riding every day; see also letter from H. B. Jones, 10 February [1866] and n. 3.

[4] Two of CD's sisters had died the previous year; Emily Catherine Langton died in February 1866, and Susan Elizabeth Darwin in October 1866 (see *Correspondence* vol. 14).

[5] Susan fell ill several months before her death (see *Correspondence* vol. 14, letter to W. D. Fox, 24 August [1866]).

[6] Fox refers to The Mount, the Darwin family home in Shrewsbury, where Susan had lived until her death (see n. 3, above). Fox was a second cousin of CD's, and had become particularly close to him while they were students at the University of Cambridge (see *Correspondence* vol. 1, and Browne 1995, pp. 94–6).

[7] Fox refers to CD's surviving sister, Caroline Sarah Wedgwood, and to her three daughters, Katherine Elizabeth Sophy, Margaret Susan, and Lucy Caroline Wedgwood (Freeman 1978).

[8] In his letter of 24 August [1866] (*Correspondence* vol. 14), CD had told Fox of his progress on *Variation*, and on the fourth edition of *Origin*.

[9] Charles Woodd Fox, Fox's second eldest son, matriculated at Christ Church, Oxford, on 13 October 1865 (*Alum. Oxon.*). Robert Gerard Fox, Fox's third eldest son, was educated at King's College, London (Boase 1894). The other two dispersed children have not been identified, but Fox later mentioned that two of his daughters were at school in London (see *Correspondence* vol. 16, letter from W. D. Fox, [before 4 November 1868]). Fox's fourth son was Frederick William Fox, aged 11.

[10] Fox refers to Henrietta Emma Darwin, who must have been visiting her uncle, Erasmus Alvey Darwin, at 6 Queen Anne Street.

[11] Freemartin or free martin: a 'hermaphrodite or imperfect female' (*OED*). Fox asked CD whether he believed in freemartins in his letter of 6 February [1863] (*Correspondence* vol. 11), noting that in his parish one had recently been born and another had been miscarried. No replies from CD on this subject have been found. In *Descent* 1: 207, CD wrote that at a very early embryonic period, both sexes of vertebrates possessed 'true male and female glands'; he also concluded that some extremely remote progenitor of the vertebrates was androgynous or hermaphroditic, but added that the sexes were separate in vertebrates long before humans appeared (*ibid.* 1: 208). See also letter to William Turner, 15 January [1867] and n. 9.

To William Turner 1 February [1867]

Down. | *Bromley.* | *Kent. S.E.*

Feb. 1st

My dear Sir

I thank you cordially for all your full information, & I regret much that I have given you such *great* trouble at a period when your time is so much occupied.[1] But the facts are so valuable to me that I cannot pretend that I am sorry that I did trouble you; & I am the less so, as from what you say, I hope you may be induced some time to write a full account of all rudimentary structures in man: it would be a very curious & interesting memoir. I shall at present give only a brief abstract of the chief facts which you have so very kindly communicated to me, & will not touch on some of the doubtful points.[2] I have received far more information than I ventured to anticipate.

There is one point, which has occurred to me, but I suspect there is nothing in it. If, however, there sh^{d.} be, perhaps you will let me have a brief note from you; & if I do **not** hear I will understand there is nothing in the notion— I have included the down on the human body & the lanugo on the fœtus as a rudimentary representation of a hairy coat.[3] I do not know whether there is any direct functional connection* between the presence of hair & the *panniculus carnosus*, but both are superficial & would perhaps together become rudimentary.[4] I was led to think of this by the places, (as far as my ignorance of anatomy has allowed me to judge) of the rudimentary muscular fasciculi, which you specify.— Now some persons can move the skin of their hairy heads, & is this not effected by the panniculus? How is it with the eyebrows? You specify the axillæ & the front region of the chest & lower part of scapulæ: now these are all hairy spots in man.[5] On the other hand the neck, and as I suppose the covering of the gluteus medius,[6] are not hairy; so, as I said, I presume, there is nothing in this notion.— If there were, the rudiments of the Panniculus ought perhaps to occur more plainly in man than in woman.—

With very sincere thanks for all that you have done for me, & for the very kind manner, in which you granted me your favour, pray believe me | My dear Sir | Yours very sincerely | Ch. Darwin

*To put the question under another point of view: is it the primary or aboriginal function of the panniculus to move the dermal appendages or the skin itself?

P.S. If the skin on the head is moved by the Panniculus, I think I ought just to allude to it, as some men alone having power to move the skin, shows that the apparatus is generally rudimentary.

Edinburgh University Library, Dc.2.96/5 folio 3
Endorsement: '1867'

[1] See letter to William Turner, 15 January [1867]. Turner's reply has not been found. John Goodsir, professor of anatomy at the University of Edinburgh, and Turner's superior, was confined to his bed from the end of 1866 and died in March 1867. Turner, who succeeded him as professor, was probably occupied with Goodsir's professorial duties as well as his own as senior demonstrator. (*DSB*.)

[2] For CD's use in *Descent* of Turner's information on rudimentary organs, see the letter to William Turner, 15 January [1867] and nn. 4–7 and 9–11. CD also cited Turner in *Variation* 2: 300, 370, and in *Expression*, p. 101 n. 18.

[3] See *Descent* 1: 25–6; see also *ibid.* 2: 375–81.

[4] The panniculus carnosus is a thin sheet of striated muscle embedded in the lowest skin layer of lower mammals; it produces local movement of the skin. In humans, only vestigial remnants remain. See Landau ed. 1986.

[5] In his missing letter, Turner evidently supplied CD with information from a paper he read to the Royal Society of Edinburgh on 21 January 1867 (Turner 1867). In this paper, Turner described the occasional presence in humans of the musculus sternalis (a chest muscle), and suggested it was closely allied to the panniculus carnosus; CD noted Turner's paper in *Descent* 1: 19. In the same letter, Turner evidently also specified areas of rudimentary muscles in the armpits (axillae) and the shoulder blades (scapulae) (see *Descent* 1: 19–20). CD also discussed the panniculus carnosus in *Expression*, pp. 101, 298.

[6] CD refers to a muscle in the hip (gluteus medius).

From Fritz Müller 2 February 1867

Desterro,
Febr. 2. 1867.

My dear Sir

I am much obliged to you for your kind letter (without date)[1] in which you ask about the number of capsules produced by the Maxillaria with the larger pods.[2] I am told by a french collector M. Gautier that it is the Maxillaria tetragona; however his names are not always to be relied upon.—[3] On large plants growing on rocks and covering often more than a square-foot you may sometimes find half a dozen or more, whilst smaller plants growing on trees yield rarely more than one or two. I think that hardly 20% of the flowers produce seed-capsules.[4] The species is interesting in some other respects also. The bifid pedicelli of the pollinia execute the same movement, you have described in Maxillaria ornithorhyncha, the right and left pollen masses are connected by an elastic tissue like that of the caudicle.[5]

Whereas in all other Vandeae, I examined (Notylia, Cirrhaea, Ornithocephalus, Polystachya, Aëranthus, Oncidium, Cyrtopodium and several other species of Maxillaria), the margins of the valves of the seed-capsules are beset with hygroscopic hairs, I could not find them in the ripe pods of this species; but there are some scattered and rudimentary hairs in the young capsules.[6]

I send you pollen-tubes taken out of a young pod of this Maxillaria showing the fringes which from the six longitudinal ribbons extend between the ovula. On comparing these pollen-tubes with those of Cattleya Leopoldi[7] you will be struck by a remarkable difference.

In Maxillaria (and with a single exception in all our Vandeae) the pollen-tubes remain fresh in their whole length; in Cattleya (as well as in all our Epidendreae) the upper part of the pollen-tubes soon becomes dry and black; this evidently is a consequence of the stigmatic chamber being shut in Vandeae, whilst it remains open and the pollen exposed to air in Epidendreae.[8]

I already told you in my last letter, that in Notylia and in Oncidium flexuosum, pollen and stigma of the same plant act, as it were, as a poison on each other; this is also the case with the curious Oncidium unicorne and with another species which seems to be nearly related to Oncidium pubes.[9]

I have had numerous racemes of a second species of Cirrhaea (perhaps the C. dependens Rchb. f.) interesting by the extreme variability of the colours of its flowers. I could not fertilize this species with fresh pollinia whilst it might be done easily after they had dried half an hour or an hour.[10] This probably will prevent the flowers being fertilized with the same plants pollen. The disk in this species is provided with a hook facilitating its being taken away.

There begins now to flower here a curious small Orchid, viz. Ornithocephalus.

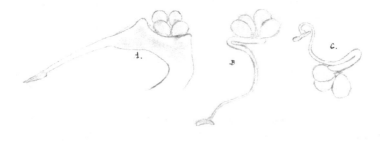

The anther-bed has a small transverse ridge on the anterior margin; the pedicellus of the pollinium passing over this ridge forms an obtuse angle; as soon as the pollinium is removed by an elastic binding of the pedicellus, this obtuse angle is transformed into a very acute one and subsequently by an hygroscopic movement, the pedicellus is curved in a very singular and elegant manner. In water the pedicellus returns to the former form.[11] The two pods of Monachanthus which I fertilized (Dec. 25) with pollen of Catasetum mentosum are now already 7 cm long and as much in circumference.[12]

The plant of Oncidium flex., on which I had fertilized some flowers simultaneously with pollen from a distinct plant of the species and with pollen of Epidendrum Zebra, having perished by an accident I have repeated the experiment on another plant; the result has been the same and I have satisfied myself that it is indeed the side of the Epidendrum-pollen which grows less and that of the Oncidium-pollen which grows more rapidly.[13]

I have made in the first weeks of January a pleasant excursion on the continent, up the river Cubatão to the German colony Theresopolis;[14] and have brought home a fine collection of living Orchids, among which a large plant probably belonging to the Catasetidae[15] You state in your Orchis-book that there are flourishing in your neighbourhood 13 species of Orchids; now on the Cubatão you may collect on the branches of a single old Cedrela-tree even a larger number of species: half a dozen Maxillariae, four or five small Pleurothallidae, a couple of Oncidia, the Leptotes bicolor etc. The beautiful Miltonia cereola was very abundant.—[16] The most striking feature of the Orchid-flora of Theresopolis was the abundance and variety of Maxillariae and the entire absence (as far I have seen) of the Cattleyae, the Amphiglott Epidendra,[17] the Brassavola, which are the dominant littoral forms. Epidendra in general seem to be much more rare than here, where I already know 13 species. . . .

Draft, incomplete[18]
Möller ed. 1915–21, 2: 109–11; DAR 70: 146

[1] Letter to Fritz Müller, [before 10 December 1866] (*Correspondence* vol. 14).

[2] In his letter to CD of 1 and 3 October 1866 (*Correspondence* vol. 14), Müller had estimated that there were 1,756,440 seeds in one *Maxillaria* capsule. CD told him he wanted to publish this fact and asked how many capsules were produced (*ibid.*, letter to Fritz Müller, [before 10 December 1866]).

[3] *Maxillaria tetragona*, native to Santa Catarina state, is now usually known as *Bifrenaria tetragona* and less often as *Cydoniorchis tetragona*. An old synonym was *Lycaste tetragona*. Hippolyte Gautier has not been further identified; Müller also mentioned him in his letter of 2 November 1866 (*Correspondence* vol. 14).

[4] In *Variation* 2: 379 n. 34, CD mentioned that the *Maxillaria* described by Müller sometimes produced six capsules; see also 'Fertilization of orchids', p. 158 (*Collected papers* 2: 155), and *Orchids* 2d ed., p. 278.

[5] For CD's description of the pollination mechanism in *Maxillaria ornithorhyncha*, see *Orchids*, p. 191. For definitions of CD's terms for orchid structures, see *Orchids*, pp. 5–8.

[6] Müller had earlier mentioned hairs on the capsules of some orchids (see letter from Fritz Müller, 1 January 1867 and n. 19).

[7] *Cattleya leopoldii*.

[8] CD did not mention the varying conditions of the pollen-tubes in Epidendreae and Vandeae in the second edition of *Orchids*.

[9] See *Correspondence* vol. 14, letter from Fritz Müller, 1 December 1866, and this volume, letter from Fritz Müller, 1 January 1867, for Müller's descriptions of the poisonous effect of same-plant pollen in *Oncidium flexuosum* and *Notylia*. CD reported Müller's finding in *Variation* 2: 134–5. See also *Cross and*

self fertilisation, pp. 330, 341. Müller published his observations on the effects of same-plant pollen in F. Müller 1868a.

[10] For Müller's earlier observation on an unnamed species of *Cirrhaea*, see the letter from Fritz Müller, 1 January 1867 and n. 17. CD mentioned Müller's observation of the second species in *Orchids* 2d ed., p. 172. Müller's published account of *Cirrhaea* is in F. Müller 1868b and was cited by CD in 'Fertilization of orchids' (*Collected papers* 2: 150).

[11] 'Anther-bed': that portion of the column under, or surrounding, the anther (Dressler 1981, p. 308 s.v. 'clinandrium'). CD included Müller's account of the movement of the *Ornithocephalus* pedicel (now stipe) in *Orchids* 2d ed., pp. 159–60. Figure 14 (*ibid.*, p. 160) was taken from Müller's enclosed sketch of this movement.

[12] See letter from Fritz Müller, 1 January 1867 and nn. 13–17.

[13] See letter from Fritz Müller, 1 January 1867 and n. 5.

[14] Müller lived in Destêrro (now Florianópolis) on Santa Catarina Island. Rio Cubatâo flows into the Atlantic off the south end of the island. Theresopolis, on the Rio Cubatâo, is now Queçaba (West 2003).

[15] CD and Müller considered the Catasetidae a subfamily within the tribe Vandeae (*Orchids* 2d ed., p. 178); for a recent classification, see Dressler 1993, pp. 168–70.

[16] See *Orchids*, p. 345. In *Orchids* 2d ed., p. 279, CD added that Müller found more than thirteen species of orchid on a single *Cedrela* tree.

[17] *Amphiglottis* is a synonym (no longer accepted) for some species of *Epidendrum*.

[18] This letter was found by Alfred Möller and published in *Fritz Müller: Werke, Briefe und Leben* (Möller ed. 1915–21). According to Alfred Möller, all of Fritz Müller's letters to CD were written in English (see Möller ed. 1915–21, 2: 72 n.); most of them have not been found. Many of the letters were later sent by Francis Darwin to Möller, who translated them into German for his book. Möller also found final drafts of some letters (*ibid.*, pp. 72 n.) and published these in their original English version. The present letter is one of these. Müller's drawings of the pollinium of *Ornithocephalus* were excised from the original letter and are in DAR 70: 146. They are included as part of the letter.

From Erasmus Alvey Darwin to Emma Darwin [before 3 February 1867?][1]

Dear Emma

There are 3 other places that Charles ought to initial, one in the same paragraph & 2 in the one below.[2]

I am very glad to hear that you are thinking of coming & shall be most happy to see you the day you propose the 4th.[3] I suppose Hensleigh is undoubtedly getting better but it a very slow process & the account that I hear every day is 'just the same' but by the end of the week he seems better.[4]

yours affec | E D

DAR 105: B122–3

[1] The date is conjectured from visits of Henrietta Emma Darwin, and CD and Emma Darwin to London (see n. 3, below), and from Hensleigh Wedgwood's poor health (see n. 4, below).

[2] The document needing signatures has not been identified.

[3] Emma Darwin's diary (DAR 242) records that Henrietta went to London on 4 February 1867, and CD and Emma on 13 February 1867; CD and Emma may have intended to visit earlier with Henrietta.

[4] Hensleigh Wedgwood's illness is also mentioned in the letter from J. D. Hooker, 4 February 1867, and the letter to J. D. Hooker, 8 February [1867].

From J. D. Hooker 4 February 1867

Kew
Feby 4/67.

Dear Darwin

How gets on the book?.[1] I send a number of the Nation, with an article on Popular Lectures &c that seems to me remarkably good. I wonder how Agassiz likes it!—[2] A. Gray[3] sends me the paper regularly.

I yesterday declined the invitation to be President of Brit: Assoc: at Norwich in 1868 with *much pain* at having to refuse so very flattering an invitation so very kindly conveyed;[4] but the fact is that I have an insuperable aversion to high places. the acceptance would have been bad dreams in anticipation for 18 months, & a downright surgical operation at the end of it!— I believe I inherit this from my father,[5] who never would put himself forward, or be put forward, & I am sure it *paid in the end*. I was also actuated by the fact that I can see no way to a good "*Address*"— I played out my Trump card at Nottingham, knowing that if I were called to be President (which I had already good reason to expect) & accepted, I was throwing away my best chance of success.[6] Lastly it would stand *terribly* in the way of my work—both Genera Plantarum, & Insular Botany.[7] Here above is a pretty dose of Egotism even from one friend to another

I am reprinting the Lecture in same type but other form for distribution—[8] the only thing I do not like & could not conscientiously consult you about, was the passage about "a Wise Providence ordering &c &c" or something of that sort. (I forget the words, it matters little)[9]

It is bosh & unscientific, but I could not resist the opportunity of turning the tables of Providence over those who think & argue the contrary of its intentions—& showing those who will have a Providence in the affair, that your's is the God one, theirs the Devils. I always felt, that if I had to print the Lecture, I should wish these passages cut out, but that this would be dishonest—so it even went forth in G. C. & now will further

What do you say to Owen's assuring me that Mammals bones, (*Deer*) are found in bogs in Mauritius.[10] he told me this himself.

Grove told me, at an excellent Phil. Club meeting, that Owen in his Dodo-paper, has taken again to himself the credit of continuity theory, & of showing the **futility** of the Type theory—[11] I wish Grove would pitch into him.

Bentham is doing Umbelliferæ for Gen. Plant, & finds that the two remarkable Umbelliferous genera of Madeira, *Monizia* & *Melanoselinum*, are only species of *Thapsia*, a Mediterranean genus, of most remarkable & exceptional habit.[12] Now this is one of those cases of genera *confined* to the Island being then created out of a continental form; the genus I suspect; not having ever existed on the continent. It appears to me that it will always be difficult to say whether a genus *that has continental allies*, is an Insular development, or an old now extinct continental genus. & the utter want of fixed system upon which genera are & *must always* be formed will always throw insuperable obstacles in the way of this enquiry— it is easy enough with regard to the Laurels & other things having no continental affinities.[13]

I dined at Murchisons[14] on Saturday, for the first time, for many years. What a charming host he makes at a small table.

I heard very poor accounts of H. Wedgwoods[15] health how is this?

The frost has done us a power of damage[16]

My wife sends her kind love to Mrs Darwin—& yourself— I took her just now to the Drawing room for the first time[17]

Ever yrs Aff | J D Hooker

Have you read the Duke's "Reign of Law", & if so is it worth reading?[18]

[Enclosure]

What I mean about Providence is this— I think & believe that all reasoning upon the subject is utterly futile—that there is no such thing in a scientific sense—but that whereas those who deal in it, hold that the theory of fixed types & creations is the only one consonant with a belief in a Providence I hold, that they are wrong & that the theory of continuity & variation is the only one consonant with the belief

DAR 102: 138–142

CD ANNOTATIONS
1.1 How ... book?.] *scored pencil*
1.1 article ... good. 1.2] *scored pencil*
2.1 I ... Assoc:] *scored pencil*
3.1 I am ... distribution—] *scored pencil*
3.3 "a Wise ... that sort.] *scored pencil*
5.1 What ... himself. 5.2] *double scored pencil*
6.2 & of ... theory— 6.3] *double scored pencil*
9.1 I heard ... this?] *double scored pencil*
11.1 I took ... time 11.2] *scored pencil*
13.1 Have ... reading?] *double scored pencil*
End of letter: 'Spectator— Abortive organs'[19] *pencil*

[1] Hooker refers to *Variation*, the manuscript of which CD had sent to his publisher in December 1866 (see *Correspondence* vol. 14). See also letter to J. D. Hooker, 15 January [1867] and n. 10.

[2] The author of 'Popularizing science' focused on a number of Louis Agassiz's lectures and publications to illustrate the 'distinct dangers' of popularisation (Anon. 1867, p. 33); one danger cited was the temptation to present only one side of a controversy. The author gave an example from recent lectures by Agassiz in Boston; under the guise of describing the natural history of Brazil, Agassiz claimed to have refuted CD's transmutation theory, but did not precisely state CD's position. The lectures mentioned may have been those given before the Lowell Institute in September and October 1866 (see Lurie 1960, p. 353). Agassiz had recently published an article in the *Atlantic Monthly*, a popular periodical (J. L. R. Agassiz 1866).

[3] Asa Gray.

[4] The British Association for the Advancement of Science met at Norwich, 19 to 26 August 1868 (*Report of the thirty-eighth meeting of the British Association for the Advancement of Science held at Norwich*). The president always gave an address.

[5] William Jackson Hooker.

6 Hooker gave a well-received lecture on insular floras at the annual meeting of the British Association in Nottingham in August 1866 (see *Correspondence* vol. 14, letter to J. D. Hooker, 30 August [1866]).

7 Hooker had been collaborating with George Bentham on *Genera plantarum* (Bentham and Hooker 1862–83) since 1860; the second of seven parts appeared in 1865, and the third in October 1867. Hooker had occasionally mentioned his intention of writing a book on insular botany and general geographic distribution (see, for example, *Correspondence* vol. 13, letter from J. D. Hooker, 1 January 1865), and had considered this further after his lecture on insular floras (see n. 6, above, and *Correspondence* vol. 14, letters from J. D. Hooker, [22 November 1866] and 4 December 1866).

8 Hooker refers to the type used when his lecture 'Insular floras' was published in the *Gardeners' Chronicle* (J. D. Hooker 1866a). When the lecture was reprinted as a pamphlet he made changes that took account of some of CD's criticisms offered in his letters of 9 January [1867], 15 January [1867], 21 January [1867], and 29 January [1867]. Hooker had been concerned about the type-size that the *Gardeners' Chronicle* would use for the article (*Correspondence* vol. 14, letter from J. D. Hooker, 28 September 1866). For a transcription of his pamphlet, see Williamson 1984.

9 See enclosure. Towards the end of the lecture published in the *Gardeners' Chronicle*, Hooker wrote (J. D. Hooker 1866a, p. 75):

> By a wise ordinance it is ruled, that amongst living beings like shall never produce its exact like; that as no two circumstances in time or place are absolutely synchronous, or equal, or similar, so shall no two beings be born alike; that a variety in the environing conditions in which the progeny of a living being may be placed shall be met by variety in the progeny itself. A wise ordinance it is, that ensures the succession of beings, not by multiplying absolutely identical forms, but by varying these, so that the right form may fill its right place in Nature's ever varying economy.

He retained the paragraph in his pamphlet (Williamson 1984, p. 76). Hooker discussed his 1866 Nottingham lecture at length with CD in July and August 1866 (see *Correspondence* vol. 14). See also n. 8, above.

10 In his paper 'Account of the late discovery of dodo's remains in the island of Mauritius', George Clark noted that remains of a deer were also found in the marsh where the dodo's remains were; he added that he sent Richard Owen and Alfred Newton all the bones he found (G. Clark 1866, pp. 144, 146). Owen described the dodo in R. Owen 1866, but the only other animal remains he mentioned as being found with the dodo bones were the carapaces and skulls of tortoises.

11 Hooker refers to William Robert Grove, and to the Philosophical Club of the Royal Society of London. On Owen's supposed claim of priority to the 'continuity theory' (transmutation of species), see *Correspondence* vol. 14, letter to J. D. Hooker, 31 May [1866] and n. 11. In 'On the osteology of the dodo (Didus ineptus *Linn.*)' (R. Owen 1866), Owen claimed that there were affinities between the dodo, an extinct flightless bird, and members of the Columbidae (pigeons and doves) that lived or had lived on Mauritius. He argued that through the 'long course of successive generations' with food available on the ground and no predators, a pigeon could cease using its wings, and develop a heavy body and stronger legs, and that this was how the dodo originated (R. Owen 1866, p. 70). In support of his argument, he cited Lamarck 1809 and Georges Louis Leclerc, comte de Buffon (R. Owen 1866, p. 80), but not CD. In R. Owen 1866, Owen criticised the 'type-form' described in Strickland and Melville 1848, that is, the notion that the 'Creator' assigned to each class of animals 'a definite type or structure' that did not change (*ibid.*, p. 81); however, in the 1840s, Owen had espoused the related notion of a vertebrate archetype (see Rupke 1994, pp. 193–204, and Camardi 2001). For CD's comments on the type theory, see *Correspondence* vol. 4, letter to H. E. Strickland, [19 February 1849], and *Correspondence* vol. 5, letter to T. H. Huxley, 23 April [1853]; see also *Correspondence* vol. 4, Appendix II.

12 See Bentham and Hooker 1862–83, 1: 930, for Bentham's description of *Thapsia*. Hooker mentioned *Monizia edulis* and *Melanoselinum* when noting how odd it would be if some of the plants on Madeira were found on a British island or mountain (see J. D. Hooker 1866a, p. 7, and letter to J. D. Hooker, 9 January [1867] and n. 6). Hooker did not mention *Thapsia* in the published versions of the lecture.

[13] In his letter to Hooker of 15 January [1867], CD had suggested that most of the genera, though not the species, that were distinctive to Atlantic islands had been derived from Europe and had since become extinct there. See also *Origin*, pp. 106–7, 397–9, 403–6, and letter from J. D. Hooker, 20 January 1867. Hooker mentioned the laurels in his paper on insular floras as one of the groups of the Atlantic island type that were 'contradistinguished from European' (J. D. Hooker 1866a, p. 7; see also p. 50).

[14] Roderick Impey Murchison.

[15] Hensleigh Wedgwood.

[16] Hooker was director of the Royal Botanic Gardens, Kew. For an account of the destructive frost, see Allan 1967, p. 217.

[17] Frances Harriet Hooker had given birth to Reginald Hawthorn Hooker on 12 January (see letter from J. D. Hooker, [12 January 1867]).

[18] *The reign of law* (G. D. Campbell 1867) was written by George Douglas Campbell, duke of Argyll (see also letter from T. H. Huxley, [before 7 January 1867]).

[19] CD's annotation was for his letter to Hooker of 8 February [1867].

From Lydia Ernestine Becker 6 February 1867

Manchester Ladies' Literary Society. | 10 Grove s^t | Ardwick

Feb. 6. 1867.

My dear Sir

I return you—with more thanks than I know how to express, the two papers which you were so good as to entrust to my care.[1] Will you have the kindness to cause me to be informed of their arrival—having once lost a book-post packet I shall feel a little anxious till I hear they are again in your hands—and this induces me to give you the little extra trouble involved in registering the packet—for which I must apologise.

I have transcribed portions of them, and made large copies of the diagrams— I hope this was not wrong—without your permission, but I thought, as they were printed—I might do so without impropriety.

The arrangements in *Lythrum* are indeed most marvellous. It sets one wondering whether different sized stamens in the same flower can ever be quite without meaning, and if there is any difference in the action of the pollen of the long and short stamens in didynamous and tetradynamous flowers.[2] In the N. O. *Geraniaceae*[3] it seems as if there might be some transition going on—for in *Geranium* each alternate stamen is smaller, and in the allied genus *Erodium* the alternate stamens have become sterile. Can it be possible that this genus was once dimorphic, and one of the female forms having by any means become exterminated, the corresponding set of stamens have shed away? If one of the forms of Lythrum were to disappear—two sets of stamens would be made useless to the species, and it is conceivable that they might then gradually become abortive.[4]

I obeyed your directions about the paper on Climbing Plants and the insight into their extraordinary and regular movements was a new revelation to all of us.[5] I made large copies of the diagrams and dived into my herbarium for specimens of each class of climbers, bringing up enough to make a goodly show. Luckily a collection of ferns from the islands of the South Pacific recently presented to me contained a specimen of one named in your paper *Lygodium scandens*. Till I read it I had never dreamed of twiners in this class, as none of our British ferns have the habit,[6] but as the "march of intellect" seems to be the order of the day, even in the vegetable world, there is no telling what they may accomplish in time!

Our society appears likely to prosper beyond my expectations the countenance you have afforded has been of wonderful service, and I do hope that by becoming useful to its members it may prove in some degree worthy of the generous encouragement you have given us.

The ladies who had the privilege of listening to the paper desire to express their thanks to you for it, which I hope you will be pleased to accept.

Believe me to be | yours gratefully | Lydia E. Becker.

DAR 160: 115

[1] Becker refers to 'Climbing plants' and 'Three forms of *Lythrum salicaria*'. Becker had asked CD for a published paper of his that would be suitable for reading at the first meeting of the Manchester Ladies' Literary Society (see *Correspondence* vol. 14, letter from L. E. Becker, 22 December 1866).

[2] Didynamous plants have two long stamens and two short, while tetradynamous plants have four long stamens in two pairs and two short; both terms, which CD did not use, signified classes in the Linnaean classification system. *Lythrum* has three forms, one with six long and six short stamens, one with six long and six mid-length stamens, and one with six short and six mid-length stamens (see 'Three forms of *Lythrum salicaria*'). CD and Becker had earlier corresponded on the possible dimorphism in *Lychnis diurna* (see *Correspondence* vol. 11).

[3] The 'natural order' Geraniaceae (Lindley 1853, p. 484).

[4] CD discussed whether or not dimorphism and trimorphism were related to a gradual separation of the sexes, that is, to dioeciousness, in 'Three forms of *Lythrum salicaria*', pp. 194–6 (*Collected papers* 2: 126–8).

[5] CD's letter to Becker with the directions has not been found (see n. 1, above). Becker presented 'Climbing plants' to the inaugural meeting on 30 January 1867 of the Manchester Ladies' Literary Society, of which she was president (Blackburn 1902, p. 31); for the portion of her address preceding her presentation of the paper, see *ibid.*, pp. 31–9. For more on Becker, see Shteir 1996, pp. 226–31.

[6] CD mentioned *Lygodium scandens* in 'Climbing plants', pp. 14, 22–3. For his acquisition of this twining fern, see *Correspondence* vols. 11 and 12.

To W. D. Fox 6 February [1867]

Down Bromley Kent
Feb 6th

My dear Fox

It is always a pleasure to me to hear from you, & old & very happy days are thus recalled.[1] This is rather a joyful day to me, as I have just sent off the M.S for two huge volumes (I grieve the Book is so big) to Printers on Domestic Animals &c &c but my book will not appear, even if completed, before next November, as Murray has strong prejudice against publishing except during Spring & Autumn.[2] I am utterly in darkness about merit of my present book; all that I know is that it has been a most laborious undertaking. Of course a copy will be sent to you.—

It is true indeed that Death has been busy with us, & it is astonishing to me that I sh^d. have survived my two poor dear sisters.[3] The old House at Shrewsbury is on sale, but has as yet found no purchaser, & I daresay will not soon.— All the furniture was sold by Auction, having been bequeathed to the Parkers, who had become like Susan's children.[4]

Caroline & Erasmus are fairly well for them; but this is not saying much for them, especially for the latter, who does not often leave the House.[5] I am so sorry to hear so poor an account of yourself; with your active habits being confined must be a terrible deprivation. You are quite right about riding; it does suit me admirably, & I am very much stronger; yet I never pass 12 hours without much energetic discomfort.[6] But I am fairly well content, now that I am no longer quite idle.— Poor Bence Jones has been for months at death's door, & was quite given up; but has rallied in surprising manner from inflammation of Lungs & heart-disease.[7] My wife is fairly well but suffers much from repeated headachs,[8] & the rest of us are well.— I hope you will get all right with returning Spring.

My dear old friend | Believe me; | Yours affectionately | Ch. Darwin

Postmark: Bromley FE 7 67
Endorsement: 'C. Darwin | Feb 8. 1867'
Christ's College Library, Cambridge (Fox 147)

[1] See letter from W. D. Fox, 1 February [1867].

[2] CD sent the manuscript of *Variation* to John Murray in December 1866 (see *Correspondence* vol. 14). In early February, he sent for the manuscript again so that he could mark sections for smaller type in order to reduce the size of the book (see letters to John Murray, 8 January [1867] and 29 January [1867]). On Murray's reluctance to publish before autumn, see also the letter to W. B. Tegetmeier, 6 January [1867] and n. 3.

[3] CD refers to his late sisters, Emily Catherine Langton and Susan Elizabeth Darwin (see letter from W. D. Fox, 1 February [1867] and n. 4).

[4] Susan had lived at The Mount, the Darwin family home in Shrewsbury; following her death on 3 October 1866, CD's surviving siblings, Erasmus Alvey Darwin and Caroline Sarah Wedgwood, travelled to Shrewsbury to sort out some of the family belongings in the house (see *Correspondence* vol. 14, letter from E. A. Darwin, 11 October [1866]). The remaining contents were put on sale from 19 to 24 November 1866 (*Shrewsbury Chronicle*, 16 November 1866; see also *Correspondence* vol. 14, letter from J. D. Hooker, [22 November 1866]). The house was let and eventually sold (see letter from Salt & Sons, 17 July 1867). CD's eldest sister, Marianne, married Henry Parker (1788–1856); after the Parker parents died, their grown-up children, Robert, Henry, Francis, Charles, and Mary Susan, had stayed at The Mount when in Shrewsbury (Freeman 1978).

[5] Fox had asked about CD's surviving sister, Caroline, in his letter of 1 February [1867]. The invalidism of CD's elder brother, Erasmus, is discussed in B. Wedgwood and Wedgwood 1980, p. 233. Erasmus was also depressed following Susan's death in October 1866; Susan had been his favourite sister (see n. 4, above, B. Wedgwood and Wedgwood 1980, p. 288, and Browne 2002, p. 267).

[6] See letter from W. D. Fox, 1 February [1867] and n. 3.

[7] Henry Bence Jones, a physician who had treated CD in 1865 and 1866 (see *Correspondence* vols. 13 and 14), became dangerously ill in September 1866 and had been close to death in January 1867 (Kyle 2001, p. 16). For more on Jones's recovery, see the letter from H. B. Jones to Emma Darwin, 1 October [1867].

[8] Emma Darwin noted headaches in her diary for 6, 7, and 8 February 1867 (DAR 242).

From William Henry Kinnaird Gibbons 7 February 1867

My dear Sir

Should your memory not have been impaired by your eminent literary successes my name will not (I think) be unfamiliar to you—![1]

I should indeed be ungrateful if I could forget *yours* as I believe that I am mainly indebted to your late fathers great medical skill for the prolongation to a very advanced age of the lives of my late revered parents with whom I resided at *Harley* when in the country) until their deaths.[2] I purchased some years ago a landed property of 500 acres here upon which I am occasionally resident but I have now also a House in London

I have sometimes thought of a drive to Down to see you and should probably have accomplished this but that I did not know whether you were generally resident there or what might be your engagements. I should be glad however of an opportunity of conversing with you upon the very interesting and profound subject upon which your mind has been engaged and in reference to which your researches have conferred such benefits upon man.

It has occurred to me that as you have thought so deeply upon the *"origin of species"* you must have given your attention also to the great question of mans **future destiny** after the separation of the *thinking principle* (or mind) from the body! and that consequently you may have investigated the phenomena of *Spiritualism?*

Having myself paid a good deal of attention to this subject I feel a very deep interest in it and from an extensive experience of the several phases of the Spiritual evidences I have arrived at the conclusion that what I have witnessed is not explicable upon any *other assumption* than that the *minds* of the *departed* are somehow exercising an *active agency* by which we are influenced!

The rapidity with which these opinions are spreading throughout Society must (I think) soon lead to some more careful enquiry into the subject than it has yet received[3]

The interesting work of Professor Gregory on animal magnetism The preface of De Morgans Volume, "From matter to Spirit" and some other publications upon Spiritualism seem to leave us without much *definite* information upon this profound question and should *you* have formed any opinions upon it I shall feel greatly indebted to you for it.[4]

I have just finished reading the Duke of Argylls new work *"The Reign of law"* (in which he takes some liberties with your name) and I perceive that in his last Chapter he attaches great importance to the spiritual phenomena *"which are so much ignored by the men of science"*![5]

If you have not read the work you will find it worth your attention.

Hoping that you are in the enjoyment of that greatest blessing—good health | I am | My dear Sir | Yours very faithfully | W H K Gibbons

The Old Lodge | East Grinstead | Sussex | Feb 7th 1867

DAR 165: 36

[1] Gibbons had lived until the late 1850s in Harley, Shropshire, a village ten miles south-east of Shrewsbury; his father had been rector of Harley (*Post Office directory of Gloucestershire, with Bath, Bristol, Herefordshire, and Shropshire* 1856). CD and his family were also social acquaintances of Gibbons and his family (*Shrewsbury Chronicle*, 20 September 1825 and 22 September 1826); W. H. K. Gibbons is probably the Mr Gibbon mentioned in the letter from Caroline and Susan Darwin, 2 [January 1826] (*Correspondence* vol. 1). See also n. 2, below.

[2] Gibbons refers to Robert Waring Darwin, John Gibbons, who died in 1858 aged nearly 90, and Helen Gordon Gibbons, who died in 1855 in her early 80s.

[3] For CD's recent and positive reaction to an article critical of spiritualism ([Tyndall] 1864c), see *Correspondence* vol. 13, letter to J. D. Hooker, 7 January [1865] and n. 14. The efforts of John Tyndall and others to denounce spiritualism are discussed in Oppenheim 1985, pp. 327–30. For studies of Victorian spiritualism, see L. Barrow 1986, A. Owen 1989, and Fichman 2004, pp. 139–210.

[4] Gibbons refers to William Gregory's *Letters to a candid enquirer on animal magnetism* (W. Gregory 1851). Sophia Elizabeth De Morgan wrote *From matter to spirit. The result of ten years' experience in spirit manifestations* ([De Morgan] 1863); her husband, Augustus De Morgan, a logician and mathematician and not a confirmed spiritualist, wrote the preface (see Oppenheim 1985, pp. 335–6, and Fichman 2004, pp. 178–80).

[5] CD had not yet read G. D. Campbell 1867, but had read some of the earlier writings of George Douglas Campbell, the duke of Argyll (see letter to J. D. Hooker, 8 February [1867]). Thomas Henry Huxley had recently mentioned that he had heard that Campbell criticised both him and CD in G. D. Campbell 1867 (see letter from T. H. Huxley, [before 7 January 1867] and n. 16). In G. D. Campbell 1867, p. 428, Campbell wrote: 'if the methods and conditions of Physical inquiry were applied in a really philosophical spirit to Spiritual Phenomena, the influence of Science would be more powerful than it is'; the words Gibbons quotes are not in G. D. Campbell 1867.

To Fritz Müller 7 February [1867][1]

Down. | Bromley. | Kent. S.E.

Feb 7

My dear Sir

I feel very guilty at not having sooner thanked you for your letter of Dec 1. which contained much valuable information.[2] The reason has been that I have been very busy in making alterations in my M.S. for my book on domestic animals &c which is at last in the printer's hands.[3] I am very much obliged for the seeds of Cordia &c which are planted & if I can make them flower will be highly interesting to me.[4] Your letter told me much on many points of value to me. You have communicated to me many more cases than any two or three botanists put together.

I have quoted your evidence on the self-sterility of Ocidium flexuosum;[5] it is said to be a native of Brazil, but I much wish to know whether it is a native of your district.

I have many striking cases of similar self-sterility with Orchids, but have hitherto always attributed them to cultivation under unnatural conditions & this makes your case very interesting to me.[6] This is a very dull letter, but I did not wish to defer any longer thanking you.

Believe me my dear Sir | yours very sincerely | Ch. Darwin

P.S. Would you be so good as to tell me whether I ought to direct to you as Professor or Dr Fritz Müller

LS
British Library (Loan 10: 12)

[1] The year is established by the relationship between this letter and the letter from Fritz Müller, 1 April 1867.

[2] See *Correspondence* vol. 14, letter from Fritz Müller, 1 December 1866.

[3] CD had sent the manuscript for *Variation* to William Clowes & Sons the previous day (see letter to W. D. Fox, 6 February [1867] and n. 2).

[4] Müller may have included seeds of *Cordia*, a genus he thought was dimorphic, with his letter of 1 December 1866 (*Correspondence* vol. 14); he had sent specimens of *Cordia* earlier in that year, and had promised to send seeds (see *ibid.*, letters from Fritz Müller, 1 and 3 October 1866). CD discussed *Cordia* in *Forms of flowers*, pp. 117–18, 253.

[5] For Müller's earlier discussions on the self-sterility of *Oncidium flexuosum*, see *Correspondence* vol. 14, letter from Fritz Müller, 1 December 1866 and n. 8, and this volume, letters from Fritz Müller, 1 January 1867 and nn. 1, 2, and 5, and 2 February 1867 and n. 9. CD reported Müller's findings in *Variation* 2: 134–5.

[6] See *Variation* 2: 133–4; see also *Correspondence* vol. 12, letter to Daniel Oliver, 18 March [1864] and n. 7. CD discussed his conclusions on this subject in *Cross and self fertilisation*, pp. 340–7.

To J. D. Hooker 8 February [1867]

Down Bromley Kent
Feb 8th

My dear Hooker

I am heartily glad that you have been offered the Presidentship of the B. Assoc. for it is a great honour, & as you have so much work to do I am equally glad that you have declined it.[1] I feel, however, convinced that you would have succeeded very well; but if I fancy myself in such a position it actually makes my blood run cold. I look back with amazement at the skill & taste with which the D. of Argyll made a multitude of little speeches at Glasglow.[2] By the way I have not seen the Duke's book,[3] but I formerly thought that some of the articles which appeared in periodicals were very clever, but not very profound. One of these was reviewed in the Saturday Review some years ago; & the fallacy of some main argument was admirably exposed, & I sent the article to you, & you agreed strongly with it.[4] Now I have forgotten this counter-argument & I know I shall be humbugged by the Duke, if I reread him as I suppose I must. There was the other day a rather good review of the Duke's book in the Spectator, & with a new explanation, either by the Duke or Reviewer (I could not make out which) of rudimentary organs; viz that economy of labour & material was a great guiding principle with God (ignoring waste of seed & of young, monsters &c &c), & that making a new plan for the structure of animals was thought & thought was labour, & therefore God kept to a uniform plan & left rudiments.[5] This is no exaggeration. In short, God is a man rather cleverer than us: I wonder they did not suggest that he would suffer from indigestion, if he worked his brains too much.— I am very much obliged for the "Nation" (returned by this post): it is *admirably* good:[6] you say I always guess wrong, but I do not believe anyone, except Asa Gray could have done the thing so well. I would bet even, or 3 to 2, that it is Asa Gray, though one or two passages staggered me.

I finish my Book on "Domestic Animals &c" by a single paragraph answering, or rather throwing doubt, in so far as so little space permits on Asa Gray's doctrine that each variation has been specially ordered or led along a beneficial line. It is foolish to touch such subjects, but there have been so many allusions to what I think about the part which God has played in the formation of organic beings, that I thought it shabby to evade the question.[7] I have even received several letters on subject. One was a funny one from a lady with a whole string of questions, & when I said I could not answer one; she wrote she was perfectly satisfied & it was exactly what she expected.—[8] I overlooked your sentence about Providence, & suppose I treated it as Buckland did his own theology, when his Bridgewater Treatise was read aloud to him for correction.[9] I do not quite understand what you mean; partly from Providence meaning either simply God or hourly, providential care.—

That seems a very difficult point to conjecture on, whether an insular genus originated on the island or survived there.[10] When several allied species occur in an archipelago the probability seems that it was created there; as it shows it has long there been a varying & is a well adapted form. I forget whether the Umbellifers live on the other Atlantic Islands.[11]

Supposing that the Deer's bones are *not* those of a naturalised animal, (for certainly there was no deer, when Mauritius was discovered) it is a grand case of continental extension & of greatest value. If you see Owen, caution him, but *not from me*, about the many animals which have been there naturalised.[12]

I saw in Proof-sheet the passage in a note by Owen about ideal types; he outdoes himself in audacious impudence on this head, & makes it the ground for an attack on Huxley.[13]

Send me a copy of your Insular paper when printed, as several of us want to read it.—[14]

I told Murray not to publish my book blindly, & he has kept the M.S long & is frightened, perhaps with good reason, for I never know when I go too much into detail; & the details are to be printed in smaller type; & at last the M.S is in printer's hands.—[15] In the interval I began a chapter on Man, for which I have long collected materials, but it has grown too long, & I think I shall publish separately a very small volume, "an essay on the origin of mankind:"[16] I have convinced myself of the means by which the Races of man have been mainly formed, but I do not expect I shall convince anyone else.—[17] I wish the dreadful six-month labour of correcting press was over.—[18]

Hensleigh Wedgwood has been very ill, & is sadly pulled down, but is now recovering.—[19]

Give our very kind remembrances to M[rs] Hooker & our congratulations on her coming down stairs[20]

Ever yours affect[y] | C. Darwin

On Feb 13th we go for a week to 6 Queen Anne St.—[21] I wish there was any chance of your being in London & seeing you.—

DAR 94: 10–13
Endorsement: '/67'

[1] See letter from J. D. Hooker, 4 February 1867 and n. 4.

[2] George Douglas Campbell, the duke of Argyll, made the customary presidential address at the twenty-fifth meeting of the British Association for the Advancement of Science at Glasgow in September 1855 (G. D. Campbell 1855). CD praised his speech in his letter to W. D. Fox, 14 October [1855]. CD had attended the meeting as a vice-president of section C (geology); see *Correspondence* vol. 5. No other speeches by Campbell are noted in the *Report of the twenty-fifth meeting of the British Association for the Advancement of Science*.

[3] Hooker had asked whether CD had read Campbell's *Reign of law* (G. D. Campbell 1867; see letter from J. D. Hooker, 4 February 1867).

[4] CD refers to Campbell's review of *Orchids* in the *Edinburgh Review* ([G. D. Campbell] 1862); he commented on the article in the letter to J. D. Hooker, 18 [November 1862], and the letter to Asa Gray, 23 November [1862] (*Correspondence* vol. 10). There is an annotated copy of [G. D. Campbell] 1862 in the Darwin Pamphlet Collection–CUL. It was reviewed in the *Saturday Review* by CD's nephew, Henry Parker ([Parker] 1862). CD sent [Parker] 1862 to Hooker in late 1862 (see *Correspondence* vol. 10, letter from J. D. Hooker, [before 29 December 1862]; see also *ibid.*, letter to J. D. Hooker, 29 [December 1862], and letter from J. D. Hooker, [31 December 1862]). Campbell also published a series of essays in *Good Words* (G. D. Campbell 1865) that he later used in G. D. Campbell 1867 (see *Correspondence* vols. 13 and 14).

[5] The anonymous review of G. D. Campbell 1867 appeared in the *Spectator. A weekly review of politics, literature, theology, and art*, 5 January 1867, pp. 17–19. CD himself viewed rudimentary organs as vestiges of earlier stages in the development of species (see *Origin*, pp. 450–6); for recent discussions on this topic, see the letter from Fritz Müller, 1 January 1867 and nn. 15 and 16, and the letters to William Turner, 15 January [1867] and 1 February [1867].

[6] See letter from J. D. Hooker, 4 February 1867 and n. 2.

[7] The last paragraph in *Variation* (2: 432) ended with the following lines:

However much we may wish it, we can hardly follow Professor Asa Gray in his belief 'that variation has been led along certain beneficial lines,' like a stream 'along definite and useful lines of irrigation.' If we assume that each particular variation was from the beginning of all time preordained, the plasticity of organisation, which leads to many injurious deviations of structure, as well as that redundant power of reproduction which inevitably leads to a struggle for existence, and, as a consequence, to the natural selection or survival of the fittest, must appear to us superfluous laws of nature. On the other hand, an omnipotent and omniscient Creator ordains everything and foresees everything. Thus we are brought face to face with a difficulty as insoluble as is that of free will and predestination.

CD and Gray had long discussed natural selection and design in nature with each other (see *Correspondence* vols. 8–10).

[8] CD had recently received a letter from Mary Everest Boole enquiring about the compatibility of natural selection with various religious beliefs (*Correspondence* vol. 14, letter from M. E. Boole, 13 December 1866; see also *ibid.*, letter from M. E. Boole, 17 December [1866]), and had corresponded with Charles Kingsley on religious and scientific subjects since the publication of *Origin*. CD had also received letters on religious subjects from John Beck (*Correspondence* vol. 12, letter from John Beck, 6 October 1864) and Alexander F. Boardman (letter from A. F. Boardman, 26 January 1867 and enclosure).

[9] In his letter of 4 February 1867, Hooker discussed a statement he made in his lecture on insular floras. CD had commented on the published lecture (see letter to J. D. Hooker, 29 January [1867]

and n. 3). CD refers to William Buckland and the sixth 'Bridgewater Treatise', *Geology and mineralogy considered with reference to natural theology* (Buckland 1836); the incident has not been identified.

[10] See letter from J. D. Hooker, 4 February 1867 and n. 13.

[11] See letter from J. D. Hooker, 4 February 1867 and n. 12.

[12] Richard Owen told Hooker of deer bones found in a bog on the island of Mauritius (see letter from J. D. Hooker, 4 February 1867 and n. 10). Owen and CD were not on good terms (see also *Correspondence* vol. 14, letter to T. H. Huxley, 4 July [1866] and n. 8). CD had visited Mauritius on the *Beagle* voyage (see *Correspondence* vol. 1, letter to Caroline Darwin, 29 April 1836, and *Journal of researches*, pp. 483–6).

[13] See letter from J. D. Hooker, 4 February 1867 and n. 11. Owen expressed his opposition to the 'type-form' in R. Owen 1866, p. 81; he also referred to Thomas Henry Huxley's opposition to his proposed expansion of the exhibition rooms for the nation's natural history collections. Owen claimed that the restricted size of the National Museum proposed by Huxley would limit exhibits to 'type-forms', which Owen considered a 'metaphysical term'. See also Rupke 1994, pp. 34–46, 99–103, and *Correspondence* vol. 14, letter to W. E. Gladstone, 14 May 1866. For more on the acrimony between Owen and Huxley, see A. Desmond 1994–7.

[14] See letter from J. D. Hooker, 4 February 1867 and n. 8.

[15] See letters to John Murray, 3 January [1867], and 8 January [1867]. The return of the manuscript to CD so that he could mark up sections to be printed in smaller type was delayed by Murray's decision to have a non-scientific friend read the manuscript (see letter from John Murray, 9 January [1867]). CD had sent the manuscript to the printers two days earlier, on 6 February (see letter to W. D. Fox, 6 February [1867] and n. 2).

[16] For CD's collection of material on human descent, see Barrett 1980, H. E. Gruber 1981, and letter to Fritz Müller, 22 February [1867], n. 11. CD eventually used the material in the writing of *Descent* and *Expression*.

[17] For CD's views on the origin of human races, see *Correspondence* vol. 12, letter to A. R. Wallace, 28 [May 1864]. See also *Correspondence* vol. 13, letter from Henry Denny, 23 January 1865, and letter from F. W. Farrar, 6 November 1865.

[18] CD refers to the task of correcting the proof-sheets of *Variation*, which occupied him from 1 March to 15 November 1867 (see CD's 'Journal' (Appendix II)).

[19] Hooker had asked about Hensleigh Wedgwood's health in his letter of 4 February 1867.

[20] See letter from J. D. Hooker, 4 February 1867 and n. 17.

[21] According to Emma Darwin's diary (DAR 242), the Darwins were in London from 13 to 21 February 1867. Six Queen Anne Street was the home of Erasmus Alvey Darwin, CD's brother.

From William Turner   8 February 1867

25 Royal Crescent
Feb.y 8th.—/67

My Dear Sir

Undoubtedly the great muscle of the scalp should be regarded as a skin muscle and it is through this same muscle that the eyebrows are elevated, their depression being affected by another muscle called corrugator supercilii—[1] Those muscles which move the ear as a whole evidently belong to the same group, and the muscles of the alæ of the nose, of the lips & the sphincter muscles of the mouth and anus are apparently also of the same character—[2] Many of these from their position are evidently for the purpose of moving not only the skin, but those appendages (hairs in the human body) which are somewhat numerous in those localities.[3] There is one spot in the human body however, viz: the skin of the inner side of the palm of the hand to which a fairly marked panniculus is attached. The muscle is called palmaris brevis. But one cannot say that its use is primarily for the movement of hairs as the inner side of the palm is destitute of those appendages—

Bye the way the power not only of moving the scalp but the ears to & fro, is possessed by some persons— I have seen it more than once—[4]

In connection with the movements of the hairs, not only in the scalp but elsewhere, I may remind you that the *individual* hairs have special muscles attached to them, which are attached on the one hand to the corium immediately beneath the epidermis & on the other to the outer surface of the hair follicle— These muscles by their contraction project the individual hairs & produce the condition known as goose skin (cutis anserina), such as we experience on sudden application of cold to the surface. These muscles are of the non striped or involuntary form— They were first discovered by Kölliker & have been carefully described by Lister in Vol VI of Microscopical Journal— The panniculus carnosus of man & quadrupeds is on the other hand formed of striped or voluntary fibre—[5]

Hence it would appear that the movement of the skin & of its appendages *en masse* is effected by the agency of a more or less generally distributed voluntary muscle, whilst the movement of *individual* hairs is effected by a special involuntary muscle attached to each hair follicle.

I remain | very truly yrs | Wm Turner

N.B. In the skin of the hedge hog there is no difficulty in seeing that the 'quills' are moved by the voluntary panniculus. A special reason (defensive purposes) requiring that they should be under the influence of the will—[6]

Kölliker states that in the cat & other mammals the vibrissæ are moved by striped, i.e. voluntary, & not involuntary fibres—[7]

DAR 80: B152–3c

CD ANNOTATIONS
1.2 that ... elevated] *double scored blue crayon*
1.4 the muscles ... muscles 1.5] *scored blue crayon*
3.8 Vol VI ... Journal] *double scored blue crayon*
4.2 by ... hairs 4.3] *scored blue crayon*; *scored pencil*; '*[*animals*]* ' *added in margin, pencil*
Top of letter: 'Keep for expression' *blue crayon*

[1] See letter to William Turner, 1 February [1867]. CD discussed the scalp muscle in *Expression*, and illustrated the facial muscles in *Expression*, pp. 24–5. He discussed the muscles controlling the eyebrows in *Expression*; see, for example, pp. 148–9 and 179–80. See CD's annotations.

[2] The alae are parts of the sides of the nose. See *Expression*, pp. 24–5.

[3] See letter to William Turner, 1 February [1867]. See also nn. 5 and 6, below.

[4] CD discussed the movement of ears in humans and other mammals in *Descent* 1: 20–3. See also *Expression*, pp. 111–15.

[5] Turner refers to the arrectores pili, smooth, involuntary muscles connecting the corium, or dermis, to the hair follicle; when these muscles contract, they cause gooseflesh (*cutis anserina*). Joseph Lister discussed the cellular structure of involuntary muscles in the *Quarterly Journal of Microscopical Science* (Lister 1856); he described the arrectores pili at length in Lister 1853, where he verified Rudolf Albert von Kölliker's findings published in Kölliker 1850–4, 2 (part 1): 14, and Kölliker 1852, p. 82. CD cited Kölliker in *Expression*, p. 101, and Lister 1853 in *Expression*, pp. 101 n. 19, and 201 n. 6.

6 CD discussed movement of the 'dermal appendages' in birds, mammals, and some reptiles when under the influence of anger or fear in *Expression*, pp. 95–104 and 298. He mentioned movement of hedgehog quills by the panniculus carnosus in *Expression*, p. 101; in *Expression*, pp. 101–4 and 298, he also discussed the possible ancestral relationship of the voluntary muscles of the panniculus carnosus and the involuntary arrectores pili in mammals, noting both the role of force of habit (influence of will on involuntary muscles), and the roles of variation and natural selection.

7 See Kölliker 1850–4, 2 (part 1): 15, and Lister 1853, p. 268; see also *Expression*, p. 101.

From J. V. Carus 11 February 1867

39, Elstertrasse, Leipzig.
Feb^{ry} 11. 1867

My dear Sir,

I hereby send you the preface (very hastily translated) which I intend to give to the new edition of the translation.[1] I prefer to give this instead of any other additions or notes partly from the reasons given in the very preface, partly because I don't think the book itself the right place for any lengthy discussion.[2] You would oblige me very much indeed, if you were to give me your opinion quite frankly. As I didn't as yet send it to Stuttgart and the print of the text has scarcely begun (I saw the corrections of the three first sheets) it is plenty of time to make any alterations you should think fair or proper.[3]

The chief part of the work, the revision and correction of the book is done; now it depends on Schweizerbart to bring it out quickly. I am sorry he will bring it out in parts.[4] But as this is a matter of business I cannot meddle with it.

Could you give me perhaps a direction where I could get a specimen of the Eozoon?[5]

Believe me | My dear Sir | yours very sincerely, | J. Victor Carus

[Enclosure]
Preface of the editor.

When I was asked to lead a new edition of the translation of Darwin's book on the origin of species through the press I had first to complain that Bronn couldn't do it himself any more. Now it was not only the piety owing to the dead which made it a duty for me to revise Bronn's work, it was especially the conscience of deep obligation under which German Science stands to the late Bronn and which could not but be augmented by his introduction of the book on the origin of species to the public at large, that I welcomed the opportunity of continuing a task begun by him My first aim could only be, to correct the errors and mistakes left unnoticed here and there, and above all to insert the many important additions of the author, which are to be found in the new english edition, into this German. Of these I may mention the detailed accounts of dimorphic and trimorphic plants and animals, of mimetic butterflies, of hybridism and the sterility and fertility of hybrids and first crosses, of means of transport and so on

A singular difficulty arose to me from the publisher's wish to put my name on the title. As I acknowledge the great merits of Bronn's heartily and without any reserve, yet here only his relation to the contents of the book edited by him could lead me. His position to Darwin's theory is quite different of mine. Bronn declares in his concluding remarks, which he even joins to the text of the work as a 15. chapter, that "he was not able, to follow the doctrine" (put down in the preceding fourteen chapters) and that he thinks it "more consistant, to" insist upon the old point of view (viz. the admission of wonders), though it is not to be maintained scientifically."

So also his foot-notes are mostly debating. Now, the more I should think it unfair to fight against Bronn's opposing notes by polemising additions (which I shouldn't have done even without giving my name) the less it seems to me advisable, to append explanations or notes betraying doubt to a work equally excellent in its richness of details as in the acuteness of its combinations, and this the less as the development of the science of organic nature within the last ten's of years tends always more pressingly to a view which now Darwin has brought into such masterly form Therefore in accordance with the author I decided, to leave out Bronn's additions. I have also resisted the temptation to append notes of my own, though I might have been led to do it, for instance with regard to the value of zoological characters, to the causes of the variability, but especially with regard to the methodological necessity of the admission of the spontaneous generation I give therefore Darwin's book so as it is in its fourth english edition published towards the close of the past year with some corrections sent me by the author. For those and for many lights on moot or doubtful points I feel myself most obliged to him.

J.V.C.

DAR 161: 55, 57

CD ANNOTATION
Top of letter: 'Eozoon' *pencil*

[1] For the published preface ('Vorrede des Herausgebers'), see Bronn and Carus trans. 1867, pp. v–vi.

[2] See enclosure. The first two German translations of *Origin* included notes and an extra chapter (chapter 15) by Heinrich Georg Bronn (Bronn trans. 1860 and 1863).

[3] In the letter from E. Schweizerbart'sche Verlagsbuchhandlung (the publishing firm in Stuttgart) of 24 January 1867, Christian Friedrich Schweizerbart informed CD that he had sent the first part of Carus's manuscript to the printer.

[4] On the publisher, see n. 3, above. Bronn and Carus trans. 1867 was not published in parts.

[5] In 1864, John William Dawson identified samples taken from pre-Silurian strata in eastern Canada as fossilised Foraminifera, single-celled protists with shells; he named the species *Eozoon canadense*, the 'Dawn animal from Canada' (Dawson 1864). Further samples were sent to William Benjamin Carpenter, an expert on Foraminifera, who confirmed Dawson's interpretation (Carpenter 1864). CD added information on the discovery of *Eozoon canadense* to *Origin* 4th ed., p. 371, as substantiating his claim, made in *Origin*, p. 307, that life existed before the Silurian period (see *Correspondence* vol. 13, letter to Charles Lyell, 25 March [1865], n. 11, and Burkhardt 1988, pp. 43–5). See also *Correspondence* vol. 14, letter to J. D. Hooker, 31 May [1866]. The interpretation of the samples as pre-Silurian fossils remained controversial, however (see, for example, Carpenter 1866, and King and Rowney 1866); and by the end of the century, comparisons with similar, more recent, formations indicated that the samples were mineral in origin (see Schopf 2000).

To William Turner   11 February [1867][1]

Down.

Feb. 11,

My dear Sir

I write to ask no more questions, but merely to thank you sincerely for your last, which has explained clearly all that I wanted to know.[2] I have sent my MS. on "Domestic Animals, &c" to the printers, and it proves so terribly bulky that I have resolved not to include my Chapt. on Man; but as I have collected materials during several years on certain points (though so ignorant on other and perhaps more important subjects) in the Natural History of man I mean to publish a separate essay hereafter.[3] I mention this merely that you might not think that your labours and kindness have been thrown away,—that is as far as I am capable of making use of them.[4]

My dear Sir | Yours sincerely & obliged | Ch. Darwin.

Copy
DAR 148: 155

[1] The year is established by the reference to *Variation* having been sent to the printers (see n. 3, below).
[2] See letter from William Turner, 8 February 1867.
[3] CD refers to the manuscript of *Variation* (see also letter to W. D. Fox, 6 February [1867], n. 2). CD ultimately used his extra material in *Descent* and *Expression* (see letter to J. D. Hooker, 8 February [1867] and n. 16).
[4] For CD's use of Turner's information in *Descent* and *Expression*, see the letter to William Turner, 1 February [1867], nn. 2 and 5.

From J. D. Hooker   12 February 1867

Royal Gardens Kew

Feby 12/67

Dear Darwin

Your approval of my declining Brit. Ass Presidency was an immense *relief*—as really I was feeling rather a guilty being.—[1] Since I wrote I had a joint attack at the X club, from Huxley, Frankland, Spottiswood, Spencer & Hirst. & had much difficulty in beating off, which I did with a heavy heart, as I would fain have obliged them—[2] Besides the personal objections, I know so well, that it would altogether interrupt, for months, both "Genera Plantarum" & my "Insular work" & it would not be fair to Bentham—[3] Do my dear friend, if the subject is called up by Huxley or Tyndall, &c, your presence, back my resolution.[4] They dwelt strongly on the scientific need of it—but really I do not recognize the British Assn needs in a scientific point of view, & though I think the D. of Buccleugh a disgraceful appointment,[5] I am not sure that it is a post for the *most* scientific men of the day to aspire to; still less should it be dependent on their support

I hope to be in town on Thursday forenoon, & shall run my chance of seeing you before noon if you do not object.—[6]

Ever yours | J D Hooker

DAR 102: 143-4

[1] See letter to J. D. Hooker, 8 February [1867] and n. 1.
[2] A dining club, later known as the X Club, was established on 3 November 1864 primarily for younger men of science united by friendship and a 'devotion to science, pure and free, untrammelled by religious dogmas' (quoted in Barton 1998, p. 411; see also *Correspondence* vol. 13). The initial members included Hooker, Thomas Henry Huxley, Edward Frankland, Herbert Spencer, and Thomas Archer Hirst; William Spottiswoode joined at the second meeting in 1864. For the social and political influence of the X Club on the British Association for the Advancement of Science, and on British science, see Brock 1981 and Barton 1998. For Hooker's role in the X Club, see Bellon 2001.
[3] Hooker refers to Bentham and Hooker 1862-83, to his work on geographic distribution, and to George Bentham (see letter from J. D. Hooker, 4 February 1867 and n. 7).
[4] John Tyndall was one of the original members of the X Club (see n. 2, above).
[5] The fifth duke of Buccleuch, Walter Francis Scott, presided over the thirty-seventh meeting of the British Association for the Advancement of Science, held at Dundee in September 1867. He was not a man of science; Alfred Russel Wallace later wrote that Scott 'evidently considered it a condescension on his part to be there at all' (A. R. Wallace 1905, 2: 48; see also Brock 1981, p. 99).
[6] See letter to J. D. Hooker, 8 February [1867] and n. 21. Hooker evidently saw CD at Erasmus Alvey Darwin's house in London during his visit between 13 and 21 February (see letter from J. D. Hooker, 14 March 1867).

From John Lubbock 12 February 1867

15, *Lombard Street. E.C.*
12 Feb 1867

My dear Mr. Darwin

Should you mind signing the enclosed?
I think the work Stainton has done would fully justify his election as an F.R.S.[1]
Yours very sincerely | John Lubbock

C. Darwin Esq

DAR 170: 55

[1] CD evidently returned the enclosure, which has not been found, to Lubbock. The entomologist Henry Tibbats Stainton became a fellow of the Royal Society of London in 1867 (*DNB*).

To William Benjamin Carpenter [13-16 February 1867][1]

6 Queen Anne Street[2]
[Asking for a specimen [of Eozoon] for Professor Victor Carus of Leipzig.][3]

Incomplete[4]
Sotheby Parke Bernet, London (18 June 1979)

[1] The date range is based on the date of CD's visit to his brother's house (see n. 2, below), and on the relationship between this letter and the letter to J. V. Carus, 17 February [1867].
[2] Six Queen Anne Street, London, was the home of CD's brother, Erasmus Alvey Darwin. CD was a visitor there between 13 and 21 February (see CD's 'Journal' (Appendix II)).
[3] Julius Victor Carus asked CD for a specimen of the *Eozoon* fossils in his letter of 11 February 1867. Carpenter had described the supposed fossils following their recent discovery; see Carpenter 1864 and

Carpenter 1866, and the letter from J. V. Carus, 11 February 1867, n. 5. See also *Origin* 4th ed., pp. 371–2, and Peckham ed. 1959, pp. 514–15.

[4] The original letter is described in the sale catalogue as being four pages long, and undated.

To J. V. Carus   17 February [1867]

Down. | Bromley. | Kent. S.E.

Feb 17th

My dear Sir

I have read your Preface with care.[1] It seems to me that you have treated Bronn with complete respect & great delicacy, & that you have alluded to your own labour with much modesty. I do not think that any of Bronn's friends can complain of what you say & what you have done. For my own sake I grieve that you have not added notes, as I am sure that I should have profited much by them; but as you have omitted Bronn's objections, I believe that you have acted with excellent judgment & fairness in leaving the text without comment to the independent verdict of the reader.[2] I heartily congratulate you that the main part of your labour is over: it would have been to most men a very troublesome task, but you seem to have indomitable powers of work, judging from those two wonderful & most useful volumes on zoological literature, edited by you, & which I never open without surprise at their accuracy & gratitude for their usefulness.—[3] I cannot sufficiently tell you how much I rejoice that you were persuaded to superintend the translation of the present edition of my book, for I have now the great satisfaction of knowing that the German public can judge fairly of its merits & demerits.—[4]

I have written to Dr Carpenter for a specimen of the Eozoon & I hear from Mrs Carpenter that he is out of London, but she thinks he will be able & will have great pleasure in sending you a specimen.[5]

When I receive it, I will send it by post if not too heavy; but if too heavy by the Booksellers M^{essrs} Williams & Norgate.

with my cordial & sincere thanks, believe me | My dear Sir | Yours very faithfully | Ch. Darwin

Staatsbibliothek zu Berlin, Sammlung Darmstaedter (Carus 3)
Endorsement: '1867'

[1] CD refers to the preface to Carus's German translation of the fourth edition of *Origin* (Bronn and Carus trans. 1867). Carus enclosed an English translation with his letter of 11 February 1867.

[2] In the preface to Bronn and Carus trans. 1867, Carus referred to the two earlier editions of the translation of *Origin* by Heinrich Georg Bronn (Bronn trans. 1860, 1863).

[3] CD refers to the *Bibliotheca zoologica* (Carus and Engelmann 1861). The volumes are in the Darwin Library–CUL; the second volume is lightly annotated (see *Marginalia* 1: 160).

[4] Christian Friedrich Schweizerbart, head of the Stuttgart publishing firm E. Schweizerbart'sche Verlagsbuchhandlung, had asked Carus to translate the fourth edition of *Origin* (see *Correspondence* vol. 14, letter from E. Schweizerbart'sche Verlagsbuchhandlung, 26 October 1866). On complaints regarding the first two editions of the translation of *Origin*, and on CD's satisfaction with Carus as translator, see *Correspondence* vol. 14, letter from E. Schweizerbart'sche Verlagsbuchhandlung, 10 May 1866 and n. 10, and letter to J. V. Carus, 10 November 1866.

[5] See letter to W. B. Carpenter, [13–16 February 1867]. The letter from Louisa Carpenter has not been found.

To Edward Blyth [18 February 1867][1]

6. Queen Anne St | Cavendish Sq^{re}.— | W.
Monday

My dear Mr Blyth

I want to ask you a question about the title of some printed pages, which gave me some 20 or 30 years ago!!²

But I enjoyed our walk a month or two ago in the Zoolog. Gardens so much,³ & as I was told that you often walk there in the morning, it has occurred to me that, if disengaged, you would not object to meet me there.

I will go there next Wednesday morning at 10 oclock & go to Monkey House & wait within for quarter of an hour for the chance of your coming or being there before me.— If you know positively that you cannot meet me, perhaps you would have the kindness, to send me a line, as then, I should probably not go to the Gardens; if you *think* you can be there do not trouble yourself to write

Believe me | Yours very sincerely | Ch. Darwin

P.S. I am bound to add that my health is always doubtful, & every few days I have a bad day & cannot go anywhere,— & thus unavoidably break any engagements.⁴

McGill University Libraries, Rare Books and Special Collections Division

[1] The date is established by the relationship between this letter and the letter to Edward Blyth, 23 February [1867] (see n. 2, below). The only Monday during CD's February 1867 visit to London was 18 February.

[2] CD refers to galley proofs of pages 158 and 159 from a translation of Georges Cuvier's *Règne animal* (Cuvier 1840 or 1849; see letter to Edward Blyth, 23 February [1867] and nn. 6 and 9).

[3] CD walked with Blyth in the Zoological Gardens in Regent's Park during his visit to London in November 1866 (see *Correspondence* vol. 14, letter to Edward Blyth, 10 December [1866]).

[4] On CD's recent health, see letter to W. D. Fox, 6 February [1867] and n. 6.

To Edward Blyth [19 February 1867][1]

6 Queen Anne St | W.
Tuesday

My dear Mr Blyth

I am so unwell today that there is hardly any chance of my being able to meet you tomorrow morning at the Zoolog. gardens, so will you be so good as to consider my former note cancelled & excuse the trouble I have caused you—²

If able to do any thing tomorrow I have some other more pressing engagements which I have failed to do today.

I shall be in London probably in a few months' time,³ & I hope you will allow me to propose another meeting at the gardens.

yours sincerely | Ch. Darwin

LS
McGill University Libraries, Rare Books and Special Collections Division

[1] The date is established by the relationship between this letter, the letter to Edward Blyth, [18 February 1867], and the letter from Edward Blyth, 19 February 1867.

[2] See letter to Edward Blyth, [18 February 1867]. Emma Darwin recorded in her diary (DAR 242) for 18 and 19 February: 'C. unwell'.

[3] CD left London on 21 February 1867. He returned for a visit in June 1867. See CD's 'Journal' (Appendix II).

From Edward Blyth 19 February 1867

Feb.ʸ 19ᵗʰ. 1867—

My Dear Sir,

I am very sorry to hear of your indisposition, which has disappointed me of the pleasure of meeting you.[1] I have been much interested with the 4ᵗʰ. edition of your 'Origin of Species',[2] & have just been penning a few remarks which occurred to me, which I intended to place in your hands. As it is, I send them to you; but fear that you will regard some of them as a little fanciful.

Yours very Sincerely, | E. Blyth

[Enclosure]

Memoranda for Mʳ. Darwin—

N.B. The paging refers to the 4ᵗʰ. edit. of "Origin of Species".

About *mocking* (p. 506).[3] Among mammalia there is one very striking instance in the case of certain Malayan Squirrels (*Rhinosciurus* of Gray), which wonderfully *mimic* the appearance of the *Tupaiæ* among the *Insectivora*, which inhabit the same region. Size, form (elongated muzzle), colour, character of fur, flattened brush, and even the pale humeral line which is characteristic of the Tupayes, but found in no other Squirrel that I know of.[4] I do not, however, perceive the object in this case of mimicry—

Wallace's instance in the bird class is an extremely remarkable one,[5] but several others occur to me. The most familiar to most ornithologists will be that of certain cuckoos (*Hierococcyx*), the plumage of which is exceedingly Hawk-like both in its immature and adult phases. There are 5 or 6 races of them, all peculiar to the Indian region;[6] but though they approximate certain Sparrow-hawks in appearance, the most striking resemblance is with sundry southern hemisphere species allied to *Pernis*, but which are more widely distributed, as *Cymindis uncinatus* and *C. magnirostris* in S. America, *Baza cuculoides* in S. Africa and *B. Reinwardtii* and other species in the Malay countries, Celebes, and Australia; to the former of which they can bear no especial reference.[7] Even our common cuckoo has somewhat of a hawk-like aspect

A very remarkable instance of apparent mimicry occurs in certain other true cuckoos (*Surniculus Dicruroides* and *S. lugubris*), which bear an extraordinary resemblance to the Drongos or 'King Crows' (*Dicrurus*), both in immature and adult plumage;[8] & probably they lay their eggs in the Drongos' nests, but this has not been ascertained.

Certain Doves, as several of the *Macropygiæ*, are also remarkably cuculine in appearance, but with what object is not very manifest. Also, some of the *Graucalus* and *Campephaga* series are very *Cuculus*-like.[9]

The *Haliaëtus blagrus*, which feeds chiefly on sea snakes, is remarkably gull-like, as I have thought when seeing it skim over the waves—pure white with ashy mantle; but Gulls are rare in the Indian Ocean, where indeed there are no species of large size. The great *H. pelagicus* also strikingly reminds one of the large black-backed Gull.[10] There is a striking similarity in the barking voice of the larger Gulls and most of the Sea eagles.

The marked resemblance in facial expression of the Orang-utan to the human Malay of its native region, as that of the Gorilla to the Negro, is most striking, & what does this mean? Unless a divergence of the anthropoid type prior to the specialization of the human peculiarities, which however would imply a parallel series of at least two primary lines of human descent which seems hardly probable; & moreover we must bear in mind the singular facial resemblance of the *Lagothrix Humboldtii* (a *platyrrhine* form) to the negro, wherein the resemblance can hardly be other than accidental.[11] The accompanying diagram will illustrate what I suggest (rather than *maintain*); & about *Hylobates* or Gibbons, I am not sure that I place it right, for, upon the whole, the Gibbons approximate Chimpanzee more than they do the Orang-utan, notwithstanding *geographical position*. *Aryan* I believe to be improved *Turánian* or *Mongol*—[12]

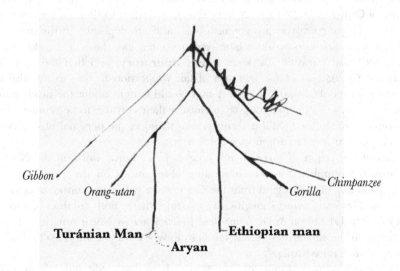

To appreciate the likeness of a Malay to an Orangutan, you should see an old Malay women chewing pâu, & note the mobility of the lips, in additional to the general expression. However to be explained, the likeness is much less decided in other races of the grand Turánian stock. We cannot call this a case of mimicry.

I find that the leaf-nosed bats of America are emphatically *platyrhine* while those of the Old World are *catarrhine*, & the *pteropodine* bats are *strepsirhine*, like the *Lemuridæ*.[13] The genus *Taphozous* bears a corresponding resemblance to *Galæopithicus*.[14] The facial expression (or physiognomy) of the larger species of *Hipposideros* recals to mind that of the Orang-utan; and the ordinary bats with simple stomach and cheek-pouches remind us of the Baboon & *Macacus* series.[15] Can we soundly interpret these resemblances, which in the cases first mentioned correspond with geographical distribution? Is there not a significant analogy between the prominent cheek-callosities of the adult male Orang-utan and the facial membranes of the horse-shoe bats?[16] I would hardly like to suggest all this in print but it may be thought over.

If different genera of *Cheiroptera* adumbrate corresponding genera of *Quadrumana*, so also it may be thought that some genera of *Insectivora* forecast some of *Carnivora*; as *Tupaia-Herpestes*; *Potamagale-Lutra*; just plausibly indicating different lines of ascent from the lower order to the higher.[17] May not the *Bruta* (i.e. *Edentata* and *Cetacea*) descend more directly from *Monotremata* and other placental mammalia from Marsupialia?

About parasitic cuckoos (p. 259).[18] The American *Coccyzi* are in the condition of the Screech-owls, in having eggs and young of different ages in the same nest.[19] Very closely allied to them are the Old World crested cuckoos (*Coccystes*), which are *parasitic*, and lay their eggs in the nests of birds about their own size or *larger*, as *C. glandarius* in those of magpies and crows.[20] *Eudynamis orientalis* constantly in those of Crows, and the eggs are very similarly marked to those of the *Corvi*; while those of *Coccystes jacobinus* are greenish-blue, and are deposited in the nests of the *Malacocerci* which also produce blue eggs, coloured like those of *Accentor modularis*. Note well that parasitic *Eudynamis* is **not migratory**; and that the eggs of the parasitic *Coccystes*, as of *Coccyzus*, are of fair proportionate size; as are also those of the parasitic *Molothrus*![21] I suspect that Gould is right about the newly hatched cuckoo starving its companions, & so causing their carcases to be removed by the proprietors of the nest. Still it seems wrong to reject the personal observations of Jenner, as you seem to admit in p 291.[22]

About the origin of dogs (p. 19), consult J. K. Lords work on the Nat. Hist. of British Columbia, for very interesting observations on the derivation of the aboriginal dog of that region from the *Canis latrans* (which he considers as identical with the Mexican coyote, though Gray separates them. Both are thorough forms of Jackal).[23] In Ld Milton & Dr. Cheadles "Journey across North America" are some remarks on the similarity of the sleigh dogs to wolves; & see also Paget's Travels in Hungary & Transylvania.[24]

If the different-sized races of European or humpless cattle derive from different wild races, as *primigenius, frontosus, longifrons*, & *trachoceros*, we should suspect the same of the still more contrasted races of domestic humped cattle, which also probably descend from more than one wild original. *N.B.* "Bráhmini cattle" is a misnomer. Any bull dedicated to Bráhma becomes a "Bráhmini bull", and the latter are generally of the smaller races so far as I have seen. I think it *probable* that *there have been* different wild races of the Zebu form, as likewise of the humpless taurines?[25]

P. 281. A nuthatch does not hammer *downwards* at an object, like a titmouse, or a nutcracker (*Nucifraga*), but delivers the blow *forward* with a swing of the body, if I mistake not; so different an action that one is not likely to pass into the other. I can hardly imagine a titmouse striking in the manner of a nuthatch![26]

P. 417. Recent observations have shewn that there is a much greater community of species than was supposed on the two sides of the Isthmus of Panamá; on which subject consult Günther.[27] The absence of corals on the western side is believed to be due to the existence of a cold current, as of course you know.

p. 556. Those curious South American birds, the *Palamedea* and *Chauna*, are most closely allied to the spur-winged geese of Africa (*Plectropterus*), and in reality, are true *Anatidæ*, though not web-footed; in the *semi-palmate* geese of Australia (*anseranas melanoleuca*), there is a decided approach in the shape of the bill to the screamers; and I cannot think how any naturalist can look at a living *Palamedea* without perceiving at once that it is *non-palmate goose!*[28] A. Newton[29] agrees with me in this opinion. The *Parra* group is quite distinct, & its affinity is with the snipe and plover series, & not at all with the *Rallidæ*, Anatomy, plumage, eggs, chicks, &c &c.[30]

P. 13 Have you ever *examined* the additional toe so constant in the Dorking fowls, as also in the Chinese *Shang-hai* bird?[31] How many phalanges has it? A sixth finger or toe is exceedingly apt to be transmitted by generation in human beings, but being taken off soon after birth, & the fact kept secret, we do not generally hear of it; though I have heard of some remarkable instances, wherein almost every child of a large family has inherited this redundancy.[32]

For the *highly predatory* and *carnivorous* habits of the Weka Rail of New Zealand (*Ocydromus australis*), vide 'Ibis', 1862, p. 103.[33] Some of the New Zealand birds seem hardly to have yet acquired sufficient distrust of man (*ibid.* p. 105–6), as Petroica alighting on the hand, and *Carpophaga* allowing to have a snare be placed on its neck; the latter ascribed to "stupidity", whereas I should rather say *non-experience*. The extraordinary familiarity of the Canada jay (**Perisoreus** *canadensis*) is worthy of notice. *Vide* Lord and others.—[34] Note *diving* habits of common Cape Petrel (Daption *capensis ibid* p. 99).[35]

DAR 160: 209, 209/1 & 2, DAR 47: 190, 190a, DAR 80: B99–99a, DAR 205.11: 138, DAR 48: A75

CD ANNOTATIONS[36]
Enclosure:
1.1 *Memoranda* ... eagles. 7.6] 'Mimicry in Mammals & Birds' *blue crayon*; 'E. Blyth | Feby 1867'
8.1 The marked ... mimicry. 9.4] '*Man* | E. Blyth | Feb 21/67'
12.1 About parasitic ... p 291. 12.14] 'Cuckoos | Instinct. laying eggs in other nests (very good) | Feb. 22/67/'
15.1 P. 281 ... 99). 19.8] 'Abnormal Habits in Birds for Transitions.[37] | E. Blyth | Feb. 22/1867'

CD note:[38]

I w^d have sent you Origin, if I s^d have thought you w^d have cared for it. | Dom. A. | I did not know about Orang & Malay but I have [*del illeg*] to consider Vogts view, founded on Gratiolets remark on brain. | (Mimicry) (*Bats*) (Cuckoos most valuable remarks) Dogs I have used the [*inf*] | I regretted since I did not see in Lond, but during both last days was confined to house. | *Sexual Selection.*— too many questions to ask | *Title of paper*. I find [*del*] I more in your paper, even when you are not writing on subject.— Even in last L. W. summer plumage of gulls & plover *nuptial* | Spur-winged sexual

[1] See letter to Edward Blyth, [19 February 1867].

[2] The fourth edition of *Origin* was published in November 1866 (*Publishers' Circular*).

[3] CD made a number of additions to his section on mimicry or 'Analogical resemblances' in *Origin* 4th ed., pp. 502–6; see also Peckham ed. 1959, pp. 663–70.

[4] John Edward Gray listed '*Rhinosciurus Tupaioides*', the sharp-nosed squirrel, in J. E. Gray 1843, p. 195; it is now known as *Rhinosciurus laticaudatus*, the long-nosed squirrel (Nowak 1999, p. 1288). Nowak 1999, pp. 244–9, classifies the taxonomically controversial *Tupaia* and Tupaiidae (tree shrews) as the only genus and family in the order Scandentia. Though CD did not add Blyth's information to later editions of *Origin*, when discussing the mimicry most common in insects (see n. 5, below), he changed his comment that no cases of mimicry were known in 'larger animals' to read 'larger quadrupeds' (Peckham ed. 1959, p. 669).

[5] In *Origin* 4th ed., p. 506, CD wrote that the known cases of mimicry were all in insects except for Alfred Russel Wallace's one case of mimicry in birds. Though CD included no details, Blyth evidently knew of Wallace's case of an oriole (*Mimeta bouruensis*) mimicking a honeysucker (*Tropidorhyncus bouruensis*) on the Island of Bouru (or Buru); see A. R. Wallace 1863, pp. 26–8.

[6] Blyth refers to the hawk-cuckoos. *Birds of the world* 4: 548–50 lists six species of hawk-cuckoos in the genus *Cuculus*; the genus is still called *Hierococcyx* by some sources. For Blyth on the classification of the family Cuculidae, see Blyth 1842–3.

[7] Blyth refers to the honey-buzzards (*Pernis*). When Blyth mentions '*Cymendis uncinatus* and *C. magnirostris*', he is referring probably to the hook-billed kite (*Chondrohierax uncinatus*), and possibly to the roadside hawk (*Buteo magnirostris*), both of South and Central America; see Peters *et al.* 1931–87, 1: 285–6, 361–2. *Baza cuculoides*, the African cuckoo-hawk, is now named *Avecida cuculoides*, and the Pacific baza now includes a subspecies named *A. subcristata reinwardtii* (*Birds of the world*).

[8] Two species of *Surniculus* in the family Cuculidae are called drongo-cuckoos; *S. dicruroides* is now *S. lugubris dicruroides* (*Birds of the world* 4: 569). However, see also Davies 2000, p. 266. *Dicrurus* is a genus in the drongo family, Dicruridae.

[9] *Macropygia* is a genus of cuckoo-doves in the family Columbidae (pigeons and doves); see *Birds of the world* 4: 143–6. The family Campephagidae, or cuckoo-shrikes, includes *Coracina*, *Campephaga*, and under some classifications, *Graucalus*.

[10] Blyth refers first to what is now called *Haliaeetus leucogaster*, the white-bellied sea-eagle (see *Birds of the World* 1: 121–3). He also refers to *H. pelagicus*, Steller's sea-eagle, and to *Larus marinus*, the great black-backed gull. Blyth had been curator of the Museum of the Asiatic Society of Bengal, Calcutta, India from 1841 to 1862 (*DNB*). For more on Blyth's career, see Eisely 1979 and Brandon-Jones 1995.

[11] There are now two recognised species of *Lagothrix* (woolly monkeys), both found in South America; *L. lagotricha* and *L. flavicauda* (Nowak 1999, pp. 538–40). The Platyrrhini are the New World monkeys, but Blyth is probably using 'platyrrhine' to mean flat-nosed. For CD's recent discussions on the origin of human races, see *Correspondence* vols. 12 and 13. For discussion of nineteenth-century western perspectives on the human races, see, for example, Stocking 1982, S. J. Gould 1997, and Graves 2002, pp. 37–73.

[12] Chimpanzees are found in Africa and gibbons in south-east Asia; orang-utans are now limited to Sumatra and Borneo. For discussions of other mid-nineteenth-century classifications of primates, including humans, see Stanton 1960 and S. J. Gould 1997. On modern classifications of primates, see Nowak 1999, pp. 490–3. The term Turanian is now applied to any of the peoples speaking Ural-Altaic languages, but in the mid-1860s the term was usually applied to people who spoke any of

the Asian languages, and were generally nomadic rather than agricultural (see *OED*, and Stocking 1987, pp. 58–9). For a contemporary source on the development of languages and human races, see C. Lyell 1863, pp. 454–70. CD did not include a genealogical chart of human descent in *Descent*, but see *Descent* 1: 185–213, 2: 385–96 for his conclusions.

[13] Blyth refers first to the American leaf-nosed bats (Phyllostomidae; Nowak 1999, pp. 350–3). For the Old World leaf-nosed bats, see n. 15, below. Blyth is also probably referring to the Family Pteropodidae, the Old World fruit bats. The terms platyrrhine, catarrhine, and strepsirrhine are usually used to refer to primates. Platyrrhine: flat-nosed, widely spaced nostrils facing outwards (New World monkeys); catarrhine: nostrils close together, oblique, and directed downwards (Old World apes and monkeys) (*OED*). Some classifications include the Strepsirrhinae. Strepsirrhine: having a moist rhinarium and a cleft upper lip bound to the gum (e.g., lorises and lemurs; Allaby ed. 1999). See also Fleagle 1999, pp. 82–4, 136–7.

[14] Blyth refers to *Taphozous*, tomb bats in the family Emballonuridae (Nowak 1999, pp. 307–9), and to *Galeopithecus*, flying lemurs and colugos (now called *Cynocephalus* in the order Dermoptera); see *ibid.*, pp. 250–2.

[15] *Hipposideros* is the genus of Old World leaf-nosed bats (Nowak 1999, pp. 333–7). Blyth also refers to the macaques (*Macaca*).

[16] Horseshoe bats (Rhinolophidae) have a 'nose-leaf expansion of the skin surrounding the nostrils'; see Nowak 1999, pp. 329–32.

[17] Blyth refers to the bat order (Chiroptera), and to what older classifications called the order of Quadrumana (monkeys, apes, baboons, and lemurs). He also refers to the insectivore *Tupaia* (tree shrew), the carnivore *Herpestes* (mongoose), and the insectivore *Potamogale* (giant African water shrew or giant otter shrew), and the carnivore *Lutra* (Old World river otter).

[18] In *Origin* 4th ed., p. 259, and in earlier editions, CD speculated on how the parasitic European cuckoo acquired its habit of laying eggs in other birds' nests.

[19] Blyth refers to members of the genus *Coccyzus*; for example, *C. americanus*, the yellow-billed cuckoo, often lays eggs at intervals throughout the breeding season (*Birds of the world* 4: 595–6).

[20] *Coccystes* is a synonym of *Clamator*. *Clamator glandarius* is the great spotted cuckoo of Europe and Africa. Some *Clamator* species are called crested or pied cuckoos (*Birds of the world* 4: 547–8). See Peters *et al.* 1931–87, 4: 12, and Davies 2000, pp. 98–108, 263.

[21] The common koel or Asian koel, *Eudynamys scolopacea*, now includes a subspecies *E. s. orientalis*; it is known to lay eggs in the nests of *Corvus* (crow) species (*Birds of the world* 4: 570). The pied cuckoo or Jacobin cuckoo is now called *Clamator jacobinus* (see n. 20, above), or *Oxylophus jacobinus*; in India, its eggs are bluish (*Birds of the world* 4: 547). They evidently resemble eggs of a babbler genus, *Turdoides*; *Malacocircus* and *Malacocercus* are synonyms for some *Turdoides* species (Peters *et al.* 1931–87, 10: 331). *Accentor modularis* is the hedge-sparrow. In *Origin* 4th ed., pp. 258–9, CD further emphasised his point, made in *Origin*, p. 217, that laying eggs at intervals would be difficult for a migratory bird since extended hatching time would require a bird to nest longer. He also discussed whether parasitic birds tended to lay eggs in nests containing eggs similar in appearance (*Origin* 4th ed., p. 261; see also *Correspondence* vol. 8, letter to W. B. Carpenter, 6 January [1860] and n. 6). CD discussed the parasitic habits of *Molothrus*, the cowbird, in *Journal of researches* 2d ed., pp. 52–4.

[22] In *Origin* 4th ed., pp. 261–2, CD added a discussion of John Gould's belief that the nestlings of the parent bird starved to death rather than being ejected from the nest by the young cuckoo (J. Gould 1873, vol. 3, text with plates LXVII and LXVIII); J. Gould 1873 was first issued in twenty-five parts from 1862 to 1873; part 4, issued by 1865, included the plates and texts on cuckoos (see *Ibis* n.s. 1 (1865): 98–9). Edward Jenner had observed the nestling cuckoo ejecting the eggs or young of the foster-parent (Jenner 1788, p. 225). CD did not explicitly refer to Jenner in the fourth edition of *Origin*; however, he included the 'young cuckoo ejecting its foster-brothers' as an example of 'small consequences of one general law, . . . namely, multiply, vary, let the strongest live and the weakest die' on page 291. CD did mention Jenner's 'celebrated paper' in the manuscript of his 'big book' on species (see *Natural selection*, p. 508 n. 2). For recent conclusions on cuckoo nesting behaviour, see Davies 2000.

[23] Blyth refers to *The naturalist in Vancouver Island and British Columbia*, in which John Keast Lord observed that the 'true Indian dog' was a tame coyote, but only in inland areas where there were no opportunities for breeding with imported dogs (Lord 1866, 2: 218–25). CD cited Lord 1866 in *Variation* 1: 22–3; both Lord and CD noted that the coyote was *Canis latrans*. For John Edward Gray's classification of the genus *Canis*, see J. E. Gray 1843, pp. 57–9.

[24] Blyth refers to William Wentworth Fitzwilliam (Viscount Milton), Walter Butler Cheadle, and their book, *The north-west passage by land* (Fitzwilliam and Cheadle [1865]). In J. Paget 1839, 2: 18, John Paget mentioned that the Hungarian shepherd-dog looked much like a wolf; see *Variation* 1: 24.

[25] CD made minor changes to *Origin* 4th ed., p. 19, that slightly strengthened his assertion that European cattle and the Indian humped cattle 'descended from a different aboriginal stock'; he noted facts communicated to him by Blyth (see Peckham ed. 1959, p. 92). CD also cited Blyth on Indian cattle in *Origin* 4th ed., p. 301 (Peckham ed. 1959, p. 434). CD did not use the term 'Bráhmini' of cattle in *Origin* or *Variation*. In *Variation* 1: 79–83, he referred to 'Zebus' in India, with the specific name of *Bos indicus*; CD cited Blyth on the humped cattle twice in these pages.

[26] In *Origin* 4th ed., p. 212, CD noted that he had seen titmice hammering yew seeds to break them 'like a nuthatch'. On page 281, he suggested that natural selection might preserve each slight variation of the titmouse's beak, until it had one 'as well constructed for this purpose as that of the nuthatch'.

[27] In *Origin* 4th ed., p. 417, CD wrote, 'No two marine faunas are more distinct, ... than those of the eastern and western shores of South and Central America'. Blyth probably refers to Albert Charles Lewis Günther's recent paper on fishes of the states of Central America; an account was given in *Proceedings of the Zoological Society of London* (1866): 600–4. See also Günther 1864–6, and *Correspondence* vol. 14, letter from J. D. Hooker, 14 December 1866 and n. 10. In *Origin* 5th ed., p. 424, CD noted Günther's conclusion that thirty per cent of the fishes were the same on each side of the isthmus of Panama, leading naturalists to believe that the isthmus was formerly open.

[28] In *Origin* 4th ed., pp. 556–7, CD wrote, 'How strange it is ... that upland geese, which never or rarely swim, should have been created with webbed feet. ... But on the view of each species constantly trying to increase in number, with natural selection always ready to adapt the slowly varying descendants of each to any unoccupied or ill-occupied place in nature, these facts cease to be strange, or perhaps might even have been anticipated.' CD's comment led Blyth to mention other apparently related geese with webbed, semi-webbed, or non-webbed feet. Blyth refers here to the screamers, now the family Anhimidae, of South America; these include two species of *Chauna*, and *Anhima cornuta*, the latter of which is also called *Palamedea cornuta* (*Birds of the world* 1: 534–5). The screamers do not have webbed feet. *Plectropterus gambensis*, in the family Anatidae, is also web-footed. *Anseranas melanoleuca* (a semi-palmate goose) was Gould's name for what is now called *A. semipalmata*, the magpie goose (*Birds of the world* 1: 575). All birds mentioned have hooked bills.

[29] Alfred Newton.

[30] The 'Parra group' were birds known as jacanas, many of which belonged to the genus *Parra*. Jacanas had formerly been placed in the family Rallidae, but by the mid-1860s, some systematists placed them in a new family, Parridae (see Blyth 1866–7, p. 170). The family Parridae has since been renamed Jacanidae, which in modern classification is often included in the order Charadriiformes with snipes (family Scolopacidae) and plovers (family Charadriidae), while the family Rallidae is in the order Gruiformes (see *Birds of the world* 3: 108–12, 276–7, 289–91).

[31] In *Origin* 4th ed., p. 13, CD began discussing 'inheritable deviations of structure'. CD referred several times to the fifth toe of the fowls Blyth mentions, in addition to that of some sub-breeds, in *Variation*, noting that it indicated a long period of selection (*ibid.* 1: 260, 2: 202, 238).

[32] Blyth had earlier equated the extra toe in fowls to the supernumerary digit in dogs and the sixth finger in human beings (see *Correspondence* vol. 5, letter from Edward Blyth, [1–8 October 1855] and n. 8). CD collected information on polydactylism in 1863 (see *Correspondence* vol. 11, especially letter to T. H. Huxley, 27 June [1863]). In concluding his discussion of extra digits in humans and other animals, CD suggested the case was 'one of reversion to an enormously remote, lowly-organised, and multidigitate progenitor' (*Variation* 2: 17).

[33] Blyth refers to an article in *Ibis* (J. F. J. von Haast 1862); Julius von Haast described the weka (now *Gallirallus australis* in the family Rallidae) as omnivorous (p. 103).

[34] Blyth refers to the 'Totoara, the New Zealand Robin' (the Toutouwai, *Petroica australis*), and to the pigeon, '*Carpophaga novæ zelandiæ*' that Haast described as 'so stupid' as to sit on a branch until caught by a snare (J. F. J. von Haast 1862, pp. 105, 106). CD discussed the acquired fear of humans in *Origin* 4th ed., pp. 253–4; see also *Journal of researches* 2d ed., pp. 398–401. For Lord's description of the Canada jay, see Lord 1866, 2:151–2.

[35] Blyth refers to the Cape petrel's diving behaviour as described in Layard 1862; the scientific name is now *Daption capense*. For CD's comments on another diving petrel, see *Origin* 4th ed., p. 213.

[36] CD cut up much of Blyth's enclosure to place the data in folders for the different subjects discussed.

[37] A section entitled 'On the origin and transitions of organic beings with peculiar habits and structure' in *Origin* 4th ed., pp. 207–18, mentioned several of the birds and mammals Blyth discusses here; however, CD did not add the enclosed information on the habits or structures to later editions.

[38] Most of the content of CD's note is a reminder for his letter to Blyth of 23 February [1867]. CD refers to Carl Vogt and Louis Pierre Gratiolet.

To King's College, London   22 February [1867]

Down. | Bromley. | Kent. S.E.

Feb 22.

Sir

I should be very much obliged if you w^d send me any printed document by which I c^d learn something about the course of studies at King's College for the sake of one of my sons.[1]

I wish for instance to know whether the students are allowed any choice in their particular lines of study, & whether they are compelled to reside in the College.

I should therefore be much obliged if you c^d send me any paper that w^d give me such information.

I beg leave to remain Sir | your obedient servant. | Ch. Darwin

LS
Endorsement: '1867 | C. Darwin | 22 Feb'
King's College Library, University of London: file KA/IC/D45

[1] CD refers to Leonard Darwin, who was 17 years old and attending Clapham Grammar School. No further correspondence with King's College has been found. Ultimately, Leonard attended the Royal Military Academy, Woolwich.

To the Linnean Society   22 February [1867][1]

Down. | Bromley. | Kent. S.E.

Feb. 22

My dear Sir

I send by this post a paper which has been sent to me by a gentleman at the Cape of Good Hope with the request that I would communicate it, if I thought fit, to the Linn. Soc.[2] On a subject on which I feel so much interest I do not feel capable of judging whether it is worthy of being read & published by the Society. The paper describes a new species of that extraordinary genus Bonatea, & gives some new &

curious details on its fertilization; but the paper is too diffuse & perhaps gives too many extraneous remarks. It will be difficult for the referee, should the paper be read & published, to determine what illustrations sh^d be added;[3] but perhaps the Council may think that enough or too much has already been published on the fertilzation of Orchids.[4]

My dear Sir | yours sincerely | Ch. Darwin

LS
Linnean Society of London, SP1249

[1] The year is established by the reference to Weale 1867 (see n. 2, below), which was read at the Linnean Society in March 1867.

[2] James Philip Mansel Weale enclosed a manuscript with his letter of 9 January 1867; the manuscript (now in the Linnean Society archives, SP1249) was later published as Weale 1867. Weale lived in Bedford, Cape Colony.

[3] The paper was read at the Linnean Society on 7 March 1867, and was published in the *Journal of the Linnean Society (Botany)*; one illustration showing four figures was included (see Weale 1867, p. 473). On the publishing of the paper, see the letter to the Linnean Society, 9 December [1867], and the letter to J. P. M. Weale, 9 December [1867]. An annotated copy of Weale 1867 is in the Darwin Pamphlet Collection–CUL.

[4] Papers on orchids appearing in the *Journal of the Linnean Society (Botany)* following the publication of *Orchids* and 'Three sexual forms of *Catasetum tridentatum*' included Trimen 1863 and 1864, Crüger 1864, and Moggridge 1864; the first three of these were communicated to the Linnean Society by CD.

To Fritz Müller   22 February [1867]

Down Bromley Kent
Feb 22

My dear Sir

Your last letter of Jan. 1. is more valuable to me even than some of your previous ones.[1] The fact about the own-pollen being poisonous is quite extraordinary; I will quote your remarks & explanation after giving your former facts.[2] Can the cause of the decay be due to parasitic cryptogams?[3] I sh^d be very much obliged to you if you w^d. inform me soon whether *Oncidium flexuosum* is a native of your district.[4] These observations of yours will be a most valuable addition to my discussion on self-impotent plants.[5] There never was a more curious case than that of the rudimentary condition of the organs in Catasetum. It explains the fact, which I have been assured of, that Catasetum in some countries not rarely produces seed-capsules.[6] Your facts also about the sucking in of the pollen-masses & of the dispersal of the seeds in Gesneria are all quite new to me.[7] I hope you keep a record of all your miscellaneous observations, for you might thus hereafter publish a wonderful book.

Although you have aided me to so great an extent in many ways, I am going to beg for any information on two other subjects. I am preparing a discussion on "sexual selection", & I want much to know how low down in the animal scale sexual selection of a particular kind extends.[8] Do you know of any lowly organized animals, in which the sexes are separated and in which the male differs from the female in arms of offence, like the horns & tusks of male mammals, or in gaudy plumage & ornaments as with birds & butterflies? I do not refer to secondary sexual characters by which the male is able to discover the female, like the plumed antennæ of Moths, or by which the male is enabled to seize the female, like the curious pincers described by you in some of the lower crustaceans.[9] But what I want to know is how low in the scale sexual differences occur which require some degree of self-consciousness in the males, as weapons by which they fight for the female, or ornaments which attract the opposite sex. Any differences between males & females which follow different habits of life w^d have to be excluded. I think you will easily see what I wish to learn. A priori it w^d never have been anticipated that insects w^d have been attracted by the beautiful colouring of the opposite sex, or by the sounds emitted by the various musical instruments of the male Orthoptera.[10] I know no one so likely to answer this question as yourself. & sh^d be grateful for any information however small.

My second subject refers to expression of countenance, to which I have long attended, & in which I feel a keen interest; but to which unfortunately I did not attend when I had the opportunity of observing various races of Man.[11] It has occurred to me that you might without much trouble make a **few** observations for me in the course of some months on Negros, or possibly on native S. Americans; though I care most about Negros. Accordingly I enclose some questions as a guide & if you c^d answer me even one or two I sh^d feel truly obliged. I am thinking of writing a little essay on the origin of Mankind, as I have been taunted with concealing my opinions; & I sh^d do this immediately after the completion of my present book. In this case I sh^d add a chapter on the cause or meaning of Expression.[12]

With gratitude for all your great kindness & sincere admiration of all your powers of observation I remain | my dear Sir yours very | sincerely C. Darwin

PS. | You must not give yourself any great trouble about these questions, but possibly you might in the course of a few months be able to observe for me one or two points.

I have sent copies to other quarters of the world

an answer within 6 or 8 months w^d be in time.—[13]

If you kept the subject occasionally before your mind, an opportunity of observing some few cases, such for instance as (4) or (5) or (13) &c would almost certainly occur.—

But you must not plague yourself on a subject which will appear trifling to you, but has, I am sure, some considerable interest.

[Enclosure]
Queries about Expression

(1) Is astonishment expressed by the eyes & mouth being opened wide & by the eyebrows being elevated?

(2) Does shame excite a blush, when the colour of the skin allows it to be visible?

(3) When a man is indignant or defiant does he frown, hold his body & head erect, square his shoulders & clench his fists?

(4) When considering deeply on any subject or trying to understand any puzzle does he frown, or wrinkle the skin beneath the lower eyelids?

(5) When in low spirits are the corners of the mouth depressed, & the inner corner or angle of the eybrows raised by that muscle which the French call the 'grief' muscle?

(6) When in good spirits do the eyes sparkle with the skin round & under them a little wrinkled & with the mouth a little extended?

(7) When a man sneers or snarls at another is the corner of the upper lip over the canine teeth raised on the side facing the man whom he addresses?

(8) Can a dogged or obstinate expression be recognized, which is chiefly shewn by the mouth being firmly closed, a lowering brow & slight frown?

(9) Is contempt expressed by a slight protrusion of the lips & turning up of the nose with a slight expiration?

(10) Is disgust shewn by everted lower lip, slightly raised upper lip with sudden expiration something like incipient vomiting?

(11) Is extreme fear expressed in the same general manner as with Europeans?

(12) Is laughter ever carried to such an extreme as to bring tears into the eyes?

(13) When a man wishes to shew that he cannot prevent something being done, or cannot himself do something does he shrug his shoulders, turn inwards his elbows, extend outwards his hands & open the palms?

(14) Do the children when sulky pout, or greatly protrude their lips?

(15) Can guilty or sly or jealous expressions be recognized? tho' I know not how these can be defined.

(16) As a sign to keep silent is a gentle hiss uttered?

(17) Is the head nodded vertically in affirmation, & shaken laterally in negation?

Observations on natives who have had little communications with Europeans would be of course the most valuable, tho' those made on any natives w^{d} be of much interest to me. General remarks on expression are of comparatively little value. A definite description of the countenance under any emotion or frame of mind w^{d} possess much more value, & an answer to any one of the foregoing questions w^{d} be gratefully accepted.

Ch. Darwin
Down Bromley Kent

LS(A)
Endorsement: 'Received March 24/67 | Answered April 1'
British Library (Loan 10: 13)

[1] See letter from Fritz Müller, 1 January 1867.

[2] See *Correspondence* vol 14, letter from Fritz Müller, 1 December 1866, and this volume, letter from Fritz Müller, 1 January 1867 and nn. 1, 5, and 8–10.

[3] CD may be recalling John Scott's observation of fungal threads on the pollen masses of *Bletia* (see *Correspondence* vol. 11, letter from John Scott, [1–11] April [1863]). In *Variation* 2: 134, CD noted that the discoloration and decay were not caused by parasitic cryptogams, which were observed by Müller only once (see letter from Fritz Müller, 1 April 1867).

[4] CD also asked this question in his letter to Müller of 7 February [1867].

[5] See n. 2, above. CD discussed self-impotent plants in *Variation* 2: 131–40. See also *Cross and self*

fertilisation, pp. 329–47. In *Origin* 5th ed., p. 333, CD wrote: 'With plants, so far is cultivation from giving a tendency towards sterility between distinct species, that in several well-authenticated cases ..., certain plants ... have become self-impotent, whilst still retaining the capacity of fertilising and being fertilised by, other species.'

6 On the rudimentary nature of some *Catasetum* organs, see the letter from Fritz Müller, 1 January 1867 and nn. 15 and 16. CD had received letters regarding the production of seed capsules in *Catasetum* in Trinidad (see *Correspondence* vol. 11, letter from H. F. Hance, 10 May 1863, and letter from Edward Bradford, 31 July 1863; see also *Correspondence* vol. 12, letter to Daniel Oliver, 18 March [1864], and 'Fertilization of orchids', p. 154 (*Collected papers* 2: 151)). See also *Orchids* 2d ed., p. 197 n.

7 See letter from Fritz Müller, 1 January 1867 and nn. 17–19.

8 CD had briefly discussed sexual selection in *Origin*, pp. 87–90, 197–200. He had also discussed sexual selection with Alfred Russel Wallace in 1864, with Charles Lyell in 1865, and with James Shaw in 1866 (see *Correspondence* vols. 12, 13, and 14). CD was currently collecting material on sexual selection that would ultimately appear in *Descent*.

9 For a discussion of secondary sexual characters, see *Origin*, pp. 150–8. In *Für Darwin* (F. Müller 1864, pp. 12–17), Müller discussed the chelae in species of crustacea that had two male forms. In one form the chelae were larger; the chelae were used to clasp females (see Dallas trans. 1869, pp. 20–6). Darwin cited Müller on these crustacea in *Descent* 1: 328 *et seq.*

10 CD discussed the colouring of insects throughout *Descent* 1: 361–423; he discussed the auditory apparatus of Orthoptera in *ibid.* pp. 352–61.

11 CD kept notes made during 1837, 1838, and 1839 on human descent and on expression in his M and N notebooks, and in 'Old and useless notes' (see Barrett 1980; see also H. E. Gruber 1981, p. 39); he also made careful observations of his children's expressions (see *Correspondence* vol. 4, Appendix III). He made occasional observations of the expressions of indigenous peoples while on the *Beagle* voyage (see *Journal of researches*, and R. D. Keynes ed. 1988).

12 CD did not explicitly discuss human descent in *Origin*; for objections to what his correspondents took to be his opinion on the topic following its publication, see, for example, *Correspondence* vol. 8, letter from Leonard Jenyns, 4 January 1860, and letter from W. H. Harvey, 24 August 1860. More recently, CD had been deeply disappointed by Charles Lyell's *Antiquity of man* (C. Lyell 1863), and had disagreed on 'minor' points with Alfred Russel Wallace's essay on humans and natural selection (A. R. Wallace 1864b; see, respectively, *Correspondence* vol. 11, and *Correspondence* vol. 12, letter to A. R. Wallace, 28 [May 1864]). He had only recently decided to write a separate essay on humans instead of devoting a chapter to it in *Variation* (see letter to William Turner, 11 February [1867]). Ultimately, he published *Expression* separately from *Descent*.

13 CD sent handwritten queries about expression to a number of correspondents at the end of February 1867. CD had earlier sent similar queries on expression to the Falkland Islands (see *Correspondence* vol. 8, letter to Thomas Bridges, 6 January 1860) and had also evidently sent questions in late 1866 (see *Correspondence* vol. 14, letter to B. J. Sulivan, 31 December [1866]; see also, this volume, letter from B. J. Sulivan, 11 January 1867 and n. 3). For more on CD's queries about expression, see Freeman and Gautrey 1970 and 1975, and Freeman 1977. On CD's research for *Expression*, see Browne 1985 and Ekman ed. 1998. Müller's replies, written on 5 October 1867 and sent to CD, have not been found; see letter from Fritz Müller, [8 October 1867], n. 2. See also *Expression*, pp. 268–9.

To J. P. M. Weale 22 February [1867]

Down. | Bromley. | Kent. S.E.
Feb 22.

My dear Sir

I am very much obliged for your letter & the paper on the Bonatea which I have read with much interest.[1] Several of the points which you describe are new to me, but I will not here discuss them. I have forwarded your paper to the Linnean

Soc., but as the Society has published so much on the fertilization of orchids & as it is rather in arrear I do not feel sure that it will publish your paper.[2]

I have in fact no means of judging on this head; but should your paper not be published you must not infer from this that the Society does not think it of value. If not published you could if you thought fit reclaim it. The difficulty I foresee will be the expense of engravings; but I hope I may be altogether mistaken in my mistrust.[3]

Your observations on Asclepias will I have no doubt be new & curious; I am nearly sure that Robt Brown many years ago told me that he did not understand how the pollen masses were retained by the stigmas which do not emit viscid matter.[4] I have not heard anything about Dr Brown & his conduct as botanist in your colony.[5] I am much obliged for your very kind & flattering expressions with respect to what I have been able to do in natural History.[6] It is a most serious drawback to me that I am very seldom able to go to London or to see any of my fellow-workers in Natural History owing to my constant state of ill-health; I thus lose much pleasure & profit.

With my best wishes for your continued success in Natural History I remain | my dear Sir | yours very faithfully | Ch. Darwin

LS
Postmark: FE 22 67
American Philosophical Society (326)

[1] Weale enclosed a manuscript on the structure and fertilisation of *Bonatea* with his letter of 9 January 1867.
[2] See letter to the Linnean Society, 22 February [1867].
[3] The paper (Weale 1867) was eventually published by the Linnean Society (see letter to the Linnean Society, 22 February [1867] and n. 3).
[4] See letter from J. P. M. Weale, 9 January 1867 and n. 6. Robert Brown (1773–1858) published on the Asclepiadaceae in Robert Brown 1831. Brown's statement is not recorded in a letter; however, CD often saw Brown in London before and after the *Beagle* voyage (see *Autobiography*, pp. 103–4).
[5] Weale wrote about the colonial botanist, John Croumbie Brown of Cape Town, in his letter of 9 January 1867.
[6] See letter from J. P. M. Weale, 9 January 1867 and n. 5.

To Edward Blyth 23 February [1867][1]

Down. | Bromley. | Kent. S.E.
Feb 23.

My dear Mr Blyth

I have been very much interested by your remarks on the "Origin"; many of which are quite new to me, such as those on mimicry.[2] I knew already some few of the other facts which you mention. I am thinking of writing a short essay on Man & have consequently been much struck with your remarks on the Orang. Do you know C. Vogt's nearly similar remarks on the origin of Man from distinct Ape-families, founded on Gratiolet's observations on the brain? I think you cannot object to my cautiously alluding to your observation on the similarity of the Orang & Malay &c: I think the similarity must be accidental, & I would confirm this by your observation on the S. American genus with respect to the Negro.[3] I do not know what to think about the almost parallel case of Bats. If I had known that you w^d have cared for a copy of the new Edit. of the Origin assuredly I w^d have sent you one: you will of course receive my book on "Dom. Animals &c" whenever published.[4] I regret much that I did not meet you in London, but during the two last days I was unable to leave the house.[5]

You gave me long since two printed pages Royal 8vo with a black line round the page with notes in very small type, it contains some excellent remarks on sexual plumage of birds evidently by yourself. Please to tell me the title that I may refer to it.[6]

I have picked up more facts on sexual characters, even when you are not discussing the subject from your writings than from those of any one else. Thus in the last N^o of "Land & Water" there is a notice which I am sure must be written by you, in which you indicate that the summer plumage of Gulls, Plovers &c is nuptial plumage common to both sexes like that confined to the Drake. This is a new idea to me, if I understand you rightly.— But I presume that you admit that winter plumage, as with the Ptarmigan, may be acquired for a special end.[7]

How I wish that you w^d always sign your name to whatever you write. Can you guide me to any papers on sexual differences, especially in colour, in Mammals? I have picked out two cases by you on Bos & Antelopes in the "Indian Field".[8]

Can you tell me whether the canine teeth differ in the sexes in the true Carnivora? But I must not ask any more questions except one:—Do you still maintain the law on sexual plumage in Birds as given in the sheets above referred to? Yarrell gives a rather different law, dependent on there being an annual change in plumage.[9]

But I have written to you at quite unreasonable length— pray forgive me & believe me yours very sincerely— | Ch. Darwin

LS(A)
McGill University Libraries, Rare Books and Special Collections Division

[1] The year is established by the relationship between this letter and the letter from Edward Blyth, 19 February 1867.

[2] For Blyth's remarks on the fourth edition of *Origin*, see the letter from Edward Blyth, 19 February 1867.

[3] For CD's intended publication on humans, see the letter to William Turner, 11 February [1867] and the letter to Fritz Müller, 22 February [1867] and n. 12. For Blyth's comments on the orang-utan and on human resemblances to apes and monkeys, see the letter from Edward Blyth, 19 February 1867 and nn. 11 and 12. In *Lectures on man* (C. Vogt 1864, pp. 464–6), Carl Vogt discussed the three 'ape-types', the orang-utan, chimpanzee, and gorilla, and ways in which their features approached human features; in making these comparisons, he quoted Louis Pierre Gratiolet's studies of cerebral structure. CD discussed Vogt's idea in *Descent* 1: 230; he cited C. Vogt 1864 and referred to Gratiolet's conclusions without citing their source, Gratiolet [1854]; CD did not allude to Blyth's views. CD

viewed the conclusions of Vogt and Gratiolet as unlikely (see *Descent* 1: 230–1). An annotated copy of C. Vogt 1864 is in the Darwin Library–CUL (see *Marginalia* 1: 824).

[4] In his letter of 19 February 1867, Blyth sent comments on the fourth edition of *Origin*, and discussed resemblances between particular bat genera and primate groups. CD also refers to *Variation*, which was published in January 1868.

[5] CD stayed at Erasmus Alvey Darwin's home in London from 13 to 21 February 1867 (see letter to J. D. Hooker, 8 February [1867], n. 21); before falling ill, he had hoped to walk with Blyth in the Zoological Gardens, Regent Park (see letters to Edward Blyth, [18 February 1867] and [19 February 1867]).

[6] CD also referred to these pages in his letter to Blyth of [18 February 1867]. They were galley proofs of a portion of *Cuvier's animal kingdom* translated from Georges Cuvier's *Règne animal* by Blyth (Cuvier 1840 or 1849 (new edition), pp. 158–9; these are pages 146 and 147 in Cuvier 1859 (reprint of the new edition)). The pages, annotated by both CD and Blyth, are in DAR 84.1: 179–80.

[7] CD marked a passage mentioning the nuptial plumage of plovers, sandpipers, and some gulls and terns in the 16 February 1867 issue of *Land and Water*, p. 84; the annotated clipping entitled 'Arrivals in the zoological gardens' is in DAR 84.1: 182 and is not signed. CD wrote 'Blyth' at the bottom. He cited the article in *Descent* 2: 84 n. 77. CD discussed birds with similar plumage in both genders, even when moulting twice a year, in *Descent* 2: 80–6, with references to plovers, birds closely related to sandpipers, and the ptarmigan; see also *ibid.*, pp. 180, 208–19.

[8] CD made a general reference to Blyth's observations of feral humped cattle (*Bos indicus*) in the *Indian Field* 1858, in *Variation* 1: 79 n. 33. CD cited Blyth, but not an article in the *Indian Field*, on colour differences in antelopes, in *Descent* 2: 288.

[9] See n. 6 above. Blyth added editorial notes describing various sexual plumages according to different groups of species in Cuvier 1840 and 1849, pp. 158–9; in the last paragraph of the note on p. 159, he discussed the juvenile plumage in birds that undergo a double moult. In DAR 84.2: 2, CD listed '*Blyths Laws* (corrected)'; these include three laws of the plumage of juvenile birds in relation to the plumage of the adults. CD also refers to the third of three laws listed in William Yarrell's *A history of British birds* (Yarrell 1843, 1: 159); this states that when adult birds assume breeding plumage different from winter plumage, the young birds' plumage is intermediate 'in the general tone of its colour' between the winter and breeding plumage of the parent birds; his first two laws are on juvenile plumage in relation to sexual adult plumage. The first volume of Yarrell 1843 is in the Darwin Library–CUL and is heavily annotated, including the pages on sexual plumage (see *Marginalia* 1: 883–5). CD also wrote out six laws of plumage in his notes (see DAR 84.2: 3–4). In *Descent* 2: 186–223, CD listed the 'classes of cases or rules under which the differences and resemblances, between the plumage of the young and the old, of both sexes or of one sex alone, may be grouped'.

To Paul Rohrbach 23 February [1867][1]

Down. | Bromley. | Kent. S.E.

Feb 23—

Dear Sir

I am much obliged to you for your kindness in sending me your Memoir on Epipogium, which has interested me much, as the structure & manner of fertilization of the genus are quite new to me. You seem to have worked out the details of the plant with great care. I have long suspected that the classification of Orchids w^d require considerable modification, & this seems to be confirmed by some of your remarks.[2]

With my renewed thanks & best wishes for your continued success in y^r researches in Natural History I remain Dear Sir yours faithfully & obliged | Ch. Darwin

LS
Staatsbibliothek zu Berlin, Sammlung Darmstaedter (Dokum. Sig. 37.24)

[1] The year is established by the probable publication of Rohrbach 1866 after the paper was awarded
  a prize on 4 June 1866 (see n. 2, below).
[2] There is a lightly annotated copy of Rohrbach's memoir on *Epigogium* (Rohrbach 1866) in the Darwin
  Pamphlet Collection–CUL. An archival note attached to the letter indicates that Rohrbach's paper
  was a submission to a competition; he was awarded the prize on 4 June 1866. CD referred to
  Rohrbach's 'admirable memoir' in 'Fertilization of orchids', p. 155 (*Collected papers* 2: 152). In *Orchids*
  2d ed., p. 103, CD cited Rohrbach's observation that the structure and manner of fertilisation of *E.
  gmelini* were similar to those of *Epipactis*, and concluded that the genera were allied, contrary to John
  Lindley's classification.

To Alfred Russel Wallace 23 February 1867

> *Down, Bromley, Kent, S.E.*
> *February* 23, 1867.

Dear Wallace,—

I much regretted that I was unable to call on you, but after Monday I was
unable even to leave the house.[1] On Monday evening I called on Bates[2] and put a
difficulty before him, which he could not answer, and, as on some former similar
occasion, his first suggestion was, "You had better ask Wallace." My difficulty is,
why are caterpillars sometimes so beautifully and artistically coloured? Seeing that
many are coloured to escape danger, I can hardly attribute their bright colour in
other cases to mere physical conditions. Bates says the most gaudy caterpillar he
ever saw in Amazonia (of a Sphinx) was conspicuous at the distance of yards from
its black and red colouring whilst feeding on large green leaves.[3] If anyone objected
to male butterflies having been made beautiful by sexual selection, and asked why
should they not have been made beautiful as well as their caterpillars, what would
you answer? I could not answer, but should maintain my ground.[4] Will you think
over this, and some time, either by letter or when we meet, tell me what you think?
Also, I want to know whether your *female* mimetic butterfly is more beautiful and
brighter than the male?[5]

When next in London I must get you to show me your Kingfishers.[6]

My health is a dreadful evil; I failed in half my engagements during this last visit
to London.—[7] Believe me, yours very sincerely, | C. Darwin.

Marchant ed. 1916, 1: 178

[1] CD had been at 6 Queen Anne Street in London, the home of his brother, Erasmus Alvey Darwin,
  from 13 to 21 February (see 'Journal', Appendix II).
[2] Henry Walter Bates.
[3] In a note dated 18 February 1867, CD wrote that Bates had observed a caterpillar of the sphinx moth
  (family Sphingidae). CD had been considering this general question as early as 1861 (see DAR 81: 5).
  CD discussed colourful caterpillars and gave more detail of Bates's observation in *Descent* 1: 415–16.
  Wallace and Bates had travelled together in the Amazon River basin (*DNB*).
[4] For CD's interest in sexual selection, see the letter to Fritz Müller, 22 February [1867] and n. 8. See
  also letter from Alfred Newton, 21 January 1867 and n. 8. The first extant letter from CD to Wallace

concerning CD's belief in the importance of sexual selection is the letter to A. R. Wallace, 28 [May 1864] (*Correspondence* vol. 12).

[5] In his letter of 19 November 1866 (*Correspondence* vol. 14), Wallace told CD that he had a metallic-blue female *Diadema* that mimicked a metallic-blue *Euploea*; he also wrote that the male *Diadema* was dusky brown. In his paper 'On reversed sexual characters in a butterfly' (A. R. Wallace 1866a, p. 186), Wallace suggested the name *Diadema anomala* (now *Hypolimnas anomala*, the Malayan eggfly); he pointed out that it was the female of the species that mimicked the male *Euploea midamus* (blue-spotted crow). CD mentioned the case in *Descent* 1: 413. CD also discussed mimetic butterflies with Wallace when he was in London later in November 1866 (see letter to Alfred Newton, 19 January [1867]).

[6] CD and Wallace evidently had a discussion on the differences between the sexes in kingfishers when CD was in London in November 1866 (see letter to A. R. Wallace, 29 April [1867] and n. 6). In 1862, Wallace had returned from eight years in the Malay Archipelago and had since organised his extensive collections (see *Correspondence* vol. 12, letter from A. R. Wallace, 2 January 1864, and A. R. Wallace 1905, 1: 385–405). Wallace's kingfishers were displayed in his Bayswater exhibition (see letter from A. R. Wallace, [19 June 1867] and n. 3, and A. R. Wallace 1905, 1: 404).

[7] See also letter to Edward Blyth, [19 February 1867].

From Edward Blyth   24 February 1867

24/2/67—

My dear M[r]. Darwin,

I wish that you could write me word that your bodily health is less precarious, & that the cheerful influences of the coming spring are likely to have a beneficial effect. The remarks of mine which you refer to on the sexual plumage of birds, I cannot recal to mind, & perhaps therefore you would not mind sending them to me, to be returned when I have looked them over.[1] It is true that in the bustards the seasonal adornment in the breeding season is peculiar to the male sex, but I want more information about the ruff of the houbara bustards. In the Bengal floriken and likh (*Otis deliciosa* & *aurita*), the seasonal change is considerable, & confined to the male, which is *smaller* than the female. In the little bustard (*O. tetrax*) the sexes are *alike in size*, & the male only undergoes great seasonal change of colouring.[2] In *O. tarda* the male is $\frac{1}{3}$ larger, & he alone shows the long moustachial plumes, besides having the gular bag, which I have recently seen (unequivocally developed) in a fresh specimen. The large bustards of the division *Eupodotis* have also the male $\frac{1}{3}$ larger, & no other secondary sexual difference that I know of, except the gular bag in *O. Edwardii*, as stated by Sykes & Sir Walter Elliot.[3] Now in all the plover and sandpiper series (with one exception) the seasonal change of colouring is common to the two sexes. The exception is the ruff, which alone has the male larger, for in the others I think the male is constantly smaller, & sometimes very conspicuously so, as in *Numenius lineatus*, and *Limosa ægocephala*. The seasonal adornment of the ruff is most remarkable (& I am not sure whether or not it is analogous to that of the ruffed bustards—*Houbara*); but most reeves also, undergo a certain amount of change, with sometimes a slight indication of the frill, the feathers composing which being not much longer than the rest.[4] Now the ruff is polygamous; & I suspect that the Kora (*Gallicrex cristatus*) is also polygamous amongst the *Rallidæ*. Here again the male is $\frac{1}{3}$ larger, & undergoes a remarkable seasonal change of colouring, which the female does not, besides the development of the frontal caruncle.[5]

2. *E. cristatus,*
   ♂ breeding season.

1. do. non-breeding time.

In this bird the male becomes very deeply tinged with dusky cinereous in the breeding season, & the frontal caruncle is coral-red; after breeding the latter shrinks into a small flat acuminated shield, & the bird moults back into the olivaceous plumage of the female, assuming the dusky colouring afterwards by a change of colour in the same feathers. From Wolf's figures of the weka rail of N. Zealand (*Ocydromus*), I am led to suspect that this species undergoes a similar change of hue, & probably in the male only.[6] In the gulls and terns that assume a breeding plumage, the latter is common to both sexes, *e.g.* the hoods of the *Xema* gulls, and the black *pileus* of many terns. In *Hydrochelidon indica* (Stephens, *hybrida*, Pallas, *[leucoparius]*, Tem.), there is a much greater amount of seasonal change in both sexes, and the abdominal region becomes black, whereas in *Sterna melanogaster* the same part is permanently black; in the black terns (*Hydrochelidon fissipes* & *leucoptera*) the seasonal change attains its maximum among the *Laridæ*; & about the minimum, where there is any change, in *L. canus*.[7]

You ask if I can think of any true *carnivora* amongst the mammalia in which there is a sexual diversity in the *canines*.[8] Very decidedly in the *walrus*, on the extreme limit of the order; & you should look to the southern hemisphere seals (*Otaria*, &c). I have an *impression* that there may be a difference in the sea-elephant (*Morunga*), but it may only be proportionate to the size of the skull. The proboscis of *Morunga* is, if I mistake not, *said to be* a sexual distinction, & how about the hood of the monk-seal? The only secondary sexual difference I can remember among the terrene *carnivora* is the mane of the lion.[9]

In hollow-horned ruminants, there are never upper canines; in the camels & llamas these are much larger in the males; & in the cervine ruminants peculiar (whenever they occur at all) to the males, in which they attain so remarkable a development in the muntjacs, the chevrotains, and the musks (the two latter being hornless).[10]

In the anthropoid apes the canines are more developed in the male sex, but hardly so (at least disproportionally) among other *Quadrumana*. One of the most remarkable instances is that of the narwhal, & remember also *Mastodon* ohioticus.[11]

Most (if not all) of the deer shed two coats twice annually, having a distinct summer and winter vesture, generally very different in colouring; & I can recal no marked *sexual* difference of colouring, though the males of many are darker when fully adult, as C. axis, & *C. Duvauceli*.[12] But in many the males have a considerable nuchal ruff, & the larynx is tumid at the rutting season.

In the *bovine* and *antilopine* series (*i.e.* the sheath-horned ruminants generally) the coat, I think, is *never* shed more than once in the year, & there is no seasonal enlargement of the larynx, for that of *Antilope gutturosa* is tumid *permanently*. In several the adult males are black or blue-grey where the female is brown, & in such species the castrated male resembles the female in colour, *e.g.* for certain, *Ant. cervicapra*, *Portax pictus*, & *Bos sondaicus*. In the first the black colour disappears after the rutting season, & in fact indicates that the animal is *must*, as termed in India.[13] The sable antelope (*Aigoceros niger*) has also a permanently black male; & probably the new *Cobus maria*, from what Sir H. Baker told me; & I feel satisfied that the long-lost *Aigoceros leucophæus* (as supposed) is no other than adult male *A. equinus*, coloured like mature male Nil-gai![14]

Returning to birds, among what I must call the *Parridæ* (which are allied to the plovers and not to the rails), the *Hydrophasianus sinensis* undergoes, alike in both sexes, an extraordinary change of *plumage* in spring; whereas the *Metopidius indicus* (& its immediate congeners doubtless) undergoes *no vernal change*; but the juvenile plumage is very different from the adult, & has been erroneously considered the winter dress.[15] In *Hydrophasianus* there is scarcely any difference between the juvenile and the mature winter plumage. This is a remarkable difference in birds otherwise so nearly allied.

I must see C. Vogt's paper you allude to.[16] I want to get at good descriptions & figures of the *unimproved* races of domestic animals, especially of sheep just now. Bentley wants me to publish in a volume my essays on Wild Types, which I should like well enough to do, in a rather more scientific manner; but I do not wish to interfere in any way with your promised volume, & would rather supply you with any information I may have.[17] As for sheep, I now see clearly—more so than when I penned the article on sheep now publishing—that there must have been two wild types of the European races, one a moufflon very like the Corsican, if not identical with it; the other a lost race, long-tailed with horns spiring in a double circle. The latter would be an immediate prototype of the old *heath* race, the former of the diminutive short-tailed sheep with crescentic horns, as the genuine old Highland, the Shetland, and cognate races.[18] Can you help me at all in this enquiry? The horns upon a stuffed head of a Shetland ram now with Leadbeater are uncommonly moufflon-like, & so is a skull marked Indian in B.M.,[19] which may however have been brought from India without being *Indian*! Do you happen to remember the description in some French work (20 years or more ago) of *Ovis arkar*, from the mountains bordering on the Caspian? Blasius figures the horns, but without giving dimensions, or referring to the original description, which I decidedly remember having read in French, but I cannot now find out where.[20] You will have seen my essay on origin of goats, since writing which I have come to know the true *C. caucasica*, which is widely different from *ægagrus*, & akin to *pyrenaica*.[21]

Yours ever truly, | E. Blyth—

P.S. I will write again on secondary sexual differences in mammalia.

AL incomplete?
DAR 83: 34, 150–1, DAR 84.1: 26–7, 138

CD ANNOTATIONS
1.1 I wish ... over. 1.5] *crossed blue crayon*
1.5 It is true] *after opening square bracket blue crayon*
1.21 but most reeves ... change, 1.22] *scored blue crayon*
2.2 after breeding ... *L. canus*. 2.14] *crossed pencil*
2.5 From Wolf's ... male only. 2.7] *enclosed in square brackets and crossed, blue crayon*
3.1 You ask ... *C. Duvauceli*. 6.4] *crossed blue crayon*; 'Deer' *blue crayon*
6.4 But in ... season. 6.5] *double scored blue crayon*
9.1 I must ... *pyrenaica*. 9.21] *crossed blue crayon*

CD note:
Blyth abstract of letter on Sexual S.
Sheet I. on Bustards nuptial plumage confined to males. In plover & sandpipers common to both sexes, except in Ruffs and Reeve, but in latter female does undergo some change, with trace of frill.— case of do in likewise [*interl*] polygamous Rallidæ
Sheet II. seasonal changes in Gulls & Terns common to both sexes.
Sheet II. p. 2. Canines in Walrus*—perhaps seals [*interl*]— Do not differ in terrene carnivora—
— — p. 3. Canines in Ruminants
No canines in Hollow-Horned Ruminants [vary] in Antelope— Larger in ♂ Camel & Guanaco
Narwhal & Mastodon ohioticus
Not much in other monkeys
Sheet III. p. 4. some male Deer have nuchal ruff & larynx increase in size during breeding season. no change of colour during breeding season in Deer
— — p. 4 sexual colouring of Antelopes
Sheet IV. Case of allied genera, in one of which birds undergo great seasonal change *in both sexes [*interl*] & not in other.—

[1] See letter to Edward Blyth, 23 February [1867] and nn. 6 and 9.
[2] CD had inquired about a notice in the 16 February 1867 issue of *Land and Water*, p. 84, evidently written by Blyth (see letter to Edward Blyth, 23 February [1867] and n. 7); in the notice, Blyth had discussed the breeding plumage of the 'Bengal floriken (*Sypheotides deliciosa*)' and a bustard at the the Zoological Gardens, Regents Park, London. Blyth refers to the houbara bustard (now *Chlamydotis undulata*), the Bengal florican (now *Houbaropsis bengalensis*), and the lesser florican or likh (now *Sypheotides indica*). The little bustard is now *Tetrax tetrax*. See also Jerdon 1862–4, 2: 606–26. For systematic changes in the family Otidae, see *Birds of the World* 3: 240–73. For CD's discussion of moulting and plumage in 'certain bustards', see *Descent* 2: 81, 83.
[3] Blyth refers to the great bustard (*Otis tarda*), and the great Indian bustard, now *Ardeotis nigriceps* (Peters *et al.* 1931–87, 2: 220 and *Birds of the world* 3: 264). Blyth also refers to William Henry Sykes and Sykes 1834, pp. 639–40; Elliot's publication on bustards has not been identified.

[4] For the sexual plumage of plovers and sandpipers, see the letter to Edward Blyth, 23 February [1867] and n. 7. Blyth also refers to the ruff (now *Philomachus pugnax*); the female was sometimes called the reeve. The ruff is now in the family Scolopacidae with sandpipers, godwits (*Limosa* species), and curlews and whimbrels (*Numenius* species); plovers are now in the family Charadriidae (*Birds of the World*). In *Descent* 1: 270, CD stated that the ruff ('*Machetes pugnax*'), unlike other waders ('Grallatores') showed strong differences in sexual plumage. *Numenius lineatus* is now named *N. arquata*, the Eurasian curlew (*Birds of the world* 3: 504). *Limosa aegocephala*, the black-tailed godwit, is now *L. limosa*. For additional comments on sexual differences in waders, see the letter from Alfred Newton, 21 January 1867 and n. 5.

[5] On the possible polygamy of ruffs, see *Descent* 1: 270. The watercock, known also as the kora, is now *Gallicrex cinerea* (*Birds of the world* 3: 194). In *Descent* 2: 80, CD mentioned the large red caruncle on 'one of the rails, *Gallicrex cristatus*'. In *ibid.*, p. 41, CD cited Blyth on the size difference in *G. cristatus*.

[6] An illustration of the weka (*Ocydromus australis*; now *Gallirallus australis*) is included in *Zoological sketches by Joseph Wolf* (Sclater ed. 1861–7).

[7] The monospecific genus *Xema* now includes four subspecies (*Birds of the world* 3: 621). Blyth's *Hydrochelidon* species, the terns, are now named *Chlidonias*, and include *C. hybridus indicus* and *C. leucoptera* or *leucopterus*. *Sterna melanogaster*, the black-bellied tern, is now *S. acticauda*. The black tern is now *C. nigra*. See *Birds of the world* 3: 663–4. In some classifications, the family Laridae includes gulls and terns. *Larus canus* is the mew. In *Descent* 2: 228–9, CD noted that in some smaller gulls, 'or sea-mews (Gavia)', and in some terns, the heads of the adults turned darker during the breeding season.

[8] See letter to Edward Blyth, 23 February [1867].

[9] Blyth probably refers to *Otaria flavescens*, the South American sea lion. For differences in the proboscis and teeth between the sexes in the elephant seal (*Mirounga*), see Nowak 1999, p. 877. For CD's comments on sexual differences in lions and in marine mammals, see *Descent* 1: 268, 2: 241–2, 266–70, 277–8. See also n. 11, below.

[10] Blyth refers to the sizes of canine teeth in some members of the order Artiodactyla. Hollow-horned ruminants, the Bovidae, include antelope, cattle, bison, buffalo, sheep, and goats; see Nowak 1999, p. 1135. Camels and llamas, also in the Artiodactyla like Bovidae, are not considered ruminants (*ibid.*, p. 1072). Of the cervine ruminants (now the superfamily Cervoidea), Blyth refers to the muntjac (*Muntiacus*), the chevrotains (family Tragulidae), and the musk deer (*Moschus*); see *ibid.*, pp. 1096–8, 1081–4, 1809–91. For a contemporary discussion on relationships among the Ruminantia, see *Zoological Record* 3 (1866): 35–7. See also *Descent* 2: 257–8.

[11] For CD on the canines of the anthropoid apes, see *Descent* 1: 144, 156; on the narwhal, see *Descent* 2: 242. Though CD did not mention the large mastodon tusks in *Descent*, he referred to the sexual differences in elephant tusks several times.

[12] Blyth refers to the axis deer (*Cervus axis*, now *Axis axis*) and the swamp deer or barasingha (*C. duvaucelii*); see Nowak 1999, pp. 1100–2, 1106–7. CD cited Blyth's information on darker adult males in the axis deer in *Descent* 2: 290.

[13] The family Bovidae now also includes antelope (see n. 10, above). *Antilope gutturosa* is a synonym of *Procapra gutturosa*. Blyth also refers to the blackbuck (*Antilope cervicapra*), the nilgai, or nilgau (*Portax picta*; now *Boselaphus tragocamelus*), and the banteng (*Bos sondaicus*; now *B. javanicus*). See Nowak 1999, pp. 1145–1203. For CD's discussion of the sexual differences in colours of ruminants, including *Portax picta* and the banteng, see *Descent* 2: 287–90.

[14] The sable antelope is now *Hippotragus niger*. Blyth also evidently refers to James Murie's recent description of an antelope that Murie thought was allied with or identical to *Cobus sing-sing* (now *Kobus* species). Blyth probably refers to Samuel White Baker, who was also mentioned in Murie 1867, pp. 3, 7. The 'long-lost *Aigoceros leucophæus*' (now known as *Hippotragus leucophaeus* or the blue buck) was hunted to extinction in the early nineteenth century. *Aigoceros equinus*, the roan antelope, is now also in the genus *Hippotragus*. See Nowak 1999, pp. 1174–5. Blyth also refers to the nilgai (see n. 13, above).

[15] The family Parridae is now called Jacanidae (see letter from Edward Blyth, 19 February 1867 and n. 30). *Hydrophasianus sinensis*, the pheasant-tailed jacana (now *H. chirurgus*), and *Metopidius indicus*, the bronze-winged jacana, were the two Indian species belonging to the family Parridae.

[16] Blyth refers to Carl Vogt and C. Vogt 1864 (see letter to Edward Blyth, 23 February [1867] and n. 3).

[17] Blyth presumably refers to the publisher Richard Bentley. CD cited Blyth extensively in his chapters on domestic animals in *Variation*.

[18] The article Blyth refers to was published in the 2 March 1867 issue of *Land and Water*, p. 134; he published a second article on sheep, confirming his conviction that there were two European wild types, in the 9 March 1867 issue of *Land and Water*, pp. 156–7. Both articles were under the heading 'Wild types and sources of domestic animals'; they were published under Blyth's pseudonym, 'Zoophilus'. CD cited these articles, as well as Blyth's article on *Ovis* (Blyth 1841), in *Variation* 1: 94. For earlier comments by Blyth on sheep descent, see *Correspondence* vol. 5.

[19] Blyth refers to John Leadbeater, a taxidermist, and to the British Museum.

[20] Johann Heinrich Blasius's drawing of *Ovis arkal* is in Blasius 1857, p. 469, figs. 243 and 244. The original description has not been identified. *Ovis arkal*, the Transcaspian urial, is considered a subspecies of *O. vignei* in Lydekker 1898. See also Blyth's article in the 2 March 1867 issue of *Land and Water*, p. 134.

[21] Blyth refers to his article 'Wild types and sources of domestic animals', which appeared in the 2 February 1867 issue of *Land and Water*, pp. 37–8. CD's annotated copy of the article is in DAR 83: 118g. CD cited the article in *Variation* 1: 101, n. 96. Blyth also refers to three species of *Capra*, now called, respectively, the West Caucasian tur, the wild goat, and the Spanish ibex (see Nowak 1999, pp. 1220–7).

From A. R. Wallace 24 February [1867][1]

9, St. Mark's Crescent | N.W.
Feb. 24th.

Dear Darwin

I saw Bates a few days ago & he mentioned to me this difficulty of the catterpillars.[2] I think it is one that can only be solved by special observation. The only probable solution I can imagine is something like this Catterpillars are very similar in form & there are hundreds of species that are only to be distinguished by *colour*.

Now great numbers are protected by their green colours assimilating with foliage or their brown colours resembling bark or twigs. Others are protected by prickles and long hairs—which no doubt render them distasteful to birds, especially to our small birds which I presume are the great destroyers of catterpillars. Now supposing that others, not hairy, are protected by a disagreeable taste or odour, it would be a positive advantage to them never to be mistaken for any of the palatable catterpillars, because a slight wound such as would be caused by a peck of a bird's bill almost always I believe kills a growing catterpillar. Any gaudy & conspicuous colour therefore, that would plainly distinguish them from the brown & green eatable catterpillars, would enable *birds* to recognise them easily as a kind not fit for food, & thus they would escape *seizure* which is as bad as being *eaten*.[3]

Now this can be tested by experiment, by any one who keeps a variety of insectivorous birds. They ought as a rule to refuse to eat and generally refuse to touch gaudy coloured catterpillars, & to devour readily all that have any protective tints. I will ask Mr. Jenner Weir of Blackheath about this, as he has had an aviary for many years & is a very close and acute observer, & I have no doubt will make the experiment this summer.[4]

When our discussion on Mimicry took place a most interesting little fact was mentioned by Mr. Stainton. After *mothing* he is accustomed to throw all the common species to his poultry & once having a lot of young turkeys he threw them a quantity of moths which they eat greedily, but among them was one common *white moth* (*Spilosoma menthastri*) One of the young turkeys took this in his beak, shook his head & threw it down again, another ran to seize it and did the same, and so on, the whole brood in succession rejected it! Mr. Weir tells me that the larva of this moth is hairy & is also rejected by all his birds, which sufficiently accounts for the insect being *very common*.[5] But what is still more curious, another moth much *less common* (*Diaphora mendica*) has the *female* also white, (although the male is quite different) and might at night be easily mistaken for the other! So here we have a case of British mimicry exactly analogous in all its details to that of the *Heliconidæ* & *Danaidæ*; and it is particularly valuable because it is a *direct proof* that Lepidoptera *do* differ in flavour, & that certain flavours *are* distasteful to birds.[6]

My female mimetic butterfly *is* much more beautiful than the male, being metallic blue while the male is dull brown.[7] I sometimes doubt whether sexual selection has acted to produce the colours of *male butterflies*. I have thought that it was merely that it was advantageous for the females to have less brilliant colours, & that colour has been produced merely because in the process of infinite variation *all colours* in turn were produced. Undoubtedly two or three male butterflies do often follow a female, but whether *she* chooses between them or whether the strongest & most active gets her is the question.[8] Cannot this also be decided by experiment? If a lot of common butterflies were bred, say our "brimstone" or better, the *"orange tip"*, & the males and females separated & then a certain number of the males *discoloured* by rubbing the wings carefully;—and we were then to turn out a female along with a coloured and a discoloured male into a room or greenhouse, would the female always or in the majority of cases choose the best coloured male.? A series of experiments of this kind carefully carried out would I think settle the question. I will suggest these two classes of experiment at the next meeting of the Entomological & perhaps some country residents may be induced to carry them out.[9]

I hope you will take care of your health, & not work too hard when you get a little better.

I often wish I lived in the country, & was able to carry out some of these most interesting observations but I do not know whether I shall be able to manage it.

Yours very faithfully | Alfred R. Wallace

C. Darwin Esq.

DAR 82: A19–21

CD ANNOTATIONS

1.1 I saw ... *very common.* 4.9] *crossed pencil*

1.1 I saw ... of catterpillars. 2.4] 'Mr Mansell Weales larvæ cases like Mimosa horrida'[10] *added in margin blue crayon*

4.2 After ... it! 4.7] 'white moth' *added in margin blue crayon*

4.9 But what ... the other! 4.11] *scored red crayon*

5.1 My female ... were produced. 5.6] *crossed pencil*

Top of letter: 'Mimetic' *blue crayon*; 'Caterpillars' *red crayon*

[1] The year is established by the relationship between this letter and the letter to A. R. Wallace, 23 February 1867.

[2] Henry Walter Bates had advised CD to pose his question on colourful caterpillars to Wallace (see letter to A. R. Wallace, 23 February 1867).

[3] In *Descent* 1: 416, CD paraphrased Wallace's remarks and quoted from the similar statement made by Wallace at the 4 March 1867 meeting of the Entomological Society of London (see *Transactions of the Entomological Society of London* (*Journal of the Proceedings*) 3d ser. 5 (1865–7): lxxx).

[4] Though there is no further extant correspondence in 1867 regarding these experiments, John Jenner Weir conducted them in the summers of 1867 and 1868 (see *Correspondence* vol. 16, letters from J. J. Weir, 24 March 1868 and 31 March 1868). See also letter from A. R. Wallace, 10 March 1869 (*Calendar* no. 6651). In 1869 and 1870, Weir read a paper describing his experiments with insectivorous birds to the Entomological Society (Weir 1869–70); see *Descent* 1: 417.

[5] Wallace refers to a discussion held at the Entomological Society on 3 December 1866 (see *Transactions of the Entomological Society of London* (*Journal of the Proceedings*) 3d ser. 5 (1865–7): xliv–xlviii); Henry Tibbats Stainton's case of the *Spilosoma* and Weir's comment were recorded on p. xlv. The general discussion on mimicry and CD's theory was continued from the Entomological Society meeting of 19 November 1866, reported in *ibid.*, pp. xxxvi–xli (see also *Correspondence* vol. 14, letter from A. R. Wallace, 19 November 1866). CD reported the case in *Descent* 1: 398 n. 16, and Wallace in [A. R. Wallace] 1867a, p. 25. A synonym for *Spilosoma menthastri* (white ermine moth) is *S. lubricipeda*, and the caterpillar is the woolly-bear.

[6] In his study of the Amazonian Heliconidae and their mimics, Bates argued that natural selection accounted for the phenomenon of mimicry; both male and female Heliconidae are brightly coloured, and Bates surmised that they were protected by a secretion or odour unpalatable to predators, and that their mimics were protected by their resemblance to Heliconidae (Bates 1861, pp. 502–15). Wallace had discovered cases where only the females of some species of *Diadema* (now *Hypolimnas*) mimicked species of *Euploea* of the family Danaidae, while females of several species of *Pieris* resembled *Heliconius* species of the family Heliconidae (see A. R. Wallace 1866a, p. 186; see also n. 7, below). CD mentioned Wallace's consideration of *Diaphora mendica* in *Descent* 1: 398 n. 16, giving the earlier generic name of *Cycnia*. See also *ibid.*, pp. 412–14. *Diaphora mendica* is the muslin moth.

[7] See letter to A. R. Wallace, 23 February 1867 and n. 5.

[8] CD had asked Wallace how he would defend the view that the bright colours of male butterflies were due to sexual selection (see letter to A. R. Wallace, 23 February 1867).

[9] Wallace refers to *Gonepteryx rhamni* (brimstone butterfly), and to *Anthocharis cardamines* (orange-tip butterfly). In *Descent* 1: 409, CD stated that the male brimstone butterfly had probably acquired his bright colours as a result of sexual selection, but did not mention experiments. There is no mention in the transactions of the meeting of the Entomological Society of 4 March 1867 of Wallace's asking for the experiments with discoloured males to be carried out. He did, however, ask for observations on which caterpillars were eaten by birds (see *Transactions of the Entomological Society of London* (*Journal of the Proceedings*) 3d ser. 5 (1865–7): lxxx, and *Descent* 1: 417.

[10] CD refers to James Philip Mansel Weale's observation of caterpillars mimicking thorns (see letter from J. P. M. Weale, 9 January 1867 and n. 7). In *Descent* 1: 416–17, CD referred to Weale's observation in his discussion of Wallace's opinion regarding conspicuous colour in caterpillars.

From Benjamin Dann Walsh [25 February 1867][1]

insects; in Stainton's *Entom. Annual* (1861, p. 39) you will find good proof that worker wasps can & do generate worker wasps.[2] The demonstration is simple. A nest containing a single female & several workers is in early spring deprived of the female; & it is found that the building of fresh cells & the production of fresh workers therein goes on as successfully as if the mother-female had remained in the nest. With regard to your ⟨   ⟩

⟨   ⟩ which Dr Velie assures me never builds a nest for itself, & the books say the same? As with your Cuckoo, the other species belonging to the same genus have no such parasitic habits.[3]

I enclose you a copy of a recent Lecture by Agassiz, the marked portions in which I thought would interest you. I suspect he has mistaken the deposits left by floating Ice-bergs for true Glaciers. His theory about Glaciers moving **on level ground** might do for high northern latitudes[4] ⟨   ⟩

Editing the *Practical Entomologist* does undoubtedly take up a good deal of my time, but I also pick up a good deal of information of real scientific value from its correspondents.[5] Besides, this great American nation has hitherto had a supreme contempt for Natural History, because they have hitherto believed that it has nothing to do with the dollars and cents. After hammering away at them for a year or two, I have at last succeeded in touching the 'pocket nerve' in Uncle Sam's body, and he is gradually being galvanised into the conviction that science has the power to make him richer. ⟨   ⟩

⟨   ⟩ You cannot have the remotest conception of the ideas of even our best-educated Americans as to the pursuit of science. I never yet met with a single one who could be brought to understand how or why a man should pursue science for its own pure and holy sake.

AL incomplete
Darwin Library–CUL (bound with Siebold 1857), *ML* 1: 248–9

[1] The date is established by the relationship between this letter and the letter to B. D. Walsh, 23 March [1867].

[2] Walsh referred to Henry Tibbats Stainton and to an article on Hymenoptera in the *Entomologist's Annual* (F. Smith 1861); the article included a summary of an account of the deposition of fertile eggs by worker wasps presented in Stone 1860, pp. 7263–4.

[3] The bird that Jacob W. Velie referred to has not been identified. For a discussion of parasitic nesting habits in birds, for CD's discussion of the parasitic instinct of cuckoos in *Origin*, and on his additions to the topic in the fourth edition of *Origin*, see the letter from Edward Blyth, 19 February 1867 and nn. 18–22. Walsh may be responding to additions in the fourth edition of *Origin*; CD had sent a copy to him in late 1866 (see *Correspondence* vol. 14, letter to B. D. Walsh, 24 December [1866]). The species ('your cuckoo') that Walsh refers is the common cuckoo (*Cuculus canorus*); the genus *Cuculus* now contains sixteen species. For modern cuckoo systematics, see *Birds of the world* 4: 508–607.

[4] The enclosure has not been found. Walsh may be referring to a lecture given by Louis Agassiz at the Cooper Institute in New York in the winter of 1866 and 1867 and published in the *New York Herald Tribune* (see J. L. R. Agassiz 1867 and E. C. Agassiz 1885, 2: 645); Agassiz also gave lectures at the Lowell Institute in September and October 1866 (Lurie 1960, p. 353). Agassiz returned in August 1866 from an expedition to Brazil, during which he claimed to have seen evidence of glaciation in the basin of the Amazon River (see *Correspondence* vol. 14).

[5] In his letter of 24 December [1866], CD had expressed his sympathy with Walsh over the hard work that he thought was involved in being an editor of a journal (*Correspondence* vol. 14).

From Trübner & Co 26 February 1867

London. | 60, Paternoster Row. ⟨E.C.⟩
Febry. 26⟨th 1867⟩

Messrs Trübner & Co present their respectful compliments to Charles D⟨arwin⟩ Esq and beg to inform him, that a russian Correspondent of theirs Mr Kowalewsky is desirous of translating into russian Mr Darwin's new work "Domesticated Animals".[1] Mr Kowalewsky offers to pay £1 p. sheet for the advantage of early sheets.

L (damaged)
DAR 169: 70

[1] The references are to Vladimir Onufrievich Kovalevsky and to *Variation*.

To A. R. Wallace 26 February [1867]

Down. | Bromley. | Kent. S.E.
Feb 26

My dear Wallace

Bates was quite right, you are the man to apply to in a difficulty. I never heard any thing more ingenious than your suggestion & I hope you may be able to prove it true.[1] That is a splendid fact about the white moths: it warms one's very blood to see a theory thus almost proved to be true.[2] With respect to the beauty of male butterflies, I must as yet think that it is due to sexual selection; there is some evidence that Dragon-flies are attracted by bright colours; but what leads me to the above belief is, so many male Orthoptera & Cicadas having musical instruments. This being the case the analogy of birds makes me believe in sexual selection with respect to colour in insects.[3] I wish I had strength & time to make some of the experiments suggested by you;[4] but I thought butterflies w^d not pair in confinement; I am sure I have heard of some such difficulty.

Many years ago I had a dragon-fly painted with gorgeous colours but I never had an opportunity of fairly trying it.

The reason of my being so much interested just at present about sexual selection is that I have almost resolved to publish a little essay on the Origin of Mankind, & I still strongly think (tho' I failed to convince you, & this to me is the heaviest blow possible) that sexual selection has been the main agent in forming the races of Man.[5]

By the way there is another subject which I shall introduce in my essay, viz expression of countenance; now do you happen to know by any odd chance a very good-natured & acute observer in the Malay Arch. who you think w^d make a few easy observations for me on the expression of the Malays when excited by various emotions? For in this case I w^d send to such person a list of queries.[6]

I thank you for your *most interesting* letter & remain | yours very sincerely | Ch. Darwin

Endorsement: '1867'
LS(A)
British Library (Add 46434, f. 76)

[1] Henry Walter Bates had advised CD to ask Wallace about colourful caterpillars (see letter from A. R. Wallace, 24 February [1867] and n. 2). When he described his enquiry in *Descent* 1: 416, CD mentioned that Wallace had 'an innate genius for solving difficulties'.

[2] See letter from A. R. Wallace, 24 February [1867] and nn. 5 and 6.

[3] Wallace was not convinced that the bright colours of some male butterflies were a result of sexual selection (see letter from A. R. Wallace, 24 February [1867] and n. 8). CD eventually concluded in *Descent* 1: 399, 403–4, that bright colours in male butterflies and some male moths were generally a result of sexual selection rather than for protection. For CD and Bates's earlier consideration of bright colours and sexual selection in butterflies, see also *Correspondence* vol. 9, letter to H. W. Bates, 26 March [1861], and letter from H. W. Bates, 28 March 1861 and n. 7. He reported the belief that the colours of dragonflies served as a sexual attraction in *Descent* 1: 361–2, and discussed the sounds made by cicadas and Orthoptera in *ibid.*, pp. 350–61.

[4] For Wallace's suggested experiments, see his letter of 24 February [1867].

[5] For CD's intended publication on human descent, see also the letter to Edward Blyth, 23 February [1867] and n. 3. CD had tried in 1864 to convince Wallace that sexual selection had been a powerful influence in forming the human races (see *Correspondence* vol. 12, letter to A. R. Wallace, 28 [May 1864], and letter from A. R. Wallace, 29 May [1864]).

[6] Ultimately, CD published *Expression* separately from *Descent*. See also letter to Fritz Müller, 22 February [1867] and n. 13.

To Julius von Haast   27 February [1867][1]

Down Bromley Kent

My dear Dr Haast

I have thought that you might know some Missionary, protector[2] or colonist who associates with the Natives any where in N. Zealand & who w^d at your request oblige me by making a few observations on their expression of countenance when excited by ⟨va⟩rious emotions. Perhaps you might have some opportunity yourself of observing. I should be most grateful for any, however small, information, & I enclose some queries for this purpose. You must not take much trouble but I believe you will aid me if you can. I have sent copies of these queries to various parts of the world for I am greatly interested on the subject.[3]

I hope your geological investigations continue to be as interesting as they have hitherto been.[4]

Believe me | my dear Dr Haast— | yours sincerely | Ch. Darwin

Feb 27th.—

[Enclosure]

Queries about Expression.

1. Is astonishment expressed by the eyes and mouth being opened wide and by the eye-brows being raised?

2. Does shame excite a blush, when the colour of the skin allows it to be visible?

3. When a man is indignant or defiant does he frown, hold his body and head erect, square his shoulders and clench his fists?

4. When considering deeply on any subject, or trying to understand any puzzle, does he frown, or wrinkle the skin beneath the lower eyelids?

5. When in low spirits, are the corners of the mouth depressed, and the inner corner or angle of the eyebrows raised by that muscle which the French call the "Grief muscle"?

6. When in good spirits do the eyes sparkle, with the skin round and under them a little wrinkled and with the mouth a little drawn back?

7. When a man sneers or snarls at another, is the corner of the upper lip over the canine teeth raised on the side facing the man whom he addresses?

8. Can a dogged or obstinate expression be recognized, which is chiefly shewn by the mouth being firmly closed, a lowering brow and a slight frown?

9. Is contempt expressed by a slight protusion of the lips and turning up of the nose, with a slight expiration?

10. Is disgust shewn by the lower lip being turned down, the upper lip slightly raised, with a sudden expiration something like incipient vomiting?

11. Is extreme fear expressed in the same general manner as with Europeans.

12. Is laughter ever carried to such an extreme as to bring tears into the eyes?

13. When a man wishes to show that he cannot prevent something being done, or cannot himself do something, does he shrug his shoulders, turn inwards his elbows, extend outwards his hands, and open the palms?

14. Do the children when sulky, pout or greatly protrude the lips?

15. Can guilty, or sly, or jealous expressions be recognized?—though I know not how these can be defined.

16. As a sign to keep silent, is a gentle hiss uttered?

17. Is the head nodded vertically in affirmation and shaken laterally in negation?

Observations on natives who have had little communication with Europeans would be of course the most valuable, though those made on any natives would be of much interest to me.

General remarks on expression are of comparatively little value.

A definite description of the countenance under any emotion or frame of mind would posess much more value, and an answer within 6 or 8 months or even a year to any *single* one of the foregoing questions would be gratefully accepted.

Memory is so deceptive on subjects like these that I hope it may not be trusted to.

Down, Bromley, Kent
Ch Darwin

March

[1] The year is established by the relationship between this letter and the letter from Julius von Haast, 12 May – 2 June 1867.

[2] Protector (of aborigines): an official post established in 1840 to look after Maori interests *vis-à-vis* Europeans and abolished in 1846 (*Dictionary of New Zealand English*).

[3] See letter to Fritz Müller, 22 February [1867], n. 13.

[4] Haast was provincial geologist for Canterbury province, New Zealand, and mentioned his geological work briefly in his letters to CD of 17 July 1866 and 8 September 1866 (*Correspondence* vol. 14).

To Robert Swinhoe [27 February 1867][1]

[Enclosure][2]

Queries about Expression.

1. Is astonishment expressed by the eyes and mouth being opened wide and by the eyebrows being raised?

2. Does shame excite a blush, when the colour of the skin allows it to be visible?

3. When a man is indignant or defiant does he frown, hold his body and head erect, square his shoulders and clench his fists?

4. When considering deeply on any subject, or trying to understand any puzzle, does he frown, or wrinkle the skin beneath the lower eyelids?

5. When in low spirits, are the corners of the mouth depressed, and the inner corner or angle of the eyebrows raised by that muscle which the French call the "Grief Muscle?"

6. When in good spirits do the eyes sparkle, with the skin round and under them a little wrinkled and with the mouth a little drawn back?

7. When a man sneers or snarls at another, is the corner of the upper lip over the canine teeth raised on the side facing the man whom he addresses?

8. Can a dogged or obstinate expression be recognised, which is chiefly shown by the mouth being firmly closed, a lowering brow and slight frown?

9. Is contempt expressed by a slight protrusion of the lips and turning up of the nose, with a slight expiration?

10. Is disgust shown by the lower lip being turned down, the upper lip slightly raised, with a sudden expiration something like incipient vomiting?

11. Is extreme fear expressed in the same general manner a with Europeans?

12. Is laughter ever carried to such an extreme as to bring tears into the eyes?

13. When a man wishes to show that he cannot prevent something being done, or cannot himself do something, does he shrug his shoulders, turn inwards his elbows, extend outwards his hands, and open the palms?

14. Do the children when sulky, pout or greatly protrude the lips?

15. Can guilty, or sly, or jealous expressions be recognised?—though I know not how these can be defined.

16. As a sign to keep silent, is a gentle hiss uttered?

17. Is the head nodded vertically in affirmation and shaken laterally in negation?

Observations on natives who have had little communication with Europeans would be of course the most valuable, though those made on any natives would be of much interest to me.

General remarks on expression are of comparatively little value.

A definite description of the countenance under any emotion or frame of mind would possess much more value.

Memory is so deceptive on subjects like these that I hope it may not be trusted to.

Notes and Queries on China and Japan 1 (1867): 105

[1] The date is established by the mention in the letter from Robert Swinhoe, 5 August 1867, of a letter to Swinhoe (now missing) dated 27 February, and an enclosure on expression. See also letter to Fritz Müller, 22 February [1867] and n. 13.

[2] The letter and its original enclosure, CD's queries on expression, have not been found. Swinhoe had the queries published in *Notes and Queries on China and Japan*, 31 August 1867; the transcript has been made from that publication. The queries were headed by the following paragraph:

Signs of emotion amongst the Chinese &c.— The following Queries have been addressed to me by a friend in England. He wishes them to be applied to the expression displayed under various emotions by the Chinese or by any other outlandish race. Some of your readers may find leisure to record their observations on this subject in *Notes and Queries*. I give my Querist's own words:—

Beneath the queries was printed, 'Amoy, July, 1867. R.S.' No replies were published.

To J. P. M. Weale 27 February [1867][1]

Down. | Bromley. | Kent. S.E.
Feb. 27th,

Dear Sir

Since writing to you about a week ago,[2] it has occurred to me that you w^d. perhaps grant me a great favour; namely to forward, & back up with your own influence, the enclosed queries on Expression to any capable & trustworthy observer, who associates with Hottentots or Caffres. I am sending these queries to all parts of the world, for I am much interested in the subject, & sh^d. be grateful for *any* however small information.[3]

Anyone who w^d. keep the subject before his mind for 2 or 3 months, would easily observe some of the points.—[4] Pray excuse me begging this favour & do what you can to aid me.—

Believe me | Yours very faithfully | Ch. Darwin

[Enclosure]

Queries about Expression

(1) Is astonishment expressed by the eyes & mouth being opened wide, & by the eyebrows being raised?

(2) Does Shame exite a blush when the colour of the skin allows it to be visible? Especially how far down the body does blush extend?

(3) When a man is indignant or defiant does he frown, hold his body & head errect, square his shoulders & clench his fists?

(4) When considering deeply on any subject or trying to understand any puzzle, does he frown, or wrinkle the skin beneath the lower eyelids?

(5) When in low spirits are the corners of the mouth depressed & the inner corner or angle of the eyebrows raised & contracted by that muscle which the french call the grief muscle?

(6) When in good spirits do the eyes sparkle, with the skin round & under them a little wrinkled & with the mouth a little drawn back in the corners?

(7) When a man sneers or snarls at another, is the corner of the upper lip over the canine teeth raised on the side facing the man whom he addresses?

(8) Can a dogged or obstinate expression be recognised, which is chiefly shown by the mouth being firmly closed, a lowering brow & a slight frown?

(9) Is contempt expressed by a slight protrusion of the lips & turning up of the nose with a slight expiration?

(10) Is disgust shown by the lower lip being turned down, the upper lip slightly raised, with a sudden expiration something like incipient vomiting?

(11) Is extreme fear expressed in the same general manner as with Europeans?

(12) Is laughter ever carried to such an extreme as to bring tears into the eyes?

(13) When a man wishes to show that he cannot prevent something being done or cannot himself do something, does he shrug his shoulders, turn inwards his elbows, extend outwards his hands & open the palms?

(14) Do the children when sulky pout or greatly protrude the lips?

(15) Can guilty, or shy,[5] or jealous expressions be recognised—though I know not how these can be defined?

(16) As a sign to keep silent is a gentle hiss uttered?

(17) Is the head nodded vertically in affirmation, & shaken laterally in negation

Observations on natives who have had little communication with Europeans would be of course the most valuable, though those made on any natives would be of *much* interest to me.

General remarks on expression are of comparatively little value.

A definite description of the countence under any emotion or frame of mind would possess much more value; & an answer within 6 or 8 months or even a year to any *single* one of the foregoing questions would be gratefully accepted.

Charles Darwin

Down, Bromley, Kent | 1867

P. S. Memory is so deceptive in subjects like these that I hope it may not be trusted to.

University of Virginia Library, Special Collections, Darwin Evolution Collection (3314)

---

[1] The year is established by the date of the enclosure.

[2] Letter to J. P. M. Weale, 22 February [1867].

[3] Weale lived in Cape Colony (now the Eastern Cape province of the Republic of South Africa). In the nineteenth century, the term 'Caffre' or Kafir was usually used to refer to some groups of the Xhosa people of south-eastern Africa, while 'Hottentot' was usually used to refer to peoples of south-western

Africa (the Khoikhoi); for nineteenth-century uses of the terms 'Hottentot' and 'Caffre', see Stocking 1987, Dubow 1995, and S. J. Gould 1997.

[4] Weale enclosed one set of replies to the queries with his letter of 7 July 1867. See also letter from M. E. Barber, [after February 1867].

[5] 'Shy' is a copyist's error for 'sly' (see Appendix IV).

To Ferdinand von Mueller 28 February [1867][1]

Down Bromley Kent

My dear Sir

I have thought that you would grant me a favour. Perhaps you know some Missionary or Protector of the Aborigines, or some acute colonist in the far interior who w^d take a little trouble to oblige you.[2] In this case will you have the kindness to forward soon the enclosed Queries, & beg your correspondent to make a few observations for me, as any opportunity may occur, on the expression of the aborigines under the several emotions specified. If you c^d obtain answers to even 1, 2, or 3 of the questions you would do me a considerable service, for I am at present much interested on this subject.[3]

I hope you will excuse my begging this favour from you & I remain my dear Sir | yours sincerely | Ch. Darwin

Feb 28th

[Enclosure]

Queries about Expression

(1) Is astonishment expressed by the eyes & mouth being opened wide, & by the eyebrows being raised?

(2) Does shame excite a blush, when the colour of the skin allows it to be visible?

(3) When a man is indignant or defiant does he frown, hold his body & head erect, square his shoulders & clench his fists?

(4) When considering deeply on any subject, or trying to understand any puzzle, does he frown, or wrinkle the skin beneath the lower eyelids?

(5) When in low spirits are the corners of the mouth depressed, & the inner corner or angle of the eyebrow raised by that muscle which the French call the "Grief muscle?"

(6) When in good spirits do the eyes sparkle, with the skin round & under them a little wrinkled & with the mouth a little drawn back in the corners?

(7) When a man sneers or snarls at another, is the corner of the upper lip over the canine teeth raised on the side facing the man whom he addresses?

(8) Can a dogged or obstinate expression be recognized—which is chiefly shewn by the mouth being firmly closed, a lowering brow, & a slight frown?

(9) Is contempt expressed by a slight protusion of the lips, & turning up of the nose, with a slight expiration?

(10) Is disgust shewn by the lower lip being turned down, the upper lip slightly raised, with a sudden expiration something like incipient vomiting?

(11) Is extreme fear expressed in the same general manner as with Europeans?

(12) Is laughter ever carried to such an extreme as to bring tears into the eyes?

(13) When a man wishes to shew that he cannot prevent something being done, or cannot himself do something, does he shrug his shoulders, turn inwards his elbows, extend outwards his hands, & open the palms?

(14) Do the children when sulky, pout, or greatly protrude the lips?

(15) Can guilty, or sly, or jealous expressions be recognized—tho' I know not how these can be defined?

(16) As a sign to keep silent, is a gentle hiss uttered?

(17) Is the head nodded vertically in affirmation, & shaken laterally in negation?

Observations on natives who have had little communication with Europeans w^d be of course the most valuable; tho' those made on any natives would be of much interest to me.

General remarks on expression are of comparatively little value?

A definite description of the countenance under any emotion or frame of mind would possess much more value; & an answer within 6 or 8 months or even a year to any *single* one of the foregoing questions w^d be gratefully accepted.

Memory is so deceptive on subjects like these that I hope it may not be trusted to.

Down Bromley Kent. | 1867 | Ch. Darwin

LS(A)
DAR 92: A33–5

[1] The year is established by the date of the enclosure.

[2] On the development of the term 'protector of Aborigines' in the various colonies or states of Australia from the 1830s, see *Encyclopaedia of Aboriginal Australia*. Mueller was the director of the Botanical Gardens, Melbourne, Australia.

[3] For replies to the queries sent to Mueller, see the letter from F. A. Hagenauer to Ferdinand von Mueller, [12 September 1867], and the letter from Samuel Wilson to Ferdinand von Mueller, 12 November 1867. See also letter to Fritz Müller, 22 February [1867] and n. 13.

From Mary Elizabeth Barber [after February 1867][1]

Answers to some of the queries about expression in the Native races of S. Africa[2]

No. 1. By the Kafir and Fingoe tribes[3] astonishment is expressed by a serious look and by placing the right hand upon the mouth at the same time uttering the word Mawo! which means wonderful

No. 2. I have never observed a blush of any kind upon the dark colord skins of either Kafirs or Fingoes.

No. 3. I have never seen a Kafir or Fingoe clench his fists, they do not fight with their fists.

No. 11. Yes, in very much the same way as in Europeans. I have often heard it said of Kafirs and Fingoes that they were pale with rage or fear.

No. 12— Yes. they frequently laugh until the tears run down their cheeks, especially the women.

No. 13. I have never seen a Kafir or Fingoe shrug his shoulders or extend outwardly the palms of his hands.

No. 14. When Kafirs or Fingoes are sulky their lips are protruded and eyes cast down.

No. 15— Yes. their faces are very expressive and a guilty look can easily be detected.

No. 16— No. the sign used by Kafirs and Fingoes to keep silent is by gently waving the right hand backwards and forwards just below the face, on the right hand side, with the hand open and the palm turned slightly downwards, while the expression of the face is very serious.

No 19— I am not quite sure about the head being nodded vertically in affirmation but I have never seen it shaken lateraly in negative

My observations only apply to the Kafir and Fingoe tribes, with the other numerous races I have had no intercouse and know nothing of their manners and customs.

M. E. Barber

DAR 160: 40

CD ANNOTATION
End of letter: Sent through M[r] J. P. Manson Weale.[4]

[1] The date is established by the relationship between this letter and the letter to J. P. M. Weale, 27 February [1867] (see CD annotation, and n. 4, below).

[2] For the questions, see the enclosure to the letter to J. P. M. Weale, 27 February [1867]. In 1867, Barber was residing on a farm near Grahamstown, Cape Colony (now the Eastern Cape province of the Republic of South Africa; see Gunn and Codd 1981, p. 87). CD referred to Barber's replies in *Expression*, pp. 22, 269, 289. See also Shanafelt 2003, pp. 829–30.

[3] For the term 'Kafir', see the letter to J. P. M. Weale, 27 February [1867] and n. 3. 'Fingoe' refers to refugee groups of indigenous South Africans who were driven from Natal into the Eastern Cape Province by colonial authorities and missionaries in the 1820s (*OED* and Shanafelt 2003, p. 830). For nineteenth-century western perceptions of native South Africans, see Dubow 1995. See also Stocking 1987.

[4] Barber may have received the list directly or indirectly from James Philip Mansel Weale. He did not mention her as one of the recipients in his letter to CD of 7 July 1867; however, he did mention giving the list to Barber's brother, James Henry Bowker, and Bowker may have passed it on to Barber.

From Alfred Newton 1 March 1867

10 Beaufort Gardens | S.W.
1 March 1867.

My dear Sir,

On Tuesday last I met in a birdstuffer's shop at Brighton an intelligent young gentleman by name Booth.[1] He told me that last summer he had opportunities of studying the breeding habits of the Dotterel (*Eudromias morinellus*) in Scotland

and volunteered—without any leading question—the information that the cocks "looked after the young", and that the hens seemed to care very little about their offspring.

The Dotterel as you no doubt are aware is one of the species in which the hens are much more brilliantly coloured than the cocks.[2]

Believe me | Y^{rs} very truly | Alfred Newton

I asked M^r. Booth, (after he had told me what I have mentioned) whether he had taken the trouble to ascertain the sexes of the birds he killed by dissection, & he said he had done so, & shewed me a very dingy looking cock bird that he had obtained while anxiously "looking after" its young.

DAR 84.1: 28–9

CD ANNOTATIONS

2.1 The ... cocks. 2.2] *scored pencil*
Top of letter: 'Males Birds duller than female | Dottrell.' *pencil*
End of letter: 'Blyth in letter ['says' *del*] speaks of those species of Turnix in [*interl*] which the adult [*interl*] females only have a more or less black in front Quotes Jerdon on native testimony that male Turnix alone incubates & tends young—[3] If there were many females & only few males in *species not polygamous* the most beautiful females w^d be selected.—[4] I think I say that females very few with Polyborus N Zelandiæ.'[5] *ink*; Blyth ... N Zelandiæ.] *crossed pencil*; If ... *polygamous*] *double scored ink*

[1] Edward Thomas Booth.
[2] CD discussed differences between the sexes of the dotterel plover (now *Caradrius morinellus*), including their breeding habits, in *Descent* 2: 203–4. He mentioned Newton's observations and 'those of others' (*ibid.*, p. 204 n. 20). CD had recently exchanged letters about birds in which the female was more brightly coloured and the male cared for the young with Newton and Edward Blyth (see letter to Alfred Newton, 23 January [1867], and letter from Edward Blyth, 24 February 1867).
[3] See letter to Alfred Newton, 4 March [1867] and n. 5.
[4] See letter to Alfred Newton, 4 March [1867] and n. 6.
[5] See letter to Alfred Newton, 4 March [1867] and n. 3.

From Edward Blyth [2–30 March 1867][1]

⟨½ *page*⟩ which ⟨⅔ *line*⟩ do so ⟨⅔ *line*⟩ chanced to meet ⟨*two words*⟩, and his impression is that neither Chim nor Orang shew anything of the kind, the movements of their very protrusile lips being quite different from those of the human being—[2] I have been writing about the yak, & bring to notice some very interesting facts. 1stly, this animal does not lie down or[3] ⟨½ *page*⟩ ⟨   ⟩ably to whence the probability of the lláma & alpáca having derived from the guanáco & vicuña respectively.[4] You will have seen what I have said of the sheep, but I now incline more than I have there expressed myself to the opinion that the Corsican moufflon answers to the conditions required for it to be considered the true wild origin of the small short-tailed domestic sheep with crescent horns, as the old Highland and Shetland sheep, but certainly not of the various larger races with long tail & double flexure of horn, as the Dorset, &c—[5]

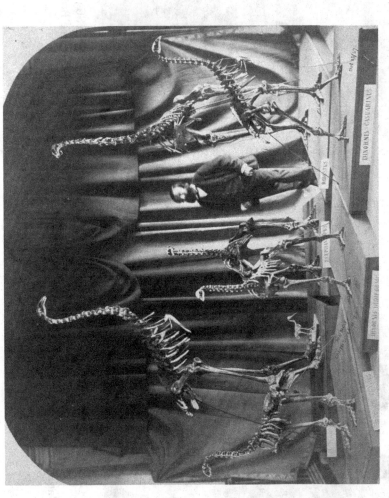

Julius von Haast and the Glenmark moas in the Canterbury Provincial Council Building, 1867
Photograph by D. L. Mundy
Canterbury Museum, Canterbury, New Zealand, ref. 7558

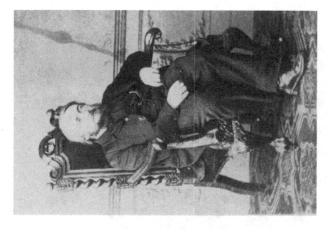

Bernard Peirce Brent, 1863
Courtesy of Richard Peirce Brent

Frederick Geach and Alfred Russel Wallace,
Singapore, 1862
© The Natural History Museum, London

I had indeed totally forgotten the paper to which you referred me in the old Magazine of Natural History.[6]

You will see in my remarks on the yak some curious facts on the seasonal shedding or non-shedding of the coat in wild as compared with domestic animals of the same genus, if not species,[7]

Yours Sincerely, | E. Blyth

Incomplete
DAR 160: 208

CD ANNOTATIONS
1.7 I now … conditions 1.9] *double scored brown crayon*
1.10 crescent horns,] *underl brown crayon*
1.11 long tail] *underl brown crayon*
1.11 double … Dorset, 1.12] *underl brown crayon*
End of letter: 'Blyth' *blue crayon*

[1] The date range is established by the publication dates of Blyth's first article on sheep and his article on the yak, in the 2 March 1867 and 30 March 1867 issues of *Land and Water* (see nn. 5 and 7, below).

[2] CD and Blyth had discussed facial resemblances between humans and other primates, as well as their possible genealogical relationships (see letter from Edward Blyth, 19 February 1867 and nn. 11 and 12, and letter to Edward Blyth, 23 February [1867] and n. 3).

[3] Blyth's article on the yak in *Land and Water*, 30 March 1867, pp. 237–8, appeared under the heading 'Wild types and sources of domestic animals'. Blyth wrote that the yak lies down and rises not like other ruminants but like the horse. CD mentioned the yak several times in *Variation* but did not refer to Blyth.

[4] In his 30 March 1867 article in *Land and Water* (pp. 237–8), Blyth wrote that if the domestic yak was derived from the wild yak, then it was likely that the llama and alpaca derived, respectively, from the guanaco and vicuña. CD wrote a similar comment regarding llamas, alpacas, guanacos, and vicuñas in *Variation* 2: 208. See Nowak 1999, pp. 1072–8, for modern classification in the family Camelidae.

[5] Blyth probably refers to his first article on sheep in *Land and Water*, 2 March 1867, p. 134; he published a second article on sheep in the 9 March 1867 issue, pp. 146–7, in which he emphasised the idea that the Corsican moufflon was the ancestor of small, short-tailed domestic sheep. See the letter from Edward Blyth, 24 February 1867 and n. 18, for Blyth's similar comment on the origin of domestic sheep and for CD's use of the information.

[6] No letter from CD to Blyth containing such a reference has been found. Blyth himself published a number of articles in the 1830s and 1840s in the *Magazine of Natural History*; CD cited Blyth 1837, an article on seasonal changes in fur and feathers, a number of times in *Descent* 2: 183–238.

[7] Blyth wrote that most wild ruminants shed their coats at least once a year, but that the domestic yak did not; instead, the under-growth of wool rose in 'felted masses' to the surface of the coat in the spring (*Land and Water*, 30 March 1867, pp. 237–8).

From A. R. Wallace 2 March [1867][1]

9 St. Mark's Crescent | N.W.
March 2nd.

Dear Darwin

I am very glad you like my notion about the catterpillars It is a kind of "forlorn hope", but fortunately it can be easily tested.[2]

I dare say you are right about sexual selection in butterflies, but I still think that *protective adaptation* has *kept down* the colours of the females, because the Heliconidæ and Danaidæ are almost the only groups in which the females are generally *equally brilliant* with the males.[3]

I can tell you several persons in the East who would I think observe "expression" for you. The best is Mr. Charles Johnson Brooke acting Rajah of Sarawak author of "Ten Years in Sarawak". Address him as

C— J— B— Esq,
Rajah Mudah
Sarawak, Borneo.[4]

He has grand opportunities, as he sees Malays, Dyaks, & Chineese under all kinds of excitements, in *war* in *hunting*, in *law suits* and under every occasion of daily life.[5] He would also I have no doubt send copies of your questions to some of the Missionaries and deputy governors in the interior.

Another person who would I am sure do the same for you is Mr. F. F. Geach, a young Cornish mining engineer, engaged in Tin & Copper mining in the interior of Malacca;—address, care of Mess[rs]. Paterson Simons and Co. *Singapore.*[6]

If you would send me a copy of your questions I sh[d]. like to see how far I could answer them from memory.

I certainly cannot yet see my way to any action of sexual selection in forming the races of man.[7] Stealing wives from other tribes for instance is a *very common* practice, & it would I imagine tend to check any selective action. Youth is almost the only thing a savage cares about, and the handsomest & finest women very often become prostitutes & leave few or no offspring. The women certainly don't choose the men, & the men want chiefly in a wife, a *servant*. Beauty is I believe a very small consideration with most savages, as it is very rare to find a woman so *plain* as not to leave as many or more offspring than the most beautiful.[8] This of course is a delicate subject to go into.

My present impression is, that the distinctive characters of human races are almost wholly due to *correlation* with *constitutional adaptations* to climate soil food & other *external conditions*. You must have facts of which I am quite ignorant,—& at all events your essay will be most welcome & is sure to be valuable.[9]

Believe me Dear Darwin | Yours very faithfully | Alfred R. Wallace—

DAR 85: A98

CD ANNOTATIONS
1.1 I am ... tested. 1.2] *crossed pencil*
2.1 but ... males. 2.4] *scored brown crayon*
3.1 I can ... memory. 6.2] *crossed pencil*
6.1 If ... memory. 6.2] '(*See insects other side*)' *added pencil*
7.1 I certainly ... *servant.* 7.6] 'just as we marry heiresses & yet beauties marry most'[10] *added pencil*
8.1 My present ... valuable. 8.4] *crossed pencil*
Top of letter: 'Keep' *pencil*; '& insects' *brown crayon*
Bottom of p. 2 'Sex. Selection | Wallace | Man' *pencil*

[1] The year is established by the relationship between this letter and the letter from A. R. Wallace, 24 February [1867].

[2] For Wallace's hypothesis on caterpillar mimicry, and his proposed experiment for testing it, see his letter of 24 February [1867]. CD praised Wallace's idea in his letter of 26 February [1867].

[3] In his letter of 26 February [1867] CD had argued that sexual selection accounted for colour differences in male and female butterflies; see the letter from A. R. Wallace, 24 February [1867] and n. 6 for Wallace's thoughts on sexual selection, and for more on the Heliconidae and Danaidae. For more on CD's and Wallace's diverging views of sexual selection, see Kottler 1980 and Bajema ed. 1984, pp. 110–255; see also Fichman 2004, pp. 262–8.

[4] CD had mentioned his queries on human expression in his letter to A. R. Wallace, 26 February [1867]. Wallace's references are to Charles Anthoni Johnson Brooke, and to Brooke 1866. Sarawak, now a Malaysian state, was a dependency of the Brunei Sultanate on the island of Borneo until James Brooke became governor as raja of Sarawak in 1841; he left Sarawak in 1863, naming his nephew Charles Brooke as his heir in 1867 (*ODNB*). Charles Brooke succeeded him in 1868. 'Rajah Mudah': heir apparent. For more on Charles Brooke's position as raja, see Payne 1986. See also Baring-Gould and Bampfylde 1989.

[5] The Dyaks or Dayaks are the indigenous, generally non-Muslim, people of parts of Borneo while the Malays and Chinese are more recent immigrants to the island.

[6] Frederick F. Geach was working for Paterson Simons & Co., the first British company to speculate in Malayan tin mining on a large scale (see Wong 1965, pp. 33–5). For tin mining in the state of Malacca (now Meleka), see also Turnbull 1972. Geach had earlier been hired by the Portuguese to open a copper mine in Timor (A. R. Wallace 1869, pp. 147–9).

[7] See letter to A. R. Wallace, 26 February [1867] and n. 5. Wallace maintained this view when he expanded his earlier essay on the origin of human races (A. R. Wallace 1864b) in A. R. Wallace 1870, pp. 303–31. For CD's and Wallace's differing views of the importance of sexual selection in the origin of the human races, see Kottler 1985, pp. 420–4, and Fichman 2004, pp. 266–70.

[8] In *Descent* 2: 343, CD suggested that the view that male 'savages' were indifferent to the beauty of women did not agree with the care women took in ornamenting themselves; he presented European opinions of how men of different peoples considered beauty in women in *Descent* 2: 344–54. CD mentioned the effective enslavement of women among 'savages' as a practice that could counteract sexual selection (see *Descent* 2: 358, 366). For Wallace's views on peoples of the Malayan archipelago, see A. R. Wallace 1864c and A. R. Wallace 1869.

[9] For CD's 'essay' on human descent, see letter to A. R. Wallace, 26 February [1867].

[10] CD touched on the roles of beauty and wealth in human sexual selection in *Descent* 1: 170, and 2: 356, 371. See also *Correspondence* vol. 12, letter to A. R. Wallace, 28 [May 1864] and n. 20.

From E. A. Darwin 3 March 1867

3 March 1867

Dear Charles

I am not quite sure that this is not a duplicate & perhaps the other is within my will— It is only to tell you that I promised Mark that you would continue the payment of the £20 per annum after my death. The house I give him rent free—[1]

yours | Eras Darwin

C R Darwin Esq
Payable 1st May & Nov

DAR 105: B57–8

¹ Mark Briggs had been the Darwin family coachman at The Mount, Shrewsbury (*Correspondence* vol. 1). He lived with the family until Susan Elizabeth Darwin's death in October 1866, after which he lived in a cottage near The Mount (*Emma Darwin* (1904) 2: 13).

From Fritz Müller 4 March 1867

Desterro, Brazil,
March 4. 1867.

My dear Sir

I am very much obliged to you and thank you cordially for Bentham's and Hooker's Genera Plantarum, which I received in due time and which have already been very serviceable to me.¹ The "conspectus generum"² and the list of the abnormal and allied forms at the head of each order facilitate extraordinarily the task of finding the name of any unknown genus. I was surprised at seeing that the authors never had an opportunity of examining the seeds of some of our most common and conspicuous plants, such as Schizolobium and Norantea.³ I should be very glad, if I could satisfy any wish, which the authors might have respecting to our flora.

I have also to thank you for your kind letter of Jan. 1ˢᵗ, and will now first answer the questions, you ask in this letter.⁴ As to *Adenanthera pavonina*, the only tree, I know, stands in a garden; but I had not even suspected, that it had been planted there, and still less, (as we are here very rich in Mimoseæ) that the species had been introduced from India.—⁵ The *Oncidium flexuosum* is an *endemic* species and is even the most common species of that genus.—⁶ There is now flowering another common species of *Oncidium*, perhaps the *O. micropogon Rchb. f.* in which pollen and stigma of the same individual plant have the same deletery action on each other, which they have in *O. flexuosum, unicorne* and *pubes(?)*.— This is also the case with a species of *Gomeza* R. Br. (Rodriguezia Lindl.), and with a small, but extremely pretty *Sigmatostalix* Rchb. f. (S. tricolor n. sp?).⁷ The same plants of Gomeza and of the several species of Oncidium on which I ascertained this fact, were fertile with pollen of other plants of the species. Of Sigmatostalix I have but one flowering plant.

*[dried specimen]*⁸

Flower of *Gomeza*, split longitudinally
3 days after fertilization with own pollen.

Another allied Orchid, which is even placed in the same genus (Odontoglossum) with Gomeza by Mr. Reichenbach, the *Aspasia lunata*, is fertile with own pollen;⁹ I had a single flower, which being fertilised with its own pollen, is yielding a seed-capsule.

I have now had several ears of the *Notylia* (pubescens?) of which I could not fertilize the few flowers, I observed last year. The stigmatic slit, extremely narrow, when the flower expands, widens gradually in the course of the next days and 2 or 3 days afterwards fertilization is effected rather easily.¹⁰ Even during the first day I sometimes succeeded in introducing *dry* pollenmasses. The stigma has room only for one pollen mass, as is also the case with *Ornithocephalus* so that in Notylia each pollinium may fertilize two and in *Ornithocephalus* four flowers. (In some Vandeæ with 4 pollen masses the anterior pair covers the posterior so completely, that the latter can touch the stigma only after the removing of the former; and thus each pollinium is apt to fertilize two flowers; so it is, for instance, with *Dichæa* and with the most beautiful of all our Vandeæ, the *Zygopetalum maxillare*).—[11] As in our other *Notylia*, the same individual plants pollen soon becomes blackish-brown in the stigmatic chamber, whilst pollen of any other plant of the species remains fresh, emits tubes ec.—

$$\left(\tfrac{15}{1}\right)$$

Decbr. 23/66.

Pollinia of a *Dichæa* (with dark-blue lip)
a. anterior pollenmasses, concealing the smaller posterior ones.
b. posterior pollen masses, after fertilizing a flower with the anterior ones.
c caudicles of the removed anterior pollen masses.

All the *Epidendreæ*, I hitherto tried, are fertile with own pollen; but from some experiments on *Epidendrum cinnabarinum* I suspect, that they will be less fertile with own, than with a distinct plants pollen. From several flowers, fertilized (Decbr. 20) with their own pollen, I obtained two pods, (ripe Febr. 19 & 20), the seeds of each of which weighed 5 grains. An ear of a second plant, the flow⟨ers⟩ ⟨of which were⟩ fertilized (Decbr. 20) with pollen of a ⟨   ⟩ same plant yielded two pods also ⟨   ⟩ seeds weighing 5.5 and 6 grains.

A second ear of the same plant ⟨was fertilized with⟩ pollen of a distinct plant of the species (Decbr. ⟨   ⟩) ⟨   ⟩ pod (ripe febr. 17), the seeds of which weighed 12,5 grains—more than those of both the pods fertilised with pollen of the same plant.

Three capsules of a third plant, fertilised (Decbr. 21) with pollen of a distinct plant, (ripe febr. 17), contained 26 grains of seeds; (each capsule, on an average, 8.7 grains).—

Lastly—and this is rather curious, a pod of a fourth plant, fertilised (Decbr. 21) with pollen *of a distinct species* (*Ep. Schomburgkii?*), (ripe febr. 17), was larger than all the other pods and its seeds weighed 14 grains! A second pod of Ep. cinnabarinum fertilized (Jan. 18) with pollen of Ep. Schomburgkii, is also much larger, than several pods, fertilized (febr. 17) with pollen of the own species.

Among 300 seeds from a pod fertilised with the same plants pollen only 86 seemed to be good, while at least 9/10 appeared to be so in the pods fertilised by pollen of a distinct plant of the species or of Ep. Schomburgkii.

Fertilisation with own pollen, at least in Orchids, seems to have much analogy with illegitimate unions of dimorphic plants or crossing of distinct species. it might be interesting to compare the offspring of plants fertilised with own pollen with hybrids and the illegitimate offspring of dimorphic plants.[12] May not the individual plants of some species, which were found to be quite sterile with own pollen, have been the offspring of flowers fertilized with own pollen?

⟨*dried specimen excised*⟩ I enclose a dimorphic *Rubiaceæ*, probably a *Diodia*. (I have not yet examined ripe fruits, which furnish the main distinction between this genus, Borreria and Spermacoce)[13] There is a small, but as far, as I have seen, constant, difference in the size of the pollen-grains, those of the short-styled flowers being larger.[14]

I had promised you to try some experiments on the fertilisation of *Scaevola*; but all the plants, which at several occasions I have brought home, have perished and from all the seeds, I planted, I did not obtain a single plant. In some other cases I have also utterly failed in transplanting into my garden plants growing in the loose sand of the sea-shore.—[15]

The copies of the paper on climbing plants, which you have been so good as to send me, have not yet arrived.[16]

Permit me again to thank you cordially for your great kindness and Believe me, dear Sir, | very sincerely and respectfully yours | Fritz Müller.

Incomplete
DAR 142: 102

CD ANNOTATIONS
1.1 I am ... India.— 2.5] *crossed brown crayon*
1.5 I was ... flora. 1.8] *scored brown crayon*
1.8 our flora.] *underl brown crayon*
2.5 The *Oncidium* ... capsule. 3.4] *crossed blue crayon*
2.11 The same ... species. 2.13] *scored blue crayon*
4.1 I have ... maxillare).— 4.11] *crossed red crayon; scored brown crayon;* '(Orchids.)' *added, brown crayon, square brackets in original*
4.11 As in ... ec.— 4.14] *enclosed in square brackets, red crayon; crossed blue crayon*
5.1 All the] *after opening square bracket, red crayon*
5.1 All the ... Schomburgkii 9.3] *crossed red crayon*
10.1 Fertilisation ... pollen? 10.6] *crossed brown crayon*
11.4 being larger 11.5] *before closing square bracket brown crayon*
12.1 I had ... yours 14.2] *crossed brown crayon*
On cover: 'June 2ᵈ | Dimorphic Plants | (Orchids [*del*]) | *It is not [*interl*] *Diodæa* (Rubiaceæ) but Borreria. Rubiaceæ | Rubiaceæ' *pencil*[17]

---

[1] In his letter of 23 August [1866] (*Correspondence* vol. 14), CD had offered to send Müller a copy of the first two parts of *Genera plantarum* by George Bentham and Joseph Dalton Hooker (Bentham and Hooker 1862–83), which was published in seven parts. The first part appeared in 1862, the second

in 1865. CD reported sending the volumes in his letter of [late December 1866 and] 1 January 1867 (*Correspondence* vol. 14).

2 'Conspectus generum': overview of genera (Latin).

3 For the descriptions of *Norantea* and *Schizolobium*, see Bentham and Hooker 1862–83, 1: 181, 569.

4 See *Correspondence* vol. 14, letter to Fritz Müller, [late December 1866 and] 1 January 1867.

5 Müller had sent CD seeds which he described as coming from 'a tree, probably belonging to the Mimoseae' (*Correspondence* vol. 14, letter from Fritz Müller, 1 and 3 October 1866). CD sent some of the seeds to Hooker, who identified them as seeds of *Adenanthera pavonina*, a tree native to India (see *Correspondence* vol. 14, letter to Fritz Müller, [late December 1866 and] 1 January 1867 and n. 14).

6 In his letter of [late December 1866 and] 1 January 1867 (*Correspondence* vol. 14), CD had asked whether *Oncidium flexuosum* was endemic to Brazil. CD later asked whether the species was native to the area where Müller lived, noting that he had previously attributed similar cases of self-sterility to cultivation under unnatural conditions (see letter to Fritz Müller, 7 February [1867] and n. 6).

7 Müller never published the name *Sigmatostalix tricolor*; however, the species he refers to is probably *S. radicans* (a synonym of *Ornithophora radicans*), which is native to Santa Catarina (Pabst and Dungs 1975–7, 2: 199).

8 Müller attached a number of plant specimens to this letter; none of the other surviving specimens is labelled.

9 Müller refers to Heinrich Gustav Reichenbach. *Gomeza* is now *Gomesa*.

10 Müller first described the case of *Notylia* in his letter of 2 August 1866 (*Correspondence* vol. 14); he had initially supposed that the plant was male because he could not get pollinia to adhere to the stigma and the ovules seemed rudimentary, but later found seed capsules and hypothesised that the ovules matured some time after the flower opened. See also *Correspondence* vol. 14, letter to Fritz Müller, 25 September [1866] and nn. 3–6). CD added Müller's observation on the widening of the stigmatic slit to *Orchids* 2d ed., p. 172.

11 The genera *Notylia*, *Ornithocephalus*, *Dichaea*, and *Zygopetalum* were all included within the tribe Vandeae (Lindley 1853; Endlicher 1836–42). For an updated classification of these genera, see Dressler 1993, pp. 272–3.

12 CD had told Müller of his plan to carry out experiments comparing the growth rates of 'plants raised from seed fertilized by pollen from the same flower & by pollen from a distinct plant' (see *Correspondence* vol. 14, letter to Fritz Müller, 25 September [1866]). CD's earlier work on dimorphic and trimorphic plants had suggested that fertility was dependent on whether pollen from the same or different form of plant was used; he had referred to the crosses of plants of the same form as 'homomorphic' unions and later as 'illegitimate' (see 'Dimorphic condition in *Primula*', p. 87 (*Collected papers* 2: 55), and 'Three forms of *Lythrum salicaria*', p. 187 (*Collected papers* 2: 121)).

13 The genus *Borreria* is now subsumed within the genus *Spermacoce* (Mabberley 1997).

14 CD had noted differences in pollen-grain size in his experiments with dimorphic and trimorphic plants (see 'Dimorphic condition in *Primula*', pp. 78–9 (*Collected papers* 2: 46), 'Two forms in species of *Linum*', p. 75 (*Collected papers* 2: 98), and 'Three forms of *Lythrum salicaria*', pp. 170–3 (*Collected papers* 2: 106–9)).

15 In a letter of 5 November 1865 (*Correspondence* vol. 13), Müller had mentioned a *Scaevola* growing on the east coast of the island of Santa Catarina, Brazil. In his letter to Fritz Müller, [9 and] 15 April [1866] (*Correspondence* vol. 14), CD wrote, 'I have long wished some one to observe the fertilization of Scævola' and reported his own observations. No letter from Müller promising to try experiments with *Scaevola* has been found, but in a letter to his brother Hermann of 29 October 1866 (Möller ed. 1915–21, 2: 97–8), Müller described the pollination mechanism of the specimens he observed in the wild, confirming CD's observations (see also a similar description in F. Müller 1868a, pp. 114–15).

16 The reference is to Müller's paper on climbing plants, copies of which had been sent by CD (F. Müller 1865; see *Correspondence* vol. 14, letter to Fritz Müller, [late December 1866 and] 1 January 1867 and n. 15).

17 CD wrote the name '*Diodæa*' (*sic* for *Diodia*) on the envelope, which probably contained the now

missing specimen that Müller sent. CD evidently amended the name on the envelope after receiving Müller's letter of 2 June 1867, in which the identification of the specimen was corrected.

To Alfred Newton 4 March [1867]

Down. | Bromley. | Kent S.E.

Mar 4—

My dear Sir

Very many thanks about the Dotterel, & I am pleased to hear of this additional evidence.[1] I have looked to Swinhoe's papers, but the case does not seem very conclusive.[2] After writing to you I remembered that the female of the carrion-hawk of the Falkland I's (formerly called Polyborus N. Zealandii) is very much brighter coloured than the male, as I ascertained (Zoolg. Voyage of Beagle: Birds) by dissection; I have written to the Missionaries there about its nidification & if I receive any answer, will inform you.[3] The other day I thought I had got a case at the Zoolog Gardens in the Casuarinus Galeatus, in which the female has the finest & brightest caruncles &c; but Sclater tells me it w^d be rash to trust to the comparison of a single pair, & he tells me that the male ostrich has the finest plumes.[4]

With my best thanks | I remain my dear Sir | yours very sincerely | Charles Darwin

P.S. Mr Blyth tells me that according to Jerdon, the natives say the male Turnix alone incubates & attends to young—[5]

There is another consideration which might lead to the females being the most beautiful, viz if they were the more numerous than the males & the species were not polygamous, for in this case the more beautiful females w^d be selected.—[6]

LS(A)
Endorsement: 'C. Darwin. March 4/67.'
DAR 185: 89

[1] See letter from Alfred Newton, 1 March 1867.
[2] Newton had suggested that CD look at some recent papers of Robert Swinhoe's for information on male *Turnix* sitting on eggs (see letter from Alfred Newton, 21 January 1867 and n. 7).
[3] The last extant letter to Newton from CD is that of 23 January [1867]. See also annotation to letter from Alfred Newton, 1 March 1867. No letter to the Falkland Islands asking about the bird has been found, nor has a later letter to Newton on the bird's nidification. CD had noted that female Falkland Islands carrion hawks were more brightly coloured than the males in *Birds*, p. 16 (see also *Ornithological notes*, pp. 237–8, and R. D. Keynes ed. 2000, pp. 210–12). He gave this information again in *Descent* 2: 205–6, but added that nothing was known about the species's incubation habits. The carrion hawk CD observed was probably *Phalcoboenus australis*, the striated caracara (*Birds of the world* 2: 250).
[4] CD refers to the Zoological Gardens, Regent's Park. He had been to London from 13 to 21 February 1867 (see 'Journal' (Appendix II)). CD also refers to Philip Lutley Sclater; no letters have been found between them on cassowaries. In *Descent* 2: 204, CD noted that the female was larger and more brightly coloured in the common cassowary (*Casuarius galeatus*, now *C. casuarius*, the southern cassowary). CD cited Sclater on the plumes of the male ostrich in *Descent* 2: 205.
[5] The letter, or fragment of a letter, in which Blyth mentioned *Turnix* (button quail) has not been found; Newton also informed CD of publications describing the incubation of eggs by males in

Turnix (see n. 2, above). In *The birds of India* (Jerdon 1862–4, 2: 597) Thomas Claverhill Jerdon discussed the incubation of eggs by males in *T. taigoor*, the 'black-breasted bustard-quail'. CD noted this information and quoted other portions of Jerdon's description in *Descent* 2: 201–2. An Indian subspecies of the barred button quail is now known as *T. suscitator taigoor* (see *Birds of the world* 3: 54).

[6] For more on CD's consideration of sexual selection in birds when females were more colourful and more numerous that males, see *Descent* 2: 207–8.

To Frederic William Farrar 5 March 1867

Down
March 5, 1867

My dear Sir

I am very much obliged for your kind present of your lecture.[1] We have read it aloud with the greatest interest and I agree to every word. I admire your candour and wonderful freedom from prejudice; for I feel an inward conviction that if I had been a great classical scholar I should never have been able to have judged fairly on the subject.[2] As it is, I am one of the root and branch men,[3] and would leave classics to be learnt by those alone who have sufficient zeal and the high taste requisite for their appreciation. You have indeed done a great public service in speaking out so boldly. Scientific men might rail for ever, and it would only be said that they railed at what they did not understand. I was at school at Shrewsbury under a great scholar, D[r]. Butler;[4] I learnt absolutely nothing, except by amusing myself by reading and experimenting in chemistry.[5] D[r]. Butler somehow found this out and publicly sneered at me before the whole school, for such gross waste of time; I remember he called me a Pococurante, which, not understanding, I thought was a dreadful name. I wish you had shown in your lecture how Science could practically be taught in a great school; I have often heard it objected that this could not be done, and I never knew what to say in answer.

I heartily hope that you may live to see your zeal and labour produce good fruit; and with my best thanks, I remain, my dear Sir | Yours very sincerely | Charles Darwin

Copy
DAR 144: 41

[1] CD refers to Farrar's lecture at the Royal Institution of Great Britain, 'On some defects in public school education' (Farrar 1867). CD's copy has not been found.

[2] Although he was a master at Harrow school and a classical scholar, Farrar was critical of the prevailing system of English education, which was based on Greek and Latin. He advocated the teaching of science, as well as reforms in classical education that would eliminate verse composition and the memorisation of abstruse grammatical rules (Farrar 1867).

[3] Root and branch men: those supporting a reform involving the total abolition of some existing institution (*OED*).

[4] Samuel Butler was headmaster of Shrewsbury School when CD was a pupil there.

[5] In his *Autobiography*, p. 46, CD wrote about his early interest in chemistry, as a result of which he was given the nickname 'Gas' by his schoolmates (see also *Correspondence* vol. 1).

To W. B. Tegetmeier 5 March [1867]

Down. | Bromley. | Kent. S.E.
March 5th.

My dear Sir

I write on the bare & very improbable chance of your being able to try, or get some *trustworthy* person to try, the following little experiment. But I may first state, as showing what I want, that it has been stated that if the 2 long feathers in the tail of the male widow-Bird at the C. of Good Hope are pulled out, no female will pair with him.[1]

Now when 2 or 3 common cocks are kept I want to know if the tail-sickle feathers & saddle feather of one, which had succeeded in getting wives, were cut & mutilated & his beauty spoiled, whether he would continue to be successful in getting wives. This might be tried with drakes or peacocks, but no one w^d be willing to spoil for season his peacock. I have no strength or opportunity of watching my own poultry, otherwise I w^d try it.— I would very gladly repay all expences of loss of value of the poultry &c— But as I said I have written on the *most* improbable chance of your interesting anyone to make the trial or having time & inclination yourself to make it.— Another & perhaps better mode of making the trial w^d. be turn down to some hens 2 or 3 cocks, one being injured in its plumage.

I am glad to say that I have begun correcting proofs.[2]

I hope that you received safely the skulls which you so kindly lent me.—[3]

My dear Sir | Yours sincerely | Ch. Darwin

Endorsement: '1867—'
Archives of The New York Botanical Garden (Charles Finney Cox collection) (Tegetmeier 102)

[1] CD wrote in *Descent* 2: 120: 'the female widow-bird (*Chera progne*) disowns the male, when robbed of the long tail-feathers with which he is ornamented during the breeding-season.' He noted that he read this observation of Martin Karl Heinrich Lichtenstein's in Rudolphi 1812, p. 184; an annotated copy of Rudolphi 1812 is in the Darwin Library–CUL (see *Marginalia* 1: 716–18). *Chera progne* is now known as *Euplectes progne*, the long-tailed widow bird.

[2] CD refers to *Variation*, which was published in 1868 (Freeman 1977). He received the first proof-sheets on 1 March 1867 (CD's 'Journal' (Appendix II)).

[3] On CD's return of the last skulls of fowl that Tegetmeier had lent him for engravings in *Variation*, see the letter to W. B. Tegetmeier, 6 January [1867].

To George Robert Waterhouse 5 March [1867?][1]

Down, | Bromley. | Kent. S.E.
Mar 5

My dear Waterhouse

Will you have the kindness to answer me a question. In the Museum there is a Specimen of a slim Abyssinian wolf, described by Rüppel in his Wirbelthiere von Abyssinien.

In my notes I have called this Canis Simensis, which I think must be wrong;[2] Gervais calls it C. Sinus; will you kindly tell me which is right?[3]

When last in the Museum I had intended calling upon you, but I staid so long talking in the working room that I expended all my strength.[4]

Believe me my dear Waterhouse | yours very sincerely | Ch. Darwin

LS

The Natural History Museum, London (LIS-Archives, DF100/9 Palaeontology Department Keeper's correspondence)

[1] The year is conjectured on the supposition that the letter concerns a possible correction to the proof-sheets of *Variation*. CD received the first proof-sheets on 1 March 1867 (see 'Journal' (Appendix II)). *Canis simensis* (see n. 3, below) is mentioned in the first chapter of *Variation*. The letter also suggests that CD may have visited London recently; CD visited London in February 1867 (see n. 4, below). His most recent visit to London in a February before this was in 1863.

[2] CD refers to Eduard Rüppell, and to *Neue Wirbelthiere zu der Fauna von Abyssinien gehörig* (Rüppell 1835–40). The description of *Canis simensis*, now the Ethiopian wolf, is in Rüppell 1835–40, 1: 39; it is illustrated on plate 14. Waterhouse worked in the geology department of the British Museum.

[3] The 1843 *List of the specimens of Mammalia in the collection of the British Museum* records the Abyssinian wolf as 'Canis Simensis', referring to Rüppell 1835–40, plate 14 (Gray 1843, p. 58). CD underlined 'Canis Sinus' in his annotated copy of Paul Gervais's *Histoire naturelle des mammifères* in the Darwin Library–CUL, and wrote 'C. Sinensis Sinensis' in the margin (Gervais 1854–5, 2: 58; see *Marginalia* 1: 325–7). No reply to CD's query from Waterhouse has been found, but CD cites Rüppell and refers to *Canis simensis* in *Variation* 1: 33, n. 51.

[4] CD had visited London from 13 to 21 February 1867 (see 'Journal' (Appendix II)).

From John Traherne Moggridge   6 March [1867][1]

March. 6

St. Roch | Mentone

Dear Mʳ. Darwin

though I have lost several of my Ophrys plants which were marked & drawn last year, I have already watched five of them through all their stages: thus far the inference is that colour, markings & shape of lip, & pubescence—, which form the principal characters of difference, vary but little from year to year—[2]

One plant brought up a monstrous flower in which the petals & sepals were strangely combined; but even here the colour & markings remained almost un-changed.—

In number 19. there was a curious inversion of the position of last year's markings.

Last year all the six lower flowers of the spike from this root, had a marking as at fig. 1. ⟨   ⟩ at fig. 2.;

⟨   ⟩ ones of Fig. 1.[3]

I am working now at the varieties of **Viola** *odorata* which are very numerous, & appear to be in several stages of advancement & fixity.[4]

I have also commenced a set of observations on *Thymus vulgaris* & I find that it differs from our **Thymus** *serpyllum* in presenting on separate plants several forms, in which the stamens are found in different stages of suppression.[5]

Thus the perfect hermaphrodite form is as represented at Fig. 1.; the next stage which I have as yet observed is as at Fig. 2 where the anthers though useless are still prominent & of a special texture; the third degree is that in which the plant has the anther-cells reduced to two minute, transparent lobes terminating the filament & similar to it in texture, as at Fig. 3.—[6] — The style always remains perfect ⟨   ⟩

I find that **Rhamnus** *Alaternus* is functionally diœcious though the organs are always complete in number, & not very markedly reduced in form.—[7]

The other day I heard a curious bit of information from the late Lord Brownlow's gardner[8] to the effect that when gardners wish to obtain seedling dwarf geraniums they always take the necessary pollen from the pair of shorter stamens of a Geranium (Pelargonium) flower; the offspring of this union being smaller than those bred from the longer stamens.—[9]

If there is no error here the statement may be suggestive.—

We fully purpose to return to Mentone next autumn & intend to leave our present quarters on our homeward route early in May next.—[10]

You know how gladly either commissions or hints for work will be recieved by y[rs] very sincerely | J. Traherne Moggridge—.

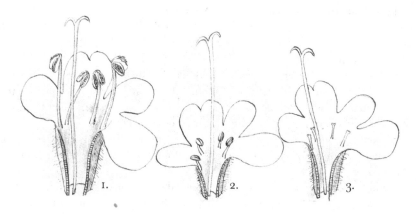

Thymus vulgaris flowers cut open & greatly magnified[11]

Incomplete
DAR 109: A90–1, DAR 111: B47

CD ANNOTATIONS
1.1 though ... fixity. 5.2] *crossed blue crayon*

9.1 The other … recieved by 12.1] *crossed blue crayon*

'Thymus' diag.: labels crossed ink

Diag. 3: diagonal cross with point in each segment added pencil

Above diag.: 'N. B. To be reduced to one half of present scale.' *ink* 'Fig. 14' *blue crayon*[12]

Below diag.: 'Sketch by Mr T. Moggridge' *ink*

CD note:

'March 19th 1867 Have seen one specimen of the Thyme, as in drawing— in Hermaphrodite flowers the pistil is evidently much shorter. He has sent me heads of a whole large plant, in which all the flowers were female with long protruding pistil, with anthers of proper shape [*altered from* 'size'] & well-sized, but containing very little pollen & this pollen all bad & grains of very unequal size— What a gradation.'[13] *ink*; 'in which … well-sized,' *scored red crayon*

[1] The year is established by the date of CD's note (see n. 12, below).

[2] In his letter of 13 October [1865] (*Correspondence* vol. 13), CD had suggested that Moggridge mark his late spider orchids (*Ophrys arachnites*) so that he could compare the flowers produced in different seasons and check for variation.

[3] CD excised a portion of the letter, but the fragment has not been found.

[4] CD was interested in the small unopened (cleistogamic) flowers of *Viola*. His notes on *V. odorata* are in DAR 111. He discussed the species in *Forms of flowers*, pp. 317–18, 336.

[5] Both *Thymus vulgaris* and *T. serpyllum* were examples of plants that CD later referred to as 'gynodiœcious' because they possessed hermaphrodite and female, but no male flowers (see *Forms of flowers*, pp. 299–303).

[6] The figure numbers refer to the enclosed sketches. CD included Moggridge's diagrams in his discussion of *Thymus vulgaris* in *Forms of flowers*, p. 302.

[7] CD discussed some *Rhamnus* species, but not *R. alaternus*, in *Forms of flowers*, pp. 293–7, 307–08. His notes on the genus are in DAR 109: A41–3, 50, and DAR 111. CD's son William Erasmus Darwin had made several observations on *R. cathartica* for CD in the summer of 1866 (see *Correspondence* vol. 14).

[8] John William Spencer Brownlow Egerton Cust, the second Earl Brownlow, had died in Mentone on 20 February 1867 (*Burke's peerage* 1868). His gardener has not been identified.

[9] In 1863, CD had questioned a similar statement made by Donald Beaton in the *Journal of Horticulture* about the production of dwarf plants in *Pelargonium* (see *Correspondence* vol. 11, letter to Isaac Anderson-Henry, 20 January [1863] and n. 10, and letter from Isaac Anderson-Henry, 26–7 January 1863).

[10] Owing to chronic ill health, Moggridge spent most winters at Mentone (now Menton), a town on the French Riviera near the Italian border (R. Desmond 1994).

[11] The sketches are reproduced at approximately 70 per cent of their original size.

[12] These are CD's instructions to the printer. In the printed version of *Forms of flowers*, p. 302, the diagrams appear as figure 15.

[13] CD evidently added this note after receiving Moggridge's letter of 15 March [1867], which contained the plant specimens. For CD's earlier work on *Thymus*, see *Correspondence* vol. 12, letter to W. E. Darwin, 14 May [1864] and n. 8.

From F. W. Farrar 7 March [1867][1]

Harrow.
March. 7.

My dear Sir,

Your kind approval gratifies me more, I think I may candidly say, then would the approval of any living man.[2] I have reason to hope that real work is being done, in the direction of relieving our present educational system from its extraordinarly fantastic & stationary condition. I am filled with sorrow & indignation when I think of the mere *paralysis* of intellectual power w^{h} it produces—not in boys of genius, for genius is a fire w^{h} calcines any amount of superimposed rubbish—but in boys of fine manly minds & average intellect. I allude especially to the fetish-worship of Latin Verse, with w^{h} both at the University & in the Public Schools I think that we c^{d} wage war to the knife. I shall be happy if my Lecture (w^{h} of course involves me in some unpleasantnesses) hastens the death of so irrational a system.[3]

The *constructive* side is really not difficult. I send, by book post, (but only at present, *privately*) a copy of the Report of a Sub Committee of the Brit. Assoc^{n} on this subject, for w^{h} I moved at the Nottingham meeting.[4] The only members of the Committee were Professors Tyndall & Huxley, Mr. Griffith (who after Easter is coming to be a regular Science master here at Harrow), & Mr. Wilson of Rugby.[5] This Report will probably come before you when it has been adopted by the Council, as I expect it will be tomorrow. It is not yet revised & corrected.

I am, dear Sir, with sincere respect, | Most faithfully your's | Frederic W. Farrer.

DAR 164: 38

[1] The year is established by the relationship between this letter and the letter to F. W. Farrar, 5 March 1867.

[2] CD had expressed his agreement with Farrar's position on public school education in Farrar 1867 (see letter to F. W. Farrar, 5 March 1867).

[3] See letter to F. W. Farrar, 5 March 1867 and n. 2.

[4] The reference is to 'Report of the committee appointed by the Council of the British Association for the Advancement of Science to consider the best means for promoting scientific education in schools', later published in the *Report of the thirty-seventh meeting of the British Association for the Advancement of Science held at Dundee in September 1867*, pp. xxxix–liv. CD's pre-publication copy has not been found. The Nottingham meeting of the British Association had been held in 1866.

[5] Farrar refers to John Tyndall, Thomas Henry Huxley, George Griffith, and James Maurice Wilson. Griffith is not named as a member of the committee in the published version of the report (see n. 4, above). On the movement to introduce science into English public schools, see White 2003, pp. 75–81.

To A. R. Wallace 7 March [1867][1]

Down. | Bromley. | Kent. S.E.

Mar 7

My dear Wallace

The addresses which you have sent me are capital, especially that to the Rajah; & I have despatched two sets of queries. I now enclose a copy to you & sh^{d} be very glad of any answers; you must not suppose the P.S. about memory has lately been inserted; please return these queries as it is my standard copy.[2] The subject is a curious one, I fancy I shall make a rather interesting appendix to my Essay on Man.[3]

I fully admit the probability of "protective adaptation" having come into play with female butterflies as well as with female birds.[4] I have a good many facts which make me believe in sexual selection as applied to man, but whether I shall convince any one else is very doubtful.[5]

Dear Wallace | yours very sincerely | Ch. Darwin

LS
British Library (Add 46434 ff. 20–20v)

[1] The year is established by the relationship between this letter and the letter from A. R. Wallace, 2 March [1867].

[2] In his letter of 2 March [1867], Wallace sent two addresses for CD, including that of Charles Anthoni Johnson Brooke of Sarawak. CD's letters and queries about human expression, including the enclosure for Wallace, have not been found. For the most recent extant queries about expression that CD sent to a correspondent, see the enclosure to the letter from Ferdinand von Mueller, 28 February [1867].

[3] *Expression* was ultimately published in 1872, a year after *Descent.*

[4] See letter from A. R. Wallace, 2 March [1867 and n. 3.

[5] See letter from A. R. Wallace, 2 March [1867] and nn. 7–10.

From Francis Trevelyan Buckland 9 March 1867

Salmon Fisheries Office, | *4, Old Palace Yard, Westminster,* | *S.W.*
March 9 | 1867.

My Dear M[r] Darwin.

I have been asking all over the country for an Otter hound for you[1] I was last week in Tiverton Devonshire—& made enquiries— I send you the reply from my friend a sporting parson as good a fellow as ever walked. Please return his note as I wish to put it into Land & Water of which I trust you approve[2]

Yours ever | Frank Buckland

When the weather gets warmer I should be so much obliged if you would come & see my Museum at the Horticultural.[3] I get no help—but I copy the Salmon & will not be beaten— Read my "Crow" article in todays-paper[4]

DAR 160: 362

[1] In his letter of 2 October 1866 (*Correspondence* vol. 14), CD had asked Buckland to publish an enquiry on the webbing between the toes of the otter-hound (see also letter to *Land and Water*, [2 October 1866]). For a more detailed enquiry regarding the feet of otter-hounds and their swimming abilities, see *Correspondence* vol. 12, letter to T. C. Eyton, 29 December [1864?]. For CD's published comments on otter-hounds, see *Variation* 1: 39–40.

[2] The enclosure has not been found. A short article, 'Otter hunting', signed Brown Willy, was published in *Land and Water*, 20 July 1867, p. 652. The article included only general information on otter hunting. 'Brown Willy' has not been identified.

[3] Buckland had set up an exhibit on pisciculture at the Royal Horticultural Society's London gardens south of Kensington Gore (now occupied by the Royal Albert Hall and Imperial College); the exhibit included casts of salmon and oysters, models of fisheries, stocked ponds, and eventually a salmon run. Most of the exhibit was later moved to the nearby South Kensington Museum. See Bompas 1885, pp. 149, 153, 198, and B. Elliott 2004.

[4] In *Land and Water*, 9 March 1867, p. 153, Buckland replied to a query about specimens of 'crow oysters' or *Anomia ephippium*, also known as saddle-back oysters. Buckland noted that they were not true oysters, and that *A. ephippium* (family Anomiidae) appeared to be replacing true oysters in some environments. He added: 'In fact there seems to be going on a continual "struggle for existence" between the oysters and the "crows," a point in submarine economy to which I would much like to call the attention of Mr. Darwin.'

From Henry Walter Bates 11 March 1867

Royal Geographical Society | *15, Whitehall Place, S.W.*

March 11 1867

My dear Mr Darwin

I promised when you were here to look up a few cases of sexual ornamentation &c in insects & send the particulars to you.[1] Here they are.

1. Gay-coloured butterflies.

The tropical American genus Epicalia is a good case. The males are amongst the most gaudy of all butterflies; the females are generally very plain. I know both sexes of 12 species: in one both sexes are similar in pattern & colours & this pattern is that of the great majority of the females of the genus; in 9 other species the two sexes are so different that Entomologists formerly placed them in separate genera, but the male of one of the 9 is totally different in colours & pattern from the other 8 males although as gaily coloured as they; in the remaining 2 species both males & females are gaily coloured but males more so than females.[2]

I think this case will interest you; the fact of the females of 10 species being of the same type of colouration gives us a clue to the ancestral type, from which the males have diverged by sexual selection, & this type is that of both sexes in several allied genera found in various parts of the world. The fact, again, of 2 species having females gaily-coloured & very different from the females of the rest of the species points to the inheritance of gay colours being partaken of by both sexes in the two species whilst in the rest male has inherited male colours & female female. The males & females in this genus do not frequent separate haunts as in many other genera; but the females fly slower & nearer the ground than the males. The facts afforded by Epicalia are not isolated in Entomology— other genera & natural groups of species within genera afford similar illustrations.[3]

In some genera when the males are much more gaily coloured than their females the males are immeasurably more numerous than the females & spend most of their days in the open sunlight whilst their obscurer partners are confined to the shade of the woods (Genera, Catagramma, Eunica, Megistanis &c).[4] In other cases the females have clearly diverged from the Common type of coloration while the males have remained unchanged in this respect; groups of species of Pieris come under this category.[5]

2 Armature of males.

I have looked through my collection of horned genera of Lamellicorn beetles; viz, Copris, Phanæus and Onthophagus and find 5 specimens with their cephalic horns broken or edge of clypeus chipped. As we generally select perfect specimens in collecting this is only an indication of what might be found if attention was drawn to the subject. The male horns & thoracic bosses are so wonderfully developed in many species that they must have been drawn out by a long course of Natural Selection & therefore must be of some use to the species; but no one has yet recorded, to my knowledge, a case of males fighting: true, the species come out of their holes only at night.[6]

Excessive variability in size of horns & bosses I find is the rule, but not a rule without exceptions. For instance, the most wonderfully horned species of Onthophagus in the world viz. O. rangifer of the Zambezi region, shows in ab.[t] 100 specimens I have inspected (collected at random by a non-Entomologist) shew no imperfectly developed males; the female is without armature in this species.[7]

The rule in Onthophagus is, however, a gradual degradation of horns &c. from fully-developed males down to males so degenerate that they are scarcely to be distinguished from females.

Another fact is that the species in Onthophagus cannot be naturally classed according to the horns of the male: in each natural group of species there are species with one cephalic horn & others with a pair, like the bull.

In Phanæus & Copris there are species with horned females scarcely to be distinguished from degenerate males.[8]

If you wish to ask me any more special questions about these matters, I shall be glad to try to answer them

Yours sincerely | H W Bates

Wallace brought forward your enquiry regarding gay caterpillars before the Entomological Society last Meeting & several practical men are looking out for explanations.[9]

DAR 82: A36–9, A46–7

CD ANNOTATIONS

2.1 1. Gay-coloured ... plain. 3.2] 'Very beautiful but not much contrast of colours.' *added pencil*

3.2 the females ... plain.] *altered to* 'the females are generally very much plainer.' *pencil*

3.2 I know ... & colours 3.3] *scored red crayon*

3.3 in one ... & colours] *double scored red crayon*

3.3 one] *underl red crayon*; 'A' *added red crayon*

3.3 in one ... & colours] '(male less gaudy)' *added pencil*; 'gay colours or plain' *interl after* 'colours' *red crayon*

6.1 2 ... beetles; 7.1] *crossed pencil*; 'H. W. Bates' *added pencil*

7.2 Copris, ... Onthophagus] 'I have looked through my collection of Horned genera of Lamellicorn Beetles viz' *added pencil*

7.7 & therefore ... species;] *scored blue crayon*

8.2 Onthophagus 8.3] *underl blue crayon*

8.3 rangifer] *underl blue crayon*

9.1 The rule ... females. 9.3] *scored blue crayon*

10.1 Another ... bull. 10.3] *scored blue crayon*

Top of letter: 'If gay, colours transferred to female: if plain the male has retained primordial colour of ancestor.' *ink*; '[London]' *red crayon*

[1] CD had called on Bates the evening of Monday 18 February 1867 during his recent visit to London (see letter to A. R. Wallace, 23 February 1867); CD wrote notes on their discussion of lamellicorn beetles, *Epicalia* (a butterfly genus), and the sphinx moth (see DAR 81: 14–15). CD had also been corresponding with Alfred Russel Wallace about whether the colour of butterflies and their larvae was due to sexual selection or to the natural selection of protective coloration (see letter from A. R. Wallace, 2 March [1867] and nn. 2 and 3).

[2] The genus *Epicalia* has been subsumed within the genera *Catonephele* and *Nessaea* in the family Nymphalidae, whose classification is still debated; recently the genera *Catonephele* and *Nessaea* were placed in a subfamily Biblidinae (see Wahlberg *et al.* 2003). Bates described an *Epicalia* species in Bates 1863, 2: 52.

[3] CD included Bates's information on *Epicalia* in *Descent* 1: 388–9. A draft for CD's discussion in *Descent* is in DAR 81: 182. For CD's explanation of the variation in colours in *Epicalia* and some other Lepidoptera, see *Descent* 1: 399–415, 419–20.

[4] Without referring to these three particular genera, Bates described these habitat preferences and the relative numbers of male and female butterflies in Bates 1863, 2: 227–8. See *Descent* 1: 309.

[5] *Pieris*, or whites, are in family Pieridae. CD referred to the brightly coloured females in some species of Pieridae in *Descent* 1: 413.

[6] Lamellicorn beetles (family Scarabaeidae) include several horned genera. CD included figures of a male and female of species of *Copris*, *Phanaeus*, and *Onthophagus* in *Descent* 1: 369. He discussed the question of whether males used their horns in fighting, noting Bates's failure to find sufficient evidence (*ibid*, p. 371). CD's report of Bates's similar information from 18 February 1867 is in DAR 81: 14. For Bates's later publication on his lamellicorn collections, see Bates 1886–90.

[7] CD discussed the 'excessive variability' of horns in species of lamellicorn in *Descent* 1: 370–1; he also mentioned Bates's comment on *Onthophagus rangifer*, but added that Bates's later research contradicted his original finding. The beetle known as *O. rangifer* is now usually considered to belong to the genus *Proagoderus*, or alternatively to the subgenus *Proagoderus* of the genus *Onthophagus* (personal communication, Clarke Scholtz).

[8] CD reported Bates's information on the number of cephalic horns in *Onthophagus*, and also noted that males of *Phanaeus lancifer*, as well as males of other *Phanaeus* and *Copris* species, had horns only slightly less developed than the females in *Descent* 1: 370.

[9] Bates refers to Wallace's query regarding colourful caterpillars made at the 4 March 1867 meeting of the Entomological Society of London (see letter from A. R. Wallace, 24 February [1867] and nn. 2–4 and 9); see also *Descent* 1: 417. CD had initially asked Bates why he thought some caterpillars were conspicuous (see letter to A. R. Wallace, 23 February 1867).

From A. R. Wallace   11 March [1867][1]

9, St. Mark's Crescent | N.W
March 11th.

Dear Darwin

I return your queries but can not answer them with any certainty. For the Malays I should say Yes. to 1. 3. 8. 9. 10. & 17. and No. to 12. 13. and 16.[2] but I cannot be *certain* in any one. But do you think these things are of much importance? I am inclined to think that if you could get good direct observations you would find some of them often differ from tribe to tribe, from island to island and sometimes from village to village. Some no doubt may be deep-seated, and would imply organic differences but can you tell beforehand which these are. I presume the Frenchman shrugs his shoulders whether he is of the Norman Breton or Gaulish stock. Would it not be a good thing to send your List of queries to some of the Bombay & Calcutta papers as there must be numbers of Indian judges & other officers who would be interested & would send you hosts of replies.

The Australian papers & N. Zealand might also publish them & then you would have a fine basis to go on.

Is your essay on Variation in Man to be a supplement to your volume on Domesticated Animals & cultivated Plants?[3] I would rather see your second volume on "The Struggle for Existence &c." for I doubt if we have a sufficiency of fair & accurate facts to do any thing with Man.[4] Huxley I believe is at work upon it.[5]

I have been reading Murray's volume on Geog. Dist. of Mammals. He has some good ideas here and there but is quite unable to understand Natural Selection, and makes a most absurd mess of his criticism of your views on Oceanic Islands.[6]

By the bye what an interesting volume the whole of your materials on that subject would I am sure make.

Yours very sincerely | Alfred R. Wallace—

P.S. I mentioned the *Catterpillar* question at the Ent. Soc. on Monday & think we shall have observations made this summer. Many members seemed to think that known facts favoured my view.[7]

Larvæ of *Cucullia verbasci* &c. often swarm for sp. of *verbascum* are very showy and conspicuous and never seem to be eaten by birds. The larvæ of *Callimorpha jacobeæ*, are a similar case.[8]

ARW.

DAR 106: B24, B45; DAR 82: A22

CD ANNOTATIONS
1.3 I am ... village. 1.6] *crossed blue crayon*
1.6 Some ... go on. 2.2] *crossed pencil*

CD *note*:
    I may say when thinking of Beauty of Butterflies—beauty of Caterpillar occurred to me that anyone might say—I applied to M[r] Wallace & ['suggested' *del*] gave me the following very curious suggestions, which he will investigate   a peck from a bird w[d] be as injurious as to be eaten—have paramount importance, for caterpillar to be recognized.[9]—on your principle that classical writers recommended shepherds to keep white sheep-dog not to be killed for wolves.—[10]
    The striping or banding w[d] follow from previously coloured marks or from differences in the tissues. It is indisputable that very many imitate leaves &c &c— Bates in cases.— Species of acacia &c[11]

[1] The year is established by the relationship between this letter and the letter to A. R. Wallace, 7 March [1867].
[2] The list of queries about expression that CD sent Wallace has not been found; see letter to A. R. Wallace, 7 March [1867] and n. 2. For a version of the questions similar to those sent to Wallace, see Appendix IV. Wallace had travelled in the Malay Archipelago from 1854 to 1862.
[3] CD referred to his work on human expression as an appendix to his 'Essay on Man' in his letter to Wallace of 7 March [1867]. *Variation* was published in 1868; CD had originally planned to present his work on humans in a final chapter of *Variation* (see letter to J. D. Hooker, 8 February [1867] and n. 16). *Expression* was published in 1872, a year after *Descent*.
[4] CD had intended to publish an expanded version of his theory following the publication of *Origin* in 1859 (see, for example, *Correspondence* vol. 7, letter to A. R. Wallace, 25 January [1859] and n. 11).

'Struggle for existence' is the title of chapter three in *Origin*. *Variation* was planned as the first of three related works; in *Variation* 1: 8, CD stated that, in a 'second work', he would discuss the variation of organisms in a state of nature, the struggle for existence, and the difficulties opposed to the theory of natural selection. The third work was to have dealt with the applicability of the theory of natural selection.

[5] Thomas Henry Huxley wrote a series of papers between 1865 and 1871 examining variations in humans, and considering divisions between human groups (see Di Gregorio 1984, pp. 160–84).

[6] Andrew Murray made critical comments on CD's theory in *The geographical distribution of animals* (A. Murray 1866, pp. 4–14); he also discussed CD's ideas on the dispersal of plant and animals to oceanic islands in *ibid.*, pp. 15–22, and CD's theory of the formation of coral islands in *ibid.*, pp. 25–7.

[7] See letter from H. W. Bates, 11 March 1867 and n. 9.

[8] Wallace refers to the mullein moth (*Cucullia verbasci*) and the cinnabar moth (*Callimorpha jacobeae*, now *Tyria jacobaeae*).

[9] CD was evidently considering the discussion that he later published in *Descent*; see *Descent* 1: 416, and letter from A. R. Wallace, 24 February [1867] and n. 3.

[10] No comment by Wallace on this topic has been found; however, in *Variation* 1: 24, CD mentioned the recommendation of Columella that white sheep dogs should be used since they would not be mistaken for wolves.

[11] CD refers to Henry Walter Bates, with whom he had also discussed coloration in caterpillars (see letter to A. R. Wallace, 23 February 1867); in Bates 1861, pp. 508–9, Bates noted that many caterpillars resembled twigs and other objects, while some butterflies resembled leaves or bark. James Philip Mansel Weale had told CD of caterpillars that mimicked *Acacia horrida* thorns (see letter from J. P. M. Weale, 9 January 1867, and CD's annotation to the letter from A. R. Wallace, 24 February [1867]).

From Benjamin Clarke 12 March 1867

2 Mount Vernon | Hampstead | N.W.
March 12 67.

Dear Sir,

I have for a long time had in view the addition, to my work on the Natural System of Botany,[1] of a new arrangement of the Classes of Zoology on principles closely analogous to those of the botanical Tables of that work, so as to combine the two systems into one system of Botany & Zoology.

By this combination I hope & even expect that they will materially assist in proving each other correct & also explain each other & so facilitate to students the study of both, there being many students in the present time who are initiated in Botany & Zoology nearly at the same time.

The Table of the new arrangement of the Classes of Zoology is now ready for the press & I write to request the favour of your subscription for the work with the zoological addition the price of which to those who have subscribed for the botanical work, & therefore to you, will be only $10/– & you will receive it carriage free.

The botanical part of the work is the same as that you have already received but the book will have a different title i.e. On Systematic Botany & Zoology & will contain 2 or 3 pages of letterpress or possibly more explanatory of the Zoological

Table & also 1 or 2 pages of botanical Addenda.[2] The botanical Addenda will I believe materially assist in proving the botanical arrangement, especially as regards the flowering plants, to be correct.

D.[r] Gray of the British Museum[3] & other Zoologists will favour me with their opinions of the Zoological Table before it is printed off.

The botanical arrangement I have every reason to believe is quite successful, especially in its principles, & I have not heard of a single objection to it, so I am encouraged to proceed.[4]

I beg to enclose a notice of progressive development believing it will interest you. It is presented as an addendum to the Tables in which progressive development is the leading principle, although there is no notice of that subject in the letterpress of the work.[5]

I remain dear Sir | truly yours | Benj.[n] Clarke.

Charles Darwin Esq.[e]

[Enclosure]

On a mode of producing Varieties by Pruning.

About eight years since I commenced two experiments one on red wheat & the other on white for the purpose of producing a six-set variety, i.e. six grains in a spikelet, three or four being the usual number in cornfields.[6] Before the plants flowered all the stems were cut away except one & as soon as the ear was protruded the upper half was cut off, & all the fresh shoots at the base of the stems were removed about once a week or fortnight till the ears were ripened.

The result was that the six-set was produced & in the third year a seven-set spikelet, which as far as I know has not been observed by agriculturists & it seemed very probable that eight-set could be produced.

The experiment which had only economical purposes in view was then discontinued because it was found that cutting off the upper half of the ear for only three years had the very singular effect of occasioning the upper half of the ear in the red wheat to be nearly barren & shrivelled, & of converting the upper half of that in the white into an unnaturally dense spike approaching that of Phalaris Canariensis;[7] it was in this that the seven-set was produced. Would not this account for varieties with short spikes? A plant growing in barren ground would have unusually short spikes which in time might perhaps become a permanent variety.

One suggestion may be interesting. If the central bud of a Sun-flower were cut off when the plant was about a foot high, & those of the lateral branches which sprouted afterwards when they became a few inches long & the plant was then left to flower, I quite expect that a corymbous[8] variety of the Sun-flower would be produced having many small heads in the place of one, which could be propagated by seed, & the corymbs increased in density by continuing the pruning.

This leads to the inference that progressive development depends in great measure on the condition of the parents at the time of fertilization;— whatever that condition may be it is unalterably stamped on the offspring. But it appears to me that this is not the only cause of progressive development. Spontaneous generation although it has been regarded as unphilosophical is nevertheless it appears to me rendered most probable by Dactylium oogenum,[9] & if so it is in all probability a very common occurrence. This being the case, spontaneous alteration at the time of generation, not depending on the condition of the parents, would perhaps be a phenomenon not so unlikely to occur as otherwise might be supposed. The production of a nectarine on a peach tree I regard as an instance of this spontaneous alteration,[10] & at Hampstead there is an Elm tree one arm of which has an apparently different leaf, I think considerably smaller, & much denser branches than the rest of the tree, which I suppose to arise from one of the buds having spontaneously a new mode of growth.[11]

If then buds at the time of generation can take on a new mode of growth, of course embryos can do the same, & here we should have a prolific source of new varieties.

I therefore attribute progressive development to two causes which act either separately, or, not improbably, sometimes in conjunction.

DAR 161: 157/1, 158

[1] Clarke refers to *A new arrangement of phanerogamous plants* (Clarke 1866), a taxonomic work in which Clarke presented a natural system of botanical classification based on the position of the ovaries. Although CD's name does not appear on the list of original subscribers to the privately published work (see Clarke 1866), he evidently bought a copy; he recorded a payment in his Account books–cash account (Down House MS) of £1 under the heading 'Botanical book Clarke' for 26 December 1866. CD's copy has not been found.

[2] Clarke's new work, *On systematic botany and zoology*, was published in 1870 (Clarke 1870); CD recorded a payment in his Account books–cash account (Down House MS) of 10s. under the heading 'Syst. Bot. & Zoolʸ B. Clarke' for 20 October 1870. CD's copy is in the Darwin Library–Down.

[3] Clarke refers to John Edward Gray, who was the keeper of the zoological collections at the British Museum.

[4] For CD's view on the lack of criticism of Clarke 1866, see the letter to J. D. Hooker 17 March [1867].

[5] The tables in Clarke 1870 are arranged to show a genealogical relationship among the different classes, but this relationship is not noted in the accompanying text.

[6] Red and white wheat are varieties of *Triticum aestivum*, common bread-wheat (Peterson 1965, p. 15).

[7] The reference is to *Phalaris canariensis* or canary grass, which has spike-like panicles that resemble those of club wheat (*Triticum compactum*).

[8] Corymbous: i.e. corymbose.

[9] *Dactylium oogenum* is a species of anamorphic fungus (that is, a fungus persisting in an asexual state). Clarke evidently believed its occurrence was the result of spontaneous generation. On the Victorian debates over spontaneous generation, see Strick 2000.

[10] CD noted in *Variation* 1: 340 that there was considerable evidence of peach trees producing nectarines by bud-variation. CD defined bud-variation as including 'all those sudden changes in structure or appearance' that occasionally occurred 'in full-grown plants in their flower-buds or leaf-buds' (*Variation* 1: 373).

[11] In *Variation* 1: 390–4, CD considered certain cases of bud-variations that gave rise to different forms of growth on a single plant as consequences of reversion.

To A. R. Wallace [12–17] March [1867][1]

Down. | Bromley. | Kent. S.E.
March

My dear Wallace

I thank you much for your two notes.[2] The case of Julia Pastrana is a splendid addition to my other cases of correlated teeth & hair, & I will add it in correcting the proofs of my present volume.—[3] Pray let me hear in course of summer if you get any evidence about the gaudy caterpillars. I sh^d. much like to give (or quote if published) this idea of yours, if in any way supported, as suggested by you.[4] It will, however, be a long time hence, for I can see that sexual selection is growing into quite a large subject, which I shall introduce into my essay on man, supposing that I ever publish it. I had intended giving a chapter on man, in as much as many call him (not *quite* truly) an eminently *domesticated* animal; but I found the subject too large for a chapter. Nor shall I be capable of treating the subject well, & my sole reason for taking it up is that I am pretty well convinced that sexual selection has played an important part in the formation of races, & sexual selection has always been a subject which has interested me much.[5]

I have been very glad to see your impression from memory on the expression of Malays.[6] I fully agree with you that the subject is in no way an important one: it is simply a "hobby-horse" with me about 27 years old; & *after* thinking that I would write an essay on man, it flashed on me that I could work in some "supplemental remarks on expression,"— After the horrid tedious dull work of my present huge & I fear unreadable book, I thought I would amuse myself with my hobby-horse.[7] The subject is, I think, more curious & more amenable to scientific treatment, than you seem willing to allow. I want anyhow to upset Sir C. Bell's view, given in his most interesting work "the anatomy of Expression" that certain muscles have been given to man solely that he may reveal to other men his feelings. I want to try & show how expressions have arisen.—[8]

That is a good suggestion about newspapers;[9] but my experience tells me that private applications are generally most fruitful.— I will, however, see if I can get the queries inserted in some Indian paper.— I do not know name or address of any other papers.—

I have just ordered, but not yet received Murray's book: Lindley used to call him a blunder-headed man.—[10] It is very doubtful whether I shall ever have strength to publish the latter part of my materials.[11]

My two female amanuenses are busy with friends, & I fear this scrawl will give you much trouble to read.—[12]

With many thanks | Yours very sincerely | Ch. Darwin

British Library (Add 46434 ff. 80–83v)

[1] The beginning of the date range is established by the relationship between this letter and the letter from A. R. Wallace, 11 March [1867]. The end of the date range is established by CD's evident

receipt of A. Murray 1866 by 17 March (see letter to J. D. Hooker, 17 March [1867], and n. 10, below).

[2] See letter from A. R. Wallace, 11 March [1867]. The other letter from Wallace has not been found.

[3] Wallace's letter about Julia Pastrana, the woman deformed by excessive hairiness and an enlarged jaw, has not been found. In a section on 'Correlated variation of homologous parts' (*Variation* 2: 322), CD commented on Pastrana, mistakenly writing that she was Spanish. He also wrote that her upper and lower jaws contained a double set of teeth, giving her face a 'gorilla-like appearance' (*ibid.*, p. 328). It was later found that her features resulted from an enlargement of the gums, rather than extra teeth. See Browne and Messenger 2003.

[4] See letter from A. R. Wallace, 11 March [1867] and n. 7.

[5] See letter from A. R. Wallace, 11 March [1867]; Wallace had asked whether CD's work on humans would be a supplement to *Variation*. CD had discussed the role of sexual selection in the formation of the human races in his letter to A. R. Wallace, 28 [May 1864] (*Correspondence* vol. 12).

[6] See letter from A. R. Wallace, 11 March [1867] and n. 2.

[7] See letter from A. R. Wallace, 11 March [1867]. CD made notes on human expression as early as 1838 (see, for example, *Notebooks*, Notebook M; see also Barrett 1980). CD also refers to *Variation*, which he had been working on since 1860 (see 'Journal' (Appendix II)).

[8] CD read Charles Bell's *Essays on the anatomy of expression in painting* (Bell 1806) in 1840 (see CD's reading notebooks, *Correspondence* vol. 4, Appendix IV). CD acquired the third, enlarged edition of Bell 1806, *The anatomy and philosophy of expression as connected with the fine arts* (Bell 1844), in November 1866 from his brother, Erasmus Alvey Darwin (inscription to CD's annotated copy of Bell 1844 in the Darwin Library–CUL; *Marginalia* 1: 47–9). CD later wrote that he had not agreed with Bell that some muscles had been 'specially created for the sake of expression' (*Autobiography*, p. 132). He included a critique of Bell's ideas in *Expression*. There is an earlier note by CD on Bell 1806 and expression in *Notebooks*, Notebook C, C243 (Barrett 1980, p. 194). For more on CD's and Bell's views, see Browne 1985, Ekman 1998, pp. 7–9, 144, and Hartley 2001.

[9] In his letter of 11 March [1867], Wallace suggested that CD send his list of queries on human expression to foreign newspapers.

[10] See letter from A. R. Wallace, 11 March [1867] and n. 6. CD refers to Andrew Murray, A. Murray 1866, and John Lindley. CD's Account books–cash account (Down House MS) records payment for books from Quaritch, a London bookseller, on 13 March 1867; these may have included A. Murray 1866 (see letter to J. D. Hooker, 17 March [1867], and n. 15).

[11] See letter from A. R. Wallace, 11 March [1867] and n. 4.

[12] CD often dictated letters to his wife, Emma, or daughter Henrietta. Emma recorded in her diary (DAR 242) that Sarah Elizabeth Wedgwood and Georgina Tollet were visiting from 11 to 21 March 1867.

From J. D. Hooker 14 March 1867

Kew

March 14/67.

Dear Darwin

Hurra, I am right again; it was not A Gray that wrote the article on Agassiz— how glorious a discovery that there is another man in the world who could write such an article: by Jove it does warm the cockles of one's heart.[1]

Per contra I am in a state of deep dejection, having been persuaded by all the Botanists I respect to accept the nomination for election to the Presidentship of Brit Assn. in 1868 at Norwich.[2] You may well pity me. However in for a penny in for a pound & if I am in good health & keep at the time I will do my very best.

What have you been about since I last saw you in Queen Anne St.?[3] I have been plodding at Gen. Plant.[4]

Here is a wonderful discovery, Naudin has sent me seeds of *Chamærops humilis* fertilized by the Date Palm & by Jove they are altogether unlike Chamerops seeds in shape texture size & consistency & exactly half way to Dates![5] Young plants have been raised & these appear intermediate.

Vertical sections.

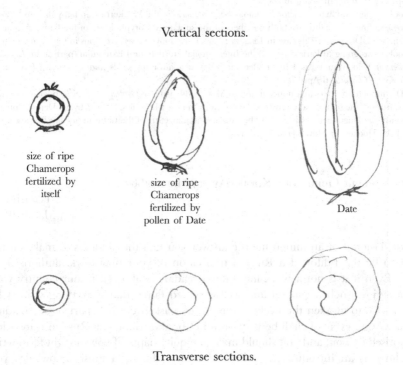

size of ripe
Chamerops
fertilized by
itself

size of ripe
Chamerops
fertilized by
pollen of Date

Date

Transverse sections.

Is this not a wonderful experiment & results.?

    Ever yrs aff | J D Hooker

    I go to Paris end of month for Jury work at Exhibn.—[6]

DAR 102: 145–6

CD ANNOTATIONS
1.1 Hurra, … Gen. Plant. 3.2] *crossed red crayon*
End of letter: 'G. Chronicle 67 p. 264 M. Denis speaks of the fruit of the Hybrid [*interl*] palm as intermediate & not as result of direct action of fertilisation.'[7] *ink*

[1] CD suspected that Asa Gray had written an article, 'Popularizing science', that was critical of Louis Agassiz (Anon. 1867; see letter from J. D. Hooker, 4 February 1867 and n. 2, and letter to J. D. Hooker, 8 February [1867]).

[2] Earlier, Hooker told CD he had declined the invitation to be president of the British Association for the Advancement of Science (see letter from J. D. Hooker, 4 February 1867).

[3] CD had visited London from 13 to 21 February 1867 (CD's 'Journal' (Appendix II)), staying at the house of his brother, Erasmus Alvey Darwin, at 6 Queen Anne Street.

[4] *Genera plantarum* (Bentham and Hooker 1862–83) was published in seven parts. The second part appeared in 1865, the third in 1867.

[5] Hooker refers to Charles Victor Naudin. *Chamaerops humilis* (the Mediterranean fan palm) and *Phoenix dactylifera* (the date palm) are both members of the subfamily Coryphoideae of the family Arecaceae (Palmae). CD reported this case in *Variation* 1: 399, but removed it as erroneous in the second edition.

[6] Hooker was going to Paris to attend the International Horticultural Exhibition (part of the *Exposition universelle* held from 1 April to 31 October 1867) as a juror for seeds and saplings of forest trees (*Gardeners' Chronicle*, 6 April 1867, p. 348).

[7] CD's annotation refers to a notice in the 16 March 1867 issue of *Gardeners' Chronicle*, p. 264, reporting that fruits of the hybrid palm grown at Hyères (a town near Marseilles, France) by M. Denis (Alphonse Amaranthe Dugomier Denis) were to be exhibited at a meeting of the Linnean Society. See also letter to J. D. Hooker, 17 March [1867] and n. 2.

From Vladimir Onufrievich Kovalevsky   15 March 1867

S Petersbourg

3/15 March 1867[1]

Sir

M[r]. Truebner transmitted me the answer you had the kindness to make on my offer to be the Editor of a Russian translation of your new work, authorised by you.[2] Even if it is not a work for general reading, still, even in our country your name enjoys such a general and well deserved fame, that everything written by you is sure to awaken the liveliest interest, at least in the best part of our reading public. At all events I shall be very proud to be the editor of a Russian translation authorised by you, and you should make me quite happy if you thought fit to write two lines as an Introduction or Preface to my edition.[3] I must aknowledge you that I am, besides being an editor, also a little naturalist now and still more so *in spe*,[4] so that you will make me doubly happy by authorising me to be the Russian editor of your new work and adressing a few lines to this translation.— You can be quite sure that the translation of the work will be made most carefully, as I shall make it myself under the guidance of my brother who is lecturer of Zoology in the University of S Petersburg, but now temporarily absent on an excursion to Triest.[5] I take the boldness to send you some of his Memoirs, perhaps some of his studies and it may be a little wild conceptions about the affinity of the Ascidian and vertebrate types by their history of developement shall, however lightly, interst you, whose most true and steady follower he professes to be.[6]

In finishing my letter, I shall beg your pardon for my bad English, and pray you, if you shall have a moment time, to inform me directly, or trough M. Truebner is there any hope that the first part of the work will be finisched this year, will it be illustrated and what approximative dimensions (in pr. sheets)[7] it will have.—

Leaving the pecuniary arrangement entyrely to you, and if you will have the kindness to adress your resolution to M[r]. Truebner, I shall wait with the utmost impatience of a naturalist and of your admirer, for the first printed sheets of your work which you had the kindness to promise to send me as soon as you receive them from the printer

I am | Sir | Your most true and impatient admirer | W. Kowalewsky

My adress. | *S Petersbourg.* | coin de la Petite Morskaja | A Gowchovaja m Mitkoff. N 19.

DAR 169: 71

[1] Kovalevsky gives both the Julian (3 March) and Gregorian (15 March) calendar dates.

[2] Nicholas Trübner's publishing firm had approached CD with an offer from Kovalevsky to translate *Variation* (see letter from Trübner & Co, 26 February 1867). CD's reply to Trübner has not been found.

[3] No preface was added to the Russian edition of *Variation* (V. O. Kovalevsky trans. 1868–9).

[4] 'In spe': hopefully, in the future (Latin).

[5] Alexander Onufrievich Kovalevsky was curator of the zoological cabinet and privat-dozent at St Petersburg University (*DSB*). Trieste, now an Italian city, situated on the gulf of Trieste at the head of the Adriatic sea, was capital of Küstenland province, Austria, in 1867 (*Columbia gazetteer of the world*).

[6] Kovalevsky sent seven papers by his brother on the anatomy and embryology of marine invertebrates, all written in German (A. O. Kovalevsky 1865a and 1865b, 1866a–d, and Ovsjannikov and Kovalevsky 1866; CD's copies of these papers are in the Darwin Pamphlet Collection–CUL). In one paper, on the embryology of simple ascidians (A. O. Kovalevsky 1866b), Kovalevsky noted similarities between the larval stages of ascidians and amphioxus, suggesting an ancestral link between invertebrates and vertebrates. CD's copy of the paper is heavily annotated.

[7] Pr. sheets: i.e. proof-sheets.

From J. T. Moggridge 15 March [1867][1]

S[t]. Roch.
March 15

Dear M[r]. Darwin

the enclosed box contains a very remarkable form of **Thymus** *vulgaris*, having its anthers of the *same form* as those of the Hermaphrodite form— The anthers contain bad pollen— The whole of a large plant—having 12 expanded heads of flowers—was in in this condition, except that in two or three cases flowers had but two instead of four lobes to the corolla.— This plant is marked with wool, & I enclose four other heads from four separate plants.—[2]

I also enclose a few racemes of **Rhamnus** *Alaternus*— I will observe this plant more, but as yet I find no pollen in the anthers of the female flowers—[3]

Each raceme of Rhamnus in the box is from a distinct bush.—

Many thanks for your letter received March 13—[4] | y[rs]. very sincerely | J. Traherne Moggridge.

DAR 171: 204

[1] The year is established by the relationship between this letter and CD's note on the the letter from J. T. Moggridge, 6 March [1867].

[2] For CD's comments on the specimens of *Thymus vulgaris* sent by Moggridge, see his note on the letter from J. T. Moggridge, 6 March [1867].

[3] In a note dated 19 March 1867, CD wrote that Moggridge had sent him flowers of *Rhamnus alaternus* and commented, 'they are simply dioiceous, in nearly same state as common Holly' (DAR 109: A43). See letter from J. T. Moggridge, 6 March [1867] and n. 7.

[4] CD's letter to Moggridge has not been found.

To George Moultrie Salt   16 March [1867]

Down. | *Bromley.* | *Kent. S.E.*
Mar 16.

Dear Sir

I am much obliged for your note & return the paper signed. When you receive the money will you be so good as to place it to my acc^t at the Union Bank (*Charing Cross Branch*).[1]

Dear Sir | yours very faithfully | Ch. Darwin

LS
Endorsement: '16 Mch 1867'
Rachel Salt (private collection)

[1] The paper that CD signed has not been identified, but a deposit into CD's account of £699 1 *s*. 2 *d*. is recorded for 20 April 1867 with the notation, 'Sale of Part of Land at Shrewsbury (Capital)' (CD's Account books–banking account (Down House MS); square brackets in original). CD had banked with the Charing Cross branch of the Union Bank of London since 1853 (see *Correspondence* vol. 5, letter to John Higgins, 11 April [1853]).

To J. D. Hooker   17 March [1867][1]

Down.
Mar 17

My dear Hooker

It is a long time since I have written, but I cannot boast that I have refrained from charity towards you, but from having lots of work. I am so much obliged to you for telling me about the palm seeds. I have got a whole string of cases equally, or perhaps more curious, but yours is infinitely the most valuable from having been observed by first rate judges.[2] The subject is of paramount importance for my beloved Pangenesis.[3] Now c^d you obtain permission for me from Naudin to insert some such paragraph as the enclosed.[4] My book will not appear till next Nov. but I sh^d have to insert the passage in the proof sheet in about a month's time. Help me if you can. Are they seeds or nuts? Correct the word if they are not seeds.

It is great news about the presidentship; I am very sorry for it, tho' you seem to keep up your spirits.[5] You ask what I have been doing; nothing but blackening proofs with corrections.[6] I do not believe any man in England naturally writes so vile a style as I do. The only fact which I have lately ascertained, & about which I dont know whether you w^d care, is that a great excess of, or very little pollen produced not the least difference in the average number, weight, or period of germination in the seeds of Ipomœa.[7] I remember saying the contrary to you & Mr Smith[8] at Kew. But the result is now clear from a great series of trials. On the other hand seeds from this plant, fertilised by pollen from the same flower, weigh less, produce dwarfer plants, but indisputably *germinate quicker* than seeds produced by a cross between two distinct plants.[9]

In your paper on Insular Floras (p. 9) there is what I must think an error, which I before pointed out to you; viz you say that the plants which are wholly distinct from those of nearest continent are often *very common*, instead of very rare.[10] Etty,[11] who has read your paper with great interest, was confounded by this sentence. By the way I have stumbled on two old notes, one that 22 species of European birds occasionally arrive as *chance* wanderer, to the Azores, & secondly the trunks of American trees have been known to be washed on shores of Canary isl^d, by gulf stream, which returns southward from the Azores.[12]

What poor papers those of A. Murray are in G. Chronicle: what conclusions he dreams from a single Carabus & that a widely ranging genus![13] He seems to me conceited: you & I are fair game geologically, but he refers to Lyell, as if his opinion on a geological point was worth no more than his own.—[14] I have just bought, but not read a sentence of, Murray's big book, second-hand for 30^s, **new**, so I do not envy the publishers.[15] It is clear to me that the man cannot reason.—

I have had a very nice letter from Scott at Calcutta:[16] he has been making some good observations on the acclimatisation of seeds from plants of same species, grown in different countries; & likewise on how far European plants will stand the climate of Calcutta; he says he is astonished how well some flourish, & he maintains, if the land were unoccupied, several could easily cross, spreading by seed, the Tropics from N. to South; so he knows how to please me, but I have told him to be cautious, else he will have Dragons down on him.—[17]

I was going to have asked you what sort of a man Benj. Clarke was (I bought his book out of kindness) but I see A. Gray calls him "that ass". He is now going to publish analogous views on Animals, & I have subscribed.[18] He tells me that no single person has or can object to his views on plants; I suspected that perhaps no one noticed them.[19] He tells me a wonderful story of the effects of an inherited mutilation from cutting off the upper half of the ears of wheat for only 3 generations, which I cannot believe; & I have told him no one would believe it, unless he repeat & rerepeats his experiment.—[20]

Farewell my dear old friend | C. Darwin

DAR 94: 13a–e

[1] The year is established by the relationship between this letter and the letter from J. D. Hooker, 14 March 1867.

[2] CD refers to seeds that had recently been sent to Hooker by Charles Victor Naudin from a fan palm pollinated by a date palm (see letter from J. D. Hooker, 14 March 1867 and n. 5).

[3] In *Variation* 1: 397–403, CD discussed cases in which the pollen of one plant, when applied to another species or variety, affected the shape, colour, or flavour of its fruit. In the chapter on pangenesis, CD concluded, 'We here see the male element affecting and hybridising not that part which it is properly adapted to affect, namely the ovule, but the partially-developed tissues of a distinct individual' (*Variation* 2: 365). For more on the development of the theory of pangenesis, see *Correspondence* vols. 13 and 14.

[4] The enclosure has not been found, but CD added a description of the fruit and seeds, citing Naudin's report to Hooker, in *Variation* 1: 399.

[5] CD refers to Hooker's decision to accept the presidentship of the British Association for the Advancement of Science for 1868 after initially turning down the invitation (see letters from J. D. Hooker, 4 February 1867 and 14 March 1867).

[6] CD had received the first proof-sheets of *Variation* on 1 March 1867 (CD's 'Journal' (Appendix II)).

[7] CD began a series of experiments with *Ipomoea* in 1866; at the end of a note dated 1866 he commented, 'too much pollen?' (DAR 78: 71). In *Cross and self fertilisation*, pp. 24–5, CD described experiments with *I. purpurea* to ascertain whether fertility rates were affected by the amount of pollen used and concluded, 'flowers fertilised with little pollen yielded rather more capsules and seeds than did those fertilised with an excess; but the difference is too slight to be of any significance' (*ibid.*, p. 25).

[8] John Smith was curator of the Royal Botanic Gardens, Kew (R. Desmond 1994).

[9] CD's notes on the germination of seeds from cross-pollinated and self-pollinated plants of *Ipomoea purpurea*, dated between 1 and 10 March 1867, are in DAR 78: 77.

[10] In his letter to Hooker of 21 January [1867], CD had pointed out the apparent misprint of 'commonest' for 'rarest' in the latest instalment of Hooker's paper on insular floras in the *Gardeners' Chronicle* (J. D. Hooker 1866a); in Hooker's pamphlet version of the paper, he changed 'the commonest of all' to 'very common' (see Williamson 1984, p. 70).

[11] Henrietta Emma Darwin.

[12] The notes referred to have not been found. In a letter to Hooker of 5 August [1866] (*Correspondence* vol. 14), CD mentioned looking for a 'note about Birds being blown to the Azores from Europe', but had not found it on that occasion.

[13] Andrew Murray's paper, 'Dr. Hooker on insular floras', appeared in two parts in *Gardeners' Chronicle* (16 and 23 February 1867, pp. 152, 181–2; A. Murray 1867). Murray cited the case of a species of beetle of the family Carabus (now Carabidae), *Aplothorax burchelli*, found only on St Helena. Murray argued that it showed the greatest affinity to species found in Switzerland and Asia Minor, and further, that as the species could not have arrived by means of occasional transport, it indicated a connection between Europe and the coast of Africa that might have extended to St Helena (*ibid.*, p. 182).

[14] Murray made two references to Charles Lyell's geological theories, in both cases associating them with CD's views. He argued that Lyell and CD were unduly cautious in not allowing for the former existence of land-bridges between Europe and the Americas because of the great depth and width of the Atlantic ocean (A. Murray 1867, p. 152).

[15] CD refers to Murray's recently published book, *The geographical distribution of mammals* (A. Murray 1866). CD's annotated copy is in the Darwin Library–CUL (see *Marginalia* 1: 624). In 1864, Murray had produced a prospectus for the work, and asked CD for suggestions on the content, but CD had been sceptical of Murray's ability (see *Correspondence* vol. 12, letter from Andrew Murray, 31 October 1864, and letter to J. D. Hooker, 3 November [1864]).

[16] See letter from John Scott, 22 January 1867.

[17] CD's response to Scott's letter of 22 January 1867 has not been found.

[18] CD refers to Benjamin Clarke and Asa Gray. The source of Gray's comment has not been identified, but Hooker frequently passed on letters he received from Gray to CD. CD had bought a copy of Clarke's work on plant taxonomy and Clarke had recently written to tell CD of his plan to produce

a similar work including animals (Clarke 1866 and Clarke 1870; see letter from Benjamin Clarke, 12 March 1867 and nn. 1 and 2).

[19] See letter from Benjamin Clarke, 12 March 1867.

[20] See enclosure to letter from Benjamin Clarke, 12 March 1867. CD's reply to Clarke's letter has not been found.

From Friedrich Hildebrand 18 March 1867

Bonn
March 18th | 1867.

Dear and honoured Sir

you must excuse me not having thanked you before for sending me the copies of my notice on Corydalis,[1] but I was waiting for a little treatise of mine to be finished to send you a copy of it. I have ventured to show how right you are when you say that nature abhors perpetual self-fertilisation.[2] I heard that there has appeared a new edition of your "Origin of Species",[3] perhaps you have said in it more about selffertilisation and intercrossing than in the first, but I suppose that you did not enter in details and I hope that my little book will not appear quite useless for the public, though I do not believe that you will find much in it that you have not known before. I cannot send you as yet the paper on Aristolochia, it is very annoying that Pringsheim goes on so slowly with his Jahrbücher.[4]

As there is a want of a good German journal with critics of botanical litterature we are going to fill up this want and I have promised to look out for all that belongs to the flowers of plants, therefore you would oblige me very much if you would let me know *occasionaly* of any paper that is published in England about this matter or tell the authors to send me a copy that I shall most gladly return with those that are wanted of my papers.[5] I send you two copies of my little work, perhaps you will be so kind to give one of them to somebody, who takes an interest in the matter.[6]

I hope that you are in good health and remain dear Sir | yours | respectfully | Hildebrand

DAR 166: 206

[1] CD had corresponded with Hildebrand on the pollination mechanism of *Corydalis cava*, and had communicated Hildebrand's paper 'On the necessity for insect agency in the fertilisation of *Corydalis cava*' to the International Horticultural Exhibition (see *Correspondence* vol. 14, letter to Friedrich Hildebrand, 16 May [1866] and n. 3). The paper was published in *International Horticultural Exhibition* 1866, pp. 157–8 (Hildebrand 1866a). A copy is in the Darwin Pamphlet Collection–CUL. Hildebrand later wrote an expanded version in German (Hildebrand 1866–7b).

[2] Hildebrand refers to his monograph *Die Geschlechter-Vertheilung bei den Pflanzen und das Gesetz der vermiedenen und unvortheilhaften stetigen Selbstbefruchtung* (Sexual division in plants and the law of avoidance and disadvantage of perpetual self-fertilisation; Hildebrand 1867a). CD's annotated copy is in the Darwin Library–CUL (see *Marginalia* 1: 378–9). On the title page of the work, Hildebrand included the following quotation, attributed to CD:

Nature tells us in the most emphatic manner that she abhors perpetual selffertilisation.

No hermaphrodite fertilises itself for a perpetuity of generations.

See *Orchids*, p. 359; see also *Origin*, p. 97.

³ The reference is to the fourth edition of *Origin*. Hildebrand may refer to the German translation (Bronn and Carus trans. 1867).

⁴ Hildebrand's paper 'Ueber die Befruchtung von Aristolochia Clematitis und einiger anderer Aristo-lochia-Arten' (On the fertilisation of *Aristolochia clematitis* and a few other species of *Aristolochia*; Hildebrand 1866–7a), was published in *Jahrbücher für wissenschaftliche Botanik*, a journal founded and edited by Nathanael Pringsheim (*NDB*).

⁵ The journal Hildebrand hoped to start was evidently not established. The first German journal devoted to reviewing botanical literature was the *Botanischer Jahresbericht*, first published in 1874, but Hildebrand was not one of the editors (Frodin 2001, p. 9).

⁶ CD sent the second copy of Hildebrand 1867a to Daniel Oliver (see letter to Friedrich Hildebrand, 20 March [1867]).

To John Murray   18 March [1867]

Down. | *Bromley.* | *Kent. S.E.*

March 18ᵗʰ

My dear Sir

The Compositors have *invented* & placed the title, as on enclosed paper above the red line, at head of my Introduction. Now I think this sounds a better title than the advertised one.¹ Have the kindness to consider the question & inform me. Does it signify another title having been advertised? If you thought it worth while, do consult Sir C. Lyell,² as he has such good judgment & I have always found him so very kind, that I feel sure he would advise me.

I can form no sort of opinion about the number of copies to be printed off, & this a point which you will soon have to decide, as I have received two Revises.—³

I presume you will not object to my having two sets of clean sheets, one for Germany & the other to Russia for translation; though when the publishers see how big a book it is, their courage may fail.⁴ If translations are made, I suppose you would allow stereotypes of the wood-blocks to be made.

Be so kind as to answer these questions & believe me, my dear Sir | Yours sincerely | Ch. Darwin

Endorsement: '1867. March 18'
John Murray Archive

¹ The enclosure has not been found. CD's book, eventually published as *The variation of animals and plants under domestication*, had been advertised as early as 1865 under the title 'Domesticated Animals and Cultivated Plants, or the Principles of Variation, Inheritance, Reversion, Crossing, Interbreeding, and Selection under Domestication' (*Publishers' Circular*, 1 August 1865 p. 386; see *Correspondence* vol. 13, letter to John Murray, 2 June [1865] and n. 3).

² Charles Lyell's geological works were published by Murray.

³ According to CD's 'Journal' (Appendix II), the first proof-sheets of *Variation* arrived on 1 March 1867.

⁴ For more on the proposed German translation of *Variation*, see the letter to E. Schweizerbart'sche Buchhandlung, [19 March 1867]. CD had received an offer via Trübner & Co from Vladimir Onufrievich Kovalevsky, to translate *Variation* into Russian (see letter from Trübner & Co, 26 February 1867, and letter from V. O. Kovalevsky, 15 March 1867).

John William Salter (seated)
The standing figure may be Henry Hicks.
© The Natural History Museum, London

William Boyd Dawkins, *circa* 1870
With permission from Derbyshire County Council:
Buxton Museum and Art Gallery

Albert Günther, 1867
Photograph by Samuel Fry & Co.
© The Natural History Museum, London

Calotes nigrilabris
Günther 1864, plate 14
By permission of the Syndics of Cambridge
University Library

From John Murray 19 March [1867][1]

50^A. *Albemarle S!* | *W.*

Mar 19

My Dear Sir

I am enclined to think, with you, that the new Title of your work is an improvement & I propose to adopt it in future.[2]

I have been giving much consideration to the question of the number to be printed & have altered my mind on this point. I do not think there will be any eventual loss if I print 1500 Copies & I propose with your assent to go to press with that number.—[3]

Of course this—like all other publishing undertakings is a lottery—but the Scientific *must study* your book & the unlearned will dip into it & pick out portions at least suited to interest them— This I gather from the proof sheets you shall certainly have the sets of fair sheets. please instruct Clowes to that effect[4]

At least if your old German publisher will have nothing to say to you, I can assure you of another candidate at Jena, who tells me he is ready to adventure[5]

I am | My Dear Sir | Yours very sincerely | John Murray

Charles Darwin Esq

DAR 171: 347

[1] The year is established by the relationship between this letter and the letter to John Murray, 18 March [1867].

[2] See letter to John Murray, 18 March [1867] and n. 1.

[3] Murray had initially proposed printing only 750 copies (see letter from John Murray, 28 January [1867]).

[4] CD had asked for proof-sheets for the German and Russian translations of *Variation*. Murray refers to the printers William Clowes & Sons.

[5] No letter concerning *Variation* from a German publisher other than E. Schweizerbart'sche Verlagsbuchhandlung has been found in the John Murray Archive (Virginia Murray, archivist).

To E. Schweizerbart'sche Verlagsbuchhandlung [19 March 1867][1]

Dear Sir

I have received the press proof of my new book, which will probably be entitled The V. of A. & P. under D. It will be in 2 Vols 8vo. with 43 wood-blocks.—[2] Whether so large a work will be worth translating into German I cannot judge. I have had two applications from G. publishers for clean sheets for translation, & one offer on payment.[3] But I would give up any claim for payment if I could get a good translator. Do you continue to wish to bring out a translation, & could you persuade Prof. V. C. to translate the work?[4] If you inform me that he will translate it, subject to your resolving to have a Translation, I would send you clean sheets as soon as half-a dozen have been printed off. But please to observe on Prof V. Carus being the translator You would have to negotiate with Mr Murray for stereotypes of the wood-blocks. I sh^d wish for an answer before much of the work *is printed* so that if you decline I might offer early sheets to the other applicants.—

Hoping that my offer may interest you, I remain, Dear Sir | Yours faithfully | C. D.—

A draft S^t
DAR 96: 33r

[1] The date is established by the reference to this letter in the letter from E. Schweizerbart'sche Verlagsbuchhandlung, 22 March 1867.

[2] CD refers to *Variation*.

[3] CD had received an offer to translate *Variation* from Rudolf Oldenburg, the business manager for the Munich branch of the firm J. G. Cotta'schen Buchhandlung (Bosl ed. 1983; see *Correspondence* vol. 14, letter from Rudolf Oldenbourg, 28 October 1866). More recently, CD's publisher, John Murray, had received an offer from a publisher in Jena (see letter from John Murray, 19 March [1867] and n. 5).

[4] Julius Victor Carus had recently completed the German translation of the fourth edition of *Origin* (Bronn and Carus trans. 1867; see letter to J. V. Carus, 17 February [1867]).

To Friedrich Hildebrand   20 March [1867][1]

Down. | Bromley. | Kent. S.E.
March 20th.

My dear Sir

I am much obliged for your new work which I see from the woodcuts will contain very much matter new to me & of great interest.[2] But I am at present so much overworked & am so poor a german scholar that I shall not be able to read it just yet, but I know it will be a real pleasure to me when I am able.[3] I have sent the second copy to Prof. Oliver of Kew who reads german easily & formerly often reviewed books & perhaps still does so.[4] I first thought of Prof. Asa Gray of Cambridge Massachusetts U.S. but I am not sure that he reads german, otherwise he would certainly notice it in the J. of Science.[5] From turning over the pages of your book I suspect that it is very like a long chapter which I have sketched out & intended to write, but which perhaps I never should have finished & certainly could not have done it nearly as well as you.[6] Perhaps you might like to hear that I have raised all three forms of Oxalis speciosa & that their mutual fertility, as far as I have tried follows exactly the same law as with Lythrum.[7] I congratulate you on the completion of your new work which I fully believe will be extremely interesting to all botanists. There is nothing of consequence on your subject in the new edit. of the "Origin".[8] I will do what I can but I fear I shall not be able to help you with information for your new journal which I hope may be successful.[9]

My dear Sir, yours sincerely | Ch. Darwin

LS
Courtesy of Eilo Hildebrand (photocopy)

[1] The date is established by the relationship between this letter and the letter from Friedrich Hildebrand, 18 March 1867.

[2] Hildebrand had sent CD two copies of his monograph on plant sexuality (Hildebrand 1867a; see letter from Friedrich Hildebrand, 18 March 1867 and n. 2).

[3] CD was correcting proof-sheets of *Variation* (see letter to J. D. Hooker, 17 March [1867] and n. 6).

[4] Daniel Oliver had been a botany editor of the *Natural History Review*, a journal that ceased publication in 1865, and had been responsible for the bibliography of phanerogamic botany that appeared regularly in the journal. He often referred CD to German and French publications on botanical topics (see, for example, *Correspondence* vol. 12, letter from Daniel Oliver, 14 June 1864).

[5] Gray was a frequent contributor to the *American Journal of Science and Arts*.

[6] CD did publish his work on illegitimate crosses in 'Illegitimate offspring of dimorphic and trimorphic plants', which was read before the Linnean Society in February 1868. A modified form of the paper later became chapter 5 of *Forms of flowers* (pp. 188–243).

[7] The results of CD's experiments with *Oxalis speciosa* were published in *Forms of flowers*, pp. 175–8. CD's notes for his experiments with *O. speciosa*, dated between 1864 and 1868, are in DAR 109: 10–26, 108. See also 'Three forms of *Lythrum salicaria*'.

[8] CD refers to the fourth edition of *Origin*, published in November 1866. The German translation, based on this edition, appeared early in 1867 (Bronn and Carus 1867).

[9] See letter from Friedrich Hildebrand, 18 March 1867 and n. 5.

From J. D. Hooker 20 March 1867

Kew
March 20/67.

Dear Darwin

I take it very hard & not a little unkind that, after having by a series of unprecedented wriggles, got my Ins. Flora into the shape of a plausible though perhaps rather sophistic form—you (& Etty especially) should call on me to wriggle out of any *mere* contradictions of facts, or irreconcileable statements.[1]

I thought I had modified the sentence you objected to (& which did appear I grant irreconcileable) & I have not looked since;— now I must: so here goes—for a *contortuplication* of wriggles (a regular colic) if I find that I must.[2]

Meanwhile I send Naudins letter, & return your note— there is no need to refer to me at all in the matter.[3] N. puts no restriction on the communication, & no doubt it will be published long before you are in print. *Fruit* or *Drupe* is the proper word.[4] I shall see N. in Paris early in April & let you know. I shall send them to L. S.[5] tomorrow. I was not aware that you had similar cases.

I am dying to understand Pangenesis that haunts me at night. Huxley told me that he had referred you to something of the kind in Bonnet.[6] I cannot conceive a Pangenesis without a correlative *Panexodus* (the Great God Pan is not yet dead, that's clear)— What I mean is this, that if every previous attribute (infinitely subdivided) of all its ancestors, exists in an organism, any of these may come out/turn up in its progeny— but I suspect I am talking nonsense to you. I was so very blind to the force of the derivative hypothesis, that I always feel too inclined to take your views au coup de (I forget what, I am coaching up french, hard, for Paris Exposition.[7]

That poor Clark is simply mad, & has been in confinement.[8] You have grasped the reason why no one attacks his theory—which by the way he will alter to suit your tastes as much as you please. When I point out a fact or structure to him he always accepts it, however subversive of his theory whose Elasticity is delightful. Thank God he is off on Zoology[9]

I am extremely interested, as is Smith, in your Ipomœa experiment.[10]

Scott has been telling me of his success in cultivating Temperate plants in thatched houses in Calcutta, he seems to be getting on capitally & his experiments are all in your favor.[11]

Now for p. 9. of my Lecture— I cannot see the error.[12] That the plants of *no* affinity are commonest, appears carried out by the Laurels being so abundant, forming forests, in the Canaries & I think also in Madeira, & Clethera[13] is certainly very common at least I think so. It is the peculiar genera *of European affinity* that are so rare—as *Merugia, Melanoselinii,* & *Dracæna* ajari is abundant in Canaries & has no European affinity. so with *Plocama, Visnea, Bosea,* &c.[14] Please look again & let me know.

NB. This is not a wriggle.

Thanks for the note about Azorean birds.[15]

I have written to ask Sir H. Barkly about Mammals bones in Mauritius[16]

I do not attempt to read Andrew Murray in G. C. the most bumptious, conceited, muddle pated pig in print.— true he cannot reason. He was Secy. Hort. Soc. but they had to chassée him: his treatment of Lyell is refreshingly civil—[17]

I am wearying very much to see you again, but cannot conceive when.

Next week I go to Paris. More & more St Helena plants prove to have Cape affinities.[18]

I have just identified another *very* marked & peculiar genus common to Chili & New Zealand it is Griselinia of Foster, with which I find *Decostea* to be generically identical & by Jove there is a species undescribed on the top of the Organ M^{ts}![19] so go ahead— As usual it is one of those genera most difficult to preserve the seeds of—a small fleshy berry. It is not a *glacial* genus.[20]

Ever dear old Darwin | yrs aff | J D Hooker

When you write, do tell me about p. 9: and do say I was right about Asa Gray!!!![21]

DAR 102: 147–50

CD ANNOTATIONS
3.2 no doubt ... proper word. 3.4] *scored pencil*
4.4 if every ... progeny— 4.6] *scored pencil*
13.1 I am ... affinities. 14.2] *scored pencil*
17.1 When ... Gray!!!! 17.2] *double scored pencil*

[1] In his letter of 17 March [1867], CD commented that both he and his daughter Henrietta Emma Darwin had found an error in the offprint of 'Insular floras' (J. D. Hooker 1866a; see Williamson 1984 for the text of the offprint, which differed slightly from the original version printed in *Gardeners' Chronicle*).

[2] For the change in the offprint, see letter to J. D. Hooker, 17 March [1867] and n. 10.

[3] CD had asked Hooker to obtain Charles Victor Naudin's permission to cite his observation about

the fruit of a fan palm pollinated by a date palm (see letter to J. D. Hooker, 17 March [1867] and nn. 2 and 4). Hooker enclosed Naudin's letter with his letter of 23 March 1867.

[4] In his letter of 17 March [1867], CD had asked Hooker for the correct terminology to refer to the 'seeds or nuts' of the palm.

[5] Linnean Society.

[6] CD had previously clarified some aspects of his theory of pangenesis in a letter to Hooker of 4 April [1866] (*Correspondence* vol. 14). CD had sent the manuscript of a version of his theory for criticism to Thomas Henry Huxley, who had told him about similar theories propounded by the French authors Charles Bonnet and Georges Louis Leclerc, comte de Buffon (see *Correspondence* vol. 13, letter to T. H. Huxley, 12 July [1865] and n. 4).

[7] Hooker alludes to his caution in supporting CD's theory of transmutation of species, which was occasionally referred to as the 'derivation' of species (see also letter to J. D. Hooker, 29 January [1867] and n. 4). For a review of the literature on Hooker's acceptance of CD's theory, see Endersby 2002, pp. 310–22. Hooker planned to attend the Paris International Horticultural Exhibition in April (see letter from J. D. Hooker, 14 March 1867 and n. 6).

[8] Hooker refers to Benjamin Clarke (see letter to J. D. Hooker, 17 March [1867] and n. 18).

[9] Clarke, who had written a taxonomic work on plants (Clarke 1866), had begun to work on zoological taxonomy (see letter to J. D. Hooker, 17 March [1867] and n. 18).

[10] See letter to J. D. Hooker, 17 March [1867] and nn. 7 and 9.

[11] Hooker refers to John Scott. For more on Scott's acclimatisation experiments, see the letter from John Scott, 22 January 1867 and n. 4.

[12] See letter to J. D. Hooker, 17 March [1867] and n. 10.

[13] Hooker refers to *Clethra*; *C. arborea* is the species native to Madeira (Mabberley 1997).

[14] Hooker may refer to *Myrica*. He refers to *Melanoselinum*, an unidentified species of *Dracaena*, *Plocama*, *Visnea*, and *Bosea*. For endemic plant species in the Canary Islands, see Bramwell 1976.

[15] CD had passed on information from a note about birds from Europe being occasionally blown to the Azores in his letter to Hooker of 17 March [1867]. The note has not been found.

[16] Henry Barkly was the governor of Mauritius (*DNB*). Hooker's letter to him, dated 18 February 1867, is in the Hooker deposit–CUL (Ms Add 9537; see letter from J. D. Hooker, 4 February 1867 and n. 10, and letter to J. D. Hooker, 8 February [1867]).

[17] In his letter to J. D. Hooker, 17 March [1867], CD discussed Murray's review in *Gardeners' Chronicle* of J. D. Hooker 1866a (Murray 1867), and criticised Murray's treatment of Charles Lyell. Murray was assistant secretary to the Royal Horticultural Society from 1860 to 1865 (*Proceedings of the Royal Horticultural Society* 5 (1865): 1).

[18] In 'Insular floras', Hooker had commented, 'St. Helena, though 1000 miles nearer to South America than is any part of the African coast, contains scarcely any plants that are even characteristic of America' (J. D. Hooker 1866a, p. 50; see also Williamson 1984, p. 70). The Cape of Good Hope is the southernmost part of South Africa.

[19] The Serra dos Orgãos, or Organ Mountains, are a coastal range in central Rio de Janeiro state, Brazil (*Columbia gazetteer of the world*). The genus *Griselinia*, named by Johann Reinhold Forster and Georg Forster, is found in the littoral zone in New Zealand, Chile, Argentina, and Brazil. Hooker was working on the third part of the first volume of *Genera plantarum* (Bentham and Hooker 1862–83); this part appeared in 1867. In Bentham and Hooker 1862–83, 1: 951, Hooker listed *Decostea* as a synonym of *Griselinia*.

[20] Hooker evidently held that the presence of *Griselinia* in south-eastern Brazil could not be accounted for by its being the remnant of an earlier cold period. In 1866, CD, Hooker, Charles Lyell, and Charles James Fox Bunbury had discussed the character of the flora of the Organ mountains and the possibility of glacial action there (see *Correspondence* vol. 14).

[21] Hooker refers to page 9 of the offprint of J. D. Hooker 1866a; see above, nn. 1 and 2. Hooker had earlier informed CD that Asa Gray was not the author of the article 'Popularizing science' (Anon. 1867; see letter from J. D. Hooker, 14 March 1867 and n. 1).

To John Murray 20 March [1867][1]

Down. | Bromley. | Kent. S.E.

March 20th.

My dear Sir

The new title is fixed.[2] Many thanks about the clean sheets. I will not forget about the other german publisher.[3] With respect to the number of copies I am in that useful frame of mind that when you propose a small number I wish for more & now that you propose 1500 I am frightened so my opinion goes for nothing.[4]

My dear Sir | yours sincerely | Ch. Darwin

LS
John Murray Archive

[1] The year is established by the relationship between this letter and the letter from John Murray, 19 March [1867].

[2] CD refers to *Variation*; see letter from John Murray, 19 March [1867] and n. 2.

[3] Murray had agreed to send extra sets of proof-sheets for the German and Russian translations of *Variation*, and had reported that another German publisher was interested in bringing out a German translation of *Variation* (see letter from John Murray, 19 March [1867] and n. 5).

[4] See letter from John Murray, 19 March [1867] and n. 3.

To J. D. Hooker 21 March [1867]

Down

Mar 21

My dear Hooker

Many thanks for yr pleasant & very amusing letter.[1] You have been treated shamefully by Etty[2] & me, for now that I know the facts, the sentence seems to me quite clear. Nevertheless as we have both blundered it w[d] be well to modify the sentence something as follows. "whilst *on the other hand* the plants which are related to those of distant continents, but have no affinity with those of the mother continent, are often very common".[3] I forget whether you explain this circumstance but it seems to me very mysterious.

You have not sent Naudin's letter which I sh[d] be glad to see as according to the Gard. Chron. it is the fruit of the hybrid itself & not of the mother plant which is intermediate in character; & if so I do not care much about the case.[4] I must say one w[d] on Pangenesis viz. that it by no means implies that "every previous attribute of all the ancestors exists in an organism," but I fear my dear Pang. will appear bosh to all you sceptics.[5] You were quite right about Asa Gray, & as seems invariably the case I quite wrong.[6]

Do always remember that nothing in the world gives us so much pleasure as seeing you here whenever you can come.

I chuckle over what you say of And. Murray, but I must grapple with his book some day.[7] yours affectionately | Ch Darwin

Endorsement: '/67'
LS

DAR 94: 13f–g

1 See letter from J. D. Hooker, 20 March 1867.
2 Henrietta Emma Darwin.
3 For the wording of the passage in J. D. Hooker 1866a and CD's comments, see the letters to J. D. Hooker, 21 January [1867] and n. 1, and 17 March [1867].
4 See letter from J. D. Hooker, 14 March 1867 and nn. 5 and 7. Hooker enclosed Charles Victor Naudin's letter with his letter of 23 March 1867. CD reported Naudin's observation of fruit growing on a *Chamaerops humilis* pollinated by a date palm in *Variation* 1: 399. For the importance of the example to CD's theory of pangenesis, see the letter to J. D. Hooker, 17 March [1867] and n. 3.
5 See letter from J. D. Hooker, 20 March 1867. In his discussion of reversion in *Variation* 2: 398, CD stated that although there were a vast number of active and dormant gemmules, there must be a limit to their number, and that those long dormant would be more liable to perish.
6 See letter from J. D. Hooker, 20 March 1867 and n. 21.
7 For Hooker's comment on Andrew Murray, see the letter from J. D. Hooker, 20 March 1867. CD refers to Murray's book, *The geographical distribution of mammals* (Murray 1866).

From E. Schweizerbart'sche Verlagsbuchhandlung[1] 22 March 1867

E. Schweizerbart, Commissions- und Verlagsbuchhandlung. Buchdruckerei

Stuttgart

22 März 1867

Verehrtester Herr!

Ihr freundliches Schreiben vom 19^{en} veranlasst mich sofort Ihnen Antwort zu geben.[2] Empfangen Sie meinen aufrichtigen Dank für den neuen Beweis Ihrer Gewogenheit, die Sie für mich haben, indem Sie mir in uneigennützigster Weise die Aushänge Bogen Ihres grösseren Werkes, das Sie unter dem Titel: *The variation of animals* & *Plants under Domestication"* in diesem Augenblicke unter der Presse haben, Behufs einer deutschen Uebersetzung anbieten.[3]

Mit allem Vergnügen nehme ich diesen Antrag an und werde noch heute an Herrn Prof Carus[4] schreiben, um ihn zu ersuchen, die deutsche Ausgabe davon zu besorgen; ich zweifle nicht daran, dass er sehr gerne diesem Auftrage sich unterziehen werde und wird er desshalb sich wohl selbst an Sie wenden.

Wegen der Holzschnitte werde ich mich demnächst an den Herrn Murray selbst wenden und möchte nur bitten dass Sie demselben gelegentlich sagen er möchte mir von den Holzschnitten zu Ihrem Werke **gute** Stereotypen machen lassen.[5]

Der Druck der neuen Auftrage von Origin deutsch, schreitet rasch vor und kann die erste Lieferung (12 Bog) in einigen Tagen verschickt werden, wenn das Ganze fertig, erhalten Sie nach Wunsch die Exemplare.[6]

Für die liebenswürdige Uebersendung Ihrer Porträte habe ich Ihnen auch noch meinen Dank zu sagen; ich war nicht wenig überrascht statt des mir bekannten jüngeren Bildes in der anziehenden form einen älteren Herrn mit grauem Barte zu erhalten.[7]

Ich sandte Herrn Carus eine hier gemachte Copie des grossen Bildes, worauf dieser mir schrieb, dass ihm das Kleinere das er von Ihnen erhalten (vermüthlich das mit der Stühllehne) fast besser gefalle, es ist das *ganz Profil*.

Es will uns scheinen dass das frühere Bild manchen besser zusagen würde; ich erlaube mir daher die Frage, ob Sie vielleicht Werth darauf legen, dass ein neueres Bild gegeben werde, woraus die mit Ihrem Aeusseren vorgegangene Veränderung ersichtlich ist; *das jenige Bild, das Sie mir bezeichnen, werde ich geben*, bitte daher nur es mir zu bestimmen.[8]

Mit gröster Hochachtung und Verehrung | Ihr ganz ergebenster | E Schweizerbart

DAR 177: 75

CD ANNOTATION
Verso of last page: '*Any Photographs* | Murray— to make *good* stereotypes' *pencil*

[1] For a translation of this letter, see Appendix I. It was written by Christian Friedrich Schweizerbart, the head of E. Schweizerbart'sche Verlagsbuchhandlung, who used the signature E. Schweizerbart in business communications.

[2] See letter to E. Schweizerbart'sche Verlagsbuchhandlung, [19 March 1867].

[3] Schweizerbart refers to *Variation* (see letter to E. Schweizerbart'sche Verlagsbuchhandlung, [19 March 1867]).

[4] Julius Victor Carus.

[5] John Murray, CD's publisher, provided Schweizerbart with electrotypes of the illustrations for *Variation*; Murray's ledger records a payment of £10 from Schweizerbart in September 1867 (John Murray Archive).

[6] Carus had recently finished translating the fourth edition of *Origin* (Bronn and Carus trans. 1867; see letter to J. V. Carus, 17 February [1867]). No presentation list for Bronn and Carus trans. 1867 has been found. CD's copy is in the Darwin Library–Down.

[7] The image used in the first two German editions of *Origin* (Bronn trans. 1860 and Bronn trans. 1863) was a reproduction of a photograph taken around 1857 (DAR 225: 175; reproduced as the frontispiece to *Correspondence* vol. 8).

[8] CD had sent Carus a photograph of himself with his letter of 10 November 1866 (*Correspondence* vol. 14). The photograph Carus received was one made by CD's son William Erasmus Darwin in 1864, reproduced as the frontispiece to *Correspondence* vol. 12 (Carus photograph album, Smithsonian Institution Archives, Record Unit 7315). An engraving made from this photograph was used for the frontispiece to Bronn and Carus trans. 1867. CD may have sent Schweizerbart a later photograph, taken by Ernest Edwards in 1865 or 1866 (see *Correspondence* vol. 13, letter from E. A. Darwin to Emma Darwin, 25 [November 1865] and n. 3).

From Alexander Shaw to E. A. Darwin 22 March 1867

40, West Abbey Road, | (Kilburn) N.W.
22nd March 1867

Dear Mr Darwin

I send you by my Sister Lady Bell's wish, a copy of this Essay, written now some years ago. It is avowedly on the Nervous System: but it contains a view of the development of the Animal Kingdom, in illustration of Sir Charles Bell's Classification of the Nerves, which she seems to desire to commend to your notice.[1]

From observing the distinctions in the origins and course of the Nerves, in Man, Sir Charles Bell was led to divide all those of the Brain and Spinal Cord into two Classes—one common to the lowest and highest animals—the other gradually introduced in correspondence with the successive changes of structure that the Respiratory organs—confined, at first, to the simple office of oxygenating the blood— undergoes in adapting it, in the highest animals, to the perfectly distinct office of producing Vocal Sounds.[2]

The way in which I have traced, through the Animal Series, the rise and progress of the different sets of parts subject to the two classes of nerves, had not been thought of by Sir Charles Bell. But the train of observation has conducted me to the conclusion—that the supremacy of the organization of Man depends on the Structure of his Mouth, rather than on the perfection of his Hands: or, otherwise, that the nobility of his frame is derived from his possessing, in his Face, a perfect organ of Voice, Speech, and Expression, in correspondence with his great endowment, his Mind.[3]

I am | Very faithfully Y^{rs} | Alex. Shaw.

Erasmus Darwin Esqr.

DAR 177: 145

[1] Shaw refers to Marion Bell, the widow of Charles Bell. Alexander Shaw wrote several essays related to the work of his brother-in-law, Charles Bell; the book sent by Shaw may have been *An account of Sir Charles Bell's classification of the nervous system* (Shaw 1844). Shaw 1844 was also published as an appendix ('On the nervous system') to *The anatomy and philosophy of expression as connected with the fine arts* (Bell 1844, pp. 231–58). See also letter from E. A. Darwin, 22 [March 1867], n. 3.

[2] Shaw discussed Bell's designation of two classes of nerves, the 'original class' that served the oxygenation of the blood, and the 'respiratory class' in Bell 1844, pp. 256–8. For CD's opposition to Bell's views on expression, see the letter to A. R. Wallace, [12–17] March [1867] and n. 8.

[3] Bell's *The hand: its mechanism and vital endowments as evincing design* (Bell 1833) was written as the fourth Bridgewater Treatise; the Bridgewater Treatises were eight works on natural theology published between 1833 and 1836 under the terms of a bequest by Francis Henry Egerton, the eighth earl of Bridgewater (*EB* and Topham 1998). CD had read Bell 1833 in 1839 (see CD's reading notebooks, *Correspondence* vol. 4, Appendix IV). For Shaw's discussion of the significance of the mouth in human development from lower animals, see Shaw 1844 or Bell 1844, pp. 256–8.

From E. A. Darwin 22 [March 1867][1]

22

Dear Charles

Lady Bell[2] having mentioned to me that Shaw had written a paper from Bells papers I asked her to get me a copy, & in consequence Mr Shaw has sent this little book I dont know whether it will be any use or not[3]

Yours affec | ED

DAR 105: B56

[1] The month and year are established by the relationship between this letter and the letter from Alexander Shaw to E. A. Darwin, 22 March 1867.

[2] Marion Bell.

[3] See letter from Alexander Shaw to E. A. Darwin, 22 March 1867 and n. 2. In November 1866 Erasmus had given CD a copy of Bell 1844 that included Shaw's 1844 essay as an appendix (inscription in CD's annotated copy of Bell 1844, Darwin Library–CUL; see *Marginalia* 1: 47–9). Neither Shaw 1844, nor any other separate publication of Shaw's, has been found in the Darwin Archive–CUL.

To [?] 22 [March? 1867][1]

> *Down.* | *Bromley.* | *Kent. S.E.*
> Friday 22$^\mathrm{d}$

Dear Sir

I am greatly obliged for your most kind offer, which, however, I will decline, as I am a good deal out of health, & as I have passed in a work which is now printing the part about Dogs.[2]

You are, also, probably not aware that this place is six miles of hilly road from Bromley.[3]

With my very sincere thanks for your kind offer, I remain | Dear Sir | Yours faithfully & obliged | Charles Darwin

Fitzwilliam Museum, Cambridge

[1] The recipient has not been identified. The month is conjectured, and the year is established, by the printing of *Variation* having begun before March 1867, and the fact that the discussion of dogs appeared in the first chapter (see n. 2, below). In 1867, the 22nd fell on a Friday in February, March, and November.

[2] CD received the first proof-sheets of *Variation* on 1 March 1867 (CD's 'Journal' (Appendix II)); by 26 March he had had proofs of 180 pages (see letter to V. O. Kovalevsky, 26 March [1867]), which would have included the section on dogs (*Variation* 1: 15–43).

[3] The closest railway station to CD's home was in Bromley.

From J. D. Hooker 23 March 1867

> Kew
> March 23/67.

Dear Darwin

After I had closed my last I remembered that Bentham had to take Naudins letter to Linnean— I enclose it herewith—[1] I find figured a very oblong fruited variety of the *Chamerops*, which lessens the wonder slightly: but I find no description of this oblong variety hitherto. The odd thing of Naudins fruit is that it further tastes like Date. I cannot help suspecting that it will prove fruit of a hybrid, though as you say you have equally curious cases.[2] Date & Chamærops belong to same tribe of Palms, though one is Fan-leaved & the other Feather-leaved.[3]

I see you "smell a rat", in the matter of insular plants that are related to those of distant continent being common"— Yes, my beloved friend, let me make a clean breast of it.— *I only found it out after the Lecture was in print*! & by Jingo it has played the very devil with me ever since.[4] I have been waiting ever since to "think it out" & write to you about it coherently. I thought it best to *squeeze* it in, any how or where, rather than leave so curious a fact unnoticed. I am glad you are the only one who has twigged it & it's importance. I next must work out what proportion of the Trees of these Insular Flora are European, for these non European things are principally trees—that have not been "improved off the face of the Islands"— "more Darwinians" & it is rather a puzzle why trees have not got across, why none of the Cupuliferæ[5] are in the Islands. but the whole thing wants working out & it is a good lode to follow, that I shall keep attending to.

Thanks for the better wording of the sentence.[6]

I like your candid knuckle down about A Gray,[7] it "warms my stomach" as they say in the East (or West?) It is like "a drop o' Gin of a morning"—to me—

A capital French collector is going to the Morocco Mts. (if he can get there) which may throw some light on the Madeiran &c Flora.[8]

Ever yrs | J D Hooker

[Enclosure][9]

Paris,
9 Mars 1867—

Cher Monsieur Hooker,

Je ne saurais vous dire combien votre lettre m'a rendu heureux. Je vous en remercie, ainsi que du charmant portrait que vous avez eu la courtaisie d'y inseré et qui va tenir une place distingué dans ma collection de célébrités contemporaines. Je vous adresse aussi mes félicitations ainsi qu'à Mme Hooker, de l'accroisement de votre famille.[10]

J'ai beaucoup médité ce que vous me dites, dans votre lettre, de la structure de différentes cucurbitacées. Cette famille contient encore bien des points obscurs, mais qui s'éclaircirsait, je l'espère, quand on pourra les étudier sur des sujets vivantes.

Vous recevrez, en même temps que cette lettre, une petite boîte renfermant des graines de ce que j'appelle provisoirement *Microphœnix decipiens*.[11] Ces graines, ou plutot ces fruits ressemblent assez à des petites dattes, et peut-ëtre pas sans raison. Leur histoire vous causera, je crois, quelque étonnement.

Ces prétendues dattes soit tout simplement des fruits de *Chamærops humilis*, var. *arborescens*, qui ont été *fécondés par le pollen du Dattier* (Phœnix). L'expérience a été faite à Hyères, par M. Denis,[12] grand amateur d'horticulture, qui a quantité de Palmiers adultes dans son jardin. Il y a 2 ans, il a eu l'idée de secouer des régimes mâles de dattier, en fleurs, sur les régimes femelles, pareillement en fleurs, d'un chamærops, et, chose remarquable à noter, les fruits du chamærops ainsi fécondés sont devenus *deux fois plus gros* que dans les cas de fécondation normale, et sensiblement plus allongés, plus dactyliformes en un mot. J'ai eu soin de faire semer de ces graines à Hyères, l'année dernière (1866), et les jeunes plantes qui en sont sorties donnant des signes très manifestes d'Hybridité, dans la forme de leurs feuilles pennifrondes, et non plus palmatifrondes, comme celles des Chamærops leur mère.

J'ai recommandé à M. Denis de recommencer son expérience en 1866; il l'a fait, et le même phènomène qu'en 1865 a reparu. Au mois de décembre dernier, étant à Hyères, j'ai vu, sur l'arbre, les fruits fécondés par le Dattier— J'en ai pris quelques uns, et je suis heureux de vous en envoyer, pensant qu'il vous sera agréable d'observer un hybride d'un nouveau genre, c'est-à-dire obtenu entre deux plantes de genres très différents, et même classés dans deux sections distinctes par M. de Martius.—[13] Ce fait sera modifier la théorie de l'hybridation, telle qu'on su la figure aujourd'hui.

Vous m'intéressez beaucoup, Cher Monsieur Hooker, par ce que vous me dites des effets du froid sur quelques plantes. Je savais le Jubæa très rustique, pourtant je ne l'aurais pas cru capable d'un tour de force comme celui que vous m'apprenez. A Montpellier, il résiste à tout, même à des froids de -12° centigr., sous couverture. Je l'estime du même degré de rusticité que le *Cham. humilis*, qui vient parfaitement dans toute la région de l'olivier, mais qui ne peut pas en sortir impunément. Le *Cham. Fortunei*[14] est, à mon avis le plus rustique de tous; il est le seul qu'on réussisse à conserver *sans abris* à Bordeaux et dans les Landes.

A propos de plantes gelés, il y aurait, *Selon moi*, une grande réforme à faire dans la météorologie horticole anglaise, et vraiment, cher Monsieur, vous rendriez service à bien du monde en prenant ici l'initiative. Cette réforme consisterait à *Substituer le thermomètre centigrade* au thermom. de Fahrenheit. Ce dernier est réellement si mal conçu, que ceux même qui ont l'habitude de s'en servir ne réussissent pas toujours à se rendre compréhensibles. J'en trouve à tout instant la preuve dans le *Gardener's Chron.* quand je lis les: Effects of the late frosts—The late Winter &c.[15] Les horticulteurs n'ayant pas du formule commune pour indiquer le degré de froid dont ils ont à parler, il en résulte une manière arbitraire de s'exprimer, qui fait que très souvent ils ne se comprennent pas les uns les autres. Toutes ces obscurités disparaitraient si ces MM. voulaient adopter le Centigrade, dont le point de départ (le zèro) coïncide avec un phénomène naturel, la température de la glace fondante, la température zèro, qui n'est ni le chaud ni le froid. Rien de plus facile que de se faire comprendre, quand on dit: -6; -2; +3 &c. D'ailleurs ce thermomètre est tout aussi bien milligrade que Centigrade, et même dix milligrade si on veut.

Je serai vraiment heureux de vous revoir, Cher Monsieur;[16] nous causerons alors du jardin d'expérimentation qui est en projet. En attendant, agréez l'assurance de mes sentiments les plus sincères et les plus affectueux

votre dévoué confrère   Ch Naudin

DAR 102: 151–3; Royal Botanic Gardens, Kew, DC 143: 643

[1] See enclosure, and letter from J. D. Hooker, 20 March 1867 and n. 3. Hooker refers to Charles Victor Naudin, George Bentham, and the Linnean Society.

[2] CD had asked Hooker to clarify whether the intermediate fruit that Naudin had sent was from the mother plant (*Chamaerops humilis*) or the hybrid plant produced by the cross (see letter to J. D.

Hooker, 21 March [1867] and n. 4). CD was only interested in the first case, which he interpreted as demonstrating the direct action of the male plant on the female (see letter to J. D. Hooker, 17 March [1867] and n. 3).

3 Hooker evidently refers to the classification of *Phoenix* and *Chamaerops* given in Lindley 1853, p. 139, in which both genera belong to the suborder Corypheae of the order Palmaceae. Later, in Bentham and Hooker 1862–83, 3: 921 and 924, these genera were placed in the tribes Phoeniceae and Corypheae respectively. For the modern classification, see the letter from J. D. Hooker, 14 March 1867, n. 5.

4 CD had questioned a statement in Hooker's paper on insular floras (J. D. Hooker 1866a) regarding the commonness of plants having no affinity with those of the 'mother' continent (see letters to J. D. Hooker, 17 March [1867] and n. 10, and 21 March [1867], and letter from J. D. Hooker, 20 March 1867).

5 The Cupuliferae were an order that included the modern families Betulaceae and Fagaceae (see Bentham and Hooker 1862–83, 3: 402–3, and Mabberley 1997).

6 CD had suggested re-wording a sentence in Hooker's pamphlet on insular floras, but no new version was ever printed (see letter to J. D. Hooker, 21 March [1867] and n. 3).

7 In his letter of 21 March [1867], CD admitted he had been mistaken in supposing Asa Gray to be the author of the article 'Popularizing science' (Anon. 1867; see letter from J. D. Hooker, 14 March 1867 and n. 1).

8 The French collector has not been identified. Morocco's principal mountain ranges include the High Atlas, Middle Atlas, Anti Atlas, and Rif (*Columbia gazetteer of the world*).

9 For a translation of the enclosure, see Appendix I.

10 Naudin refers to Frances Harriet Hooker and Reginald Hawthorn Hooker.

11 The name was later published in *Revue Horticole* (1885): 513–14.

12 Alphonse Amaranthe Dugomier Denis. See also letter from J. D. Hooker, 14 March 1867 and n. 7.

13 Carl Friedrich Phillip von Martius.

14 *Chamaerops fortunei* is a synonym of *Trachycarpus fortunei*.

15 The winter of 1866 to 1867 was unusually cold. In an editorial in *Gardeners' Chronicle* for 26 January 1867 (p. 73), readers were asked to record 'the results of the recent severe weather'. In the issue for 2 March 1867 (pp. 210–11), several replies were published. Some writers referred to degrees below freezing, others to degrees below zero, and one gave both the temperature and its relation to freezing point ('10°, or 22° below freezing'; *ibid.*, p. 211).

16 Hooker was going to Paris to attend the International Horticultural Exhibition (see letter from J. D. Hooker, 14 March 1867 and n. 6).

From Hermann Müller 23 March 1867

Lippstadt,
march 23th 1867.

My dear Sir,

You have so kindly accepted of my inconsiderable first attempts at spreading your theories of infinite consequence among bryologists, that I feel compelled not only to thank you sincerely for your encouraging reply, but also to ask your advice with respect to my further activity in the above mentioned direction.[1]

After being employed, while adhering for many years to the premises of the Linnean School, in researches about insects, phenogams and mosses I was for the first time induced by your masterly work on the origin of species to a serious examination and thereby gradually to a torough rejection of my Linnean point of view.[2] Innumerable phenomena of animal and vegetable life which I had by the bye observed as a diligent collector without being able to get to an understanding of or even to an universal interest in them appeared to me as it were magically illumined in their causal connection under the light of your theorie so that nothing in my life had made upon me a deeper impression nor given me a more lasting satisfaction than the study of your masterwork.

Although I am fully aware of the slight importance of the small memoirs which I have published till now uper mosses for the whole theory of species, I nevertheless did not keep them back, as you said yourself: "Whoever is led to believe that species are mutable will do good service by conscientiously expressing his conviction; for only thus can the load of prejudice by which this subject is overwhelmed be removed".[3]

In future my endeavours will be directed towards contributing for my part to determine clearly by a close examination of several mosses the various gradations occuring between species and varieties as well as to state the gradations of advantageous properties in mosses—particularly of their hygroscopic organs—and to gain an understanding of their origin trough natural selection.

For next summer, however, I think I could not easily find a higher enjoyment and at the same time a better preparation for the researches intended than in repeating your charming observations on the fertilisation of Orchids by insects, as far as the Westfalian Flora offers any opportunities to it and in devoting my attention in general to the fertilisation of flowers by insects. As I have made for many years a thorough study of the indigenous plants and insects I do not think the task to difficult for me. Besides Sprengel "das entdeckte Geheimniss der Natur", your own work on Orchids and several epistolary communications of my brother's (Fritz Müller in Desterro) nothing on this subject has come to my notice.[4] If, therefore, any other observations should have been published in that line of late, I take the liberty of asking you to let me kindly know them.

A lovely plant, equally subject to fertilisation by insects is at present blowing in its pot before my window. As I think it possible that nothing has been published yet on that mechanism, I venture upon adding herewith for you both figure and description[5]

Hoping that you will accept of these lines with the same benevolence as you did with respect to my small publications on mosses, I remain, | dear Sir, | Yours most respectfully | H. Müller.

Description of the mechanism of the flowers in Lopezia miniata

The numerous little flowers distinguished by their vivid red colour are fixed upon long horizontally projecting pedicels in consequence of which their sepals and petals come to lie almost in a vertical plane and offer no landing place to an insect. The flowers consist of a globular ovarium 4 narrow sepals, 4 broader petals alternating with them, 2 stamina the one of which is so transformed that instead of the anthers it bears a petal like lamina which is in the middle line folded up upwards and a pistil.

Of the sepals two are lying in a vertical line, both the others are not rectangular to these, but form with the one above an angle of about 60 degrees. Of the petals the two superior narrower ones have their stalk shaped basis in a somewhat oblique direction upwards and are then suddenly bending in a right angle like a knee, so that their lamina stands vertical or bent a little backwards. Just on the knee which is mostly projecting in front there appears a small green shining spot secreting such a copious quantity of a liquid clear as whater and sweet as sugar that not only the knee itself is covered with a drop of nectar but also on the basis of the both inferior petals a still mor abundant quantity of the nectar is gathered. The inferior petals which ought to stand in the middle of the angle between the inferior and the lateral sepals are so strongly bent upwards at their basis, that their lamina attains a still higher position than the lateral petals At the same time the inferior margin of these petals bends so much upwards that by this double incurvation their basis is fitted for collecting the overflowing drops of the nectary.

Thus much all flowers are identic, yet with regard to the structure of the anthers and the pistill they are to be distinguished in two kinds which, it seems, appear irregularly mingled.[6] With the one kind of mostly female flowers the straightly projecting pistil forms the landing place for the insects when they come flying. The real anther which seems to contain only shrivelled up grains of pollen, lies bent back as far as to be pressed against the upper sepal, likewise the transformed anther lies bent back and is pressed against the lower sepal

With the other mostly male flowers the anthers of the upper stamens which contains a copious quantity of good pollengrains are firstly closed round by the clasped or folded up lamina of the transformed anther and open in that confinement by two longitudinal splits of the upper side which is so covered with pollengrains. A small pistill with a shrivelled up stigma is embraced by the hollowed bases of the two stamina. The folded up lamina of the transformed anther offer consequently here the only landing place for approaching insects. By pushing against them from above with a short hair or by the tread of an insect upon them (as I have sometimes observed it with Tipula,[7] the only insects now in my room) the two anthers hitherto united spring asunder with an elastic impulse so that in the lapse of about 1 second the transformed anther sweep through about 60 degrees and is pressed against the lower sepal, whilst in the same time the upper turns upwards through about 30 degrees and puts itself in a horizontal position or slightly bent upwards. The hair with which I tried was in every case laid on with pollen grains

Insects which by their flying on cause that elastic springing asunder, are certainly laid on with pollen at their legs or at their underside and when they land then on the pistill of a female flower, part of that pollen will indubitably remain sticking at the prominent stigma

[Enclosure]

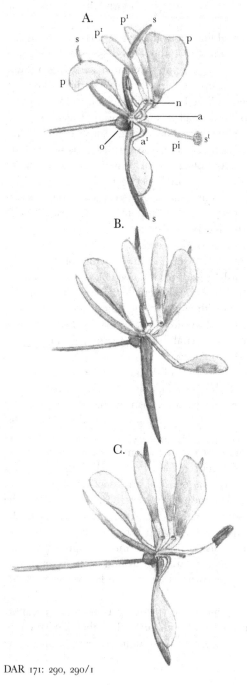

Lopezia miniata

A   Oblique view of perfect female flower, when fully expanded.

o. ovarium   s sepals
p. petals   p¹ upper petals
n nectary
a anther
a¹ transformed anther
pi pistill   sᵗ stigma

B.   Oblique view of a perfect male flower when first opened

The transformed lower anther encloses the real upper anther The small pistill with a shrivelled up stigma, is embraced by the hollowed bases of the two anthers

C   Oblique view of a perfect male flower showing the two anthers after springing asunder and pistill between them

DAR 171: 290, 290/1

CD ANNOTATIONS
2.5 Innumerable ... masterwork. 2.10] *scored red crayon*
10.12 At the ... nectary. 10.14] *scored red crayon*

[1] Müller had sent CD one paper on the distribution of Westphalian mosses, and another describing features of mosses that were supportive of CD's theory (H. Müller 1866a and 1866b). CD's copies of the papers are in the Darwin Pamphlet Collection–CUL; H. Müller 1866b is annotated. No letter from CD to Müller concerning the papers has been found.

[2] Müller refers to the system of classification advocated by the Swedish naturalist Carl von Linné. For the underlying principles of the Linnean system, see Larson 1971 and Stafleu 1971. Müller also refers to *Origin*.

[3] The quotation is from *Origin*, p. 482.

[4] The reference is to Christian Konrad Sprengel and his book, *Das entdeckte Geheimniss der Natur im Bau und in der Befruchtung der Blumen* (The secret of nature discovered in the structure and fertilisation of flowers; Sprengel 1793). Müller also refers to *Orchids*. In a letter to Hermann Müller of 11 February 1867, his brother Fritz suggested that he try several experiments with orchids related to topics that Fritz had been discussing in correspondence with CD (see Möller ed. 1915–21, 2: 111–16).

[5] Müller refers to *Lopezia miniata*. Friedrich Hildebrand had recently described the same mechanism in *L. coronata* (see Hildebrand 1866b, pp. 75–6, and Hildebrand 1867a, pp. 22–3).

[6] Hildebrand had recently described the flowers of *Lopezia* as dichogamous and had coined the term 'protandrisch' (protandrous), since the anthers matured before the stigma (see Hildebrand 1867a, pp. 17–18).

[7] *Tipula* is a genus of dipterous insects, commonly known as craneflies.

To B. D. Walsh 23 March [1867][1]

Down. | Bromley. | Kent. S.E.
March 23[d]

My dear Sir

I am a good deal overworked with the tedious labour of correcting the proofs of my new book, but I cannot resist sending you a few lines to thank you for pleasant & kind letter of Feb 25th—[2] I quite agree with what you say about the high value of disseminating scientific knowledge,—perhaps as useful as mere discovery—only *you* should remember that there are many who can do the one & only a few who can discover original truths.—[3] Thanks for Lecture of Agassiz.[4] *Lyell* does not believe a word about glacial action of any kind in lowlands of Brazil. If anyone had attempted to prove, even in a temperate region, former glacial action, & had owned he could find no striæ, he w[d]. have been laughed at.—[5]

I believe that the theory that glaciers are propelled by alternate freezing & thawing has been given up by all physicists.—[6]

I did not know about the workers' wasps breeding.—[7]

My new Book will interest you or any one else very little: it will not appear till November & will be entitled "The Variation of Animals & Plants under Domestication".—[8]

With many thanks for your letter—

Believe me, yours very sincerely | Ch. Darwin

Field Museum of Natural History, Chicago (Walsh 9)

[1] The year is established by the reference to *Variation* (see nn. 2 and 8, below).

[2] See letter from B. D. Walsh, [25 February 1867]. CD had received the first proof-sheets of *Variation* on 1 March 1867 (CD's 'Journal' (Appendix II)).

[3] Walsh had defended his decision to become editor of *Practical Entomologist* on the grounds that it would promote the pursuit of science in America (see letter from B. D. Walsh, [25 February 1867]).

[4] CD refers to Louis Agassiz and the lecture on glaciers in J. L. R. Agassiz 1867. See letter from B. D. Walsh, [25 February 1867] and n. 4.

[5] Agassiz advanced the hypothesis that the Amazon valley had been covered by a huge glacier in J. L. R. Agassiz 1866. CD had discussed Agassiz's claims regarding glacial action in Brazil with Charles Lyell and other correspondents (see *Correspondence* vol. 14).

[6] For a summary of contemporary views on the causes of glacier movement, see C. Lyell 1867–8, 1: 369–74. For CD's earlier discussion of the topic, see *Correspondence* vol. 6, letter to T. H. Huxley, 17 January [1857].

[7] See letter from B. D. Walsh, [25 February 1867] and n. 2.

[8] *Variation* was published in January 1868 (Freeman 1977).

To J. D. Hooker 24 [March 1867]

Down
24th

My dear H.

I return Naudin's letter, which I have been **very** glad to see: it is a clear case of the direct & immediate action of the pollen on the mother-plant, & rejoices my heart, for I look at such cases as unintelligible on any common view of the act of impregnation.—[1]

Etty & I admire your coolness in blowing us up for not understanding your inserted new case & difficulty, & we admire still more your candour in letting the rat out of the bag.[2] I do not envy your wriggles in making the case harmonise with other facts;—but it must be done & I do not doubt you will succeed.—

Müller counted above 13 sp. of orchids growing on one tree![3]

You said you did not know of violet on Peak of Teneriffe, I enclose scrap which if useless cannot signify.—[4]

Fritz Müller has just sent me seeds of a dimorphic Plumbago, about which I bothered you:[5] he has sent me seeds of a *climbing* Lobelia & of another kind which grows 10 ft high!!! I shall be curious to see this: I have also plants now growing of a dimorphic Cordia from S. Brazil: I never know whether any such things are worth offering you for Kew.—[6] For instance I have several (& had more, but threw away) species of Oxalis from C. of Good Hope.—[7] Here is a piece of good luck, a plant of Cyrtopodium of R. Brown,—the genus next to Catasetum—is coming into flower with me.—[8]

your affect | C. Darwin

Endorsement: 'March | /67'
DAR 185: 92

[1] See enclosure to letter from J. D. Hooker, 23 March 1867. CD had been uncertain about the usefulness of Charles Victor Naudin's case of intermediate fruit produced by a cross between a fan palm and a date palm for his theory of pangenesis (see letter from J. D. Hooker, 23 March 1867 and n. 2).

[2] Both CD and Henrietta Emma Darwin had queried a passage in Hooker's essay on insular floras on the commonness of plants having no affinity with those of the 'mother' continent (J. D. Hooker 1866a); see letters to J. D. Hooker, 17 March [1867] and 21 March [1867], and letters from J. D. Hooker, 20 March 1867 and 23 March 1867.

[3] See letter from Fritz Müller, 2 February 1867.

[4] The 'scrap' has not been found. See letter from J. D. Hooker, 20 January 1867 and n. 2.

[5] CD had asked Hooker for seed of *Plumbago* in his letter of 24 December [1866] (*Correspondence* vol. 14). Hooker sent two plants, but CD wanted only the seed (see letter to J. D. Hooker, 29 January [1867] and n. 2).

[6] CD later sent several plant specimens to Hooker for the collection at the Royal Botanic Gardens, Kew (see letter from J. D. Hooker, [14 September 1867] and n. 2).

[7] Roland Trimen had sent CD bulbs of *Oxalis* from the Cape of Good Hope, South Africa, in 1864 (see *Correspondence* vol. 12, letters to Roland Trimen, 13 May 1864 and 25 November 1864).

[8] The orchids *Cyrtopodium andersonii* and *C. punctatum* were on CD's list of hothouse plants (DAR 255: 8; see *Correspondence* vol. 11, Appendix VI). *Cyrtopodium* was named by Robert Brown (1773–1858). Both *Catasetum* and *Cyrtopodium* were placed in the subtribe Catasetidae in Lindley 1853, p. 182.

From Benjamin Clarke 25 March 1867

2 Mount Vernon | Hampstead N.W.
March 25/67.

Dear Sir,

Please accept my best thanks for the very liberal manner in which you have patronised my forthcoming work[1] & as you take so much interest, which I am glad to find, in my experiments on wheat I enclose four ears from the plants to which I referred in my note two of the red & two of the white as they had not been destroyed having kept them among my specimens.[2]

The experiment with the red wheat appears quite satisfactory;—the two ears grew on the same plant, one of them being cut off before or at the time of flowering & consequently has no seeds. I find it is more depauperated at the upper part than I described it, the flowers being abortive & it must have been precisely in the same condition *before it emerged from its sheath*. The other is 6-set[3] in the first spikelet the two largest grains having been taken out, & I feel sure the depauperated one would from the size of the flowers have been 4- or 5-set.

I think I distinctly recollect ears of white wheat more congested at the upper part than the two I have sent, but these being the largest the others were cut away,—the upper part of one of them has been broken off which has in some degree altered its appearance;—3 or 4 of the spikelets are 7-set, but the uppermost grains are small.

While writing the short account I sent you a suggestion occurred to me which I think will interest you. Supposing the male inflorescence of Zea Mays were all removed as soon as it could possibly be got at by pressing open the sheath, & the female flowers fertilized by pollen from another plant & this operation repeated year after year, there would it appears to me probably be produced in 5 or 6 years, a variety of Indian Corn with only a very few male flowers.[4] The consequence of this would most probably be, that the quantity of female flowers & consequently seed, would be increased by a third or nearly doubled, & if so the increased produce of the plant might, in such a country as the United States, amount perhaps to a million annually, or even more as the annual imports into this country are over three millions of quarters;—in 1863 the computed value of the imports was over £4.000.000. Why should not Indian Corn be as docile under the pruning knife as Red Wheat is?

Farmers would have no difficulty in trying the experiment. They could set apart a field for seed to sow again & remove the flowers (as early as possible) from every other plant in a row, year after year. The vigour of the male flowers might, by gardeners be increased, by breaking off all the smaller ones as soon as possible,—*of course not cutting off the spikes of females*. Doubtless you have a Green House & your gardener could very well attend to the directions given. In looking at Indian Corn it used to occur to me that it was misfortune that it produced so many male flowers at the expense it seemed to me of the quantity of seed produced.

I beg to enclose a note on Hemp[5] which I believe will interest you. Being the son of an agriculturist I am naturally interested in Turnips & believe I have discovered a remedy for the Turnip Fly,[6] which consists in sowing three kinds of seed, one naked as usual, one covered with thick gum & one covered with resin varnish;—the seeds are thrown into flour so that they do not stick together while drying;—the covered seeds do not come up till after a heavy rain & then would grow so fast that they would escape the Fly, so I am assured by a farmer.

I did not see the nectarine on a peach tree,—it was advertised as exhibited at one of the shows of the R.H.S. at Kensington.[7]

The Elm referred to is a tree growing on the left hand side on a green just before you come to the Heath, being 20 or 30 yards I should think from the road side, & stands on the south side of a large green mound surrounded with iron railings, which is the Water Company's reservoir very recently established.[8] The arm being out of reach I could not procure any leaves. There is no appearance of disease & it may be 3 or 4 inches in diameter. As a small tree it would be more oramental than the common Elms.

If you have an opportunity of writing a few lines of favourable review for my botanical work which you have received, its sale may be accelerated.[9] I relinquished the practice of the medical profession some time since in consequence of the injuries my health had sustained.[10] I have the pleasure of seeing your name on the list of contributors to the Royal Society Relief Fund for £100,[11] a practical proof of your readiness to assist literary men who have met with reverses.

I remain dear Sir, | Yours very truly | Benjn. Clarke.

Charles Darwin Esq^e.

P.S. Should you wish to send any of my communications to any Journal accompanied with some comments of your own you are perfectly at liberty to do so.[12]

The ears of corn are sent in hope you may make some use of them & therefore need not be returned,—I have taken out all the grains I want. They were grown in rather large flowerpots in rich earth & daily care taken with regard to moisture. I suppose that if the male flowers were removed from one plant of Indian corn & the females from another so as to make it diœcious either a gigantic or more productive variety would result.

[Enclosure]

On varieties of Cannabis sativa produced by Pruning.—
Suggestions for the improvement of Turnips

A few years since I grew a female plant of Cannabis in a rather large flower pot with rich mould, taking daily care to keep it in good condition as to moisture, for the purpose of ascertaining whether it would produce embryos by parthogenesis.[13] As well as I can recollect it produced seeds 1 or 2 in most of the axils of each pair of leaves but certainly none on the branches. I then cut about the upper third or half of the main stem off to see what the effect would be as to the production of seed. The upper branches then bore several seeds, but this was not satisfactory as regards parthogenesis because a male Hemp plant about that time began to flower.

However notwithstanding the uncertainty about the seeds produced after the male plant flowered, I sowed the seed & in the third generation somewhere about one third of the seeds, perhaps 6 or 7 of them, produced tricotyledonous plants[14] the pruning having been continued. Selecting one of these to raise it turned out to produce regularly 3 leaves in a whorl.

I also reared from the same parcel of seed two other female plants but whether they had two or three cotyledons I do not recollect but I well recollect that they had leaves opposite in pairs as Hemp usually grows. Both these plants proved to be monœcious, one of them producing two well developed male flowers, having anthers large as usual, in the axils of the first pair of leaves, & the other, one male flower also in the axil of one of the first leaves.[15] These male flowers appeared to me more surprising than the tricotyledonous embryos.

One inference deduced from this this experiment is that the production of seed by the aid of pruning would have the same effects on the generality of plants, as the cultivation of Wild Wheat (Ægilops) & Wild Oats has on those cereals. I have long entertained this view & often wished I knew a Turnip seed grower (a class of men who are known to give every attention to the selection of their plants) that I might explain to him my proposition for improving Turnips, expecting that bulbs two or three times the usual size could soon be produced, differing from ordinary Turnips as the Shaddock does from the Orange.

My proposition is, that as soon as all the flowers become sufficiently distinct so that it may be guessed which will be the largest, all the smaller ones up to 9 out of 10 or 19 out of 20 should be broken off, & the remaining 10th or 20th be allowed to flower & fruit. The experiment might be carried still further by cutting away the stamens of the flowers on the lower branches & the ovaries of those on the upper part of the inflorescence as soon as it was practicable to do it. The pollen of the upper flowers would then fall on the ovaries of those beneath them, the juices of which would not be exhausted, or rather lessened, by the production of anthers, at least not to the usual extent.

If Beet continues to be cultivated in France for the manufacture of sugar such an improvement would be still more desirable.[16]

DAR 161: 157, 159

[1] Clarke refers to his proposed new work, *On systematic botany and zoology* (Clarke 1870; see letter from Benjamin Clarke, 12 March 1867 and n. 2).

[2] For a description of Clarke's experiment, see the letter from Benjamin Clarke, 12 March 1867.

[3] Clarke refers to the number of grains in the spikelet.

[4] *Zea mays* is monoecious, with male flowers located at the top of the stalk, while female flowers are grouped together lower down on an ear or cob (Mangelsdorf 1974, pp. 4–5). In *Variation* 1: 320–3, CD discussed variability in maize, noting the occurrence both of male flowers among female and more rarely of female flowers among male, as well as reports of hermaphrodite flowers.

[5] *Cannabis sativa* (see enclosure).

[6] The insect commonly referred to as turnip fly was *Haltica nemorum* (now *Phyllotreta nemorum*), a beetle of the family Chrysomelidae that feeds on the young leaves of the turnip (*EB* 9th ed.).

[7] In the enclosure to his letter of 12 March 1867, Clarke had mentioned a nectarine produced on a peach tree. CD evidently asked for the source of this report in his reply to Clarke's letter, which has not been found. The report of the tree exhibited at a Royal Horticultural Society show in Kensington has not been found.

[8] In the enclosure to his letter of 12 March 1867, Clarke described an elm tree, with anomalous leaves on one branch, that he saw in Hampstead Heath, north-west London. The West Middlesex Water Company was responsible for the water supply in this area (*Post Office London directory* 1867).

[9] CD had recently bought a copy of Clarke 1866 (see letter from Benjamin Clarke, 12 March 1867, n. 1). No review of the work by CD was published.

[10] Although he trained as a physician at Winchester, Clarke practised for only a few months (*Journal of Botany* 28 (1890): 84).

[11] The Royal Society Scientific Relief Fund was set up in 1859; funds were raised by subscription (*Record of the Royal Society of London*).

[12] CD did not send the notices Clarke enclosed in this letter or the letter of 12 March 1867 to any journal.

[13] *Cannabis sativa* is a dioecious plant of the family Cannabaceae (Mabberley 1997). Parthenogenesis has not been reported in the species, which is wind-pollinated (Bócsa and Karus 1998).

[14] *Cannabis sativa* is normally dicotyledonous.

[15] Monoecious varieties of *Cannabis sativa* have the same flower structure as the normal dioecious plants but the location of the inflorescences is different (Bócsa and Karus 1998, p. 35).

[16] For the development of the sugar beet (*Beta vulgaris*) in France, see Winner 1993, pp. 15–18.

From John Lubbock 25 March 1867

25 Mar. 67

My dear M[r]. Darwin

I do not know whether you have read M[c]Lennans "Primitive marriage".[1]

He refers the curious practise of Exogamy to the prevalence among certain tribes of female infanticide.

I should have thought that the objection to marriage between near relations might have had much to do with it.[2]

Can you tell me whether we have any evidence that any animals have an instinctive repugnance to breeding in sin.

Believe me always | Very sincerely | Yours | John Lubbock

C Darwin Esq

DAR 170: 56

[1] The reference is to John Ferguson McLennan's *Primitive marriage: an inquiry into the origin of the form of capture in marriage ceremonies* (McLennan 1865).

[2] The terms exogamy (the custom of marrying outside a clan or group) and endogamy (the custom of marrying within a clan or group) were coined by McLennan (*OED*; see McLennan 1865, pp. 48–9, 53). McLennan surmised that female infanticide resulted in a shortage of women in a tribe, and the consequent need to capture women for marriage from other groups (McLennan 1865, pp. 138–41). For Lubbock's later discussion of exogamy and endogamy, see Lubbock 1870, pp. 92–113.

To Fritz Müller 25 March [1867][1]

Down. | Bromley. | Kent. S.E.
March 25th.

My dear Sir

Your last letter received two days ago contained a multitude of curious facts about orchids.[2] The case of the little orchid of which the long pedicelli performs such curious movements is very interesting. I once suspected that the pollen when partially dry acted best, but was quite unable to prove it; I daresay the final cause is what you suggest.[3] Hildebrand of Bonn has just published a book which I have not read but which seems to me very good on the fertilisation of plants. I think it would perhaps interest you so I will get an additional copy & send it.[4] Many thanks for your answer about the Maxillaria.[5] The seeds of the Plumbago which you have sent are a treasure to me.[6] I shall also be curious to see the two lobelias.[7] The little bulbs of your semi-dioceus oxalis are growing well, except those of the flowers with semi foliaceous stamens & none of these grew; the others will not flower this year. Your Cordias have germinated well, but as I suppose they are trees I fear I never shall see them in flower.[8] I have dispatched two copies of yr. paper on Climbing Plants as directed & will with pleasure send any others.[9] Your brother (as I now know him to be) sent me some time ago some papers on mosses in which he has worked out with great care the variability of certain forms.[10] I am working very hard at correcting proofs of my new book & the corrections are very heavy.[11] I fear that neither you nor anyone will care much about this book which has cost me much more labour, I suspect, than it is worth.

I am extremely glad to hear that you like Häckel's book;[12] it is so long & the german rather difficult that I have been able to read only small portions. He seems to me a singularly clear thinker with great powers of methodical arrangement, but

I have not met with much that seems actually new. I have, however, no right to judge. I liked the man so much that I do hope his book will be very successful.[13]

My dear Sir, believe me | yours very sincerely | Charles Darwin

P.S. What can be the cause of flowers like y^r *Echites* catching insects?[14]

LS
British Library (Loan 10: 14)

[1] The year is established by the relationship between this letter and the letter from Fritz Müller, 2 February 1867.

[2] See letter from Fritz Müller, 2 February 1867.

[3] Müller had described the movement of the pedicel (now called stipe) of *Ornithocephalus* after the removal of the pollinium. He suggested that the initial movement was elastic, but that its final position was caused by hygroscopic movement, noting that it resumed its former position if put in water (see letter from Fritz Müller, 2 February 1867 and n. 11).

[4] Friedrich Hildebrand had sent CD two copies of his book on plant sexuality, Hildebrand 1867a, one of which CD sent to Daniel Oliver (see letter to Friedrich Hildebrand, 20 March [1867] and nn. 2 and 4).

[5] Müller had given CD an estimate of the number of seed capsules produced by *Maxillaria tetragona* (letter from Fritz Müller, 2 February 1867 and nn. 2 and 4).

[6] Since Müller mentioned finding a dimorphic *Plumbago* in a letter of [2 November 1866] (*Correspondence* vol. 14), CD had been trying to obtain seeds of this genus from Joseph Dalton Hooker for his crossing experiments with dimorphic plants (see *Correspondence* vol. 14, letter to J. D. Hooker, 24 December [1866], and this volume, letter to J. D. Hooker, 29 January [1867]).

[7] In a letter to J. D. Hooker of 24 [March 1867], CD reported that Müller had sent seeds of a climbing *Lobelia*, and of another kind which grew to ten feet in height.

[8] CD had received bulbils of *Oxalis* sent by Müller in October 1866 and seeds of *Cordia* in early February 1867 (see *Correspondence* vol. 14, letter to Fritz Müller, [before 10 December 1866] and n. 2, and this volume, letter to Fritz Müller, 7 February [1867] and n. 4). Many species of *Cordia*, a member of the family Boraginaceae, are trees (Mabberley 1997).

[9] The reference is to Müller's paper on climbing plants, which had been sent to the Linnean Society by CD (F. Müller 1865; see also *Correspondence* vol. 13, letter from Fritz Müller, [12 and 31 August, and 10 October 1865]). Müller had probably given CD instructions for sending copies of the paper in his letter of 2 February 1867, of which only an incomplete copy is extant. In a letter to his brother, Hermann Müller, of 11 February 1867, Müller wrote that CD would send Hermann a copy of the paper (Möller ed. 1915–21, 2: 111).

[10] Müller had mentioned his brother's work on mosses in 1866 (see *Correspondence* vol. 14, letter from Fritz Müller, 2 August 1866). For the papers sent by Hermann Müller to CD, see the letter from Hermann Müller, 23 March 1867, n. 1.

[11] CD had begun correcting proof-sheets of *Variation* at the beginning of March (see letter to W. B. Tegetmeier, 5 March [1867] and n. 2).

[12] Müller's comments on Ernst Haeckel's *Generelle Morphologie* (Haeckel 1866) were evidently in a now missing part of his letter of 2 February 1867. In his letter to Hermann Müller of 11 February 1867, Müller wrote that he thought Haeckel's book was very important, although he disagreed with many details and disliked Haeckel's introduction of a large number of Greek neologisms (Möller ed. 1915–21, 2: 115–16).

[13] Haeckel had visited CD in October 1866 (*Correspondence* vol. 14, letter from Ernst Haeckel, 19 October 1866).

[14] *Echites* is a genus of the family Apocynaceae. Müller's comments on *Echites* were evidently in a now missing part of his letter of 2 February 1867. In his letter to Hermann Müller of 11 February 1867, Müller wrote that the flower of an *Echites* growing in his garden had trapped a small bee by its

proboscis, which was inserted between two anthers, and that the bee subsequently died (Möller ed. 1915–21, 2: 115). CD had earlier suggested that Müller should try to observe flowers that caught insects by their proboses, and mentioned that this occurred in some of the Apocyneae (a subfamily of Apocynaceae; see *Correspondence* vol. 14, letter to Fritz Müller, [9 and] 15 April [1866] and n. 9).

From Asa Gray 26 March 1867

Cambridge [Massachusetts]
26th March, 1867

Dear Darwin

This is to acknowledge yours of Feb, 28—[1]

You see I have *printed* your queries—privately—50 copies—as the best way of *putting* them where useful answers may be expected.[2] Most of them will go into the hands of agents of the Freedmen's bureau, etc—[3] Others to persons I or Wyman may know & rely on I wish I had them sooner. My crony Wyman has been 2 months in Florida[4]—but will be home again before I could send to him

I *did not write* the article in the *Nation* on Popular Lecturing—tho, it contains so many things I have *said* over and over—that it startled me.[5] Then it *hits so many nails square on the head* that I should think it could be written only in Cambridge or hereabouts—

It is generally supposed to be written by a person in New York but I suspect a person near by here— —only suspect.

There is a short capital, quiet hit at Agassiz in a later number of the Nation[6]— which Hooker may have sent you.

Yes Magnolia-seeds hang out a-while, in autumn—finally stretch & break the threads of spiral-vessels. Whether birds eat them I dont know. They look enticing & have a pulpy coat— are bitter & spicy[7]

In haste ever yours | A Gray

Shall I send you more of these circulars?

I shall send to Indian-people too.[8]

DAR 165: 157

CD ANNOTATIONS[9]
Top of letter: 'Clean sheets not worth sending' *pencil*
Between signature and postscript: '(German & Russian Edition)' *ink, del ink*; 'Experiments— — [*illeg*] published— exotic plants *I suspect* | Potato experiments— Proof sheets— Trübners | oxalis like Lythrum.' *ink*

[1] CD's letter has not been found.
[2] CD evidently enclosed a handwritten copy of his queries about expression with his letter to Gray of 28 February 1867; neither the questionnaire nor the letter have been found. A version of the questionnaire was published in 1868 in the *Annual report of the Smithsonian Institution . . . for the year 1867*, p. 324, under the title 'Queries about expression for anthropological inquiry'; this version, which contains some American spellings, may be an edited version of the queries printed by Gray (see Freeman and Gautry 1975, pp. 259–60). This questionnaire is also published in *Collected papers* 2: 136–7.

[3] The United States Bureau of Refugees, Freedmen, and Abandoned Lands, known as the Freedmen's Bureau, was established by the US Congress in 1865 to provide medical and educational aid to the freed African-Americans following the Civil War (*EB*).

[4] Jeffries Wyman was an ethnologist and comparative anatomist. Wyman travelled extensively for the sake of his health, and in order to expand his collections; he travelled to Florida eight times between 1852 and 1874 (*ANB*).

[5] Joseph Dalton Hooker had forwarded the *Nation* article, 'Popularizing science' (Anon. 1867), to CD after it had been sent to him by Gray; CD praised the article and guessed that Gray had written it (see letter from J. D. Hooker, 4 February 1867, and letter to J. D. Hooker, 8 February [1867]).

[6] Gray probably refers to a paragraph in the *Nation*, 7 March 1867, p. 182, reporting on the last of Louis Agassiz's six lectures on South America given at the Cooper Institute in New York City in February 1867 (J. L. R. Agassiz 1867). His lectures and publications on the subject purported to refute CD's theory. See also letter from B. D. Walsh, [25 February 1867] and n. 4. The *Nation* noted that Agassiz 'treated his opponents like a gentleman, and would have treated them like a philosopher had he stated their position clearly, which he did not'.

[7] CD was interested in whether the bright colour of some seeds attracted birds and so aided in seed dispersal (see *Origin* 4th ed., pp. 430–2). Fritz Müller had been sending information on this point; see *Correspondence* vol. 14, letters from Fritz Müller, 2 August 1866, and 1 and 3 October 1866. In his letter of 1 December 1866 (*Correspondence* vol. 14), Müller had told CD about the brightly-coloured seeds of *Talauma*, a genus closely related to *Magnolia*. See also letter from Thomas Belt, 12 January 1867 and n. 5.

[8] See n. 2, above. Gray sent the queries on expression to Joseph Trimble Rothrock, who had recently been among native people of British Columbia; he also probably sent the queries to Spencer Fullerton Baird (see letter from J. T. Rothrock to Asa Gray, 31 March 1867 and n. 1, and letter from George Gibbs, 31 March 1867).

[9] CD's annotations are notes for his letter to Asa Gray, 15 April [1867].

From J. D. Hooker 26 [and 27] March 1867[1]

Royal Gardens Kew
March | 26/67

Dear Darwin

I am quite crazy for seeds of plants, & you never can go wrong in sending me *rejectamenta* of seeds plants & orchids— the drain on us is terrific & we loose many things through change of hands & various causes affecting *too big* an establishment. The climbing Lobelia* must be a fine thing— Pray let me have a plant or two if it germinates.— The oxalis seeds will be most acceptable.[2]

Thanks for the *Viola* case, but what I understood you to say was, that a *boreal* violet was found in the Peak. I knew of this one.— it is a Mediterranean form I believe.[3]

We are very anxious about our Baby,[4] which after 3 months thriving was seized with Convulsions last Sunday, which continue every 2 to 3 hours though not so violent as at first. It has no head symptoms, no fever or sickness, & it is attributed to too much vegetable food. It takes it's food well, & sleeps well meanwhiles.

I do not go to Paris till about 14th. April.[5] I am glad to say.

Your Willy[6] was here on Sunday looking very well indeed & very agreeable.

Ever yrs aff | J D Hooker.

*No doubt a Siphocampylus (convolvulaceus?) of which there are various scandent species[7]

Wednesday no improvement in the Baby.

DAR 102: 154–5

[1] The date is established by the written date and the postscripts: in 1867, 27 March was a Wednesday.
[2] In his letter to Hooker of 24 [March 1867], CD had reported receiving from Fritz Müller seeds of several plants, including a climbing *Lobelia* and another kind that grew to ten feet in height (see below, n. 7); he had wondered whether it was worth his offering these and other seeds and plants to the Royal Botanic Gardens, Kew.
[3] See letter to J. D. Hooker, 24 [March 1867] and n. 4.
[4] Hooker's youngest child was Reginald Hawthorn Hooker, who had been born on 12 January 1867.
[5] Hooker previously told CD he was going to Paris at the end of March (see letter from J. D. Hooker, 14 March 1867 and n. 6).
[6] William Erasmus Darwin.
[7] *Siphocampylus*, like *Lobelia*, belongs to the subfamily Lobelioideae of the family Campanulaceae (Mabberley 1997).

To V. O. Kovalevsky 26 March [1867][1]

Down. | Bromley. | Kent. S.E.
March 26th.

Dear Sir

I have much pleasure in answering your questions as far as I can. My work will be full sized octavo in two vols, each containing, I conjecture, about 500 pages; but the second vol. will be thicker than the first & I hope more interesting.[2]

There are 42 wood cuts. If you determine on a translation you had better negociate directly with Mr. Murray of Albemarle St London for stereotypes.[3] I cannot ask him to give you them gratis as he publishes at his own risk & pays me. I have already the first proofs of 180 pages & I suppose the whole work will be completed in from 4 to 6 months.[4] The first 6 clean sheets will probably be printed off in 2 or 3 weeks & I will send them to you by post.

I am much obliged for your brother's memoirs which I have no doubt will interest me:[5] his assistance will be of great importance to you in the translation.

With respect to an introduction to your edition I really do not know what I could say.[6] You might yourself state that you received from me early sheets, & that the translation had my concurrence.

When you have occasion to write again, be so kind as to inform me whether there has been a russian translation of my "Origin of Species";[7] if not I still think that this would be a better book for you to undertake. I am aware that a russian translation of Dr. Rolle abstract of my views has appeared.[8]

Dear Sir | yours faithfully | Ch. Darwin

LS(A)
Institut Mittag Leffler

[1] The year is established by the relationship between this letter and the letter from V. O. Kovalevsky, 15 March 1867.

[2] Kovalevsky was planning to translate *Variation* into Russian (see letter from V. O. Kovalevsky, 15 March 1867 and n. 2). The two volumes of *Variation* had 411 and 486 pages, respectively.

[3] CD's publisher, John Murray, recorded the sale of electrotypes of the woodcuts for *Variation* to Kovalevsky for £10 in September 1867 (John Murray Archive).

[4] CD had begun correcting proof-sheets of *Variation* at the beginning of March (see letter to W. B. Tegetmeier, 5 March [1867] and n. 2). *Variation* was published in January 1868 (Freeman 1977).

[5] Kovalevsky had sent several papers by his brother, Alexander Onufrievich Kovalevsky (see letter from V. O. Kovalevsky, 15 March 1867 and n. 6).

[6] In his letter of 15 March 1867, Kovalevsky had asked CD to write a short introduction to the Russian edition of *Variation*.

[7] The first Russian translation of *Origin* was Rachinskii 1864.

[8] CD refers to Friedrich Rolle and Rolle 1863; Russian translations of the book appeared in 1864 and 1865 (Vladimirskii trans. 1864 and Usov trans. 1865).

To John Lubbock   26 March [1867]

Down

March 26th

My dear Lubbock

I have not read or heard of the book on "Primitive Marriage" nor do I know what "exogamy" means; so that I am not a little in the dark.—[1]

I do not think any evidence has been published of an instinctive repugnance to close intermarriage with animals. I have received some private accounts of such feeling with domestic animals, but they were so few, that I have not thought it prudent to give them.[2]

Indirectly the end must, I think, be largely gained by the wandering of the young males, & their expulsion from the herd in social animals, by the old males.[3] I heartily wish I could have given you any better information. In my new book I have a chapter on Interbreeding & give all the evidence which I have.[4]

It is a long time since we have met & if Mahomet does not come to the mountain, the mountain must come some Sunday to Mahomet.[5]

Yours affect^{ly} | C. Darwin

Endorsement: '/67'
DAR 263: 65 (English Heritage 8820 6509)

[1] See letter from John Lubbock, 25 March 1867 and nn. 1 and 2. CD later read John Ferguson McLennan's book (McLennan 1865), an annotated copy of which is in the Darwin Library–CUL (see *Marginalia* 1: 559–61). CD referred to his ideas several times in *Descent*; for CD's discussion of exogamy related to sexual selection, including the views of McLennan and Lubbock, see especially *Descent* 2: 358–65. On contemporary criticism of McLennan's ideas, see the introduction in McLennan 1970, pp. xxxviii–xl. See also Jann 1996.

[2] See letter from John Lubbock, 25 March 1867. CD refers to his book on domestic animals and plants, *Variation*. In *Descent* 2: 361–2, CD noted the breeding habits of several different primates.

[3] CD included this idea in *Descent* 2: 362–3.

[4] CD refers to chapter 17 in *Variation* (*Variation* 2: 114–44).

[5] CD alters the proverb attributed to Francis Bacon: 'If the mountain will not come to Mohamet, Mohamet must go to the mountain' (Speake ed. 2003, p. 210). The most recent recorded meeting of Lubbock and CD was on 2 October 1866 (see *Correspondence* vol. 14, letter to J. D. Hooker, 2 October [1866]).

From Andrew Smith 26 March 1867

16 Alexander Square
26 March 1867

My Dear Darwin

Until the time you called, when I was confined to bed, I was under the sorrowful belief that I was never more either to see or hear from you.[1] Under those circumstances you will easily imagine how much pleased I was to see your hand writing and find that you had me still in remembrance[2] Had I been in the enjoyment of my ordinary health when your letter reached me you would soon have had something in return but seeing my condition was then and has been ever since very different you will I know ascribe my delay to what has really caused it and not to indifference.

Now that we have a prospect of some improvement in the weather I hope to soon get out of doors and gain what I very much want—strength.[3] I see you are still as active and enthusiastic as ever and you may rest assured that if I can contribute any thing to you you will have it the moment I can venture a visit to my library which I am told is very cold as there has scarcely been a fire in it since December last. I have no doubt but that I will find among my notes something in regard to the use, which male animals, in South Africa, make of their Horns.[4] I recollect although I cannot give the full particulars at present that on one occasion we found the skeletons of two Gnu's *Catoblepas* Gnu joined together by the manner in which they had entangled their horns no doubt when *fighting*— They fight desperately especially about the time they are courting the females.[5] They are doubtless weapons both of offence & defence as I trust I will, when I get a little stronger, be able to prove to you from my notes which I dare not yet attack I strongly suspect will not find it very easy to give you any thing satisfactory as to what kind of women savage men prefer—[6] so far as the Hottentot is concerned I can with certainty say he regards a woman with huge *posteriors* as first rate[7] and he some time ago used to value highly such of the females as had very lengthened Nympæ but now he rather views these ugly developements as undesirable if not as deformities I have been told of some who had them so elongated as that they were able during sexual intercourse to encircle the mans loins and fix him by them until the appetite of both was thoroughly satisfied. I have never seen what would admit of any thing like that being effected still I have seen them pretty long.[8] Now you must reconcile this speciality in the Hottentot's formation and let me know what brought it into existence. You must not say it is artificial seeing it comes without their using weights to bring it into existence and as a proof that it is not a formation of their manifacture it still begins to appear at puberty though there is nothing they would not now do to prevent it. In regard to the large posteriors I may mention that I once came in communication with a woman more than ordinarily gifted in that way and it was all but an impossibility for her to get on her feet when she was sitting unless where she could avail herself of some slope of the ground— Where such was open to her she had to work herself round till her back was directed up the slope and if it was considerable she rose with tolerable facility.— This Lady was esteemed a beauty and was the mother of several children—[9]

I find I have just done enough to let me know I am an invalid you shall however, please God, have more when I am stronger

Yours most faithfully | Andrew Smith—

DAR 85: A103–5

CD ANNOTATIONS
1.1 Until … Horns. 2.8] *crossed blue crayon*
2.8 I recollect … deformities 2.19] *enclosed in square brackets blue crayon*
2.8 I recollect … attack 2.13] *crossed pencil*
2.15 so far … Nympæ 2.18] *scored blue crayon*
2.18 Nympæ] 'h' *interl after* 'p' *pencil*
2.19 been told … prevent it. 2.27] *crossed blue crayon*
2.27 In regard] *after opening square bracket blue crayon*
2.32 This lady … children— 2.33] *scored blue crayon*

[1] CD evidently called on Smith during his most recent visit to London from 13 to 21 February 1867 (see 'Journal' (Appendix II)).
[2] The letter to Smith has not been found.
[3] Smith had resigned from his post as director-general of the army and ordnance medical departments in 1858 due to poor health (*ODNB*).
[4] In his research on sexual selection, CD was curious about the horns of male mammals and whether they were used in fighting for females.
[5] In a discussion of the 'law of battle' in *Descent* 2: 240 n. 3, CD mentioned Smith's observation of the gnu skeletons. *Catoblepas gnu* (then the common gnu: see A. Smith 1849, Mammals, text accompanying plate 38), or *Connochaetes gnou*, is now known as the white-tailed gnu or black wildebeest; see Nowak 1999, 2: 1184. See also J. E. Gray 1852, pp. 119–22, for a contemporary description of gnu species. Smith had been stationed in South Africa from 1820 to 1836 as an army surgeon, and was an authority on South African zoology.
[6] CD had also discussed sexual selection as a means of forming the human races with Alfred Russel Wallace (see, for example, letter to A. R. Wallace, 26 February [1867] and n. 5, and letter from A. R. Wallace, 2 March [1867]).
[7] CD recorded Smith's certainty on this point in *Descent* 2: 345. For nineteenth-century European perceptions of Khoikhoin women and of steatopygia, see Qureshi 2004.
[8] CD placed Smith's information on the lengthened nymphae (inner labia) in a footnote written in Latin in *Descent* 2: 345 n. 53. Earlier in the century, Georges Cuvier had lent scientific authority to travellers' accounts when he examined the cadaver of an African woman and described what he called the 'Hottentot apron' (Cuvier 1817 in Fausto-Sterling 1995, pp. 33–6). On the fascination of nineteenth-century European naturalists with the genitalia of African women, including elongated labia, see also Schiebinger 1994, pp. 164–8.
[9] CD included Smith's account of this woman in *Descent* 2: 346.

To William Erasmus Darwin 27 [March 1867][1]

Down.
27th

My dear old William

The jacket has just come & is splendid. Your present & thinking of me has pleased me much. I do not suppose I shall use it much this Spring, but it will be the very comfort of my life next winter. I thank you heartily.—

I have not yet read the Duke & Heaven knows when I shall get the time, but I am inclined to agree with you from all that I have heard.[2]

Mamma has several times declared that the Duke did not understand the Origin, but I pooh-poohed her, & as it seems very unjustly.[3] We have been amused at how a Duke, as you say, looks at the Creator. I must try & read the book before long.—[4]

I am working very hard at proof-sheets & odious dull work it is & will be for the next 4 or 5 months. I fear the book is by no means worth the confounded labour.—[5]

Horace keeps much the same; he had a fever-fit today, which has disappointed me, as he escaped yesterday.[6] I fear the fever will now run on for another week & if so he will be much reduced.— Poor little fellow he is very patient & sweet.—

Goodnight— your present has pleased me much | Your affectionate Father | C. Darwin

Hooker's Baby, I hear, causes much anxiety—[7]

DAR 210.6: 121

[1] The month and year are established by the reference to *The reign of law* (see n. 2, below), the reference to Horace Darwin's fever (see n. 6, below), and by CD's concern for the health of Joseph Dalton Hooker's baby; the baby's health improved in early April (see n. 7, below, and letter from F. H. Hooker, [6 April 1867]).

[2] CD had received other queries and comments on *The reign of law*, by George Douglas Campbell, duke of Argyll (Campbell 1867; see letter from T. H. Huxley, [before 7 January 1867], letter from J. D. Hooker, 4 February 1867, and letter from W. H. K. Gibbons, 7 February 1867). No letter from William on the subject has been found.

[3] Joseph Dalton Hooker had also stated that Campbell did not understand *Origin* or *Orchids* (see *Correspondence* vol. 10, letter from J. D. Hooker, [15 and] 20 November [1862]).

[4] Campbell argued that the natural laws outlined in *Origin* were set in motion and controlled by God, and that God was therefore still controlling all life through the laws, or the 'Forces of Nature', and their relationships to each other through the 'principle of adjustment'. He wrote, for example (Campbell 1867, pp. 126–7):

> It is, indeed, the completeness of the analogy between our own works on a small scale, and the works of the Creator on an infinitely large scale, which is the greatest mystery of all. Man is under constraint to adopt the principle of Adjustment, because the Forces of Nature are external to and independent of his Will. They may be managed, but they cannot be disobeyed. . . . How imperious they are, yet how submissive! How they reign, yet how they serve!

CD read Campbell 1867 in June 1867 (see letter to Charles Lyell, 1 June [1867] and letter to Charles Kingsley, 10 June [1867]).

[5] CD began correcting the proof-sheets of *Variation* at the beginning of March 1867 and finished 15 November 1867 (see 'Journal' (Appendix II)).

[6] According to Emma Darwin's diary (DAR 242), Horace, their youngest son, returned to Down from school on 16 March 1867 with a fever, and began taking quinine on 1 April. For 27 March, Emma recorded 'fever at 4'.

[7] Hooker had just written that Reginald Hawthorn Hooker was suffering periodic convulsions (see letter from J. D. Hooker, 26 [and 27] March 1867).

From Maxwell Tylden Masters 28 March 1867

Gardeners' Chronicle | *& Agricultural Gazette Office,*
| *41, Wellington Street, Strand, W.C.*
March 28 *1867*

My dear Sir

I venture to send you the enclosed though I fear I may add to your vexations by so doing—[1] You will see at the end a reference to yourself—[2] The lip was smashed so I cut it away— the pollen masses **I** placed on the stigmas—[3]

—I hope you received your copy of the Congress Report and also D[r]. Hildebrand's— if you have not already forwarded the latter or have any inconvenience in so doing I can forward it from hence without troubling you.[4]

Faithfully yrs | Maxwell T. Masters

[Enclosure]

As the sexes of Orchids form a subject of considerable interest, I beg to forward you the accompanying specimens of Cypripedium insigne. Of this I have several plants, all however originally derived from the same piece, but in spite of numerous attempts, I have uniformly failed to fertilise the flowers. The seed-vessel swells and the flower fades as usual, but no seed is produced. It appears to me that my plant produces a male flower only, and is not hermaphrodite. Have any others of your correspondents made a similar observation? I enclose a flower of Cypripedium insigne and two barren seed-vessels, to which the pollen of C. barbatum and C. venustum was applied this year.[5] To prove that the pollen masses of the plant in question are good, I send also a seed-vessel of C. barbatum, fertilised with the pollen of one of the same flowers of C. insigne, and which is full of seeds. A.D.B.

DAR 96: 34–5, *Gardeners' Chronicle*, 6 April 1867, p. 350.

[1] The original enclosure, a letter to the *Gardeners' Chronicle* from 'A.D.B.', has not been found; the version included above is from the text published in the *Gardeners' Chronicle*, 6 April 1867, p. 350. See also *Collected papers* 2: 134. A.D.B. has not been identified.

[2] The published letter from A.D.B. did not mention CD (see n. 1, above); however, the *Gardeners' Chronicle* editors attached the following note:

> The specimens forwarded appeared on examination to be perfectly formed as regards their stamens and pistils, but perfectly destitute of ovules. On forwarding them to Mr. Darwin, that gentleman kindly favoured us with the following remarks. EDS.

A.D.B.'s letter with the editors' remark, and CD's reply are in *Gardeners' Chronicle*, 6 April 1867, p. 350 (*Collected papers* 2: 134–5). For a draft of CD's reply, see the letter to M. T. Masters, [28 March

– 5 April 1867]. The editors of the *Gardeners' Chronicle* in 1867 were Maxwell Tylden Masters and Thomas Moore (*DNB*).

[3] Masters presumably included a flower of *Cypripedium insigne* (lady's-slipper orchid), cutting away part of the damaged labellum (the 'slipper'), and placed its own pollen on the stigmas.

[4] Masters had evidently sent a copy of the report, *International Horticultural Exhibition 1866*; CD was a vice-president of the committee of the botanical congress, and Masters was the secretary (see *Correspondence* vol. 14, letter from M. T. Masters, March 1866). Masters also evidently sent a copy for CD to forward to Friedrich Hildebrand (the report contained a paper by Hildebrand, Hildebrand 1866a). CD had forwarded the manuscript of Hildebrand's paper to Masters the previous year (see *Correspondence* vol. 14, letter to Friedrich Hildebrand, 16 May [1866], and this volume, letter from Friedrich Hildebrand, 18 March 1867 and n. 1).

[5] CD had observed several species of *Cypripedium*, including *C. insigne*, *C. barbatum*, and *C. venustum*; however, he had not performed crossing experiments with them (see *Orchids*, pp. 270–6).

To M. T. Masters [28 March – 5 April 1867][1]

The explanation of the sterility of the seed-capsules of the Cypripedium sent to me I have little doubt lies in the circumstance of their having been fertilised by pollen taken from the same plant or seedling.[2] I now know of a long series of cases in which various orchids are absolutely sterile when fertilised by their own pollen, (proved however to be as itself effective) but which can be easily fertilised by pollen taken from other individuals of the same species, or from other and quite distinct species.—[3] These facts strike me as most remarkable under a Phy. point of view; & all point to the necessity with plants of regular or occasional [union] between distinct individuals of the same species.—

If these remarks are of the least value to you, with respect to the communication you can of course use them[4]

[Many thanks] for note.— I received Report & I forward to Hildebrand copy of his paper.—[5]

I am at present so busy that I have no time to carefully examine the interesting specimen which you have kindly sent me & which I return. From the remarkable fact, lately ascertained by D[r] H. that with many orchids the ovules do not become developed until many weeks or even months after the pollen-tubes have penetrated the stigma, it is not a little difficult to ascertain whether an orchid [four words illeg] [whether] the abortion of the female organs has started.[6] Of course there is no difficulty in ascertaining the rudimentary condition of the pollen.

Adraft
DAR 96: 34–5

[1] The date is established by the relationship between this letter and the letter from M. T. Masters, 28 March 1867, and by the fact that a finished version of the first paragraph of the letter was published in the *Gardeners' Chronicle*, 6 April 1867 (*Collected papers* 2: 134–5).

[2] With his letter of 28 March 1867, Maxwell Tylden Masters had enclosed a letter from A.D.B. to the *Gardeners' Chronicle* and included specimens. Though A.D.B.'s letter only mentions sending barren seed-capsules of *Cypripedium insigne* flowers crossed with flowers from other species, the specimens may have been labelled differently, or else CD assumed that the author had also placed pollen from *C. insigne* stamens on the stigma of the same flower, and on the stigmas of flowers from the same plant.

[3] CD refers to cases reported to him by John Scott, Fritz Müller, and others; he reported on them in *Variation* 2: 133–5. See *Correspondence* vol. 12, letter from John Scott, 28 March 1864, *Correspondence* vol. 14, letter from Fritz Müller, 1 December 1866, and this volume, letters from Fritz Müller, 1 January 1867 and 2 February 1867; see also letter to Fritz Müller, 7 February [1867] and n. 6.

[4] See letter from M. T. Masters, 28 February 1867, n. 2.

[5] See letter from M. T. Masters, 28 February 1867 and n. 4; the references are to the report, *International Horticultural Exhibition* 1866, Friedrich Hildebrand, and Hildebrand 1866a.

[6] CD refers to data presented in Hildebrand 1863 and 1865; there are lightly annotated copies of these papers in the Darwin Pamphlet Collection–CUL. CD reported Hildebrand's results in *Variation* 1: 402–3, 'Fertilization of orchids', p. 153, and *Orchids* 2d ed., p. 172. See also *Correspondence* vol. 11, letter from Friedrich Hildebrand, 16 July 1863, and *Correspondence* vol. 14, letter to Fritz Müller, 25 September [1866].

From H. W. Bates 29 March 1867

Royal Geographical Society | 15, Whitehall Place, S.W.

March 29 1867

My Dear Mr Darwin

I was interrupted in the examination of horned-beetles required to answer your last question by the arrival of a large addition to my collection of them from Paris.[1] The development of horns & excrescences & the sexual differences in this respect are so wonderfully diversified that I think of tabulating the species & furnishing you with results, if I think they are worth your having. Meantime I will try to answer your other questions.

First, the poser of the Lady Darwinian.[2] It is a very fair question & ought to be answered. There are no doubt great gaps in the gradation of forms, as now existing, between the ordinary white Pierid Leptalis & its highly specialised congeners which mimic Heliconidæ. But we must not exaggerate the width of these gaps, nor think that all the existing links are represented on the plates to my Linnæan paper.[3] There are a good many species in collections bridging over the differences between the extreme forms in the genus. Some of these do not wear the livery of any Heliconids, although the wings are elongated like them, departing in this character from the Pierid type. They are forest insects, fly slowly & are rare; how they escape extermination I cannot say. This much however may be said that the undersurface of their wings is coloured & marked something like a dead leaf & this would probably stand them in some stead, as their wings are closed in repose, & they fly very little. (These species are Leptalis Licinia, Psamathe, Kollari Nemesis &c).[4] Amongst the Leptalids mimicking Heliconids there is a very considerable diversity of coloration, for the various species are adapted to almost all the extreme forms of Heliconidæ. I have shown also in my paper that one Leptalis mimics not a Heliconid but a member of a totally different group, which nevertheless has a general resemblance in form & colours to the Heliconidæ.[5]

Now let us put all these facts together. We see that the weak, struggling Leptalids manage to escape extermination by various disguises; some perhaps more effective than others, for I ought to have mentioned that one Leptalis, at least, is an abundant insect (L. Eumelia, a perfect imitation of a very common Ithomia, I. Eurimedia).[6] There can be no difficulty, therefore, in understanding how at all previous periods in the history of the group, the species of Leptalis have found some means of escaping extinction, even though the forms had not reached the extreme divergence from the family type which they have now attained. It must be remembered that the Leptalids are inhabitants of humid forests where none, or scarcely any, of their Family allies live. The proper places for Pieridæ are open sunny grass lands; the Leptalidæ therefore, in intruding into the forest, encountered unusual difficulties & it became a necessity that they should become greatly modified if they were to maintain their ground at all. First their wings became elongated— this I believe was totally independent of adaptation to the long winged Heliconids—it may be connected with relaxation of muscular development—for many forest genera of butterflies are long-winged. Being long-winged & sporting into various colours—not at first the gayer colours of Heliconids—accidental resemblances of their varieties to some other object, no matter what, saved them from extinction. This continued for countless generations & in various parts of the great Tropical forest until the astonishing mimicry of Heliconidæ was brought about.[7]

Another question you asked was whether any female Vanessa or Machaon group of Papilio existed differing in colours from its partner & from the usual gay colouring of its genus.—[8]

I believe none is known. If you extended the search to the next allied genus (to Vanessa) Junonia, I could give you a string of cases of sexual disparity in colour. There are all grades of disparity among the species of Junonia (1) Male & female alike.— (2) male & female same colours but male brighter (3) male & female different in colours but a certain resemblance betraying their relationship (4) male & female so different that no one would judge them to belong to one & the same species.[9]

Yours sincerely | H W Bates

I know of no case of male monkeys fighting together.[10]

DAR 205.10: 95 (Letters)

CD ANNOTATIONS

1.1 I was ... questions. 1.6] *crossed pencil*

2.1 First, ... about. 3.20] *enclosed in square brackets brown crayon*

4.1 Another ... species. 5.6] *crossed blue crayon*

5.3 Male & female alike.— 5.4] *underl red crayon*; '& both beautiful??' *added ink*

5.4 but male brighter] 'male' *double underl red crayon*; '?' *added red crayon*; *underl & '?' del pencil*; 'brighter' *double underl pencil*; 'I read this lighter correct [illeg]' *added pencil*

5.5 (4) ... different 5.6] 'both beautiful' *interl ink, underl red crayon*; '?' *added red crayon*

Top of letter: '*Please return* | Keep | *Mimetic* Last Page | Sexual Selection' *pencil*; 'All used except **Mimetic**' *blue crayon*

---

[1] The letter to Bates to which this is a reply has not been found. Bates had recently been investigating the colour of male and female butterflies, and the horns and sexual differences of lamellicorn beetles,

to reply to earlier questions of CD's regarding sexual selection (see letter from H. W. Bates, 11 March 1867). For Bates's published descriptions of his collection of lamellicorn beetles, see Bates 1886–90.

[2] In the missing letter, CD had sent a question from Georgina Tollet (see letter to H. W. Bates, 30 March [1867] and n. 3).

[3] Tollet evidently asked about the extreme difference between the white *Leptalis nehemia* with shorter and broader wings, and the colourful *Leptalis* species with elongated wings. One of the plates in Bates's paper on the Amazonian Heliconidae, published in the *Transactions of the Linnean Society of London* (Bates 1861), showed the white *L. nehemia* as well as several more colourful *Leptalis* species and five Heliconidae. *Leptalis nehemia* is now *Pseudopieris nehemia* (Lamas 2004). Most species in the genus *Leptalis* in the family Pieridae are now in the genera *Dismorphia* or *Enantia*. (for a recent checklist of the genera of the family Pieridae, see Braby 2005). See also letter from H. W. Bates, 11 March 1867 and n. 5.

[4] Bates refers to what are now often known as *Enantia melite melite*, *E. lina psamathe*, and *Lieinix nemesis*. *Leptalis kollari* is a synonym of *E. lina psamathe* (Lamas 2004).

[5] Bates noted the variability in colour among *Leptalis* species in Bates 1861, pp. 504–6, and also referred to a *Leptalis* imitating a *Stalachtis* species; *Stalachtis* is a genus of the family Riodinidae (see Bates 1861, p. 504).

[6] In Bates 1861, p. 539, Bates described *Ithomia eurimedia* (now *Aeria eurimedia*), and mentioned that *Leptalis melia* (now *Dismorphia melia*) was often found with it.

[7] Bates discussed how a mimetic species may develop through natural selection in Bates 1861, pp. 511–15. However, in considering the first stage of a *Leptalis* species coming to mimic Heliconidae species, he also wrote: 'In what way our *Leptalis* originally acquired the general form and colours of *Ithomiæ* I must leave undiscussed' (*ibid.*, p. 513).

[8] Bates refers to the genus *Vanessa* and to related forms of *Papilio machaon* (swallowtails).

[9] CD mentioned these differences in *Junonia* in *Descent* 1: 389. His draft for this discussion is in DAR 81: 183.

[10] CD was interested in what he called the 'law of battle', when males fought over females, and the related secondary sexual characteristics; on the 'law of battle' in mammals, see *Descent* 2: 251–68. He mentioned the fighting of some male primates, but not monkeys, in *ibid.*, pp. 324–5.

To J. D. Hooker   29 [March 1867]

Down
Friday 29th

My dear Hooker.

Do let us have a line in a few days to tell us how the Baby is, for we are anxious to hear.[1] It is wonderful what convulsions infants will endure without any harm to their constitutions.

I will remember to save all plants for you from foreign plants, when I have done experimenting with them.[2]

Yours affect. | C. Darwin

DAR 94: 18

[1] In his letter of 26 [and 27] March 1867, Hooker had written about Reginald Hawthorn Hooker's illness.

[2] See letter from J. D. Hooker, 26 [and 27] March 1867 and n. 2.

From John Lubbock 29 March 1867

15, Lombard Street. E.C.
29 Mar. 67

My dear M^r. Darwin

Thanks for your information.[1]

It is an age since we have met. I should be delighted to see you any Sunday, or will come up next time I am at home.[2]

Would you like to have a look at McLennan's book?[3]

Believe me | Yours very sincerely | John Lubbock

C. Darwin Esq

DAR 170: 57

[1] See letter to John Lubbock, 26 March [1867].

[2] CD had suggested that he and Lubbock meet on a Sunday; Lubbock spent much of the week at his house in London. The Lubbock family estate, High Elms, bordered the grounds of Down.

[3] No answer to Lubbock's letter has been found, but see the letter to John Lubbock, 26 March [1867] and n. 1.

To Hermann Müller 29 March [1867][1]

Down, Bromley, Kent
Mar: 29.

Dear Sir

Your letter has been a great pleasure to me.[2] To believe that I have interested and in any degree aided an energetic and skilful naturalist like yourself is in truth the highest reward possible to me.—

—I sent you a few days ago a paper on climbing plants by your brother, and I then knew for the first time that Fritz Müller was your brother.[3] I feel the greatest respect for him as one of the most able naturalists living, and he has aided me in many ways with extraordinary kindess.—

Yours very faithfully | Ch. Darwin

Copy incomplete
DAR 146: 428

[1] The year is established by the relationship between this letter and the letter from Hermann Müller, 23 March 1867.

[2] See letter from Hermann Müller, 23 March 1867.

[3] CD refers to F. Müller 1865 (see letter from Hermann Müller, 23 March 1867).

To William Ogle 29 March [1867][1]

Down. | Bromley. | Kent. S.E.
Mar 29.

Dear Sir

I am much obliged for your great kindness in writing to me.[2] The subject of inheritance interests me greatly, & I have to treat of it in a book which I am now printing.[3]

I am glad to hear about the inheritance of deficient phalanges, but as several cases are recorded I do not feel *as yet* sure whether I will quote it.[4]

The case of the twins strikes me as extremely interesting, & I sh^d much like to quote it on y^r authority; but I sh^d much wish to know, if you will not object to the trouble of writing again to me, whether the little finger which was crooked occurred on the same (right or left) hand in the 2 children; & 2^{ndly} did the misplaced tooth occur in the first or second dentition & what tooth was it? Was it also the same tooth in the 2 children?[5]

I hope that you will excuse me asking these questions & accept my sincere thanks for y^r kindness.

I beg leave to remain | my dear Sir | yours faithfully | Charles Darwin

LS
DAR 261.5: 1 (English Heritage 8820 5899)

[1] The year is established by the reference to the printing of *Variation* (see n. 3, below).
[2] The letter from Ogle has not been found.
[3] CD discussed inheritance at length in *Variation*, for which he had been correcting proof-sheets since early March 1867 (see CD's 'Journal' (Appendix II)).
[4] CD discussed missing fingers in a section on inheritance limited to one sex in *Variation* 2: 73.
[5] No reply from Ogle has been found. In *Variation* 2: 523, CD specified that the little fingers on both hands of each twin were crooked, and that the misplaced tooth was the second bicuspid of the second dentition in the upper jaw.

From W. B. Tegetmeier 29 March 1867

The Field. | *346, Strand,* | *London, W.C.*
March 29 *1867*

My dear Sir

I delayed day after day replying to yours of the 5th. thinking I might have something special to communicate but now write a general answer— I know nothing about the Case of the Widow bird—[1]

But as far as regards polygamous birds do not think the statement has any foundation whatever—[2]

A Game cock closely trimmed for fighting is as unlike an ordinary male fowl as a bird can well be, yet he is always well received by the hens and as in these birds the male always forces the female if reluctant I do not think it is likely that he would be ill received—

If two or more cocks are turned down the hens will receive anyone willingly but the strongest and most valiant chastises the others if he perceives them pursuing the hens.— certainly I think appearance has but little to do with the reception of the cock in fowls— I have just turned down a polish bird (crested) with some white cochin hens removed the same day from white cochin cocks the advances of the polish male (though totally different in colour tail crest and general appearance) were received

I cannot personally try the experiment in any of my runs as they only contain a single cock each—but I will do what I can elsewhere—

Have you any objection to my enquiring in my own name in the Field. I might get some useful information—

I send you a few slips of a letter Mr Wallace published last week in the Field thinking, if you wished to circulate the queries they might save writing copies[3]

I shall be happy to make any experiments in my power

do [you] not think dying a white male pigeon magenta colour which is easily done and seeing whether his wife knows him, would be of any bearing on the question[4]

Believe me | Yours truly | W B Tegetmeier

DAR 84.1: 30–1

CD ANNOTATIONS
1.1 I delayed ... bird— 1.3] *crossed pencil*
5.1 I ... copies 7.2] *crossed pencil*

[1] See letter to Tegetmeier, 5 March [1867] and n. 1.
[2] In *Descent* 2: 117, CD noted that he had received long letters concerning the 'courtship of fowls' from Tegetmeier and two other observers, and that none of these believed that females preferred particular males because of the beauty of their plumage. In *Descent* 1: 269, CD wrote that widow-birds were evidently polygamous.
[3] A letter from Alfred Russel Wallace to the *Field*, 23 March 1867, p. 206, requested information for himself and CD on the preferences of insectivorous birds for particular caterpillars; he asked readers who kept birds to offer them a variety of caterpillars, and observe their choices. For the first responses to a similar request from Wallace following his correspondence on the subject with CD, see the letter from A. R. Wallace, 11 March [1867] and n. 7.
[4] See letter to W. B. Tegetmeier, 30 March [1867] and n. 6.

To H. W. Bates 30 March [1867][1]

Down. | Bromley. | Kent. S.E.
March 30

Dear Bates

Would not the tabulating the Horned Beetles be very troublesome:[2] if not I certainly sh^d like to hear the result. But in truth it would be a pity for you to waste or take up much time over the job, for some general remarks would do very well for my object.

Your remarks in answer to my lady-friend (Miss Tollet daughter of late Mr Tollet of Betley Hall)[3] are interesting & fairly satisfactory; but it would have been better if it could have been stated what "other objects" they first mocked; or if it could be shown that some species mocked dull-coloured Heliconidæ, for then as the latter gained their splendid colours so would the mockers.—[4] Not that I feel a shadow of doubt about the truth of your theory— it must be true.[5] Wallace told me in a letter of the pretty case of the white moth & the young Turkeys.[6]

I suppose you have, of course, seen his letter to the Field; but I enclose a couple of copies.—[7]

Many thanks about Junonia— whenever I go to B. Museum, I will ask to see the genus & will look at the differences & similarities in the sexes.—; it seems a capital case. You have indeed given me most valuable information:—[8]

Dear Bates | Yours very sincerely | Ch. Darwin

I have just finished hearing read aloud your Amazon Book, & liked it better 2^{d} time even than 1st time.[9]

I shall send your letter to Miss. T, as she begged me to do.

Cleveland Health Sciences Library (Robert M. Stecher collection), FF7

[1] The year is established by the relationship between this letter and the letter from H. W. Bates, 29 March 1867.

[2] See letter from H. W. Bates, 29 March 1867.

[3] See letter from H. W. Bates, 29 March 1867. Emma Darwin recorded in her diary (DAR 242) that Georgina Tollet visited Down House from 7 to 21 March 1867. Georgina's father, George Tollet, had resided at Betley Hall, Staffordshire (Freeman 1978). Georgina, a childhood friend of Emma, had read and commented on a manuscript of *Origin* (see *Correspondence* vol. 7).

[4] CD mentioned this idea in *Descent* 1: 412. Bates had speculated about the early stages of mimicry in the development of *Leptalis* butterfly species that came to resemble Heliconidae species (see letter from H. W. Bates, 29 March 1867 and n. 7).

[5] CD had praised Bates's paper on mimicry (Bates 1861), and had written a positive review of it (see *Correspondence* vol. 11, and 'Review of Bates on mimetic butterflies'). CD added a discussion of Bates's discovery of mimicry in butterflies to the fourth edition of *Origin* (see *Origin* 4th ed., pp. 503–6).

[6] For the case of turkeys rejecting a moth, see the letter from A. R. Wallace, 24 February [1867] and nn. 5 and 6. This case had been discussed at the 3 December 1866 meeting of the Entomological Society of London; in *Descent* 1: 411, CD cited the report of this meeting as confirming Bates's hypothesis that the Heliconidae were protected from attack by birds by a secretion or odour.

[7] See letter from W. B. Tegetmeier, 29 March 1867 and n. 3.

[8] See letter from H. W. Bates, 29 March 1867 and n. 9.

[9] For CD's earlier praise of *The naturalist on the river Amazons* (Bates 1863), see *Correspondence* vol. 11.

To W. B. Tegetmeier 30 March [1867][1]

Down. | Bromley. | Kent. S.E.
March 30

My dear Sir

I am much obliged for your note & shall be truly obliged if you will insert any questions on subject.[2] That is a capital remark of yours about the trimmed Game cocks & shall be quoted by me.[3] Nevertheless I am still inclined from many facts strongly to believe that the beauty of the male bird determines choice of female with wild Birds, however it may be under domestication. Sir R. Heron has described how one pied Peacock was extra attractive to the Hens.—[4] This is a subject which I must take up as soon as my present book is done.—[5]

I shall be most particularly obliged to you if you will die with magenta a pigeon or two. Would it not be better to die the tail alone & crown of head; so as not to make too great a difference: I shall be very curious to hear how an entirely crimson pigeon will be received by the others as well as his mate.[6]

With cordial thanks for your never failing kindness— | Yours sincerely | Ch. Darwin

PS. Many thanks for the printed Queries.[7]

P.S. Perhaps the best experiment for my purpose would be to colour a young unpaired male & turn him with other pigeons & observe whether he was longer or quicker than usual in mating.

Archives of The New York Botanical Garden (Charles Finney Cox collection) (Tegetmeier 103)

[1] The year is established by the relationship between this letter and the letter from W. B. Tegetmeier, 29 March 1867.

[2] In his letter of 29 March 1867, Tegetmeier had offered to publish a query in the *Field* as to whether female birds bred with male birds missing ornamental feathers. No such query has been found in the *Field* between March and May 1867.

[3] See letter from W. B. Tegetmeier, 29 March 1867; CD noted Tegetmeier's opinion about the trimmed gamecock in *Descent* 2: 117.

[4] In *Descent* 2: 117, CD wrote: 'some allowance must be made for the artificial state under which [fowls] have long been kept' after recording the opinion that female birds did not prefer more ornamental males. CD cited cases of female choice in both wild birds, and those that were domesticated or confined, in *Descent* 2: 113–24. He quoted Robert Heron's case of the pied peacock from Heron 1835, p. 54, in *Descent* 2: 120. For more recent discussions of sexual selection, see Mayr 1972, and J. L. Gould and Gould 1997.

[5] CD was correcting proofs of *Variation*.

[6] See letter from W. B. Tegetmeier, 29 March 1867. CD recorded in *Descent* 2: 118 that Tegetmeier stained some of his pigeons magenta at CD's request, but that they were not much noticed by the other birds. However, Tegetmeier evidently did not do this for some time (see letter to W. B. Tegetmeier, 4 August [1867], and *Correspondence* vol. 16, letter to W. B. Tegetmeier, 21 February [1868]).

[7] See letter from W. B. Tegetmeier, 29 March 1867 and n. 3.

From George Gibbs 31 March 1867

Smithsonian Institution | Washington.
Mch 31. 1867

Charles Darwin, Esq.
Dear Sir,

Professor Baird has shown me your circular, "Queries about expression".[1] After twelve years residence among the Indians of the North West Coast of America, I find to my surprise, that I can only answer one of them positively—[2] The Indians of Puget's Sound, a branch of the Sélish family, whose color is of a rather light shade of sienna (even where of unmixed blood,) certainly *do* blush from shame or anger, and the darkening of the skin is palpable.[3] As to the other points I will not pretend to answer until I can observe with the certainty you desire.

One point however, not touched upon you, I will mention, and that is that they frequently, if not always, indicate *direction* by throwing the head back and protruding the chin, instead of with the finger.

A peculiarity which, though hardly coming under your apparent limits, is noticeable, that in designating the height of a human being, as a child, the hand is held edgewise; in the case of an animal, flatwise as we hold it.[4]

Very respectfully | Your obt servt | George Gibbs

DAR 165: 37

[1] Spencer Fullerton Baird was the assistant secretary and curator of the Smithsonian Institution (*ANB*); since the early 1860s, Gibbs had been organising manuscripts for publication under the auspices of the Smithsonian (see *ANB*, Stevens 1873). Asa Gray may have sent a printed copy of CD's queries about expression to Baird (see letter from Asa Gray, 26 March 1867 and n. 2). For a later, printed, version of the questionnaire, see Appendix IV. A version of CD's list of queries was published in the *Annual Report of the Board of Regents of the Smithsonian Institution for the year 1867*, p. 324 (see *Collected papers* 2: 136–7).

[2] Gibbs lived in Oregon, and in Washington Territory, from 1848 to 1860; during much of that time he studied languages of north-western native Americans (Stevens 1873).

[3] For one version of CD's query on blushing, see the second question in the printed questionnaire, Appendix IV; for CD's variations of this query, see Freeman and Gautrey 1972. In 'Tribes of western Washington and northwestern Oregon', Gibbs wrote that the people of Puget Sound were a western branch of the 'Selish or Flatheads', and were usually mentioned as 'the Niskwalli nation' (now Nisqually; see Gibbs 1877, p. 169; see also pp. 178–9). The Salish language family comprises twenty-three languages spoken over a large area of north-western North America; the language spoken in Puget Sound is Lushootseed (Czaykowska-Higgins and Kinkade 1998).

[4] CD did not cite Gibbs or record his observations in *Expression*.

From J. D. Hooker 31 March 1867

Kew
March 31/67.

Dear Darwin

We have no hopes for our pretty little baby whose fits increase in number & duration—[1] Happily it suffers little, the nerves of motion exclusively being affected. This suspense is very painful.

M[rs] Hooker[2] is pretty well

I keep her in bed next door to the child.

Ever dear old Darwin | Yrs affec | J D Hooker

Percival Wright was here today on his way to the Seychelles to spend some months collecting—[3]

DAR 102: 156

[1] Hooker reported that Reginald Hawthorn Hooker had been suffering from convulsions (see letter from J. D. Hooker, 26 [and 27] March 1867, and letter to J. D. Hooker, 29 [March 1867]).

[2] Frances Harriet Hooker.

[3] Edward Perceval Wright spent six months in the Seychelles Islands in 1867, and brought back a collection of plants and animals (*DNB*).

From Joseph Trimble Rothrock to Asa Gray 31 March 1867

McVeytown [Pennsylvania]

Mch 31st 1867

My Dear Doctor

Some of Mr Darwins questions I feel safe in answering, Such only I attempt Any reply to.[1] all parties add that the Answers apply especially to the Atnahs and Espyox tribes on the Nasse River of North So Western N.Aa.[2] Neither tribe had much previous acquaintance with the Whites.— in fact the Espyox had not previous to my going among them seen half a dozen whites. You will observe I found the indians much like other men. I must say I nowhere saw the ideal, taciturn immovable indian of Mr Coopers conception.[3]

Question	1.	Yes
"	2	—
"	3	Many of them do, Not all
"	4	Yes
"	5	—
"	6	Yes
"	7	Sometimes. Generally in fact. Among the Siceanees[4]—a tribe adjoining the Atnahs
Question	8	Yes, decidedly[5]
"	9	—
"	10	—
"	11	Exactly So.
"	12	Laughter is often excessive, tho I know of but one instance in which tears were shed from that cause, real or feigned grief produce them often enough from the women.
Question	13	—
"	14	Yes, just as a white child[6]
"	15	Yes, and these knowing indians, look for these signs
"	16	—
	17	Vertical nod is usual, lateral not so common though I have seen it.

Mr Darwin may depend on the correctness of these Answers, I am sure of them

On my return home from the North West, I was unable to reconcile my views with those of the late Dr Morton, but disliked to adopt those opposed to his. I knew him to be an authority on the subject, and therefore kept the matter in a mental status quo. He says that "sixteen years of almost daily comparisons have only confirmed him in the opinions announced in his Crania Americana, that all the American nations are of one race (excepting the Esquimaux) and that this race is peculiar and distinct from all others."[7] Col Hamilton Smith says it is vain to assert that all American races excepting the Esquimaux have sprung originally from one stock.[8]

If by this Smith means that Indians differ from one another as much as Europeans do, I at once adopt his view. J Aitken Meigs of Philad in a paper Read before the Philadelphia Acad. of Nat. Sci. May 1866 to my mind completely upsets Mortons opinion, and is I think in the main correct. If M[r] D. has not this paper I am sure he would be interested in it.[9] Meigs is unable to understand why the Stickine river indians and some of those farther north do not always flatten the infant skull. The reason is plain, these tribes are warlike, and come south among the flat-headed tribes for slaves. Finding their captives regard the compressed skull as a mark of aristocratic blood; they (the Northern tribes) in contempt flatten the heads sometimes of their own female children, and usually the heads of both sexes when born in slavery. Hence the deformed cranium among them is the mark of inferiority and the sign of a slave.[10] Should there happen to be any other facts in my possession relating to the ends, which would interest you or Mr D. I beg that you will command them. Did you receive about 150 Species of NW. plants from me some months ago? I sent them by Adams Express.[11]

Yours Always | J. Trimble Rothrock

I shall soon have some answers from Wyman—[12] from memory— And he will send copies of your queries South to trusty persons

A.G.

DAR 176: 218

CD ANNOTATIONS
1.2 Atnahs] 'Atnaks?' *added ink*
1.3 Espyox] 'Espyox(?)' *added ink*
1.3 Nasse] 'Nasse?' *added ink*
Top of letter: 'Keep 2[d] Page for Descent of Man[13] | cause of not flattening the Head' *blue crayon*

---

[1] Rothrock refers to CD's queries about expression. CD evidently sent a copy of the queries to Asa Gray with a letter of 28 February 1867, but that letter has not been found (see letter from Asa Gray, 26 March 1867). In his letter of 26 March 1867, Gray told CD that he recently had fifty copies of the queries printed. Rothrock probably had a handwritten copy of the queries. For a later, printed, version of the queries, see Appendix IV. Rothrock had been in British Columbia as a member of the Western Union Telegraph Expedition from 1865 to 1866 (*DAB*).

[2] The Nass river of western British Columbia, Canada, flows about 236 mi (380 km) south-west through the Coast Mountains to the Pacific Ocean north of Prince Rupert. The name 'Atnah' is recorded as a synonym for the Shuswap (Secwepemc) nation (Hodge ed. 1910, 2: 561), but their territory was in south-central and eastern British Columbia. 'Espyox' may refer to the Kispiox (Gitanspayaxw) tribe of the Gitksan nation, who had territories on the upper Nass (Sterritt *et al.* 1998, p. 99). Much of the

land around the Nass river was the territory of the Nisga'a nation. For more on tribal boundaries in this area, see Sterritt *et al.* 1998.

3 Rothrock refers to the characterisation of Native Americans in the novels of James Fenimore Cooper (for more on Cooper's portrayal of Native Americans, see House 1965, pp. 47–71).

4 'Siceanee': probably the Sekani; the Long Grass band of the Sekani had territory near the headwaters of the Nass river, and were known to intermarry with the Kispiox (Sterritt *et al.* 1998, pp. 54–5). CD noted Rothrock's information on sneering in *Expression*, pp. 252 and 260.

5 In *Expression*, p. 232, CD cited Rothrock on the expression of determination or obstinacy.

6 In *Expression*, p. 233, CD noted the affirmative reply about pouting in children, but did not cite Rothrock by name.

7 Samuel George Morton had written a monograph, *Crania americana* (S. G. Morton 1839), in which he argued that there was only one distinct American race, based on an analysis of human skulls. The statement quoted by Rothrock was made in an article, 'Some observations on the ethnography and archæology of the American aborigines' (S. G. Morton 1846, p. 7).

8 Charles Hamilton Smith argued in *Natural history of the human species* that several migrations of people from different Old World origins to America had occurred (C. H. Smith 1848; for the statement paraphrased by Rothrock, see p. 251). 'Esquimaux' refers to Inuit people.

9 In a paper reviewing the ethnological literature on American races and reporting on his examination of the skulls in Morton's collection (Meigs 1866), James Aitken Meigs challenged Morton's view of a single cranial type for all Americans. CD's annotated copy of an offprint of Meigs 1866 is in the Darwin Pamphlet Collection–CUL.

10 Meigs described the 'Stikanes or Cowitchins' (Cowichan tribe or Quw'utsun' people) of Vancouver Island as practising a form of head-binding resulting in a conical head shape (Meigs 1866, p. 212). Rothrock evidently thought Meigs was referring to people from the area of the Stikine River (Tahltan nation) in north-western British Columbia. For more on the practice of slavery among Native American tribes in British Columbia, see Hodge ed. 1910, 2: 598; for the practice of head-flattening, see *ibid.*, 1: 96–7, 465.

11 Adams Express Company.

12 The postscript to CD was added by Asa Gray. No record of answers from Jeffries Wyman to the queries has been found, nor was Wyman mentioned in *Expression*.

13 CD used Rothrock's answers to the queries about expression in *Expression*, which had originally been planned as a part of *Descent* (see letter to A. R. Wallace, 26 February [1867] and nn. 5 and 6).

To Stephen Paul Engleheart? [April 1867?][1]

Down
Tuesday Evening

My dear Sir

As Leonard will go to school tomorrow (Wednesday) afternoon will you be so kind as to let us have the note, properly addressed, about the sling.—[2] His arm keeps quite comfortable.—

Your's very faithfully | & obliged | Ch. Darwin

Provenance unknown

1 The date is conjectured from a reference to 'Lenny's arm' in Emma Darwin's diary (DAR 242) for 22 April 1867. Stephen Engleheart was the village physician in Down and was frequently consulted by the Darwin family (Freeman 1978).

2 Leonard Darwin was a pupil at Clapham Grammar School in South London (M. Keynes 1943).

From Fritz Müller 1 April 1867

Desterro, Brazil,
April 1. | 1867.

My dear Sir.

The last French Steamer brought me your two letters of Feb. 7 and 22, for which I am much obliged to you.[1]

As to your question about the sexual differences of lower animals, I can hardly give you any information.[2] Among higher Crustacea I know of only one instance of males distinguished by bright colours from their modest females.

In a little *Gelasimus* of our coast the posterior half of the cephalothorax assumes often, in the *adult male*, a pure white colour. This white colour may, in a few minutes, change into dirty grey or even black, while at the same time the colours of the anterior half lose much of their brilliancy. The cephalothorax of the female is commonly of a nearly uniform greyish brown. This little gelasimus uses to run about in the sunshine and so is more able to exhibit its beauty to the female than are most other Crustaceans. The number of the males seems to be much larger than that of the females; just now, when I went to catch some specimens, I found a dozen males and only two females. I suspect, that the large pincers may serve to this and perhaps to some other male crabs for fighting for the female; at least, when several males are emprisoned in a glass, they often kill or mutilate each other.—[3] Among lower Crustaceans some species of *Sapphirina Thomps.* offer a most interesting case in point. The males are known since the time of Cook for the brilliancy of their colours, while the females are colourless according to *Gegenbaur* and *Claus*. In these little Copepods also the colours are changing (See the enclosed extract from Claus' book.).—[4] From Gerstäcker's Textbook of Zoology I see that the males of some spiders (Sparassus smaragdulus &c) are distinguished by different and more gaudy colours.—[5] Among Annelids some Syllideæ with alternate generations show very marked sexual differences, so great, that the asexual form, the male and the female were placed in three distinct genera (Autolytus Grube, Diploceræa Gr. and Sacconereis J. Müll.)[6] Some of the sexual forms have very large eyes and I have found some of these little worms having much prettier colours than is usual with Annelids. Perhaps these colours here also may be due to sexual selection.—

As to the expression of countenance of Negros, I shall keep the subject before my mind and may perhaps be able to answer some of your questions; but I have here no opportunity of observing native S. Americans.[7]

Oncidium flexuosum is a native of our district, and so are the other Orchids of my garden, all of which I collected either on our island or on the neighbouring part of the continent.[8]

I but once saw ramifying filaments of a parasitic cryptogam causing the decay of Orchid-pollen, when I had placed *old* pollen-masses of *Lælia purpurata* on the stigma of *Brassavola fragrans*.— In the case of self-impotent Vandeæ I never saw a trace of cryptogams, which might be considered as the cause of the decay.—[9]

To the list of self-impotent Orchids, in which pollen and stigma of the same plant are poisonous to each other, I can add a small *Oncidium [dried specimen]* (probably undescribed) from Theresopolis[10] in which own pollen after three days' stay in the stigmatic chamber was decayed and dark brown. This *Oncidium* grows in company of *O. unicorne* and though the flowers, as you see, are quite different, (belonging even to a different section of the genus), the pseudobulbs and leaves of the two species are almost undistinguishable. Perhaps this strange ressemblance may be due to mimicry.— I am now experimenting on the last (as to its flower-time), but not least of our Oncidia, the *Oncidium crispum*, which seems to be an exception among our other self-impotent species of that genus; at least own pollen, placed on the stigma March 23, I found to be quite fresh as well as the stigma and pollen-tubes, when I examined a flower this morning.[11]

As to the fertility of *Epidendrum cinnabarinum* being much lessened by self-fertilization, I can now give you one more striking case. I fertilized (Jan. 17) six flowers of a raceme with pollen of the same flowers and obtained six pods, the seeds of which weighed: 1.5— 2.— 1.5— 2.— 2,5— 2.5 grains; most of the seeds were bad.— On the same day I fertilized six flowers of a second raceme of the same plant with pollen of a distinct plant and obtained 5 pods, the seeds of which weighed 5.— 6.— 5.— 5.—5 grains, but few seeds being bad.—[12]

At reading Dr Hildebrand's paper on *Corydalis cava*, which you have been so kind as to send me, I was struck by the circumstance that among the 29 seed-capsules obtained by crossing distinct plants, the seeds of which were counted, none had 4 seeds, while about half the number had from 5 to 7, and the other only from 1 to 3 seeds. May not this species be functionally dimorphic, the poor seed-capsules being the result of illegitimate unions?—[13]

⟨*Drawing or specimen excised*⟩ Here is another dimorphic *Rubiacea*;[14] the pollen-grains of the short-styled plant are larger (about $\frac{1}{12}$ ^{mm} diam.), than those of the long-styled (about $\frac{1}{16}$^{mm} diam.). The plant has beautiful blue Fruits.

During the last months I have been making some experiments on the crossing of different species of Orchids; and I hope that these experiments, if continued on a larger scale, will lead to some interesting results. Some points may already be worth mentioning. It seems that much more widely distinct species of Orchids may be successfully crossed, than is usual with plants. Thus I have (as yet unripe) pods of Cattleya elatior var Leopoldi ♀ and Lælia purpurata ♂;—of Cattleya elatior var Russelliana, ♀, and Epidendrum cinnabarinum ♂;—of Brassavola fragrans Lem. ♀ and different species of Epidendrum ♂;—of Oncidium flexuosum ♀ and Cyrtopodium (Andersoni?) ♂;—of Notylia ♀ and Ornithocephalus ♂!—of Sigmatostalix ♀ and Ornithocephalus ♂!—of the small Oncidium from Theresopolis ♀ and Sigmatostalix ♂;—of Oncidium flexuosum ♀ and Gomesa ♂;—of Gomesa ♀ and Onc. flexuosum ♂ etc.—

But the case, which most surprised me, is the following: A Frenchman, Mr. Gautier,[15] bade me to cross for him, in his garden, Zygopetalum maxillare and Miltonia cereola. I did so, without expecting any result; but both the species have now fine swelling pods!— There appears to be no close systematic affinity between the two genera; at least Miltonia is much more closely allied to Oncidium than to Zygopetalum. Mr. Reichenbach even cancels the genus Miltonia, uniting it with Oncidium;[16] but I repeatedly fertilized several species of Oncidium with pollen of Miltonia, without obtaining a single pod.— The only point in which Zygopetalum maxillare and Miltonia cereola closely ressemble each other is the *colour of the labellum!*—

March 6, when I was fertilizing a *Cattleya* (probably a var. of C. elatior) with pollen of some other species of the genus and of Epidendrum, I had just at hand some pollinia of *Oncidium micropogon* and placed them on the stigma of one of the flowers of the Cattleya. Now this flower withered *sooner* than the other fertilized flowers, and has yielded a fine pod, which is now 36^{mm} long, while the other pods are about 46^{mm} long. Probably the pollen-tubes of Oncidium will not be able to fertilize the ovules of Cattleya and I think the pod will wither, when the time of fertilization arrives, but even so the case is curious.— If horticulturists contrived to raise Orchids from seeds, this facility with which widely distant species are crossed, would be a highly valuable quality.—

There is with Orchids the same unaccountable ⟨*one-third line excised*⟩ fertility of reciprocal crosses, as with other ⟨*one-third line excised*⟩ certain species seems to have a great fertilizing ⟨*one-third line excised*⟩ *Epidendrum cinnabarinum*, these same species are ⟨*one-third line excised*⟩ with difficulty. I fertilized with pollen of ⟨*one-third line excised*⟩

Epidendrum Zebra n. sp?[17]	6	flowers,	⟨ ⟩
—— fragrans	2		⟨ ⟩
—— raniferum?	2		I.
—— vesicatum	2		2
Cattleya elatior	2		I.
Brassavola fragrans.	4.		3.
	18		13.

On the other hand I fertilized Epidendrum cinnabarinum with pollen of

Epidendrum Zebra	17	flowers, which yielded	0 pod
— fragrans	4		0. —
— raniferum	37		1 —
— vesicatum	6		0. —
Cattleya elatior	2.		0.
Brassavola fragrans	16.		0.
	82		1.

The easity, with which fertilisation is effected and the number of seeds produced is not proportional. Epidendrum Zebra is easily fertilized with Ep. cinnabarinum, but the pods (not yet ripe) will be very poor. On the contrary 14 flowers of Epid. cinnabarinum fertilized with pollen of Ep. Schomburgkii(?) yielded only two pods, but one of these had more seeds than any pod fertilized with pollen of the own species, and also the other was very rich.—

The sterile unions between different species behave in most cases very differently from those of the self-impotent species. First discolours and withers the pedicel of the germen and this discolouring goes slowly on in an upwards direction and even, when the germen falls off, pollen and stigma are generally quite fresh. Generally the germen will fall off 1 or 2 weeks after fertilization; but a germen of *Cirrhæa dependens?* fertilized with *Stanhopea* Jan. 24 fell off only March 11; stigma, pollen and pollen tubes were fresh.— Sometimes the pollen grows brownish; but then this discolouring generally begins at the outer surface of the pollen masses, and but in very few cases at the inner surface, where it touches the stigma, whilst the latter is *invariably* the case, when a self-impotent Oncidium, Gomesa, Sigmatostalix or Notylia is fertilized with own pollen.—[18]

Hoping that this letter will find you in good health, Believe me, dear Sir, | very truly and respectfully yours | Dr Fritz Müller.

P.S. I forgot to thank you for the copies of my paper on Climbing plants.[19]

[Enclosure][20]

Claus, die freilebenden Copepoden. Leipzig, 1863; pg. 35;[21]

Endlich mag an diesem Orte der merkwürdige Farbenschiller besprochen werden, welcher an der äusseren Körperbedeckung einiger *Saphirinen*-Männchen bereits älteren Forschern bekannt war. Schon *Anderson* beobachtete die Erscheinung auf *Cook*'s letzter Reise an seinem *Oniscus fulgens*, einer Thierform, die offenbar mit *Banks*'s *Carcinium opalinum* und *Thompson*'s *Sapphirina* identisch ist . . . . *Meyen*, welcher *Sapphirinen* in der Nähe der Azoren beobachtete, beschreibt dieselben als äusserst bewegliche Plättchen mit beständigem Farbenwechsel.[22] In die Tiefe hinabgesunken zeige sich das Thier mit dem glänzendsten Violettroth, das einen purpurnen Kern einschliesse, eingefangen aber verliere es die Farben, welche nur durch die Brechung der Lichtstrahlen an der spiegelnden Oberfläche des Körpers hervorgerufen würden. An der letztern aber entdeckte er auf der Rückenseite 4-, 5- oder 6-seitige Schilder, die in stumpfen Kanten aneinander gereiht, eine glänzende Fläche zeigten; durch diese dachte sich Meyen wie durch aneinander gereihte Prismen das Licht gebrochen und so bei jeder Bewegung des Thieres den Farbenwechsel erzeugt. Eine bei weitem detaillirtere Beschreibung der erwähnten Farbenerscheinungen erhalten wir durch *Gegenbaur* (Müller's Archiv 1858 p. 67).[23] Auch dieser Forscher verlegt ihren Sitz in jene polygonalen . . Felder, die er auf eine zusammenhängende Schicht von platten Zellen unter der Cuticula, *auf die Matrix der Chitinhülle*, zurückführt. "Beim Weibchen", sagt unser Forscher, "ist der

Zellinhalt während des Lebens durchaus hell, das Männchen dagegen lässt im Leben beinahe dieselben Erscheinungen an jenen Zellen unter dem Mikroskope erkennen, wie man sie am frei lebenden Thiere beobachten kann. Bei durchfallendem Lichte sowohl als bei auffallendem ist der Wechsel des Farbenspieles von Zelle zu Zelle zu beobachten und während im letzteren Falle nur Metallglanz funkelt, so ist bei ersteren neben dieser Erscheinung noch ein dioptrisches Farbenspiel sichtbar. Oft grenzt sich eine Zelle von der benachbarten mit grösster Schärfe durch Farbe oder Metallschimmer ab, erscheint gelb, roth oder blau mit den verschiedensten Nüancirungen von einer Farbe in die andere übergehend, jedoch ohne alle Mittelfarben, ohne Grün, Violett oder Orange. Die beiden ersten Farben kommen jedoch bei dem katoptrischen Phänomen vor, bei welchem Blau die erste Rolle spielt. Betrachtet man die Erscheinung an einer einzelnen Zelle, so findet man den Übergang von Blau in Roth ohne die Mittelfarbe dadurch zu Stande kommen, dass an einem Theile der Zelle, etwa in einer Ecke derselben, das Blau erblasst, fast grau wird und dann plötzlich an dieser Stelle ein rother Saum auftritt, der breiter werdend über die Zelle in dem Maasse sich ausdehnt, als das Blau gewichen ist, so dass alsbald die ganze Zelle blau erscheint. Dasselbe gilt vom Gelb. Die Qualität der Farbe einer Zelle ist völlig unabhängig von den benachbarten Zellen. So erscheinen gelbe mitten im Roth, rothe mitten im Blau. Doch kann auch die Erscheinung auf benachbarte Zellen überschreiten; vom Rande einer blauen Zelle geht Blau auf die Nachbarzelle über, die eben noch roth war, und so dehnt sich zuweilen eine Farbe über eine grosse Strecke aus. Zuweilen tritt plötzlich in einer und derselben Zelle ein farbloser Fleck auf, in der Mitte oder am Rande, grösser oder kleiner, während der übrige Theil noch in voller Farbe prangt. Verwandelt man jetzt das durchfallende Licht in auffallendes, so leuchtet der Fleck in vollem Metallglanze, während die übrigen vorher und nachher gefärbten Partien dunkel sind. Die Zeiträume, innerhalb welcher diese Phänomene verlaufen, sind verschieden lang, oft wechselt in einer Secunde die Farbe dreimal, oft währt eine Farbe mehrere Secunden lang. Mit dem Tode des Thierchens, wo sich der feinkörnige Inhalt jedesmal gegen die Mitte hin zusammendrängt, ist die ganze Erscheinung erloschen."

Ich habe die ganze Beschreibung *Gegenbaur's* citirt, weil meine eigenen das Phänomen des Farbenwechsels bestätigenden Beobachtungen nicht bis in die von *Gegenbaur* erforschten Details für die Aufeinanderfolge der Farben eingedrungen sind. Indess kann ich mich nicht in allen Stücken mit *Gegenbaur* einverstanden erklären. Zunächst bilden nach meinen Beobachtungen die polygonalen Felder allerdings eine unter der Cuticula gelegene Schicht, aber sie sind 1, keine Zellen 2, liegen sie nur unter der Rückenfläche, 3, konnten sie im weiblichen Geschlechte nicht gesehen werden. Dass sie keine einfachen Zellen des Matricalepithels sind, geht nicht nur aus ihrer Grösse hervor, welche bei *Sapph. auronitens* 0,08 mm, bei *Saph. fulgens* 0,1 mm. im Durchmesser beträgt, also mit den kleinen Zellen der Matrix anderer Copepoden gar nicht verglichen werden kann, sondern vor Allem aus dem Verhalten der Begrenzung. Die polygonalen Platten sind nicht von einer festen Membran umgeben, sondern zeigen sehr feingezackte Umrisse. Man hat es mit dünnen Platten einer feinkörnigen Substanz zu thun, mit Platten, welche durch suturenartig in einandergreifende Ränder begrenzt sind, und (bei *Sapph. nitens*) häufig äusserst dichte und zarte Streifen ähnlich wie gewisse Lepidopteren-Schuppen darbieten. Kerne, wie sie *Gegenbaur* für eins der drei von ihm gezeichneten Felder abbildet, habe ich niemals deutlich und regelmässig beobachtet und ich kann mich auch aus diesem Grunde nicht dazu verstehen, die Felder für Zellen zu halten. Weit eher entsprechen dieselben ganzen Complexen von verschmolzenen und veränderten Zellen der Matrix, für die ich keine zweite tiefere Lage eines Epithels nachweisen konnte. Ferner habe ich hervorzuheben, dass der Farbenschimmer keineswegs mit dem Tode des Thieres erlischt, der nur den wunderbaren Wechsel der Farben, die Veränderung derselben Theile von Blau in Roth &c. aufhebt. Der goldgrüne Metallglanz (*S. auronitens*) sowohl als das grünlich violette Farbenspiel (*S. fulgens*) finde ich an einigen seit Jahren in diluirter Glycerinlösung aufbewahrten Formen prachtvoll erhalten. Auch bei *Sapphirinella mediterranea* treten unter der Cuticula die nämlichen polygonalen feinstreifigen Felder auf und zeigen bei auffallendem Lichte einen schwach violetten, bei durchfallendem einen blassgelblichen Schimmer.

Eine Erklärung der besprochenen Farbenerscheinungen wage ich nicht im Detail auszuführen. Von einem Vergleiche der polygonalen Tafeln mit aneinandergereihten Glasprismen, die das Licht in die Spectralfarben zerlegen, kann natürlich keine Rede sein, vielmehr haben wir es mit Interferenzerscheinungen zu thun, welche ihren Sitz in dem feinkörnigen zuweilen wie in Sprüngen und Rissen zerspaltenen Gefüge der Tafeln haben. Vollkommen dunkel aber bleiben die höchst merkwürdigen Farbenveränderungen während des Lebens in den einzelnen polygonalen Feldern, die, wenn auch nicht dem Willen des Thieres unterworfen, doch von Vorgängen des Stoffwechsels abhängig zu sein scheinen, in denen auch die Ursache für das Leuchtvermögen der *Sapphirinen* zu suchen hat.

l.c. pg. 152. Das Männchen von *Sapphirina nigromaculata Claus* hat keinen Farbenschimmer.

Incomplete
DAR 110: B111–12; DAR 81: 167

CD ANNOTATIONS
1.1 The last … selection.— 3.23] *crossed ink*
4.1 As … Americans. 4.3] *crossed pencil*
5.1 *Oncidium* … bad.— 8.7] *crossed blue crayon; enclosed in square brackets brown crayon; 'See last page' added ink*
9.1 At … unions?— 9.6] *crossed brown crayon*
17.1 The sterile … plants. 19.1] *crossed red crayon; after opening square bracket red crayon*

[1] Letters to Fritz Müller, 7 February [1867] and 22 February [1867].
[2] In his letter to Müller of 22 February [1867], CD had asked whether Müller knew of any 'lowly organized' animals in which the male differed from the female in 'arms of offence' or ornamentation.

[3] *Gelasimus* or *Uca* is the genus of fiddler crabs. CD cited Müller for this information in *Descent* 1: 333 and 336. Müller wrote a short paper on colour change in crabs that includes a description of this *Gelasimus* (F. Müller 1881). The species Müller observed was probably *Uca leptodactyla* (see Crane 1975, pp. 304–7).

[4] CD used this information in *Descent* 1: 335–6. Müller refers to Captain James Cook, Carl Gegenbaur, and Carl Friedrich Claus. With his letter Müller enclosed a handwritten extract from Claus's *Die freilebenden Copepoden* (Claus 1863, pp. 35–7); CD cited page 35 in *Descent* 1: 336. Gegenbaur discussed *Sapphirina* in Gegenbaur 1858, Claus in Claus 1863, pp. 149–53. See also n. 22, below.

[5] Müller refers to Adolph Gerstaecker; *Sparassus smaragdulus* (now *Micrommata virescens*) is discussed in Gerstaecker 1863, p. 338, and *Descent* 1: 337.

[6] Müller refers to the family Syllidae (marine worms). See Gerstaecker 1863, p. 439. See also *Descent* 1: 327, where CD repeats much the same information.

[7] CD had enclosed a handwritten copy of his questionnaire on expression with his letter to Müller of 22 February [1867]. Destêrro (now Florianópolis) is on Santa Catarina Island off the coast of south-eastern Brazil.

[8] CD had asked whether *Oncidium flexuosum* was native to Müller's part of Brazil in his letters to Müller of 7 February [1867] and 22 February [1867]; see also letter from Fritz Müller, 4 March 1867.

[9] See letter to Fritz Müller, 22 February [1867] and n. 3. CD noted Müller's sole observation of a parasitic cryptogam in *Variation* 2: 134.

[10] For Müller's most recent report of self-impotent (self-incompatible) orchids whose pollen and stigma were poisonous to each other, see his letter of 4 March 1867. Müller had travelled to Theresopolis in January 1867 (see letter from Fritz Müller, 2 February 1867 and n. 14).

[11] CD reported Müller's observations of *Oncidium* species in *Variation* 2: 134. For Müller's earlier discussions of *Oncidium* species, see *Correspondence* vol. 14, letter from Fritz Müller, 1 December 1866, and this volume, letters from Fritz Müller, 1 January 1867, 2 February 1867, and 4 March 1867.

[12] Müller most recently mentioned experiments on *Epidendrum cinnabarinum* in his letter to CD of 4 March 1867. CD reported his results in *Variation* 2: 134.

[13] Müller refers to Friedrich Hildebrand and to Hildebrand 1866a. CD mentioned this paper in his letter to Müller of 22 and 25 September [1866]; Hildebrand had concluded from his experiments that flowers of *Corydalis cava* were sterile with their own pollen, and most fertile with the pollen of any other individual plant of the species.

[14] Müller had sent CD a specimen from the family Rubiaceae with his letter of 4 March 1867.

[15] Hippolyte Gautier has not been further identified (see also letter from Fritz Müller, 2 February 1867 and n. 3).

[16] Müller refers to Heinrich Gustav Reichenbach and possibly to Reichenbach 1852. For Müller's earlier discussions of *Epidendrum*, see *Correspondence* vol. 14, letter from Fritz Müller, 1 December 1866, and this volume, letters from Fritz Müller, 1 January 1867, 2 February 1867, and 4 March 1867. See also *Variation* 2: 134.

[17] See letter from Fritz Müller, 1 January 1867 and n. 4.

[18] CD paraphrased the information in this paragraph, and cited Müller, in *Variation* 2: 135.

[19] CD had received copies of Müller's paper on climbing plants on 1 January 1867, and sent Müller eleven of them (*Correspondence* vol. 14, letter to Fritz Müller, [late December 1866 and] 1 January 1867). The paper was composed of parts of three letters from Müller to CD, edited by CD, and published in the *Journal of the Linnean Society (Botany)* (F. Müller 1865; the paper is also published in *Correspondence* vol. 13 as the letter from Fritz Müller, [12 and 31 August, and 10 October 1865]).

[20] For a translation of the text of this enclosure, see Appendix I.

[21] In *Descent* 1: 335–6, CD cited Claus's description of colour change in *Sapphirina*.

[22] William Anderson's identification of *Oniscus fulgens* is discussed in the account of Cook's third voyage to the Pacific Ocean from 1776 to 1780, during which Cook was killed (Cook and King 1784, 2: 257). John Vaughan Thompson's account of *Sapphirina* is in Thompson [1828–34]; Franz Julius Ferdinand Meyen's is in Meyen 1834, pp. 153–5. Claus also refers to Joseph Banks.

[23] Claus refers to the *Archiv für Anatomie, Physiologie und wissenschaftliche Medizin*, which was edited by

Johannes Peter Müller between 1834 and 1858. The lengthy quotation given by Claus is from Gegenbaur 1858, pp. 66–8, with some brief omissions.

From Hermann Müller 1 April [1867][1]

Lippstadt
April 1.

Dear Sir

I am very much obliged to you for your extraordinary kindness in having sent me the paper on climbing plants by my brother and moreover your own work on this object, from which I had only read an abridgement in the botanical journal Flora, likewise for your bountiful communications on the papers published lately on the fertilization of flowers.[2] The lecture of Hildebrand's book "Geschlechtervertheilung", sent me several days ago by Prof. Hanstein, has convinced myself that this department of observation, in which I intended to betake myself, is reaped almost as thoroughly as the fertilization of Orchids has been by your admirable work. Some few important details would possibly be found in examining in this regard our indigenous flowers, but of decisive importance would chiefly rest only the trial of breeding for several generations plants of the small always closed flowers of Lamium amplexicaule and of other similar species.[3]

I congratulate with my brother for having so abundant an opportunity and so perfect an ableness in discovering new interesting facts of some importance for your theorie.

With respect to Subularia I am sorry to say that it does not grow in Westfalia.[4]

Of Pyrola two species (minor and rotundifolia) grow near Lippstadt and I shall not neglect to look on whether one of them is dimorphic and to advertise you.[5]

With Lopezia at this day at first I have begun, reminded by you and by Hildebrand's book, to mark the individual flowers in order to watch more closely the revolution of the single parts. Indoubtedly L. miniata will behave as L. coronata observed by Hildebrand.[6]

With my sincere thanks I remain | My dear Sir | Yours most respectfully | H Müller

DAR 171: 289

[1] The year is established by the relationship between this letter and the letter to Fritz Müller, [late December 1866 and] 1 January 1867 (*Correspondence* vol. 14).

[2] CD had sent Müller a copy of Fritz Müller's paper on climbing plants (F. Müller 1865; see *Correspondence* vol. 14, letter to Fritz Müller, [late December 1866 and] 1 January 1867, and this volume, letter to Hermann Müller, 29 March [1867]). CD had also sent a copy of his own paper, 'Climbing plants'. A summary of 'Climbing plants' was published in the German journal *Flora* in June 1866 (*Flora* n.s. 24 (1866): 241–52). Müller had asked about recent literature on the pollination of plants by insects in his letter of 23 March 1867. The letter in which CD sent this information has not been found.

[3] Müller refers to Friedrich Hildebrand, to Hildebrand's *Die Geschlechter-Vertheilung bei den Pflanzen* (Sexual division in plants; Hildebrand 1867a), to Johannes von Hanstein, and to CD's *Orchids*. In his letter to CD of 23 March 1867, Müller had offered to repeat CD's observations on the pollination of orchids

by insects, using plants available to him in Westphalia. He discussed *Lamium amplexicaule* briefly in
H. Müller 1873, pp. 312–13.

[4] The letter in which CD enquired about *Subularia* has not been found. *Subularia* was thought to flower
under water with the corolla closed, which would make crossing with another individual impossible
(see *Natural selection*, pp. 62–3 and n. 2, and *Correspondence* vol. 6). In a letter to Asa Gray, 18 June
[1857] (*Correspondence* vol. 6), CD wrote: 'Podostemon & Subularia under water (& Leguminosæ) seem
& are strongest cases against me.' See also *Forms of flowers*, pp. 311–12.

[5] The letter in which CD made this enquiry has not been been found. In a letter to C. C. Babington,
20 January [1862] (*Correspondence* vol. 10), CD wrote that he had read that *Pyrola* was dimorphic,
and complained that he himself would never be able to see or experiment on it. In *Forms of flowers*,
p. 54, CD wrote: 'from other statements it appeared probable that Pyrola might be heterostyled, but
H. Müller examined for me two species in North Germany, and found this not to be the case'.

[6] Hildebrand had discussed *Lopezia coronata* in Hildebrand 1867a, pp. 22–3; CD remarked that the
case was new to him in his letter to Friedrich Hildebrand, 20 April [1866]. Müller described the
structure and pollination mechanism of *L. miniata* in his letter to CD of 23 March 1867. He thought
he had identified male and female flowers; however, Hildebrand had observed that in *L. coronata*,
the male parts of the flower matured first, and were replaced by the female parts when they had
withered away. The flowers were therefore dichogamous and it was impossible for them to pollinate
themselves.

From V. O. Kovalevsky 2 April 1867

S Petersburg
March 21/2 Ap. 1867[1]

Sir

I am very grateful for the very kind answer You gave to my letter;[2] in order to
occupy as little as possible of your valuable time, I shall be very short and precise
in my answer.

I certainly determine on a translation of Your new work, but in reference to
woodcuts I could make them as well here, by the artists of the Academy, the more
so as I presume that Mr Murray will be a little tickled, as an editor, to give to
somebody else stereotypes of a work which he has not already finished himself, but
if in some short lines, which I expect in answer, you shall give me the permission
to apply to M^r. Murray, I shall do so, and if his charches for stereotypes are not
much higher than the woodcuts made here, I shall certainly be very glad to receive
them beforehand.[3] All I shall ask You, dear Sir, is to inform Mr Murray, that you
have had already the kindness to stipulate with me and accept my propositions, so
that, in the very improbable case he should receive a similar proposition, not to
accept it, or at least to inform me and give me a little preference.[4] For my part, I
have made similar conditions, in absence of a litterary treaty, with some continental
writers, as M^r. Ch. Vogt, Rosmässler, Billroth[5] and other, for receiving early printed
sheets of some of their works, and shall be very happy to have the same advantage,
over other editors who do not like to honour the right of litterary property, also in
English books.

It was very foolish to ask You for a special Introduction, not Knowing the public
you speak to You shall certainly be at a loss to say something.—[6]

Your former book, the "Origin of Species" is translated and printed some three years ago, but I understand that the translation is made not from the original but from the German translation of Mr Bronn, the late Prof. at Heidelberg, and with his remarks.[7]

Mr Truebner informed me of the conditions You would accept, but I think it shall be better to inform me about it directly, so that I could make a remittance on some London house, for the first half of the first volume.[8]

Dear Sir | Yours faithfully | W. Kowalewsky

DAR 169: 72

CD ANNOTATION

Top of letter: 'Wood-cut price of | Payment to me' *red crayon*

[1] Kovalevsky gives both the Julian (21 March) and Gregorian (2 April) calendar dates.

[2] Kovalevsky refers to his letter to CD of 15 March 1867 and CD's letter to him of 26 March [1867].

[3] Kovalevsky refers to his planned translation of *Variation* (V. O. Kovalevsky trans. 1868–9). In his letter of 26 March [1867], CD had advised Kovalevsky to negotiate directly with his publisher, John Murray, about stereotypes of the woodcuts.

[4] CD mentioned Kovalevsky's concern about the cost of the stereotypes in his letter to John Murray, 10 April [1867]. No letter concerning Kovalevsky's position as preferred translator has been found.

[5] Kovalevsky refers to Carl Vogt, Emil Adolph Rossmässler, and Theodor Billroth. Kovalevsky published translations of Vogt 1863 in 1864; of Rossmässler 1863 in 1867; and of Billroth 1863 in 1866 (Davitashvili 1951).

[6] Kovalevsky had asked CD to write a short introduction or preface to his translation; see letter from V. O. Kovalevsky, 15 March 1867, and letter to V. O. Kovalevsky, 26 March [1867].

[7] CD had asked whether there had been a Russian translation of *Origin* in his letter of 26 March [1867]. The first Russian translation of *Origin* was Rachinskii 1864. Kovalevsky also refers to Heinrich Georg Bronn, professor of natural science at Heidelberg University, who died in 1862, and to Bronn trans. 1860. Bronn had added editorial comments throughout the text and an additional final chapter of remarks challenging CD's theory.

[8] Nicholas Trübner's publishing company published Kovalevsky's translation of *Variation* (V. O. Kovalevsky trans. 1868–9). In his letter of 15 March 1867, Kovalevsky had asked CD to write to Trübner about financial arrangements; neither CD's letter to Trübner nor his reply to Kovalevsky has been found (see letter from V. O. Kovalevsky, 24 April [1867]). See also CD's annotation.

From Isaac Anderson-Henry 3 April 1867

Woodend, Maderty, | Crieff.
April 3/67

My dear Sir

Your very gratifying letter of the 31st ulto has reached me here this morning[1]

I shall have much pleasure in sending your letter and backing your request to Mr Traill, tho I know him but slightly. In fact, I do not know where his place "Aberlady Lodge" is; but I return tonight to my place (Hay Lodge Trinity) where I will learn from a neighbour who knows him, his proper address.[2] Aberlady is some 17 miles to the East—yet a Lodge with that name may be in the neighbourhood of Edin^gh. as I rather think it is, tho on looking the directory I find only one "Robt

Traill" with the addition "Vulcan Foundry Admiralty Street" which I think must be the same.[3]

Like you I was much struck with M[r] Traills remark, as long before I ever took to crossing & in very early life,—boyhood I believe,—I was told, I think by a Gardener, that a hybrid was produced by inserting one eye, such as a potatoe's or a barley corn within the other & the united growth made the hybrid! I never tried the experiment believing it to be a *myth*—and if *you* have tried & failed, I fear it is[4]

I am glad to hear again from you. Knowing your herculean labours I felt unwilling to intrude on your most valuable time with my letters[5]

You are pleased to regard my small testimony as more eulogistic than deserved. Allow me to assure you that it was amply borne out by the Society—and withal, it is but one small leaf added to the Chaplet which the whole scientific world has conferred.[6]

The hybrid pod on Rhod[n] Dalhousiæ was truly a monster of its kind—[7] Some of its seeds I sent to Dr Hooker & to other friends—but it went the same way, so far as I heard of it, with them as with me.

I enclose a print of the whole transactions of the Bot[l] Societys Meeting— a Communication I had from Professor Jameson of Quito may interest you—[8]

Believe me | My dear Sir | Ever faithfully Yours | Is: Anderson Henry

DAR 159: 67

[1] CD's letter to Anderson-Henry of 31 March 1867 has not been found.

[2] Joseph Dalton Hooker had sent CD a page from the *Farmer*, apparently giving details of a meeting of the Botanical Society of Edinburgh at which Robert Trail made some remarks about hybrid potatoes (see letter to J. D. Hooker, 4 April [1867], and letter from Robert Trail, 5 April 1867). The page of the *Farmer* has not been found in the Darwin Archive–CUL. Anderson-Henry usually wrote to CD from his Hay Lodge, Edinburgh, address; Madderty, near Crieff, is about ten miles west of Perth.

[3] A Robert Traill, iron founder, is listed at 9 Admiralty Street and 1 Regent Street in the *Post Office Edinburgh directory* 1863–4. Robert Trail later wrote to CD from Aberlady Lodge, Drem (letter from Robert Trail, 5 April 1867); Drem is about three miles east of Aberlady. CD had written to Trail on 1 April; see letter from Robert Trail, 5 April 1867.

[4] CD gave information supplied by Trail about producing potato hybrids in *Variation* 1: 395–6; he remarked that he had repeated the experiments on a large scale, but with no success (see also letter to J. D. Hooker, 4 April [1867]).

[5] The last extant correspondence between CD and Anderson-Henry is from 1863 (*Correspondence* vol. 11).

[6] The report of the meeting of the Botanical Society of Edinburgh in the *Farmer* evidently included Anderson-Henry's eulogy of CD; see letter to J. D. Hooker, 4 April [1867], and Anderson-Henry 1867a.

[7] In Anderson-Henry 1867a, p. 112, Anderson-Henry mentioned the large seed-pod resulting from a cross between *Rhododendron dalhousiae* and *R. nuttallii*. CD described this fruit in *Variation* 1: 400.

[8] There is a lightly annotated offprint from the *Farmer*, 20 and 27 March 1866, giving details of the meeting of the Botanical Society of Edinburgh on 14 March, in the Darwin Pamphlet Collection–CUL. It includes a communication from William Jameson, professor of chemistry and botany at the University of Quito, Ecuador, on the Compositae of the Andes. The communication is also reproduced in the *Transactions of the Botanical Society [of Edinburgh]* 9 (1866–8): 115–18. Much of the communication concerns the medicinal effects of *Chuquiraga insignis*.

From J. D. Hooker 3 April 1867

Kew
April 3^d/67.

Dear Darwin

We have had a terrible fight with the grim destroyer, but I do hope that we have pulled baby out of his clutches for the present.[1] Up to the present hour (10 am) he has had no fits for 12 hours—& is taking his Asses milk freely & looking brighter. I quite suppose it was a case of defective nutrition, the nervous system especially not being nourished—since he was weaned, though the child looked so well & plump, & took its food so freely, that no one guessed that anything could be the matter— I kept at him day & night, craming in nutrient by oil rubbing, keeping a lump of bacon in his mouth during the fits, which lasted 5, 8, & in one case nearly 10 hours, without intermission, & during which he took no other food. The affection was confined to nerves of motion— the last was on Monday, when he lay 5 hours, violently jerked twice every second, with perfect rhythmic regularity, each jump sufficient to have tossed him up in the air: the muscles all clenched as if with strychnine, except those of the face, which were all in motion—& the pulse at 156— No doctor had seen anything like it & I have had 5 at him—& I am sure that Sibson[2] was right, & that it is a mere case of defective nutrition.—that his blood is, like his mother's, deficient in red globules, or quality of these.

Mrs Hooker is all right again.[3]

Ever yr affec | J D Hooker

DAR 102: 157–8

[1] Hooker's youngest child, Reginald Hawthorn Hooker, who had been born on 12 January 1867, had been ill at the end of March (see letters from J. D. Hooker, 26 [and 27] March 1867 and 31 March 1867).

[2] Francis Sibson.

[3] In his letter of 31 March 1867, Hooker wrote that his wife, Frances Harriet Hooker, was 'pretty well', but staying in bed.

To J. D. Hooker 4 April [1867]

Down. | Bromley. | Kent. S.E.
April 4th

My dear Hooker

We both heartily rejoice that Mrs Hooker & your anxieties about your poor Baby are over.— It must have been very distressing. I never heard of anything like such convulsions. Thank you much for your two letters.—[1]

We have had a little uneasiness, now quite over, about Horace who came from School with intermittent fever, which lasted a fortnight & has made him very thin & has brought back his indigestion & we shall have to keep him here for a month more at least.[2]

You have done me a **very great** service in sending me the page of "The Farmer": I do not know whether you wish it returned; but I will keep it unless I hear that you want it. Old I. Anderson-Henry passes a magnificent but rather absurd eulogium on me, but the point of such *extreme* value in my eyes is Mr Traill's statement that he made a mottled mongrel, by cutting eyes through & joining two kinds of potatoes: I have written to him for full information & then I will set to work on a similar trial.[3] It would prove, I think, to demonstration that propagation by buds & by the sexual elements are essentially the same process, as Pangenesis in the most solemn manner declares to be the case.—[4]

I do hope that you will have no return of anxiety.—

My dear old Friend | Yours affect^ly | C. Darwin

Endorsement: '/67'
DAR 94: 19–20

[1] Hooker refers to Frances Harriet Hooker and Reginald Hawthorn Hooker. Reginald had been ill in late March but had recovered; Hooker wrote news of him in his letters to CD of 31 March 1867 and 3 April 1867.

[2] According to Emma Darwin's diary (DAR 242), Horace Darwin, their youngest son, returned to Down from school on 16 March 1867 with a fever, and began taking quinine on 1 April. He was attending Clapham Grammar School (CD's Classed account books (Down House MS)). There is no record of when he went back to school.

[3] The page of the *Farmer* that Hooker sent to CD evidently contained a report of the meeting or part of the meeting of the Botanical Society of Edinburgh on 14 March 1867. The page has not been found in the Darwin Archive–CUL. Isaac Anderson-Henry sent an offprint of the whole proceedings reprinted from the *Farmer*, 20 and 27 March, but it does not include Robert Trail's remarks. CD's letters to Anderson-Henry and Trail have not been found, but see the letter from Isaac Anderson-Henry, 3 April 1867, and the letter from Robert Trail, 5 April 1867. CD mentioned Trail's information in *Variation* 1: 395–6; he said he had repeated Trail's experiments without success.

[4] CD published his 'Provisional hypothesis of pangenesis' in chapter 12 of *Variation* (*Variation* 2: 357–404); he had discussed it with Hooker in 1866 and during March 1867 (see *Correspondence* vol. 14, letter to J. D. Hooker, 4 April [1866?], and letter from J. D. Hooker, [6? April 1866], and this volume, letter from J. D. Hooker, 20 March 1867, and letter to J. D. Hooker, 21 March [1867]). CD thought pangenesis could explain both sexual and asexual reproduction, as well as reversion and the regrowth of body parts (see *Correspondence* vol. 13, letter to T. H. Huxley, 27 May [1865], n. 7).

To John Murray 4 April [1867][1]

Down Bromley Kent
Ap. 4th

My dear Sir

Many thanks.— Herr Schweizerbart is the man.— But please take *no* steps about the stereotypes till you hear from me again; as I said to him I w^d. not agree about Translation until I knew that a certain Prof. V. Carus w^d. undertake this & that I have not yet heard.[2]

My dear Sir | Yours sincerely | Ch. Darwin

John Murray Archive

[1] The year is established by the relationship between this letter and the letter to E. Schweizerbart'sche Buchhandlung, [19 March 1867].

[2] The letter to which this is a reply has not been found, but evidently concerned the German translation of *Variation*. CD had offered to send clean proof-sheets of *Variation* to the German publisher Christian Friedrich Schweizerbart on condition that the translation was made by Julius Victor Carus, and had advised Schweizerbart to negotiate with Murray about stereotypes of the illustrations (see letter to E. Schweizerbart'sche Buchhandlung, [19 March 1867]). Schweizerbart wrote to CD on 22 March 1867 accepting his terms and saying that he had written to Carus to ask whether he would accept the commission.

To Thomas Blunt 5 April [1867][1]

Down. | Bromley. | Kent. S.E.
April 5th

My dear Sir

I write merely a line to thank you for your kind note.— We knew that your partnership was dissolved, but added the name from habit.—[2]

I most sincerely congratulate you on the success of your son in his scientific studies. Science runs in your blood.—[3]

The loss of my poor sister, Susan, has been a great one to many; & now after so many years there is no longer any connection between our family & Shrewsbury.[4]

With every good wish | Believe me | My dear Sir | Yours sincerely | Ch. Darwin

Houghton Library, Harvard University, bMS Am 1631 (95)

[1] The year is established by the reference to Susan Elizabeth Darwin's death; see n. 4, below.

[2] CD's letter to Blunt and Blunt's reply have not been found. *Post Office directory* entries suggest that the partnership of Blunt and Salter was dissolved sometime between 1863 and 1870; Blunt continued in business as a chemist with his son, Thomas Porter Blunt (*Post Office directory of Gloucestershire, with Bath, Bristol, Herefordshire, and Shropshire* 1863, 1870).

[3] Thomas Porter Blunt took a first class degree in natural sciences at Oxford University in December 1864 (*Oxford University calendar* 1865).

[4] Susan died in October 1866; she had lived at the Darwin family home, The Mount, in Shrewsbury. After her death, the house was sold. (See *Correspondence* vol. 14.)

From J. V. Carus 5 April 1867

Leipzig,
April 5th. 1867

My dear Sir,

You will have heard from Schweizerbart that I am willing to translate your new work, the more so, as it is my full conviction, that the progress of biology depends on the firm proving of your theory.[1] I am so very much occupied just now and within the next twelve months, that I should feel exceedingly obliged if you would kindly tell me, at what rate your work will be published. I have to complete my handbook of Zoology, then the historical Commission of the Munic Academy asked me to write the History of Zoology, which is to be finished Easter 1868.[2] The way in which Schweizerbart will publish the translation will partly depend on the form of the publication of the original. Now can you tell me, if the two volumes will come out at the same time, or one after the other. How long will they be? When will the print begin? The simplest way would be, I think, to send me directly the sheets. I shall do then what I can to keep pace with the English edition I trust you will kindly allow me to ask you directly on doubtful cases

You will have seen Haeckel's book on general Morphology.[3] I am very sorry that he has not been more moderate. So, as it is, it will do great mischief. There is of course no doubt that it is exceedingly well meant, and it contains capital observations on morphological questions. But by the many personal and quite unnecessary remarks, by the immoderate sharpness of many of his criticisms he weakens the effect. If he had written twenty sheets instead of two big volumes he would have done more and really good. His personal friends, especially Gegenbaur[4] and myself, tried in vain to mitigate his fury; and for what I know the critics and reviews he will laugh at, if they don't agree entirely with him One cannot even say, that this ought to be the way to promote a new theory or to further any new point of view. Instead of a true philosophy (I am sure he doesn't know Kant,[5] and I can prove it) he introduces a way of reasoning which is almost worse than the so called natural philosophy of yore. There is only one man, to whose judgement he would subdue; that is yours. He would not only mind very much what you tell him about his book, but your opinion will be no doubt the only one which will make some impression on him. He likes to consider himself as martyr of the new monistic theory of the world.[6] If you were to tell him that he did you a bad service I dare say he would be cured. I like him so very much that I am sorry for him. He is such an enthusiastic man that even his best friends have no influence over him. The last day, I stayed with Gegenbaur in Jena[7] and we both were convinced that one word of you would do more than long disputations with us.

Believe me | My dear Sir | Yours very faithfully, | J. Victor Carus

P.S. | Schweizerbart tells me that he has written to Mr Murray about the clichés.[8]

DAR 161: 58

[1] No letter from Christian Friedrich Schweizerbart, head of the German publishing firm E. Schweizerbart'sche Buchhandlung, confirming Carus's willingness to translate *Variation* has been found; however, Schweizerbart had told CD that he had written to Carus to ask him whether he was willing (see letter from E. Schweizerbart, 22 March 1867).

[2] Carus and Adolph Gerstaecker's *Handbuch der Zoologie* (Carus and Gerstaecker 1863–75) was published in two volumes, the first volume in 1863, the second in two parts in 1868 and 1875. Carus's *Geschichte der Zoologie* (Carus 1872) was published in Munich in 1872 as part of the Historical Commission of the Royal Academy of Science's series, *Geschichte der Wissenschaften in Deutschland* (History of science in Germany).

[3] Carus refers to Ernst Haeckel and Haeckel 1866. CD's annotated copy of Haeckel 1866 is in the Darwin Library–CUL (see *Marginalia* 1: 355–7). For an indication of CD's initial opinions, see the letter from T. H. Huxley, [before 7 January 1867] and n. 3, and the letter to Ernst Haeckel, 8 January 1867.

[4] Carl Gegenbaur.

[5] Immanuel Kant. For Kant's influence on biology see Grene and Depew 2004.

[6] Monism, as propounded by thinkers like Haeckel, asserts the fundamental unity of organic and inorganic nature, and abolishes traditional divisions between plant and animal, animal and human (see Tort 1996).

[7] Gegenbaur was a professor of anatomy at Jena.

[8] Carus refers to John Murray and to arrangements for the German publisher Schweizerbart to acquire stereotypes of the illustrations for *Variation* for Carus's translation.

To J. D. Hooker 5 April [1867]

Down
Ap. 5th

My dear H.

C. Nägeli writes to me that he has worked hard for 4 years on Hieracium to show causes & manner & steps of variation,—hybridism &c &c, & I sh^d. think from a very long letter that his results were valuable.[1] Now he wants me to get for him a complete set of British forms & will give in exchange a large set of German & Alpine forms.— Is there not some one in England who has made a special study of this genus?— If you cannot advise me better, should I apply to Babington (if he is well again) or to H. C. Watson??[2]

I am sorry to trouble you,—mere name & address w^d suffice.—[3]

Yours affect | C. Darwin

[Enclosure 1]

In Annales du Museum Tom. X. p. 471— A new genus of Umbelliferæ is described, Drusa, said to be confined to the Canary Isl^d.— the pericarp is furnished with elegant stellate hooks (see Plate at p. 456). Now this seed must have been developed or formed when mammals existed to transport it, ie not in Canaries.[4] Do you at all agree??

Copied from ancient notes of mine.—

[Enclosure 2]

(I possess old good note from Lowe on the very different soil of P. Santo compared with Madeira (& on proportion of endemic plants in P. Santo) so that some plants which grow in latter will not grow in P. Santo.—)[5]

(You remember, of course, Watson's paper in Hooker J. of Bot. on the changes which some Azorean plants underwent when cultivated in England.—)[6]

Endorsement: 'Replied 8.IV/67'
DAR 94: 14–16

[1] The letter from Carl Wilhelm von Nägeli, which was written on 31 March 1867 (see letter to C. W. von Nägeli, [after 8 April 1867]), has not been found. Nägeli published on *Hieracium* (hawkweeds) in a number of articles in the *Sitzungberichte der königlichen bayerischen Akademie der Wissenschaften zu München* 1866.

[2] Hooker had told CD of Charles Cardale Babington's illness in his letter of 4 December 1866 (*Correspondence* vol. 14); CD also refers to Hewett Cottrell Watson.

[3] CD's enquiry was passed to Daniel Oliver by Frances Harriet Hooker; see letter from Daniel Oliver, 8 April 1867.

[4] The text and plate mentioned are in A. P. de Candolle, 'Mémoire sur le *Drusa*, genre nouveau de la famille des Ombellifères', *Annales du Muséum d'Histoire Naturelle* 10 (1807): 466–71.

[5] The references are to Richard Thomas Lowe, and to the islands of Madeira and Porto Santo. See *Correspondence* vol. 6, letter from R. T. Lowe, 12 April 1856.

[6] The reference is to Watson's 'Supplementary notes on the botany of the Azores', in the *London Journal of Botany*, edited by William Jackson Hooker (*London Journal of Botany* 6 (1847): 380–97). CD referred to Watson's observations in *Origin*, p. 140. See also *Natural selection*, p. 126. The sections beginning 'In Annales . . .' and 'Copied from . . .' are on separate sheets of paper. They have been assigned to this letter on the basis of their position in the Darwin Archive–CUL.

From Robert Trail 5 April 1867

Aberlady Lodge | Drem
5th April 1867

Sir

I received yours of the 1st Instant last night and in reply beg to say that in listening to the very interesting and instructive paper by M[r] Anderson Henry on the subject of Hybridization of Plants it brought to my recollection the experiment which I had made on the Potato a great many years before.[1] I am afraid however that all I have to say on the subject will be very meagre and of very little use in a scientific point of view.—

The experiment was made with two sorts which are now supposed to have perished with the Potato disease.— They were at the period extensively cultivated in Forfarshire where I then was and were called the old Blue and the large American the latter a very coarse potato The sets were cut with a very sharp knife and only one eye was left which was divided as near as possible through the middle and carefully joined to the eye of the other and kept in its place by a piece of bass.— There might be about Sixty sets so treated.— The results were very various. Some produced Potatoes all white, some all blue; Some with part all white and part all blue and a few probably four or five had the produce regularly mottled.—[2] The only difference that I observed from the rest of the crop was that the Potatoes were much smaller in size though not less numerous

The two sorts experimented on, regularly produced the one white and the other blue flowers. Not having then any idea that the subject was of any interest I did not pay any attention to the colours of the flowers where the varieties seem to have united and in consequence of removal from the house and garden the experiments were not repeated as I had no opportunity at that time of continuing them and I had totally lost sight of the subject till I heard M[r] Henrys valuable Paper

The foliage of the blue Potato was of a much darker green than the other.— If there are any further particulars which you would wish to know it will give me great pleasure to supply them.— I am scarcely a Botanist but have often amused myself producing new varieties of florists flowers by crossing

   I am | Sir | Yours Faithfully | Robert Trail

DAR 178: 175

[1] CD's letter to Trail of 1 April 1867 has not been found. Isaac Anderson-Henry had read a paper on hybridisation before the Botanical Society of Edinburgh on 14 March (Anderson-Henry 1867a). CD had read of Trail's remarks in a report in the *Farmer* (see letter from Isaac Anderson-Henry, 3 April 1867, n. 2).

[2] CD reported Trail's remarks in *Variation* 1: 395–6, commenting that the potatoes that turned out part white and part blue, or mottled, afforded 'clear evidence of the intimate commingling of the two varieties'. This was an example of one of the types of reproductive phenomena that he hoped his provisional hypothesis of pangenesis would account for (*Variation* 2: 357–404).

From Frances Harriet Hooker [6 April 1867][1]

My dear M[r]. Darwin—

   Joseph went off to Paris yesterday morning at 7.AM.—in company with D[r]. Thomson—so your letter must wait awhile for an answer—[2] I do not know when he will return— I am happy to tell you that baby is very much better—recovering fast, in fact.— D[r]. Withecombe said this morning that he only wants food & nursing—[3] You & M[rs]. Darwin[4] will be glad to hear this.

   Poor Smith & his wife are in trouble about their baby, who is ill of inflammation on the chest—dying, I fear— it is not expected to live through the day.—[5]

   We were very sorry to hear of Horace's being ill—[6] I trust you will soon have no further cause for anxiety.

   Believe me | Y[rs]. affect[ly]. | F H Hooker

Kew. W. | Sat[y].—

DAR 102: 159–60

[1] The date is established by the relationship between this letter and the letter to J. D. Hooker, 4 April [1867]. In 1867, 6 April was a Saturday.

[2] Joseph Dalton Hooker was attending the Paris International Horticultural Exhibition as a juror for seeds and saplings of forest trees. Thomas Thomson, who had travelled in the Himalayas with Hooker and collaborated with him on various publications (*DNB*), was associate juror for hothouses and horticultural implements. (*Gardeners' Chronicle*, 6 April 1867, p. 348.) Hooker answered CD's letter of 4 April [1867] on 13 April.

[3] Reginald Hawthorn Hooker had recently recovered from an illness (see letter from J. D. Hooker, 3 April 1867 and n. 1). John Rees Withecombe was a medical practitioner in Richmond, Surrey.

[4] Emma Darwin.

[5] John Smith was curator of the Royal Botanic Gardens, Kew; his wife was named Mary. The child has not been further identified. J. D. Hooker sent CD news of the baby's death in his letter of 13 April 1867.

[6] CD mentioned the illness of his son Horace in his letter of 4 April [1867].

From Daniel Oliver 8 April 1867

Royal Gardens Kew
8 April 67

My dear Sir,

Mrs Hooker has put your note into my hands referring to Naegeli & *Hieracia.*[1]

The difficulty about them is this that hardly anyone has a good set of the peculiar *Highland* forms excepting M[r] James Backhouse Jr of York & the *few* to whom he has communicated specimens & I doubt if he would undertake to make up a set for Naegeli.[2]

M[r] Baker of the Kew Herbarium has a fair collection of British Hieracia (he has himself published a fasciculus of Teesdale or No. Yorks. species) & he would be glad either to *lend* Naegeli his collection entire—or to make him up a set of duplicates so far as his stores allow. He has a good set of French critical species but these prob[y]. N. has already from Jordan &c.[3]

Baker thinks neither Babington nor Watson likely to be able to help—at any rate in the way of *giving* N. a set.[4]

I sh[d]. have mentioned that D[r] Hooker is at Paris—a Juror[5]

Ever sincerely yrs | Dan[l] Oliver

DAR 173: 33

[1] Oliver refers to Frances Harriet Hooker, Carl Wilhelm von Nägeli, and the letter to J. D. Hooker, 5 April [1867].

[2] Backhouse published his *Monograph of the British Hieracia* in 1856 (Backhouse 1856). Nägeli wanted to exchange a set of German and Alpine species of *Hieracium* (hawkweeds) for a complete set of British species (see letter to J. D. Hooker, 5 April [1867]).

[3] John Gilbert Baker was an assistant in the Kew herbarium; he published on the botany of North Yorkshire, including *Hieracium*, in Baker and Nowell 1854 and Baker 1863. Oliver also refers to the French botanist Alexis Jordan.

[4] CD had suggested applying to Charles Cardale Babington or Hewett Cottrell Watson; see letter to J. D. Hooker, 5 April [1867].

[5] Joseph Dalton Hooker was a juror for seeds and saplings of forest trees at the Paris exhibition (*Gardeners' Chronicle*, 6 April 1867, p. 348).

From Carl Vogt[1] 8 April 1867

Genève
ce 8 Avril 1867

Monsieur et très-honoré Maître!

Je recois à l'instant même de mon éditeur et ami, Mr. J. Ricker, libraire à Giessen (Hesse) une lettre dans laquelle il m'informe, que vous allez publier un ouvrage en deux volumes sous le titre "Domesticated animals and cultivated plants etc.[2] Mr. Ricker me demande, si je ne suis pas disposé à traduire ce livre en allemand et dans le cas où vous n'auriez pas encore disposé du droit de traduction en faveur d'une autre personne, si vous vouliez nous confier à lui et à moi, comme éditeur et traducteur, une édition allemande? Mr. Ricker ajoute: "Veuillez demander à Mr. Darwin s'il demande des des droits d'auteur pour la traduction allemande et quelle serait la somme demandée par lui et, en outre, si son éditeur cède les clichés des illustrations qui doivent orner l'ouvrage et à quel prix?"[3]

Lors de ma dernière visite en Angleterre il y a quelques années, Mr. Wallace[4] et autres me disaient, que vous étiez trop souffrant et que je devais par conséquent renoncer au projet d'aller vous voir. Je me réjouis donc de vous voir rétabli au point de pouvoir songer à la publication d'un grand ouvrage, car c'est vous, Monsieur, qui avez ouvert des voies nouvelles à la science— c'est vous dont nous nous disons avec fierté les disciples.

Je ne vous aurai pas adressé ces lignes si votre ancien traducteur, Mr. Bronn, n'avait malheureusement succombé à de longues souffrances.[5] Vous connaissez sans doute mon nom comme homme de science— j'ai aussi quelques titres comme traducteur des "Vestiges of the natural history of creation", des Lecons de Mr. Huxley etc.[6]

Agréez, Monsieur et cher maitre, l'expression de mon attachement devoué | C Vogt

Prof. C. Vogt | Pleinpalais | *Genève* (Suisse)

DAR 180: 10

[1] For a translation of this letter, see Appendix I.

[2] Vogt was born in Giessen, studied there until 1835, and taught zoology there between 1846 and 1848, when he went to Geneva (Judel 2004). Franz Anton Ricker, Vogt's friend and publisher, took over his brother Joseph's publishing and bookselling business in Giessen in 1835 after his brother's death, keeping the name J. Ricker'sche Buchhandlung (*Hessische Biographien* 3: 301). Vogt also refers to *Variation*.

[3] CD was in the process of arranging for the German translation of *Variation* to be translated by Julius Victor Carus and published by E. Schweizerbart'sche Buchhandlung; see letter to John Murray, 4 April [1867].

[4] Alfred Russel Wallace.

[5] Heinrich Georg Bronn, who translated the first and second German editions of *Origin* (Bronn trans. 1860 and 1863), and *Orchids* (Bronn trans. 1862), died in 1862.

[6] Vogt was professor of geology at Geneva, and had applied CD's theory of descent to humans in his *Vorlesungen über den Menschen* (C. Vogt 1863). His translation of *Vestiges of the natural history of creation* ([Chambers] 1844) was published in 1851 (C. Vogt trans. 1851); his translation of Thomas Henry Huxley's *On our knowledge of the causes of the phenomena of organic nature* (T. H. Huxley 1863b) was published in 1865 (C. Vogt trans. 1865).

To Carl Wilhelm von Nägeli [after 8 April 1867][1]

Thanks for your long & valuable letter of March 31 on morphological laws &c.[2]
Your lecture on Hieracium, very valuable & instructive.—[3] I have received 2 or 3 days ago Photograph, which I am very glad to possess & your several memoirs

though I have not yet had time to read.—[4] Delay in answering your letter **About specimens**—[5]

"Copy of Origin".[6]

Munich. C. Nägeli.

Adraft
DAR 173: 33v

[1] The date is established by the fact that this letter was drafted on the back of the letter from Daniel Oliver, 8 April 1867.

[2] Nägeli's letter of 31 March 1867 has not been found, but see the letter to J. D. Hooker, 5 April [1867].

[3] The lecture has not been identified or found in the Darwin Pamphlet Collection–CUL, but see the letter to J. D. Hooker, 5 April [1867], n. 1.

[4] The photograph has not been found in the Darwin Archive–CUL. Nägeli published ten papers in the *Sitzungsberichte der königl. bayer. Akademie der Wissenschaften zu München* (1866), including three on *Hieracium* and four on hybridisation. They have not been found in the Darwin Pamphlet Collection–CUL.

[5] Nägeli wished to exchange a set of German and Alpine forms of *Hieracium* for a set of British forms; CD had written to Joseph Dalton Hooker for the name of someone wishing to make the exchange (letter to J. D. Hooker, 5 April [1867]), and in Hooker's absence CD's letter was answered by Daniel Oliver (letter from Daniel Oliver, 8 April 1867).

[6] CD may have offered to send a copy of the fourth edition of *Origin*, which was published in 1866.

To John Murray 10 April [1867][1]

Down. | *Bromley.* | *Kent. S.E.*
Ap 10

My dear Sir

Will you have a set of Stereotypes made for Herr Schweizerbart: he told you in his note how to forward them & how he w^d repay you. I am anxious that they sh^d be good stereotypes; & would it not be a good plan to have a proof struck from each sent to me for inspection?[2]

I should be *particularly* obliged if you w^d let me hear the cost of the whole set as soon as you know what it amounts to, for the Russian translator wants a set, but first wishes to know the cost, as he thinks he could get them done very cheaply in Russia—but I should prefer his having the originals.[3]

I am *sincerely* sorry to say that my corrections are frightfully heavy, though I had this part of the MS. copied & thought I had corrected it well.[4]

My dear Sir | yours very sincerely | Ch. Darwin

LS
John Murray Archive

[1] The year is established by the relationship between this letter and the letter from V. O. Kovalevsky, 2 April 1867.

[2] CD refers to stereotypes of the illustrations for *Variation*, which Murray was publishing. Christian Friedrich Schweizerbart was the director of the German firm that was publishing the German translation of *Variation*. See also letter from E. Schweizerbart'sche Buchhandlung, 22 March 1867, letter to John Murray, 4 April [1867], and letter from J. V. Carus, 5 April 1867. The letter from Schweizerbart to Murray has not been found in the John Murray Archive, but there is a ledger entry under *Variation* that reads: 'Sept 18th 1867 Schweizerbart £10'.

[3] The Russian translator was Vladimir Onufrievich Kovalevsky; see letter from V. O. Kovalevsky, 2 April 1867.

[4] CD was correcting the proof-sheets of *Variation* (CD's 'Journal' (Appendix II)).

To J. V. Carus 11 April [1867][1]

> *Down. | Bromley. | Kent. S.E.*
> Ap 11

My dear Sir

First for my own book: I am quite delighted to hear that you will undertake the translation. I am sorry to say my book is very large, much larger I fear than it is worth. A considerable part is a compilation, but it contains a good many original observations. I advised Schweizerbart not to undertake the publication until he had submitted a large part to some competent judge; but he seems resolved & I am of course glad of it.[2]

It consists of 2 Vols. *large* 8vo. with I fear *at least* 500 pages in each. The 2 Vols will be published together next Nov.[3] I have already corrected 250 pages & I suppose the first clean sheets will soon be printed off, & they shall be sent direct to you. I have written to Murray about the stereotypes.[4] I fear from what you say about y^r many works in hand, that there must be a considerable delay in the German edition.[5]

With respect to Haeckel's book, I admire & quite sympathize with all that you say about him with so much kindness & frankness. I agree with you that the book is too large.[6] It is, also, I believe bad policy to speak so positively as he does about any disputed theory.

I am so poor a German scholar that I have read here & there only portions, but these seem to me to be put with admirable clearness, force & method. With respect to the severity of his criticisms I am in a difficulty, for I do not know German well enough to perceive the sting of all his remarks, & these passages I have merely skimmed over. Nevertheless I have stumbled on some severe & contemptuous remarks on diff^t authors; & Fritz Müller has written to me from Desterro to the same general effect with you.[7] Hence I can, without mentioning any names, say that I have heard from several quarters of the severity of his criticisms, & I will speak strongly how injudicious I think this is, for my experience leads me to agree with you that severe strictures do no good, but only cause bitter anger.[8] I fear however that my remonstrance will be of no use, for when a man has once taken to writing in letters of blood & is conscious of his own power, he seldom, as far as I see, is ever cured of this bad practice.

With sincere respect I remain | my dear Sir yours very truly | Ch. Darwin

LS(A)
Staatsbibliothek zu Berlin, Sammlung Darmstaedter (Carus 4)

[1] The year is established by the relationship between this letter and the letter from J. V. Carus, 5 April 1867.

[2] Carus had confirmed his willingness to undertake the German translation of *Variation* in his letter of 5 April 1867. CD's letters to Christian Friedrich Schweizerbart, the publisher of Carus's translation, have not been found; a draft letter to E. Schweizerbart'sche Buchhandlung of [19 March 1867] does not mention submitting the book to a reviewer.

[3] The first volume of *Variation* contained 411 pages, and the second 486, excluding preliminaries. They were published in January 1868 (Freeman 1977).

[4] CD refers to John Murray, the English publisher of *Variation*; see letter to John Murray, 10 April [1867].

[5] Carus had written that he would be very much occupied for the next year; see letter from J. V. Carus, 5 April 1867.

[6] Carus had criticised the aggressive tone and the length of Ernst Haeckel's *Generelle Morphologie* (Haeckel 1866) in his letter to CD of 5 April 1867.

[7] The letter to CD in which Müller criticised Haeckel 1866 has not been found; however, Müller remarked in a letter to his brother, Hermann Müller, of 11 February 1867 (Möller ed. 1915–21, 2: 115–16), that the book would stir up controversy.

[8] Carus had suggested CD write to Haeckel explaining that his combative style did CD no service (letter from J. V. Carus, 5 April 1867).

To Ernst Haeckel   12 April [1867][1]

Down. | Bromley. | Kent. S.E.
Ap 12.

My dear Sir

I hope you have returned home well in health, & that you have reaped a rich harvest in natural science.[2] I have been intending for some time to write to you about your great work, of which I have lately been reading a good deal.[3] But it makes me almost mad with vexation that I am able to read imperfectly only 2 or 3 pages at a time. The whole book w^d be infinitely interesting & useful to me. What has struck me most is the singular clearness with which all the lesser principles & the general philosophy of the subject have been thought out by you & methodically arranged. Your criticism on the struggle for existence offers a good instance how much clearer your thoughts are than mine.[4]

Your whole discussion on dysteologie has struck me as particularly good.[5] But it is hopeless to specify this or that part; the whole seems to me excellent. It is equally hopeless to attempt thanking you for all the honours with which you so repeatedly crown me. I hope that you will not think me impertinent if I make one criticism: some of your remarks on various authors seem to me too severe; but I cannot judge well on this head from being so poor a German scholar. I have however heard complaints from several excellent authorities & admirers of your work on the severity of your criticisms.[6] This seems to me very unfortunate for I have long observed that much severity leads the reader to take the side of the attacked person. I can call to mind distinct instances in which severity produced directly the opposite effect to what was intended. I feel sure that our good friend Huxley,[7] though he has much influence, w^d have had far more if he had been more moderate & less frequent in his attacks. As you will surely play a great part in science, let me as an older man earnestly beg you to reflect on what I have ventured to say. I know that it is easy to preach & if I had the power of writing with severity I dare say I sh^d triumph in turning poor devils inside out & exposing all their imbecility. Nevertheless I am convinced that this power does no good, only causes pain. I may add that as we daily see men arriving at opposite conclusions from the same premises it seems to me doubtful policy to speak too positively on any complex subject however much a man may feel convinced of the truth of his own conclusions. Now can you forgive me for my freedom? Though we have met only once I write to you as to an old friend, for I feel thus towards you.

With respect to my own book on Variation under domestication I am making slow, but sure progress in correcting the proofs. I fear that it will interest you but little, & you will be struck how badly I have arranged some of the subjects which you have discussed. The chief use of my book will be in the large accumulation of facts by which certain propositions are I think established. I have indulged in one lengthened hypothesis, but whether this will interest you or any one else, I cannot even conjecture.[8]

I hope before long you will write to me & tell me how you are & what you have been doing & believe me my dear Häckel yours very sincerely | Ch. Darwin

LS
Ernst-Haeckel-Haus (Bestand A-Abt. 1–52/13)

[1] The year is established by the reference to CD's working on the page-proofs of *Variation* (see CD's 'Journal' (Appendix II)).

[2] Haeckel had spent from November 1866 to March 1867 travelling and doing research on Tenerife and Lanzarote (see Haeckel 1867; see also letter from Ernst Haeckel, 12 May 1867).

[3] CD had received a copy of Haeckel's *Generelle Morphologie* (Haeckel 1866) in late 1866 (see letter to Ernst Haeckel, 8 January 1867). There is an annotated copy in the Darwin Library–CUL (see *Marginalia* 1: 355–7).

[4] CD wrote in his copy of Haeckel 1866, 2: 239: 'good criticism on my term of struggle for existence— says ought to be confined to struggle between organisms for the same end—all other cases are dependance—Misseltoe depends on apple' (*Marginalia* 1: 356).

[5] Haeckel discussed dysteleology ('the study of functionless rudimentary organs in animals and plants': *Chambers*) in Haeckel 1866, 2: 266–85; these pages are annotated in CD's copy in the Darwin Library–CUL (see *Marginalia* 1: 356–7). CD cited Haeckel for his discussion of rudimentary organs in *Descent* 1: 17.

[6] CD had received letters criticising Haeckel 1866 from Julius Victor Carus and Fritz Müller (see letter from J. V. Carus, 5 April 1867, and letter to J. V. Carus, 11 April [1867] and n. 7).

[7] Thomas Henry Huxley and Haeckel corresponded with one another; see Uschmann and Jahn 1959–60.

[8] CD refers to *Variation*, and to his 'provisional hypothesis of pangenesis' (*Variation* 2: 357–404).

To Carl Vogt 12 April [1867][1]

Down. | *Bromley.* | *Kent. S.E.*

Ap 12.

My dear Sir

I thank you sincerely for your very obliging letter. I look at it as a very great honour that a naturalist whose name I have respected for so many years should be willing to undertake the translation of my book.[2] But Herr Schweizerbart, who published the Origin of Species applied to me some time ago, & as he had persuaded Prof. Victor Carus to make the translation, I have agreed to his proposal.[3] The book, I am sorry to say, is very large, viz 2 vols. large 8vo, with I suppose at least 500 pages in each vol. Prof. Carus, though he has undertaken the translation informs me that he has much work on hand, & it is possible (though not probable) that when he hears (& I wrote to him on the subject yesterday) of the size of the book, & that several sheets will be printed immediately & sent to him, he may wish to give up the task.[4] In that case nothing w^d give me higher satisfaction than that Schweizerbart sh^d arrange with you, if that be possible, for a translation; for I have often heard of the fame of your excellent translations.[5] My present work I greatly fear is of much greater length than value. Its publication has been long delayed owing to ill-health from which I still suffer though in a less degree. The entire book will be published next November, & then I will do myself the pleasure of sending you a copy.[6] Permit me to add that I have lately read with extreme interest the English translation of your Lectures on Man.[7]

With the most sincere respect & with my best thanks I remain my dear Sir | yours very faithfully | Charles Darwin

P.S. I should very much like to possess a photograph of you if you will send me one; & I enclose one of myself in case you w^d like to have it.—[8]

LS
Bibliothèque Publique et Universitaire de Genève, Ms fr. 2188, f. 300–1

[1] The date is established by the relationship between this letter and the letter from Carl Vogt, 8 April 1867.

[2] Vogt had offered to prepare the German translation of *Variation* in his letter of 8 April 1867.

[3] CD refers to Christian Friedrich Schweizerbart, head of the firm that published the German translations of *Origin* (Bronn trans. 1860 and 1863), and Julius Victor Carus.

[4] See letter from J. V. Carus, 5 April 1867, and letter to J. V. Carus, 11 April [1867].

[5] See letter from Carl Vogt, 8 April 1867 and n. 6.

[6] See CD's 'Journal' (Appendix II). *Variation* was published in January 1868. Vogt's name appears on CD's presentation list for the first English edition of *Variation* (DAR 210:11. 33).

[7] There is a lightly annotated copy of Vogt's *Lectures on man* (C. Vogt 1864), a translation of C. Vogt 1863, in the Darwin Library–CUL (*Marginalia* 1: 824). See also letter to Edward Blyth, 23 February [1867] and n. 3.

CD may have sent a copy of a photograph of himself taken by Ernest Edwards during CD's visits to London in November 1865 or April 1866 (see *Correspondence* vol. 13, letter from E. A. Darwin to Emma Darwin, 25 [November 1865] and n. 3), or a copy of one taken by his son William Erasmus Darwin in 1864 (see *Correspondence* vol. 14, letter from W. E. Darwin, 8 May [1866] and n. 10). Vogt's photograph has not been found in the Darwin Archive–CUL.

From J. D. Hooker 13 April 1867

Kew
April 13 /67

Dear Darwin

That certainly is a curious observation of Traill's, but do you know I have little faith in I. A. Henrys soundness— I put several questions to him regarding his paper & I send his answer— he is a nice liberal & enthusiastic fellow, but not sufficiently exact I suspect— I do hope that the Potato case is a true one.[1]

I was greatly pleased with what I saw at Paris, & I think the Exposition is most unjustly abused in the English papers;[2] it will be far the finest thing of the kind when finished.

How curious is Horace having intermittent fever—[3] I hope it will not prostrate him too much. Our baby is fairly well & troubled only with a few muscular twitches—at times— Mrs Hooker was in Town with me yesterday for the first time for many months— I hope she will not knock herself up now, as the Governess has gone away for a month's holiday, & the boys come home next week. Charlie will go to his Grandmamas at Norwich, & I am going to ask Mrs Darwin if I may bring Willy to Down on Saturday 20th. till Monday.[4] One bed is sufficient, as he is a quiet-sleeper. I know she will not hesitate to say *no*, if in any way inconvenient.

Only fancy poor Smith has lost his infant, (the contemporary of mine), during my visit to Paris! & I only returned in time for its burial— it was a magnificent huge baby—. inflam. of lungs—[5]

I saw Huxley—well—yesterday. He Lubbock & I go to Brittany on 24th.[6]

If any of your boys go to Paris at Easter, there is my room which is mine for this month at their service, & there are others in same house, clean new & nice, at 3 frs. a night.—& they need not spend more than 7 fr. a day, in living like fighting-cocks, if they go the right way about it.

Ever Yr aff | J D Hooker.

DAR 102: 161–2

[1] Robert Trail's claim to have made a mottled hybrid potato by joining the eyes of two different types of potato had been reported in an account of a meeting of the Botanical Society of Edinburgh in the *Farmer*, sent to CD by Hooker; see letter to J. D. Hooker, 4 April [1867] and n. 3. The letter to Hooker from Isaac Anderson-Henry, whose paper on hybridisation inspired Trail's remarks, has not been found. For Anderson-Henry's paper, see Anderson-Henry 1867a.

[2] *The Times* published articles on the Paris exhibition, criticising in particular the failure to have the exhibits ready for the official opening on 1 April 1867, in articles on 1, 3, and 4 April 1867. See letter from J. D. Hooker, 14 March 1867 and n. 6.

3 CD mentioned his son Horace's ill health in his letter to Hooker of 4 April [1867].

4 Reginald Hawthorn Hooker, born in January 1867, had been ill, and Frances Harriet Hooker, Hooker's wife, had also suffered poor health; see letter from J. D. Hooker, 3 April 1867 and nn. 1 and 3. Hooker also refers to his elder sons William Henslow Hooker (aged 14) and Charles Paget Hooker (aged 11), and to his mother, Maria Hooker, who went to live in Norwich in 1867 (Allan 1967, p. 224). The governess has not been identified. J. D. Hooker and William Hooker did not visit Down; see letter to J. D. Hooker, 25 [April 1867].

5 Frances Harriet Hooker mentioned the illness of John Smith's baby in her letter to CD of [6 April 1867].

6 Hooker spent some time in Brittany with John Lubbock and Thomas Henry Huxley, exploring megalithic monuments (L. Huxley ed. 1918, 2: 89).

From J. D. Hooker [to W. E. Darwin?]¹   [13 April? 1867]²

Kew
Saty

Dear Darwin

Adoxa is a standing puzzle, which I have not myself examined except once with you at Down?³ It is the subject of various memoirs of which Profr Oliver⁴ refers me to the following—

Wydler in Flora 1850— p. 433.

— " — in Bot. Zeit. 1844. p 657.

Decaisne in Ann. Sc. Nat. ser ii vi. 72. under *Helwingia*.

There are also good dissertions in *Nees Gen. Plant. Floræ Germanicæ.*—⁵

These books are all in the Linnæan Library.⁶

We hope that the baby is now out of danger though it still has starts & twitches.⁷

The piece of Elers. pottery was I think at Browns an antiquity shop in Wardour street, on the right as you go up towards Oxford St., & some 5 doors from the latter.⁸

Ever sincerely yrs | Jos D Hooker

I was greatly pleased with Paris Exhibition though not yet half finished.⁹

DAR 186: 48

CD ANNOTATION
Verso of last page of letter: 'Victor Hugo | Shirt'

1 William Erasmus Dariwn is conjectured to be the addressee because of the existence of another letter from Hooker to CD on 13 April 1867 containing similar information, and because of annotations, possibly by W. E. Darwin, on the letter that are similar to notes in W. E. Darwin's botanical sketchbook (DAR 186: 43).

2 The date is conjectured from the similarity of the references to Hooker's baby and the Paris exhibition to those in the letter from J. D. Hooker, 13 April 1867. In 1867, 13 April was a Saturday.

3 There are some annotations about *Adoxa*, possibly by W. E. Darwin, on this letter.

4 Daniel Oliver often provided CD with bibliographic references.

5 Hooker refers to Wydler 1850 and 1844, Decaisne 1836, and Nees von Esenbeck 1835[-61].

[6] CD regularly borrowed books from the library of the Linnean Society.

[7] See letter from J. D. Hooker, 13 April 1867 and n. 4.

[8] John Philip and David Elers produced high-quality red stoneware using innovative techniques in Staffordshire between about 1690 and 1700; their work was known to Josiah Wedgwood I, CD's grandfather, whose pottery works were also in Staffordshire (G. Elliott 1998). There was an antique furniture warehouse run by William Brown at 14 Wardour Street (*Post Office London directory* 1867).

[9] See letter from J. D. Hooker, 13 April 1867 and n. 2.

From J. V. Carus 15 April 1867

Leipzig,
Apr. 15th. 1867

My dear Sir,

Many thanks for your kind letter which I received last night.[1] I answer it at once because I think it fair to ask you frankly which translator you would prefer for your work. After consideration I see I can manage it to have the first volume ready for November, so that the first half of the translation can be published at the same time with the whole of the original⟨.⟩ But as C. Vogt offered himself to do it, it depends, at least as far as the publishers will arrange it, on your will.[2] C. Vogt has most certainly a greater name than I. But by his unmeasured satyrical and I am sorry to say sometimes quite cynical extravagances he lost a good deal of the influence, his judgement could otherwise still have. Although I am not impartial, yet I don't think for the reason just mentioned that he is the fit person to introduce your work to the German public.

Just now he travels over Germany and delivers lectures like a travelling preacher, but not "ad propagandam fidem", but in favour of the materialism in the absurdest form.[3] I trust you understand me. I should most gladly desist from translating your work if you find an abler man to do it. But on the other hand I should be sorry for your work's sake, if it should be associated with the name of a man, who would contrast by his fighting and scoulding manners most singularly with the sober and earnest tenor of a book full of observations. I know, many of our younger naturalists are losing that feeling of scientific decency which according to ⟨my⟩ opinion is utterly indispensable with a true mind of research. Yet I should not think that I am standing alone with the judgment given above. Now you will decide and let me kindly know your will

I am very happy to see that you agree with me about Haeckel's book.[4]

Believe my dear Sir | Yours very faithfully | Prof J. Victor Carus

DAR 161: 59

[1] Carus refers to CD's letter to him of 11 April [1867].

[2] CD had planned to publish the two volumes of *Variation* in November 1867 (see letter to J. V. Carus, 11 April [1867]). It was eventually published on 30 January 1868 (CD's 'Journal' (Appendix II)). Carl Vogt had also offered to translate *Variation* (see letter from Carl Vogt, 8 April 1867), and CD had written to him on 12 April [1867], suggesting that if Carus was too busy to make the translation, Vogt might do it instead. For more on Vogt, see F. Gregory 1977 and Montgomery 1988.

[3] *Ad propagandam fidem*: for the propagation of the faith (Latin). For more on Vogt's controversial lectures, given in various European countries between 1867 and 1869, see W. Vogt 1896, pp. 177–84.
[4] Carus refers to Ernst Haeckel's *Generelle Morphologie* (Haeckel 1866); see letter from J. V. Carus, 5 April 1867, and letter to J. V. Carus, 11 April [1867].

To Asa Gray 15 April [1867][1]

Down. | Bromley. | Kent. S.E.
April 15

My dear Gray

How good you have been to take so much trouble about the Expression-queries.— I wish I had thought earlier of having them printed, for in that case I might have sent a dozen to each of my few correspondents, as it is I can think of no one to send them to, so do not want any more.[2]

By the way I have just thought of Thwaites in Ceylon & will send him a couple.[3]

I have been lately getting up & looking over my old notes on Expression, & fear that I shall not make so much of my hobby-horse, as I thought I could: nevertheless it seems to me a curious subject, which has been strangely neglected.[4]

I have seen no one for months (but Hooker I rejoice to say will be here next Saturday)[5] & have no news.— I am plodding on heavily correcting, & trying to make an atrociously bad style a little better, my book "on the Variation of Animals & Plants under Domestication": I would offer to send you clean sheets, but I do not think you would care to receive them. There is not much about plants, & what there is, is almost all mere compilation; it will be a fearfully big book in two vols. & I shall be the next 5 or 6 months merely correcting the press: it is enough to make one curse one's fate in being an author.—[6]

I manage to get a little amusement by some of my experiments.— I have proved that the trimorphic species of Oxalis behaves in exactly the same complicated manner in regard to their fertilisation as Lythrum.—[7] I am going on with my trials of the growth of plants raised from self-fertilised & crossed seeds, & begin now to suspect that the wonderful difference in growth & conststitutional vigour occurs only with exotic plants which have been raised by seed during many generations in England, but which are not properly visited by insects & so have been rarely crossed.—[8]

I have just heard of a case which has interested me hugely, & which I am inclined to believe is true; namely that by cutting the tubers of differently coloured potatoes through the eye, & joining them, you can make a hybrid or mongrel. I am repeating this experiment on a large scale, for it seems to me, if true, a wonderful physiological fact.[9]

Here is a long prose all about my own doings.

Farewell with many thanks | Yours most sincerely | Ch Darwin

Gray Herbarium of Harvard University (97)

[1] The year is established by the relationship between this letter and the letter from Asa Gray, 26 March 1867.

[2] Gray had had copies of CD's questionnaire on expression printed and had sent some to CD (see letter from Asa Gray, 26 March 1867). CD may have begun sending out a standard list of questions in December 1866 (see *Correspondence* vol. 14, letter to B. J. Sulivan, 31 December [1866]). CD had sent out a number of questionnaires in February 1867.

[3] CD did not send questionnaires to George Henry Kendrick Thwaites, director of the Peradeniya Botanic Gardens in Ceylon (Sri Lanka), until 1868, when he sent him a printed copy (see *Correspondence* vol. 16, letter to G. H. K. Thwaites, 31 January [1868]).

[4] CD's M and N notebooks (*Notebooks*), written between 1838 and 1840, and concerned in particular with human beings, contain notes on expression. See Barrett 1980 for additional notes. See also Browne 1985. *Expression* was published in 1872.

[5] CD refers to Joseph Dalton Hooker; see letter from J. D. Hooker, 13 April 1867.

[6] CD revised the page-proofs of *Variation* between March and November 1867 (CD's 'Journal' (Appendix II)).

[7] CD and Friedrich Hildebrand had worked simultaneously on the trimorphism of *Oxalis*; see *Forms of flowers*, pp. 169–83, and *Correspondence* vol. 14, letter from Friedrich Hildebrand, 11 May 1866, and letter to Friedrich Hildebrand, 16 May [1866]. CD's notes on *Oxalis* are in DAR 109 and 111. For CD's work on *Lythrum*, see 'Three forms of *Lythrum salicaria*' and *Correspondence* vols. 10–12.

[8] For CD's report of the start of these experiments, see *Correspondence* vol. 14, letter to Asa Gray, 10 September [1866]. In *Cross and self fertilisation*, CD gave the examples of *Lathyrus odoratus* (sweetpea) and *Pisum sativum* (common pea), which had been self-fertilised for many generations in Britain, and which insects rarely succeeded in pollinating, stating that their size and vigour could be increased by a cross with different stock (*Cross and self fertilisation*, pp. 157–63, 439).

[9] See letter from Robert Trail, 5 April 1867. CD mentioned Trail's information in *Variation* 1: 395–6; he said he had repeated Trail's experiments without success.

To J. D. Hooker 15 [April 1867]

Down
15th

My dear Hooker

Hurrah, Hurrah,—we will find a bed for you & Willy even if we turned some one out, not that we shall have to do anything of the sort.—[1]

My opinion of old I. A. Henry is just the same as yours.[2] Most of his experiments are failures & those that have succeeded are not recorded with sufficient exactitude. He has, however, elsewhere recorded with more detail his curious case of the ovaria of Rhododendron *directly* affected by foreign pollen, like your Chamærops–Date-palm case.—[3]

We are very glad at so good an account of the Baby;[4] as we shall see you so soon, Hurrah again, I will write no more.—

Some of our Boys will be hugely tempted by your most kind offer of your room in Paris—[5]

Yours affect[ly.] | C. Darwin

P.S. I fear we shall not be able to send the carriage for you on Saturday as we expect Lizzie home from Germany, & George has promised to meet with pony-

carriage a friend; but you had better enquire at "Railway Inn" near to Station.—[6]
We can send you back on Monday.—

Endorsement: April /67
DAR 94: 21–2

[1] Hooker planned to visit Down with his son William Henslow Hooker from 20 to 22 April 1867; however, the visit did not take place (see letters from J. D. Hooker, 13 April 1867, and letter to J. D. Hooker, 25 [April 1867]).

[2] Hooker gave his opinion of Isaac Anderson-Henry in his letter of 13 April 1867.

[3] CD reported Anderson-Henry's cross of *Rhododendron dalhousiae* with the pollen of *R. Nuttallii*, as a case of foreign pollen increasing the size of the ovary, in *Variation* 1: 400, citing Anderson-Henry 1863. Anderson-Henry also discussed the case in Anderson-Henry 1867a; there is a lightly annotated offprint of this paper from the *Farmer*, which Isaac Anderson-Henry sent with his letter of 3 April 1867, in the Darwin Pamphlet Collection–CUL. In his letter of 14 March 1867, Hooker reported a case of 'seeds' of a *Chaemerops humilis* crossed with a date-palm that were more like dates than like the regular 'seeds' of *Chaemerops*. Hooker's 'seeds' were fruits (drupes; see *Variation* 1: 399).

[4] CD refers to Reginald Hawthorn Hooker; see letter from J. D. Hooker, 13 April 1867.

[5] See letter from J. D. Hooker, 13 April 1867.

[6] CD refers to his children Elizabeth and George Howard Darwin. George's friend was probably Richard Paul Agar Swettenham, a fellow undergraduate at Trinity College, Cambridge; Emma Darwin's diary (DAR 242) for Saturday 20 April 1867 mentions 'Mr Swettenham'. Hooker would have been travelling to Bromley Station.

From John Murray 17 April [1867][1]

50^A, Albemarle S^t | W.
Ap. 17

My Dear Sir

The cost price of a set of Electrotypes from the woodcuts of your new work is £4" 10" — I think Mr Schweitzererbart sh^d pay £10" as for the Russian Edition, you can fix the price as you please—[2] I will insure that Mr Clowes makes good Electros—[3]

I can send those for Schweizerbart as he directs.

I remain | My Dear Sir | Yours faithfully | John Murray

Chas Darwin Esq

DAR 171: 348

[1] The year is established by the relationship between this letter and the letter to John Murray, 10 April [1867].

[2] In his letter to Murray of 10 April [1867], CD had asked Murray to have a set of stereotypes of the illustrations to *Variation* made for the publisher Christian Friedrich Schweizerbart for the German translation; he had enquired about the cost on behalf of the Russian translator, Vladimir Onufrievich Kovalevsky.

[3] Murray refers to William Clowes, a London printer.

From Carl Vogt[1] 17 April 1867

Genève
ce 17 Avril | 1867.

Monsieur et cher Maître!

Ci-jointe, mirant votre désir, une photographie.[2] Je pourrais presque dire avec Goethe:

Da hast Du, weil Du's willst, mein garstig Gesicht,—
Aber meine Liebe, die siehst Du nicht![3]

La votre m'a rappelé, d'une manière étonnante, un défunt ami, Théodore Parker de Boston.[4]

J'ai envoyé votre le⟨tt⟩re à Mr. Ricker— il tâchera de se consoler[5]

Bientôt, j'aurai le plaisir de vous envoyer un mémoire sur les microcéphales ou hommes-singes, fruit de mes études de l'année passée pour lesquelles la guerre me faisait des loisirs.[6] J'arrive à la conclusion, que cette conformation anormale est un atavisme, qui ramène vers le point de départ des deux souches, homme et singes—mais que ce point de départ n'est plus représenté dans la création actuelle. Je pense avoir élucidé la question d'une manière satisfaisante—autant que le permettent les faits à notre disposition.

Votre livre nouveau traitant des animaux domestiques, vous aurez sans doute remarqué le beau mémoire de Mr. Nathusius sur le cochon domestique et ses métamorphoses. C'est extrêmement important, puis qu'il démontre, d'une manière peremptoire, les changements survenus dans la conformation de la tête et surtout du groin pas suite de la domesticité.[7]

Agréez, Monsieur, l'assurance de ma considération très distinguée. | Votre devoué | C Vogt

Mr. Ch. Darwin.

DAR 180: 11

[1] For a translation of this letter, see Appendix I.

[2] CD asked for Vogt's photograph in his letter to Vogt of 12 April [1867]; it has not been found in the Darwin Archive–CUL.

[3] Vogt slightly modified a quotation from a poem by Johann Wolfgang von Goethe, 'Das garstige Gesicht':

's ist ungefähr das garst'ge Gesicht—
Aber meine Liebe siehst du nicht.
[This is nearly the most ugly face—
But you cannot see my love upon it.]

See Goethe 1988, p. 413.

[4] CD sent his photograph with his letter to Vogt of 12 April [1867]. Parker, an American Unitarian theologian, social reformer, and writer, died in 1860.

[5] Franz Anton Ricker was Vogt's friend and publisher; he had hoped to publish a translation of *Variation* by Vogt. In fact, *Variation* was translated by Julius Victor Carus and published by Christian Friedrich Schweizerbart (Carus trans. 1868).

[6] Vogt refers to his *Mémoire sur les microcéphales* (C. Vogt 1867); there is an inscribed, annotated copy in the Darwin Library–CUL (see *Marginalia* 1: 824–6). CD thanked Vogt for it in his letter to Vogt of

7 August [1867]. Vogt assembled a collection of microcephalic skulls at the meeting of the natural history section of the Institut Genevois in June 1866 (C. Vogt 1867, p. 4), near the time of the Seven Weeks War between Prussia, Austria, and allied German states in June and July 1866 (*EB*).

[7] Vogt refers to *Variation*, and to Hermann Engelhard von Nathusius and Nathusius 1864. There is a heavily annotated copy of Nathusius 1864 in the Darwin Library–CUL (see *Marginalia* 1: 630–5); CD frequently cited this and other works by Nathusius in *Variation*. On changes in the shape of the skull in domestic pigs, see *Variation* 1: 71–3.

To J. V. Carus   18 April [1867][1]

Down. | *Bromley.* | *Kent. S.E.*

April 18th

My dear Sir

My letter has given you a false impression.— The wish never for a moment crossed my mind that Vogt should translate my book in preference to you; but I thought it *possible* from what you said of your other great undertakings in hand, that you might wish to give up the translation. In my letter to Prof. Vogt in thanking him for his wish I said that I thought it possible, *but **not** probable*, that you might, when you *heard* of *the size of my book*, wish to give it up, & in that case I sh^d. feel gratified by his undertaking the Translation. As you are not frightened at the undertaking, I sh^d. be *most truly* grieved that there should be any change.— I am sorry that my letter has given you the trouble of writing to me.—[2]

I am surprised that I have not yet received any clean sheets; but my printers, I know, sometimes have the habit of not beginning to print off, until the whole volume is set up in type.[3]

With cordial thanks for all your kindness, I remain | My dear Sir | Yours sincerely | Ch. Darwin

I wrote two days ago to Häckel.—[4]

I have been rather surprised at receiving an application for clean sheets for a Russian Translation of my Book—[5]

Staatsbibliothek zu Berlin, Sammlung Darmstaedter (Carus 31)

[1] The year is established by the relationship between this letter and the letter from J. V. Carus, 15 April 1867.

[2] CD refers to his letter to Carl Vogt, 12 April [1867], in which he suggested that Vogt might translate *Variation* into German if Carus was too busy. Carus had offered to make the translation in his letter of 5 April 1867, asking at the same time how long *Variation* would be and when it would be published; CD answered his questions in his letter of 11 April [1867].

[3] In his letter of 5 April 1867, Carus had asked CD to send him proof-sheets to translate.

[4] CD wrote to Ernst Haeckel on 12 April [1867]. Carus had asked CD to persuade Haeckel that his combative style did CD no service (see letter from J. V. Carus, 5 April 1867).

[5] The publishers Trübner & Co. wrote to CD on 26 February 1867, asking for proof-sheets of *Variation* on behalf of Vladimir Onufrievich Kovalevsky, who wanted to prepare a Russian translation. See also letter from V. O. Kovalevsky, 15 March 1867.

To W. B. Tegetmeier 20 April [1867]

Down. | Bromley. | Kent. S.E.

Ap 20.

My dear Sir

Many thanks for your offer, but I have the pamphlet; it is clever but has not changed my conclusions.[1]

I forget whether I have made any alterations about the pigeons in the Origin (excepting about a crossed bird about which I blundered) but to make sure, I send by this post the revises of the last Edit. which are correct excepting perhaps here & there a word;[2] but I must particularly beg you to be so kind as to return the sheet as it is of consequence to me.

My dear Sir | yours very sincerely | Ch. Darwin

LS

Endorsement: '1867'

Archives of The New York Botanical Garden (Charles Finney Cox collection) (Tegetmeier 104)

[1] The letter to which this is a reply has not been found; the pamphlet has not been identified.

[2] CD refers to the revised page-proofs of the fourth edition of *Origin*, which was published in 1866. By the 'crossed bird about which I blundered', CD may refer to a passage in the first edition that was altered in the third and later editions. The original passage read (*Origin*, p. 25):

> I crossed some uniformly white fantails with some uniformly black barbs, and they produced mottled brown and black birds; these I again crossed together, and one grandchild of the pure white fantail and pure black barb was of as beautiful a blue colour, with the white rump, double black wing-bar, and barred and white-edged tail-feathers, as any wild rock-pigeon!

In the third edition (*Origin*, p. 26), it read:

> I crossed some white fantails, which breed very true, with some black barbs . . . ; and the mongrels were black, brown, and mottled. I also crossed a barb with a spot, which is a white bird with a red tail and red spot on the forehead, and which notoriously breeds very true; the mongrels were dusky and mottled. I then crossed one of the mongrel barb-fantails with a mongrel barb-spot, and they produced a bird of as beautiful a blue colour, with the white croup, double black wing-bar, and barred and white-edged tail-feathers, as any wild rock-pigeon!

See also Peckham ed. 1959, p. 100.

From J. T. Moggridge 22 April [1867][1]

S.t Roch.

Ap. 22.

Dear M.r Darwin

I send by private hand a few plants of Orchis *intacta* Link— These are the best I can get but are rather far gone—[2]

This plant remains but a very short time in flower—

I have not been able to find any insect visiting the flowers; but as the pollen-grains are sufficiently loose & light to adhere to the dissecting needle when gently introduced within the hood of the flower, I think that a minute insect would probably bear away a grain or two in like manner.—

In rather advanced flowers the pollen-grains seem to lie quite loose round & on the stigmatic surface, & any moist object—or even a dry one as the needle—carries away pollen-grains—[3]

We propose to leave Mentone on May 6th. & to be in London (adress care of Nevil Maskelyne Esq[r]. 112 Gloucester Terrace | Hyde Park.) on May 18th.

If any plant is wanted from here which I can bring you, pray write & tell me— Y[rs]. most sincerely | J. Traherne Moggridge.

DAR 171: 211

[1] The year is established from the address, and from a botanical note made by CD concerning the specimens sent by Moggridge. Moggridge, who regularly spent the winter in the south of France, wrote to CD from St Roch, Mentone, on 9 November [1866] (see *Correspondence* vol. 14), 6 March [1867], and 15 March [1867]. In earlier years, he wrote from other addresses. CD made a note dated 1868: 'Orchis intacta from M[r] Moggridge in Italy— flowered under net and almost every flower produced five Pods' (DAR 70: 91). (Mentone is very close to the border between France and Italy.) Since *Orchis intacta* flowers in late winter and early spring, the note must have been made no earlier than the year after CD received the specimens.

[2] CD mentioned Moggridge's sending plants of *Orchis* or *Neotinea intacta* in 'Fertilization of orchids', p. 143 (*Collected papers* 2: 140–1), and in *Orchids* 2d ed., p. 27.

[3] In 'Fertilisation of orchids', p. 143 (*Collected papers* 2: 140–1), CD concluded that *Neotinia intacta* was adapted for both self-fertilisation and crossing.

To Fritz Müller 22 April [1867][1]

Down. | Bromley. | Kent. S.E.

Ap 22.

My dear Sir

I am very sorry your papers on climbing plants never reached you.[2] They must be lost, but I put the stamps on myself, & I am sure they were right. I despatched on the 20th all the remaining copies, except one for myself. Your letter of Mar 4th contained much interesting matter,[3] but I have to say this of all your letters.

I am particularly glad to hear that Oncidium flexuosum is endemic, for I always thought that the cases of self-sterility with orchids in hot-houses might have been caused by their unnatural conditions.[4] I am glad also to hear of the other analogous cases, all of which I will give briefly in my book that is now printing.[5] The lessened number of good seeds in the self-fertilized Epidendrons is to a certain extent a new case.[6] You suggest the comparison of the growth of plants produced from self-fertilized & crossed seeds; I began this work last autumn & the result in some cases has been *very* striking, but only as far as I can yet judge with exotic plants which do not get freely crossed by insects in this country. In some of these cases it is really a wonderful physiological fact to see the difference of growth in the plants produced from self-fertilized & crossed seeds, both produced by the same parent-plant; the pollen which has been used for the cross having been taken from a distinct plant that grew in the *same* flower-pot.[7] Many thanks for the dimorphic rubiaceous plant.[8] Three of your Plumbagos have germinated, but not as yet any of the lobelias.[9]

Have you ever thought of publishing a work which might contain miscellaneous observations on all branches of natural history, with a short description of the country & of any excursions which you might take. I feel certain that you might make a very valuable & interesting book, for every one of your letters is so full of good observations. Such books, for instance Bates's Travels on the Amazons,[10] are very popular in England.

I will give your obliging offer about Brazilian plants to D.[r] Hooker, who was to have come here to day, but has failed.[11] He is an excellent good fellow, as well as naturalist: He has lately published a pamphet, which I think you w[d] like to read, & I will try & get a copy & send you.[12]

Yours most sincerely. | C. Darwin

LS(A)
British Library (Loan 10: 15)

[1] The year is established by the relationship between this letter and the letter from Fritz Müller, 4 March 1867.

[2] CD told Müller in his letter of [before 10 December 1866] (*Correspondence* vol. 14) that he was sending copies of Müller's paper on climbing plants (F. Müller 1865; also printed in *Correspondence* vol. 13 as the letter from Fritz Müller, [12 and 31 August, and 10 October 1865]); Müller mentioned that the copies had not arrived in his letter of 4 March 1867.

[3] See letter from Fritz Müller, 4 March 1867.

[4] Müller had written that *Oncidium flexuosum* was completely infertile with its own pollen and fertile with the pollen of any other plant of the same species in his letter of 1 December 1866 (*Correspondence* vol. 14). CD had asked Müller whether this *Oncidium* was endemic to Brazil (see *ibid.*, letter to Fritz Müller, [late December 1866 and] 1 January 1867). CD reported Müller's findings in *Variation* 2: 134.

[5] In his letter of 4 March 1867, Müller listed some orchids whose pollen and stigma had the same deleterious effect on one another as did those of *Oncidium flexuosum*; CD named some of these in *Variation* 2: 135. CD was checking the proof-sheets of *Variation*. See also *Correspondence* vol. 14, letter from Fritz Müller, 1 December 1866, and this volume, letters from Fritz Müller, 1 January 1867 and 2 February 1867.

[6] In his letter of 4 March 1867, Müller recorded the production of seeds in *Epidendrum cinnabarinum* after pollination with own pollen, pollen of a distinct plant of *E. cinnabarinum*, and pollen of *E. schomburgkii*. CD reported the *E. cinnabarinum* results in *Variation* 2: 134.

[7] See letter to Asa Gray, 15 April [1867] and n. 8.

[8] See letter from Fritz Müller, 4 March 1867.

[9] In his letter to J. D. Hooker, 24 [March 1867], CD mentioned that Müller had sent him seeds of a dimorphic *Plumbago* and of two types of *Lobelia*, one of which was a climber. CD thanked Müller for them in his letter of 25 March [1867].

[10] Bates 1863.

[11] Müller had expressed surprise that Joseph Dalton Hooker and George Bentham had not been able to examine the seeds of some common Brazilian plants described in their *Genera plantarum* (Bentham and Hooker 1862–83), and had offered to 'satisfy any wish' of theirs regarding Brazilian plants (letter from Fritz Müller, 4 March 1867). Hooker had planned to visit CD from 20 to 22 April 1867 (see letter from J. D. Hooker, 13 April 1867).

[12] CD refers to a privately printed version of Hooker's article on insular floras in the *Gardeners' Chronicle* (J. D. Hooker 1866a). The text of the privately printed version, which had some changes from the *Gardeners' Chronicle* version, is reproduced in Williamson 1984. Müller thanked CD for the paper in his letter of 17 July 1867.

From Francis Parker 22 April 1867

39, Watergate Street, | *Chester.*
22nd. April | 1867

Dear Uncle Charles

You are probably aware that Aunt Susan left each of your younger children a Legacy of £*100*—[1] on the other side I send you an extract from her Will, and I now enclose you a Bankers order for £*600*— I think it will be more convenient that you should receive the Legacies and invest the amount in one sum and you can pay it to each as they come of age—

Perhaps you will have the goodness to write me a letter stating that you have received the amount of their Legacies on their behalves

I am | Yours very sincerely | Francis Parker

Charles Darwin Esq[re].
Down Bromley Kent.

Extract from the Will of the late Miss Susan Elizabeth Darwin dated 1st. November 1865—

"I give and bequeath to my nephews George Howard Darwin, Francis Darwin Leonard Darwin, and Horace Darwin and my nieces Henrietta Emma Darwin and Elizabeth Darwin (the younger sons and daughters of my Brother Charles Robert Darwin) One hundred pounds each"—

I send you in a separate cover the Undermentioned Legacy receipts for signature—

	Value
Geo. Howard Darwin ..	100
Francis Darwin	100
Leonard Darwin	100
Horace Darwin	100
Henrietta Emma Darwin	100
Elizabeth Darwin.......	100

Cha[s]. Rob[t]. Darwin
(Specific Legacy ... a portrait of Miss Darwins father[2]—valued at £4—)
Henrietta Emma Darwin
(a Silver Tea Urn—valued at £28·2·6)—

DAR 174: 19

[1] Susan Elizabeth Darwin, CD's sister, died in October 1866 (see *Correspondence* vol. 14). Henry Parker (1827/8–92) and his brother Francis were the executors of her will (Susan Elizabeth Darwin's will, Probate Registry, York).
[2] Robert Waring Darwin.

From Carl Vogt[1] 23 April 1867

Genève
ce 23 Avril | 1867.

Monsieur et très-honoré maitre!

Quelques jours après avoir reçu votre lettre, dont j'ai donné connaissance à Mr. Ricker, j'ai rencontré Mr. le Colonel Moulinié, dont vous connaissez sans doute le nom.[2] C'est un de mes plus anciens élèves, je puis dire— il a fait un excellent travail sur les larves des trématodes et de plus il a soigné la traduction de mes "Lecons sur l'homme" en français.[3] Je lui parlai de votre nouveau livre.[4] "Pardieu, me dit-il, ce serait quelque chose pour moi! Ma mère est, comme vous savez, anglaise,[5] je parle l'anglais comme ma langue maternelle et je n'ai rien à faire dans ce moment, de sorte que je pourrai me vouer corps et âme à une traduction du livre de Mr. Darwin, qui sera meilleure, j'éspère, que celle que Mlle. Royer a faite de son livre sur l'espèce."[6] Mr. Moulinié a donc écrit à notre éditeur, Mr. Reinwald,[7] 15 rue des S[ts]. Pères à Paris et je me fais un plaisir de vous envoyer ci-jointe la lettre de mon ami Reinwald en copie.

Il va sans dire que ce serait pour moi un véritable plaisir que de pouvoir revoir le travail de Mr. Moulinié et d'y ajouter une préface.[8] Si donc vous et votre éditeur n'aviez pas encore disposé du droit de la traduction française et que les prétentions, dont parle Mr. Reinwald N°. 3 ne fussent pas trop élevées, je vous prierai de favoriser ces deux amis. Quant aux question 1 et 2 que fait Mr. Reinwald, je lui ai transmis les renseignemens que contenait votre dernière lettre.[9]

Quant aux prétentions relatives au droit de traduction, je crois que Mr. R. a raison, d'après ce que je sais moi-même. La France est, en géneral, le plus mauvais marché pour des livres scientifiques, que l'on puisse imaginer— les romans et les text-books pour les écoles et collèges, voilà ce qui se vend— les livres scientifiques sont les plus mauvaises spéculations des libraires français.

Agréez, Monsieur et cher maitre, l'assurance de mon parfait dévouement | C Vogt

[Enclosure]

Par⟨is⟩
⟨   ⟩ Avril 1867

Monsieur le Colonel Moulinié Gene⟨v⟩e, 15 Rue du Montblanc
Monsieur

J'ai eu l'honneur de recevoir votre lettre du 22 par laquelle vous me faites une proposition qui me convient parfaitement.

Je crois comme vous que le nouveau livre de Mr. Darwin *"Domesticated Animals and Cultivated Plants"*,[10] pourr⟨ait⟩ être utilement traduit en français. Le livre aura, à mon ⟨avis⟩ du succès, s'il n'est pas trop cher. Il s'agirait donc a⟨vant⟩ tout de savoir:

1[o]. De quel étendue sera cette publication
2[o]. Si elle sera accompagnée de gravures sur ⟨bois⟩ ou fer acier
3[o]. À quelle conditions l'auteur ou l'éditeur ⟨   ⟩ voudraient vous céder le droit

de traduction en lang⟨ue française⟩ avec exclusion de toute autre traduction en Fran⟨ce⟩ ⟨ ⟩ et en Suisse.

Je ne ferai naturellement des démarches à ⟨ ⟩ ⟨ ⟩ lorsque vous m'y aurez autorisé. Je pense *[même]* ⟨ ⟩ mieux que cette demande fut addressée directe⟨ment⟩ ⟨ ⟩ Darwin plutot qu'à Mr Murray[11] (vie⟨ ⟩ ⟨ ⟩ anglais ont souvent ses prétentions trop élevées ⟨ ⟩ du continent) et encore que cette demande ⟨ ⟩ faite par un homme de lettres que par un libraire.

Si Mr. Vogt était en relation avec l'auteur anglais et s'il voulait nous permettre de le mêler à la negociation, comme par exemple, par la promesse de donner à notre traduction une préface ou une collaboration quelconque, nous serons a peu près certains de réussir au près de l'auteur.

Mais avant tout, il faut savoir a quoi on s'engage. J'attends donc vos renseignements à cet égard et je ne doute pas que les préliminaires une fois réglées, nous nous entendrons parfaitement ensemble sur le reste.[12]

Veuillez agréer encore mes rémerciements pour votre bienveillante communication et croire que je serai enchanté si nous ⟨ ⟩ rénouer nos relations et les faire fructifier. | J'ai l'honneur d'être Monsieur votre tout dévoué ser⟨viteur⟩ | C Reinwald

DAR 180: 12; DAR 176: 90

[1] For a translation of this letter, see Appendix I.
[2] Vogt refers to CD's letter to him of 12 April [1867], to his friend the publisher Franz Anton Ricker, and to Jean Jacques Moulinié.
[3] Vogt refers to Moulinié 1856 and Moulinié trans. 1865. Moulinié trans. 1865 was a translation of C. Vogt 1863.
[4] *Variation.*
[5] Moulinié's mother has not been further identified.
[6] Clémence Auguste Royer translated the third English edition of *Origin* (Royer trans. 1862); she published a second French edition, also based on the third English edition, in 1866 (Royer trans. 1866). For CD's opinions of these translations, see *Correspondence* vols. 10 and 13. See also J. Harvey 1997.
[7] Charles-Ferdinand Reinwald.
[8] Vogt did write the preface to Moulinié's translation of *Variation* (Moulinié trans. 1868).
[9] See letter to Carl Vogt, 12 April [1867].
[10] *Variation.*
[11] Reinwald refers to John Murray, CD's publisher.
[12] CD's response has not been found, but see the letter from C.-F. Reinwald, [May 1867].

From V. O. Kovalevsky 24 April [1867][1]

S Petersburg
12/24 April.[2]

Dear Sir

I received to day Your letter of 20 Apr. in answer to my writing of Ap. 2, and am happy to inform You, that I find the price at £10 a very reasonable one and beg Mr. Murray to make a set of stereotypes for me and to deliver them, on payement, to *Mr. Trüebner* 60. Paternoster Row, who is my commissione for English Books, and will forward the stereotypes to me.[3] Demanding You, dear Sir, a free pardon for my importunity I shall entreat You, in case M^r. Murray intends to set the whole I volume in type, before going to print, to send me duplicates of the last corrected proofs (Abdrücke der letzten correcturen).— The extreme kindness You showed me, make me bold even to ask You such a condesention, because as I shall receive stereotypes ready made, the last proof sheets will do as well as clean printed ones.

Dear Sir | Yours faithfully | W. Kowalevsky

P.S. Beeing a sportsman, I killed last week a very fine big black Russian bear; will You allow me to send the skin, made in form of a carpet with the heaad whole and paws, as a present and a natural specimen to You. You shall make me very happy in accepting the offer.

DAR 169: 73

CD ANNOTATION[4]
Top of letter: 'viâ France or Belgium' *ink*

[1] The year is established by the relationship between this letter and the letter from V. O. Kovalevsky, 2 April 1867.
[2] Kovalevsky gives both the Julian (12 April) and Gregorian (24 April) calendar dates.
[3] In his letter of 2 April 1867, Kovalevsky had asked permission to approach CD's publisher, John Murray, to ask for stereotypes of the illustrations to *Variation*, which Kovalevsky was planning to translate into Russian. CD's letter of 20 April has not been found. Kovalevsky also refers to Nicholas Trübner, a bookseller in Paternoster Row, London.
[4] CD's annotation is a note for his letter to Kovalevsky of 2 May [1867].

To J. D. Hooker 25 [April 1867][1]

Down
25th.

My dear Hooker

I was very much grieved at your not appearing on Sunday. I sincerely hope you did not fail on account of the baby.[2]

You were so kind as to offer your bed in Paris to our boys.[3] Is the offer yet open? & for how long? Can you also tell us where they cd get additional beds & how live as cheaply as you said.

I sent to Fritz Müller in S. Brazil yr Genera Plantarum—which he has found extremely useful. He says he observes that you have not examined the seeds of Schizolobium & Norantea & he says he should have much pleasure in supplying you with any specimens of S. Brazilian plants.[4]

Can you spare me a copy of y^r. Insular Floras for I should like to send him one from you as I am sure he would appreciate it.[5]

It was a very g^t. disappointment to me not seeing you on Sunday.

Yrs affectionately | C. Darwin

LS
Endorsement: 'April.'
DAR 94: 23–4

[1] The year is established by the relationship between this letter and the letter from J. D. Hooker, 13 April 1867.

[2] Hooker had planned to visit Down from Saturday 20 to Monday 22 April; Hooker's son Reginald Hawthorn Hooker, born in January 1867, had been ill (see letter from J. D. Hooker, 13 April 1867).

[3] See letter from J. D. Hooker, 13 April 1867. Hooker was attending the Paris exhibition as a juror (*Gardeners' Chronicle*, 6 April 1867, p. 348).

[4] See letter from Fritz Müller, 4 March 1867. CD had offered to send Fritz Müller a copy of the first two parts of Hooker and George Bentham's *Genera plantarum* (Bentham and Hooker 1862–83) in his letter of 23 August [1866] (*Correspondence* vol. 14).

[5] See letter to Fritz Müller, 22 April [1867] and n. 12.

From Thomas Rivers 26 April 1867

Bonks Hill, Sawbridgeworth.
April 26/67

My Dear Sir/

Pardon me for giving you trouble

Enclosed I send a root of a sort of wild oat grass, Bromus?, from the limestone hills of California this is N⁰. 1—said to grow on all the hills in N.W. America. N⁰. 2 is a root of a very peculiar & distinct variety of barley which came from N⁰. 1, at which you will I have no doubt shake your head. The most curious fact is that several roots of barley all differing from our English barley came up in the same bed of oat grass.[1] The transmutation of a genus seems almost incredible but I have seen so many changes that I have ceased to doubt strongly. I hope to be able to show you both barley & wheat from the root of grass enclosed N⁰. 1*

again apologising for troubling you with the rather wild ideas of my old age | I am My Dʳ Sir | Yʳˢ. ever truly | Thoˢ. Rivers

*This kind of grass has been domesticated some ten years or so not here but in New South Wales where it was taken by a gold digger from California & brought thence by my brother[2] who was much struck with its vigour in that dry climate

DAR 176: 170

[1] For a similar letter from Rivers, see *Correspondence* vol. 14, letter from Thomas Rivers, 14 October 1866.

[2] Rivers's brother has not been identified.

From A. R. Wallace 26 April [1867][1]

9, St. Mark's Crescent, N.W.
April 26th.

Dear Darwin

I have lately hit upon a generalization connected with sexual characters which pleases me very much, and I make no doubt will interest you. I have been writing a popular article on "Mimicry" & allied phenomena in which I have tried to bring together all the groups of facts which bear upon the subject.[2] While doing this I have become more than ever convinced of the powerful effect of "*protective resemblances*" in determining and regulating the development of colour,—more especially in the *females* of *insects*. Following out this view I impute the absence of brilliant or conspicuous tints in the female of Birds (when it exists in the male) almost entirely to this *protective* adaptation because in Birds, the female while *sitting* is much more exposed to attack than the male. This is I think pretty well *demonstrated* by the wonderful case of the *Phalaropus fulicarius*.[3] In this bird the sexes are alike in winter plumage, but in the summer plumage the *female* is much the most gaily coloured, having a black head, dark wings & reddish back, while the *male* is uniform brownish with dusky spots; the usual sexual differences being exactly reversed— Now strange to say the sexual habits are exactly reversed also, the *male alone* sitting on the eggs!!

The genus *Turnix* is another example, the males being always *smaller*, & often *less brightly coloured* than the females;—and the males certainly do also sit on the eggs if they do not do it exclusively.

Now these facts led me to consider why it is that in a number of groups of conspicuously coloured birds the sexes are *alike*, or at least *equally conspicuous*, contrary to the more general rule; and I was immediately led to the very simple reason that in these cases *sexual selection* had acted unchecked in both sexes, because the habits of the species were such that the female was not more exposed during incubation than the male— Hence the law,—that where birds nidificate in *holes in the ground*, or *in holes in trees*, or build *covered nests*, the females will generally be as gaily coloured as the males.

This is very generally true. For example

    Kingfishers—nests in holes in the earth
        sexes alike,—or females quite as conspicuous.
    Bee Eaters. . . do. do.
    Rollers . . . do. do.
    Woodpeckers nests in holes in trees.
        females very conspicuous, head often wh. spotted.
    Parrots and Perroquets.[4] nests in holes.
        sexes generally alike and very conspicuous
  Fam.
    Icteridæ. Hungnests.[5] *covered nests*
        sexes, generally alike, and very conspicuous,
          black red and yellow colours.
  Genus.
    Pardalotus .. nests in holes, or dome shaped.
        sexes alike, or females equally conspicuous.
    Tits .. *nests concealed in holes* or covered,
        sexes alike— very ornamental & conspicuous.
    Wagtails— nests well concealed
        sexes nearly alike.

and there are many more equally curious examples which I shall bring together perhaps in a paper for the Linnæan Soc^y.[6]

I think this proves that the primary action of *sexual selection* is to produce colour pretty equally in *both sexes*, but that it is checked in the *females* by the immense importance of *protection*, and the danger of conspicuous colouring. Of course this rule will not apply always, as there are many unknown causes affecting both the habits & the colouration of animals. When a bird is strong and has few enemies or is too large and numerous to build covered nests, as in rooks, jays & such like birds, the females remain conspicuous and unprotected.

The case of *female* birds being in several cases more *brightly coloured* than the males when they do *not incubate*, and almost always *as brightly coloured* when incubation is performed in *perfect concealment* proves I think that the male admires gay colours in the female as well as the female in the male,—and that the *direct cause* of the prevalent dull colours in the female, is solely their *danger*; & does not at all shew that the males have no *taste in colour*, which would be the natural inference if sexual selection alone had produced *all* the phenomena.

This case of the birds is exactly parallel to that of insects. The *objects of mimicry* have the sexes alike or equally bright coloured, for both are equally protected; witness Heliconidæ, Danaidæ, Carabidæ, Malacoderms,[7] Eumorphidæ[8] Bees & wasps &c. And *Coccinellidæ*, which though not *mimicked* are certainly protected since they are refused by birds, have the sexes alike.

I shall be glad to hear how far you agree with these views and what objections occur to you.

Lyell has some splendid facts about the swimming powers of pigs *at sea*, which quite explains their wide diffusion in the East beyond all other placental land Mammals.[9]

Hoping you are well Believe me | Yours very sincerely | Alfred R Wallace

C. Darwin Esq.

DAR 84.1: 32–5

CD ANNOTATION

On cover: 'Humming Birds?'[10] *pencil, underl ink* | 'About the colours of young Birds— why sh^d. young Kingfisher be bright, because Hen lays in concealed nest.— When transmitted to both sexes is transmitted to early age'[11]

---

[1] The year is established by the relationship between this letter and the letter from A. R. Wallace, 1 May 1867.

[2] Wallace refers to his paper 'Mimicry and other protective resemblances among animals', published in the *Westminster Review* in July 1867 ([A. R. Wallace] 1867a). CD discussed [A. R. Wallace] 1867a extensively in *Descent* vol. 2.

[3] Now *Phalaropus fulicaria*, the red phalarope.

[4] Parroquet is an alternative spelling of parakeet (*Chambers*).

[5] Hangnests or hangbirds: (*Chambers*), members of the family *Icteridae*.

[6] Wallace wrote about the effect of birds' nests on their coloration in the *Journal of Travel and Natural History* (A. R. Wallace 1868–9).

[7] Malacoderms: '(of a beetle) of the former group Malacodermata (or Malacodermi), comprising soft-bodied species such as soldier beetles and fireflies (now in the superfamilies Cantharoidea, Cleroidea, and Lymexyloidea)' (*OED*).

[8] The Eumorphidae are now a subfamily, Eumorphinae, of the Endomychidae or handsome fungus beetles (Downie and Arnett 1996, 2: 1066).

[9] Wallace refers to a passage in the tenth edition of Charles Lyell's *Principles of geology* (C. Lyell 1867–8, 2: 356), where Lyell recounted an anecdote about a pig that swam from a ship to a shore 'many miles distant'. In early 1865, CD and Wallace had discussed the distribution of pigs in East Asia; see *Correspondence* vol. 13, letters to A. R. Wallace, 29 January [1865] and 1 February [1865], and letter from A. R. Wallace, 31 January [1865].

[10] In *Descent* 2: 168, CD used humming-birds as a counter-example to Wallace's argument, pointing out that all species built open nests, yet the sexes were alike in the most gorgeous species, and in the majority the females were brightly coloured.

[11] CD's annotation may be a note for his letter to Wallace of 29 April [1867]. In *Descent* 2: 187, CD wrote: 'When the adult male resembles the adult female, the young of both sexes in their first plumage resemble the adults, as with the kingfisher, many parrots, crows, hedge-warblers.'

From Charles Loring Brace 29 April 1867

Hastings on | Hudson | N.Y.
April 29—1867

Charles Darwin Esq
My dear Sir

Permit me to introduce to you as a correspondent my friend & neighbor Rob't S. Rowley Esq, who first called my attention to that passage in D[r] Well's Essay on the Party-colored Female, to which you allude in the preface of your New Edition.[1] Your acknowledgement should be to him— — I have written an article, defending the religious idea of your Hypothesis, which I hope to send you—[2]

Prof Agassiz has been attacking the theory this winter very earnestly, but in too *ad captandum* a style to raise his reputation with scientific men—[3]

I think you will find that Mr Rowley has some ingenious suggestions to offer—

Trusting that your health is much better, believe me dear Sir | Yours sincerely | C. L. Brace

DAR 160: 272

[1] In the Historical sketch in *Origin* 4th ed., pp. xiv–xv, CD added a reference to William Charles Wells's 'Account of a white female, part of whose skin resembles that of a negro' (Wells 1818), and acknowledged Brace for drawing his attention to it. In the same connection in *Origin* 5th ed., pp. xvi–xvii, he mentioned both Rowley and Brace. Robert S. Rowley has not been further identified. Brace had evidently sent the reference to Wells 1818 to CD before August 1866 (see letter from Asa Gray, 7 August 1866). No correspondence between CD and Rowley has been found.

[2] The article has not been identified; it has not been found in the Darwin Archive–CUL. However, for Brace's views on Darwinism and religion, see a later article of his, 'Darwinism in Germany', in the *North American Review* 90 (1870): 284–99.

[3] Since his return from his expedition to South America in 1866, Louis Agassiz had been propagating what he took to be the anti-Darwinian findings of the expedition, in a series of six lectures given at the Cooper Institute in New York City in February 1867 (J. L. R. Agassiz 1867). His lectures and publications on the subject, including also his and Elizabeth Cabot Cary Agassiz's *A journey in Brazil* (J. L. R. Agassiz and Agassiz 1868), were aimed at a popular audience (see Lurie 1960, pp. 351–7). *Ad captandum vulgus*: 'to catch the rabble; to tickle the ears of the mob' (H. P. Jones ed. 1900).

To A. R. Wallace 29 April [1867][1]

Down. | Bromley. | Kent. S.E.

Ap. 29.

Dear Wallace

I have been greatly interested by your letter, but your view is not new to me.[2] If you will look at p. 240 of 4th Ed. of Origin you will find it very briefly given with two extreme examples of the Peacock & Black grouse.[3] A more general statement is given at p. 101 or at p. 89 of the 1st Ed., for I have long entertained this view, though I have never had space to develope it.[4] But I had not sufficient knowledge to generalize as far as you do about colouring & nesting. In your paper perhaps you will just allude to my scanty remark in the 4th Ed, because in my Essay upon Man I intend to discuss the whole subject of sexual selection, explaining as I believe it does much with respect to man.[5] I have collected all my old notes & partly written my discussion & it w[d] be flat work for me to give the leading idea as exclusively from you. But as I am sure from your greater knowledge of ornithology & Entomology that you will write a much better discussion than I c[d], your paper will be of great use to me.

Nevertheless I must discuss the subject fully in my essay on man. When we met at the Zoolog. Soc. & I asked you about the sexual differences in kingfishers I had this subject in view; as I had when I suggested to Bates the difficulty about gaudy caterpillars which you have so admirably, (as I believe it will prove) explained.[6] I have got one capital case (genus forgotten) of a Mexican bird in which the female has long tailed plumes & which consequently builds a different nest from all her allies.[7] With respect to certain female birds being more brightly coloured than the males, & the latter incubating I have gone a little into the subject & cannot say that I am fully satisfied.[8] I remember mentioning to you the case of Rhynchæa, but its nesting seems unknown.[9] In some other cases the difference in brightness seemed to me hardly sufficiently accounted for by the principle of protection.

At the Falkland I's there is a Carrion hawk in which the female (as I ascertained by dissection) is the brightest coloured, & I doubt whether protection will here apply; but I wrote several months ago to the Falklands to make enquiries[10]

The conclusion to which I have been leaning is that in some of these abnormal cases the colour happened to vary in the female alone, & was transmitted to females alone, & that her variations have been selected through the admiration of the male.—[11]

It is a very interesting subject but I shall not be able to go on with it for the next 5 or 6 months, as I am fully employed in correcting dull proof sheets;[12] when I return to the work I shall find it much better done by you than I c[d] have succeeded in doing.

With many thanks for your very interesting note | believe me dear Wallace | yours very sincerely | Ch. Darwin

It is curious, how we hit on the same ideas.—

I have endeavoured and show in my M.S. discussion that nearly the same principles account for young birds *not* being gaily coloured, in many cases,—but this is too complex a point for a note.—[13]

[Enclosure]

Down.—
Ap 29th.

Postcripts

My dear Wallace

On reading over your letter again & on further reflexion, I do not think (as far as I remember my words) that I expressed myself *nearly strongly* enough on the value & beauty of your generalisation, viz that all Birds, in which the female is conspicuously or brightly coloured, build in holes or under domes. I thought that this was the explanation in many, perhaps most cases, but do not think I sh[d]. ever have extended my view to your generalisation.— Forgive me troubling you with this. P.S.

yours | C. Darwin

LS(A)
British Library (Add 46434, f. 84)

[1] The year is established by the relationship between this letter and the letter from A. R. Wallace, 1 May 1867.

[2] CD refers to the letter from A. R. Wallace, 26 April [1867], and to Wallace's view that the brightness of the plumage of female birds was influenced by whether they sat on eggs in a covered or an open situation.

[3] In *Origin* 4th ed., pp. 240–1, in a paragraph on how the beauty of many male animals was acquired as a result of selection by the females, CD wrote:

> We can sometimes plainly see the proximate cause of the transmission of ornaments to the males alone; for a pea-hen with the long tail of the male bird would be badly fitted to sit on her eggs, and a coal-black female capercailzie would be far more conspicuous on her nest and more exposed to danger than in her present modest attire.

[4] In *Origin*, p. 89, in a section on sexual selection, CD wrote:

> I strongly suspect that some well-known laws with respect to the plumage of male and female birds, in comparison with the plumage of the young, can be explained on the view of plumage having been chiefly modified by sexual selection, acting when the birds have come to the breeding age or during the breeding season; the modifications thus produced being inherited at corresponding ages or seasons, either by the males alone, or by the males and females; but I have not space here to enter on this subject.

The same passage appears in *Origin* 4th ed., p. 101.

[5] Wallace had written a paper on mimicry for the *Westminster Review* ([A. R. Wallace] 1867a), and published an article on bird coloration and nesting habits in the *Journal of Travel and Natural History* (A. R. Wallace 1868–9). Wallace cited the fourth edition of *Origin* in general in [A. R. Wallace] 1867a, but not specifically in connection with the nesting habits of birds, which he discussed on pp. 38–9. In A. R. Wallace 1868–9, Wallace wrote (pp. 77–8):

> The sexual differences of colour and plumage in birds are very remarkable and have attracted much attention; and in the case of polygamous birds have been well explained by Mr Darwin's

principle of sexual selection ... but this theory does not throw any light on the causes which have made the female toucan, bee-eater, parroquet, macaw and tit, in almost every case as gay and brilliant as the male.

CD's material on sexual selection for his 'Essay upon Man' was published as *Descent* in 1871: CD had first intended to publish the material as a chapter in *Variation* but later decided to publish it separately (see letter to J. D. Hooker, 8 February [1867] and n. 16). Sexual selection among animals, including humans, was discussed in the second part of the first volume and in the second volume of *Descent*.

[6] CD had visited London from 13 to 21 February 1867 (CD's 'Journal' (Appendix II)), but evidently did not see Wallace then (see letter to A. R. Wallace, 23 February 1867). He was also in London from 22 to 29 November 1866 (*Correspondence* vol. 14, Appendix II), and visited the Zoological Gardens (*Correspondence* vol. 14, letter to Edward Blyth, 10 December [1866]). Henry Walter Bates had referred CD to Wallace for an answer to the question why some caterpillars were brightly coloured; see letter to A. R. Wallace, 23 February 1867, and letter from A. R. Wallace, 24 February [1867].

[7] The Mexican bird (if it existed) has not been identified. CD may have been thinking of *Menura superba*, a lyre-bird of Australia, in which both sexes had long tails, and built a domed nest, which CD said was an anomaly in so large a bird (see *Descent* 2: 164–5).

[8] In the section on colour and nidification among birds in *Descent* (*Descent* 2: 167), CD wrote that in one bird species that built open nests, the male sat on the eggs and was brightly coloured.

[9] See letter to Alfred Newton, 19 January [1867]. CD discussed *Rhynchaea* (now *Rostratula*), the painted snipe, in *Descent* 2: 202–3, pointing out that the female was more brightly coloured than the male, and that there was reason to believe that the male sat on the eggs.

[10] CD had given this information in *Birds*, p. 16, and repeated it in *Descent* 2: 205–6. CD's letter to the Falkland Islands has not been found. See also letter to Alfred Newton, 4 March [1867] and n. 3.

[11] See also *Descent* 2: 207–8.

[12] CD was correcting the proof-sheets of *Variation* (see 'Journal' (Appendix II)).

[13] In *Descent*, CD argued that most juvenile birds were dull coloured, and that many males and some females acquired bright colours through sexual selection. These colours, being acquired late in life, tended not to be transmitted to the opposite sex, and any tendency for them to be transmitted to the same sex would have been eliminated because it was dangerous for young birds to be brightly coloured. (See *Descent* 2: 196, 200, 222.)

To Charles Kingsley 30 April [1867][1]

Down. | Bromley. | Kent. S.E.
Ap. 30

My dear Mr Kingsley

I fully agree with you on the importance of giving to a certain extent a scientific character to Fraser's Mag.; & I do assure you that it would have given me sincere pleasure to have assisted you in any way.[2] But at present I really cannot: I am daily knocked up by correcting proof sheets & the printers are a dozen sheets a head of me, so that I greatly doubt whether I can finish the book by November when Murray wants it.[3] Hence it would be ruin to me to stop work for a week or fortnight, & I could not write an article in less time than this. I have so much unpublished matter half completed that I have often vowed I w^d never write miscellaneous articles, but I sh^d have been much tempted under other circumstances to have broken my vow at your request.

So I hope that you will forgive me & believe me | my dear Mr Kingsley | yours very sincerely | Ch. Darwin

LS

B. C. Guild (private collection)

[1] The year is established by the reference to the proof-sheets of *Variation* (see n. 3, below).

[2] No letter in which Kingsley invited CD to contribute to *Fraser's Magazine* has been found. Kingsley himself was a contributor to *Fraser's*; the main emphasis of the magazine was politics, religion, and social conditions, and it was an 'organ of progressive thought'. From May to August 1867 Kingsley ran *Fraser's Magazine* during the absence of the editor, James Anthony Froude. (*Wellesley index* 2: 303–14.)

[3] CD was working on the page proofs of *Variation* (see 'Journal' (Appendix II)). No letter from John Murray, CD's publisher, asking that the book be finished by November has been found, but see the letter to W. B. Tegetmeier, 6 January [1867].

From Peter S. Robertson 30 April 1867

Peter S. Robertson & Co. | *Nurserymen* | *and* | *Seedsmen* | *at* | *33 St Andrew Square* | *and* | *Trinity Nursery.* | *Edinburgh*

30 April *1867*.

Ch. Darwin Esqr | Down Bromley, Kent.

Sir,

My first attempt to cross the different varieties of Borecole was in 1859—[1] for this purpose I planted near each other two plants each of the following Garden varieties, 1 Delaware, a purplish medium tall, curled sort— 2nd. Russian a very curled variety beautifully purple & not above 9 inches in height,

3^d. Purple Scotch, a tall growing kind, with leaves nearly plain, greenish purple—

4th. Proliferous Purple,

5th. Perennial Kale, a purplish tall cut leaved suffruticose sort.

6th Pale Green triple curled German Greens

7 Plain Green German Greens very little curled—

On growing the seeds the following grew from each of these sorts— I found the seeds of No 5. germinate only about 10 per cent & not one of them seemed in any way different from the parent, only that they were weaker & did not survive the following winter

No 4. grew well, all the produce more or less proliferous but the colours were all shades from the Purple of the parent to nearly green & several were whitish, blanched like, & very unequal in size, there was no white Borecoles near them when the plants were in flower—*

No 2. The produce was very similar to the parent, only that about one tenth part were paler in colour or stronger as if crossed with No 1— or No 3.

No—1—3—6—&7. All produced a large proportion different from the parents, & appeared to have mutually affected each other greatly, a few plants in each kind were true to the type, & a great many plants were so different that one could hardly believe them to be Borecoles, some were nearly as plain as Cabbages, others had long narrow leaves like Brocoli but coarser looking—& a good many closed (hearted) like a Cabbage in autumn & seemed very like a new Garden sort called Cottagers Kale

I had about one acre of No 6. growing within a quarter of a mile of where these plants were planted for crossing & the whole produce of this acre was affected to the extent of about 1 per cent with monstrous crosses also— some grew a pretty dark purple.

I did not find a single proliferous plant grow from any of them but from No 4.

In 1861, I procured seeds of all the variegated & curious leaved kinds I could, & had them planted among each other in 1862 to see if they would cross & produce plants with white & purple on the same root, & I have twice since done the same thing, but have always found, that the varieties 6 & 7, sport into white or yellowish white & the Purple kinds into magenta, rose, or pale lilac, the variegated plants of both produce some selfs & I find that No 6. when grown of a very pure white ceased to produce seeds, only malformed siliquae with rudimentary seeds. I am of opinion that the whites & magentas do not cross.

I find that the plants do not show either their true colour or character till about November, during that month and December the young leaves produced are often very pure, delicate in texture & of intense beauty, both the magentas & whitish varieties stand winter frost & snow well, the colours being improved thereby—but in March & April the whitish varieties are very much sooner damaged by frost than the purples, and intense sunlight destroys the fine [tail] of both very quickly about the time they begin to throw up the seed stems—

In growing the seeds in spring in the open air I find that the coarser crosses most removed from the finest forms in appearance always grow strongest & first, so that parties cultivating them for the flower Garden should reject the strongest & carefully transplant the weakly plants, as from them the finest forms are got.—

I will be glad to write you any explanation you further require of these remarks—

I am—Sir | Your most obt Servant | Peter S. Robertson

*In successive growings of this variety it becomes quite destitute of leaves, nothing but the thread like ribs remain, which are very beautiful, & resemble some sea weeds.—

DAR 76: B49–51

CD ANNOTATION
Top of letter: 'Edinburgh' *pencil*

[1] No letter from CD to Robertson has been found, but one might have been inspired by a series of references in the *Gardeners' Chronicle* to Robertson's work on variegated kale or borecole (*Gardeners' Chronicle*, 9 February 1867, p. 130; 9 March 1867, p. 236; 23 March, p. 294). In the issue for 23 March, Robertson was quoted as saying that white and magenta borecoles did not readily cross, but that one plant of either would 'taint an acre of pure Green or common purple German greens' (*ibid.*, p. 294). A clipping from page 294 including this quote is associated with this letter in the Darwin Archive–CUL. CD discussed the cabbage family in *Variation* 1: 323–6, but did not cite Robertson.

From Charles-Ferdinand Reinwald [May 1867][1]

⟨*half page missing*⟩ of the wood-cuts from Mrs. Murray for the price of £10.—and to let you have six copies of our translation.[2]

I am highly gratified by this kind proposal which I accept with pleasure for my part, and hope that Prof. Carl Vogt will not only give his advice to our french translator Col. Moulinié, but allow me to put his name on the title piece as being done under his direction.[3]

⟨*half page missing*⟩ honoured to be in ⟨*3 or 4 words*⟩ of your highly deserved reputation and celebrity and beg to believe me

Dear Sir | Your most obedient Servant | C Reinwald

C. Reinwald, libraire editeur
15. Rue Des Saints-Pères. Paris

Incomplete
DAR 210.11: 35

[1] The date is established by the relationship between this letter, the letter from Carl Vogt, 23 April 1867, and the letter from J. J. Moulinié, 2 June 1867.

[2] John Murray was the publisher of *Variation*. This letter evidently concerned arrangements for the publication of a French translation of *Variation* by Reinwald's firm. See letter from Carl Vogt, 23 April 1867.

[3] Reinwald refers to Carl Vogt and Jean Jacques Moulinié. The words 'Préface de Carl Vogt' appear on the title page of Moulinié trans. 1868.

From A. R. Wallace 1 May 1867

9, St. Mark's Crescent
May 1st. 1867

Dear Darwin

I was afraid you had rather misunderstood my letter on first reading it; for I assure you I never for a moment imagined that any of the more obvious facts connected with *sexual selection* (which is *altogether* your own subject) could have been new to you. The remarkable coincidence, of so many of the birds which have females coloured as *gayly* as the males, making their nests so that the female is *concealed* during incubation, while almost all in which the female differs remarkably from the male in colour build *exposed* and *uncovered* nests;—appeared to me to get over one great difficulty in the way of explaining the origin of colour in birds; and as it was so new & interesting to me I thought it might not possibly have occurred to you.[1]

There are some exceptions which I cannot yet explain, but this is to be expected, for we cannot but suppose that *many* different causes have favoured or checked the developement of colour at different times. The exceptions are not I think numerous enough to upset the rule.

This view is I think also interesting as explaining the absence of much sexual difference of colour among mammals or reptiles, in which the sexes are not very differently situated as regards danger from enemies.

The mode of nidification in birds is no doubt *primarily* dependent on their structural peculiarities and their general habits (on which subject I have a paper written ten years ago,) and we may therefore conclude that the mode of nidification of Kingfishers Toucans &c. has been the acting cause in determining or permitting the action of sexual selection on the female bird. In other cases however it is quite possible, that the colour being first produced by sex! slectⁿ. has led to the modification of the nest for safety, as in the Australian finches which make *domed nests* while our European species make open ones.[2]

On powerful and pugnacious birds, such as crows and hawks I do not expect the principle of protection has acted much in modifying colour.

I enclose you a copy of my notes on the subject, which I beg you to make what use of you like, in your proposed essay. I will merely allude to the subject in my paper on "mimicry", which is finished & sent to the "Westminster" to see if they will publish it.[3] As you are going to treat fully the whole subject of "sexual selection" I hope you will not call it an *"Essay on Man"*.[4] I had thought of a short paper on *"The connexion between the colours of female birds & their mode of nidification"*,—but had rather leave it for you to treat as part of the really *great* subject of *"sexual selection"* which combined with *"protective resemblances"* and *"differences"* will I think when thoroughly worked out explain the whole colouring of the Animal Kingdom.[5]

Believe me Dear Darwin | Yours very faithfully | Alfred R. Wallace—

DAR 84.1: 36–7

[1] See letter from A. R. Wallace, 26 April [1867], and letter to A. R. Wallace, 29 April [1867]. Wallace had suggested that the dull colouring of many female birds was related to their need for protection while sitting on exposed nests.

[2] Wallace discussed how the structure and habits of birds, as opposed to their instincts, determined how they built their nests in A. R. Wallace 1867b; in A. R. Wallace 1868–9 he went on to argue that whether a nest was open or covered influenced the colour of the female bird. In A. R. Wallace 1868–9, p. 83, he wrote: 'When the confirmed habit of a group of birds was to build their nests in holes of trees, like the toucans, or in holes in the ground, like the kingfishers, the protection the female thus obtained, during the important and dangerous time of incubation, placed the two sexes on an equality as regards exposure to attack, and allowed "sexual selection" to act unchecked in the development of gay colours and conspicuous markings in both sexes.'

[3] Wallace refers to [A. R. Wallace] 1867a, which was published in the *Westminster Review* in July 1867. The need for protection of birds sitting on eggs is discussed in *ibid.*, pp. 33–4. There is a heavily annotated copy of the paper in DAR 133: 13. CD discussed [A. R. Wallace] 1867a extensively in the second volume of *Descent*.

[4] See letter to A. R. Wallace, 29 April [1867] and n. 5.

[5] Wallace published two papers on birds' nests, A. R. Wallace 1867b and A. R. Wallace 1868–9 (see n. 2, above). There is a lightly annotated copy of A. R. Wallace 1867b in DAR 133: 12. CD discussed Wallace's arguments in *Descent* 2: 166–80.

To V. O. Kovalevsky 2 May [1867][1]

Down. | Bromley. | Kent. S.E.
May 2.

Dear Sir

I have bespoken the stereotypes & desired Mr Murray to send them when ready to Messrs Trübner.[2] I sent yesterday by post some revised proof sheets, which will keep you employed for some time. You will find the whole of the first vol. dull, but I hope the 2nd vol. is somewhat more interesting.[3]

I thank you cordially on my own part & in the name of the ladies of my family for your very very kind offer of the bears skin, which it will be a great gratification to me to possess.[4] It will be the safest plan to address it to

> C Darwin Esqr
> 6 Queen Anne St
> Cavendish Square
> London.[5]

When you next have occasion to write be so good as to inform me whether the proof sheets had better be send via France or Belgium. The postage is cheaper by France. Also please to tell me whether it is necessary to give the whole of the address to you which you gave me.

Dear Sir | yours very faithfully | Ch. Darwin

LS
Institut Mittag-Leffler

[1] The year is established by the reference to arrangements for Kovalevsky's translation of *Variation* into Russian. *Variation* was published in January 1868, and Kovalevsky's Russian translation (V. O. Kovalevsky trans. 1868–9) was first issued in parts starting in 1867 (Freeman 1977).
[2] CD refers to his publisher, John Murray, and to Kovalevsky's commissioner for English books, Trübner & Co. See letter from V. O. Kovalevsky, 24 April [1867].
[3] CD refers to *Variation*. The first volume of *Variation* discussed details of variation in particular animals and plants; the second volume included more general and theoretical discussions on topics including inheritance, hybridity, variation, and CD's hypothesis of pangenesis.
[4] See letter from V. O. Kovalevsky, 24 April [1867].
[5] CD gave the address of his brother, Erasmus Alvey Darwin.

From John Maurice Herbert 3 May 1867

Rocklands, | Ross.
3 May 1867

My dear Darwin

I think dear old Dawes was a friend of yours— If so, and you consider him worthy of a memorial, do you feel disposed to subscribe? The enclosed papers will show you that we have made a beginning. The form of the Memorial has not yet been decided upon. The Dean's death has been a sad blow to me— He was my dearest & most intimate friend in this neighbourhood—[1]

I suppose you are still busy at the old work—writing huge books, and gaining fresh heaps of immortality— I was glad to hear from Prof.^{r} Tyndal some few months back that D^r. Bence Jones had made a new man of you—[2]

I sincerely hope, my dear old friend, that the restoration is complete and permanent— Oh that you were frisky enough to pay us a visit in the course of the Summer— Nothing w^d give Mrs Herbert & me greater pleasure than to receive you & Mrs Darwin.[3]

Pray give her our kindest regards, & accept the same from y^{rs} always | J M Herbert

C. Darwin Esq

DAR 166: 184

[1] Richard Dawes, dean of Hereford, died on 10 March 1867; he had been a fellow at Downing College, Cambridge, when CD was an undergraduate at Christ's and Herbert at St John's (*Alum. Cantab.*, *DNB*). Dawes had lived at the Deanery in Hereford; Herbert lived sixteen miles from Hereford at Goodrich, near Ross, Herefordshire (*Post Office directory of Gloucestershire, with Bath, Bristol, Herefordshire, and Shropshire* 1863). Dawes's memorial was an altar tomb in the north-east transept of Hereford Cathedral, paid for by public subscription (Havergal 1881). The enclosures have not been found.

[2] Herbert refers to John Tyndall, professor of natural philosophy at the Royal Institution of Great Britain and the Royal School of Mines (*ODNB*), and the physician Henry Bence Jones. Jones had been treating CD since July 1865; CD believed Jones's treatment had been beneficial (see *Correspondence* vol. 13).

[3] Herbert refers to his wife, Mary Anne Herbert, and to Emma Darwin. For an indication of the friendly relations between the Darwins and the Herberts, see *Correspondence* vol. 3, letter to J. M. Herbert, [3 September? 1846], and *Correspondence* vol. 6, letter to J. M. Herbert, 2 January [1856].

From Jean Jacques Moulinié 3 May 1867

Geneva
3 May 1867.

Sir,

I just receive from Professor Vogt communication of your letter to him of 30th April last, in which you grant me in so favourable terms the right of the french translation of your actual publication on "Domesticated animals and cultivated plants"; an honourable favour for which I beg you, Sir, to receive here, the expression of my deep gratitude.[1]

Like my russian and german co-translators, allow me, sir, not to agree with the opinion you seem to have of your first volume, owing to the nature of its contents;[2] I believe on the contrary that the numerous facts, observations and documents you must have abundantly and laboriously collected, and of which I suppose the first volume contains the exposal, besides their importance as the ground-work of the whole edifice, the basis of the theoretical views and conclusions of the second part, will by themselves and in their particulars, have a high interest for the reader, so gradually brought down from positive facts to the natural consequences they lead to.

I shall therefore be very happy to receive the first sheets of your publication as soon as you will judge convenient to send them, they will be welcome.

I remain Sir, yours very truly | and respectfully | J. J. Moulinié

15. rue du Mont Blanc | Geneva.

DAR 171: 266

[1] CD's letter to Carl Vogt has not been found. Vogt was directing the French translation of *Variation* by Moulinié (see letter from C.-F. Reinwald, [May 1867]).

[2] The German translation of *Variation* was by Julius Victor Carus (Carus trans. 1868); the Russian by Vladimir Onufrievich Kovalevsky (Kovalevsky trans. 1868–9). CD had recently suggested to Kovalevsky that he would find the first volume dull (see letter to V. O. Kovalevsky, 2 May [1867] and n. 3). No extant earlier letters to the translators mention this opinion of CD's.

To Ludwig Rütimeyer 4 May [1867][1]

> Down. | Bromley. | Kent. S.E.
>
> May 4

My dear Sir

I am very much obliged to you for your never failing kindness in sending me so many of your works, which I have long esteemed as of the highest value.[2] I am now correcting the proof sheets of a work on the variation of domestic animals & plants, & in this I have a short chapter on pigs, cattle &c; it contains no original matter & is compiled largely from your works; but this chapter has now been printed, & therefore I shall not be able to improve it from any facts contained in the second part of your work on cattle which I have just received.[3]

Prof. Victor Carus is going to translate my book into German & I will send you a copy when it is published, which I suppose will be towards the close of this year; but I fear there will not be much in it that will interest you.[4]

With my sincere thanks | believe me my dear Sir | yours very faithfully | Ch. Darwin

LS

Elizabeth Rütimeyer (private collection)

[1] The year is established by the reference to the printing of *Variation*, which had started to be set up in type by March 1867 (see letter from John Murray, 19 March [1867]); it was published in January 1868 (Freeman 1977).

[2] There are a number of works by Rütimeyer published in 1867 or earlier in the Darwin Archive–CUL, all inscribed by the author. In the Darwin Library–CUL are Rütimeyer 1861 (heavily annotated; see *Marginalia* 1: 718–26) and Rütimeyer 1863 (annotated; see *Marginalia* 1: 718). In the Darwin Pamphlet Collection–CUL are offprints of Rütimeyer 1862 (annotated on the back, 'not read'), Rütimeyer 1865a (lightly annotated), Rütimeyer 1865b, Rütimeyer 1866 (annotated), and Rütimeyer 1867a–d; Rütimeyer 1867d is annotated.

[3] CD refers to chapter 3 of *Variation*, on pigs, cattle, sheep, and goats; Rütimeyer is cited extensively throughout. The second part of Rütimeyer's work on cattle was Rütimeyer 1867b.

[4] Julius Victor Carus's translation of *Variation* (Carus trans. 1868) was published by July 1868 (see letter from Eduard Koch, 21 July 1868, *Calendar* no. 6284). CD sent a presentation list for the translation to

his German publisher in November 1867 (see letter to E. Schweizerbart'sche Verlagsbuchhandlung, 22 November [1867]); Rütimeyer's name was on it.

To A. R. Wallace 5 May [1867][1]

Down. | Bromley. | Kent. S.E.

May 5

My dear Wallace

The offer of your valuable notes is *most* generous, but it w^d vex me to take so much from you, as it is certain that you c^d work up the subject very much better than I could. Therefore I earnestly & without any reservation hope that you will proceed with y^r paper, so that I return y^r notes.[2]

You seem already to have well investigated the subject. I confess on receiving yr note that I felt rather flat at my recent work being almost thrown away, but I did not intend to shew this feeling.[3] As a proof how little advance I had made on the subject, I may mention that though I had been collecting facts on the colouring, & other sexual differences in mammals, your explanation with respect to the females had not occurred to me. I am surprized at my own stupidity, but I have long recognized how much clearer & deeper your insight into matters is than mine. I do not know how far you have attended to the laws of inheritance, so what follows may be obvious to you. I have begun my discussion on sexual selection by shewing that new characters often appear in one sex & are transmitted to that sex alone, & that from some unknown cause such characters apparently appear oftener in the male than in the female. Secondly characters may be developed & be confined to the male, & long afterwards be transferred to the female. 3^{rdly} characters may arise in either sex & be transmitted to both sexes, either in an equal or unequal degree.[4] In this latter case I have supposed that the survival of the fittest has come into play with female birds & kept the female dull-coloured.[5] With respect to the absence of spurs in female gallinaceous birds, I presume that they w^d be in the way during incubation; at least I have got the case of a German breed of fowls in which the hens were spurred, & were found to disturb & break their eggs much.[6]

With respect to the females of deer not having horns, I presume it is to save the loss of organized matter.[7]

In y^r note you speak of sexual selection & protection as sufficient to account for the colouring of all animals, but it seems to me doubtful how far this will come into play with some of the lower animals, such as sea anemones, some corals &c &c—[8]

On the other hand Häckel has recently well shewn that the transparency & absence of colour in the lower oceanic animals, belonging to the most different classes, may be well accounted for on the principle of protection.[9]

Some time or other I sh^d like much to know where y^r paper on the nests of birds has appeared, & I shall be extremely anxious to read y^r paper in the West. Rev. Your paper on the sexual colouring of birds will I have no doubt be very striking.[10]

Forgive me, if you can, for a touch of illiberality about y^r paper & believe me yrs very sincerely | Ch. Darwin

LS(A)
British Library (Add 46434 f. 89)

[1] The year is established by the relationship between this letter and the letter from A. R. Wallace, 1 May 1867.

[2] See letter from A. R. Wallace, 1 May 1867 and n. 3.

[3] See letter from A. R. Wallace, 26 April [1867], and letter to A. R. Wallace, 29 April [1867].

[4] See *Variation* 2: 71–5; see also *Descent* 1: 282–5.

[5] See *Origin* 4th ed., pp. 240–1.

[6] See *Variation* 1: 256. The information was from Bechstein 1789–95.

[7] CD's views on horns, or the lack of them, in female deer, were developed further during his work on *Descent*, the projected work on sexual selection, particularly among humans, that he had been discussing with Wallace. In *Descent* 2: 243 he wrote: 'No doubt with female deer the development during each recurrent season of great branching horns ... would have been a great waste of vital power, on the admission that they were of no use to the females.'

[8] See *Descent* 1: 322.

[9] CD refers to Ernst Haeckel's *Generelle Morphologie* (Haeckel 1866), 1: 241–3. This section is annotated in the copy in the Darwin Library–CUL (see *Marginalia* 1: 355–7). See *Descent* 1: 323.

[10] See letter from A. R. Wallace, 1 May 1867 and nn. 2 and 3.

To J. M. Herbert 7 May [1867][1]

Down. | Bromley. | Kent. S.E.
May 7th

My dear Herbert

It has given me great pleasure to receive a letter from you after the interval of so many years.[2] I saw a good deal of poor Dawes when I was an undergraduate & was charmed with him, but I have never seen him excepting once since that period.[3] His subsequent career every one must honour & I will send by this post a subscription of 2 guineas to the treasurer for his memorial.[4] It has often amused me to reflect how little a man sometimes knows himself, or his friends know him, for I well remember Ramsay of Jesus[5] chaffing Dawes that he was a man who w^d go through life laughing & doing nothing else, & Dawes himself & all of us quite agreed that this was true.

Thank you for your kind enquiries about my health; I had two or three bad years but am now much better, & am able to work; but Tyndall gave too flourishing an account, for I am never well for the whole day; this prevents me going any where, & we cannot accept your cordial invitation, which I sh^d in every way have much enjoyed.[6] Whenever you & Mrs Herbert[7] (to whom we both send our very kind remembrances) happen to be in London & have a little spare time, we should be extremely glad if you would pay us a visit.

Believe me my dear old friend | yours very sincerely | Ch. Darwin

LS(A)
American Philosophical Society (327)

252 *May 1867*

[1] The year is established by the relationship between this letter and the letter from J. M Herbert, 3 May 1867.

[2] Letter from J. M. Herbert, 3 May 1867. The last known correspondence with Herbert is a letter to Herbert of 18 November [1856] (*Correspondence* vol. 13, Supplement), evidently a response to a letter from Herbert.

[3] See letter from J. M. Herbert, 3 May 1867, n. 1. CD spoke to Richard Dawes in or before July 1853 (see *Correspondence* vol. 5, letter to W. D. Fox, 17 July [1853] and n. 6).

[4] See letter from J. M. Herbert, 3 May 1867 and n. 1. CD recorded a payment of £2 2s. to 'Dawes memorial' on 6 May 1867 (CD's Account books–cash account (Down House MS)).

[5] Marmaduke Ramsay was a fellow of Jesus College, Cambridge, while CD was an undergraduate at Christ's.

[6] CD refers to John Tyndall. See letter from J. M. Herbert, 3 May 1867.

[7] Mary Anne Herbert.

From Nicholas Trübner 7 May 1867

London. | 60, Paternoster Row E.C.
May 7 *1867*

Dear Sir,

A correspondent of mine, Mr. Weisse of Stuttgart,[1] is very desirous of publishing a *German* translation of the work upon which you are now engaged. In case you have not already arranged with a German Translator I shall be very glad if you will permit me to negotiate with you on behalf of my friend.[2]

I have the honour to be | Dear Sir | yours most respectfully | N Trübner

Charles Darwin Esq

DAR 178: 194

[1] Mr Weisse has not been further identified.

[2] Trübner refers to *Variation*. CD had already accepted the offer of Julius Victor Carus to translate the work; it was published by E. Schweizerbart'sche Buchhandlung (see letter to J. V. Carus, 11 April [1867]).

From Julius von Haast 12 May – 2 June 1867

Christchurch N Z.
12 May 1867

My dear Mr Darwin

I have had the pleasure of receiving your valued letter of the 27th of February, in which you enclose some queries about Expression, which I consider of very high importance for generalisations concerning the human species, if, as I hope you are able to obtain sufficient material.[1] As far as I am aware, nobody has ever tried to use thus psychological expressions in comparing them with each other & perhaps with those of the higher Mammalia for ethnological studies and I need scarcely tell you with what impatience I am expecting the results of your investigations on that important subject. I had at once your queries copied & sent them to the following Gentlemen.

1.) Fenton Chief Judge of the Native Court in Auckland

2). Rev^d. W. Colenso, the capital botanist in Napier Hawkes Bay

3) W. Buller FL.S. ornothologist, a Native Judge in Wanganui

4) The Revd I. Stack a Maori Missionary in Kaipoi; the first three in the northern, the last in this or South Island. No 2 & 3 are great Darwinians & 1 & 4 also very liberal minded men & I have no doubt, that they will furnish you with a great deal of reliable & valuable material.[2]

I do not know if you have written to Dr Hector in Wellington,[3] but in any case I shall have two other copies prepared and send him one & the other to M^r W^m. Mantell FGS also a very clever Maori scholar & excellent observer.[4]

Unfortunately I shall not be able to furnish you with any observations of mine, as I am not within reach of the natives, but I am certain that these Gentlemen will be able to give all the necessary information. I told them to send their answers either direct to you or to me for transmission.—[5] Hooker's second volume of the Handbook of the N. Z. flora[6] came by last mail & you may imagine with what avidity I went through it— The two volumes form a most excellent compact work and which will be of far greater use to the Colony, than we can at present conceive, as it will awake the minds of many colonists to look round them in their country of adoption & read the book of Nature, instead of solely money-grubbing. I think one of the greatest triumphs of modern Science has been to popularise it and thus open the lofty halls of knowledge also to those, whose ordinary occupations would formerly have prevented them from entering.

I am expecting your new work[7] with great impatience from my London agent & trust that your health will allow you to work for many years without interruption In a few weeks I shall have the pleasure of sending you another little work of mine, which will make you acquainted with the head waters of the Rakaia, one of our large alpine rivers.[8]

With best wishes believe me my dear Mr Darwin | ever yours very sincerely | Julius Haast.

June 2

Three of my correspondents have already written & will answer your questions most conscienciously.—

Ch^s. Darwin Esqre
FRS etc etc
Down Bromley Kent.

DAR 166: 11

CD ANNOTATIONS
1.1 I have ... subject 1.8] *crossed blue crayon*
3.1 Unfortunately ... rivers. 4.5] *crossed blue crayon*

[1] See letter to Julius von Haast, 27 February [1867].

[2] Haast refers to Francis Dart Fenton, chief judge of the Native Land Court in Auckland; the missionary William Colenso of Hawke's Bay; Walter Lawry Buller, resident magistrate in Wanganui; and James West Stack, whose Christchurch mission house was built on land gifted from the Kaiapoi Reserve (*DNZB*). Buller was elected a fellow of the Linnean Society in 1858 (*List of the Linnean Society of London*). In a letter to Haast dated 20 May 1867, Stack agreed to provide answers to CD's questions, and commented that he believed that the biblical account of creation and CD's theory would one day be reconciled (Reed ed. 1935, pp. 63–4).

[3] Haast refers to James Hector, director of the meteorological department of the New Zealand Institute, of the Colonial Museum, and of the botanical garden in Wellington, New Zealand (*DNZB*). There are no extant letters to or from Hector in this period.

[4] Haast refers to Walter Baldock Durrant Mantell, fellow of the Geological Society of London since 1858 (*List of the Geological Society of London*).

[5] Stack's replies to CD's questionnaire are enclosed in the letter from Julius von Haast, 4 December 1867; no replies from the other observers have been found in the Darwin Archive–CUL. CD wrote to Mantell with further enquiries, possibly in 1869 (see letter to W. B. D. Mantell, [1869?] (*Calendar* no. 6520)), and cited him for the information given on ideas of female beauty in *Descent* 2: 369. Only Stack is cited in *Expression* for giving information on the New Zealand Maori population (*Expression*, p. 20).

[6] J. D. Hooker 1864–7.

[7] *Variation.*

[8] Haast explored the headwaters of the Rakaia in 1866 (H. F. von Haast 1948, pp. 463–75). His *Report on the headwaters of the Rakaia* was printed in 1866 by direction of the Provincial Government, Christchurch (J. F. J. von Haast 1866; see *Correspondence* vol. 14, letter from Julius von Haast, 17 July 1866). The report has not been found in the Darwin Pamphlet Collection–CUL; see also the letter from Julius von Haast, 4 December 1867.

From Ernst Haeckel[1]   12 May 1867

Jena
12. Mai 1867

Mein theurer, hochverehrter Freund!

Erst vor wenigen Tagen von meiner Reise zurückgekehrt, fand ich Ihren lieben Brief vom 12. April vor, sowie die neue Ausgabe Ihres epochemachenden Werkes.[2] Für Beide sage ich Ihnen meinen herzlichsten Dank. Leider ersehe ich aus Ihrem Briefe, dass Sie einen langen Brief nicht erhalten haben, welchen ich schon vor mehreren Monaten an Sie geschrieben hatte, und welchem auch eine Beilage an unseren Freund Huxley eingelegt war. Bei der grossen Unordnung, welche die liederliche spanische Post auf den canarischen Inseln *beherrscht*, kann ich mich allerdings nicht darüber wundern.[3] Dagegen habe ich Ihren ersten, nach Madeira gesandten Brief, richtig erhalten, wenn auch erst 2 Monate später, in Lanzarote.[4]

Erlauben Sie nun zunächst, theurer Freund, dass ich Ihnen nochmals meinen *herzlichsten* Dank wiederhole für die überaus freundliche Aufnahme, die mir in Ihrem lieben Hause zu Theil wurde.[5] Jenen Tag, den ich bei Ihnen verleben durfte, und auf den ich schon so lange vorher gehofft hatte, wird mir immer unvergesslich sein.

Ich kann Ihnen nicht sagen, welche ausserordentliche Freude Sie mir gemacht haben, als Sie mir erlaubten, Sie zu besuchen, und welche grosse Satisfaction es für mich war, persönlich den Naturforscher kennen zu lernen, der als der Reformator der Descendenz Theorie und als der Entdecker der "natural Selection", einen grösseren Einfluss auf die Richtung meiner Studien und die Thätigkeit meines Lebens gehabt hat, als alle anderen. Nochmals Ihnen und Ihrer lieben Familie meinen herzlichsten, innigsten Dank.

Hoffentlich geht es mit Ihrer Gesundheit jetzt besser, und Sie können wieder Ihre reichen Kentnisse und Ihre gedankreiche Naturbetrachtung zum Fortschritte unserer Wissenschaften verwenden. Als ich auf meiner Reise das wunderschöne Klima von Madeira und von den canarischen Inseln kennen lernte, habe ich oft gewünscht, dass Sie bei uns wären, um Ihre angegriffene Gesundheit dauernd zu befestigen. Ich selbst bin dort wieder ganz zu meiner alten Kraft gelangt und habe mich von den Anstrengungen, die mir die Arbeit der letzten Jahre auferlegte, vollständig erholt.

Dass Ihnen meine generelle Morphologie im Ganzen gefallen hat, ist mir eine sehr grosse Genugthuung, und *Ihr Lob* ist für meine Anstrengungen der *höchste Lohn.* Besonders erfreut haben mich aber Ihre aufrichtigen *kritischen* Anmerkungen über die Schattenseiten des Werkes, weil ich daraus ersehe, dass Sie mich der *aufrichtigsten Freundschaft* für würdig erachten. Gewiss ist Ihr Tadel über die allzugrosse Härte meiner kritischen Angriffe und die Bitterkeit meiner Polemik an sich sehr richtig.[6] Auch meine besten hiesigen Freunde, vor Allen Prof. Gegenbaur,[7] haben mich sehr desshalb getadelt. Ich kann zu meiner Entschuldigung nur sagen, dass ich im vorigen Sommer und Winter, als ich das Buch hier in meiner trüben Einsamkeit schrieb, ausserordentlich bitter gestimmt und nervös gereizt war. Auch hatte ich zu grossen Aerger über die unverschämten und dummen Angriffe Ihrer Gegner, als dass ich sie so ungestraft hätte können hingehen lassen. Meiner Person habe ich durch diese Haltung jedenfalls sehr geschadet, und ich bin schon von vielen Seiten nicht weniger heftig angegriffen worden. Dies ist mir aber *ganz gleichgültig*, da mir an dem Ansehen meiner Person und der Achtung der Zeitgenossen sehr wenig liegt. Möge meine vielen Feinde immerhin mein Werk sehr angreifen; das trägt nur zu seiner Verbreitung bei; ob sie mich dabei schelten und verläumden, ist mir ganz gleichgültig.

Sehr leid sollte es mir aber thun, wenn ich durch meine allzu harten Angriffe der guten Sache, für welche wir gemeinschaftlich kämpfen, geschadet haben sollte. Mein Freund Gegenbaur, den ich wie einen Bruder liebe, ist dieser Ansicht, und war über meine Litterarischen Extravaganzen sehr ungehalten. Auch Sie, hochverehrter Herr, scheinen diese Befürchtung zu theilen.

Ich gestehe, dass ich in diesem Punkte nicht ganz Ihrer Ansicht bin, und dass ich Ihre Sorge für etwas zu gross halte. Ihre Befürchtung würde richtig sein, wenn man vor der Majorität der heutigen Naturforscher eine selbstständige und unparteiische Haltung und eine nach beiden Seiten hin gerechte Beurtheilung zu erwarten hätte. Dies ist aber leider *nicht* der Fall. Vielmehr sehen wir, dass die grosse Mehrheit kein eigenes selbstständiges Urtheil, keine scharfe und klare Überzeugung von dem Rechte der Wahrheit besitzt. Und da glaube ich, ist es nothwendig, dass man laut und deutlich die Wahrheit sagt, und die Schwächen der Gegner nicht schont.

Wie ich glaube, handelt es sich jetzt bei uns um eine *radikale Reform* der ganzen Wissenschaft, zu welcher Sie, hochverehrter Herr, durch Ihre mechanische causale Begründung der Descendenz-Theorie den ersten Anstoss gegeben haben. Eine solche Reformation, die überall mit ungeheuern Hindernissen und Vorurtheilen zu kämpfen hat, ist aber noch niemals durch sanfte Worte und mit wohlwollender Überredung durchgekämpft worden. Vielmehr sind überall energische Angriffe und schonungslose Stösse nöthig gewesen, um das alte Gebäude der befestigten Irrthümer zu zertrümmern. Wie in jedem Kampfe, so hat auch hier der kühne Angreifer grosse Vortheile voraus, und ich habe es daher für besser gehalten, selbst schonungslos anzugreifen, als hinterher von meinen übelwollenden Gegnern überfallen zu werden.

Wenn ich sehe, wie ungerecht und falsch man so vielfältig Ihr grosses Werk beurtheilt, wie man Sie selbst verläumdet hat, so verschwindet alle meine Achtung von dem grossen Publicum der Naturforscher. Ihre ausserordentliche *Bescheidenheit* hat man für *Schwäche*, und ihre bewunderungswürdige *Selbstkritik* für Mangel an fester Überzeugung ausgegeben. Gewiss haben Sie bei guten, verständigen und denkenden Männern dadurch nur gewonnen. Aber leider sind diese in der Minorität. Wenn ich statt dessen die Sache weit schärfer und polemischer angefasst habe, so kann ich in der That kaum glauben, der Sache selbst auf die Dauer geschadet zu haben. Jedenfalls hoffe ich zur Verbreitung Ihrer grossen Ideen und Ihrer bahnbrechenden Reform etwas beitragen, und eine Anzahl von wichtigen, darauf bezüglichen Fragen bestimmt formulirt und auf den offenen Markt gebracht zu haben, deren offene Diskussion den Gegenstand nur fördern und die Wahrheit an das Tageslicht bringen kann. Auch ich vertraue, wie Sie, hochverehrter Herr, vor Allem auf den empfänglichen und vorurtheilsfreien Sinn der *jüngeren* Naturforscher, und zweifle nicht, dass wir unter der jetzt aufblühenden Generation schon eine sehr viel grössere und lebendigere Theilnahme finden werden. Schon in den wenigen Tagen, die ich wieder hier bin, habe ich dies vom Neuen empfunden. Ich habe sogleich meine Vorlesungen über die Descendenz-Theorie wieder begonnen,[8] und empfinde mit lebhaftem Vergnügen die warme, gegen früher noch gesteigerte Theilnahme, welche die 〈　〉hsame akademische Jugend unserer Sache entgegen trägt.

Ich werde übrigens in Zukunft Ihre wohlwollenden väterlichen Rathschläge, für die ich Ihnen aufrichtig danke, berücksichtigen, und werde meiner allzu eifrigen und leidenschaftlichen Feder einen strengeren Zügel anlegen. Sie erweisen mir die aufrichtigste Freundschaft theurer Herr, wenn Sie mich auch künftighin so offen auf die Fehler meiner Arbeiten aufmerksam machen.

Über meine Reise werde ich Ihnen demnächst einen gedruckten Bericht zusenden.[9] Ich war nur kurze Zeit in Madeira, nur 14 Tage in Teneriffa (wo ich den Pic bestieg)[10] aber 3 Monate in *Lanzarote*, wo ich vorzüglich die Entwickelung der herrlichen *Siphonophoren* und die atlantischen *Radiolarien* studirt habe.[11] Im März war ich 14 Tage in Marocco 14 Tage in Gibraltar. Die erste Hälfte April reiste ich durch Sevilla, Cordova, *Granada* (!) Madrid; die zweite Hälfte April benutzte ich, um die merkwürdige Exposition universelle in Paris zu studiren.[12] Meine Gesundheit war in der ganzen Zeit vortrefflich, und ich bin mit den Resultaten der Reise sehr zufrieden.

Vielleicht komme ich noch in diesem Jahre (im Herbste) nach England, und es würde mich sehr freuen, wenn ich Sie dann in besserer Gesundheit wieder sehen dürfte. Inzwischen nehmen Sie, theurer Herr, nochmals meinen herzlichsten Dank für Ihre aufrichtige Freundschaft entgegen, und die Bitte, mir dieselbe auch fernerhin zu erhalten.

Mit der Bitte, M^{rs} und Miss Darwin[13] meine ehrerbietigste Empfehlung zu sagen, von Herzen Ihr treu ergebener | Ernst Haeckel

DAR 166: 44

[1] For a translation of this letter, see Appendix I.

[2] Haeckel had spent from November 1866 to March 1867 travelling and doing research on Tenerife and Lanzarote; he also visited Morocco, Portugal, and the Spanish mainland (see Haeckel 1867). Haeckel also refers to the letter from CD of 12 April [1867], and the fourth edition of *Origin*. Haeckel's name appears on the presentation list for this book (see *Correspondence* vol. 14, Appendix IV).

[3] Neither the letter to CD nor the letter to Thomas Henry Huxley has been found. For transcripts of the extant correspondence between Haeckel and Huxley, see Uschmann and Jahn 1959–60. Haeckel wrote to Huxley on 12 May 1867 (Uschmann and Jahn 1959–60, pp. 11–12; the letter is at the Imperial College of Science and Technology), mentioning the enclosure to the letter to CD that went astray. The Canary Islands, which include Tenerife and Lanzarote, are governed by Spain (*EB*).

[4] Haeckel presumably refers to the letter from CD of 8 January 1867.

[5] Haeckel had visited CD at Down House in October 1866 (see *Correspondence* vol. 14). On their meeting, see also A. Desmond and Moore 1991, pp. 539–40.

[6] See letter to Ernst Haeckel, 12 April [1867] and n. 6.

[7] Carl Gegenbaur.

[8] Haeckel gave lectures on CD's theory of descent both to students at Jena University and to the general public, but these seem to have been given in the winter (see Uschmann 1959, p. 45; see also *Correspondence* vol. 14, letter from Ernst Haeckel, 11 January 1866 and n. 14). His lectures given in the winter of 1867 to 1868 were published as Haeckel 1868 (see Uschmann 1959, p. 45), which was translated as Haeckel 1876.

[9] Haeckel sent a preliminary report on his travels to CD with his letter of 28 June 1867; this report has not been found in the Darwin Archive–CUL. A report was also published on 12 September 1867 in the *Jenaische Zeitschrift für Medicin und Naturwissenschaft* (Haeckel 1867).

[10] The highest point of the Pico de Teide on Tenerife is 3770 m (*Times atlas*). For Haeckel's account of the climb, see Uschmann 1984, pp. 92–6.

[11] Haeckel published his research on siphonophores in 1869 (Haeckel 1869).

[12] See also letter from J. D. Hooker, 13 April 1867.

[13] Emma and Henrietta Emma Darwin.

To J. D. Hooker [12] May [1867][1]

Down
Sunday | May 13

My dear Hooker

We are both thoroughly ashamed of ourselves for having given you so much trouble about the Paris lodgings.[2]

If you can come here on the 25th it will give us real delight. It is a long time since I have seen you; but if the day sh[d] prove very bad or unexpected business occur, do not fash yrself & give us up at once.[3] Tho' mind you will then be doubly in our debt & come you shall & must.

The address is

    D[r] Fritz Müller

    Desterro

    Brazil

    via Bordeaux

(9[d] for ¼ oz. 1/6 for ½ oz.)[4]

I sent to him y[r] "Ins. Floras."[5]

yours affectionately | Ch Darwin

(base forgery)[6]

L(S)(A)
Endorsement: '/67.'
DAR 94: 25

[1] In 1867, 12 May was a Sunday.

[2] Hooker had booked a room in Paris for the the month of April, while he was attending the Paris exhibition as a juror (*Gardeners' Chronicle*, 6 April 1867, p. 348); in his letter of 13 April 1867, by which time he had returned temporarily to Kew, he offered it to CD's sons. CD wrote on 15 [April 1867] that some of them would be 'hugely tempted'; see also letter to J. D. Hooker, 25 [April 1867]. There is no record in Emma Darwin's diary of any of the children going to Paris in April or May 1867. George Howard Darwin visited the Paris exhibition in June (see letter from G. H. Darwin, [3 June 1867] and n. 7.

[3] 'Fash': trouble or bother (Scottish; *Chambers*). Hooker did not come to Down on 25 May; see letter to J. D. Hooker, [21 May 1867].

[4] Hooker may have wanted Fritz Müller's address in connection with the dispatch of plants from Brazil to Kew; see letter to Fritz Müller, 22 April [1867].

[5] J. D. Hooker 1866a. See letter to J. D. Hooker, 25 [April 1867].

[6] Emma Darwin forged the signature after beginning to sign her own name. The words '(base forgery)' are in CD's hand.

From V. O. Kovalevsky 14 May 1867

S Petersburg
2/14 May 1867.[1]

Dear Sir

It is with the fullest acknowledgment of my importunity and with a real repentence of taking away Your precious time by my futile correspondence that, at last, I determined on writing You this letter, but dear Sir, I hope You will forgive me when I say You that all this time, from the receipt of Your very kind message of 2 May, I was quite in a fever of expectation and till this day (14 May, eight days after Your letter) I have *not received the proofs* You had the goodness to send me.[2] All printed matters send to Russia must be *sous bande*, that is, open at two ends, and if this precaution is neglected, the message will go as a parcel which are very often lost.

I have received very much book-parcels from M^r. Truebner[3] and all generally came in five days and very regularly, not one was lost, imagine then what a unhappy occurence that is, that the most dear of all the parcels should be lost by the bad management of the Continental Post Offices I will patiently(?!!) wait for the printed sheet and pray You, dear Sir to put them simply in ordinary letter covers (as You send Your letters), one sheet in one cover, and send them without prepaying or putting any stamps, direct on my adress; or deliver them to M^r. Truebner whom I shall request to send the sheets as letters and register them.

My adress can be shortened thus: *Petite Morskaja, m. Mitkoff* (via Belgium)[4]

The bear skin will go to London by one of the first steamers, but this year the river is covered till yet with ice, and the snow is laying man-high in the woods.[5]

Yours | very faithfully | W Kowalewsky

DAR 169: 74

[1] Kovalevsky gives both the Julian (2 May) and Gregorian (14 May) calendar dates.

[2] See letter to V. O. Kovalevsky, 2 May [1867]. Kovalevsky refers to the proof-sheets of *Variation*, which he was translating into Russian.

[3] Trübner & Co. were Kovalevsky's commissioners for English books (see letter from V. O. Kovalevsky, 24 April [1867]).

[4] See letter to V. O. Kovalevsky, 2 May [1867].

[5] Kovalevsky had offered to send CD a bearskin (see letter from V. O. Kovalevsky, 24 April [1867]). St Petersburg is at the mouth of the Neva; the Neva was generally frozen between November and April, and was sometimes unnavigable for longer because of ice floes (*EB*).

From J. W. Salter 14 May 1867

8 Bolton Road | St Johns Wood.
May. 14/67.

Dear Mr Darwin

You bade me apply to you if I wanted help. And this generosity has prevented me till I have tried every other means—[1] Those who should have aided me for kin's sake would do nothing when I left the Survey—except old Mr Sowerby and he could not.[2] I can borrow no money— I cannot sell my English Botany nor carry it on—& serious & frequent illness prevents me doing half that I ought, even in my precarious occupation.[3] And not one of my monied friends but are themselves in difficulties four men who would gladly aid me—and who had thousands, are now borrowing money.— And now I have my family engaged in schoolkeeping—my house partly let, & retrenchment in every sort— I am still heavily in debt, & have no means of getting out of it.[4]

I know not what I can do for you in return for any aid, unless your library wants the English Botany.— it is the only thing in my power. Pray let me send it you if you have it not.[5]

I dont think I should have written now, had not a dear sister, as poor as myself, been in the last stage of illness—[6] I shall probably have to follow her to the grave directly—& you know what all this involves.

Aid me, if you can, & command me in any way in return.

I am, Dear Sir | Yours gratefully | J W Salter

C. Darwin Esq[r].

DAR 177: 13

[1] Salter had asked CD for financial help in his letter of 31 December 1866 (*Correspondence* vol. 14), and CD recorded payments to him of £10 on 1 January 1867 and £5 on 19 May (CD's Account books–cash account (Down House MS); see also letter from J. W. Salter, 4 January [1867]). No letters from CD to Salter have been found.

[2] Salter had resigned his post as palaeontologist to the Geological Survey in 1863; he had earlier been apprenticed to the naturalist and scientific illustrator James de Carle Sowerby, and was also his son-in-law (*ODNB*).

[3] After resigning from the Geological Survey, Salter worked at local museums arranging their palaeozoic invertebrates (*ODNB*). On Salter's work on the 'English Botany', see the letter from J. W. Salter, 4 January [1867] and n. 6; on Salter's illness, see Secord 1985.

[4] See *Correspondence* vol. 14, letter from J. W. Salter, 31 December 1866. Salter and his wife, Sarah, had seven children (*ODNB*); on his wife's keeping a school, see also J. Secord 1985, p. 62 and p. 73 n. 26.

[5] There are copies of the five volumes of the *Supplement to the English botany* (W. J. Hooker and Sowerby [*et al.*] 1831–63) in the Darwin Library–CUL. Volumes 3 and 5 are in the form of monthly parts. A printed sheet at the back of the last part of volume 5, dated June 1865, contains an offer by Salter to make up incomplete editions. See also letter from J. W. Salter, 18 June 1867.

[6] Salter's sister has not been further identified; two had survived to maturity (Secord 1985, p. 62).

From V. O. Kovalevsky 15 May 1867

Dear Sir

The same day when my lamentable letter was send to You I received the 5 proof sheets You had the goodness to send me on the 2 May. But still I have been patient enough and waited ten long days after the receipt of Your letter, it was impossible for me to bear this longer and so I wrote You my letter (even registered it) which You no doubt received yesterday.[1]

I thank You dear Sir most cordially for all the kindness You show to me, believe me I shall always remember it.

For the future I shall beg You to put the sheets in letter covers and send them on my adress without prepaying (non affranchie) it is the safest way,— prepaid letters are often lost but unprepaid always reach in Russia their destination

Believe me | Dear Sir | Yours very faithfully | W. Kowalewsky

May 3/15 1867.[2]

P.S. I am also the editor of the large and beautiful work of *Brehm Thierleben*, I translated the two big volumes myself and am now translating the III vol. *Birds*, as

You certainly are aware in this work nearly all the varieties of *Dogs* & *Cats* etc you mention in Your work are represented by very good & true woodcuts, as I have the tsereotypes of Brehm's work, I think to print them on separate sheets & put them at the end of the volume. So the reader will have the woodcuts of all the varieties & species You mention in Your work, and the striking facts You mention in your book will be much better understood and appreciated, I dont think that such a Plan can be objected to, because, as the woodcuts shall be printed separately from the text, every one is at liberty if wishing to conserve the book in the form of the original, has only to tear the additional sheets and to trow them away. I shoul like very much Your opinion on the subject, but in case You have not seen the work I mention, I pray You to do so.— it is: *Brehm Illustrirtes Thierleben* 1863–1868.[3]

DAR 169: 75

[1] Kovalevsky refers to proof-sheets of *Variation*, which he was translating into Russian. See letter to V. O. Kovalevsky, 2 May [1867], and letter from V. O. Kovalevsky, 14 May 1867.

[2] Kovalevsky gives both the Julian (3 May) and Gregorian (15 May) calendar dates.

[3] Alfred Edmund Brehm's *Illustrirtes Thierleben* (Brehm *et al.* 1864–9) was issued in parts between 1863 and 1869 (*NUC*). Kovalevsky's Russian translation was published between 1866 and 1870 ([Kovalevsky] trans. 1866–70).

To V. O. Kovalevsky 16 May [1867][1]

Down. | Bromley. | Kent. S.E.
May 16th.

Dear Sir

I write merely to say that I despatched this morning four clean sheets. I am glad to say that the 1st. Vol. will contain only about 412 pages & the 2nd. Vol. about 460 but large portions are in small type.[2] I have given orders for the stereotypes.[3]

Dear Sir, | yours faithfully | Ch. Darwin

P.S. I forgot to say in my last letter that I shall make no charge for right of Translation to you: circumstances have led me to charge nothing for German Translation & therefore it could not be fair in me to make any charge for other Translations.—[4]

When you have occasion to write please inform me whether I must use present long address.—

Mittag Leffler Institut

[1] The year is established by the relationship between this letter and the letter to V. O. Kovalevsky, 2 May [1867].

[2] CD refers to *Variation*. The first volume is 411 pages long, and the second 486, not including preliminaries. For the decision to print portions of *Variation* in small type, see the letter from John Murray, 9 January [1867] and n. 3.

[3] See letter to V. O. Kovalevsky, 2 May [1867]; see also letters from V. O. Kovalevsky, 2 April 1867 and 24 April [1867].

[4] The German translation of *Variation* was by Julius Victor Carus. See letter to E. Schweizerbart'sche Verlagsbuchhandlung, [19 March 1867]).

To J. J. Moulinié 16 May [1867][1]

Down. | *Bromley.* | *Kent. S.E.*
May 16th.

Dear Sir—

I have the pleasure to inform you that I despatched this morning four clean sheets.[2] I am very glad to say that my book does not turn out so large as was expected. The Printers estimate that the 1st. Vol will contain 412 pages & the 2nd. Vol 460 pages but large portions are in small type.[3] In accordance with a letter from M. Reinwald I have given the order for the casting of the stereotypes[4]

Pray believe me, dear Sir | with much respect | Yours very faithfully | Ch. Darwin

LS
Bibliothèque Publique et Universitaire de Genève, Ms. suppl. 16, f. 8–8vº.

[1] The year is established by the relationship between this letter and the letter from J. J. Moulinié, 3 May 1867.
[2] Moulinié was to translate *Variation* into French (see letter from J. J. Moulinié, 3 May 1867).
[3] See letter to V. O. Kovalevsky, 16 May [1867] and n. 2.
[4] CD refers to Charles-Ferdinand Reinwald, possibly to Reinwald's letter of [May 1867], and to stereotypes for the illustrations for *Variation* (see also letter to V. O. Kovalevsky, 16 May [1867] and n. 3).

From J. D. Hooker 17 May 1867

Kew
May 17/67

Dear Darwin

I find I must not go to Down tomorrow, having sent Smith to the country for his health, which causes me some anxiety. I must put it off till the Gooseberry Season!—[1]

I go again to Paris at end of month, a good holiday it makes, though the only rest I get is in the Theatres— Still the show is very interesting & I have an awful deal to learn in the matter of the plants.[2]

I hear that Wallace & Mueller of Victoria are the most likely candidates for gold medal for Biology, & am puzzled a little to decide, but have so very high an opinion of Wallace that I incline to him— his work is so very good though less than one could have wished.[3]

Do let me know soon how your health is. & how your book gets on.[4]

We are all pretty well & my wife getting about a little.[5]

Ever Yr affec | J D Hooker

[Enclosure]

Mauritius

15th April 1867

My dear Dr Hooker

Thanks for your Letter of 18th. Feb^y, and for its enclosures. I read with the deepest interest your Lecture on Insular Floras.[6] It is a very valuable contribution towards the investigation of that most puzzling subject. I am glad to see that you incline a little to the theory of submerged continents; for the supposition that currents, oceanic or aerial, and birds, and fishes, have conveyed all the original progenitors of the Plants found on such Islands as those of New Zealand or Kerguelen, has always struck me as quite insufficient to account for the facts of the case.

Even as regards my favourite Ferns, which from the extreme lightness of their spores may be distributed for vast distances by the Winds, it is difficult to understand why certain genera are so carried to certain places and not to others close by, affording similar climate & conditions.

When once placed in such isolated spots their departure from the original type, and consequent formation of new species is readily conceivable.

I wish I could give you more information about the Seychelles, but you will be able to learn a good deal from Mr Edward Newton, brother to Professor Alfred Newton of Cambridge, who has just gone on leave of absence from home & spent a month at the group for the purpose of collecting Birds, on his way to England.[7] He writes me that he was disappointed on the whole, as he found both their Fauna and Flora exceedingly limited.

To the latter cause he attributes the great deficiency of Insects which he states to be very remarkable. He saw but four species of Butterflies—of which one was a common Mauritius one. To the paucity of insects and flowering plants again, he attributes the small number of Birds, of which he got but few new ones for his Collection.[8]

Does not this tend rather to discountenance a Continental origin? Unless we may assume that the smallness of the different Islets and their distance from each other has led to wholesale extinction of species.

With regard to their Granite formation, and Indian Flora, betokening former connection in that direction, I am unable to judge as to the second point but as to the first I imagine that Granite though occasionally protruding in Southern India and Ceylon is not a prominent feature in the geological structure of either, and it must be remembered that it is on the contrary largely developed in the North of Madagascar much nearer at hand. I have been told by those who have seen both that its characters are very similar.

I never heard of any indigenous Mammals being found at Seychelles, at least no quadruped.[9]

The large Bat or Flying Fox is there but it is common here & in Madagascar also.

They have not even our little insectivorous quadruped the Tendrec (Ericulus spinosus) the only one pretending to be indigenous to Mauritius, though as it is eaten as a delicacy by the Blacks I suspect it may have been introduced from Madagascar.[10] As to our laying claim to fossil bones of Deer, I doubt—saving Professor Owen's opinion—whether they are of older date than the first Colonization of the Island when it is on record that the Portuguese turned loose Deer, Monkeys, &c &c. I am not quite sure to which discovery you allude, but it is probably the recent one when the Bones of Deer, Pigs, & Monkeys, were found in the swamp at Mahébourg in which the remains of the Dodo were embedded.[11]

I did not see any animals bones from this spot, nor find any when I visited it, and I have no idea what of size or description those of the Deer were— Mr Clark in his Paper on the discovery (vide Ibis— April 1866) states—"The Deer's bones only were found in juxtaposition so as to make it probable that the animal had died on the spot on which they were found"[12] Now as the Deer are every year hunted and driven from the Woods, often wounded to die in Cornfields & elsewhere, if these Bones were those of the large species which is very near the Ceylon Elk or Jamboc Deer though I believe Mr Newton has identified it with a Java variety, there can be little question as to where they are derived. If on the other hand they are as I think I heard from Mr Newton of a small species, they may be those either of the Gazelle which was introduced many years ago by the French from Senegal, or of the Indian Axis Deer which has also often been tried here.[13]

It would be curious to know in either case whether they seem identical with a former determination of semi fossilized Deer's bones by Professor Owen from a Cave near Black River sent to him with some remains of the Tortoise by the late Dr Ayres under the impression that they might be those of the Dodo. This was I believe in 1860 or 61.[14]

Mr Edward Newton can give either yourself or Dr Owen full details as I called his attention to the subject.

I hope soon to have an opportunity of becoming better acquainted with the Seychelles Flora, as a travelling companion of Mr Newtons, Mr Nevill remained there for two months after the former had left and is expected back by next Steamer. I lent him my [Boxes Straps] &c, & he promised to lay down all he could, though beyond the Ferns he knows but little of Botany, Shells being his speciality.[15]

I was anxious that Dr Meller should avail himself of an opportunity which now presents itself for spending 6 months at Mahé Dr Brookes the Government Medical Officer there having applied for leave to go home—[16] He writes to me however that his health is so bad, that he could be of no good there, & that it would only be tantalising to him to go there as an invalid. He has suffered terribly from Dysentery during all the Summer, and I almost fear that the heat of Seychelles might be too much for him. I have not seen him for many weeks, as he is quite unable to attend to business. In fact I think the opiates to which he sometimes has recourse are nearly as bad as the disease.

Mr Ward writes to me that failing Meller he hopes to secure the services of a Dr Wright (or Bright, for he does not write distinctly) who is coming out to the islands in June on a Scientific Mission.[17] Do you know anything of this? I trust he will carry out your wishes of a thorough exploration.

Pray thank Mr Baker for the Copy of his Paper on Hymenophylæ read before the Linnean Society. I hope the Trichomanes Barklianum will stand, but I have an uncomfortable suspicion that it may prove the same as Trichomanes cuneatum Poir. in the Bourbon Catalogue, though Dr Meller could get no specimen of that & I have never met with it. The very general description given by D^r [Borias] would suit any simple leaved species and he says nothing of the venation.[18]

I enclose another species from Bourbon which is not in the Catalogue & which they told Meller was new & undescribed. I am sorry Mr Baker had not got it when he was dealing with the question, but I concluded Dr Meller had sent it home.

We are now going over our Bourbon specimens, & will set aside any that may be interesting to you at Kew.

I regret to hear of a hitch in the publication of the Synopsis but trust you will succeed in having reasonable remuneration for his work secured to Mr Baker. Will you ask the Publisher to put down my name for two copies of the work.[19]

Your letter has provoked rather a long story, but you must bear the consequences of your kindness in writing to me, | very truly Yrs | Henry Barkly

DAR 102: 163–4; Royal Botanic Gardens, Kew, DC 188: 125

CD ANNOTATION
Verso of last page of letter: 'The early accounts expressly state no mammal or Batrachian aboriginal[20]—& numbers turned out. The rusa of Java is one.— *Pigs & Monkeys*' ink

[1] CD had been expecting Hooker to visit Down on 25 May (see letter to J. D. Hooker, [12] May [1867]). John Smith was the curator of the Royal Botanic Gardens, Kew. Hooker's visit to Down during the gooseberry season was something of a tradition; see *Correspondence* vol. 13, letter from J. D. Hooker, [after 17 June 1865] and n. 6.

[2] Hooker was attending the Paris exhibition as juror for seeds and saplings of forest trees (*Gardeners' Chronicle*, 6 April 1867, p. 348).

[3] Hooker refers to the Royal Medal of the Royal Society of London. Two were awarded each year 'to each of the two great divisions of Natural Knowledge' (*Record of the Royal Society of London*, p. 349). Hooker had also considered whether Alfred Russel Wallace might be awarded a Royal Medal in 1864 (see *Correspondence* vol. 12, letter from J. D. Hooker, 26[–8] October 1864 and n. 18). Ferdinand von Mueller was director of the Botanical Gardens in Melbourne (*Aust. dict. biog.*). Neither Wallace nor Mueller won a Royal Medal in 1867; Wallace won one in 1868.

[4] Hooker refers to *Variation*.

[5] Frances Harriet Hooker's health had been intermittently poor since the birth of her sixth child in January.

[6] There is a transcript of Hooker's letter to Henry Barkly of 18 February 1867 in CUL (MS Add 9537). Hooker enclosed his paper on insular floras (probably his pamphlet version of J. D. Hooker 1866a; see Williamson 1984) and a paper on new species of Hymenophyllaceae (Baker 1866).

[7] Edward Newton was auditor-general at Mauritius (*Modern English biography*). Alfred Newton was professor of zoology and comparative anatomy at Cambridge University (*ODNB*). Hooker had suggested in his letter to Barkly of 18 February 1867 (see n. 6, above) that an exploration of the Seychelles would add vastly to knowledge of insular floras.

[8] Edward Newton gave an account of the birds of the Seychelles, including five new species, in Newton 1867a and 1867b.

[9] In his letter to Barkly of 18 February 1867 (see n. 6, above), Hooker had said that he supposed the Seychelles were granite and had an Indian flora, and remarked that he and CD thought them relics of an old continental coastline; he asked whether they had any indigenous mammals.

[10] The tenrec (now *Tenrec ecaudatus*) is thought to have been introduced to Mauritius from Madagascar (Nowak 1999).

[11] In his letter to Barkly of 18 February 1867 (see n. 6, above), Hooker had written:

Since delivering the discourse, I am informed that Mammalian bones (Deer of a small species) have been found in bogs in Mauritius, is it so? & if so are the species certainly indigenous or can they be of animals introduced during an early period of the the colony? I saw Owen the other day, who assured me that the discovery of the bones was a fact.

The reference is to Richard Owen. See also letters from J. D. Hooker, 4 February 1867 and 20 March 1867.

[12] The reference is to George Clark's paper on his discovery of dodo remains in a marsh near Mahébourg, Mauritius (G. Clark 1866); the quotation is from p. 145.

[13] By 'Ceylon Elk or Jamboc' Barkly probably meant the sambar, *Cervus unicolor* (subgenus *Rusa*; see *EB* s.v. Ceylon, and Nowak 1999). Alfred Newton, the editor of *Ibis*, the periodical in which Clark's paper was published, added a note to Clark's remarks about an earlier discovery of antlers of the deer that still existed on Mauritius in the marsh near Mahébourg, saying that these deer had been identified by Edward Blyth as *Cervus rusa*, a species introduced from Java (*Ibis* 4 (1862): 92). The axis deer (*Axis axis*) is native to the Indian subcontinent (Nowak 1999).

[14] Philip Burnard Ayres, superintendent of quarantine on Mauritius, died in 1863 (*Modern English biography*). No published remarks by Owen on the Black River cave specimens have been found.

[15] Geoffrey Nevill published two papers on the land-shells of Mauritius and the Seychelles (Nevill 1868 and 1869).

[16] Charles James Meller was director of the botanic gardens, Pamplemousses, on Mauritius; Mahé is the largest of the Seychelles Islands, which were a dependency of the colony of Mauritius (see McCracken 1997, pp. 46–9). J. H. Brooks was medical officer on the Seychelles between 1858 and 1883; in 1868 he was on leave and his place was taken by P. Vaudagne, who has not been further identified. (*Colonial Office list*.)

[17] Swinburne Ward was civil commissioner on the Seychelles (*Colonial Office list* 1867). Edward Perceval Wright spent six months in the Seychelles in 1867, and brought back a collection of plants and animals (*ODNB*); he did not take the post of medical officer (see n. 16, above).

[18] Barkly refers to John Gilbert Baker, assistant in the Kew Herbarium, and Baker 1866 (see n. 6, above). *Trichomanes barklianum* is one of the species listed in Baker's paper. Ile Bourbon is now Réunion Island, and is 130 miles south-west of Port Louis, Mauritius; it was a French colony (*EB*). The Bourbon catalogue, the second *Trichomanes* species mentioned, and Dr *[Borias]* have not been identified.

[19] Hooker had written in his letter to Barkly of 18 February (see n. 6, above): 'M\`r\` Baker has nearly brought the Synopsis Filicum to a completion, but we are rather in a deadlock by the Publisher proposing to pay Mr Baker *nothing* for his work, except the work should "pay"—thus repudiating the contract made with my father, to pay £150 for the work.' The reference is to W. J. Hooker and Baker 1868, published by Robert Hardwicke.

[20] See n. 13, above, and letter to J. D. Hooker, [21 May 1867].

From Camille Dareste[1]   19 May 1867

Lille;
19 Mai 1867.

Monsieur

Vous n'avez peut-être pas oublié que vous avez reçu, il y a cinq ans, un travail d'un Professeur d'histoire naturelle de Lille, sur la production artificielle des monstruosités. La lettre que vous lui avez écrite, et qu'il conserve précieusement, lui prouvait que vous aviez bien voulu prendre connaissance de ce travail et le lire avec quelqu'intérêt.[2] Cette lettre a été pour lui, ou plutôt pour moi, car c'est de moi qu'il s'agit ici, le meilleur de tous les encouragements; car je le dis avec beaucoup de regret, mes travaux sont, pour le moment, fort peu appréciés en France. Un seul homme les avait bien compris, c'etait Is. Geoffroy Saint-hilaire. Depuis sa mort, on n'y a, malheureusement, attaché qu'une attention très distraite.[3]

Depuis que je vous ai écris, j'ai continué mes travaux avec une grande persévérance, et j'ai pu assister au développement embryogènique de presque toutes les formes de la monstruosité. Ce travail sera, je l'espère, terminé dans un an, et je pourrai alors en entreprendre la publication.[4] Jusqu'à présent, j'ai dû me contenter d'en faire connaître les principaux résultats dans les comptes rendus de l'Académie des Sciences de Paris;[5] et ces notes n'ayant pas été tirées à part, je n'ai pas pu vous les faire remettre. Quand ce premier travail sera terminé, je pourrai, je l'espère, arriver à déterminer les conditions de la production des monstres, et produire ainsi à volonté, telle ou telle anomalie. Je possède déjà un certain nombre d'indications à ce sujet; et j'ai tous lieu de croire que je pourrai les completer. Je vous tiendrai alors au courant de tous ce que j'aurai réussi sur un sujet, qui se rattache trop à vos propres études pour ne pas vous intéresser un peu.[6]

En attendant, je vous adresse un petit travail qui se rattache directement à une question que vous avez vous même observée. Elle est relative aux boeufs *niata* de l'Amérique du Sud.[7] J'ai eu occasion d'étudier un de ces Animaux qui eu né spontanément à Lille, d'une vache flamande ordinaire, et qui reproduisait exactement les Caractères des *niata*. Si cet animal avait pu arriver à l'âge de la puberté, il aurait pu donner à ces descendants les caractères qui le distinguaient. Ce fait est parfaitement en rapport avec vos travaux sur l'espèce; j'ai donc pris le parti de vous l'adresser.[8]

Et à ce propos je dois vous dire que les conclusions que j'en ai voulu tirer relativement à la formation de la race *niata* ont été contredites assez vivement en France. Il y a des personnes, assez aveuglées par des fausses théories pour ne pas vouloir admettre l'existence de la race *niãta*, et pour rejeter entièrement votre témoignage.[9] Je serais heureux si vous pouviez le donner de nouveau d'une manière bien evidente.[10]

J'ai eu occasion egalement de constater dans la race des poules Du Saxes, la production de deux individus qui portaient le caractère céphalique et cérébral des poules polonaises. Une hernie du cerveau, qui plus tard se revêt d'une enveloppe osseuse résultant de l'ossification de la fontanelle.[11]

Veuillez m'excuser, Monsieur, de v(ous) interrompre en vous parlant de mes travaux: mais je crois qu'ils se lient aux vôtres par des liens très étroits J'espère toujours que mes études me conduiront à quelques considérations nouvelles sur la formation des espèces, et que mon travail pourra peut-être un jour prendre place dans la Science, après votre livre, quoiqu' à une bien grande distance (*longo sed proximus intervallo*)[12]

C'est l'une des pensées qui me soutient la place dans la carrière souvent ingrate que j'ai entreprise.

Recevez je vous prie, l'expression de mon admiration et de mon respect. | Camille Dareste

Professeur à la Faculté des Sciences de Lille.

DAR 162: 43

[1] For a translation of this letter, see Appendix I.

[2] Dareste wrote to CD on 8 February 1863, enclosing a copy of Dareste 1863; CD replied in a letter dated 16 February [1863] (see *Correspondence* vol. 11). There is a lightly annotated copy of Dareste 1863, in which Dareste provides an account of monstrosities artificially induced in chicken eggs, in the Darwin Pamphlet Collection–CUL.

[3] Isidore Geoffroy Saint-Hilaire died in 1861; he had worked on teratology, like his father, Etienne Geoffroy Saint-Hilaire, whose work influenced Dareste (Tort 1996).

[4] Dareste published a book summarising his research in teratology in 1877 (Dareste 1877).

[5] Dareste published articles on teratology regularly in the *Comptes rendus hebdomadaires des séances de l'Académie des sciences.*

[6] CD cited Dareste a number of times on teratology in the second volume of *Variation*, and commended his work in *Descent* 2: 338.

[7] Niata: 'An abnormally small variety of cattle, found in South America' (*OED*). CD discussed niata cattle in *Journal of researches* 2d ed., pp. 145–6.

[8] There is a copy of Dareste 1867, *Rapport sur un veau monstreaux*, in the Darwin Pamphlet Collection–CUL. Dareste suggested that monstrous forms, under the right conditions, could originate new races.

[9] Dareste had been involved in an exchange of papers with André Sanson in the *Comptes rendus hebdomadaires des séances de l'Académie des sciences.* Sanson denied the existence of a race of niata cattle in South America; Dareste had cited CD as one of his authorities. See *Comptes rendus* 64 (1867): 423–6, 669–70, 743–5, 822–4, 1101–3. See also *Correspondence* vol. 11, letters from Armand de Quatrefages, [28 March –] 11 April 1863 and n. 4, and 19 May [1863] and n. 4.

[10] CD discussed niata cattle in *Variation* 1: 89–91; see letter to Camille Dareste, 23 May 1867.

[11] See Dareste's first paper in *Comptes rendus* (see n. 9, above): 'Mémoire sur le mode de production de certaines races d'animaux domestiques'. CD discussed the skulls of Polish and other crested breeds of fowl in *Variation* 1: 262–5.

[12] *Longo sed proximus intervallo*: 'The next, but after a long interval. A poor second' (H. P. Jones ed. 1900).

From Isaac Anderson-Henry 20 May 1867

Hay Lodge, Trinity, | *Edinburgh.*
May 20/67

My dear Sir

Happening to pass thro' Perth last week and having an hour to spare I visited the Nursery of which Mr Brown, who communicated to Dr Neill the extraordinary results of a graft I alluded to in the Paper I lately wrote on hybridisation, I thought I might learn some particulars of it from my friend Mr Turnbull of Belwood the Head of the existing firm now an old gentleman.[1] I have not been disappointed, (tho I missed Mr Turnbull) as you will see by the enclosed letter I have from his nephew—of which make any use you please[2]

I have got another instance of monster pods in *Arabis blepharophylla* a new North american sp.* of which the seeds were sent me by Dr Hooker, resulting from being crossed with *Arabis Soyeri*—the 2 pods so obtained being about twice the size of the normal pods. They are yet quite green however[3]

Have you ever seen a singular book I fell in with lately the "*Telliamed*" of M. Maillet in which he treats "of the origin of men & Animals". My copy, a Translation, is dated 1750 If you have not seen, and should wish to see it, I will gladly send it. It stoutly asserts the fact of men having *tails* & gives instances[4]

Very faithfully yours | Is: Anderson Henry

Charles Darwin Esqre FRS. &c. &c

[Enclosure]

Nursery & Seed Warehouse, | *26. George Street.* | *Perth*
16 May *1867*

Isaac Anderson Henry Esq | Edin^h
My Dear Sir

My Uncle was sorry he did not see you and hopes you will have more leisure the next time you look in.

With regard to the Ash, he says, it is an event of half a century ago. Mr Brown and he were on a Botanical excursion in the Highlands, being the time they discovered the *Menzesea Cærulea*,[5] and on their way back, as they were looking over a Glen about three miles west from Kenmore they noticed the branch of an Ash entirely yellow which they took away with them and budded it on the common Ash at Perth Nurseries. as far as I can ascertain from him none of the buds grew, and memory fails him in recollecting much about it, only the fact that the operation communicated the disease or blotch to the stocks on which it had been budded, and it has been grown since that time, and annually grafted and catalogued under the name of Blotched Breadalbane Ash.[6] It had been further experimented upon by grafting Weeping Ash on the same stocks and it communicated to the Weeping Ash the blotch also.— We will have to presume that the piece of bark introduced with the bud did attach or grow, although the eye did not push.[7] this is a circumstance that takes place often in budding. I have seen the bud of a variegated Holly lie dormant, or blind as we term it, for a number of years and grow afterwards.— It is fortunate you made the enquiry, as it fixes the fact of the inoculation of the disease by budding in this case, and perpetuated through a long series of years.

as you appear much interested in the fact I have forwarded to your address a plant of the Breadalbane Blotched Ash.

I am | Yours truly | John Anderson

DAR 159: 68, 68a

CD ANNOTATION
Enclosure 2.10 Blotched Breadalbane Ash] *underl red crayon*

[1] Anderson-Henry refers to Robert Brown (*c.* 1767–1845), a Perth nurseryman, Patrick Neill, a well-known horticulturalist, and Archibald Turnbull, also a Perth nurseryman. Turnbull had inherited property at Belwood, near Perth (*Florist and Pomologist* (1875): 48). Anderson-Henry's paper (Anderson-Henry 1867a) was read before the Botanical Society of Edinburgh in March, and published in the society's *Transactions* for 1867. There is a lightly annotated offprint of the paper from the *Farmer*, which Isaac Anderson-Henry sent with his letter of 3 April 1867, in the Darwin Pamphlet Collection–CUL. Anderson-Henry, discussing changing the colour of a calceolaria by crossing, added, 'I communicated the result to Dr Neill, who, I remember, felt great interest in it, instancing something of a like nature produced by grafting operations, communicated to him by Mr Brown, of Perth' (Anderson-Henry 1867a, p. 105).

[2] Turnbull's nephew was John Anderson Anderson (R. Desmond 1994).

[3] In Anderson-Henry 1867a, p. 112, Anderson-Henry mentioned the large seed pod resulting from a cross between *Rhododendron dalhousiae* and *R. nuttallii*. He discussed his cross between *Arabis blepharophylla* and *A. soyeri* in Anderson-Henry 1867b; there is a copy of this paper in the Darwin Pamphlet Collection–CUL. CD mentioned the *Arabis* cross in *Variation* 1: 400. Anderson-Henry refers to Joseph Dalton Hooker.

[4] Anderson-Henry refers to Maillet 1750, *Telliamed: or, conversations between an Indian philosopher and a French missionary, on the diminution of the sea, the formation of the earth, the origin of men and animals, and other curious subjects.* 'Men with tails' are discussed on pp. 246–53. In it, Benoît de Maillet gives a number of examples, some of which he claims to have seen himself, and argues that humans with tails are a different species from humans without, and that the tail is not due to chance or the effect of the mother's imagination upon the foetus. On the *Telliamed* and its influence, see Carozzi trans. and ed. 1968.

[5] Brown discovered *Menziesia* (or *Phyllodoce*) *caerulea* near Aviemore, Strathspey (J. E. Smith 1824–36).

[6] Kenmore, near where the branch was found, is at the head of Loch Tay, in the Breadalbane region of the Scottish Highlands.

[7] 'The eye did not push': that is, the bud did not produce a shoot. See *OED*.

To V. O. Kovalevsky 20 May [1867][1]

Down. | *Bromley.* | *Kent. S.E.*
May 20

Dear Sir

I write merely a line to acknowledge your note of May 14 & to say how sorry I am that the proof sheets were lost.[2] But you will have received I hope by this time the clean sheets & my note.[3] They were sent via France, but I hope this will not cause a second loss. In both cases the sheets were sent open at the 2 ends.[4]

Dear Sir | yours very faithfully | Ch. Darwin

LS
Institut Mittag-Leffler

[1] The year is established by the relationship between this letter and the letter from V. O. Kovalevsky, 14 May 1867.

[2] CD had sent proof-sheets of *Variation* to Kovalevsky for his Russian translation. See letter from V. O. Kovalevsky, 14 May 1867.

[3] CD sent revised sheets on 1 May (see letter to V. O. Kovalevsky, 2 May [1867]), and clean sheets with his letter of 16 May [1867]. Kovalevsky received the revised sheets on 14 May and the clean sheets shortly after (see letters from V. O. Kovalevsky, 15 May 1867 and [after 24? May 1867]).

[4] See letter from V. O. Kovalevsky, 14 May 1867.

To Ernst Haeckel 21 May [1867][1]

Down. | Bromley. | Kent. S.E.
May 21

Dear Häckel

Your letter of the 18th has given me great pleasure, for you have received what I said in the most kind & cordial manner.[2] You have in part taken what I said much stronger than I had intended. It never occurred to me for a moment to doubt that your work with the whole subject so admirably & clearly arranged, as well as fortified by so many new facts & arguments, w[d] not advance our common object in the highest degree.—

All that I think is that you will excite anger & that anger so completely blinds every one that your arguments w[d]. have no chance of influencing those who are already opposed to our views. Moreover I do not at all like that you towards whom I feel so much friendship sh[d] unnecessarily make enemies, & there is pain & vexation enough in the world without more being caused. But I repeat that I can feel no doubt that your work will greatly advance our subject, & I heartily wish it c[d] be translated into English for my own sake & that of others. With respect to what you say about my advancing too strongly objections against my own views, some of my English friends think that I have erred on this side; but truth compelled me to write what I did, & I am inclined to think it was good policy.

The belief in the descent theory is slowly spreading in England, even amongst those who can give no reason for their belief. No body of men were at first so much opposed to my views as the members of the London entomolog. Soc; but now I am assured that with the exception of 2 or 3 old men all the members concur with me to a certain extent.[3] It has been a great disappointment to me that I have never rec[d] your long letter written to me from the Canary I.s. I am rejoiced to hear that your tour which seems to have been a most interesting one has done yr health much good.[4] I am working away at my new book, but make very slow progress & the work tries my health, which is much the same as when you were here.[5]

Victor Carus is going to translate it,[6] but whether it is worth translation I am rather doubtful

I am very glad to hear that there is some chance of your visiting England this autumn & all in this house will be delighted to see you here.

Believe me my dear Häckel | yours very sincerely | Charles Darwin

LS
Ernst-Haeckel-Haus (Bestand A-Abt. 1-52/14)

[1] The year is established by the relationship between this letter and the letter from Ernst Haeckel, 12 May 1867.

[2] See letter from Ernst Haeckel, 12 May 1867. CD was mistaken about the date. CD had commented on Haeckel's *Generelle Morphologie* (Haeckel 1866) in his letter to Haeckel of 12 April [1867].

[3] CD was himself a founder member of the Entomological Society of London. He had written before about the supposed hostility of entomologists to his views: see, for example, *Correspondence* vol. 8, letter to H. W. Bates, 22 November [1860], and *Correspondence* vol. 11, letter to Charles Lyell, 17 March

[1863]. On the attitude of members of the society to CD's views, see Poulton 1901. A member of the society who consistently opposed CD's theory of descent was John Obadiah Westwood (see Neave *et al.* 1933, p. 131).

[4] See letter from Ernst Haeckel, 12 May 1867.

[5] CD refers to *Variation*. Haeckel had visited CD at Down House in October 1866 (see *Correspondence* vol. 14).

[6] The German translation of *Variation* was by Julius Victor Carus (Carus trans. 1868)

To J. D. Hooker [21 May 1867][1]

Down
Tuesday

My dear Hooker

I suppose you are now in France. I am very sorry you c^d not come here, but remember that you are bound by honour to two visits here.[2]

We intend to be in London in the early part of June for a week & shall perhaps see you but that must not count.[3] I am very glad to hear about Wallace & the gold medal. Every thing that I have read of his gives me the highest idea of his extraordinary talents. I cannot of course judge about Müller, but he must have done much good work if he is to beat Wallace.[4] Thanks for the enclosed letter; I am glad to see that Barkely takes the same view that I did about the bones of the deer. I remember distinctly in one of the old voyages the express statement that no quadruped inhabited the island. With respect to Barkly's belief in continental extensions, the argument which always brings me round to my old belief again is that you must extend the continental theory to every single island, as far as I know, in every ocean.[5]

I am getting on very slowly with my book for I have unparalleled power of expressing myself badly, so that I doubt whether it will be finished by Nov. & this half breaks my heart & injures my stomach[6]

yours affectly | Ch. Darwin

LS
DAR 94: 26–7

[1] The date is established by the relationship between this letter and the letter from J. D. Hooker, 17 May 1867. In 1867, the first Tuesday after 17 May was 21 May.

[2] Hooker had evidently cancelled two visits to Down; one on 20 April (see letter from J. D. Hooker, 13 April 1867 and n. 4), and one on 18 May (see letter from J. D. Hooker, 17 May 1867). Hooker was attending the Paris Exhibition as a juror (*Gardeners' Chronicle*, 6 April 1867, p. 348).

[3] The Darwins visited London from 17 to 24 June 1867 (see CD's 'Journal' (Appendix II)), but did not see Hooker (see letter to J. D. Hooker, [23 June 1867]).

[4] CD refers to Alfred Russel Wallace, Ferdinand von Mueller, and the Royal Medal of the Royal Society of London. See letter from J. D. Hooker, 17 May 1867 and n. 3.

[5] See enclosure to letter from J. D. Hooker, 17 May 1867, and letter to J. D. Hooker, 8 February [1867]. For CD's views on continental extensions, see, for example, *Correspondence* vol. 6, letter to Charles Lyell, 25 June [1856], and letters to J. D. Hooker, 19 July [1856] and 30 July [1856] and n. 3,

and *Origin*, pp. 357–8. Hooker and CD had disagreed on the subject: see, for example, *Correspondence* vol. 11, letter to J. D. Hooker, 5 March [1863] and n. 17.
[6] CD was working on the proof-sheets of *Variation*; he finished on 15 November (see 'Journal' (Appendix II)).

To Isaac Anderson-Henry 22 May [1867][1]

Down, Beckenham, Kent. S.E.
May 22nd.

My dear Sir,

I am very much obliged to you for the case of the grafted ash I presume I may keep the note for a few weeks and will then return it if required. M[r]. Rivers gave me a very nearly similar case; and in correcting the press of my present book I shall be glad to add this new case.[2]

I am sorry to give you any trouble but if the Arabis soyeri has much larger pods than those of the other Arabis (the name of which I cannot read) I should be very much obliged to you if you would inform me; but if I do not hear I shall understand that the large pod is a mere monstrosity and not *directly* caused by the pollen of a species having been used which itself naturally produces a large pod.[3]

You are so kind as to offer to lend me Maillet's work which I have often heard of but never seen. I should like to have a look at it, and would return it to you in a short time. I am bound to read it, as my former friend and present bitter enemy Owen, generally ranks me and Maillet as a pair of equal fools.[4]

With very many thanks for all your kindness. | I remain, my dear Sir, | Yours very faithfully. | Charles Darwin.

Copy
DAR 145: 4

[1] The year is established by the relationship between this letter and the letter from Isaac Anderson-Henry, 20 May 1867.
[2] For the case of the grafted ash, see the letter from Isaac Anderson-Henry, 20 May 1867. The similar case, also involving an ash, is discussed in *Correspondence* vol. 11, letter to Thomas Rivers, 7 January [1863] and n. 3. CD gave both cases in *Variation* 1: 394; he was correcting the page-proofs of *Variation* at this time.
[3] The name of the other *Arabis* was *Arabis blepharophylla*. See letters from Isaac Anderson-Henry, 20 May 1867, 24 May 1867, and 25 May 1867. CD was interested in the direct action of the male element upon the female as a phenomenon to be explained by his hypothesis of pangenesis (*Variation* 2: 365–6).
[4] CD refers to Maillet 1750; see letter from Isaac Anderson-Henry, 20 May 1867. Richard Owen mentioned Benoît de Maillet several times in his review of *Origin* ([Owen] 1860): in particular he said that exceptions to the hypothesis of transmutation were 'so many and so strong, as to have left the promulgation and advocacy of the hypothesis, under any modification, at all times to individuals of more imaginative temperament; such as Demaillet in the last century, Lamarck in the first half of the present, Darwin in the second half' (p. 503). There is a copy of Maillet 1750 in the Darwin Library–Down.

From Albert Gaudry[1]   22 May 1867

Monsieur,

Je vous demande la permission de vous envoyer une note sur un reptile ancien qui semble intermèdiaire entre les vrais reptiles triasiques et les poissons dévoniens.[2] Je saisis cette occasion pour vous remercier de la lettre que vous avez bien voulu m'envoyer, il y a quelques mois.[3] J'ai été très honoré qu'un naturaliste aussi éminent que vous attache de l'intérêt à mes recherches sur les transitions des animaux fossiles. Ce n'est pas sans raison qu'on me range parmi vos admirateurs, car si je ne partage pas toutes vos vues pour l'explication des transformations des êtres, du moins ces transformations me paraissent chaque jour plus probables, et votre livre sur l'origine des espèces aura puisamment contribué à les mettre en relief.[4] Je ne connais pas d'étude plus belle que celle de la filiation des êtres qui se développe pendant la durée indéfinie des âges, et je m'efforcerai d'apporter des preuves de cette filiation tirées de la paléontologie. Quant aux explications des transformations, j'évite de m'en occuper, parcequ'un sujet si difficile peut être abordé seulement par un naturaliste tel que vous, ayant une expérience consommée et une science très vaste. Je vous avouerai même qu'en étudiant l'embryogénie et en voyant qu'il y a bien des causes dont Dieu seul a le secret, je pense que, dans l'évolution des espèces comme dans l'embryogénie des individus, il y a des causes que les plus beaux génies ne sauraient découvrir.

Votre lettre m'a trouvé au Mt Léberon, près de Cucuron (Vaucluse) dans un gisement fort riche où on rencontre la même faune qu'à Pikermi. Je n'ai pas encore dégagé les ossemens que j'ai rapportés; mais ce que j'entrevois déjà me fait penser que les animaux du Mt Léberon présentent de si légères différences avec ceux de Grèce qu'il est difficile de ne pas admettre entre eux des liens d'une réelle parenté, quoiqu'il n'y ait pas identité.[5]

Je serais bien heureux, Monsieur, si votre santé et vos grands travaux vous donnaient la possibilité de venir à Paris; je vous montrerais au jardin des plantes des fossiles qui me semblent de curieux intermédiaire.

Je vous prie d'agréer l'expression de mes sentiments les plus respectueux. | Albert Gaudry

12 Rue Taranne. Paris.
22 Mai 1867.

DAR 165: 15

CD note:[6]
I quite agree about filiation of species, in a century to come there will be splendid work— I also agree how much unknown in Embryology & causes of each variation—utterly unknown [*after del illeg*]— | My book in French | I am aware that few follow your views & Most of your leading men are bitterly opposed | **Paris**—your papers

---

[1] For a translation of this letter, see Appendix I.

[2] Gaudry refers to his 'Mémoire sur le reptile (*Pleuracanthus Frossardi*) découvert par M. Frossard à Muse (Saône-et-Loire)' (Gaudry 1867a). There is an inscribed and scored offprint of this article in the Darwin Pamphlet Collection–CUL. Gaudry also enclosed a second paper; see letter to Albert Gaudry, 27 May [1867] and n. 2. The name *Pleuracanthus* had already been assigned to a genus of beetles: most members of *Pleuracanthus* (freshwater sharks) have been reassigned to the genus *Xenacanthus*.

[3] See *Correspondence* vol. 14, letter to Albert Gaudry, 17 September [1866].

[4] After excavations at Pikermi in Attica in 1855 and 1860, Gaudry had reconstructed several skeletons of new species that appeared to be intermediate between species already known. He believed that such intermediates were evidence for transmutation, but differed from CD in believing that these transmutations were the result of continuing creation by God. See Gaudry 1862–7, 1: 365–70. Gaudry had sent CD an extract from this work in 1866 (see *Correspondence* vol. 14, letter to Albert Gaudry, 17 September [1866]).

[5] Gaudry published on the fossils of Mont Léberon (now Montagne de Lubéron) in the south of France in Gaudry *et al.* 1873. He made comparisons with findings at Pikermi and other sites on pp. 75–98.

[6] CD's note is for his letter to Albert Gaudry, 27 May [1867].

To Camille Dareste 23 May 1867

> Down. | Bromley. | Kent. S.E.
> May 23rd. 1867

Dear Sir

I thank you for your very kind letter, & for the present of your pamphlet.[1] Whether or not many persons in France are at present interested in your subject of Teratology I feel thoroughily convinced that the time will come when your labour & that of all the few others who have worked on this subject will be highly valued. Therefore I am glad to hear that you intend to publish a book on this subject. I have read the whole & often consulted I. Geoffroy St Hilaire's work;[2] but no doubt you will be able to correct some erroneous views & add much matter. It will be a great advance if you can explain the precise cause of even a few monstrosities.

I am now printing a book on "The Variation of Animals & Plants under Domestication", & in this book I add a little to what I have previously said about the Nata Oxen; but I am sorry to say this part is already printed, so that I shall not be able to allude to your instructive case. Hermann von Nathusius states that in Germany calves have occasionally been born with exactly the same structure as you describe.[3]

My book is being translated into french by Colonel Moulinié of Geneva & when it is published I will do myself the pleasure of sending you a copy; but I suppose this work will not appear till towards the close of the year.[4]

With sincere respect | believe me, dear Sir | yours faithfully & obliged | Charles Darwin

LS
Jean-Louis Fischer (private collection)

[1] See letter from Camille Dareste, 19 May 1867 and n. 8. The pamphlet was Dareste 1867.

[2] CD probably refers to Isidore Geoffroy Saint-Hilaire's *Histoire générale et particulière des anomalies de l'organisation chez l'homme et les animaux* (Geoffroy Saint-Hilaire 1832–7). There is an annotated copy in

the Darwin Library–CUL (see *Marginalia* 1: 306–16); according to CD's reading notebooks (*Correspondence* vol. 4, Appendix IV), he read it in March 1845.

[3] CD discussed niata cattle in *Variation* 1: 89–91; for his earlier comments, see *Journal of researches* 2d ed., pp. 145–6. He also refers to Hermann Engelhard von Nathusius. He cited this information from Nathusius 1864, 1: 104–5, in *Variation* 1: 89.

[4] CD refers to Jean Jacques Moulinié. His translation appeared in 1868 (Moulinié trans. 1868).

From Isaac Anderson-Henry 24 May 1867

Hay Lodge, Trinity, | Edinburgh.
May 24/67

My dear Sir

I have just a minute before starting for a run to the Country to say in reply that I have found the X^d pods of Arabis *blepharophylla larger* in proportion to its normal ones or those of its male parent A. *Soyerii* than I stated—[1] Particulars tomorrow

I have sent off Maillet. Please keep him as from me[2]

Ever yours very faithfully | Is Anderson Henry

DAR 159: 69

[1] Anderson-Henry had crossed *Arabis blepharophylla* with *A. soyeri*, and obtained two pods 'about twice the size of the normal pods' (letter from Isaac Anderson-Henry, 20 May 1867). CD had asked him to confirm that the pods were larger than those of *A. soyeri* (letter to Isaac Anderson-Henry, 22 May [1867]).

[2] Anderson-Henry refers to Maillet 1750 (see letter to Isaac Anderson-Henry, 22 May [1867] and n. 4).

From J. D. Hooker 24 May 1867

Kew
May 24/67

Dear Darwin

I do not go to Paris till next Saty or Sunday (tomorrow week)[1] I am so glad that Barkly's letter pleased you—[2] I always feel a happier & a better man when I have added a mite to your knowledge.

I do not share your objection to continental extension because you must extend it to all Islands in every ocean— It was the old objection to Lyellism that it must be applied to the most horrid convulsions,— the highest Mountains & so forth. I forget how you account for the chalk in the middle of Great continents except by deep ocean there— & the Limestone fossils at 19000 ft in the Himalaya surely show a much greater continental change since the Silurian than you admit.[3]

Do not break your heart over your book whatever you do.[4]

My friend Hodgson, who has elected to live in a state of benighted ignorance saw somewhere a request of Agassiz for information on Domesticated animals, & asked me to give the accompanying (sent by post) to him in Paris!.— H. does not know that Agassiz is in America!—& I have told him that it is better in your hands.[5]

I shall not fail to come at Gooseberry season[6]

Ever Yr aff | J D Hooker

DAR 102: 165–6

[1] Hooker refers to 1 or 2 June 1867. He had told CD that he would be returning to Paris at the end of May (see letter from J. D. Hooker, 17 May 1867 and n. 2).

[2] The reference is to Henry Barkly. See enclosure to letter from J. D. Hooker, 17 May 1867, and letter to J. D. Hooker, [21 May 1867].

[3] See letter to J. D. Hooker, [21 May 1867] and n. 5.

[4] Hooker refers to *Variation*. See letter to J. D. Hooker, [21 May 1867].

[5] Hooker and Brian Houghton Hodgson had met in Darjeeling, India, in 1848 (R. Desmond 1999, pp. 107–8). The enclosure has not been identified, but see the letter to J. D. Hooker, 26 [May 1867] and n. 2. Louis Agassiz emigrated from Switzerland to the United States in 1846 (*DAB*). The request for information on domesticated animals has not been identified.

[6] Hooker's visit to Down during the gooseberry season was something of a tradition. See, for example, *Correspondence* vol. 13, letter from J. D. Hooker, [after 17 June 1865].

From V. O. Kovalevsky [after 24? May 1867][1]

Dear Sir

I am not quite sure that You have received my last letter in which I informed You, that I have received the Revise sheets, and some days afterwards, the clean sheets.— I asked Your opinion as to the propiety of adding at the end of the Book, on separate Plates, all the species and varieties of animals You mention in Your work, taken from *Brehm's Thierleben*, all the casts of woodcuts from this work beeing in my possesion.—[2] I may inform You that at least I send yesterday the bear You have so kindly accepted and probably in two weeks You will receive it.[3]

The casts from the woodcuts of Your work arrived some five days ago,[4] they are made very well, so that I have now only one very impatient desire to receive some more sheets of your work.— As the casts are all in my possetion, be so kind, my Dear Sir, if only possible, to send me revise sheets without awaiting clean ones. Your kindness spoilt me, and so I am growing importunous and wishing for more.

Believe me | Yours very faithfully | W. Kowalewsky

P.S. If you put the revise sheets in ordinary letter-covers and send them *without prepaying* it will be the safest and shortest way to forward them.—

DAR 169: 69

[1] The date is conjectured from the relationship between this letter and the letter from V. O. Kovalevsky, 15 May 1867. Kovalevsky would probably have waited at least eight days for a reply to his letter of 15 May, since he knew that it took about four days for a letter to reach him from Down; see letter from V. O. Kovalevsky, 14 May 1867.

[2] Kovalevsky had written to CD on 15 May 1867 to say that he had received proof-sheets of *Variation* and to suggest adding to his Russian translation woodcuts from Alfred Edmund Brehm's *Illustrirtes Thierleben* (Brehm *et al.* 1864–9).

[3] See letter from V. O. Kovalevsky, 14 May 1867 and n. 5.

[4] See letters to V. O. Kovalevsky, 2 May [1867] and 16 May [1867].

From Isaac Anderson-Henry 25 May 1867

Hay Lodge, Trinity, | *Edinburgh.*
May 25/67

My dear Sir

Referring to the hasty note I wrote you yesterday[1] I now beg to send annexed exact measurements of the X^d pods of Arabis blepharophylla x A Soyeri being $1\frac{1}{16}$ inch by $\frac{1}{8}$th inch broad the girth being $\frac{6}{16}$ths—and of the *normal* pod of the same A. blepharophylla being $\frac{12}{16}$ths of an inch long & $\frac{1}{14}$ inch broad the *girth* being $\frac{4}{16}$ths. The pod of the *A Soyeri* is somewhat narrower than this last & of the same length. I enclose leaves of each species, the larger being of A. blepharophylla. The hybrid pods threaten still further to out distance the normal pods.[2] I have another instance of a similar result—in a very different—indeed a *hard* wood tribe of plants, but the seed pods are not yet sufficiently developed to state any particulars[3]

I hope the Book reached you in safety. No one could justly place you or your Books on the same platform with M. Maillet and his Book in which last there is no doubt much nonsense—tho there is no denying ability to the man whose credulity seems to have been shamefully practised upon[4]

But my servant calls for Letters and I fear I am too late for post. So with every good wish I remain | Yours most faithfully | Is: Anderson Henry

(1) Seed pod of *Arabis blepharophylla* x *Arabis Soyeri*
($1\frac{1}{16}$) $1\frac{1}{16}$ inch long by $\frac{1}{8}$th inch broad

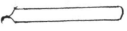

(2) Normal pod of *A blepharophylla*
$\frac{12}{16}$ inch long by $\frac{1}{14}$ inch broad

(3) *A. Soyeri* normal pod, same length as N° 2 but scarcely quite as broad
    Girth of No (1) —— $\frac{6}{16}$ inch
    D° of No (2) —— $\frac{4}{16}$

DAR 159: 70

[1] Letter from Isaac Anderson-Henry, 24 May 1867.
[2] See letter from Isaac Anderson-Henry, 24 May 1867 and n. 1.
[3] The plants to which Anderson-Henry refers have not been identified. For an account of his crossing experiments, see Anderson-Henry 1867b.
[4] Anderson-Henry had sent CD a copy of Maillet 1750 (see letter from Isaac Anderson-Henry, 24 May 1867). See also letter from Isaac Anderson-Henry, 20 May 1867 and n. 4.

To J. D. Hooker 26 [May 1867][1]

Down
Sunday 26

My dear Hooker

I write a line only to thank you for Hodson's paper, but I have it in print & it is a very good one; therefore I return it to you.[2]

It was foolish in me to say a word about continental extensions, for I said it so briefly that I think you misunderstand me; we can however fight that battle when we meet.[3] Remember the gooseberries[4]

yours affecty | Ch Darwin

Endorsement: '1867'
L(S)
DAR 94: 17

[1] The month and year are established by the relationship between this letter and the letter from J. D. Hooker, 24 May 1867.
[2] See letter from J. D. Hooker, 24 May 1867. CD refers to Brian Houghton Hodgson. The only paper by Hodgson now in the Darwin Pamphlet Collection is an annotated and inscribed offprint of Hodgson 1847, 'On the tame sheep and goats of the sub-Himaláyas and of Tibet'. CD cited this work in *Variation* 1: 95, 96, 102, and 2: 301; he cited other papers by Hodgson less frequently.
[3] See letter to J. D. Hooker, [21 May 1867], and letter from J. D. Hooker, 24 May 1867.
[4] See letter from J. D. Hooker, 24 May 1867 and n. 6.

To Fritz Müller 26 May [1867][1]

Down. | Bromley. | Kent. S.E.
May 26

My dear Sir

I thank you much for your information on sexual differences with the pretty little sketch of the male crab, & for the great trouble which you have taken in copying the long extract from Claus,[2] All such facts interest me much. I shall also be very glad for answers to *any* of my questions on the expression of negros.[3] Your additions about the self-sterile orchids are particularly valuable, & I shall give an abstract of all the information which you have so kindly given me.[4] The comparison of the pollen with that when species are crossed makes the facts much more curious.[5] The analogous cases which have been observed in Europe I have always attributed to the unnatural condition under which the orchids were grown; but it now appears that this is an error; I fear however that I shall not be able to alter the place in my book where I give these facts.[6] As Dr Hildebrand in experimenting on Corydalis used the pollen from several individuals & always with the same result I do not think the species can be dimorphic.[7]

Only 3 plants of your Plumbago are alive, but they are doing well. Hundreds of y^r Gesnera have germinated.[8] Kuhn announces in the Bot. Zeitung that he is going to publish a work on dimorphic plants of all kinds, but as far as I can judge, he does not experiment, & therefore will not interfere with me.[9]

I dare say your observations in crossing orchids will be very interesting; they already illustrate most of the leading laws; but I fear their interest will be greatly lessened by the crossed seeds not germinating. One single man in Europe has found out how to make these seeds germinate, & he keeps it a secret in his trade of nurseryman.[10] He also has made some strange crosses between *distinct* genera, & these hybrids have flowered.

Dr Hooker tells me that they have in vain tried at Calcutta to make the seeds of hybrids germinate;[11] yet American orchids growing in the Bot. Garden there have spontaneously sown themselves and grown on adjoining trees.

I presume this strange difficulty in making the seeds germinate explains their astonishing number, which you and others have shewn.[12] I am not able to do much this summer in experimenting as all my time is taken up in getting my book thro' the press;[13] it progresses very slowly & is I fear hardly worth the great labour it costs me.

With *cordial* thanks for your never failing kindness I remain my dear Sir | yours very sincerely | Ch. Darwin

LS
British Library (Loan 10: 16)

[1] The year is established by the relationship between this letter and the letter from Fritz Müller, 1 April 1867.

[2] See letter from Fritz Müller, 1 April 1867. CD refers to Carl Friedrich Claus.

[3] See letter to Fritz Müller, 22 February [1867], and letter from Fritz Müller, 1 April 1867. CD was seeking material for *Expression*.

[4] In 'Fertilization of orchids', p. 154 (*Collected papers* 2: 150), CD referred to Müller's work on the structure and manner of cross-fertilisation of orchids in Brazil but stated that he had not had 'space or time to give an abstract of his many discoveries'. CD cited Müller frequently in the second edition of *Orchids* for information that Müller had sent him.

[5] See letter from Fritz Müller, 1 April 1867. Müller had discovered that when he tried to self-pollinate self-sterile species, the pollen placed on the stigma became discoloured and decayed, whereas when a cross was attempted between distinct plants of the same species or of different species, the pollen stayed fresh (see also *Variation* 2: 134–5).

[6] CD discussed Müller's findings in *Variation* 2: 134–5.

[7] CD refers to Hildebrand 1866a. For Müller's suggestion that *Corydalis* might be dimorphic, see the letter from Fritz Müller, 1 April 1867.

[8] Müller had sent CD seeds of *Plumbago* (letter to J. D. Hooker, 24 [March 1867]), possibly with his letter of 2 February 1867; he sent seed of *Gesneria* with his letter of 1 January 1867.

[9] Max Kuhn made this announcement in his article 'Einige Bemerkungen über Vandellia und den Blüthenpolymorphismus' (*Botanische Zeitung*, 1 March 1867, p. 67). There is an annotated copy of this issue in the Darwin Pamphlet Collection–CUL.

[10] CD probably refers to John Dominy, a gardener for James Veitch & Son, who produced the first known artificial orchid hybrid, between *Calanthe furcata* and *C. masuca*, in 1856. Orchid hybrids were produced only at Veitch & Son's nursery for fifteen years. See Shephard 2003, pp. 127–9; see also *Correspondence* vol. 10, letter from J. D. Hooker, [23–5 March 1862] and n. 3.

[11] See *Correspondence* vol. 11, letter to John Scott, 24 March [1863], and letter to Hermann Crüger, 25 May [1863] and n. 8.

[12] Müller had estimated that a seed capsule of *Maxillaria* contained well over one and a half million seeds (see *Correspondence* vol. 14, letter from Fritz Müller, 1 and 3 October 1866; see also 'Fertilization of orchids', p. 158 (*Collected papers* 2: 155), and *Orchids* 2d ed., p. 278).

[13] CD was working on the page-proofs of *Variation*.

To George Howard Darwin 27 May [1867][1]

Down
May 27

My dear George

I have got to think that perhaps a better name c^d be made for my hypothesis than Pangenesis.[2] Cell-genesis w^d be perfect if it c^d be put into Greek. Atom-genesis or particle-gen—or tissue-gen—might do.

Now do you know any really good Classic who c^d suggest any Greek word expressing cell, & which c^d be united with genesis? The cells referred to are such as form the tissue of plants. I sh^d be sorry for a very long word, yet Partheno-gen. & Agamo-gen—have been extensively used. Perhaps I shall have to stick to Pan—

yours affectly | Ch Darwin

L(S)
DAR 210.1: 2

[1] The year is established by the relationship between this letter and the letter from G. H. Darwin, [3 June 1867].
[2] CD published his provisional hypothesis of pangenesis in *Variation* 2: 357–404; he discussed the hypothesis in letters in 1865 and 1866 (see *Correspondence* vols. 13 and 14), and finished work on the chapter concerning it on 21 November 1866 (*Correspondence* vol. 14, Appendix II). On the name pangenesis, see also the letter to T. H. Huxley, 12 June [1867].

To Albert Gaudry 27 May [1867][1]

Down. | Bromley. | Kent. S.E.
May 27

Dear Sir

I am much obliged for your kind letter, & for the present of your two memoirs. The one in the Bulletin I sh^d have naturally seen, but perhaps not that one in the Archives.[2]

I have been very glad to read this latter paper, as all inosculating forms are very interesting to me. I quite agree with what you say on the extreme interest of attempting to affiliate extinct & existing Species.[3]

With your great knowledge you will no doubt produce some valuable results, & I feel well convinced that in the course of time the most interesting genealogical tables will be constructed. I am aware that there are very few naturalists in France who at all concur with our views & therefore I presume you will meet with much opposition.[4]

I am at present printing a book "on the variation of animals & plants under domestication"; which will I believe be translated into French, & in this case I will direct the publisher to send you a copy, though I do not suppose it will possess much interest for you.[5] I am much obliged for y^r kind wish to see me in Paris; I sh^d much enjoy this but the state of my health renders it impossible.—

With very sincere respect | I beg leave to remain | Dear Sir | yours very faithfully | Charles Darwin

LS
Museo Civico di Storia Naturale, Milan

[1] The year is established by the relationship between this letter and the letter from Albert Gaudry, 22 May 1867.

[2] Gaudry had sent an offprint from the *Nouvelles Archives du Muséum d'Histoire Naturelle* of his 'Mémoire sur le reptile (*Pleuracanthus Frossardi*) découvert par M. Frossard à Muse' (Gaudry 1867a); see letter from Albert Gaudry, 22 May 1867. The other paper has not been found in the Darwin Pamphlet Collection–CUL. Gaudry published a note on the fossil reptile in the *Bulletin de la Société Géologique de France* in 1867 (Gaudry 1867b).

[3] See letter from Albert Gaudry, 22 May 1867 and n. 4. Inosculate: 'to pass into; to join or unite so as to become continuous; to blend' (*OED*). See also *Correspondence* vol. 1, letter to J. S. Henslow, [c. 26 October–] 24 November [1832] and n. 8.

[4] On the reception of CD's theory in France, see Stebbins 1988 and J. Harvey 1997. See also letter from Camille Dareste, 19 May 1867 and n. 8.

[5] CD refers to *Variation*. A translation by Jean Jacques Moulinié appeared in 1868 (Moulinié trans. 1868); Gaudry thanked CD for his copy in a letter to CD of 11 April 1868 (*Correspondence* vol. 16).

From J. T. Moggridge   29 May [1867][1]

34 Eastbourne Terr: | Paddington | W.
May 29

Dear M[r]. Darwin

On receipt of your letter I wrote to D[r]. Bornet to say that though his kind offer seeds of Draba was very tempting to you, yet on account of press of work you were obliged to decline with thanks.—[2]

Yesterday I received a letter containing the following passage:

"Since M[r]. Darwin might be interested by recieving seeds of Draba, I shall send him some as soon as the seed-harvest is completed. If he cannot make use of them this year, he may next. Besides, the mode of cultivation is most easy. The seeds are sown in pots in August & the pots buried in earth almost to the level of the ground— The seeds develop fine rosettes of leaves & open their flowers in the first fine days of the following spring—

Thus M[r]. Darwin might have time to complete his present work."—

I write this to exculpate myself, & to make it clear that I am not responsible for your recieving seeds of Draba against your will.—

D[r]. Bornet tells me now that Indigofera does not always throw off its wings as I described but that they fall readily with the keel after the emission of pollen.—[3]

This is very similar to what happens in Genista & Sarothamnus— I was misled by watching the artificial treatment of the flowers.

Y[rs]. Very sincerely | J. Traherne Moggridge.

DAR 157a: 101

CD ANNOTATIONS
1.1 On ... thanks.— 1.3] *crossed pencil*
5.1 I write ... flowers. 7.2] *crossed pencil*
Top of letter: 'Draba' pencil

[1] The year is established by the relationship between this letter and the letter from Edouard Bornet, [before 20 August 1867].

[2] Neither CD's letter to Moggridge, nor a letter from Moggridge mentioning Bornet's offer, has been found in the Darwin Archive–CUL. Bornet had sent CD seeds of *Papaver* in 1866 (see *Correspondence* vol. 14, letter to Edouard Bornet, 1 December 1866). See also letter from Edouard Bornet, [before 20 August 1867], and letter to Edouard Bornet, 20 August [1867].

[3] No letter on this subject from Moggridge has been found; however, CD and George Henslow discussed the 'curious movements' of *Indigofera* in 1866 (see *Correspondence* vol. 14, letter to George Henslow, 16 April [1866], and letter to Friedrich Hildebrand, 20 April [1866]).

From Tendler & Co. to John Murray 29 May 1867

Vienna—Wien | Graben, Trattnerhof
29/5 67

Mr. John Murray, London

We will take the liberty in asking you the permission of translating from the work:

Domesticated animals and cultivated plants
by Charles Darwin[1]

Accepting our kind offer you will very much favour us in naming the conditions by which you would consent to our propositions.

Hoping to get a favourable answer we promise to send you immediately the required amount and are Sir your | very respectfully | ppa Tendler & Com | Julius Grosser[2]

DAR 178: 87

[1] The reference is to *Variation*. All the German editions of *Variation* were translated by Julius Victor Carus and published by E. Schweizerbart'sche Verlagsbuchhandlung of Stuttgart. There is a note in ink at the foot of the letter: 'Answered & declined June 5. 1867 for J.M. | CAP'. John Murray was CD's British publisher.

[2] Julius Grosser has not been further identified.

From Frederick F. Geach June 1867

Johore[1]
June. 1867

My dear Sir,

I enclose the result of a few observations on Malays *who have not* had any communications with Europeans, and beg to assure you I feel a pleasure in making any memoranda that will be of the slightest Assistance.[2]

I feel sorry your queries did not reach me before I visited the Aborigines in Feby,—nevertheless I shall meet with the difficult parts in time, I have made a Copy of your letter for my Friend A. E. Hart, Timor, a Gentleman well known to M.ʳ Wallace and an admirer of a certain book in that Gentlemans possession I hope to get his reply within 12 mos, when I will enclose it to you.[3]

Please inform me if this mode of answering is in accordance with your 19.ᵗʰ clause,[4]

With best wishes for your improved health. | Yours very truly | F. F. Geach
C. Darwin Esq.

[Enclosure]

Peninsula of Johore.[5]
June. 1867.

Reply to Queries.[6]

N.º 1— It will be a long time before I can get an opportunity of seeing the expression.

" 2. With 3 natives to the breasts.
=I think it covers the whole body=
One Chinese = The whole body

" 3 In 3 or 4 mo.ˢ I may get a few subjects.

" 4 Corners of the mouth depressed.
Chin in wrinkles.
Frowns, Contracts eye brow.

do Frowns and looks down.

do Brow wrinkles head raised
(Orang Pahang)[7]

"

do— Frowns=

$\left.\begin{array}{c} 5 \\ 6 \\ 7 \end{array}\right\}$ I have not had an opportunity of seeing the expressions

Queries continued

8— *Yes*

$\left.\begin{array}{c} 9 \\ 10 \\ 11 \end{array}\right\}$ difficult and will take months to answer=

12. Seldom

13. You have beautifully expressed the Native actions

14— yes.

15— yes

16— require more time and better opportunities

17— yes—

1/17— yes=

All from Natives who have *NO* communication with Europeans=
F. F. Geach

DAR 165: 21

CD ANNOTATIONS
Top of enclosure: 'Singapore' *ink*; 'Malay' *pencil*; '35 Malay' *red crayon*
Enclosure, bottom of first page: 'See over' *blue crayon, circled blue crayon*
Enclosure, last paragraph: All from Natives ... Europeans=] *double scored blue crayon*

[1] The Malay State of Johor was an independent state within the British sphere of influence (Joseph Kennedy 1962, p. 101). The main town is Johor Baharu (*Times atlas*).

[2] CD had sent Geach a handwritten copy of his queries about expression after receiving his address from Alfred Russel Wallace in March (see letter from A. R. Wallace, 2 March [1867], and letter to A. R. Wallace, 7 March [1867]). This questionnaire has not been found. For a transcript of a later, printed, version of the queries, see Appendix IV. CD cited Geach a number of times in *Expression* for providing information on the Malay people and Chinese immigrants in Malacca (now Melaka). See also *Correspondence* vol. 16, letter from F. F. Geach, April 1868.

[3] Wallace had met A. E. Hart in Delli (now Dili), east Timor, in 1861. He wrote of him: 'He is what you may call a *speculative* man: he reads a good deal, knows a little and wants to know more, and is fond of speculating on the most abstruse and unattainable points of science and philosophy' (Marchant ed. 1916, 1: 79). No reply from Hart has been found in the Darwin Archive–CUL, and he is not cited in *Expression*.

[4] Geach refers to a paragraph at the end of CD's questionnaire where CD wrote: 'General remarks on expression are of comparatively little value; and memory is so deceptive that I earnestly beg it may not be trusted.'

[5] Peninsula of Johore: i.e. mainland Johore, not including Singapore.

[6] See n. 2, above.

[7] Orang Pahang: a person from Pahang. Pahang was an independent Malay State (Joseph Kennedy 1962, p. 135).

To Charles Lyell 1 June [1867][1]

Down. | Bromley. | Kent. S.E.
June 1st

My dear Lyell

I do not think H. Parker ever reviewed the Origin: perhaps you are thinking of an article on the D. of Argyll, which I remember praising, perhaps over-praising, to you.— I now enclose it, & you can look at it or not as you like, & please return it.—[2] I am at present reading the Duke & am *very much* interested by him; yet I cannot but think, clever as the whole is, that parts are weak, as when he doubts whether each curvature of beak of Humming Birds is of service to each species. He admits, perhaps too fully, that I have shown use of each little ridge & shape of each petal in Orchids, & how strange he does not extend the view to Humming Birds. Still odder, it seems to me, all that he says on Beauty, which I sh[d] have thought a non entity except in the mind of some sentient being: he might have as well said that Love existed during the Secondary or Palæozoic periods.[3]

I hope you are getting on with your Book, better than I am with mine, which kills me with the labour of correcting & is intolerably dull, though I did not think so when I was writing it.[4] A naturalist's life w[d] be a happy one, if he had only to observe & never to write.—

We shall be in London for a week in about a fortnights time, & I shall enjoy having a break-fast talk with you.—[5]

Yours affectionately | C. Darwin

American Philosophical Society (328)

[1] The year is established by the reference to G. D. Campbell 1867 (see n. 3, below).

[2] No letter from Lyell enquiring about a review by CD's nephew Henry Parker has been found. CD refers to [Parker] 1862, an article in the *Saturday Review* discussing [G. D. Campbell] 1862, a review of *Orchids* by the duke of Argyll, George Douglas Campbell. See *Correspondence* vol. 10, letter from J. D. Hooker, [before 29 December 1862] and n. 4, and letter to J. D. Hooker, 29 [December 1862]. No copy of [Parker] 1862 has been found in the Darwin Archive–CUL; there is an annotated copy of [G. D. Campbell] 1862 in the Darwin Pamphlet Collection–CUL. No discussion between CD and Lyell on [G. D. Campbell] 1862 has been found, but see *Correspondence* vol. 13, letter from Charles Lyell, 16 January 1865, and letter to Charles Lyell, 22 January [1865], for their discussion of G. D. Campbell 1864. For more on CD's view of Campbell's arguments, see the letter to J. D. Hooker, 8 February [1867].

[3] CD refers to Campbell's *The reign of law* (G. D. Campbell 1867); there is an annotated copy in the Darwin Library–CUL (see *Marginalia* 1: 17). Campbell suggested that specialised beaks did not give the greatest possible advantage to humming-birds, which had equal access to a wide range of flora, and that the rule governing the proliferation of humming-bird species had as its object rather 'the mere multiplying of Life, and the fitting of new Forms for new spheres of enjoyment' (G. D. Campbell 1867, pp. 241–2). His discussion of orchids is in *ibid.*, pp. 37–9. He argued that the phenomena of nature would never be understood except on the admission that 'mere ornament or beauty' was 'in itself a purpose, an object, and an end' (*ibid*, p. 197), and 'was not intended only for Man's admiration' (*ibid.*, p. 199).

[4] Lyell was working on the second volume of the tenth edition of his *Principles of geology* (C. Lyell 1867–8); see *Correspondence* vol. 14, letter to Charles Lyell, 1 December [1866], n. 2). CD was correcting the proof-sheets of *Variation* (see CD's 'Journal' (Appendix II)).

[5] The Darwins were in London from 17 to 24 June (CD's 'Journal' (Appendix II)). No record has been found of a meeting with Lyell.

To Daniel Oliver 1 June [1867][1]

Down. | Bromley. | Kent. S.E.
June 1st

Dear Oliver

Would you have the great kindness to name for me the enclosed plant— if you cannot make out the species, the genus w[d] suffice.

A friend sent a Woodcocks foot with 9 gr. of earth adhering to it, & this plant came up.— It has been grown in pot in greenhouse & the flowers have never opened & stamens appear rudimentary— whether this is natural, or due to the plant being properly a marsh plant, or to some other cause, I know not.—[2]

Excuse me bothering you & believe me | Yours very sincerely | Ch. Darwin

[1] The year is established by the relationship between this letter and the letter from H. G. H. Norman, 30 November 1866 (*Correspondence* vol. 14).

[2] Herbert George Henry Norman, the nephew of CD's neighbour George Warde Norman, sent CD a woodcock's leg with earth attached to it in November 1866 (see *Correspondence* vol. 14, letter from H. G. H. Norman, 30 November 1866, and letter to H. G. H. Norman, [after 30 November 1866]). There is a note in CD's Experimental notebook, DAR 157a, p. 83, that reads:

> Dec. 2^{d} 1866 Mr Herbert Norman of Oakley near Bromley sent me leg of Woodcock with tarsus coated with mud, which *when dry* weighed 8–9 grains. Planted on burnt sand, Decr 3^{d} Dec. 8th a monocot: plant, apparently a rush, has [*after del* 'or Luzula'] has germinated!!! The plant turns out *Juncus bufonius* or toad rush— — grows "commonly in marshy ground, especially on watery sandy Heaths." Sir J. E. Smith.—

In his *English flora* (J. E. Smith 1824–36, 2: 168), James Edward Smith had written of *Juncus bufonius*: 'In marshy ground, especially on watery sandy heaths, common.' Since no record of any other plants having germinated from the mud attached to the woodcock's leg has been found, the *J. bufonius* was probably the plant CD asked Oliver to identify. Oliver's reply has not been found in the Darwin Archive–CUL. CD mentioned the woodcock's foot and the *Juncus bufonius* in *Origin* 5th ed., p. 440. CD had earlier grown plants from seeds found in a ball of earth attached to a partridge's foot; see *Correspondence* vols. 11 and 12, and *Origin* 4th ed., p. 432. He was interested in such cases as examples of possible seed distribution by birds.

From J. J. Moulinié 2 June 1867

Geneva
2^{d} June. 1867.

Sir

I have received in due time the first four sheets you have been so kind to send me, and have proceeded to their translation with a great interest, which will, I doubt not, grow always deeper as the subject goes on.[1]

Mr Reinwald informs me that having received the stereotypes, he is ready for printing, and I can send him the copy, which I have read over, as agreed, with professor Vogt.[2]

Hoping to be soon enabled to continue my interesting occupation, I remain, Sir | yours very respectfully | J J Moulinié

15. rue du Mont-Blanc.

DAR 171: 267

[1] See letter to J. J. Moulinié, 16 May [1867]. Moulinié was translating *Variation* into French.

[2] Moulinié refers to Charles-Ferdinand Reinwald and Carl Vogt. See letter from C.-F. Reinwald, [May 1867].

From Fritz Müller 2 June 1867

Desterro, Brazil,
June 2^{d} | 1867.

My dear Sir

Your two kind letters of March 25th and Apr. 22^{d} arrived here a few days ago and at the same time the copies of the paper on climbing plants despatched Apr. 20th. I already told you, I think, that the formerly despatched copies also, after an unusual delay, reached me safely.[1] Many thanks for all the pain, you have taken.

I must to-day begin with correcting an error of a former letter; the Rubiaceous plant, which I sent you under the name of *Diodia*, does not belong to that genus, but is a *Borreria*, as I now see on examining ripe fruits.[2] Some species of Diodia, which I observed, are monomorphic. Here is one more dimorphic Rubiaceous plant, ⟨*specimen excised*⟩a *Coccocypselum*; I found no appreciable difference in the size of the pollen-grains of the two forms.[3] I enclose seeds of two species of that genus; having as yet seen only a single flowering plant of the second species, I do not know, whether it is dimorphic, though I hardly doubt that it will be so.[4] The plants are worth cultivating for their very pretty blue fruits.— Have you ever seen a *Danais*? I suspect that this Rubiacea also will prove to be dimorphic; Endlicher (gen. plant. n° 3267) says it is dioecious, the male (short-styled?) flower having exserted stamens and enclosed styli, whilst in the female (long-styled?) flower the stamens are said to be enclosed and the styli exserted.—[5]

I must correct also my statements on the curious terrestrial Orchid, of which I told you in my last letter.[6] I have since found myself a flowering plant, which has continued flowering in my garden for about six weeks. The flowering stem of my plant is about four (or five) years old, and four feet high, with large Veratrum-like leaves;[7] it bears its golden flowers in loose compound ears in the axillæ of the leaves of the preceding year. The pollinium consists of a large boat-shaped disk, from the back of which springs a straight long staff, lying in the furrow of the anterior side of the anther.[8]

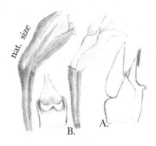

Feuerbachiae aureae F. M.
A. Upper part of columna, side-view.
B. The same, front view, after removing the pollinium.

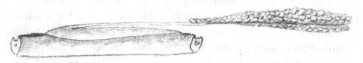

Pollinium from the side. ($\frac{15}{1}$)

(What, in the dry bud, I erroneously took for a caudiculus,[9] was the tip of this staff, wrinkled and curved). When the anther opens, the numerous small pollenmasses adhere to the staff by means of elastic threads. I had no very young buds, but as far as I have seen, the staff never adheres to the dorsal surface of the rostellum, but is *always free*, growing downwards from the tip of the rostellum. In a bud, 6mm. long, the staff had a length of 2,5mm. and a smooth surface; in a bud 16mm long it was 4mm long and the cells near its tip had grown out into short blunt warts or hairs, probably facilitating the adhesion of the pollen masses.— This seems to me to be a fine instance of the same end being gained through different means; here, as well as in Bonatea and in many Vandeæ, the pollen-masses are borne by a long stalk; but here this stalk is a long process, freely springing from the back of the disk, whilst in Bonatea it is a metamorphosed part of the pollen-masses and in the Vandeæ a part of the dorsal wall of the rostellum.

To the list of Orchids in which own pollen decays and ⟨becomes dark brown⟩[10] a few days after being placed on the stigma, ⟨I can now add two⟩ more species, viz. a second *Gomeza* (crispa) ⟨and one⟩ common *Burlingtonia* (decora Lem.?).—[11] ⟨In Oncidium crispum⟩ there seems to be, in this respect, a large amount ⟨of ind⟩ividual variability.[12] On a vigorous large plant I crossed 8 flowers, all of which are producing fine pods; 12 flowers were self-fertilized, of which 7 being dissected within 10 days after fertilization showed fresh pollen and pollen-tubes; 2 dissected a fortnight after fertilization had the pollen-tubes brown and withering; 3 are producing seed capsules; of these 2 (pollen from distinct flowers of the same plant) are equal in size to the crossed pods, 1 (pollen of the same flower) is a little smaller.— On a second vigorous plant with a very large panicle I crossed 10 flowers; all have now fine pods.— 9 flowers were self-fertilized, of which 3 had fresh pollen, when dissected 6, 7, 9 days afterwards; the germens of the rest began to discolour after 9, 10, 12, 13, 14, 15 days, and showed brown pollen, when dissected.— On a third, small but healthy, plant one flower was fertilised with the same plants pollen and is yielding a pod *much smaller*, than that of a second crossed flower; on the stigmas of two flowers I placed one own pollen-mass and one from a distinct plant; they are producing seed-capsules equal in size to the crossed one. Lastly, on two other weak plants 3 flowers were self-fertilized and perished, the germens discolouring after 9 days.

Some Orchids, which I had supposed to be self-fertile, because self-fertilized flowers yielded fine pods, seem notwithstanding to be perfectly sterile, or nearly so, with own pollen. I examined microscopically numerous seeds taken from 19 self-fertilized pods of *Epidendrum Zebra* (variegatum?)[13] and only in one of these pods I met with a small number of seeds having an embryo, which seemed to me unusually small;—all the rest contained only empty embryoless hides instead of seeds. This was also the case with two self-fertilized capsules of a *Maxillaria*. (Subgen. *Xylobium*).—[14]

Some of the pods which I had obtained by crossing distinct species of Orchids have also proved to contain only bad seeds or no seed at all; so those of *Epid. Zebra* (♀) crossed with *Ep. fragrans* and. *Ep. cinnabarinum*, and those of *Oncidium flexuosum* (♀), crossed with Cyrtopodium.[15] This latter pod grew very well in the beginning, even more rapidly, than most other pods of the same plant; fertilized Jan. 17, it measured 30mm Feb. 28, whilst a pod fertilized at the same time with *Oncid. pubes*(?), had only 23mm. But all the seeds of the latter were apparently good. In the seedless pod of *Onc. flexuosum* fertilized with pollen of *Cyrtopodium* the hygroscopical hairs on the inside of the valves, so common among Orchids, are unusually well developed.[16]

How wonderful the fact is, which you observed, of the difference of growth in the plants produced from self-fertilized and crossed seeds,[17] and what a fine gradation we now already have in the results of self-fertilization, from the less vigorous growth of the seedlings, through lessened number or total want of good seeds, to the mutual poisonous action of the same plants pollen and stigma.

I have lately begun collecting our Ferns and in a few weeks brought together about 50 species, belonging to more than 30 genera. It is a curious fact that even till the present day the first leaves (I do not know the proper English term for the *"frondes"*[18] of ferns) of many exhibit the form of *Cyclopteris*, the dominant genus in the eldest fossil fern-flora.— I enclose here two small pretty species closely ressembling each other, though belonging to distinct genera;[19] one of them, with anastomosing nerves (*Doryopteris*) is a most common species; the other, with free nerves (Pellæa?) seems to be very rare and I have *always* found it in company of *Doryopteris*.[20] The leaves of Pellæa are generally more divided, but I have often been unable to say to which of the two species a plant belonged without looking for the nerves.— A *Campyloneuron*[21] growing almost always in company of a very common *Drynaria* also ressembles to this latter so closely, that sterile plants can often hardly be distinguished without examining their nervation. I already mentioned, I think, a small *Oncidium* from Theresopolis, growing in company of, and, by its pseudobulbs and leaves, closely ressembling to, Oncid. unicorne, to which it has no close systematic affinity.—[22] Many years ago, on a voyage to the Itajahy, I was struck by a Papilionaceous plant, which at first sight, deceived by the colour of its leaves and flowers, I had mistaken for a common littoral Ipomoea, among which it flourished.

In most of these cases of two species of distinct genera or families growing in company of and closely ressembling, each other in colour, or odour, or form,—it will be very difficult to decide, whether this ressemblance is due to adaptation to the same conditions of life or to mimicry. In the case of the Leguminosa it seems to me by far more probable, that it is a mimetic plant.

I am quite unable to explain the fact of Echites catching insects.[23] I have lately planted in my garden several species of the genus, (which, besides, are very beautiful plants) in order to have an opportunity of learning something about this fact— Your *Cordia* is a small shrub, and very small plants are able to flower, so that perhaps you may see your plants flowering next year.[24]

On the Itajahy I intend to make some essays on the domestication of some of our wild mammals (Aguti, Paca, Tapir) and birds, and one of my main occupations will be cultivating plants; so, far from not caring much, as you fear, about your new book, I look for it with an intense interest.[25]

I had thought of publishing a little book containing miscellaneous zoological observations, a year or more ago; but on communicating the plan to a friend of mine I knew that it would be difficult to find a publisher for such a book in Germany.[26] So I have given up the idea.—

Are not climbing Genera generally more widely spread, than other genera of their family, as is for instance, the case with *Clematis*, or *Smilax?*—

Hoping that this letter will find you in good health, believe me, dear Sir, | very truly and respectfully yours | Fritz Müller

DAR 110: B113–14

CD ANNOTATIONS
3.1 I must ... anther. 3.8] *crossed pencil*
4.1 To ... stigma. 7.5] *crossed red crayon*
8.1 I have ... *Smilax?*— 13.2] *crossed pencil*
Top of letter: 'This first page *[alone]* on Dimorphism' *blue crayon*

[1] See letters to Fritz Müller, 25 March [1867] and 22 April [1867]. In his letter of 1 April 1867, Müller thanked CD for the copies of F. Müller 1865 that CD had sent earlier.

[2] The specimen was enclosed with Müller's letter of 4 March 1867. *Borreria* is now subsumed within the genus *Spermacoce* (Mabberley 1997).

[3] Müller evidently attached dried flowers of two forms of an unidentified species of *Coccocypselum*, which CD excised from the letter. The specimens have not been found, but CD described them in *Forms of flowers*, pp. 133–4, and noted the equal size of the pollen grains (*ibid.*, p. 250).

[4] The seeds have not been found.

[5] Müller refers to Stephan Ladislaus Endlicher and Endlicher 1836–42, p. 554. *Danais* is a genus of the family Rubiaceae, and is confined to East Africa, notably Madagascar and Mauritius.

[6] The last extant letter from Müller was that of 1 April 1867; there is no mention in it of a curious terrestrial orchid.

[7] *Veratrum* is a genus of the Melanthiaceae (Mabberley 1997).

[8] Müller never published the name *Feuerbachia aurea*, but in a letter to his brother Hermann, of 30 May 1867, he mentioned that the plant belonged to the tribe Neottieae (Möller ed. 1915–21, 2: 126). The orchid has been identified as *Corymborkis flava* (identified by Robert Dressler; see also Dressler 1993, pp. 31–3, 116–17).

[9] 'Caudiculus': caudicle.

[10] Text missing from the original letter has been reconstructed using the draft of this letter published in Möller ed. 1915–21, 2: 128. For Müller's earlier report of pollen becoming dark brown, see the letters from Fritz Müller, 1 January 1867 and 1 April 1867.

[11] *Gomeza crispa* is a synonym of *Gomesa crispa*. *Burlingtonia decora* (named by Charles Lemaire) is a synonym of *Rodriguezia decora*.

[12] Müller had earlier reported that *Oncidium crispum* might be self-fertile (see letter from Fritz Müller, 1 April 1867). CD added Müller's observations on the fertility of *O. crispum* to *Variation* 2: 134.

[13] For the identification of *Epidendrum zebra*, see the letter from Fritz Müller, 1 January 1867 and n. 4. For Müller's experiments on self-fertility in *Epidendrum* species, see also his letters of 4 March 1867 and 1 April 1867.

[14] *Xylobium*, then a subgenus of *Maxillaria* (Endlicher 1836–42, p. 197), is now a genus of the subtribe Lycastinae, while *Maxillaria* is a genus of the subtribe Maxillariinae. Both subtribes belong to the tribe Maxillarieae (Dressler 1993, pp. 173–5). CD mentioned self-fertilised *Maxillaria* capsules containing worthless seeds in *Variation* 2: 134.

[15] For more on these experiments, see the letters from Fritz Müller, 4 March 1867 and 1 April 1867. *Epidendrum fragrans* is now *Prosthechea fragrans* (Higgins 1997).

[16] Attached to the letter is a small packet containing a specimen, which Müller labelled: 'Seed-capsule of *Oncidium flexuosum*, fertilised (Jan. 17. 1867) with pollen of *Cyrtopodium*. | April 11. 67.' For earlier discussion of the presence and function of hygroscopic hairs, see the letters from Fritz Müller, 1 January 1867 and n. 19, and 2 February 1867.

[17] CD had first informed Müller of his interest in comparing the growth of plants produced from self-fertilised and crossed seeds in a letter of 25 September [1866] (*Correspondence* vol. 14; see also letter to Asa Gray, 10 September [1866] and n. 13). For CD's reports to Müller on early results of these experiments, see the letter to Müller of 22 April [1867]; see also letter to Asa Gray, 15 April [1867].

[18] 'Frondes': fronds (i.e. the leaves of ferns).

[19] The fern specimens are in DAR 142: 105.

[20] Anastomosing nerves: cross-connected nerves or veins of leaves. Free nerves: non-cross-connected nerves or veins.

[21] '*Campyloneuron*': *Campyloneurum*.

[22] See letter from Fritz Müller, 1 April 1867. *Oncidium unicorne* is a synonym for *O. longicornum*.

[23] See letter to Fritz Müller, 25 March [1867]. Müller's original statement about *Echites* was evidently contained in his letter of 2 February 1867, of which only an incomplete draft is extant. However, in a letter to his brother, Hermann, of 11 February 1867, he wrote that an *Echites* in his garden had imprisoned a small bee by its proboscis, which it had inserted between the anthers, and that the bee had died (see Möller ed. 1915–21, 2: 115).

[24] Müller had sent CD seeds of *Cordia* (see letter to Fritz Müller, 7 February [1867]), but CD had expressed doubts about whether he would ever see the plants flower, since he thought that they were trees (see letter to Fritz Müller, 25 March [1867]).

[25] Müller refers to *Dasyprocta aguti* (the agouti), *Agouti paca* (the paca), and *Tapirus terrestris* (the tapir) and to *Variation* (see letter to Fritz Müller, 25 March [1867]).

[26] See letter to Fritz Müller, 22 April [1867]. In a letter to his brother, Hermann, of 30 May 1867, Müller mentioned that a year earlier he had approached the Leipzig publisher Wilhelm Engelmann through his friend Max Johann Sigismund Schultze about publishing a collection of his natural historical observations, but that nothing had come of it (see Möller ed. 1915–21, 2: 125). Müller never published such a collection.

From G. H. Darwin [3 June 1867][1]

Tr: Coll.
Monday

Dear Papa,

I have asked a good classic about some word for Pangenesis,—he seems to thing Atomogenesis wd be all right, from the classical point of view—& wd mean the genesis of ultimate particles, litterally of particles wh: cannot be subdivided.[2] I found a word κυτταρος wh: wd make Cyttarogenesis, the word meaning the cell of a plant—but then altho' used by a good author it is rare as might be expected, & of course wd not convey any meaning to anyone who was not a good classic.[3] This man said he wd see if he cd find any more words but I have not asked him since. I do'nt think you cd get any common word for tissue or cells.

I shd say from the little notion I have about pangenesis that atomo-g. wd represent it better, unless you have particular objection to your cells or gemmules being called atoms i.e indivisible particles.— If I hear more I will write again.—

We have had some most goloptious weather since Mama &c were here—[4] Yesterday I loafed about all day. I am getting in rather a fright about the May exam wh: begins next Friday, as it is the first one in which we shall have had really hard papers & if I don't do well in it I suppose I sha'nt in the Tripos. There was a concert in the town hall on Friday & Frank[5] played in a duet with a piano; he had practised a good deal & I really thought played it very well— I was surprised to hear how loud the flute was when I was right at the further end of a very large hall.

Swettenham's eye got nearly alright at one time but he foolishly uncovered it too soon & it has relapsed again, tho' not so bad as before.—[6] I shall only be at home for two days before I go to Paris & one of them I shall have to go to London to see the dentist, more's the bore.—[7] I wish there was a little longer time betw: the end of this term & July, when I shall have to come back here.—

Yrs G H Darwin

DAR 210.2: 2

[1] The date is established by the relationship between this letter and the letter to G. H. Darwin, 27 May [1867], and by the reference to Emma's visit to George in Cambridge (see n. 4, below). In 1867, 3 June was the first Monday after 27 May.

[2] CD had asked George for an alternative term to pangenesis (see letter to G. H. Darwin, 27 May [1867]).

[3] κυττάρος is used by Aristotle in his *Historia animalium* to mean the cell of a honey-comb (Liddell and Scott comps. 1996).

[4] Emma Darwin visited Cambridge from 22 to 25 May 1867 (Emma Darwin's diary (DAR 242)). It is not known who accompanied her. Goloptious: a slang or humorous term meaning 'delightful', first appearing in print (as 'galoptious') in 1856 (*OED* s.v. 'goluptious').

[5] Francis Darwin, like George, was an undergraduate at Trinity College, Cambridge (*Alum. Cantab.*).

[6] The reference is probably to Richard Paul Agar Swettenham, an undergraduate at Trinity College, Cambridge (*Alum. Cantab.*).

[7] George visited the *Exposition universelle* in Paris; his entrance ticket, valid for the week until 24 June 1867, is in DAR 219.12: 12.

To V. O. Kovalevsky 3 June [1867][1]

Down. | Bromley. | Kent. S.E.
June 3.

Dear Sir

I enclose a sheet & have 4 others ready to send but as each sheet with the envelope weighs an ounce the postage will be very expensive to you.[2] Therefore I will wait till I hear again before sending the 4 other sheets.

With respect to what you say about illustrating my book with wood cuts from Brehm's work, it w^d I think be an excellent scheme.[3]

I have not seen Brehm's work nor do I know where I c^d see it.

Believe me dear Sir | yours very faithfully | Ch. Darwin

LS
Institut Mittag-Leffler

[1] The year is established by the relationship between this letter and the letter from V. O. Kovalevsky, 15 May 1867.

[2] Kovalevsky had asked CD to send the sheets of *Variation* for his Russian translation not pre-paid (letter from V. O. Kovalevsky, 15 May 1867).

[3] CD refers to Brehm *et al.* 1864–9 (see letter from V. O. Kovalevsky, 15 May 1867).

From Edward Cresy 6 June 1867

Metropolitan Board of Works | Spring Gardens
6 June 67—

My dear Sir,

Forgive my having so long omitted to return Dr Hooker's very interesting paper—[1] I meant & ought to have given it your daughter when she was with us—but having been disappointed in not seeing her again as I expected before she left I took it up to town & buried it among some of my other papers—[2] I hope you have not been inconvenienced by my inadvertence—

I was greatly rejoiced to hear from your daughter's report to my wife that you were better & able to work again— I sincerely hope you will keep now steadily improving & that you won't suffer from a plethora of proofs.[3]

When George comes "down" pray ask him to come & see me before he goes "up" again & I understood he meant staying to work the latter part of the long.[4]

Yours very truly | E Cresy

C Darwin Esq.

DAR 161: 248

[1] Cresy may refer to J. D. Hooker 1866a, Joseph Dalton Hooker's paper on insular floras. There are two offprints from the *Gardeners' Chronicle* of January 1867 in the Darwin Pamphlet Collection–CUL, one annotated, one lightly scored. The offprint was a slightly revised version of the paper printed in *Gardeners' Chronicle*; see Williamson 1984 for the text of the offprint.

[2] Henrietta Emma Darwin went to visit the Cresys on 10 May 1867 (Emma Darwin's diary (DAR 242)); there is a letter from her to George Howard Darwin about her visit in DAR 245: 277.

[3] CD was working on the proof-sheets of *Variation*. Cresy's wife was Mary Louis Cresy.

[4] George was an undergraduate at Cambridge University; Cresy had advised him on his career (see *Correspondence* vol. 13, letter to Edward Cresy, 7 September [1865], and letter from Edward Cresy, 10 September 1865). 'Down' and 'up': that is, not in residence, and in residence, at college; the long vacation was the summer vacation (*OED*).

From Charles Kingsley 6 June 1867

Eversley Rectory, | *Winchfield.*
June 6/67

My dear M^r. Darwin

I am very anxious to obtain a copy of a pamphlet, w^h. I unfortunately lost. It came out shortly after your 'origin of species', & was entitled "Reasons for believing in M^r. Darwins theory"—or some such words. It contained a list of phænomenal puzzles 40 or more w^h. were explicably by you & not otherwise. If you can recollect it, & tell me where I can get a copy, I shall be very glad—as I very specially want it, in your defence.[1]

I advise you to look at a wonderful article in the North British about you. It is a pity the man who wrote it had not studied a little zoology & botany, before writing about them.[2]

The Duke of Argyll's book is very fair & manly.[3] He cannot agree with you, but he writhes about under you as one who feels himself likely to be beat. What he says about the humming birds is his weakest part. He utterly overlooks *sexual* selection by the females, as one great branch of Natural selection. Why on earth are the *males* only (to use his teleological view) ornamented, save for the amusement of the females *first*?[4] In his earnestness to press the point—(w^h. I think you have really overlooked too much) that beauty in animals & plants is intended for the æsthetic education & pleasure of man, And (as I believe in my old fashioned way), for the pleasure of a God who rejoices in his works as a painter in his picture— In his hurry, I say, to urge this truth, he has overlooked that beauty in any animal must surely first please the animals of that species, & that beauty in *males* alone, is a broad hint that the females are meant to be charmed thereby—& once allow that any striking new colour w^d. attract any single female, you have an opening for endless variation. His argument that the females of each species are as distinct as the males, is naught—for a change in the embryo w^h. w^d reproduce the peculiar markings of the father, in a male w^d be surely likely to produce *some* change of markings in a female.[5]

Altogether—even the North British pleases me—for the man is forced to allow *some* Natural Selection, & forced to allow *some* great duration of the earth; & so every one who fights you, is forced to allow some of your arguments, as a tub to the whale,[6] if only he may be allowed to shun others— While very few have the honesty to confess, that they know nothing about the matter, save what you have put into their heads.

Remark that the argument of the N. British, that geological changes were more violent, & the physical energies of the earth more intense in old times, cuts both ways.[7] For if that be true—then changes of circumstance in plants & animals must have been more rapid, & the inclination to vary from *outward circumstance* greater, & also—if the physical energies of the earth were greater—so must the *physical* energies of the Animals & plants; & therefore their tendency to *sport* may have been greater; & not without a gleam of scientific insight have the legends of so many races talked of giants & monsters on the earth of old.

Yours ever faithfully | C Kingsley

DAR 169: 35

CD ANNOTATION
3.14 His argument ... naught 3.15] *scored blue crayon*

[1] CD guessed that Kingsley was referring to a short paper in the *Geologist* by Frederick Wollaston Hutton (Hutton 1861; see letter to Charles Kingsley, 10 June [1867]). There is a lightly marked copy in the Darwin Pamphlet Collection–CUL. The paper includes six objections to CD's theory and answers to them, and twenty-six questions answered by supposing that animals have descended from a common prototype.

[2] The article referred to by Kingsley, published in the *North British Review* for June 1867 ([Jenkin] 1867), was by Henry Charles Fleeming Jenkin, an engineer and critic (*Wellesley index*). The *North British Review* was published in Edinburgh and was intended to be 'liberal in politics and Christian in tone', giving due place to 'art, science, philosophy, literature, and culture in general' (*Wellesley index* 1: 663). See D. L. Hull 1973, pp. 302–50, for a copy of [Jenkin] 1867, and Hull's commentary; see also n. 6, below.

[3] Kingsley refers to G. D. Campbell 1867.

[4] Campbell had written: 'there is no connexion which can be traced or conceived between the splendour of the Humming Birds and any function essential to their life. If there were any such connexion, that splendour could not be confined, as it almost exclusively is, to one sex' (G. D. Campbell 1867, p. 243). See also letter to Charles Lyell, 1 June [1867] and n. 3.

[5] Campbell had written (G. D. Campbell 1867, pp. 250–1):

it would appear that every variety which is to take its place as a new Species must be born male and female; because it is one of the facts of specific variation in the Humming Birds, that although the male and female plumage is generally entirely different, yet the female of each Species is as distinct from the female of every other, as the male is from the male of every other.

[6] For more on Jenkin and his criticism of CD's theory, see Morris 1994 and Gayon 1998, pp. 85–102. A tub to the whale: '(to throw out) a tub to the whale, to create a diversion, esp. in order to escape a threatened danger' (*OED*).

[7] See [Jenkin] 1867, pp. 296–302.

To Charles Lyell 9 June [1867][1]

Down. | Bromley. | Kent. S.E.
June 9th

My dear Lyell

I write one line to say that certainly the variability & passage of Primrose into Cowslip must be given up.— I have this summer proved that the common oxlip is a natural Hybrid between the two; but the Bardfield oxlip, which occurs almost only in Essex (the P. elatior *of Jacquin*) is a perfectly distinct & good *[&]* third species.—[2]

I am very glad you like H. Parker's article.—[3]

I want much talk over N. British Review & to have the real pleasure of seeing you & Lady Lyell.— We shall be up, I believe, on 15th, & if you do not hear to contrary, stomacho volente I will come to your breakfast on Monday 17th.—[4]

Nothing was enclosed in your note about Primula.—[5]

Yours affect. | C. Darwin

American Philosophical Society (329)

[1] The year is established by the relationship between this letter and the letter to Charles Lyell, 1 June [1867], and by the reference to the *North British Review* (see n. 4, below).

[2] CD had been conducting experiments since 1862 to establish that the primrose (*Primula vulgaris*) and the cowslip (*P. veris*) were distinct species, and to investigate the relationship between them and the common oxlip (now *P. vulgaris* x *P. veris*) and the Bardfield oxlip (*P. elatior*). He published the results of his crosses in 'Specific difference in *Primula*', pp. 443–7, and *Forms of flowers*, pp. 55–75. His notes, dated between 1862 and 1867, are in DAR 157a: 75–7 and DAR 108. See *Correspondence* vol. 12, letter to J. D. Hooker, 10 June [1864] and n. 17, and *Correspondence* vol. 13, letter from L. C. Wedgwood, [April–May 1865?] and n. 3. Lyell included this information from CD in the tenth edition of his *Principles of geology* (C. Lyell 1867–8, 2: 324–5). He may have wanted to cite the results of observers who claimed to have raised cowslips, oxlips, and a primrose from the seed of the same plant, and thus ranked the three as varieties. CD discussed these claims in 'Specific difference in *Primula*', pp. 440–2, and *Forms of flowers*, pp. 61–2. See also *Correspondence* vol. 7, letter to Charles Lyell, 11 October [1859] and nn. 12 and 13.

[3] CD refers to his nephew Henry Parker and to [Parker] 1862. See letter to Charles Lyell, 1 June [1867] and n. 2. The letter in which Lyell commented on [Parker] 1862 has not been found.

[4] CD refers to [Jenkin] 1867, an article published in the *North British Review* for June 1867; see also letter from Charles Kingsley, 6 June 1867. He also refers to Mary Elizabeth Lyell. The Darwins were in London from 17 to 24 June (CD's 'Journal' (Appendix II)). No record has been found of a meeting with the Lyells. 'Stomacho volente': stomach willing; an invented phrase based on *Deo volente*, God willing. CD's stomach problems often prevented him from going into company; see *Correspondence* vol. 13, Appendix IV.

[5] Lyell's letter has not been found.

To Charles Kingsley 10 June [1867][1]

Down. | Bromley. | Kent. S.E.

June 10

My dear M[r] Kingsley

I have been deeply interested by your letter. I have looked through my whole large collection of pamphlets on the "Origin" & the only thing which I can find at all answering to y[r] description is that which I send by this post by Cap. Hutton. I dare say you know his name; he is a very acute observer. Please sometime return it to me.[2]

I have just finished reading the Duke's book & N. Brit. Rev.;[3] & I sh[d] very much like for my own sake to make some remarks on them, & as my amanuensis[4] writes so clearly, I hope it will not plague you. The Duke's book strikes me as very well written, very interesting, honest & clever & very arrogant. How coolly he says that even J. S. Mill does not know what he means.[5] Clever as the book is, I think some parts are weak, as about rudimentary organs,[6] & about the diversified structure of humming birds. How strange it is that he sh[d] freely admit that every detail of structure is of service in the flowers of orchids, & not in the beak of birds.

His argument with respect to diversity of structure is much the same as if he were to say that a mechanic w^d succeed better in England if he c^d do a little work in many trades, than by being a first-rate workman in one trade.[7] I sh^d like you to read what I have said upon diversity of structure at 226 in the new Ed. of Origin,[8] which I have ordered to be sent to you. Please also read what I have said (p. 238) on Beauty.) Other explanations with respect to beauty will no doubt be found out: I think the enclosed ingenious letter by Wallace is worth y^r notice.[9] Is is not absurd to speak of beauty as existing independently of any sentient being to appreciate it? And yet the Duke seems to me thus to speak. With respect to the Deity having created objects beautiful for his own pleasure, I have not a word to say against it but such a view c^d hardly come into a scientific book. In regard to the difference between female birds I believe what you say is very true; & I can shew with fowls that the 2 sexes often vary in correlation.[10] I am glad that you are inclined to admit sexual selection. I have lately been attending much to this subject, & am more than ever convinced of the truth of the view. You will see in the discussion on beauty that I allude to the cause of female birds not being beautiful; but Mr Wallace is going to generalize the same view to a grand extent, for he finds there is almost always a relation between the nature of the nest & the beauty of the female.[11]

No doubt sexual selection seems very improbable when one looks at a peacock's tail, but it is an error to suppose that the female selects each detail of colour. She merely selects beauty, & laws of growth determine the varied zones of colour: thus a circular spot w^d almost certainly become developed into circular zones, in the same manner as I have seen the black wing-bar in pigeons become converted into 3 bars of colour elegantly shaded into each other. The Duke is not quite fair in his attack on me with respect to "correlation of growth"; for I have defined what I mean by it, tho' the term may be a bad one, whilst he uses another definition: "correlation of variation" w^d perhaps have been a better term for me.[12] He depreciates the importance of natural selection, but I presume he w^d not deny that Bakewell, Collins, &c had in one sense made our improved breeds of cattle,[13] yet of course the initial variations have naturally arisen; but until selected, they remained unimportant, & in this same sense natural selection seems to me all-important.

The N. Brit. Rev. seems to me one of the most telling Reviews of the hostile kind, & shews much ability, but not, as you say, much knowledge. The R. lays great stress on our domestic races having been rapidly formed, but I can shew that this is a complete error; it is the work of centuries, probably in some cases of 1000s of years.[14] With respect to the antiquity of the world & the uniformity of its changes, I cannot implicitly believe the mathematicians, seeing what widely different results Haughton Hopkins & Thompson have arrived at.[15] By the way I had a note from Lyell this m^g who does not seem to value this article enough.[16]

Is there not great doubt on the bearing of the attraction of gravity with respect to the conservation of energy? The glacial period may make one doubt whether the temperature of the universe is so simple a question. No one can long study the Geolog. work done during the glacial period, & not end profoundly impressed with the necessary lapse of time; & the crust of the earth was at this recent period as thick as now & the force of Nature not more energetic. But what extremely concerns me, is R. statement that I require million of years to make new species; but I have not said so, on contrary, I have lately stated that the change is probably rapid both in formation of single species & of whole groups of species, in comparison with the duration of each species when once formed or in comparison with the time required for the development of a group of species—[17] with respect to Classification, it is the idea of a *natural* classification, which the genealogical explains.[18] The best bit of Review, which c^d make me modify wording of few passages in origin is I think about sudden sports, & these I have always thought, but now more clearly see, w^d generally be lost by crossing.[19] The R does not however notice, that any variation w^d be more likely to recur in crossed offspring still exposed to same conditions, as those which first caused the parent to vary.— I have moreover expressly stated that I do not believe in the sudden deviation of structure under nature, such as occurs under dom: but I weakened the sentence in deference to Harvey.—[20]

When speaking of the formation for instance of a new sp. of Bird with long beak Instead of saying, as I have sometimes incautiously done a bird suddenly appeared with a beak *[particularly]* longer than that of his fellows, I would now say that of all the birds annually born, some will have a beak a shade longer, & some a shade shorter, & that under conditions or habits of life favouring longer beak, all the individuals, with beaks a little longer would be more apt to survive than those with beaks shorter than average.[21]

The preservation of the longer-beaked birds, would in addition add to the augmented tendency to vary in this same direction.— I have given this idea, but I have not done so in a sufficiently exclusive manner.— The Reviewer w^d have left his article stronger if he had not attempted to exclusively grapple with the [*illeg*] problem of *[variation]*[22]

⟨*missing text*⟩ of facts. Pray excuse this unreasonable letter, which you may not think worth the labour of reading; but it has done me good to express my opinion on the 2 works in question, so I hope & think that you will forgive me—

With very sincere thanks for letter believe me my dear Mr Kingsley | yours sincerely | Charles Darwin

Do you know who wrote the article in N. B. Review?[23]

LS(A) incomplete & draft
American Philosophical Society (330) & DAR 96: 28–9, 32

[1] The year is established by the relationship between this letter and the letter from Charles Kingsley, 6 June 1867.

[2] CD refers to Frederick Wollaston Hutton and to Hutton 1861; see letter from Charles Kingsley, 6 June 1867 and n. 1.

[3] CD refers to G. D. Campbell 1867 and [Jenkin] 1867; see letter from Charles Kingsley, 6 June 1867.

[4] The letter is in Emma Darwin's hand, except for the last sentence. A missing section, from 'as now & the force of Nature' in the fifth paragraph to 'problem of *[variation]*' in the seventh has been supplied from a draft in CD's hand.

[5] George Douglas Campbell criticised a statement of John Stuart Mill's about human action in G. D. Campbell 1867, pp. 314–16.

[6] Campbell discussed rudimentary limbs and homology of structure in G. D. Campbell 1867, pp. 204–16. He argued that rudimentary limbs were part of a universal plan that had clearly been mentally conceived.

[7] Campbell discussed humming-birds in G. D. Campbell 1867, pp. 233–52, and CD's work on orchids in *ibid.*, pp. 37–46. He suggested that a bill adapted to probing all flowers with ease would be more advantageous than a specialised one (*ibid.*, pp. 241–2).

[8] *Origin* 4th ed.

[9] CD probably sent the letter from A. R. Wallace, 26 April [1867].

[10] See letter from Charles Kingsley, 6 June 1867 and n. 5. For more on correlated variation, see *Descent* 1: 208, 282–3, 285–6; fowls are discussed in *ibid.*, pp. 294–5.

[11] See letters from A. R. Wallace, 26 April [1867] and 1 May 1867. For CD's ongoing work on sexual selection, see, for example, the letter to Fritz Müller, 22 February [1867], and the letter to A. R. Wallace, [12–17] March [1867].

[12] On the wing-bars of pigeons and ocellated spots on peacocks' tails, see *Descent* 2: 131–5. On correlation of growth, see *Origin* 4th ed, pp. 11–12, 170–4; CD wrote: 'I mean by this expression that the whole organisation is so tied together during its growth and development, that when slight variations in any one part occur, and are accumulated through natural selection, other parts become modified' (*Origin* 4th ed., p. 170). Campbell discussed CD's theory of correlation of growth in G. D. Campbell 1867, pp. 256–62. He suggested that correlation of growth comprised two distinct notions, symmetry, which he thought had simple physical causes akin to those governing the growth of a crystal, and fitness; the latter he thought suggested 'the operation of Forces working under Adjustment with a view to Purpose' (G. D. Campbell 1867, p. 260).

[13] CD refers to Robert Bakewell, a notable stock-improver, and probably to Robert or Charles Colling, brothers who produced an improved breed of shorthorn cattle (see Trow-Smith 1959).

[14] See [Jenkin] 1867, pp. 280–6. Henry Charles Fleeming Jenkin had argued that there were physical limits to the amount of variation that could be produced in any one direction. He had used the example of a breeder selecting a feature and effecting remarkable variation in it in the first few years but reaching a limit beyond which no further variation could be achieved (a 'sphere of variation'; see *ibid.*, p. 282). He then argued that no extension of time could reverse the 'rule' governing the limits of a given sphere of variation. CD evidently interpreted Jenkin's argument as having been based on the mistaken notion that CD's own argument was that since varieties could be produced in domestic animals in a relatively short period of time, by extension new species could be created given a sufficiently long period of time (see also Gayon 1998, pp. 88–9). In *Variation* 2: 243, CD pointed out that several animals had already been domesticated in the Neolithic era.

[15] CD refers to Samuel Haughton, William Hopkins, and William Thomson (later Lord Kelvin). CD made a similar comment regarding the estimates of physical scientists in his letter to J. D. Hooker, [28 February 1866] (*Correspondence* vol. 14). Thomson held that the crust of the earth had solidified only 100 million years ago, and had criticised CD's estimate in the first edition of *Origin* of 300 million years for the denudation of the Weald (see W. Thomson 1862, pp. 391–2). In 1864, Haughton had calculated a period of 2 billion years from the formation of the oceans to the beginning of the Tertiary period, though earlier he had agreed with Thomson (see Burchfield 1990, pp. 100–1). Hopkins had sought to put geology on a firm mathematical footing but was not noted for speculations about the age of the earth (*DSB*).

[16] Charles Lyell's letter has not been found.

[17] See [Jenkin] 1867, p. 294, and *Origin* 4th ed., pp. 359–60.

[18] See [Jenkin] 1867, pp. 305–13, where Jenkin argued that the difficulty in classifying species did not need to be explained by a theory of the transmutation of species, but was a common problem in many systems of classification, and could be explained by the vast number of possible combinations

of variables that existed among living beings. See also *Origin* 4th ed., pp. 486–9, for CD's views on the 'Natural System' of classification.

[19] See [Jenkin] 1867, pp. 288–92. In *Origin* 4th ed., p. 47, CD had speculated that if a monstrous form occurred in nature and was propagated, albeit in a modified state owing to its being crossed with the ordinary form, if it was advantageous to the organism it would spread by means of natural selection. In *Origin* 5th ed., p. 49, he said that such variations would spread only under unusually favourable circumstances. CD cited Jenkin's article in *Origin* 5th ed., p. 104, for his argument on how rarely single variations in nature could be perpetuated.

[20] See *Origin* 3d ed., p. 46, and *Origin* 4th ed., p. 47. See also Peckham ed. 1959, p. 121. In the fourth edition of *Origin*, CD omitted, amongst other changes, the words added in the third edition, 'Monsters are very apt to be sterile'. William Henry Harvey had published an article in the *Gardeners' Chronicle*, 18 February 1860, pp. 145–6, reporting the apparent origination of a new species through the abnormal development of *Begonia frigida*. See also *Correspondence* vol. 8, letter to Charles Lyell, 18 [and 19 February 1860], and *Correspondence* vol. 12, letter from W. H. Harvey, 19 May 1864, n. 4. CD also discussed the nature of monstrosities in the manuscript of his 'big book' on species (see *Natural selection*, pp. 318–21).

[21] In *Origin*, p. 61, CD wrote: 'Owing to this struggle for life, any variation, ... if it be in any degree profitable to an individual of any species, ... will tend to the preservation of that individual, and will generally be inherited by its offspring.' In the fifth edition, he changed 'an individual' to 'the individuals', and 'that individual' to 'such individuals' (p. 72). See also *Journal of researches* 2d ed., pp. 379–80, for CD's comments on the gradation in the sizes and shapes of the beaks of different species of *Geospiza* on the Galápagos Islands. For a discussion of Jenkin's argument about the tendency of variation appearing in a single individual to be swamped, and CD's response, see Gayon 1998, pp. 94–102. See also D. L. Hull 1973.

[22] The text from 'as thick as now' to 'problem of [variation]' is supplied from a draft.

[23] The article was by Henry Charles Fleeming Jenkin (*Wellesley index*).

To T. H. Huxley 12 June [1867][1]

> Down. | Bromley. | Kent. S.E.
> June 12th

My dear Huxley

We come up on Saturday 15th for a week.—[2] I want much to see you for a short time to talk about my youngest Boy & the School of Mines.—[3] I know it is rather unreasonable, but you must let me come a little *after* 10 o clock on Sunday morning 16th— If in any way inconvenient, send me line to "6 Queen Anne St. W.",[4] but if I do **not** hear, I will (stomacho volente)[5] call, but I will not stay very long & spoil your whole morning as a holiday.—

Will you turn 2 or 3 times in your mind this question: What I called *pangenesis* means that each cell throws off an atom of its contents or a gemmule & that these aggregated from the true ovule or bud &c.—[6]

Now I want to know whether I could not invent a better word.

Cyttarogenesis, ie cell-genesis is more true & expressive but long.—

Atomogenesis, sounds rather better, I think, but an "atom" is an object which cannot be divided; & term might refer to the origin of atom of inorganic matter.—[7]

I believe I like *pangenesis* best, though so indefinite; & though my wife says it sounds wicked like pantheism; but I am so familiar now with this word, that I cannot judge, I supplicate you to help me.—

My dear Huxley | Yours ever most truly | C. Darwin

Imperial College of Science, Technology, and Medicine Archives (Huxley 5: 235)

¹ The year is established by the reference to the Darwins' visit to London (see n. 2, below) and the discussion of an alternative term to pangenesis (see letter to G. H. Darwin, 27 May [1867]). June 15 was a Saturday in 1867.

² The Darwins in the event went to London from 17 to 24 June (see CD's 'Journal' (Appendix II)).

³ CD's youngest son was Horace, then aged 16. Huxley was a professor at the Royal School of Mines (L. Huxley 1900).

⁴ CD gave the address of his brother, Erasmus Alvey Darwin (*Post Office London directory* 1867).

⁵ *Stomacho volente*: stomach willing.

⁶ Huxley had read CD's manuscript on pangenesis in 1865 (see *Correspondence* vol. 13, letter to T. H. Huxley, 12 July [1865]).

⁷ See letter to G. H. Darwin, 27 May [1867], and letter from G. H. Darwin, [3 June 1867].

To Charles Lyell   12 June 1867

Down
12 June, 1867.

My dear Lyell

I am not sure whether you require an answer. I can only reiterate that you had better omit whole passage.¹ I alluded to the case in 1ˢᵗ edit. of Origin, and struck it out afterwards.² I can answer any question when we meet on Monday morning.³ It is not I think, odd that Herbert & Co were deceived, for nothing was then known on reciprocal dimorphism⁴

Yours affect. | C. Darwin.

Copy
DAR 146: 326

¹ Lyell's letter has not been found. He had evidently written for further information or clarification following CD's letter to him of 9 June [1867], concerning the variability of primroses and cowslips. In the tenth edition of *Principles of geology*, Lyell mentioned that CD had in the summer of 1867 completed experiments suggesting that the common oxlip was a hybrid between the primrose and the cowslip, and that the primrose and the cowslip were themselves distinct species (C. Lyell 1867–8, 2: 324). CD probably advised omitting mention of experiments suggesting that oxlips, cowslips, and primroses could be produced from the seed of a single plant (see n. 4, below).

² In the fourth edition of *Origin*, CD omitted sentences giving the primrose and the cowslip as examples of doubtful species, and mentioning Karl Friedrich von Gärtner's unsuccessful attempts to cross the primrose and the cowslip, that appeared in the first edition of *Origin*, pp. 49 and 247.

³ The Darwins visited London from 17 to 24 June 1867 (CD's 'Journal' (Appendix II)).

⁴ CD refers to William Herbert, and probably to John Stevens Henslow and Hewett Cottrell Watson. All three are cited in 'Specific difference in *Primula*', pp. 441–2, and *Forms of flowers*, pp. 60–1, as claiming to have produced cowslips, oxlips, and primroses from the seed of a single plant. In 'Specific difference in *Primula*', p. 441, CD wrote that 'dimorphism not being formerly understood, the seed-bearing plants were in no instance protected from the visits of insects'.

To J. D. Hooker [16 June 1867]

Down
Sunday Evening

My dear Hooker

We go up tomorrow to 6 Q. Anne St for a week.— Send me a line to say whether you are, & will be, at Kew; & whether you are at all likely to be in London & to have an hour's time to spare to see us..— Or if I sh^d be supernaturally strong, might I drive down for an hour or hour & half rather early some morning.— But I am not very strong & have a good deal to do, so I doubt whether I could give myself this treat, even if you are at Kew & not worked to death.—[1]

Ever yours Affect^ly | C. Darwin

Endorsement: 'June 16/67'
DAR 94: 29–30

[1] The Darwins stayed in London until 24 June (CD's 'Journal' (Appendix II)). Hooker and CD did not meet (see letter to J. D. Hooker, [23 June 1867]). See also letter from J. D. Hooker, 18 June 1867.

From J. D. Hooker 18 June 1867

Kew
June 18/67.

Dear Darwin

I shall be in town on Thursday, & will try & hit your lunch time if your Brother will kindly allow me.[1]

I do hope you may be able to run down here too. I am turning into a Landscape Gardener, getting up cheerfully at 6 & before it, & sleeping like a Plough-boy in consequence, or rather in spite of it.—[2] I have been reading Tate's? Review in N. British,[3] & wish I was not so confoundedly lazy & I would answer it. Also I have read "Mount Sorel" & Ben D'Izzy's political life of Lord G. Bentinck[4]—& this is the sum of my acquisitions in Science polite Literature & Politics for the past 9 months— it is a blessed retrospect—for which I *am* sufficiently thankful:— bad science—bad novel,—bad political Economy— what more can a man want, that does not feed on bread alone.

Mrs Hooker will "beat you up" or "run you down" these terms being synonymous, if you will let her one day this week[5]

Ever yrs affec | J D Hooker

DAR 102: 167–8

[1] Hooker refers to Erasmus Alvey Darwin, at whose house in Queen Anne Street CD stayed when he was in London. Hooker and CD did not meet (see letter to J. D. Hooker, [23 June 1867]).

[2] Hooker was director of the Royal Botanic Gardens, Kew. On his work there after his appointment in 1865, see R. Desmond 1995, pp. 225 ff.

[3] Hooker probably refers to Henry Charles Fleeming Jenkin's anonymous discussion of CD's transmutation theory in the *North British Review* ([Jenkin] 1867). Hooker was probably guessing that the

anonymous author of the article was Peter Guthrie Tait, a physicist who had published articles in the *North British Review* (see *Wellesley index*). For CD's opinion of the article, see the letter to Charles Kingsley, 10 June [1867].

[4] Hooker refers to Anne Marsh-Caldwell's *Mount Sorel; or the heiress of the De Veres* ([Marsh-Caldwell] 1845) and Benjamin Disraeli's *Lord George Bentinck: a political biography* (Disraeli 1858).

[5] Frances Harriet Hooker did not visit (see letter to J. D. Hooker, [23 June 1867]).

From J. W. Salter 18 June 1867

8 Bolton Road | S[t] Johns Wood
June 18/67

Dear M[r] Darwin

You really must let me send you something in return for your valued aid—which I hope I shall never require again.[1] A death in our family deeply regretted, has had this effect, that it just clears me from vexing liabilities & enables me to work with quiet mind.

I did apply to the Royal Soc[y]. some time ago—and they gave me £50, a princely sum for me then. I hope to be a subs[r] to that fund some day—soon.[2]

Let me send you the Supp[t]. Eng. Bot[y]. You may not care for British plants, but few men have taught us so much of their true meaning—& the causes of the local distribution of species. I believe in you heartily, half way.

It will cost me nothing—so I shall send it. I believe few works on Botany can boast better plates—and if I can only get the means I will finish it, please God, before another two years are out.[3]

I am obliged to write: for we *never* see you at the Geol. Soc[y]. and tomorrow, Wednesday, is our last meeting for the season. Can you not come. We have few philosophers among us. Even Huxley has cut us, and Phillips seldom comes. We have only raw material except Lyell & one or two others. Think of it, & dont let it go down.[4]

There is a glut of papers on Wednesday & I shall exhibit the oldest certain fossil known—a Lingula from the *red* Cambrian rocks.[5]

Eozoon is very mythical indeed. Rupert Jones says it will pass next into the superstitious stage, & then the positive— But I am all but positive it is mineral only, & have held that from the first—no proof you will say that the opinion is just.[6]

Yours ever truly, & obliged | J. W. Salter

C. Darwin Esq[re]

DAR 177: 14

[1] Salter had applied to CD for financial assistance in May (see letter from J. W. Salter, 14 May 1867 and n. 1).

[2] No letter from CD to Salter has been found, but CD had evidently advised him to apply to the Royal Society of London for help; the society administered a scientific relief fund for the aid of 'scientific men, or their families' (*Record of the Royal Society of London*, p. 111).

[3] Salter refers to the *Supplement to the English botany of the late Sir J. E. Smith and Mr. Sowerby* (W. J. Hooker,

Sowerby [*et al.*] 1831–63), of which he was the proprietor. See letter from J. W. Salter, 14 May 1867 and nn. 3 and 5.

[4] Salter refers to the Geological Society of London, Thomas Henry Huxley, John Phillips, and Charles Lyell.

[5] Salter gave two papers at the 19 June meeting of the Geological Society: 'On some tracks of *Pteraspis* (?) in the Upper Ludlow Sandstone', and, jointly with Henry Hicks, 'On a new *Lingulella* from the Red Lower Cambrian rocks of St. Davids' (see *Quarterly Journal of the Geological Society of London* 23 (1867): 333–41).

[6] CD had evidently asked Salter's opinion of *Eozoon canadense*. See letter from J. V. Carus, 11 February 1867, n. 5. Salter presumably refers to Thomas Rupert Jones, a fellow member of the Geological Society, who was alluding to Auguste Comte's three stages of thought: the theological, the metaphysical, and the positive.

From A. R. Wallace [19 June 1867][1]

9, St. Mark's Crescent | N.W.
Wednesday.

Dear Darwin

I am very sorry I was out when you called yesterday. I had just gone to the Zool. Gardens, and I met Sir C. Lyell who told me you were in Town.[2]

If you should have time to go to Bayswater I think you would be pleased to see the collections which I have displayed there in the form of an *Exhibition* (though the public will not go to see it.).[3] If you can go with any friends I should like to meet you there, if you can appoint a time.

I am glad to find you continue in tolerable health.

Believe me | Yours very faithfully | Alfred R. Wallace—

Charles Darwin Esq.

What do you think of the Duke of Argyll's Criticisms,[4] and the more pretentious one in the last Number of the N. British Rev.?[5]

I have written a little article answering them both, but I do not yet know where to get it published.[6]

A. R. W.

DAR 106: B41–2

[1] The date is established by the reference to CD's visit and the mention of the article in the *North British Review* (see n. 5, below). CD was in London from 17 to 24 June 1867 (see CD's 'Journal' (Appendix II)). In 1867, the first Wednesday after 17 June was 19 June.

[2] Wallace refers to the gardens of the Zoological Society, Regent's Park, London, and to Charles Lyell.

[3] Wallace had arranged an exhibit of his collection of bird skins and butterflies at Thomas Sims & Co. photographic gallery, 76 Westbourne Grove, Bayswater (A. R. Wallace 1905, 1: 404–5, *Post Office London directory* 1866; see also Raby 2001, p. 194).

[4] Wallace refers to George Douglas Campbell and to G. D. Campbell 1867 (for more on Campbell's criticisms, see the letter from Charles Kingsley, 6 June 1867, and the letter to Charles Kingsley, 10 June [1867]).

[5] The unsigned article in the *North British Review* for June 1867 was [Jenkin] 1867 (see letter from Charles Kingsley, 6 June 1867 and n. 2).

[6] Wallace's article, 'Creation by law', appeared in the October 1867 issue of the *Quarterly Journal of Science* (A. R. Wallace 1867c).

From Louisa Frances Kempson to Emma Darwin   20 June 1867

Plas Maur | Penmaenmawr | Conway
Thursday | 20 June 67.

My dear Aunt Emma

Will you please to tell Uncle Charles, that I have been making enquiries in my nursery about the tears. but I can only give him hearsay evidence as I cannot see so small a thing as a tear   My nurse says that tears begin to stand in a baby's eyes when they are a few weeks old, & that they begin to run down the cheeks at about six weeks. my baby is just 4 months & the tears run down her cheeks in a piteous manner when she crys, which I am happy to say is very seldom   of course I need not say that there never was such a baby since the world began![1] but I have never seen such a happy good tempered little soul— the whole house is ⟨*2? pages missing*⟩

My private secretary is gone out boating so Amy fills his place.[2]

AL incomplete
DAR 169: 4

CD ANNOTATIONS
1.1 Will ... tears. 1.2] *crossed pencil*
1.6 of course ... is 1.8] *crossed pencil*
Top of letter: 'Mrs Kempson' *pencil*

[1] The baby was Jessie Kempson. No letter to Kempson has been found, but CD had evidently been making enquiries about babies' tears for use in his work on expression (see *Expression*, pp. 153–4).
[2] The private secretary and Amy have not been identified.

To J. D. Hooker   [23 June 1867][1]

6 Queen Anne St
Sunday

My dear Hooker

We have been very much disappointed at not seeing you & Mrs Hooker here. I suppose you got Henrietta's note saying that any day would have suited us.[2]

I have had so much to do that I have found it impossible to come to Kew.[3] Remember you are doubly pledged to come to Down some time soon. We are off home tomorrow morning early & have much enjoyed our London week—

Yours affectately | Ch. Darwin

LS
Endorsement: 'June /67'
DAR 94: 28

[1] The date is established by the endorsement and by the reference to the Darwins' visit to London (see n. 2, below). The only Sunday during their visit was 23 June.

[2] Henrietta Emma Darwin's letter has not been found. The Darwins were in London from 17 to 24 June 1866 (CD's 'Journal' (Appendix II)), and had hoped to meet Hooker and his wife, Frances Harriet Hooker (see also letter from J. D. Hooker, 18 June 1867).
[3] CD had hoped to visit the Royal Botanic Gardens, Kew (see letter to J. D. Hooker, [16 June 1867]).

To V. O. Kovalevsky 24 June [1867][1]

Down. | *Bromley.* | *Kent. S.E.*
June 24—

Dear Sir

On my return home this morning your kind & magnificent present of Brehm's Illustrirtes Thierleben had safely arrived. I am very glad to possess this work with its astonishing number of illustrations & I thank you very sincerely for it. The wood cuts will do admirably to illustrate my book.[2]

When I left London this morning your other present of the bear skin had not as yet arrived, but I dare say it soon will.[3]

I send by this post two sheets & I hope you have received all the previous ones.[4]

Believe me dear Sir | yours very faithfully | Ch. Darwin

LS
Institut Mittag-Leffler

[1] The year is established by the relationship between this letter and the letter to V. O. Kovalevsky, 3 June [1867].
[2] Kovalevsky had suggested using woodcuts from Alfred Edmund Brehm's *Illustrirtes Thierleben* (Brehm *et al.* 1864–9) to illustrate his Russian translation of *Variation* (see letters from V. O. Kovalevsky, 15 May 1867 and [after 24? May 1867]). CD had mentioned in his letter to Kovalevsky of 3 June [1867] that he did not have a copy and did not know where he could find one. Four of the six volumes of Brehm *et al.* 1864–9 (dated 1864–7) are in the Darwin Library–Down; they are annotated (see *Marginalia* 1: 69–71). See also letter from H. J. Meyer, 30 July 1867.
[3] Kovalevsky had offered to send CD a bearskin (see letter from V. O. Kovalevsky, 24 April [1867]). In a letter to Hope Elizabeth Wedgwood dated 'Summer 1867', Henrietta Emma Darwin wrote:

Did I tell you about the enormous bear I found on my arrival making it dangerous to go into the best room as he was put there for a mat & has an enormous head to tumble over & 4 large paws with sharp slightly upturned nails which tear y[r]. gown and scratch y[r]. legs? Cos if not this Kowalewski who is the Russian translator of Papa's new book, sent it as a gage d'Amour. & when he wrote to say that he was coming to England on purpose to see Papa we had to have the enormous thing in the drawing room where we endanger life & limb over it many times every evening ...

[4] CD was sending Kovalevsky the proof-sheets of *Variation*.

To A. R. Wallace [24 June 1867][1]

6 Queen Anne | St W
Monday

My dear Wallace

I return by this post the journal. Your resumé of glacier action seems to me very good and has interested my brother much and as the subject is new to him he is a better judge.[2] That is quite a new & perplexing point which you specify about the Freshwater fishes during the glacial period——[3] I have also been very glad to see the article on Lyell which seems to me to be done by some good man.[4]

I forgot to say when with you, but I then indeed did not know so much as I do now, that the sexual i.e. *ornamental* differences in fishes, which differences are sometimes very great, offer a difficulty on the wide extension of the view that the female is not brightly coloured on account of the danger which she would incur in the propagation of the species.[5]

I very much enjoyed my long conversation with you; and today we return home & I to my horrid dull work correcting proof sheets.[6]

Believe me, my dear Wallace | yours very sincerely | Charles Darwin

P.S. I had arranged to go & see your collection on Saturday evening, but my head suddenly failed after luncheon & I was forced to lie down all the rest of day.——[7]

LS(A)
British Library (Add 46434, f. 74)

[1] The date is established by the relationship between this letter and the letter from A. R. Wallace, [19 June 1867]. The first Monday after 19 June 1867 was 24 June.

[2] CD refers to the *Quarterly Journal of Science* and to Wallace's article 'Ice marks in north Wales', which appeared in the January 1867 issue (A. R. Wallace 1867d); see *Correspondence* vol. 14, letter from A. R. Wallace, 19 November 1866. While in London, CD stayed with his brother, Erasmus Alvey Darwin.

[3] Wallace had argued that the presence of so many peculiar species of freshwater fishes in some North American lakes could be explained only if these lakes were formed through glacial action (A. R. Wallace 1867d, pp. 50–1).

[4] The unsigned article, 'Sir Charles Lyell and modern geology' (*Quarterly Journal of Science* 4 (1867): 1–19), was a review of Lyell's services to geology.

[5] CD may have met Wallace on 21 June 1867, after failing to see him earlier (see letter from A. R. Wallace, [19 June 1867]). In notes for *Descent* written on that date (DAR 82: B5–6), CD mentioned Wallace's view that protective coloration in fishes could be analogous to that in birds, where the nesting parent of either sex was the least conspicuous (see letter from A. R. Wallace, 26 April [1867]). In *Descent* 2: 12–23, CD discussed colour differences in fish depending on gender, age (juvenile or adult), and spawning times. He concluded that while both sexes of some fish species might have developed protective colouring, there were no examples of females alone having colours or other characters modified for protection.

[6] CD was correcting proof-sheets of *Variation*.

[7] The reference is to an exhibition of Wallace's collection of bird skins and butterflies (see letter from A. R. Wallace, [19 June 1867] and n. 3).

From Ernst Haeckel[1]   28 June 1867

Jena
28. Juni 1867.

Mein theurer, hochverehrter Freund!

Ich beginne meinen Brief diesmal mit einer persönlichen Nachricht, von welcher ich weiss, dass Sie, bei Ihren freundschaftlichen Gefühlen gegen mich, den herzlichsten Antheil daran nehmen werden. Ich habe mich wieder verlobt und hoffe, in meiner Braut ein treues und gutes weibliches Herz gefunden zu haben, welches das Glück meines Lebens neu aufbaut. Meine Braut is *Agnes Huschke*, die jüngste Tochter des verstorbenen hiesigen Professors der Anatomie.[2] Es ist ein einfaches und natürliches, sehr verständiges und heiteres Mädchen, welches mir hoffentlich in vielen Beziehungen den Mangel meiner verstorbenen, in der That ausgezeichneten und unvergesslichen Frau ersetzen wird.[3] Wir werden Mitte August heirathen, und dann zusammen in die Alpen der Schweiz und nach Ober-Italien reisen. Ich werde daher nicht, wie ich gehofft hatte, diesen Herbst nochmals nach England kommen, und muss mir die Freude, Sie wieder zu sehen, auf das nächste Jahr versparen.

Von meiner canarischen Reise erhalten Sie beifolgend einen vorläufigen Bericht. Mit einer ausführlicheren Reisebeschreibung bin ich beschäftigt.[4] Dann werde ich zunächst die Siphonophoren bearbeiten, welche ich auf Lanzarote untersucht habe.[5] Gegenwärtig habe ich auch wieder gelegentlich die merkwürdige *Protamoeba primitiva* beobachtet, welche ich auf p. 133 in I Band meiner Morphologie beschrieben habe (in der Anmerkung).[6] Diese einfachsten aller Organismen, homogene, einfache, contractile, lebende Eiweiss-Klumpen, ohne alle Differenzirung des Stoffes, ohne alle Organe und ohne bestimmte Formen, sind in der That äusserst merkwürdig, und ich glaube, dass sie die Hypothese der Autogonie wesentlich erleichtern.[7] Nun solche ganz einfache, noch gar nicht differenzirte Organismen, bei denen der Stoff noch nicht einmal eine bestimmte Form constituirt hat, dürften als primitive, autogon entstandene Stammformen aller übrigen Organismen anzusehen sein. Da die Differenzen in der chemischen Constitution der vielleicht ursprünglich verschiedenen Moneren[8] nur sehr geringe gewesen sein können, so verliert, glaube ich, die scheinbar so wichtige Frage, ob eine oder mehrere ursprüngliche Stammformen für die verschiedenen organischen Phylen anzunehmen sind, sehr viel von ihrer principiellen Bedeutung. Es scheint mir das im Grunde sehr gleichgültig.

Mit meinem Reisebericht erhalten Sie zugleich ein Ehrendiplom von der K. K. Zoologisch-botanischen Gesellschaft in Wien.[9] Der Secretär derselben schickte mir dasselbe mit der Bitte, es an Sie weiter zu befördern, und lässt Sie zugleich ersuchen, die formelle Bestätigung des Empfanges des Diploms direct an die Zool. bot. Gesellschaft nach Wien zu melden. Derselbe Naturforscher, (Cand. jur. Aug. Kanitz[10] in Wien, Blutgasse 3), welcher sehr eifrig für unsere Sache in *Wien* und besonders in *Ungarn* wirkt, schickt Ihnen zugleich durch mich beifolgend seine Photographie[11] und bittet Sie zugleich ergebenst, ihm dafür *Ihre Photographie* mit eigenhändiger Namensunterschrift zukommen zu lassen. Sie könnten ihm dieselbe entweder direct oder durch mich schicken.

Hoffentlich meldet mir Ihr nächster Brief, hochverehrter Freund, dass Ihre Gesundheit sich mit Eintritt des schönen Sommers zunehmend bessert. Ich denke mir, dass Ihr Landsitz, an desser Anmuth ich mit Vergnügen zurückdenke, ein recht gesunder und erfrischender Sommeraufenthalt sein muss, besonders durch seine Stille und Ruhe, im Gegensatz zu dem geräuschvollen London.

Indem ich Sie bitte, Ihrer hochverehrten Frau Gemahlin meinen ergebensten Gruss zu bestellen, bleibe ich mit der vorzüglichsten Verehrung | Ihr treu ergebener | Ernst Haeckel

DAR 166: 45

[1] For a translation of this letter, see Appendix I.

[2] The reference is to Emil Huschke, who died in 1858.

[3] Haeckel's first wife, Anna Sethe, died in 1864.

[4] The enclosure has not been found. Haeckel published a report on the trip as well as an account of climbing Mount Teide in Tenerife (Haeckel 1867 and 1870).

[5] Haeckel's study of the developmental history of the Siphonophorae attempted to show the colonial nature of these animals (Haeckel 1869). Lanzarote is one of the Canary Islands.

[6] See Haeckel 1866, 1: 133–4 n. Haeckel had written earlier about a similar primitive organism (see *Correspondence* vol. 13, letter from Ernst Haeckel, 11 November 1865 and nn. 11 and 12).

[7] On Haeckel's hypothesis of autogeny, or spontaneous generation, see Haeckel 1866, 1: 179–90. For more on Victorian views on spontaneous generation, see Strick 2000.

[8] For Haeckel's definition of 'Monera' (moner), see Haeckel 1866, 1: 135. According to Haeckel, they were the primordial, anucleate ancestors of all animals (see S. J. Gould 1977, p. 170).

[9] See Appendix III.

[10] The reference is to August Kanitz. 'Cand. jur.': Candidatus juris (Latin); a law degree.

[11] Kanitz's photograph has not been found.

To Ernst Haeckel   4 July [1867][1]

Down. | Bromley. | Kent. S.E.

July 4.

My dear Haeckel

I heartily congratulate you on your approaching marriage.[2] You have my entire sympathy & I feel sure it is the wisest step which you could take; for life without a wife to love & be loved by is a poor burthen. I trust you may pass a long & happy life, & I am sure it will be an active one, & that you will do admirable work in our beloved subject of natural history. I am sorry you will not come here this autumn, but perhaps another year you will bring your wife & shew England to her, & pay us a visit in this quiet place.

I am glad you are re-examining the Protamœba for I fully agree with you on the importance of studying these lowly organized creatures; but I am rather puzzled to think what you can find to observe.[3]

Many thanks for the account of your travels, but I have not read them yet, German being as you know, no easy task to me. I have written to thank for the honour of the Diploma, & likewise to Kanitz with my photograph.[4]

I received the other day a newspaper from N. America with an abstract of a speech by Agassiz, who seems much stirred up by your book. He says he rejoices at every new work which appears on our subject, as by this means the folly of our views will the sooner be exposed & the whole subject be quickly forgotten. He is forced to admit that no one knows better than you the structure & affinities of animals, but he is very savage at your genealogical tables & says they are flatly

Ernst Haeckel and colleagues, Helgoland, September 1865
Anton Dohrn, Richard Greeff, Ernst Haeckel (standing, left to right)
Martin Salverda, Petro Marchi (seated, left and right)
Courtesy of Ernst-Haeckel-Haus, Friedrich-Schiller-Universität Jena, Germany

Federico Delpino
Malphigia 19(1904): plate 3
By permission of the Syndics of Cambridge
University Library

Mary Elizabeth Barber
Courtesy of the Hunt Institute for Botanical
Documentation
Carnegie Mellon University, Pittsburgh, PA, USA

contradicted by paleontology. It is curious that he sh^d not remember that only a few years ago he maintained that a reptile c^d not exist during the carboniferous period, & now he designates this very period as the Reptilian.[5]

Farewell my dear Haeckel with my warm wishes for your happiness in which my wife cordially joins, yours very sincerely | Charles Darwin

LS(A)
Ernst-Haeckel-Haus (Bestand A-Abt. 1-52/15)

[1] The year is established by the relationship between this letter and the letter from Ernst Haeckel, 28 June 1867.

[2] Haeckel had announced his engagement to Agnes Huschke in his letter of 28 June 1867.

[3] See letter from Ernst Haeckel, 28 June 1867 and n. 6.

[4] See letter from Ernst Haeckel, 28 June 1867. For the diploma, from the Zoological and Botanical Society in Vienna, see Appendix III. CD's letter of thanks and his letter to August Kanitz have not been found.

[5] The newspaper and the speech by Louis Agassiz have not been identified. Haeckel's book was his *Generelle Morphologie der Organismen* (Haeckel 1866); the genealogical tables, or trees, are at the end of the second volume; on these trees, see S. J. Gould 1977, pp. 76–85, 170–2 (see also letter from T. H. Huxley, [before 7 January 1867] and n. 6). Agassiz had been cautious about admitting the existence of reptiles in the Carboniferous as late as 1866; see J. L. R. Agassiz 1866–76, 1: 15, 72, 91.

From J. D. Hooker 4 July 1867

Kew
July 4th/67.

Dr Darwin

I thought I never was to get time to write to you again, day after day I have been at my wits ends with things to do, & was greatly disappointed at not seeing you in London—[1] The fact is that Smith has been away in Paris, & one of my best men at the Herbarium is very ill—[2]

Mrs Hooker has worried me into taking her to Switzerland for a fortnight & we are off tonight to return on 23^d— we shall stop in the Enghedien.[3]

A friend of mine was a run-away-boy in his youth, & if I am rightly informed 2 of his boys have run away from home. several times—for no other motive but love of roaming— is the case worth my verifying for you?[4]

I will take the first Sunday for Down that I can get free after my return— Aug. 4, I hope.[5]

Ever yrs aff | J D Hooker

I thought Lyell looking well the other day— he seems to be looking into insular floras & faunas well.[6]

The Earl of Arran (father of my next door neighbour) has twice asked to be most specially remembered to you—as Philip Gore who knew you at Monte Video & arranged for your trip in the interior (he seems a great ass, but goodnatured one)[7]

DAR 102: 169–70

[1] The Darwins had visited London from 17 to 24 June (CD's 'Journal' (Appendix II)). See also letter from J. D. Hooker, 18 June 1867, and letter to J. D. Hooker, [23 June 1867].

[2] The reference is to John Smith, curator at the Royal Botanic Gardens, Kew. The other man has not been identified. Smith may have been attending the horticultural exhibition at Paris.

[3] Hooker's wife was Frances Harriet Hooker. The Engadine is a valley of the upper Inn River in east Switzerland, renowned for its magnificent scenery and bracing climate (*Columbia gazetteer of the world*).

[4] Hooker's friend has not been identified.

[5] There is no record in Emma Darwin's diary (DAR 242) of a visit from Hooker until December 1867. However, see the letter from J. D. Hooker, [27 July 1867], and the letter to J. D. Hooker, 29 July [1867].

[6] Charles Lyell was working on the second volume of the tenth edition of *Principles of geology*; it included a new chapter, chapter 41, 'Insular floras and faunas considered with reference to the origin of species' (C. Lyell 1867–8, 2: 402–32). Hooker had read a paper on the subject of insular floras at the meeting of the British Association for the Advancement of Science in August 1866 (see J. D. Hooker 1866a).

[7] Hooker refers to Philip Yorke Gore. Gore had one surviving son and three daughters, all unmarried in 1867 (*Burke's peerage*), but no persons of the name of Gore are listed at Kew in the *Post Office London suburban directory* for 1865 or 1868. As director of the Royal Botanic Gardens, Hooker's residence was 49 The Green, Kew (R. Desmond 1995, p. 416). The neighbour probably lived at 51 The Green, a grace-and-favour residence. CD mentioned Gore in his *Beagle* diary (R. D. Keynes ed. 1988, pp. 115–16).

To A. R. Wallace   6 July [1867][1]

Down. | Bromley. | Kent. S.E.
July 6

My dear Wallace

I am very much obliged for your article on mimicry, the whole of which I have read with the greatest interest.[2] You certainly have the art of putting your ideas with remarkable force & clearness; now that I am slaving over proof sheets it makes me almost envious.[3]

I have been particularly glad to read about the bird's nests, & I must procure the Intellectual Observer;[4] but the point which I think struck me most was about it being of no use to the Heliconias to acquire in a slight degree a disagreeable taste.[5]

What a curious case is that about the coral snakes.[6] The summary & indeed the whole is excellent & I have enjoyed it much.

With many thanks | yours very sincerely. | Ch. Darwin

LS
British Library (Add 46434, f. 92)

[1] The year is established by the references to Wallace's articles on mimicry and birds' nests (see nn. 2 and 4, below).

[2] CD refers to Wallace's article 'Mimicry, and other protective resemblances among animals' ([A. R. Wallace] 1867a), published in the *Westminster Review* in July. There is an annotated offprint, inscribed by the author, in DAR 133: 13.

[3] CD was working on the proof-sheets of *Variation*.

[4] Wallace discussed birds' nests in [A. R. Wallace] 1867a, pp. 38–9. See also letter from A. R. Wallace, 1 May 1867. Wallace's article 'The philosophy of birds' nests' was published in July in the *Intellectual Observer* (A. R. Wallace 1867b); there is an annotated copy in DAR 133: 12.

[5] CD probably refers to a passage in [A. R. Wallace] 1867a, p. 20, where Wallace wrote:

> If any particular butterfly of an eatable group acquired the disagreeable taste of the Heliconias while it retained the characteristic form and colouring of its own group, this would be really of no use to it whatever; for the birds would go on catching it among its eatable allies (among whom, we suppose, it is comparatively rare), and it would probably be wounded and disabled, even if rejected, and would be as effectually killed as if it were devoured.

CD double-scored this passage in his copy.

[6] CD refers to a section in [A. R. Wallace] 1867a, pp. 31–2, where Wallace discussed the poisonous snake genus *Elaps*, including *E. corallinus* (now *Micrurus corallinus*, the Brazilian coral snake), and its non-poisonous mimics. CD scored this discussion in his copy, and cited it in *Descent* 2: 31–2.

From Asa Gray to J. D. Hooker [after 6 July 1867][1]

Dear Hooker

Send this letter to Darwin. The doings of Dionæa will delight him, and he will probably have suggestions of good experiments for Canby[2] to make. A. G.

[Enclosure]

Wilmington [Delaware]
July 6th 1867

Dear Dr. Gray,

Dr. Engelmann has sent me the sheets of the new ed. of Manual containing descriptions of Isoetes et seq., and I was surprized to find there "Thermopsis mollis". If I really wrote that I had found this plant near "Franklin, Va.", it was a gross slip of the pen, as the species I had in my mind was what I take for Baptisia villosa, Ell. a common sand barren plant further south. To make the matter sure, I send a specimen collected a few miles over in N. C. and which is the same I saw near Franklin. I hope it may not be too late to make the correction, I feel great regret & shame that my carelessness should have caused this mistake.[3]

The little Dionaea has become exceedingly interesting to me, & I have been making some experiments with a view to test its carniverous propensities.[4] On June 7th I got some raw beef and placed fragments in several of the leaves; some of these were so weak from the moving that they had lost their sensitiveness, and did not well close up on the beef. Two however did so. In one on the 9th the beef was partly dissolved, and a drop of the reddish matter resulting, hung from the lower part of the trap. This is the only instance I have seen of any fluid escaping from the leaf. Generally the leaf seems to absorb the whole. I attribute this to the meat having been dissolved with unusual rapidity. On June 18th only 11 days after the leaf was fed, the beef was entirely dissolved and apparantly absorbed by the leaf, nothing remaining but a very faint and almost imperceptable film of (perhaps) indigestable matter. The other leaf with beef, and still others with flies and other insects in, (one caught a centipede), acted in the same way but more slowly. A healthy leaf never "lets up" on any thing it has caught but retains it until it is completely digested, except the wings and horny parts of insects; and when it has finished its meal it opens, but its sensitiveness is gone and it catches no more game. In fact from what I have observed the leaf acts as a stomach, secreting a sort of gastric fluid, dissolving the food, and apparantly absorbing the resulting matter. In one leaf I placed a piece of cheese. This was partly dissolved but it was evidently not the food nature intended,—as the leaf after going through a considerable part of the operation of digestion, became sick and finally died, not however affecting the general health of the plant.

Most of the large leaves of my plants are now dead, but young ones are coming on and the plants look healthy. I have not the chemical knowledge necessary to test these matters thoroughly, nor even sufficient practice with the microscope, to examine the leaves with a view to determine whether there are any absorbent vessels. I cannot ascribe the disappearance of the leaf-food to decomposition, but a little muriatic acid will test this. You may have observed that when the leaf first closes, it does so somewhat loosely as I have attempted to show, thus;—

 a cross section

but in a little while it closes tight around the food, pressing against every part of it so that the ouline of the substance contained, may be seen on the outside of the leaf. The cross section of the leaf then appears somewhat thus;—

this pressure forces the outer edges of the leaf open and the fringes stand erect.

I wish someone would take hold of this who was fully able to elaborate it. I would gladly aid all I can in furnishing specimens, &c. At the same time the leaves of Sarracenia should be examined to see if there is not something analagous in their drowning of insects.

The enclosed specimen of Aira caryophyllea, I found the other day at Salem, N.J.

DAR 58.1: 16–17

CD ANNOTATIONS
Enclosure: 1.1 Dr. . . . mistake 1.8] *crossed blue crayon*

2.1 The little ... leaves of 4.3] *crossed pencil*
2.9 On June ... leaf was 2.10] *double scored blue crayon*
2.13 A healthy ... game. 2.16] *scored blue crayon*
Top of Enclosure: 'Dionæa' *added red crayon*

[1] The date is established by the date of the enclosure.
[2] William Marriott Canby.
[3] Canby refers to George Engelmann and to the fifth edition of Gray's *Manual of the botany of the northern United States* (A. Gray [1867]). Engelmann worked on the genus *Isoetes* for this edition (*ibid.*, p. 675). Canby's correction was not made: *Thermopsis mollis* is listed in the Addenda as having been found by Canby in Franklin, Virginia (*ibid.*, p. 679). It is described as a genus like *Baptisia*; *Baptisia villosa* is not listed in the *Manual*. It is a synonym of *Baptisia cinerea*.
[4] CD made experiments on *Dionaea* and another carnivorous plant genus, *Drosera*, between 1860 and 1862 (see *Correspondence* vols. 8–10). He decided to postpone further work until after the publication of *Variation* (see *Correspondence* vol. 10, letter to Edward Cresy, 15 September [1862]); he published *Insectivorous plants* in 1875. Canby published his findings in the *Gardener's Monthly & Horticulturist*, August 1868, pp. 229–32, and is cited by CD in *Insectivorous plants*. The 'little Dionaea' was *Dionaea muscipula*.

From J. P. M. Weale 7 July 1867

Bedford [Cape of Good Hope]
July 7th. 1867

My dear Sir,

I received your kind reply to my notes, which was very gratifying to me. I have to note some errors which I am afraid slipped into my paper, but am not quite certain.[1]

I think I wrote after *P. Zochalia* "*the only purely S. African form*". If I did it was a mistake.[2] Amongst the plants which I have noticed as presenting a deceptive resemblance to orchids is the "*Impatiens Capensis*", a plant which loves moisture & shade, & appears very attractive to small hairy Diptera with a long proboscis. I am afraid this plant got misplaced amongst the labiates, which I allow is a gross error on my part, & one which made me blush when I discovered it.[3]

My friend, M[r]. MacOwan of Grahamstown has suggested to me that it would be more correct to name the Bonatea "*Darwiniana*" as more according to the Botanical Codes.[4]

I have sent the list of Questions relative to the Native Races to D[r]. Grimmer of Colesberg, where most Hottentots & Bushmen are to be found, so far as concerns the Colony,—to the Rev[d]. Tiyo Soga, a Kafir missionary beyond the borders, to Charles Brownlee Esqre the Gurka Commissioner, to D[r]. Macarthey of the Katberg Convict Station, to Henry Bowker Inspector of the Mounted Police, & to some friends in Natal.[5]

Enclosed I send you some answers written by Christian Gaika, constable here & brother to the Chief Sandilli.[6]

I am sorry to say it is very difficult to get those, who best know the Kafirs to write replies. Intelligence is at such a low stage here, that it is the exception to find scientific enquiries treated otherwise than with contempt.

I wrote out some copies & gave them to some farmers in the neighbourhood, but they have never taken any trouble with the subject.

With respect to the muscle the French call 'Grief Muscle' I was a little puzzled, as I thought that the 'corrugator' drew the eyebrows together & downward, while 'Frontalis' raised & opened them out.[7]

With respect to both Kafir & Hottentot children the lips are much protruded when they are displeased & sulky, & I have noticed the same in Kafir adults, of both sexes.[8]

In great grief it is a common gesture amongst Kafirs to place the palms of both hands on the head.

When I have finished my observations I will send you fuller particulars, & will now come to the object of this letter.

I read the notes in the Linnean Soc: Journal on the action of the stamens in Indigofera & medicago.— I had several times tried the larger flowering leguminous plants here with pieces of Horse hair & pins, but to no effect.[9]

I have now I believe to add another order in which there are special contrivances for crossing different flowers, viz "*Polygaleæ*".[10] I am not certain, but think this is a novelty. I have tried several Polygalas at different times, but could never succeed.

While out collecting a few weeks ago I gathered on the mountain a Muraltia, which appears to me to be "*Muraltia ericæfolia*" D.C. var: β. *curvifolia*. It hardly quite agrees with Harvey's description,[11] but from the specimens I have gathered appears to vary slightly even on the mountain according to its situation. When I brought my plants home for drying I was struck with the different positions of the stamens in different flowers & with Medicago Sativa fresh in my memory I inserted a small pin, & was delighted to see the whole bundle of stamens protruded from the enveloping carina, & start up violently against the two posterior petals. My drawings were made from plants which had been sometime in water, & so the action is hardly so intensely displayed as in growing plants.[12]

The bundle of stamens is held down in the fleshy concave case of the carina, the stigma being considerably below the anthers. As far as I could see with the unassisted eye it appeared to me that in the fresh flowers a minute quantity of juice spirted up with the stamens, but before closing this I hope to visit the top of the mountain & ascertain it with a glass. As the anthers only open at the apex, & the pollen is shot out in an upward direction it seems highly probable that none of the plant's own pollen can fall on the stigma. I noticed that almost all the plants lower blossoms had been visited, & I noticed a moderate sized grey Dipterous Insect & bees buzzing round the plants. It struck me also—at a time of the year when so few plants in this neighbourhood are in flower—that its ⟨   ⟩ it appears ⟨   ⟩ with white ⟨   ⟩ resembling it in foliage ⟨   ⟩ might not be a ⟨   ⟩ perhaps be essential to it ⟨   ⟩[13] do how attractive heaths are to bees. I have made some dissections of the flower in mature & immature conditions under the microscope. In the mature specimens the stamens invariably started up on the incision being made with the knife, but in the immature ones they appeared to be held down by the closely adhering edges of the carinal case.

You will notice in my drawing that the thick bundle of spiral vessels beneath the ovary sends off a far larger quantum to the posterior petals than to the carina & that they are curved at a much sharper angle than on the other side. (I have only made very rough diagrammatic tracings of my drawings) I am inclined to think that when the pin is inserted it pushes against the two petals, & also against the base of the stamens, so that the anthers are pushed upwards through the opening of the carinal sheath, & then the spiral vessels having been thus stretched act in the manner of a steel spiral spring, & produce the action observed. As I have not concluded my dissections, I only venture this as a suggestion.

The plants appear full of seed.

I do not know whether it will strike you as it does me, but it seems to me that this is a case analogous to the apparent similarity of adaptations between the Orchids & Asclepiads.[14] In this case we have adaptations of a similar nature in two most distinct orders. The strikingly Papilionaceous resemblance in the Polygalas proper with their large & conspicuous keel & alæ, seems to me carried out in an inferior degree in the muraltias, whose analogues would seem to be Amphithaleas.[15] The Medicagos nigra & laciniata & one species of Clover have somewhat similar actions in their stamens here.

Since writing the above I have again visited the mountain. On examining the flowers with a glass I found that I was in error, & must have mistaken the pollen for the juice. Almost all the flowers on many dozens of plants had been visited by insects, & it was difficult to find a blossom with the flowers unaltered. As it was a warm day I observed bees in plenty especially a small species, which I captured in the act of loosening the stamens. I found on further examination that a slight pressure on the tips of the petals was sufficient to make the stamens leap out of their carinal case.

I am expecting shortly to leave Bedford, but whether I go to Cape Town or Natal is yet uncertain.

I have since living here often thought of your remarks on Colonies in the Voyage of the Beagle.[16] It is the exception here to meet agreeable associates & one has to fall back on books for company.

I send enclosed a small packet of Locust Dung. It has been suggested to me that certain obnoxious masses have been carried from different parts of the colony by these insects. I have my doubts on the subject, but you could more easily ascertain whether any seeds of African plants are contained in it in England than I could here.

Trusting that your health may long permit us to enjoy the fruits of your interesting labours | I am, my dear Sir | faithfully yours, | J. P Mansel Weale.

P.S. I may remark that in cases of extreme fear Kafirs are unable to control their bowels, but freely void both fæces & urine. I once saw a Kafir seized by a powerful colonist by the throat, & half choked, when the above unpleasant results occurred most freely, & I have often heard of similar cases. I have been told but have never witnessed the fact myself that Kafirs will turn pale under extreme terror.

A Kafir when he has committed any offence & is trying to screen it will look sideways towards the ground, & every now & again, when he thinks you are not observing look slyly up as if to gather what impression his story has made on you. I remember noticing a somewhat similar habit in a little Chinese girl we had in the house, when I was living in England.

[Enclosure]

Answers written by Christian Gaika, Constable at Bedford, brother of the Chief Sandilli[17]

[(1.) Is astonishment expressed by the eyes and mouth being opened wide, & by the eyebrows being raised?]

Answer to first question, Yes they do open their mouths and rise their eyebrows.

[(2.) Does Shame excite a blush when the colour of the skin allows it to be visible? Especially how far down the body does the blush extend?]

Second question　Know there is no discolour of the face visible

[(3.) When a man is indignant or defiant does he frown, hold his body and head erect, square his shoulders and clench his fists?]

Third question　yes when the indignation is much in them, but they do not square their shoulders.

[(4.) When considering deeply on any subject, or trying to understand any puzzle, does he frown, or wrinkle the skin beneath the lower eyelids?]

Fourth question　yes and some times puts his hand to his chin, and pul his beard

[(5.) When in low spirits, are the corners of the mouth depressed & the inner corner or angle of the eyebrows raised & contracted by that muscle which the french call the grief muscle?]

Fifth question　Know

[(6.) When in good spirits do the eyes sparkle, with the skin round and under them a little wrinkled & with the mouth a little drawn back in the corners?]

Sixth question　Know no signe is seen.

[(7.) When a man sneers or snarls at another, is the corner of the upper lip over the canine teeth raised on the side facing the man whom he addresses?]

Seventh question　Know he lifts his upper lip a little and shows his upper teeth and turns his head on the side of the one he is adressing

[(8.) Can a dogged or obstinate expression be recognised, which is chiefly shown by the mouth being firmly closed, a lowering brow & a slight frown?]

eigth question　yes they do when fighting

[(9.) Is contempt expressed by a slight protrusion of the lips & turning up of the nose, with a slight expiration?]

ninth question　Contempt is expressed by smiling and laughing

[(10.) Is disgust shown by the lower lip being turned down, the upper lip slightly raised, with a sudden expiration something like incipient vomiting?]

Tenth question yes, but not always.

[(11.) Is extreme fear expressed in the same general manner as with Europeans?]

Eleventh question yes the shaking of the body is much experiensed and the eyes widely opend.

[(12.) Is laughter ever carried to such an extreme as to bring tears into the eyes?]

Twelfth question yes that is their common practice.

[(13.) When a man wishes to show that he cannot prevent something being done or cannot himself do something, does he shrug his shoulders, turn inwards his elbows, extend outward his hands & open the palms?]

Thourteenth question yes they are then restless and look ashamed to keep their heads up.

[(14.) Do the children when sulky pout or greatly protrude the lips?]

Fourteenth question yes and some times showes dishonour to the one

[(15.) Can guilty, or sly, or jealous expressions be recognised—though I know not how these can be defined?]

Fifteenth question Guilt can be recognised by the eyes half opend, and the chin to the breast, and some times by the movements in the body. jealous by the distemper showen to the party

[(16.) As a sign to keep silent is a gentle hiss uttered?]

Sixteenth question yes

[(17.) Is the head nodded vertically in affirmation, & shaken laterally in negation]

Seventeenth question yes

DAR 181: 41

CD ANNOTATIONS
1.1 I received ... out. 8.3] *crossed red crayon*
9.1 With ... sexes. 9.3] *scored red crayon*
10.2 on the head.] *before closing square bracket red crayon*
12.1 I read ... succeed. 13.4] 'Polygalea' *added blue crayon*
18.4 The strikingly ... labours 23.2] *crossed blue crayon*
Enclosure:
Top: 'Gaika' *ink*; 'S. Africa' *pencil* 'Natal' *brown crayon*; '26' *brown crayon*
Left margin: 'Answers procured for me by Mr J. P. Mansel Weale, relates to Kaffirs' *ink*
Right margin: 'Introduce somewhere under Laughter case the extreme that question uncovered 〈 〉 Kaffir *[illeg]*' *pencil*
Bottom: *excised paragraph 9 pasted here: scored both sides red crayon*; 'J. P. Mansel Weale of Natal' *ink*
On cover: 'Expression—' *ink*, *del pencil* '(cf Pouting)' *pencil*, *del pencil*; 'Dichogamy in Polygala— Drawings in Portfolio of Dichogamy'[18] *ink*, *scored pencil*; '+ answers by Gaika' *pencil*, *del pencil*; 'J. P. Mansel Weale of Bedford, Natal C. of Good Hope'[19] *ink*

[1] Weale sent CD a paper on *Bonatea* in January, in the hope that it could be published in an English journal (see letter from J. P. M. Weale, 9 January 1867). CD replied to him in a letter of 22 February [1867], saying that he had forwarded it to the Linnean Society. It was read there on 7 March 1867 and published in 1869 (Weale 1867).

[2] In the manuscript of Weale 1867 (Linnean Society archives, SP1249), Weale wrote, 'Amongst the Pieridæ, P. Zochalia (according to M^r. Trimen, the only purely S. African form), is, as far as the

author's observations go, very local in its range, whilst its congeners vary much and have an almost general distribution.' The words in parentheses are crossed in ink. This sentence was not included in the version published in the *Journal of the Linnean Society*.

³ In the manuscript of Weale 1867 (see n. 2, above), Weale wrote: 'The Scrophulariaceae in many instances are found to possess an orchideous resemblance in the Karoo, where the orchids are, as far as I am aware, unknown; & in the Bedford forest I could point to many Labiates, and especially to *Impatiens Capensis* which has so close a similarity of growth to an orchid, that it might easily be mistaken for one by a casual and inexperienced observer.' This sentence was part of a section omitted in the version published in the *Journal of the Linnean Society*. *Impatiens* is a member of the family Balsaminaceae, not the Labiatae.

⁴ Weale refers to Peter MacOwan, a botanist and headmaster of the Shaw College Grammar School, Grahamstown, Cape Colony (Gunn and Codd 1981). *Bonatea darwinii* is a synonym of *Bonatea cassidea*. At the International Congress of Botany in Paris in August 1867, it was stipulated that when species were named after people, the genitive form of the noun should be used if the name was that of the person who described or distinguished the species, and an adjectival form in all other cases. It was admitted in the commentary that this rule had not been adhered to in the past and should have been proposed as a recommendation; it was later abandoned. See Candolle 1867, pp. 22, 41–2, and Candolle 1883, pp. 20, 69. MacOwan may have been referring to this distinction. Darwinii is the genitive of the noun, Darwiniana the adjectival form. No alteration to the word Darwinii was made in the manuscript (see n. 2, above).

⁵ CD had sent Weale a handwritten set of queries about expression with his letter of 27 February [1867]. W. Grimmer was district surgeon of the division of Colesberg, Cape Colony; J. McCarthy was surgeon at the Katberg Convict Station, Cape Colony (*Cape of Good Hope general directory*). The others referred to are Tiyo Soga, a Xhosa missionary working in British Kaffraria, Charles Pacalt Brownlee, civil commissioner of Stutterheim in British Kaffraria, and James Henry Bowker, commandant in the Frontier Armed and Mounted Police, Cape Colony (*DSAB*).

⁶ Christian Gaika was described as 'Kafir Interpreter and Constable' in the division of Bedford, Cape Colony, in the *Cape of Good Hope general directory*. See Shanafelt 2003, pp. 822–4, 842. Sandile was paramount chief of the Rarabe, son of Ngqika (Gaika), a Xhosa chieftain (*DSAB*).

⁷ In question 5 on his questionnaire, CD had asked: 'When in low spirits are the corners of the mouth depressed & the inner corner or angle of the eyebrows raised by that muscle which the french call the grief muscle?' (see enclosure to letter to J. P. M. Weale, 27 February [1867]). On the involvement of both the corrugator muscle and the occipito-frontalis in this expression, see *Expression*, pp. 179–80. CD called these muscles, when 'in conjoint yet opposed action', the 'grief-muscles' (*Expression*, p. 181); this was the first published usage of this phrase in English (*OED*). The phrase 'muscle de la douleur', referring to the corrugator supercilii, was used by Guillaume Benjamin Amand Duchenne in his *Mécanisme de la physionomie humaine* (Duchenne 1862, pp. 35–53). For the difference between CD's and Duchenne's understanding of the muscles used in creating this expression, see *Expression*, p. 181 n. 3; see also letter to J. P. M. Weale, 27 August [1867].

⁸ See *Expression*, p. 233.

⁹ Weale refers to Henslow 1865 and 1866, on the structure of *Medicago sativa* and *Indigofera* respectively, both published in the issue of the *Journal of the Linnean Society* (*Botany*) dated 29 November 1866. Both papers included notes from CD. Weale used horsehair and pins to try to trigger the mechanism by which the pollen was ejected.

¹⁰ Weale refers to the order Polygaleae or Polygalaceae (Lindley 1853), now the family Polygalaceae.

¹¹ *Muraltia ericaefolia* var: *curvifolia* is described in *Flora Capensis* (W. H. Harvey and Sonder 1859–65, 1: 100). Weale refers to William Henry Harvey.

¹² Weale's drawings have not been found.

¹³ This section of text (from '⟨   ⟩ it appears') is on the verso of 'With respect ... sexes.' in paragraph 9. This part of the letter has been excised from the main body and stuck on to the enclosure. Some words, where the ink has come through the paper, are legible by means of a mirror.

[14] In *Origin*, p. 193, CD had written, 'the very curious contrivance of a mass of pollen-grains, borne on a foot-stalk with a sticky gland at the end, is the same in Orchis and Asclepias,—genera almost as remote as possible amongst flowering plants'. He suggested that natural selection had sometimes modified in a similar manner parts of different organic beings, the similarity not being due to inheritance from a common ancestor (*Origin*, p. 194).

[15] The genus *Muraltia* was a member of the order (now family) Polygalaceae, like the genus *Polygala*; *Amphithalea* was listed in the suborder Papilionaceae of the order (now family) Fabaceae (Lindley 1853).

[16] See, for example *Journal of researches*, pp. 529–30, where CD commented unfavourably on the state of European society in New South Wales, Australia.

[17] CD quotes or cites Gaika's answers in *Expression*, pp. 209 (question 12), 255 (question 9), 295 (question 11), and 320 (questions 2 and 13). The questions have been transcribed from the handwritten questionnaire sent to Weale (see letter to J. P. M. Weale, 27 February [1867]), and are given here in square brackets.

[18] The drawings have not been found. CD kept a number of portfolios on different subjects; the contents have since been dispersed.

[19] Bedford was in Cape Colony, on the south coast of South Africa; Natal is a province on the north-east coast of South Africa.

From James Paget 9 July 1867

1 Harewood Place | Hanover Square | W.
July 9. 1867

My dear Darwin

I will gladly seek for answers to your questions—[1] I have begun to make as many people as possible blush in conditions favourable for inspection—[2] But, pray tell me whether it will be useful for your purpose to examine the condition of the platysma during screaming in the partial or complete insensibility produced by chloroform—.[3] This may be often observed: but the occasions of watching screaming during diseases of the brain is so rare—except in children—that it might take me years to collect facts enough for an answer to your question.[4]

Always sincerely your's | James Paget.

Charles Darwin Esq.

DAR 174: 6

[1] Paget had evidently received a copy of CD's queries about expression (see Appendix IV for a later, printed, copy of the queries). This may have been handwritten, or it may have been one of the printed questionnaires sent to CD by Asa Gray (see letter to Asa Gray, 15 April [1867]).

[2] CD asked in his later, printed, queries about expression: 'Does shame excite a blush when the colour of the skin allows it to be visible? and especially how low down the body does the blush extend?' (see Appendix IV). Most of the extant handwritten questionnaires from 1867 do not include the second part of the question. Gray's printed questionnaire (see n. 1, above) is not extant. An example of blushing reported by Paget appears in *Expression*, p. 312.

[3] This sentence is quoted in Paget's *Memoirs and letters*, p. 407 (S. Paget ed. 1901). Preceding it an undated fragment from a letter to Paget from CD is quoted: 'I am greedy for facts'. A fragment carryiing the same wording, in CD's hand, is held at the American Philosophical Society (626): on the verso are the words, 'I will ask'.

[4] CD wrote to William Erasmus Darwin in 1868: 'I have asked some London surgeons to observe about the Platysma. . . . Paget tells me he has never had an opportunity of observing since I asked him, though plenty previously.—' (*Correspondence* vol. 16, letter to W. E. Darwin, 8 April [1868]). Paget was a surgeon at St Bartholomew's Hospital, London (*ODNB*).

From E. A. Darwin   17 [July 1867][1]

17th

Dear Charles

I dont know if Mr Salt has written to you about Wynne.[2] He suggests that some sum should be paid him out of the purchase money of the Mount. I dont see what it has to do with the purchase money which is treating it as if it were Susans Estate, but such of us as liked might subscribe.[3] I for one should be willing to subscribe £10 annually, & perhaps you may as well pass this on to Caroline to see what she says and give me your ideas if you have any. Salt says that WED & the Parkers have been giving him 10/- weekly.[4]

I send you the Deed for Sig^r by this post

Yours affec. | E. D

DAR 105: B55

[1] The date is established by the relationship between this letter and the letter from Salt & Sons, 17 July 1867.

[2] The reference is to either George Moultrie Salt or William Salt of Salt & Sons, and to George Wynne, the former gardener at The Mount, Shrewsbury. See letter from Salt & Sons, 17 July 1867.

[3] Francis Parker had suggested paying Wynne out of the purchase price of The Mount, where Susan Elizabeth Darwin, CD and Erasmus's sister, had lived until her death in October 1866 (see letter from Salt & Sons, 17 July 1867). Robert Waring Darwin had left a share in The Mount to each of his children, with the proviso that his unmarried daughters should be permitted to live there (Will of Robert Waring Darwin, 27 September 1845, Department of Manuscripts and Records, National Library of Wales).

[4] Erasmus refers to his and CD's sister, Caroline Sarah Wedgwood, to CD's son William Erasmus Darwin, and probably to Henry Parker (1827/8–92) and Francis Parker, the executors of Susan's will.

From Fritz Müller[1]   17 July 1867

Itajahy, Santa Catharina,
17. Juli 1867.

. . . . . Ich muss Sie noch um Verzeihung bitten, dass ich Ihnen nicht früher für Dr. Hildebrand's Buch über die Befruchtung gedankt habe, welches Sie mir freundlichst zuschickten. Ich habe es mit grossem Interesse gelesen. Ich bekam auch Dr. Hooker's höchst interessante Schrift "Vorlesung über Insel-Floren".[2]

Die Familie der Rubiaceen scheint sehr reich an dimorphen Pflanzen zu sein; obwohl jetzt im Winter viel weniger Pflanzen blühen, wie zu irgend einer andern Jahreszeit, so bemerkte ich doch schon wieder drei dimorphe Arten: 1) eine zierliche etwa Erica ähnliche Hedyotis, 2) eine zweite Art von Suteria mit braunem Kelch und gelber Krone, von der man sagt, sie sei das Lieblingsfutter des Tapirs, 3) die

Manettia bicolor, welche hier sehr häufig ist.[3] Ich kenne zwei andere Arten von Manettia mit rothen Blüthen, von der ich vermuthe, dass sie auch dimorph ist, obwohl alle Pflanzen, die ich bisher beobachtete, langgrifflig waren.

Ich fand hier auch eine weisse trimorphe Oxalis; ich habe verschiedene kurzgriffflige und mittelgrifflige, bisher aber noch keine langgrifflige gesehen; eine einzige Pflanze hatte die Staubfäden der langgriffligen Form in Verbindung mit den Griffeln der mittelgriffligen, so dass die Griffel dieselbe Länge hatten, wie die längeren Staubfäden.[4]

Ehe ich Desterro verliess,[5] untersuchte ich die meisten Kapseln, die ich aus Kreuzungsversuchen von Orchideen erhalten hatte; die meisten waren noch unreif.[6] Beinahe alle Samen waren offenbar gut in den Kapseln von:

Cattleya intermedia ♀ und Cattl. elegans[7] ♂
Epidendrum vesicatum ♀ und Ep. raniferum ♂
  "   "   ♀ und Ep. glumaceum[8] ♂
  "   raniferum ♀ und Ep. cinnabarium ♂
  "   "   ♀ und Brassavola fragans Leon[9] ♂!
Cirrhaea saccata (?) ♀ und Cirrhaea dependens (?)[10] ♂
Oncidium micropogon ♀ und Gomeza[11] sp. ♂
Gomeza sp. ♀ und Oncidium micropogon ♂.

Es waren viel mehr gute Samen als schlechte Samen in den Kapseln von:

Cattleya elatior var. Leopoldi ♀ und Laelia purpurata ♂
  "   intermedia ♀ und Cattl. elatior var. Russeliana ♂.[12]

Ungefähr die Hälfte der Samen war offenbar gut in einer Kapsel von:

Cattleya elatior ♀ befruchtet mit Cattl. intermedia ♂.

Wenigstens $\frac{9}{10}$ der Samen waren schlecht in den Kapseln von:

Cattleya elatior ♀ gekreuzt mit Cattl. elegans ♂ und
Oncidium (aus Theresopolis)[13] ♀ gekreuzt mit Sigmatostalix ♂.

Eine ausserordentlich kleine Zahl von offenbar guten Samen wurde gefunden in den Kapseln von:

Cattleya elatior var. Russeliana ♀ und Epidendron cinnabarinum ♂
Notylia sp. ♀ und Ornithocephalus sp. ♂ (!).[14]

Alle Samen waren schlecht in den Kapseln von:

Cattleya elatior ♀ und Epidendron[15] Schomburgkii (?) ♂
Epidendron vesicatum ♀ und Epid. Schomburgkii (?) ♂
  "   "   ♀ und Epid. cinnabarinum ♂
Sigmatostalix sp. ♀ und Ornithocephalus sp. ♂.

Bei einer selbstbefruchteten Kapsel von Ornithocephalus fand ich, dass sie viel kleiner war und viel mehr schlechte Samen enthielt als eine andere Kapsel, die mit Pollen einer fremden Pflanze befruchtet war. . . . .

Was Ihre Frage über Geschlechtsverschiedenheiten bei niederen Thieren betrifft, so mag vielleicht eine unserer Amphipoden, Brachyscelus diversicolor F.M.,[16] hier zu nennen sein. Das Männchen dieser Art, welche auf einigen unserer grösseren

Acalepheen lebt (Rhizostoma cruciatum Less. = Rhacopilus cruciatus und cyanolobatus Agass. und Chrysaora Blossevillei Less. = Lobocrocis Blossevillei Agass.),[17]
unterscheidet sich nicht nur durch seine Antennen, deren erstes Paar sehr dick und
reichlich mit Riechhaaren versehen ist, während das zweite ausserordentlich lang ist
(dies zweite Paar fehlt den ♀ und den jungen ♂)—sondern auch durch seine Farbe.
Das Weibchen ist gewöhnlich von milchweisser oder mattgelblicher Farbe, das
Männchen dunkelröthlich-braun oder schwärzlich. Die Art hat aussergewöhnlich
grosse Augen, wie in der That die meisten Hyperina sie haben, und so ist es nicht
unwahrscheinlich, dass das Weibchen durch die Farbe des Männchens angezogen
wird.

Bei dieser Gelegenheit will ich hinzufügen, dass es im Meere bei Santa Catharina einen Fisch giebt, der sehr melodische Töne hervorbringt, welche auch dazu
dienen mögen, das andere Geschlecht anzulocken.[18] Die Töne klingen wie ent-
ferntes Geläute von Kirchenglocken. Ich habe sie nur an ruhigen Abenden gehört,
wenn diese Seemusikanten um einen Felsen dicht bei der Küste schwammen, aber
ich sah den Fisch nie.

Um noch einmal auf das Epidendron mit seitlichen fruchtbaren Antheren zurück-
zukommen, bei dem Pollenschläuche sogar noch vor dem Oeffnen der Blüthe
getrieben werden, so scheint die grosse Veränderlichkeit der seitlichen Antheren,
welche in einem Falle beobachtet wurde, der Ansicht günstig zu sein, dass sie erst
neuerdings erworben wurden in Folge eines zufälligen Rückschlags auf einen längst
verlorenen Character.[19]

Mit der grössten Hochachtung, glauben Sie, dass ich, werther Herr, aufrichtigst
der Ihrige bin | Fritz Müller.

Möller ed. 1915–21, 2: 130–1

[1] For a translation of this letter, see Appendix I. All Fritz Müller's letters to CD were written in English
   (see Möller ed. 1915–21, 2: 72 n.); most of them have not been found. Many of the letters were later
   sent by Francis Darwin to CD, who translated them into German for his *Fritz Müller: Werke, Briefe
   und Leben* (Möller ed. 1915–21). Möller also found drafts of some Müller letters among Fritz Müller's
   papers and included these in their original English form (*ibid.*, 2: 72 n.). Where the original English
   versions are missing, the published versions, usually appearing in German translation, have been
   used.
[2] The references are to Hildebrand 1867a and J. D. Hooker 1866a (see letters to Fritz Müller, 25 March
   [1867] and 22 April [1867]).
[3] In *Forms of flowers*, pp. 131–3, 135, CD reported Müller's information on heterostyly in *Hedyotis*, *Suteria*,
   and *Manettia bicolor*. *Suteria* is now subsumed within the genus *Psychotria*; *M. bicolor* is a synonym of
   M. luteo-rubra.
[4] CD mentioned Müller's observation of a white-flowered trimorphic *Oxalis* in *Forms of flowers*, p. 180.
[5] Müller had lived in Destêrro (now Florianópolis) from 1856, but had just returned to live on the
   Itajahy (now Itajaí) river near Blumenau, where he bought a new property in September 1867.
   Müller typically gave his address simply as 'Itajahy' until the mid-1870s, when he began to add
   'Blumenau', but only consistently gave his address as 'Blumenau' from 1878 (Möller ed. 1915–21, 3:
   94; see also West 2003, pp. 149–50).
[6] Müller had begun a series of crossing experiments with orchids earlier in the year (see letter from
   Fritz Müller, 1 April 1867).

[7] *Cattleya elegans* is a synonym of *Laelia elegans*.

[8] *Epidendrum glumaceum* is a synonym of *Prosthechea glumacea* (see Higgins 1997, p. 378).

[9] *Brassavola fragans* (named by Charles Lemaire; Möller mistranscribed 'Lem' as 'Leon') is a synonym of *B. tuberculata*.

[10] *Cirrhaea dependens* is a synonym of *C. viridipurpurea*.

[11] *Gomeza* is a synonym of *Gomesia*.

[12] *Cattleya elatior* is a synonym of *C. guttata*; the varieties mentioned by Müller are now *C. guttata* var. *leopoldii* and *C. guttata* var. *russelliana*.

[13] Now Queçaba (West 2003).

[14] Müller probably found it surprising that a cross between two fairly widely separated genera was possible. In orchids, natural hybrids are usually confined to intergeneric crosses within a subtribe, making a cross at this level rare (see Dressler 1981, pp. 147–8). *Notylia* is now a member of the orchid tribe Cymbidieae, while *Ornithocephalus* belongs to the tribe Maxillariae.

[15] 'Epidendron': a mistranscription of *Epidendrum*.

[16] For CD's query, see the letter to Fritz Müller, 22 February [1867]. The name *Brachyscelus diversicolor* was never published by Müller. For a description of all currently recognised species within the genus, see Vinogradov *et al.* 1996, pp. 488–96. Müller had discussed sexual differences in the antennae of *Brachyscelus* in F. Müller 1864, p. 53 (see also Dallas trans. 1869, pp. 78–9; CD annotated this section in his copy; see *Marginalia* 1: 609).

[17] *Rhizostoma cruciata* and *Rhacophilus cruciatus* are both synonyms of *Catostylus cruciatus* (Kramp 1961, p. 370). *Chrysaora blossevillei* is now considered to be a doubtful species (Kramp 1961, p. 324). Two species of *Chrysaora* have been recorded in Brazil, *C. lactea* and *C. plocamia* (Migotto *et al.* 2002, p. 23). It is likely, because of its size, that the species Müller refers to is *C. plocamia*.

[18] The fish has not been identified, but of thirty-four possible sound-producing species in the families Batrachoididae and Sciaenidae in Brazil, the Bocon toadfish, *Amphichthys cryptocentrus*, is the only reef-associated species known to occur at depths of a metre or less (Joseph Luczkovich, personal communication). CD mentioned musical sounds made by fish in *Descent* 2: 23, 331 (see also Pauly 2004, pp. 193–4).

[19] Müller first mentioned this orchid in a letter of 1 June 1866 that has not been found (see *Correspondence* vol. 14, letter to Fritz Müller, 23 August 1866). He gave an account of it in letters to Max Johann Sigismund Schultze (2 June 1866) and Hermann Müller (1 July 1866; the letters are reproduced in Möller ed. 1915–21, 2: 83–4, 86–9). Müller described it as a species of the orchid tribe Epidendreae with three fertile anthers, two in the outer whorl and one in the inner whorl (see Möller ed. 1915–21, 2: 87–8). He later published his observations in F. Müller 1868a, 1869, and 1870; CD added a reference to Müller's work in *Orchids* 2d ed., p. 148.

From Salt & Sons 17 July 1867

Shrewsbury
17th. July 1867

Dear Sir

   You will have been anxious to know we doubt not how the Mount house & Gardener's cottage stand as to prospect of sale & the present arrangements.—[1] The house is at present let temporarily terminable by a month's notice to M^rs^. G. B. Watton with a view to keep it aired &c at the rate of £11 a month you paying the taxes. Wynne the gardener has been allowed to retain the cottage rent free but he is quite incapacitated for work. His wife and sons have had charge of the house until M^rs^. Watton occupied it & have kept that and the grounds in order. They ask for some allowance as they cannot live in the house earning nothing and it seems reasonable something should be allowed them for work done if not for the future though of course they do not expect gardener's wages but desire to leave the matter in your hands. Their wages were £5·10/— a month.[2] M[r] F & M[r]. C. Parker[3] have been here to day & it is proposed to advertise the property for immediate sale offering the house first in one Lot, & if not sold then the whole property in four Lots. *Lot 1* the gardener's cottage, *Lot 2* the House, (described as being capable of being made into 2 good Dwellings) but excluding the kitchen garden which would be severed for a separate building lot and excluding the Hill Head Bank field above the house.

Lot 3 the Kitchen garden with an approah by S[t]. George's Church.

Lot 4 the Hill Head Bank field

Lots 1, 3, & 4 we should be sanguine of selling & the mode proposed would we think be the most likely way of selling lot 2 and would be more within the compass of an ordinary buyer.[4]

As regards Wynne M[r] F Parker suggests that if the property is sold something should be given to Wynne in one sum as finally settling with him. If not all sold you would know what is likely to be wanted from him for the future but he desires that you should express your wishes on the subject.

Both M[r] F. & M[r]. C. Parker would attend the sale for which a separate valuation would be forwarded to you & we should suggest a discretionary power as to the price of the lots should be left with them. A duplicate of this letter will be sent to each branch of the family & we shall be obliged if you will express any views that occur to you

We are Dear Sir | With Great Respect | Your obliged & faithful Servants | Salt & Sons

C R Darwin Esq

LS
DAR 177: 10

[1] The Darwin family residence, The Mount in Shrewsbury, Shropshire, was to be sold following the death of Susan Elizabeth Darwin, CD's sister, in 1866 (Freeman 1978). The house had been advertised for sale by auction on 30 November 1866 (*Shrewsbury Chronicle*, 16 November 1866), but only four portions of land near The Mount had been sold then (Harris 2004).

[2] Mrs G. B. Watton was probably Mary Anne Watton, a widow who later lived for a time at her brother's residence, Mount House, near the Mount (Census returns, Frankwell area, Shrewsbury, 1871). George Wynne was left £10 in Susan's will; a codicil added that those in her service at her death should receive half a year's wages in addition to any amount owed to them (Susan Elizabeth Darwin's will, Probate Registry, York). Wynne and his wife Margaret had three sons, John, Arthur, and Ambrose (Census returns, 1871, Frankwell area, Shrewsbury). See also the letter from E. A. Darwin, 17 [July 1867].

[3] Francis and Charles Parker, CD's nephews.

[4] In the event, all the remaining Darwin property at The Mount was bought by Edward Henry Lowe for £3450 (Harris 2004).

To Charles Lyell 18 July [1867]

Down. | Bromley. | Kent. S.E.
July 18.

My dear Lyell

Many thanks for y^r long letter. I am sorry to hear that you are in despair about y^r book; I well know that feeling but am now gettng out of the lower depths. I shall be very much pleased if you can make the least use of my present book, & do not care at all whether it is published before yours. Mine will appear towards the end of Nov. of this year; you speak of yours as not coming out till Nov. 68, which I hope may be an error.[1] There is nothing in my book about man which can interfere with you;[2] so I will order all the completed clean sheets to be sent (& others as soon as ready) to you. But please observe you will not care for the 1st vol., which is a mere record of the amount of variation; but I hope the 2nd will be somewhat more interesting tho' I fear the whole must be dull.

I rejoice from my heart that you are going to speak out plainly about species.[3] My book about Man if published will be short, & a large portion will be devoted to sexual selection, to which subject I alluded in the Origin as bearing on Man.[4] Many thanks about 6 fingered men, but that Chapter is finished.[5]

Tahiti is I believe rightly coloured; for the reefs are so far from the land & the ocean so deep that there must have been subsidence tho' not very recently: I looked carefully, & there is no evidence of recent elevation.[6] I quite agree with you versus Herschel on Volcanic I.s. W^d not the Atlantic & Antarctic volcanos be the best examples for you, as there there can be no coral mud to depress the bottom?[7] In my "Volcanic I." p 126 I just suggest that volcanos may occur so frequently in the oceanic areas, as the surface w^d be most likely to crack when first being elevated.[8] I find one remark p. 128 which seems to me worth consideration, viz. the parallelism of the lines of eruption in volcanic archipelagoes with the coast-lines of the nearest continent, for this seems to indicate a mechanical, rather than a chemical connection in both cases; ie the lines of disturbance & cracking. In my S. American Geology p. 185 I allude to the remarkable absence, *at present* of active volcanos on the E. side of the Cordillera in relation to the absence of the sea on this side.[9] Yet I must own I have long felt a little sceptical on the proximity of water being the exciting cause. The one volcano in the interior of Asia is said, I think, to be near great lakes; but if lakes are so important why are there not many other volcanos within other continents? I have always felt rather inclined to look at the position of volcanos on the borders of continents, as resulting from coast-lines being the lines of separation between areas of elevation & subsidence. But it is useless in me troubling you with my old speculations.

Rütimeyer sent me his book, but I have not even had time to cut the pages.[10]

I heartily wish you good progress with your book & remain | Yours affectionately | Ch. Darwin

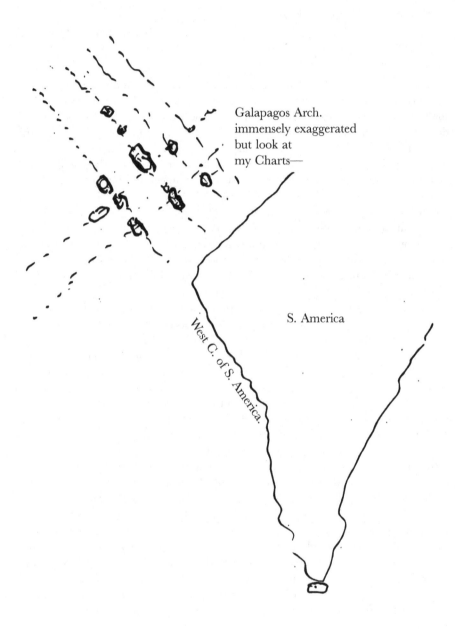

Galapagos Arch.
immensely exaggerated
but look at
my Charts—

S. America

West C. of S. America.

Endorsement: '1867'
LS(A)
American Philosophical Society (331)

[1] The letter from Lyell has not been found. Lyell was working on the tenth edition of his *Principles of geology* (C. Lyell 1867–8); the first volume was published in the first half of December 1866 and the second in the second half of March 1868 (*Publishers' Circular*). CD was working on the proof-sheets of *Variation*, which was published in January 1868 (Freeman 1977).

[2] CD had intended to publish a chapter on human descent as part of *Variation*, but decided to publish it instead as a separate work (see letter to J. D. Hooker, 8 February [1867]).

[3] Lyell devoted three largely new chapters, 35 to 37, of *Principles of geology* to natural selection (C. Lyell 1867–8, 2: 261–328). For CD's dismay regarding Lyell's discussion of 'species' in *Antiquity of man* (C. Lyell 1863), see *Correspondence* vol. 11, letter to Charles Lyell, 6 March [1863].

[4] CD's *The descent of man, and selection in relation to sex* (*Descent*) was published in 1871; it was in two volumes. CD's work on expression of the emotions, which he had intended to form one chapter of his 'essay on the origin of mankind' (see letter to Fritz Müller, 22 February [1867]), was published in 1872 (*Expression*). CD discussed sexual selection briefly in *Origin*, pp. 87–90. For CD's growing interest in sexual selection and human descent, see also the letters to A. R. Wallace, 26 February [1867] and [12–17] March [1867].

[5] CD discussed polydactylism in chapter 12 of *Variation* (*Variation* 2: 12–17).

[6] In the map (plate 3) in the front of CD's *Coral reefs*, Tahiti is coloured mostly light blue, indicating a barrier reef. Lyell may have queried whether it should not be red, indicating a fringing reef. CD pointed out in *Coral reefs* that the 'moat' between a barrier reef and the land was often filled with detritus and coral forming a type of fringing reef, as was reported to be the case with Tahiti, but he ignored this phenomenon for the purpose of colour-coding the map. Generally speaking, according to CD, barrier reefs occurred in deep water, where subsidence had taken place, and fringing reefs in shallower water, where the sea bed had remained stationary or been uplifted. See *Coral reefs*, pp. 119–20, 127, 152–3. CD discussed and rejected apparent evidence for the recent elevation of Tahiti in *Coral reefs*, pp. 138–9. Lyell did not alter his remarks on Tahiti in the tenth edition of *Principles of geology* (C. Lyell 1867–8, 2: 592), where he mentioned it as an example of a barrier or encircling reef. See also C. Lyell 1867–8, 2: 602–3, on the bearing of CD's map of coral reefs on zones of subsidence and their coincidence with lack of volcanic activity.

[7] John Frederick William Herschel had attempted to account for the proximity of chains of active volcanoes to the sea by suggesting that the carrying of solid matter off the land into the sea by rivers caused pressure on the sea-bed to increase while pressure on the land decreased; he proposed that lava erupted through the cracks where the earth's crust gave way to the opposing forces (Herschel 1866, pp. 1–46). In C. Lyell 1867–8, 2: 229, Lyell pointed out that active volcanoes rarely occurred near great river deltas, and added: 'The number, also, of active volcanos in oceanic islands is very great, not only in the Pacific, but equally in the Atlantic, where no load of coral matter or any sedimentary deposits derived from the weight of neighbouring lands can cause a partial weighting and pressing down of a supposed flexible crust.'

[8] In *Volcanic islands*, p. 126, CD asked: 'Do volcanic eruptions . . . reach the surface more readily through fissures, formed during the first stages of the conversion of the bed of the ocean into a tract of land?'

[9] In *South America*, p. 185, CD speculated that lava streams had flowed in the eastern Cordillera when the Atlantic reached that far to the west, and pointed out in a footnote the rarity of volcanic action except near the sea or large bodies of water, and the fact that there were now no active volcanoes on the eastern side of the Cordillera.

[10] See letter to Ludwig Rütimeyer, 4 May [1867].

To Henrietta Emma Darwin   26 July [1867]

July 26th

My dear Etty.—

You are a very good girl to wish for remaining slips of present chapter, but they are enormously altered & 10 folio pages of MS added, & the slips themselves have had to be cut into pieces & rearranged, so I will not send them.[1]

But for the future I shall be only too glad for you to see the slips, as well as Revises.— I will either keep, according to quantity finished, the whole of present chapter till your return, or send part to you.—

All your remarks, criticisms doubts & corrections are excellent, excellent, excellent

Yours affect[ly]. | C. D.

Endorsement: '1867'
DAR 185: 57

[1] Henrietta was reading the proof-sheets of *Variation*. She spent the second half of July visiting Devon and Cornwall; in a letter to her brother George Howard Darwin she mentioned having '11 pages of proof to do' (letter to G. H. Darwin, 23 July [1867], DAR 245: 280). CD paid Henrietta £20 for correcting the proofs (Browne 2002, p. 407). On Henrietta's contribution as editor to CD's work, see Browne 2002, pp. 347–9.

From J. D. Hooker [27 July 1867][1]

Kew
Saturday.

Dear Darwin

We returned last Monday, Mrs Hooker much the better of the trip.[2]
May I come to you on Saturday next 4th?[3]
Ever Yrs Aff | J D Hooker

DAR 102: 171

[1] The date is established by the relationship between this letter and the letter from J. D. Hooker, 4 July 1867. The Saturday before 3 August 1867 (see n. 3, below) was 27 July.
[2] Hooker had taken his wife, Frances Harriet Hooker, to Switzerland. They evidently returned a day earlier than planned; they had meant to return on Tuesday 23 July. See letter from J. D. Hooker, 4 July 1867.
[3] In 1867, 4 August was a Sunday.

To J. D. Hooker 29 July [1867]

Down
July 29

My dear Hooker

It rejoices my heart that you will come here next Saturday, & I hope nothing will prevent you, but remember should any thing occur you must not put yourself out to give me pleasure great though that will be.[1]

If you have Willy or Charley with you we should be most glad to see them.[2]
yours affectionately | C. Darwin
If you can lay your hand on Adam Bede bring it back.—[3]

LS(A)
Endorsement: '/67'
DAR 94: 143

[1] See letter from J. D. Hooker, [27 July 1867].
[2] CD refers to Hooker's sons Charles Paget and William Henslow Hooker.
[3] CD refers to Eliot 1859.

To William Bowman 30 July [1867][1]

Down. | Bromley. | Kent. S.E.
July 30

Dear Bowman

On the great principle that a man who has done one a kindness will probably do another I want to beg a little information from you.[2] Sir C. Bell says that when an infant screams violently, it closes the orbicular muscles so as to compress the eyes & prevent them becoming gorged with blood owing to the retarded circulation. He states that on opening the eyelids of a screaming infant he has actually seen the tissues become gorged with blood. He explains on this same principle of protection the firm closing of the eyes in violent coughing, sneezing &c.[3] Now as I have not elsewhere met with a similar remark I sh^d esteem it a great favour if you w^d inform me whether you have observed any thing of the kind, & believe in Sir C. Bell's statement.

Expression in animals & men is at present a hobby of mine & I think I shall probably utilize my notes made during several years.[4] This leads me to ask one other question; when any one (not short-sighted) looks intently at a distant object he generally contracts his eyebrows & the muscles going to the upper lip, which raises the cheeks & thus apparently compresses the eye slightly. Now do you suppose that these movements serve merely to contract the aperture of the eye & protect it from superfluous light, or does slight pressure aid the distinct vision of a distant object?

After long delay from ill-health I have at last printed $\frac{3}{4}$ of a book, including a chapter on inheritance, on which subject you formerly so much aided me.[5]

I hope you will forgive me for troubling you, & pray believe me yours very sincerely | Charles Darwin

LS(A)
DAR 261.11: 8 (English Heritage 8820 6060)

[1] The year is established by the relationship between this letter and the letter from William Bowman, 5 August 1867.
[2] Bowman, an ophthalmologist, had given CD information for *Variation* and had shown kindness to one of CD's sons (see *Correspondence* vol. 13, letter to William Bowman, 30 July [1865?], and *Correspondence* vol. 14, letter to William Bowman, 26 September [1866]).
[3] Charles Bell made these remarks in his *Nervous system of the human body* (Bell 1836), p. 175, and *Anatomy and philosophy of expression as connected with the fine arts* (Bell 1844), pp. 105–6. There is an annotated copy of Bell 1844 in the Darwin Library–CUL (see *Marginalia* 1: 47–9). See also *Expression*, pp. 158–9. For CD's doubt about Bell's view of the derivation of human facial muscles, see the letter to A. R. Wallace, [12–17] March [1867].

⁴ See letter to Fritz Müller, 22 February [1867], n. 11. *Expression* was published in 1872.
⁵ CD refers to *Variation*; chapters 12 to 14 are on inheritance. See also CD's 'Journal' (Appendix II). CD cited Bowman for information on inherited eye defects in *Variation* 2: 8–9, 79.

From J. D. Hooker 30 July [1867][1]

Kew
July 30th.

Dear Darwin

I hope to leave by the Victoria Train at 4.30. pm for Bromley on Saturday & shall be very glad to bring a boy with me.[2]

Indeed—*Indeed* I did return Adam Bede 2 years ago, I think that my wife took it with her to town to Q. Anne St. when you were there. You had asked me for it— if I remember right— —[3]

I wish you would return me "Tylor's early History of mankind" & my own copy of my "Himalaya Journals" with my own mss notes!—both of which I have lent ie. lost.[4]

Lyell was here yesterday very well & full of some "Insular" difficulty, that he is to propound to me.[5]

Ever yrs affec | J D Hooker

Pray remember there is no occasion to send a carriage— I can always get a fly if I should not prefer the walk.

DAR 102: 172–3

[1] The year is established by the relationship between this letter and the letter to J. D. Hooker, 29 July [1867].
[2] See letter to J. D. Hooker, 29 July [1867] and n. 2.
[3] Hooker had read *Adam Bede* (Eliot 1859) by April 1860; see *Correspondence* vol. 8, letter to J. D. Hooker, 18 [April 1860]. CD and Hooker also read and discussed several novels by George Eliot in 1865 (see *Correspondence* vol. 13). Hooker's wife was Frances Harriet Hooker; CD often visited his brother, Erasmus Alvey Darwin, at his house in Queen Anne Street, London.
[4] Hooker refers to Tylor 1865 and J. D. Hooker 1854. CD and Hooker read and discussed Edward Burnett Tylor's book in 1865 (see *Correspondence* vol. 13); CD cited it in *Variation* 2: 122 n. 22. CD cited J. D. Hooker 1854 in *Variation* 1: 259 n. 67, 307 n. 5, 356 n. 129.
[5] Charles Lyell included a chapter on insular floras (chapter 41) in the second volume of the tenth edition of his *Principles of geology* (C. Lyell 1867–8, 2: 402–32); he referred to Hooker's recent article on insular floras (J. D. Hooker 1866a; see C. Lyell 1867–8, 2: 419).

From Hermann Julius Meyer 30 July 1867

Bibliographisches Institut. | *Hildburghausen,*
den 30. July 1867

Dear Sir

You will have come in possession of the 3 volumes of Brehm's Zoological work, which we have forwarded to you accordingly to the directions of our friend Mr.

Kowalewsky in Petersburg. This day we sent the 4^(th). vol. which has just been finished. The same will be followed by the two last vols, in the course of a year.[1]

If you deem the work worth your examination, it will undoubtedly engage your interest, being the first description of animal life emanating from the principles, the discovery of which we owe to your genius.

for this reason you will find no objection, that Mr. Brehm has very often introduced your words verbaliter,[2] or at least refered to your works.

We think that to this new point of view the uncommon success of our work is due. Besides the several reprints of the original publication it is already in the course of publication in four different languages, viz. one in russian, one in french, one in italian and another in danish.[3]

We had not yet the opportunity to offer the work to an english publisher but should be very proud, if we could do so—with a recommendation from your part.

To our knowledge your theory has not yet been introduced into an english popular work of the extent of ours, and therefore an english edition of the same will not fail to answer the intentions of an english publisher and the taste of the english public, if it enjoys the favor of your authority.[4]

Leaving the matter to your kind attention we are, Sir, | your's very respectfully | The Bibliographic Institute | H J Meyer

DAR 171: 169

[1] Meyer refers to Alfred Edmund Brehm's *Illustrirtes Thierleben* (Illustrated animal life; Brehm *et al.* 1864–9) and to Vladimir Onufrievich Kovalevsky. Kovalevsky had arranged for the volumes to be sent to CD so that he could assess the suitability of the illustrations for inclusion in Kovalevsky's Russian translation of *Variation* (see letter to V. O. Kovalevsky, 24 June [1867] and n. 2). Four of the six volumes of Brehm *et al.* 1864–9 (dated 1864 to 1867) are in the Darwin Library–Down; they are annotated (see *Marginalia* 1: 69–71).

[2] 'Verbaliter': i.e. verbatim.

[3] The first Russian translation of Brehm *et al.* 1864–9 was [V. O. Kovalevsky] trans. 1866–70; the French was Gerbe trans. 1869–73; and the Italian was Branca and Travella trans. 1869–74. Several Danish works were derived from Brehm's publications; see Mariager trans. 1871 and 1873, and Mariager and Feddersen trans. 1879. On the popularity of Brehm *et al.* 1864–9, see Tort 1996.

[4] CD forwarded Meyer's letter to John Murray; see letter to John Murray, 4 August [1867]; see also *Correspondence* vol 16, letter to the Bibliographisches Institut, Hildburghausen, 8 June 1868. The first English translation of part of Brehm *et al.* 1864–9 was *Cassell's book of birds* (T. R. Jones trans. [1869–73]).

To Fritz Müller 31 July [1867][1]

Down. | Bromley. | Kent. S.E.

July 31.

My dear Sir

I received a week ago your letter of June 2. full as usual of valuable matter & specimens.[2] It arrived at exactly the right time, for I had just completed, & was enabled to correct a pretty full abstract of your observations on the plant's own pollen being poisonous. I have inserted this abstract in the proof sheets in my chapter on sterility, & it forms the most striking part of my whole chapter.[3] I thank you very sincerely for these most interesting observations, which however I regret that you did not publish independently. I have been forced to abbreviate one or two parts more than I wished—viz. about the fertility of the Epidendreae & on the advantage to the plant of the pollen being poisonous, for I was not quite sure that I understood parts of your letter.[4] The seeds which you have sent are very valuable.[5] Your letters always surprize me from the number of points to which you attend. I wish I c^d make my letters of any interest to you, for I hardly ever see a naturalist & live as retired a life as you in Brazil. The terrestrial Orchis with the pollen-staff seems very curious; my first impression was that it was allied to Spiranthes. With respect to mimetic plants I remember Hooker many years ago saying he believed that there were many, but I agree with you that it w^d be most difficult to distinguish between mimetic resemblance & the effects of peculiar conditions.[6] Who can say to which of these causes to attribute the several plants with heath-Erica-like foliage at the C. of Good Hope?[7] Is it not also a difficulty that quadrupeds appear to recognize plants more by their odour than their appearance? What I have just said reminds me to ask you a question: Sir J. Lubbock brought me the other day what appears to be a terrestrial planaria (the first ever found in the northern hemisphere) & which was coloured exactly like our dark coloured *slugs*. Now slugs are not devoured by birds like the shell-bearing species, & this made me remember that I found the Brazilian planariæ actually together with striped Vaginuli which I believe were similarly coloured.[8] Can you throw any light on this? I wish to know, because I was puzzled some months ago how it w^d be possible to account for the bright colours of the planariæ in reference to sexual selection. By the way I suppose they are Hermaphrodites?[9]

Do not forget to aid me, if in y^r power, with answers to **any** of my questions on expression, for the subject interests me greatly.[10]

with cordial thanks for y^r never failing kindness | believe me yours very sincerely | Charles Darwin

LS(A)
British Library (Loan 10: 17)

[1] The year is established by the relationship between this letter and the letter from Fritz Müller, 2 June 1867.

[2] See letter from Fritz Müller, 2 June 1867.

[3] CD described Müller's observations in *Variation* 2: 134–5, in a section headed, 'On certain hermaphrodite plants which, either normally or abnormally, require to be fertilised by pollen from a distinct individual or species'.

[4] CD refers to discussions in the letter from Fritz Müller, 1 January 1867.

[5] Müller had sent seeds of two species of *Coccocypselum* with his letter of 2 June 1867; he also sent seeds of *Oncidium flexuosum* resulting from a cross with a *Cyrtopodium*. These are still in a packet glued to the letter.

[6] See letter from Fritz Müller, 2 June 1867.

[7] James Philip Mansel Weale evidently noted this resemblance in an excised portion of his letter of 7 July 1867 (see letter to J. P. M. Weale, 27 August [1867]).

[8] John Lubbock was a neighbour of CD's; he documented his discovery in 'Note on the discovery of *Planaria terrestris* in England', *Journal of the Linnean Society (Zoology)* 10 (1870): 193–5. For CD's discovery of planarias in association with *Vaginulus* (a genus of slug) at Valparaiso, Chile, see *Correspondence* vol. 1, letter to J. S. Henslow, 24 July – 7 November 1834; CD also mentioned finding planarias in the Brazilian forest in his *Journal of researches*, pp. 30–2. See also R. D. Keynes ed. 2000. He discussed the species he found, and their similarity to *Vaginulus* in '*Planariae*'.

[9] CD later decided that the bright colours of planarias and other 'lower' animals, whether hermaphrodite or not, was not due to sexual selection (see *Descent* 1: 322).

[10] Müller's answers to CD's queries about expression have not been found, but they were evidently enclosed in the letter from Fritz Müller, [8 October 1867] (see n. 2 of that letter). CD had sent a copy of the queries with his letter to Müller of 22 February [1867].

From George Henslow [*c.* August 1867?][1]

Adderly Rectory[2] | Market Drayton | Salop

My dear Sir

I beg to thank you very much for your very interesting papers. I fortunately possess all the Vols of Gard: Chron: so I shall be able to refer to the papers at any time.[3]

I am very anxious to see your book: & indeed, ordered it some months back, thinking it was out; as I met with an advert! of it.[4] I hope also to read the new Ed^n. of the "Origin of Species"[5] when I get back to town.

Very faithfully yours | Geo Henslow

C. Darwin Esq

DAR 166: 148

[1] The date is conjectured from the address and the narrow mourning border, which also appear on the letter from George Henslow, 15 August 1867.

[2] The incumbent of Adderley Rectory from 1863 was Athelstan Corbet; he and Henslow had been contemporaries at Cambridge University.

[3] CD published numerous notes and queries in the *Gardeners' Chronicle* from 1841 onwards (see *Calendar*, pp. 583–4).

[4] Henslow probably refers to *Variation: Variation* had been advertised as early as 1865 (*Reader*, 15 April 1865, p. 427).

[5] Henslow probably refers to the fourth edition of *Origin*, which was published in November 1866 (*Publishers' Circular*).

From Charles Lyell 4 August 1867

73 Harley Street:
August 4, 1867.

My dear Darwin,—

I must write a word before starting to-morrow morning for Paris, to thank you for your last letter, and to say what a privilege I feel it to be allowed to read your sheets in advance.[1] They go far beyond my anticipations, both as to the quantity of original observation, and the materials brought together from such a variety of sources, and the bearing of which the readers of the 'Origin' will now comprehend in a manner they would not have done had this book come out first.[2] The illustrations of the pigeons are beautiful, and most wonderful and telling for you, and the comparison of the groups with natural families difficult to divide will be most persuasive to real naturalists. The rabbits are famously worked out, osteology and all.[3] The reason I have not got on faster is, that I have been correcting the press of my recast of Mount Etna, which I have reviewed twice since my former edition of fourteen years ago,[4] also the Santorin eruption of 1866, and my grand New Zealand earthquake, which produced more permanent change than any other yet known.[5] I have also had to rewrite my chapters on the 'Causes of Volcanic Heat,' the 'Interior of the Earth,' &c.[6] But all this is in the printers' hands, and I can now give myself to variation and selection.

Believe me, my dear Darwin, ever affectionately yours, | Charles Lyell.

K. M. Lyell ed. 1881, 2: 415–16

[1] Lyell went to Paris for the *Exposition universelle*, held from 1 April to 31 October 1867 (Mainardi 1987, p. 129); see K. M. Lyell ed. 1881, 2: 406. CD had arranged for the proof-sheets of *Variation* to be sent to Lyell (see letter to Charles Lyell, 18 July [1867]).

[2] CD had described *Origin* as an abstract for which he planned to give further evidence in future volumes (*Origin*, pp. 1–2). *Variation* was originally conceived as the first of these supporting volumes, and was to be followed by books on variation in nature and the principle of natural selection (see *Variation* 1: 1–14).

[3] *Variation* included twelve illustrations of pigeons and eleven of rabbits, among which were many drawings of bones. For CD's comparison of his own 'groups' of breeds of domestic pigeons with the genera of the family Columbidae, see *Variation* 1: 157.

[4] Lyell had visited Mount Etna in 1828, 1857, and 1858. He added material from a paper on the volcano (C. Lyell 1858) to the tenth edition of *Principles of geology* (C. Lyell 1867–8, 2: 1–47). The ninth edition had been published in 1853 (C. Lyell 1853).

[5] For Lyell's accounts of the eruption in the Gulf of Santorin (now Santorini) and the New Zealand earthquake of 1855, see C. Lyell 1867–8, 2: 69–71 and 82–9.

[6] See C. Lyell 1867–8, 2: 230–2, 209–13.

To John Murray 4 August [1867][1]

Down. | Bromley. | Kent. S.E.
Aug 4.

My dear Sir

I do not think that you often bring out German translations, but perhaps you will be so kind as to read over the enclosed note & *return* it to me.[2] The note explains itself, & why I feel any interest in the subject. I have not had time yet to do more than look at the book so I cannot speak of its merit; but its immense sale on the continent speaks strongly in its favour. It is illustrated *most* profusely by *excellent* woodcuts & Plates (uncoloured) which no doubt is a main attraction. It will consist of 5 thick Vols Royal 8vo.[3] The expense of getting up this book must have been prodigious. If by any chance you w[d] like to see the four vols. which I have, I c[d] send them to you.

My dear Sir | yours very sincerely | Ch. Darwin

P.S. I am making good progress with my book & have corrected the slips of exactly half the 2nd vol.[4]

LS(A)
John Murray Archive

[1] The year is established by the relationship between this letter and the letter from H. J. Meyer, 30 July 1867 (see n. 2, below).
[2] CD enclosed the letter from H. J. Meyer, 30 July 1867. Hermann Julius Meyer had asked CD to recommend Alfred Edmund Brehm's *Illustrirtes Thierleben* (Brehm *et al.* 1864–9) to an English publisher for translation.
[3] Royal octavo.
[4] CD was correcting the proof-sheets of *Variation*.

To W. B. Tegetmeier 4 August [1867]

Down. | *Bromley.* | *Kent. S.E.*

Aug 4.

My dear Sir

I am much obliged for the proofs which I have been glad to see, tho' my Chapter on Pigeons is printed off. Your new book will be I have no doubt as good as your last & more than that no one need desire.[1]

I saw with much surprize that M[r] Brent is dead, & thus an eccentric but worthy man is lost.[2] I sh[d] have liked to hear something about his death.

I hope you have not quite forgotten the experiment which you so very kindly offered to try about dying pigeons with bright colours. I feel much curiosity on this head.[3]

My book will be published in Nov. & I need not say that a copy will of course be sent to you—[4]

My dear Sir | yours very sincerely | Ch. Darwin

LS
Endorsement: '1867'
Archives of The New York Botanical Garden (Charles Finney Cox collection) (Tegetmeier 105)

[1] Tegetmeier evidently sent CD proof-sheets of his new book on pigeons (Tegetmeier 1868). CD refers to chapters 5 and 6 of *Variation* (1: 131–224), and to Tegetmeier's *Poultry book* (Tegetmeier 1867).
[2] The reference is to Bernard Peirce Brent, whom CD had once described as 'very crotchetty' (see *Correspondence* vol. 13, Supplement, letter to W. B. Tegetmeier, 8 May [1861]). Brent sent CD information on domestic birds, and is frequently cited in *Variation*.
[3] See letter to W. B. Tegetmeier, 30 March [1867] and n. 6.
[4] *Variation* was published in January 1868 (Freeman 1977). Tegetmeier's name appears on CD's presentation list for *Variation* (DAR 210.11: 33).

From William Bowman 5 August 1867

5 Clifford St
Aug. 5 1867

Dear D⟨arwin⟩

I have ⟨referred to⟩ Sir C Bells ⟨wo⟩rk ⟨in⟩ regard to the subject of your query.[1] I suppose there is no doubt that in violent sneezing &c the eyes are protected by the distinctive closing of the lids— and that when the closure is Spasmodic & strong, the vessels of the eyeballs, tending to be engorged under the strain put upon them are mechanically supported by the External Compression and all the more if the organ by a certain degree of engorgement of its ⟨own⟩ ⟨& the⟩ neighbouring vessels begins ⟨to be made more⟩ pro⟨minent for then⟩ it wou⟨ld⟩ be more ⟨exposed to th⟩e pressure of the ⟨of the orbicularis⟩ muscle. I should ⟨be in⟩cline⟨d⟩ to place little reliance on the experiment related by Sir Charles, as to opening the lids of a screaming infant & finding the eyes at first pale becoming red (if that be his meaning)—because the mechanical restraint & pressure of the fingers at the moment of inspection would be likely to modify the condition—[2] Sir C. Bells remarks are always ingenious & often full of meaning, but he was ⟨perhaps not always a rigidly⟩ exact experimenter.

As ⟨to your other question,⟩ I should ⟨say⟩ that ⟨in intent looking whether at near⟩ or distant ⟨objects the contraction⟩ of the brows & of some other parts near the eye under the influence of intense, attention given to the perceptions of the sense—is attended with (often) some aid to acuteness of sight by partial occlusion of the pupil—& may partly sustain the consensual movements of the two eyes—by giving a firmer external support while the globes are brought to binocular vision by their ⟨own pr⟩oper muscles—[3]

⟨I am⟩ delighted to hear you ⟨are⟩ likely soon to give us som⟨e more of⟩ your deep thoughts ⟨and wide gen⟩eralizations.[4] Be sure no one more appreciates your labours than myself.— I think your views the only rational ones on the grand subject of the course of life upon our planet. You seem to be on the high road to a future which yet I suppose will be ever distant from the most advanced philosophers of human mould. But with long looking towards things unseen as yet our power of sight may improve with improvement of the material organs of mental & physical sight & our race may know when it is come to somewhat higher than its at present.

Yrs ever most sincerely | W. Bowman

DAR 160: 267 (fragile)[5]

CD ANNOTATION

Top of letter: 'Frowning' *blue crayon*[6]

---

[1] The reference is to Charles Bell and to Bell 1836 and 1844. CD had asked for Bowman's opinion on some of Bell's observations on crying infants (see letter to William Bowman, 30 July [1867]). In *Expression*, p. 171, CD acknowledged Bowman for providing information on the eye.

[2] See letter to William Bowman, 30 July [1867] and n. 3.

[3] CD had asked whether contraction of the brow when looking at a distant object might limit the amount of light received by the eye, or whether the added pressure might improve vision (see letter to William Bowman, 30 July [1867]). CD quoted Bowman on this point in *Expression*, p. 227.

[4] Bowman refers to *Variation*.

[5] The content of missing or illegible text in the original letter has been supplied from a copy (DAR 261.11: 9).

[6] CD used information from this letter in discussing the origin of frowning in *Expression*, pp. 224–8.

From John Murray 5 August [1867][1]

50, Albemarle S! | *W.*
Aug[t] 5

My Dear Sir

I thank you for sending me M[r] Meyers letter from Hildburghausen & return it as desired.[2] I am sorry to say I am unacquainted with Brehm's work to wch it relates & unless you do me the further favor to lend me one or two Vols of it I shall be unable to return an answer[3]

I am glad to find that you are getting on so well with the revision of your own work & I gather from that that your health is improved[4]

Sir Ch. Lyell was rejoicing in the privilege of being allowed to read your proofs—wch gave him great satisfaction[5]

Yours Dear Sir very | sincerely | John Murray

Chas Darwin Esq

DAR 171: 349

[1] The year is established by the relationship between this letter and the letter to John Murray, 4 August [1867].

[2] CD had enclosed Hermann Julius Meyer's letter of 30 July 1867 with his letter to Murray of 4 August [1867].

[3] Murray refers to Alfred Edmund Brehm and Brehm *et al.* 1864–9, of which four volumes (of an eventual six) had appeared.

[4] CD was correcting proof-sheets of *Variation*.

[5] See letter from Charles Lyell, 4 August 1867. Murray was Lyell's publisher as well as CD's.

From Robert Swinhoe 5 August 1867

Amoy—
5 August, 1867.

My dear M[r]. Darwin

Your letter of 27 February with its enclosure on human expression has a long time been before me;[1] and has my attention— I am taking long to make observations, but it is your wish that such should not be made in a hurry. To make them as complete as I can I am enlisting the services of Surgeons, Missionaries &c The skin on a Chinaman's countenance is so tightly stretched that it is often difficult to detect the wrinkles, and it is the study of the mandarins & literati to affect a stolid look, often under the most trying circumstances. The lower classes will I expect give the best opportunities for experiment. I will do my best for you any way.

I have taken the following note for you under date 13 June. Mr. Gisbert, the Spanish Consul at Amoy,[2] visited me this afternoon. He ⟨   ⟩s me that when he was an engineer on the roads in Spain some years ago, he was fond of shooting and roaming about the country. That in the *Sierra Morena*, a Strawberry-tree (*Arbutus unedo?*) was very abundant, and bore large quantities of a red fruit like fine large red strawberries. These gave quite a glow to the woods. The spot on the mountain chain he speaks of is on the divisional line between the provinces of Seville and Badajos.[3] Under these trees Hedgehogs occurred innumerable and fed on the fruit, which the Spaniards call *Madroño*.[4] He has often seen an *Erizo* (Hedgehog)[5] trotting along with at least a dozen of these Strawberries sticking on its spines. He supposes that the Hedgehogs were carrying the fruit to their holes to eat in quiet and security, and that to procure them the Hedgehogs must have rolled themselves over the fruit which was scattered in great abundance all over the ground beneath the trees.

Let me call your attention to the *Red* domestic Pigs in the Zoological Gardens— They are the Pigs and only *domestic* animal of the Formosan Savage. I have seen white patched and white varieties, but the Red is the prevalent colour. It seems to me that these pigs are direct descendants from the wild pig of the Formosan Mountains *Sus taivanus*;[6] but I have found them breed freely with the curious long-haired Pig of North China.[7] The red mother in two cases produced each time 8 young, all *black* like their sire. There has not been time yet to see if the offspring are prolific.

The wild pig of Formosa is blackish-brown with a whitish moustache streak. The latter disappears or shows very faintly in the domestic state. The colour of these animals would seem to show that with them at least *erythrism* is the first stage in domestication as in the gold-fish; *albinism* the second; at *melanism* they have not yet arrived— In the sty they are shy and wildish.

The Chinese in Formosa import the black long-faced and curved-back breed of Amoy and have not much fancy for the savages' breed.

Dr Sclater on noting the arrival of these Pigs from Formosa unfortunately stated that they were the wild Formosan Pig, and that he therefore could not understand the fact of one of them having a *white* patch.[8] We have *also* sent the wild pig, which I hope will reach safe and give an opportunity for comparison—

I sent a *Bear* from Chefoo (N. China)[9] noting how like it was to & probably the same as the Himalayan (*Ursus tibetanus*). I suppose *Chefoo* (one of the ports open to trade) was taken for some place in Formosa, & the Society were informed that it was sent to illustrate my *Ursus formosanus* but unfortunately turned out to be *U. tibetanus*.

I have lately sent a true Formosan Bear to put the matter straight. I think no one will deny that it is a good species.[10] Insular forms of *Bears, Flying Squirrels, Jays* & many other animals, you can with almost certainty indicate as distinct from those of the adjacent main.[11]

I trust your health is fast reestablishing, and that the work we are all looking forward to will appear ere long—[12]

Your's very sincerely, | Robert Swinhoe

Charles Darwin

DAR 177: 330

CD ANNOTATIONS
1.1 Your letter ... trees. 2.14] *crossed pencil*
2.1 Mr. Gisbert,] *after opening double quotes, pencil*
2.2 visited ... He] *crossed pencil*
Top of letter: 'As you have lately published an account of Hedgeh wch carry apples *[illeg]* on their spines, you may think the following statement sent me by' *pencil, del pencil* | 'Formosa Pigs' *pencil*

[1] CD evidently sent Swinhoe a handwritten copy of his queries about expression. Neither the letter nor the original enclosure has been found. Swinhoe had the enclosure published in *Notes and Queries on China and Japan* 1 (1867): 105; this published version is reproduced in this volume as the enclosure to the (missing) letter to Robert Swinhoe, [27 February 1867].

[2] Mr Gisbert has not been further identified. See letter to *Hardwicke's Science Gossip*, [before 1 December 1867]. Amoy (now Xiamen) is located in Fujian Province at the mouth of the Amoy (now Jiulong) river (*Columbia gazetteer of the world*).

[3] The Sierra Morena is a mountain range, parts of which are in the provinces of Seville and Badajoz, in south-western Spain (*Columbia gazetteer of the world*).

[4] Madroño: the Spanish name for the tree *Arbutus unedo*.

[5] Erizo: the Spanish name for the common hedgehog, *Erinaceus europaeus*.

[6] Swinhoe had earlier described the Formosan mountain pig as *Porcula taivana* (see Swinhoe 1862, pp. 360–1), but later amended his description, calling the species *Sus taivanus* (see *Proceedings of the Zoological Society of London* (1864): 383). *Sus taivanus* is now considered a sub-species of *S. scrofa* (see Oliver ed. 1993, p. 113). Formosa (now Taiwan) has both coastal and inland mountain ranges.

[7] Swinhoe is probably referring to the Chinese subspecies now designated *Sus scrofa moupinensis* (Oliver ed. 1993, pp. 112–13).

[8] Philip Lutley Sclater was secretary of the Zoological Society of London (*DSB*). For Sclater's account of the pigs sent to London from Taiwan, see *Proceedings of the Zoological Society of London* (1866): 419.

[9] Chefoo (now Yantai) is situated on the north coast of the Shandong peninsula, China. It was opened to foreign trade in 1862 (*Columbia gazetteer of the world*).

[10] Swinhoe had initially described the Formosan black bear as probably a distinct species from the Himalayan black bear (see Swinhoe 1862, pp. 351–2), but after examining a number of skulls of both forms, he concluded that they belonged to the same species. The bear that Swinhoe had just sent to the London Zoological Society Gardens survived and, according to Abraham Dee Bartlett, could not be distinguished from the Himalayan form (see Swinhoe 1870, p. 621). The species is now known as the Asiatic black bear, *Ursus thibetanus* (Nowak 1999, pp. 681–2).

[11] Swinhoe described two species of flying squirrels native to Formosa (Swinhoe 1862, pp. 358–9). He also described the jay native to Formosa as a distinct species (Swinhoe 1863, p. 386).

[12] Swinhoe refers to *Variation*, published in January 1868.

To John Murray 6 August [1867][1]

Down
Aug 6.

My dear Sir

I enclose the 4 large vols.[2] After you have considered them will you be so kind as to return them to the enclosed address. in the same box.—

Yours sincerely | Ch. Darwin

LS(A)
S. J. Hessel (private collection)

[1] The year is established by the relationship between this letter and the letter from John Murray, 5 August [1867].
[2] CD had recommended Brehm *et al.* 1864–9, four volumes of which had been sent to him by the publisher, for translation. See letter from H. J. Meyer, 30 July 1867, letter to John Murray, 4 August [1867], and letter from John Murray, 5 August [1867] and n. 3.

To William Bowman 7 August [1867][1]

Down. | *Bromley.* | *Kent. S.E.*
Aug. 7th

Dear Bowman

Very many thanks for your kind note & clear answers to my queries, which are valuable to me.—[2] I hope to do a little more work in natural science,—much is out of the question—but I agree with you entirely that no one can say how glorious the future prospect of knowledge is.— What an instance the spectroscope is![3]

With hearty thanks for your uniform kindness— believe me | Yours very sincerely | Ch. Darwin

DAR 261.11: 10 (English Heritage 8820 6062)

[1] The year is established by the relationship between this letter and the letter from William Bowman, 5 August 1867.
[2] See letter from William Bowman, 5 August 1867 and n. 1.
[3] The spectroscope (invented in the late 1850s by Gustav Robert Kirchhoff and Robert Wilhelm Eberhard Bunsen) made possible the chemical analysis of stars as well as providing a highly accurate means of identifying the presence of even minute amounts of chemical substances (*DSB* s.v. Kirchhoff, Gustav Robert, and Bunsen, Robert Wilhelm Eberhard).

To Carl Vogt 7 August [1867][1]

Down. | *Bromley.* | *Kent. S.E.*
Aug 7.

Dear Sir

I thank you very sincerely for your kind present of your Mémoire sur les Micro-céphales, which I had intended ordering for I had received from M. de Quatrefages

Hermann Müller
Courtesy of Heiner Kresse

George Henslow
Courtesy of Paul Akerhielm

Carl Vogt
Courtesy of the Stazione Zoologica 'Anton Dohrn'
(Historical Archives, ASZN:La.9)

Vladimir Onufrievich Kovalevsky, 1869
Davitashvili 1951, p. 125
By permission of the Syndics of Cambridge
University Library

his report.[2] I have not had time as yet to read the work, but I could not resist carefully reading the Chapter on Genèse, & it has interested me extremely,[3] It is really curious how closely we have considered the same classes of facts, & have come to similar conclusions about atavism &c. The proofs of my chapter on this subject are corrected; & this I regret for your admirable illustration of the Aphis had not occurred to me, & I should have much liked to have quoted it from you.[4]

I am sure I shall feel deep interest in the whole work.

With my best thanks & sincere respect, I remain | Dear Sir | yours very faithfully | Charles Darwin

LS
Bibliothèque Publique et Universitaire de Genève, Ms fr. 2188, f. 302–3

[1] The year is established by the relationship between this letter and the letter from Carl Vogt, 17 April 1867.

[2] The references are to C. Vogt 1867, Armand de Quatrefages, and Quatrefages de Bréau 1867, a review of Vogt's work. CD's annotated copies, bound together, are in the Darwin Library–CUL (see *Marginalia* 1: 824–6). Vogt had earlier promised to send a copy of his work (see letter from Carl Vogt, 17 April 1867).

[3] CD refers to the final chapter of C. Vogt 1867, on the origin or causes of microcephaly as these relate to the laws of heredity (C. Vogt 1867, pp. 187–200). CD cited C. Vogt 1867 in *Descent* 1: 57, 121.

[4] Vogt cited aphids because they seemed to show reversion to characteristics of an earlier generation under certain environmental conditions. He concluded that the transmission of latent characters in higher animals was a similar phenomenon, although the causes were as yet not understood (Vogt 1867, pp. 191–3). In *Variation* 2: 47–61, CD discussed causes of reversion and transmission of latent characters.

To Asa Gray 8 August [1867][1]

> Down, Bromley, Kent,
> August 8.

My dear Gray,—

I have been glad to see Mr. Canby's interesting letter on Dionæa, and I thank you for sending it; but unfortunately the facts are not new to me.[2] Several years ago I observed the secretion of the "gastric juices" and the close adhesion of the two sides of the leaf when a fly was caught. I keep my notes in such an odd fashion that it would take me some time to find them. I am almost sure I ascertained the acid reaction of the secretion and its antiseptic power, but I cannot remember whether in this, or in analogous cases, I found its subsequent reabsorption.[3] This letter fires me up to complete and publish on Drosera, Dionæa, etc., but when I shall get time I know not. I am working like a slave to complete my book.[4]

Incomplete
J. L. Gray ed. 1893, 2: 557

[1] The year is established by the relationship between this letter and the letter from Asa Gray to J. D. Hooker, [after 6 July 1867].

² CD refers to a letter from William Marriott Canby to Asa Gray of 6 July 1867, in which Canby described his experiments with the insectivorous plant *Dionaea* (see enclosure to letter from Asa Gray to J. D. Hooker, [after 6 July 1867]).

³ See *Insectivorous plants*, p. 296. Most of CD's extant notes on his experiments with insectivorous plants, dated between 1860 and 1875, are in DAR 54 to 61. Notes made in September 1861 on the application of various substances to the leaves of *Dionaea* are in DAR 54: 50–61. See also *Correspondence* vol. 8.

⁴ CD was working on the proof-sheets of *Variation*.

From Seth Sutton   8 August 1867

Zoological Gardens | Regents Park
Aug. 8ᵗʰ. 67

Sir,

In accordance with your request, I beg to say that I have made the following observations respecting the animals you refered too.¹ the Chimpanzee and Orang, when they Cough or sneeze they then shut their eys. When screaming violently they *do not* shut their eys, and if suddenly alarmed they then erect their hair. At extraordinary noise, such as *Thunder—hamering* &c, &c, they become excited, and if teased or anoyed the same. When they listen or are astonished their mouths are *Closed.*²

The Anubis

When excited the hair stands erect from the back part of the neck to the loines, the other part (the rump) remaining quite smooth³

Nither the Chimpanzee nor the orang use their lips as an organ of touch when examining any strange object—

In Conclusion I beg to say if theres anything else you require notice off I shall be *pleased* to attend ⟨to it⟩—

I am Sir your obdᵗ Servant.—

S. Sutton

C. Darwin Esqʳ.

DAR 177: 322

CD ANNOTATIONS

1.1 In accordance ... refered too 1.2] *crossed pencil*
1.2 the Chimpanzee] 'T' *over* 't' *pencil*
1.3 When screaming ... eys, 1.4] *double scored red crayon,* '(good do not cry)' *pencil, underl red crayon*
1.4 and if] *after closing square bracket, pencil*
1.4 and if ... the same. 1.6] *crossed ink*
1.6 When they] *after opening square bracket, ink*
1.6 When they ... attend ⟨to it⟩— 5.2] *crossed pencil*
1.7 *Closed*] *before closing square bracket, pencil*
2.1 The Anubis] *scored blue crayon*
2.1 The Anubis ... smooth 3.2] *crossed pencil*
3.1 When excited] *underl blue crayon*
Top of letter: 'Animals crying' *pencil*

[1] CD evidently requested that Sutton make observations on expression in apes and monkeys at the London Zoological Society Gardens, where he was a keeper. CD may have made the request during his visit to London from 17 to 24 June 1867 or through his son Francis Darwin (see CD's 'Journal' (Appendix II)).

[2] CD referred to these observations in *Expression*, pp. 95, 145, and 163.

[3] Sutton refers to the Anubis baboon, *Papio anubis*. CD mentioned observing the erection of hair in an angry baboon in *Expression*, p. 96.

To B. D. Walsh 9 August [1867]

Down, Bromley, Kent

Aug. 9th.

My dear Sir

I am very much obliged for your note & for the Practical Entomologist.[1] I received your last paper & read it carefully & have just looked at it & find many passages marked, amongst others the concluding paragraph; but I am not sure that I think so much of this argument as of some others which you have advanced.[2] I must say I am very glad to hear that you are going to give up your Journal for it must have been a very heavy burden, especially of late with your wife in such a suffering state.[3] You will also now have more time for science.

With respect to the duplicate of the "Origin" I should rather like it to be sent to D^r. Leidy, the Paleontologist;[4] but if there is any one else to whom you would like to send it, pray do so. I have been working very hard at my new book & I have no brains left, so you must excuse the stupidity of this letter, & the circumstance that I cannot say positively whether I received your letter of Feb 25, but if I did receive it it is safe for future reference in one of my portfolios.[5] I do not remember ever receiving an unpaid letter from you. I am sorry to say I do not know the name of the oak gall which has spread throughout England.[6] I was much interested by the passages which you marked in the Prac: Entomol: & have one question which I should be much obliged if you would sometime answer; it is, are you sure that the Lucanidæ use their great jaws to hold the females in copulation; I always thought that they used them in fighting with other males, & I am nearly sure that this is the case.—[7]

My dear Sir | Yours very sincerely | Ch. Darwin

LS(A)
Postmark: AU10 67
Field Museum of Natural History, Chicago (Walsh 15)

[1] Walsh's letter has not been found. Walsh evidently enclosed copies of *Practical Entomologist* for May and July 1867 (see n. 7, below). The copies have not been found in the Darwin Pamphlet Collection–CUL.

[2] The reference is to the second half of a paper on gall insects (Walsh 1866). CD probably received the paper early in 1867 (see *Correspondence* vol. 14, letter to B. D. Walsh, 24 December [1866]). CD's heavily annotated copy is in the Darwin Pamphlet Collection–CUL; most of the final paragraph was scored. In it Walsh wrote: 'Surely, therefore, upon general principles, a hypothesis, which accounts clearly and satisfactorily for a great mass of phenomena, is more likely to be a correct one, than a

hypothesis which accounts for nothing, and, while it mercifully spares our Reasoning powers, draws most largely and exorbitantly upon our Faith' (Walsh 1866, p. 288).

[3] The *Practical Entomologist*, which was edited by Walsh, ceased publication with the double issue for August and September 1867 (vol. 2, nos. 11 and 12). Walsh's wife was Rebecca Walsh.

[4] Walsh evidently received two copies of the fourth edition of *Origin* (see *Correspondence* vol. 14, letter to B. D. Walsh, 24 December [1866]). Joseph Leidy had written to CD expressing support for the theory of natural selection (see *Correspondence* vol. 9, letter to Joseph Leidy, 4 March [1861]). It is not known whether Leidy received Walsh's spare copy of the book.

[5] CD was working on the proof-sheets of *Variation*. CD kept a number of portfolios, amassed over a long period of time and since dispersed. The likely contents of some portfolios were reconstructed when some of CD's papers were catalogued in 1932 (see DAR 220: 113). Only a fragment of the original letter from Walsh of [25 February 1867] has been found, suggesting that CD may have put an excised portion of the letter into a portfolio.

[6] CD had mentioned a new gall, remarkable because it attached 'not to the leaf but to twigs' in his letter to B. D. Walsh of 20 August [1866] (*Correspondence* vol. 14).

[7] The reference is to Walsh's statement that male horn-bugs (family Lucanidae) used their enlarged jaws to grasp females during mating; Walsh made this statement in a two-part paper that appeared in the issues of *Practical Entomologist* for May and July 1867 (*Practical Entomologist* 2 (1867): 88, 107). CD's copies have not been found. CD cited these pages of *Practical Entomologist* in *Descent* 1: 342, but added that the jaws were probably also used for fighting.

From Andrew Murray   12 August 1867

67 Bedford Gardens | Kensington
12 August 1867

My dear Sir

I am going to do something w^h I feel sure must give you pleasure.— I take it that you have spent more time than any other man in picking the kernels out of books of travels—and that you must be more sensible than others to the great advantage w^h a regular record of the Nat: Hist: facts scattered thro' books of travels would be of to the Naturalist—

I am going to make the somewhat desperate attempt to revive the Natural Hist^y Review—or rather to start another Journal of the same stamp, and to help to make it go I shall add travels to the subjects reviewed—and altho I shall not confine such reviews to Nat: Hist: topics, practically they will be reviewed from the Nat. Hist. point of view.—[1]

Fancy what an Ease it would have been to you to have had a Journal of Travels extending over a number of years in w^h the Nat: Hist^y points incidentally touched on in each were all buoyed off & all you had to do was to run your Eye down the Index— For me a mere list of travels in Each Country (like Engelmanns German one)[2] would have been often most valuable—

Now as one good turn deserves another, even when done not expressly for our behoof, and here I assure you the advantage w^h such a record would be of to such studies as yours was the thing that put the idea in my head, I have come to ask *the other*.— It is to point out to me from your past experience the way in w^h I should work the idea so as to be most useful— of course I have my own notions on the subject derived from my own past Experience but as yours has been so vastly more Extensive, not to speak of other good causes, I turn to you as a better guide

I propose to eliminate from this Journal all heavy original articles many of w^h. in the old N H. Review were only fitted for the pages of Scientific Transactions, & to have no laborious Essays in shape of Reviews, but simply a fair account of the contents of the book reviewed— Corresp^cc. from travellers & collectors abroad and original papers from those at home containing isolated facts or special information or new ideas with a miscellaneous corner will complete the contents

Believe me my dear Sir Yours very truly | And^W. Murray

DAR 171: 328

[1] The *Natural History Review* ceased publication with volume 5 in 1865 (see A. Desmond 1994–7, 1: 342–3). The journal that Murray started was the *Journal of Travel and Natural History*; only one volume was published, containing issues from 1868 and 1869.

[2] The reference is to the German publishing firm Wilhelm Engelmann, whose titles included many travel and natural history publications. The list of travels referred to has not been identified.

From George Henslow 15 August 1867

Adderley Rectory | Market Drayton | Salop.

Aug 15/67

My dear Sir

I have found a field here abounding in the common *Genista tinctoria* in full blossom & on examination of the opening buds; it appears that the keel *&* *wings*, have the same peculiarity of curling backwards by their claws on being touched as the Indigo which was growing in your Green-house; though in that instance *the claws of the Alæ* do not curl but only the keel.

I think of making a note of the fact, for the Linnean,[1] & should much like,—if you will kindly permit me—as you did before with reference to the structure of Broom—to describe the peculiar process of the French bean, to which you called my attention when I saw you.[2]

Of course I am presuming it has not been described; if *it has*, there would be no need to do it again.

Trusting your health improves | believe me | My dear Sir | to be ever faithfully yours | George Henslow

Ch. Darwin Esq.

DAR 166: 163

[1] Henslow published his observations on *Genista tinctoria* in the *Journal of the Linnean Society (Botany)* (Henslow 1868).

[2] There is no reference to the French bean (*Phaseolus vulgaris*) or information from CD on its structure in Henslow 1868. Henslow had mentioned CD's observations on *Phaseolus* in a paper on *Medicago sativa* (Henslow 1865). In a paper on *Indigofera*, Henslow quoted a communication from CD on the

structure of the flower in the common broom, *Cytisus scoparius* (Henslow 1866; see also *Correspondence* vol. 14, letter to George Henslow, [before 19 April 1866] and n. 2). Henslow last visited Down on 2 and 3 April 1866 (Emma Darwin's diary (DAR 242), *Correspondence* vol. 14, letter to J. D. Hooker, 4 April [1866]).

To Fritz Müller 15 August [1867][1]

Down. | Bromley. | Kent. S.E.
Aug 15th

My dear Sir

I have just received the seeds of the Adenanthera for a necklace, & it has been very kind of you to remember my little wish.[2] Mr. D. Hanbury says you have given him valuable information & specimens.[3] I know him very little, but Dr. Hooker thinks highly of him.—[4]

I know very well that I am quite unreasonable in making so many requests & asking so many questions; but there is one point, about which I am very curious & *possibly* you might obtain information for me.— Some American monkeys cry a little (Rengger & Humbolldt):[5] now when they cry from grief or pain, do they wrinkle up & close their eyes like a *Baby always does when screaming & crying*? When they cry, especially whilst young, do they scream & make loud noise?

It is possible that monkeys may be kept tame in S. Brazil. & you could ascertain this.—

I have lately received an admirable letter from your Brother on the fertilisation of Orchids: he describes, writes English & draws almost as well as you![6]

Borreria-seed are come up—[7] Plumbago, Cordia, the tall Lobelia & Gesneria all growing well.[8] Most of other seeds have failed. The Oxalis from the bulbs, flowered, but all utterly sterile, so I could make nothing out about them.[9] After I have experimented with these plants Dr. Hooker wants them for Kew.—[10]

Forgive me for being so very troublesome & believe me | Yours very sincerely | Ch. Darwin

I have written to *India* to try & find out how Adenanthera is disseminated.—[11]

British Library (Loan 10: 18)

[1] The year is established by CD's reference to his having written to India (see n. 11, below).

[2] In the letter to Fritz Müller, [before 10 December 1866], CD had written, 'These seeds would make a beautiful bracelet for one of my daughters if I had enough' (*Correspondence* vol. 14).

[3] Daniel Hanbury had asked CD for Müller's address in order to obtain information on 'pharmacological matters'; Hanbury worked on the botanical origin of drugs (see *Correspondence* vol. 14, letter from Daniel Hanbury, 1 December 1866; see also Ince ed. 1876).

[4] Joseph Dalton Hooker had travelled to Syria and Palestine with Hanbury on a botanical expedition in 1860 (L. Huxley ed. 1918, 1: 528).

[5] The references are to Johann Rudolph Rengger and Alexander von Humboldt and to Rengger 1830 and Humboldt 1814–29. CD cited their observations on crying in monkeys in *Expression*, p. 137.

[6] The letter from Hermann Müller has not been found, but see the letter to Hermann Müller, 16 August [1867].

[7] Müller had enclosed a specimen of *Borreria* with his letter of 4 March 1867. He described it as a 'dimorphic *Rubiaceæ*, probably a *Diodia*'. Müller corrected his identification of the genus after examining ripe fruits of his own specimens (see letter from Fritz Müller, 2 June 1867 and n. 2).

[8] CD had received seeds of *Cordia* sent by Müller in December 1866 (see letter to Fritz Müller, 7 February [1867]). Müller enclosed seeds of *Gesneria* with his letter of 1 January 1867. He enclosed seeds of *Plumbago* and *Lobelia* with his letter of 2 February 1867 (see letter to J. D. Hooker, 24 [March 1867], and letter to Fritz Müller, 25 March [1867]).

[9] Müller sent CD bulbils of *Oxalis* with his letter of 1 and 3 October 1866 (see *Correspondence* vol. 14, letter to Fritz Müller, [before 10 December 1866]). CD reported that the bulbs were growing in his letter to Müller of [late December 1866 and] 1 January 1867 (*Correspondence* vol. 14). See also letter to Fritz Müller, 25 March [1867].

[10] See letter to J. D. Hooker, 24 [March 1867], and letter from J. D. Hooker, 26 [and 27] March 1867.

[11] CD wrote to John Scott for information on the means by which seeds of *Adenanthera pavonina* were disseminated (the letter has not been found, but see the letter from John Scott, 24 September 1867). In *Origin* 4th ed., p. 240, CD had argued that the beauty of fruits served 'merely as a guide to birds and beasts, that the fruit may be devoured and the seeds thus disseminated'. *Adenanthera pavonina*, which had conspicuous seeds but no surrounding pulp with nutritive value, presented a problem for CD's hypothesis (see *Correspondence* vol. 14, letter to J. D. Hooker, 10 December [1866] and n. 3).

To Hermann Müller 16 August [1867][1]

Down, Bromley, Kent
Aug: 16.

Dear Sir

— — — — I was made aware by Prof. Asa Gray (either in a paper in the Amer. Journal of Science or in a letter) of my error with respect to Cypripedium.[2] By an odd chance I put an Andrena into the labellum, and saw what you describe as naturally taking place.[3] — — — —

— — — I do not doubt that this species is generally self-fertilized; and I am aware that I erred in supposing that this happened to so few species.[4] Neottia nidus avis is often self-fertilized.[5] Epipactis latifolia I find is always fertilized by wasps. (vespa)[6] — — — —

Yours very faithfully | Charles Darwin

Copy incomplete
DAR 146: 429

[1] The year is established by the reference, in the letter to Fritz Müller, 15 August [1867], to a letter from Hermann Müller containing observations on the fertilisation of orchids. Hermann began his work on orchids in the summer of 1867 (Möller ed. 1915–21, 2: 111).

[2] In *Orchids*, pp. 274–5, CD had speculated on the means by which flowers of *Cypripedium* might be pollinated, concluding that an insect would have to insert its proboscis through one of two small openings above the lateral anthers. Gray concluded, from observations of American species of *Cypripedium*, that an insect would enter a flower through the large opening on the dorsal surface of the flower, then crawl out through one of the small openings above the anthers (see *Correspondence* vol. 10, letter to Asa Gray, 10–20 June [1862] and n. 16). Gray later published his observations in the *American Journal of Science and Arts* (A. Gray 1862).

[3] In *Orchids* 2d ed., pp. 230–1, CD described his experiment with a small bee of the genus *Adrena* and referred to the observations of Gray and Müller (see also *Correspondence* vol. 11, letter to Asa Gray,

20 April [1863]). In the published version of his observations on Westphalian orchids, Müller described the pollination of *Cypripedium calceolum*, noting that the bee's path through the flower necessitated its touching the stigma before the anthers, thus ensuring cross-fertilisation (H. Müller 1868, pp. 1–3).

⁴ The reference is probably to Müller's observations on *Epipactis viridiflora* (see H. Müller 1868, pp. 7–10, and *Orchids* 2d ed., pp. 102–3). Müller observed that the flower lacked a rostellum, which in most orchids separates the anther from the fertile stigma, and so was easily self-pollinated. In *Orchids*, p. 358, CD had concluded that self-fertilisation in orchids was a 'rare event', but in the second edition he modified his view, acknowledging that some species were 'regularly or often self-fertilised' (*Orchids* 2d ed., p. 290).

⁵ In *Orchids* 2d ed., p. 290, CD included *Neottia nidus-avis* among those species capable of self-pollination, but more often pollinated by insects.

⁶ In *Orchids* 2d ed., pp. 101–2, CD noted that wasps were the only insects he had seen visiting *Epipactis latifolia*. Müller cited CD's letter informing him of this fact in his paper on Westphalian orchids (H. Müller 1868, p. 12).

From B. J. Sulivan 16 August [1867]¹

Bournemouth
Aug 16ᵗʰ.

My dear Darwin

I forgot to speak to you the other day about this shell, but I think you will like to see the account of it.² Mʳ Davidson has not quoted me correctly, as he makes me *command* Beagle. & also say that you *never* found one, which I could not know.³ What I meant was that if Solander found it in T. D. Fuego—and perhaps near Good Success Bay—as Cook was there—it would be singular if you did not find one all the time you were in T. D F.⁴ and that as my officers were dredging for four species at Falklands & never found one, it was singular that I should have got this specimen in Port William the only time I ever used the dredge—⁵ I think I once showed you the shell. I had no idea it was so rare an one until the Gentleman who was here working for the Pleon! Society⁶ at Fossil bones saw it and asked me to send it to Mʳ D as he thought it a very valuable specimen.

We returned home this day week

With kind remembrances to Mʳˢ Darwin & all your party

Believe me | very sincerely yours | B. J. Sulivan

DAR 177: 289

¹ The year is established by the reference to Davidson 1867 (see n. 2, below).

² No record of Sulivan's meeting CD in August 1867 has been found. Sulivan refers to a shell that he had sent to Thomas Davidson for identification and Davidson's notice of it, 'On *Waldheimia venosa*, Solander, *sp.*' (Davidson 1867). The genus name *Waldheimia* was replaced by *Magellania* (Doescher 1981, p. 42). The paper has not been found in the Darwin Pamphlet Collection–CUL.

³ See Davidson 1867, p. 83. Sulivan had been lieutenant on the *Beagle* from 1831 to 1836, when CD served as naturalist. The commander of the *Beagle* was Robert FitzRoy (see *Correspondence* vol. 1). Sulivan was later commander of the *Philomel*, which surveyed the Falkland Islands from 1842 to 1846 (Sulivan ed. 1896, p. 56 *et seq.*; see also n. 5, below).

⁴ Sulivan refers to Daniel Solander and James Cook. Good Success Bay (now Bahía Buen Suceso) is in the strait of Le Maire, Tierra del Fuego (see *Journal of researches*, p. 227). While on the *Beagle*, CD

visited Tierra del Fuego from December 1832 to February 1833 and from May to June 1834 (see *Journal of researches*, pp. 227–44, 263–307; see also R. D. Keynes 2002).
[5] Port William is an inlet on East Falkland Island. Sulivan reported he had dredged the specimen alive in 1843 or 1844 (Davidson 1867, p. 81).
[6] Sulivan refers to the Palaeontographical Society.

From J. D. Hooker 17 August 1867

Royal Gardens Kew
Aug 17/67

Dear Darwin

I return, by the Train to Bromley, Mary Barton & Vol. II of North & South—[1] The whole of the vraisemblable[2] of the latter falls before the Darwinian Gospel— how could such imbecile parents have such a child as Margaret? Also the denouement is too abrupt, & I have no sympathy with the Hero.[3]

As to Mary Barton it is the most horrible story I ever read— I got through the first $\frac{1}{3}$d of the book of deaths of all classes & ages of starvation, fever, & consumption spiced and garnished with Paralysis & blindness—when poor Esther[4] came on the scene & floored me— I took thereupon a sip or two of the last chapter, & being somewhat revived thereby, I managed to struggle through the rest. Esther is the blunder of the book— there was no occasion to run her hapless hopeless misery through the whole story, & thus leave a most painful impression of the whole book—a regular poisoning of the tale— It is tremendously unnatural & sensational in plot & matter but most powerful, & I suppose extremely well written— At least I could always reread bits with pleasure. The life of the Manchester work people, is I suppose pourtrayed to the life.

I also send Cunningham's letters— do not read them unless you like— they are all details—but some of them may interest you.[5]

Mirabilis not in flower yet.[6]

We go to Scotland on the evening of the 30th—via Dunfermline (near which my old Indian friend Sir J Colvile lives)—to Dundee—[7]

Ever Yr affec | J D Hooker

DAR 102: 174–5

CD ANNOTATION
Top of letter: 'Novels | Mirabilis' *pencil*; 'Letters of Cunningham' *ink*

[1] CD's home was about six miles from Bromley station. Hooker refers to two novels by Elizabeth Cleghorn Gaskell, *Mary Barton* (Gaskell 1848), and *North and south* (Gaskell 1855).
[2] Vraisemblable: i.e. vraisemblance.
[3] Hooker refers to the characters Margaret Hale and John Thornton in Gaskell 1855.
[4] Esther is Mary Barton's aunt in Gaskell 1848.
[5] Robert Oliver Cunningham was the naturalist on the *Nassau*, which was surveying in the Straits of Magellan; he sent plant specimens to Hooker (R. Desmond 1994). CD returned Cunningham's letters with his letter to Hooker of 2 September [1867].

[6] In the previous autumn, CD had asked Hooker for specimens of *Mirabilis* species other than *M. jalapa* because he wanted to try a 'curious' crossing experiment (see *Correspondence* vol. 14, letter to J. D. Hooker, 20 November [1866] and n. 6).

[7] Hooker refers to James William Colvile. The annual meeting of the British Association for the Advancement of Science was held in Dundee from 4 to 10 September 1867 (*Report of the thirty-seventh meeting of the British Association for the Advancement of Science; held at Dundee in September 1867*, p. lxxiii).

From Edouard Bornet[1] [before 20 August 1867][2]

Antibes | Alpes Maritimes

Monsieur,

Nous voici à l'époque indiquée par M. Jordan pour semer les graines de *Draba verna*. Je prends la liberté de vous en adresser 10 formes différentes que nous avons cultivées l'hiver dernier à Antibes et qui sont intéressantes à observer.[3] La culture est très facile et ne demande presque aucun soin.

M. Jordan recommande de faire le semis "vers la fin d'Août ou au commencement de Septembre, dans des pots que l'on enterre à peu près au niveau du Sol et que l'on tient assez éloignés les uns des autres, jusqu'à ce que les graines soient bien levées. Elles se développeront d'abord en belles rosettes pendant l'automne, et donneront ensuite 〈   〉 fleurs, dès les premiers beaux jours du 〈printem〉ps."

〈L〉es semis de *Papaver* ont été très-contrariés 〈   〉 le mauvais temps et ont très-mal réussi. Un seul des croisements que j'avais faits est venu à bien, c'est l'hybride du *Pap. dubium* (*modestum, Jord.*) L. fécondé par le pollen du *Pap. Rhœas* (*cruciatum, Jord.*).[4] La plante etait extraordinairement vigoureuse. Elle avait le port et les fruits du *P. dubium*, mais les fleurs etaient plus grandes, plus rouges et souvent marquées d'une large croix noire. Les rayons stigmatiques etaient violets et fortement papilleux comme dans le *P. Rhœas*. Cet hybride etait absolument sterile. Les anthères ne contenaient point de pollen. Les ovules, que j'ai examinés avec soin, m'ont paru bien conformés.

J'ai reçu l'an passé le mémoire que vous avez eu la bonté de m'envoyer et je vous prie de recevoir mes bien sincères remerciements.[5]

Veuillez, Monsieur, agréer l'assur〈ance de〉 mes sentiments respectueux | Ed. Bornet

DAR 160: 256

CD ANNOTATION
Top of first page: 'Sterility of close Specie' *pencil*

[1] For a translation of this letter, see Appendix I.

[2] The year is established by Bornet's reference to receiving the memoir sent by CD (see n. 5, below). The date is established by the relationship between this letter and the letter to Edouard Bornet, 20 August [1867].

[3] Alexis Jordan described twenty forms of *Draba verna* (now *Erophila verna*), classifying these as separate species with the genus name *Erophila*, in the first part of *Icones ad floram Europæ novo fundamento instaurandam spectantes* (Jordan and Fourreau 1866–1903). Bornet's experimental work was done in the garden of Gustave Alphonse Thuret at Antibes in the south of France.

4 *Papaver modestum* is described in Jordan 1851, p. 215; *P. cruciatum* is described in Jordan 1860, pp. 465–6. Jordan maintained that *P. rhoeas* as described by other authors corresponded to approximately twenty separate species by his account (Jordan 1860, p. 467).

5 CD had sent a copy of 'Climbing plants' to Bornet by the same post as his letter to Bornet of 1 December 1866 (see *Correspondence* vol. 14).

To Edouard Bornet   20 August [1867][1]

Down. | *Bromley.* | *Kent. S.E.*

Aug 20

Dear Sir

I thank you sincerely for your great & renewed kindness in sending me the seeds of the Draba.[2] I am so much engaged this autumn that I do not think I shall sow them & no doubt they will keep alive until next year, when they will be of great interest to me.[3]

Owing to the same cause I sowed only four lots of the seeds of the Papaver this spring & have made only a few experiments. Papaver vagum, depressum, Lecoqii came up quite true, but the seedlings of P. pinnatifidum varied a good deal.[4]

With sincere thank & my best respect I remain dear Sir | yours very faithfully | Ch. Darwin

LS

Muséum National d'Histoire Naturelle (Département Systématique & Evolution; Cryptogamie), Ms 505 fol. 388

1 The year is established by the relationship between this letter and the letter from Edouard Bornet, [before 20 August 1867].

2 Bornet sent CD seeds of ten varieties of *Draba verna* (see letter from Edouard Bornet, [before 20 August 1867] and n. 3).

3 No record of CD's performing experiments with *Draba verna* has been found.

4 *Papaver vagum* was described in Jordan 1860, pp. 458–9; it is now generally subsumed within *P. dubium*, as are *P. depressum* and *P. lecoqii*. CD recorded the results of his experiments with *P. vagum* seeds sent by Bornet in *Cross and self fertilisation*, pp. 108–9. In 1876, CD suggested experiments for John Scott to perform with varieties of *P. somniferum* (letter to John Scott, 1 July 1876, *Calendar* no. 10555).

From William Angus Knight   20 August 1867

Dundee

August 20th. 1867

Dear Sir,

I understand you are to visit Dundee during the meetings of the British association in September:[1] and I write to say that if you have made no other arrangement, it will give me much pleasure to receive you at my house.

I regret that I can only offer you a small bedroom; but I take it for granted that the gentlemen visiting Dundee at that time will be prepared to make the best of circumstances.

If you find you can accept this invitation I hope to hear from you what day I am to expect you.

My address is

Revd William Knight
17 Roseangle
Dundee.

I am, | Yours sincerely | William Knight.

Charles R. Darwin Esq.

DAR 169: 40

[1] The British Association for the Advancement of Science held its annual meeting for 1867 in Dundee from 4 to 11 September (*Report of the thirty-seventh meeting of the British Association for the Advancement of Science; held at Dundee in September 1867*, p. lxxiii). CD did not attend the meeting.

From William Boyd Dawkins 22 August 1867

Upminster, Romford
22. August. 1867.

My dear Sir,

Along with this I have taken the liberty of sending an essay on *Bos longifrons* that contains a very interesting letter of Lord Selkirks on the breeding *out* of horns in the Galloway Breed of cattle—p 177 *note*,[1] thinking that you might be interested in the clear proof of the disappearance of so pronounced a character under the selection of man in so short a time as 80 years.

There is one important fact bearing on development, which I hope to publish in the winter—the lineal descent of the recent species of Rhinoceros from the Palæo- and Paloplotheria.[2] I have taken the dentition as the standard of comparison, and I find nearly all the intermediate forms that connect the two extremes. The characters scattered among, and as I believe inherited by, the more recent species converge in the older, and the whole form a finely graduated series. My method of work has been to take the elements of form in the most differentiated teeth and express them by symbols a, b, and the like, and thus I find I can express the measure of differentiation from the Palæotherium, or Paloplotherium of the Eocene.

The carnivores approached in a similar way lead to similar results. Would you be kind enough to tell me whether you consider this method a safe one? Its practical working I find of the utmost value in determining Pleistocene, Pliocene, and Eocene mammals.

Apologising for this intrusion | I am | My dear Sir | Yours truly | W. Boyd Dawkins

Charles Darwin Esq. F.R.S.

DAR 162: 117

[1] The reference is to Dawkins 1867, p. 177 n., in which Dawkins quotes information received in a letter from the earl of Selkirk (Dunbar James Douglas) dated 6 March 1867. The paper was the second part of an essay on fossil British oxen (Dawkins 1866 and 1867); Dawkins evidently sent both parts to CD (see letter to W. B. Dawkins, 26 August [1867]). The paper has not been found in the Darwin Pamphlet Collection–CUL.

[2] Dawkins's paper, 'On the dentition of *Rhinoceros Etruscus*, Falc.', was read on 8 January 1868 and appeared in the 1 August 1868 issue of the *Quarterly Journal of the Geological Society* (Dawkins 1868a). The Eocene genera *Palaeotherium* and *Paloplotherium* are members of the order Perissodactyla (odd-toed ungulates), which also includes the family Rhinocerotidae. In modern nomenclature, *Rhinoceros etruscus* is now *Dicerorhinus etruscus*.

To Charles Lyell   22 August [1867][1]

Down. | Bromley. | Kent. S.E.
Aug 22

My dear Lyell

I thank you cordially for your two last letters. The former one did me *real* good, for I had got so wearied with the subject that I c^d hardly bear to correct the proofs, & you gave me fresh heart. I remember thinking that when you came to the pigeon chapter you w^d pass it over as quite unreadable.[2]

Your last letter has interested me in very many ways, & I have been glad to hear about those horrid unbelieving French men.[3] I have been particularly pleased that you have noticed Pangenesis. I do not know whether you ever had the feeling of having thought so much over a subject that you had lost all power of judging it. This is my case with Pangen: (which is 26 or 27 years old!) but I am inclined to think that if it be admitted as a probable hypothesis, it will be a somewhat important step in Biology.[4]

I cannot help still regretting that you have even looked at the slips, for I hope to improve the whole a good deal. It is surprizing to me & delightful that you sh^d care in the least about the plants. Altogether you have given me one of the best cordials I ever had in my life, & I heartily thank you. I despatched this m^g the French edit. The introduction was a complete surprize to me, & I dare say has injured the book in France; nevertheless with all its bad judgment & taste it shews I think that the woman is uncommonly clever.[5] Once again many thanks for the renewed courage with which I shall attack the horrid proof sheets.

Our kind love to Lady Lyell—[6] | yours affecty | Charles Darwin

You can leave the French Edit. at 6 Queen Anne st, when finished.—[7] A Russian who is translating my new Book into Russian, Kowalewsky, has been here, & says you are immensely read in Russia & many editions, how many I forget.— Six Editions of Buckle! & 4 Editions of Origin.[8]

LS(A)
American Philosophical Society (332)

[1] The year is established by the relationship between this letter and the letter from Charles Lyell, 4 August 1867.

² CD refers to Lyell's letter of 4 August 1867 and to a later letter (see n. 3, below). Lyell read proof-sheets of *Variation* that CD had sent, and praised both the text and illustrations in his letter of 4 August 1867.

³ Lyell had recently attended the Paris exhibition (see letter from Charles Lyell, 4 August 1867 and n. 1). His letter, evidently written during or after his visit to France, has not been found.

⁴ CD discussed pangenesis, his theory of hereditary transmission, in *Variation* 2: 357–404. He had written an earlier version of the essay and solicited advice on it from Thomas Henry Huxley in 1865 (see *Correspondence* vol. 13; see also Olby 1963 and Geison 1969 for more on the development of the hypothesis).

⁵ CD sent a copy of the second French edition of *Origin* (Royer trans. 1866) to Lyell. The translator, Clémence Auguste Royer, had revised her preface since the first edition (Royer trans. 1862; for more on the changes to both the preface and text, see J. Harvey 1997, pp. 76–9).

⁶ Mary Elizabeth Lyell.

⁷ The address is that of CD's brother, Erasmus Alvey Darwin, in London. CD visited London from 17 to 24 September 1867 (see CD's 'Journal' (Appendix II)).

⁸ The date of Vladimir Onufrievich Kovalevsky's visit to CD has not been determined. Two Russian editions of *Origin* had been published (Rachinskii trans. 1864 and Rachinskii trans. 1865), but Kovalevsky may have been including reprints as editions. Lyell's *Principles of geology* was translated into Russian in 1866 and *Antiquity of man* in 1864; *Manual of elementary geology* appeared in two volumes between 1866 and 1878 (*GSE*, s.v. Lyell, Charles). Henry Thomas Buckle's *History of civilisation in England* (Buckle 1857–61) appeared in Russian as part of an edition of collected works in 1861 and as a separate volume in 1863 (*GSE*, s.v. Buckle, Henry Thomas).

From J. T. Rothrock to Asa Gray   22 August 1867

New York
Aug 22nd 1867

Dear Doctor

   The names which Mr Darwin is in doubt about are the *Atnah* and *Espyox* Tribes on the Nasse River.¹ Thurber² wants an autograph of D. So do I. but as you request I have returned the coveted signature. Should there be any more items I can communicate it will give me pleasure to do so. I am overhauling Thurbers duplicate grasses ect. and enriching my own collection therefrom

   In packing those snow-shoes I was obliged to cut the tip of the heel off. There being no box in town long enough. Dr W.³ can however make a nite of it replacing the part.

   Yours Ever | J. T. Rothrock

   Please give my kindest regards and heartiest sympathy to Mrs G⁴ in her sickness. Ill be with Thurber until the 3rd or 4 of sept.⁵

DAR 176: 219

CD ANNOTATIONS
1.1 The names ... River. 1.2] *scored pencil*
1.2 Thurber ... part. 2.3] *crossed pencil*
1.3 I have ... do so. 1.4] *scored pencil*
Top of letter: '31' *brown crayon*

¹ CD evidently asked Gray to ask Rothrock to clarify the names of the two tribes on which Rothrock based his answers to CD's questionnaire for *Expression* (see letter from J. T. Rothrock to Asa Gray,

31 March 1867 and n. 2). The Nass river of western British Columbia, Canada, flows about 380 km (236 mi) south-west through the Coast Mountains to the Pacific Ocean north of Prince Rupert.
[2] George Thurber.
[3] The reference is probably to Charles Wright, who worked at the Gray Herbarium when he was not collecting plants in the field (*ANB*).
[4] Jane Loring Gray.
[5] Gray added the following note to the end of the letter: 'J. T. Rothrock (soon to be *M.D.*) The former pupil of mine who worked a little over *Houstonia* fertilisation | A. G.' Gray had sent CD Rothrock's observations on *Houstonia* (see *Correspondence* vol. 10).

To Cassell, Petter, & Galpin 24 August [1867]

> *Down.* | *Bromley.* | *Kent. S.E.*
> Aug 24.

Dear Sir

I have had much correspondence with M. Kovalevsky, & he has visited me here. I have formed a very high opinion of him, as his knowledge is extraordinarily great. He is brother of a distinguished naturalist—[1] He is preparing a translation of a book which I shall soon publish & has translated & published some very expensive German works into Russian.[2] He has made me very liberal offers for the translation of my book.[3] I should certainly on my own part trust M. Kovalevsky to almost any extent, but you will see that it is obviously impossible for me to answer for his pecuniary circumstances. As far as I can judge his circumstances appear to be good.

Dear Sir | yours faithfully | Charles Darwin

LS
Endorsement: 'C. Darwin | 24th. August 1867.'
American Philosophical Society (333)

[1] Vladimir Onufrievich Kovalevsky's brother was Alexander Onufrievich Kovalevsky.
[2] V. O. Kovalevsky was translating *Variation* into Russian. He had also translated part of Alfred Edmund Brehm's *Thierleben* ([Kovalevsky] trans. 1866–70; see letter from V. O. Kovalevsky, 15 May 1867).
[3] Kovalevsky left the financial arrangements for his translation to CD, and CD decided not to demand payment (see letter from V. O. Kovalevsky, 15 March 1867, and letter to V. O. Kovalevsky, 16 May [1867]).

From Thomas Wright 24 August 1867

> St. Margarets Terrace | Cheltenham
> 24 Aug 67

Dear M[r] Darwin

I have much pleasure in sending you by this post a copy of my paper on corals reefs present & past.[1] to so great a master on this subject as yourself I can offer little that is new. For the more I peruse your valued work on Coral islands[2] the more deeply do I become impressed with the most profound respect for its accomplished author. Still I thought the observations I had made on ancient reefs especially those in the Oolitic rocks might interest you and the inference I have drawn from your theory of subsidence may hold true of the bed of the Jurassic ocean.[3] at all events it is a new explanation of the formation of the Oolitic strata and when we reflect on the enormous lapse of time that must have taken place to allow so many different forms of polyp life to appear and depart, as shewn by our tables, we have no difficulty in understanding how a slow subsidence of the bed of the Jurassic ocean would have allowed all those tropical forms which the Oolites contain to come in and go out ere the Jurassic drop scene fell. I should like to have your thoughts on this point. I was very sorry to learn some time since that you were far from well I hope this fine genial season has restored you to health and with my very best wishes for your health & happiness

believe me always | Yours most truly | Thomas Wright

C. Darwin Esq FR.S | &c. &c &c.

DAR 181: 179

[1] Wright sent an offprint of his article, T. Wright 1866. CD's copy is in the Darwin Pamphlet Collection–CUL.

[2] The reference is to *Coral reefs*.

[3] In his paper, Wright summarised CD's theory of the formation of coral reefs (T. Wright 1866, pp. 112–18). In the section on oolitic corals, Wright concluded that the bed of the Jurassic sea was a slowly subsiding area of great extent, similar to parts of the present-day coral sea in the Indo-Pacific Ocean (*ibid.*, p. 159).

To W. B. Dawkins 26 August [1867]

Down. | Bromley. | Kent. S.E.
Aug 26

My dear Sir

I am much obliged for the present of your two papers. I have not had time as yet to do more than glance at them & refer to the case of the Galloways.[1] I had heard something of this latter case, but not in such detail. I have nearly finished printing a book on the Variation of domestic animals;[2] but I am sorry to say for my own sake that the chapter on cattle is printed off, so that I shall not be able at present to profit by your papers.[3] I have been interested by all your previous work, & I am sure I shall be so in an especial degree on the descent of the species of Rhinoceros.[4]

As far as I understand your plans of ascertaining the amount of difference between existing & extinct forms, it seems to me very good; & I feel sure that the attempt will be valuable & very interesting.[5]

Permit me to add that your letter has pleased me much, for from your previous papers I supposed that you considered species to be immutable, & as I am an advocate for mutability, the opinion of so able a judge as yourself was a great discouragement to me.

With my best thanks & sincere respect for your scientific labours I remain | My dear Sir | Yours very faithfully | Charles Darwin

W. Boyd Dawkins
Upminster | Romford

LS (photocopy)
Postmark: AU 26 67
DAR 249: 79

[1] Dawkins sent an essay on fossil British oxen that had been published in two parts (Dawkins 1866 and 1867). In the second part, Dawkins had quoted the account by Dunbar James Douglas of the disappearance of horns from Galloway cattle (see letter from W. B. Dawkins, 22 August 1867 and n. 1).

[2] CD refers to *Variation*, which was published in January 1868.

[3] CD refers to Dawkins 1867 in *Variation* 2d ed., 1: 85.

[4] Dawkins was attempting to establish the relationship between modern and extinct members of the family Rhinocerotidae by a study of changes in dentition (see letter from W. B. Dawkins, 22 August 1867 and n. 2).

[5] See letter from W. B. Dawkins, 22 August 1867.

From W. B. Dawkins 27 August 1867

Upminster, Romford
27. August 1867.

My dear Sir,

I thank you very much for your kind note. Your approval of my scheme will help very much to its completion.[1]

I am very glad indeed to hear that you take an interest in my few essays written rather by a learner than a teacher.[2] In all of them I have tried as far as possible to break down the barrier between Pleistocene and living species, by attaching weight to individual variation, and to results of living under different conditions. Thus the Pleistocene animals utterly extinct amount only to about 10 or 11 out of 56.[3]

Permit me to add that I feel unspeakably grateful to you for your book.[4] Without it I should probably have been groping in darkness, and manufacturing species in proportion to my own ignorance. All the men who are working minutely at one branch of Nat. Hist, that I know of, endorse your views as regarding their own peculiar provinces,—Dr Duncan in the corals, H. Woodward in the crustacea.[5] I dont indeed see how any naturalist can work on any other hypothesis. It is cropping up continually even in Prof. Owens works.[6] If I can add only one grain of proof to its truth, I shall feel amply rewarded for any quantity of labour.

When I was working at the hyænas, I was struck by the wonderful effect that captivity has had in modifying the shape of the skull. The prehensile character of the jaws is to a certain extent lost, and the maxillaries, and rami become shortened in a remarkable degree. Indeed for purposes of comparison I find the bones of animals from menageries absolutely useless. I mention this because it shows how very pliant even the skeletons of animals are, and how with the disuse of a muscle, its points 'd'appui'[7] become aborted.

I am | My dear Sir | Yours very truly | W. Boyd Dawkins

Charles Darwin Esq.

DAR 162: 118

[1] Dawkins refers to his research comparing the dentition of living and extinct members of the family Rhinocerotidae (see letter from W. B. Dawkins, 22 August 1867 and n. 2, and letter to W. B. Dawkins, 26 August [1867]).

[2] Dawkins had sent CD papers on fossil oxen (Dawkins 1866 and 1867; see letter from W. B. Dawkins, 22 August 1867, and letter to W. B. Dawkins, 26 August [1867]).

[3] The source of Dawkins's statement on the proportion of extinct animals has not been found. However, in a paper on the prehistoric Mammalia of Great Britain (Dawkins 1868b), there is a table listing more than fifty animals and charting their presence in Britain from prehistoric times to the recent past.

[4] The reference is to *Origin*.

[5] Dawkins refers to Peter Martin Duncan and Henry Woodward.

[6] A review in the *London Review of Politics, Society, Literature, Art, and Science*, 28 April 1866, pp. 482–3, of the first two volumes of Richard Owen's *On the anatomy of vertebrates* (Owen 1866–8) argued that Owen had made a partial admission of 'the truth of the principles of Natural Selection' (see *Correspondence* vol. 14, letter to J. D. Hooker, 31 May [1866], n. 11). For a discussion of Owen's views on evolution, see Rupke 1994, pp. 220–58.

[7] Point d'appui: point of support, fulcrum (French).

To J. P. M. Weale 27 August [1867][1]

Down. | Bromley. | Kent. S.E.

Aug 27

My dear Sir

I thank you cordially for all your kindness.[2] The case of the Muraltia, which you describe & figure so clearly is very curious; & I quite agree with you, the case is especially curious in the resemblance of the movement of the stamens to those in papilionaceous plants. I doubt whether the movement, at least in the latter, is due to irritability, nor is it a case of simple elasticity. The resemblance of your Muraltia to a heath, of which I believe there are other cases at the Cape is curious.[3] I have formerly examined, but with no great care, our English Polygola[4] & convinced myself that its fertilization depended on insects.

You have been extremely kind in taking such great trouble about expression, which is a subject that interests me to an unreasonable degree. That I sh[d] receive answers written by the brother of a Kaffir chief is a truly wonderful fact in the progress of civilization.[5]

Thank you for telling me about the children pouting,—a gesture which I hear from N. America is common to Indian children.[6] I shall be most grateful for any further trustworthy information. I believe the French are quite wrong in speaking of **a** "grief muscle"; the movement apparently results from a combined action of the upper orbicular & that part of the frontal muscle which is seated above the inner angle of the eyebrows.[7] I enclose a poor photograph of a young woman who c[d] voluntarily make this movement;[8] but the eyebrows are hardly oblique enough; the transverse wrinkles on the forehead which extend only a short distance on each side of the centre are eminently characteristic; as is a slight swelling close above the inner end of the eyebrow. I sh[d] be very glad to hear whether this expression can be seen in any savage race. The only chance w[d] be visiting a person in anxiety or grief.

When I rec[d] the locust-dung I c[d] not imagine what it was, & I might have gone on guessing till doomsday. I will try the experiment carefully, but shall be as much surprized as interested if it sh[d] prove to contain any seeds.[9]

With very sincere thanks | I remain my dear Sir | yours very faithfully | Charles Darwin

LS(A)
University of Virginia Library, Special Collections, Darwin Evolution Collection (3314)

[1] The year is established by the relationship between this letter and the letter from J. P. M. Weale, 7 July 1867.
[2] Weale had sent CD's handwritten set of queries on expression to several people in the Cape Colony (see letter from J. P. M. Weale, 7 July 1867 and n. 5.
[3] Weale's observation on the resemblance of *Muraltia* to a heath was evidently in the section of his letter that was later excised (see letter from J. P. M. Weale, 7 July 1867 and n. 13).
[4] *Polygala* and *Muraltia* are both members of the family Polygalaceae (milkworts; characterised in Lindley 1853 as an order) in the order Fabales.
[5] Weale enclosed answers to CD's set of queries on expression written by Christian Gaika, whom Weale described as a brother of Chief Sandile (see letter from J. P. M. Weale, 7 July 1867 and n. 6).
[6] CD may refer to information from Joseph Trimble Rothrock on children of tribes living in the Nass river area of western British Columbia, Canada (see letter from J. T. Rothrock to Asa Gray, 31 March 1867).
[7] CD introduced the phrase 'grief-muscles' to refer to the combined action of these muscles (*Expression*, p. 181; see letter from J. P. M. Weale, 7 July 1867 and n. 7).
[8] The enclosure has not been found, but see *Expression*, p. 180 and Plate II, figure 3 (facing p. 178) for the published version of the photograph.
[9] Weale had enclosed a packet of locust dung with his letter of 7 July 1867. For the results of CD's experiments, see the letter to Asa Gray, 16 October [1867].

From J. D. Hooker 31 August 1867

Royal Gardens Kew
Aug 31/67.

Dear Darwin

I have given up Dundee—my Mother being very ill with a severe attack of Peritonitis & Enteritis, which has called me down to Norwich, where she lives.—[1] My wife[2] is there, & I return tonight to be back here again on Monday mg.

I most stupidly have forgotten the name of your Brazil friend, to whom I want to write *badly* as he can help us much.[3]

We are quite ready for any refuse Orchids you may be good enough to send us.[4]

Ever aff Yrs | J D Hooker

Self impregnated Victoria has given a wretched yield— The results of the impregnation by Chats-worth plants are now being ripened.[5]

It is quite curious how difficult I find it to get our best Foremen to keep an accurate register—though they have your instructions *in writing*. I have now had to order that every capsule ripened be reported to me & kept for my examination!

DAR 102: 176–7

[1] Hooker refers to the meeting of the British Association for the Advancement of Science held in Dundee from 4 to 11 September 1867 (*Report of the thirty-seventh meeting of the British Association for the Advancement of Science; held at Dundee in September 1867*, p. lxxiii). Hooker's mother, Maria Hooker, had moved to Norwich in 1867 (Allan 1967, p. 224).

[2] Frances Harriet Hooker.

[3] Fritz Müller had offered to supply Hooker with specimens of Brazilian plants (see letter to J. D. Hooker, 25 [April 1867]).

[4] CD had received orchid specimens from John Traherne Moggridge (see *Correspondence* vols. 13 and 14, and this volume, letter from J. T. Moggridge, 22 April [1867]). He later sent several of the specimens to Hooker (see letter to J. T. Moggridge, 1 October [1867]).

[5] The reference is to *Victoria regia* (now *V. amazonica*) and to the garden of Chatsworth House in Derbyshire, which had a greenhouse specially built by Joseph Paxton to house specimens of the plant (R. Desmond 1994). CD had become interested in this species and the related *Euryale ferox*, which was reported to be perpetually self-pollinating (see *Correspondence* vol. 14, letter from Robert Caspary, 25 February 1866, and letter to J. D. Hooker, 1 November [1866]).

To Alfred Wrigley [September 1867][1]

My dear D^r. W.—

As experience in an Examination, of the same nature, as those to be ultimately passed, must be very useful to any young man, in curing him from nervousness & as a stimulus to exertion, I wish my son L. to try for W. this winter.[2] From what I gather, he has *small* chance of success, but he assures me that he will not feel discouraged by failure.—[3]

I may take this opportunity of informing you that for various considerations, I wish to send H. after Xmas to a private Tutor.[4] But I hope that you will not suppose that I feel dissatisfied with Clapham. On the contrary I always rejoice that my 4 sons have been under D^r P & your care;[5] & I cordially thank you for your [wonderful] K.[6] to them & myself—

Believe me my dear D^r. W | Yours very f. | C. D

P.S. My son H. has a strong desire to make greater progress in Math; therefore if you can assist him in this respect. I sh^d. feel much obliged.—⁷

AdraftS
DAR 96: 31

¹ The date is established by the relationship between this letter and the letter to Alfred Wrigley, 7 March [1868]. In the 7 March [1868] letter, CD wrote that he had written to Wrigley 'half a year ago' saying that he wished to place Horace with a private tutor after Christmas.
² CD refers to Leonard Darwin, and probably to the Royal Military Academy, Woolwich. See also the letter from Alfred Wrigley, 2 January 1868, on Leonard's success in examinations for entrance to the Military Academy at Sandhurst.
³ For more on Leonard's education and academic ability, see J. R. Moore 1977.
⁴ Horace Darwin.
⁵ Horace and Leonard were both pupils at Clapham Grammar school in south-west London; their brothers George and Francis were former pupils. Charles Pritchard was the school's former head-master, Wrigley the current one.
⁶ K.: probably kindness.
⁷ Wrigley had been professor of mathematics at the Royal Military College, Addiscombe, Surrey.

From John Brodie Innes 1 September [1867]¹

Keston
1st. Sept^r

Dear Darwin

I found out the name of the tutor I mentioned to you about whom you may think it worth while to enquire. I only knew him by reputation, not personally.

Rev^d. C. Bradley²
? Hatfield

I am not sure of his parish but it joins on to Colney Hatch—Eastward and is a short walk from the C Hatch station—³ You will find it on a map at once—

I am off to the north tomorrow night and hope to be home to lunch on Tuesday⁴

I hope Horsman⁵ will stay quietly, at least for the year he has promised: and that you will continue to like him—

Faithfully Yours | J Brodie Innes

DAR 167: 5

¹ The year is established by the reference to Samuel James O'Hara Horsman (see n. 5, below).
² Charles Bradley.
³ Colney Hatch station was in the north London suburban area (*Post Office London suburban directory*). The station is now called New Southgate.
⁴ Innes's home was at Milton Brodie, near Forres, Scotland (*DNB*).
⁵ Horsman was curate at Down for part of 1867 and 1868; he had recently arrived to take up the post (J. H. Moore 1985, pp. 470, 477). According to Emma Darwin's diary (DAR 242), Horsman visited on 2 September 1867.

To J. D. Hooker 2 September [1867][1]

Down
Sep 2

My dear Hooker

I am extremely sorry to hear about Lady Hooker for I fear she must be in a very dangerous & what I always think worse a very suffering state.[2]

The address you require is "D[r] Fritz Müller Desterro Brazil" (viâ Bordeaux.)[3] He always writes via Bordeaux so I suppose there is some advantage. He is a capital observer & most obliging man.

We rec[d] the novels all safe, & your jolly note of Aug 17. but I do not think you do justice to Mary Barton.[4] I return Tyndall which I have been particularly glad to read. How true & striking it is what he says on the power of pondering.[5] Parts of the article seem to me very obscurely written. I return Cunningham's letters;[6] they are too full of details about plants for my taste. I am very glad you are attending to the Victoria & Euryale.[7]

Will you be so kind as to get any of your assistants (but do not think of doing it yourself) to name for me the enclosed common greenhouse Mimulus. The nurseryman c[d] only tell me that he always sold the seed as "mixed vars. of Mimulus". It is of *real importance* to me to know the name, as I have experimented much on it.[8] Also ask some one (if possible) to send me a few fresh & youngish flower of the yellow var. of Mirabilis jalapa.[9]

yours affectionately | Ch Darwin

L(S)(A)
DAR 94: 33–4

[1] The year is established by the relationship between this letter and the letter from J. D. Hooker, 31 August 1867.

[2] Hooker's mother, Maria Hooker, had been diagnosed as suffering from peritonitis and enteritis (see letter from J. D. Hooker, 31 August 1867).

[3] See letter from J. D. Hooker, 31 August 1867 and n. 3.

[4] For Hooker's opinion of the novel *Mary Barton* (Gaskell 1848), see the letter from J. D. Hooker, 17 August 1867.

[5] CD refers to an article by John Tyndall on 'Miracles and special providences' that appeared in the 1 June 1867 issue of the *Fortnightly Review* (Tyndall 1867). In a section on the nature of scientific discovery, Tyndall remarked, 'There is much in this process of pondering and its results which it is impossible to analyse', and concluded that by 'a kind of inspiration' the investigator moved from 'contemplation of facts to the principles on which they depend' (*ibid.*, p. 655).

[6] CD refers to Robert Oliver Cunningham (see letter from J. D. Hooker, 17 August 1867 and n. 5).

[7] See letter from J. D. Hooker, 31 August 1867 and n. 5. CD reported on the self-fertility of *Victoria regia* and *Euryale ferox* in *Cross and self fertilisation*, p. 365.

[8] CD reported the results of his experiments with several generations of *Mimulus luteus* in *Cross and self fertilisation*, pp. 63–81. In a note (*ibid.*, p. 63), he thanked Hooker for identifying the specimens he sent to Kew. Notes on crossing experiments with *Mimulus* that CD made between 1866 and 1873 are in DAR 77: 111, DAR 78: 17–40, 42–5, and DAR 109: B12.

[9] CD had mentioned his wish to carry out crossing experiments with *Mirabilis* species in his letter to Hooker of 20 November [1866] (*Correspondence* vol. 14). In his letter of 17 August 1867, Hooker stated that *Mirabilis* was 'not in flower yet'.

From Federico Delpino[1] 5 September 1867

Illustre Signore

Pieno d'ammirazione pel grande talento della S. V. Illma, oso offerirle in omaggio due miei scritti.[2]

Se Ella si compiacerà di darvi uno sguardo, si accorgerà tosto quanta influenza abbiano avuto su di me le Sue ammirabili opere sulla variabilità delle specie e sulla fecondazione delle orchidee.[3]

Quanto alla prima opera, accettando interamente *il piano di variazione* così ingegnosamente trovato ed esposto dalla S. V., mi parve doverlo interpretare Spiritualisticamente, locché (se non sono in errore) toglie via tutte quante le objezioni sollevate contro il piano medesimo.[4]

Quanto alla seconda opera, Ella vedrà che io sono stato fortunato di ritrovare che la legge delle nozze miste mediante gl'insetti ha luogo nelle asclepiadee e in altre famiglie di piante con pari ragione che nelle orchidee.[5]

Le chieggo perdono della libertà che mi prendo, ed ho l'onore di protestarmi con profonda stima

Suo vero ammiratore e servitore | Delpino Federico

Italia | Genova | Chiavari addì 5 7.bre 1867

DAR 162: 142

[1] For a translation of this letter, see Appendix I.

[2] Delpino sent his paper on Friedrich Hildebrand's essay on the distribution of the sexes in plants and the prevention of self-fertilisation (Delpino 1867a) and another paper on the mechanisms to ensure fertilisation in seed plants (Delpino 1867b). CD's annotated copies (together with three additional pages of notes affixed to Delpino 1867a) are in the Darwin Pamphlet Collection–CUL.

[3] Delpino refers to *Origin* and *Orchids*.

[4] Delpino refers to Delpino 1867a. On Delpino's philosophical position and his attempt to reconcile Darwinian transmutation theory with his own teleological approach, based on Kantian nature philosophy, see Pancaldi 1991, pp. 107–36.

[5] See Delpino 1867b. Delpino alludes to CD's statement: 'it is apparently a universal law of nature that organic beings require an occasional cross with another individual' (*Orchids*, p. 1; see also *Origin*, p. 97). Delpino had published an earlier article on the mechanism to ensure fertilisation in asclepiads, with some considerations on final causes and CD's theory of the origin of species (Delpino 1865).

From W. B. Dawkins 7 September 1867

Upminster | Romford, Essex.

7 Sept. 1867.

My dear Sir,

I thank you very much for your suggestion, which I will carry out in the winter.[1] I find menagerie bones absolutely useless for comparison with those of the same animals of Pleistocene age.[2] Between wild individuals even of the same recent species there is a large amount of variation, and that variation M[r]. Sanford and myself have used, or rather are using to prove the identity of *Felis spelæa* with *F. leo*, in the Palæont. Soc.[3]

I am | My dear Sir | Yours truly | W. Boyd Dawkins

Charles Darwin Esq. F.R.S.

DAR 162: 119

[1] CD had encouraged Dawkins in his plans to carry out further comparative studies of the dentition of living and extinct representatives within the Rhinocerotidae (see letter to W. B. Dawkins, 26 August [1867]). CD may have made a specific suggestion about the work in a reply to the letter from W. B. Dawkins, 27 August 1867, although no such letter has been found.

[2] Dawkins mentioned changes he had observed in the skulls of captive hyenas compared with their wild counterparts in his letter to CD of 27 August 1867.

[3] Dawkins was collaborating with William Ayshford Sanford on a study of British Pleistocene Felidae (Dawkins and Sanford 1866–72). The four-part monograph was published by the Palaeontographical Society. The second part of the work, published in 1868, was on *Felis spelaea*, the cave lion (now *Panthera leo spelaea*). *Felis leo* is now *Panthera leo*.

From Thomas Rivers 9 September 1867

Nurseries, Sawbridgeworth, Herts.[1]

Sep 9/67

My Dear Sir/.

I am tempted to submit to you the result of a curious cross in peaches.

Some three years since my son wishing to raise, a fine peach with fine flowers, took some pollen from that brilliant beautiful variety the "Double Crimson Chinese peach" & fertilised carefully (ie removing the stamens from the female parent) some flowers of a variety called Leopold the 1st.[2] this is a very large melting peach giving large pale flowers. Last spring one of the seedlings raised from the crossed flowers blossomed & to our surprise gave no sign of change, for its flowers were large & pale like its female parent, this tree escaped notice till very recently when it was found full of Almond-like fruit their pulp even harder & more almond-like than those given by the male parent the Double Crimson Chinese Peach.

I enclose two fruit

N°. 1 is the seedling raised from the fertilised flowers

N°. 2 is the fruit of the Double Crimson Peach the male parent of N°. 1

The force of the male in making a fine large peach into a fruit more almond-like than itself is *to me* most surprising

Pray forgive my intrusion I am looking anxiously for your new book[3]

I am My Dr Sir | Y^rs. very truly | Thos. Rivers

DAR 176: 171

[1] The location of Rivers's nursery is followed on the letterhead by: '*Harlow Station is the most convenient for passengers. | Great Eastern Railway*'.

[2] Rivers's son, Thomas Francis Rivers, developed several new varieties of peach and other soft fruit for the commercial market (*Journal of Horticulture* 39 (1899): 161–2).

[3] Rivers refers to *Variation*; CD discussed Rivers's observations on the relationship between the peach and the almond in *Variation* 1: 338–9 (see also *Correspondence* vols. 11 and 13).

From J. V. Carus 11 September 1867

39, Elsterstrasse, | Leipzig,
Septbr. 11th. 1867

My dear Sir,

You had kindly ordered Mr Murray to send me clean sheets. To-day I venture to ask you for the same kindness. Since the 8th. of July I didn't get any sheets, when I received them up to page 224. As I agree entirely with Mr Schweizerbart that it would be the best to bring out the first volume of the translation at the same time with the original, viz. in November, I want the rest of the first volume very much.[1]

As I am writing to you, may I ask you some more questions. The Dogs give me some trouble. I cannot get clear with some English expressions. Taking *Linné*, Systema naturae (I have Gmelin's edition before me)[2] as a means of explanation: mastiff is *C. anglicus*,[3] Bull-dog is *C. molossus*, Water-dog is *C. aquaticus*. Now, Turnspit is C. *vertagus*, Dachs in German. What do you mean by "the turnspit-like German badger-hound".[4] We call the "vertagus" Dachs-hund (i.e. badger-dog). Is *hound* = C. *gallicus*? Then what is fox-*hound*, deer-*hound*? The *retriever*, is that C. *venaticus* L. Gm.? For the other names, spaniel, setter, pointer, you gave already notes to the late Mr Bronn, which I used of course.[5] But even here I am afraid to get into a scrape. You say on p. 19. (Variation &c) "our hounds *and* setters *or* spaniels" and you add (in brackets) the translation: "Jagdhund *und* Wachtelhund".[6] Therefore "*hound*" is Jagdhund (C. *gallicus* L. Gm.?), *spaniel* = C. avicularius L. Gm.? (Hühnerhund or Wachtelhund). Bronn translates "setter" with "Spürhund",[7] which I think is correct. Terrier is C. terrarius of later authors, our Pinscher, good animals of this race ought to have a black palate as far as I heard, a character which Linné gives to the King-Charles-Dog; but this is altogether another race. If you will be so kind as to give me some explanation about these matters I should be most thankful. I have here Ham. Smith's Dogs (Natur. Libr) what I mention in the case you would refer to figures[8]

Pigeons are less troublesome. I am sorry I cannot get hold of Neumeister, whose work is completely out of print.[9] Carrier is "Botentaube"; Frillback is Strupptaube. You give the last word for the "common Frillback",[10] I think I may use the same expression for the "Indian Frillback" (of course then "Indische Strupptaube). Only the "Barb" makes me some difficulty. May I not use the same word in German: "Barb' taube"? You give as translation "Indische Taube";[11] what is then an *English* Barb? Then: the Runt. Here you give two German expressions: Florentiner *and* Hinkeltaube.[12] But Runt means the whole race. "Archangel" means the colouring like Marrubium; isn't it?[13] Is dragon a race of itself or is it synonymous with

Carrier?[14] Do you know any work, where I could find such toy-forms, as Priests, Monks, Porcelains, Breasts Shields and so on?[15]

I am sorry I trouble you; but I wish to make the translation as trustworthy as possible. And here I don't see how I can do it without your kind help.

My best wishes for your health. I do hope indeed that you may be able to publish your second and third work[16]

Believe me | My dear Sir | Yours ever truly | J. Victor Carus

DAR 161: 60

CD ANNOTATIONS

2.4 *C. molossus*] *underl red crayon*
2.6 We call ... *gallicus?* 2.7] *scored red crayon*
2.7 Then ... Gm.?] *scored red crayon*
2.10 add ... Jagdhund 2.12] *scored red crayon*
2.12 C. avicularius] *underl red crayon*; 'Pl. 15' *red crayon, circled red crayon*; 'Ham Smith' *red crayon; added bottom margin*[17]
3.2 Carrier] 'Turkish' *added red crayon*
3.3 for the ... Frillback" 3.4] *double scored red crayon*
3.8 "Archangel ... Carrier? 3.10] *scored red crayon*
3.10 toy-forms ... so on? 3.11] *scored red crayon*

[1] CD had informed Carus he expected *Variation* to be published in November 1867 (see letter to J. V. Carus, 11 April [1867]). In the only extant letter to Carus between 11 April and 11 September, CD mentioned he had not yet received any clean proof-sheets (see letter to J. V. Carus, 18 April [1867]).

[2] Carus would probably have used the last edition by Johann Georg Gmelin of the *Systema naturæ* by Carl von Linné (Carolus Linnaeus) (Linnaeus 1788–93).

[3] *Canis anglicus*.

[4] See *Variation* 1: 28.

[5] Carus refers to Heinrich Georg Bronn, who had translated the first and second German editions of *Origin* (Bronn trans. 1860 and 1863). The notes referred to have not been found.

[6] See *Variation* 1: 19. CD included the German translation because he was quoting a passage from Rütimeyer 1861.

[7] See Bronn trans. 1863, p. 46.

[8] Carus refers to Charles Hamilton Smith's work on the natural history of dogs (C. H. Smith 1839–40), which was part of a series, The naturalist's library, edited by William Jardine.

[9] CD makes several references to Neumeister 1837 in *Variation*.

[10] See *Variation* 1: 155.

[11] See *Variation* 1: 144.

[12] See *Variation* 1: 142.

[13] See *Variation* 1: 157. 'Archangel' refers to a breed of pigeon, generally of a bronze or copper colour with black, blue, or white wings, characterised by a metallic sheen. *Marrubium vulgare* (horehound) has leaves with whitish or silvery hairs.

[14] See *Variation* 1: 141.

[15] See *Variation* 1: 187.

[16] In the introduction to *Variation* (*Variation* 1: 3–9), CD set forth his plan to produce three related books covering variation under domestication (i.e., *Variation*), variation in the wild, and the principle of natural selection. Only *Variation* was completed.

[17] CD's annotation refers to plate 15 of C. H. Smith 1839–40 (see letter to J. V. Carus, 16 September 1867, and n. 8, above).

From Friedrich August Hagenauer to Ferdinand von Mueller[1] [12 September 1867][2]

Aborigena Mission Station | Lake Wellington, Gippsland

Werthester Herr

Es thut mir leid dass ich nicht eher dazu gekommen bin Ihrer Bitte Genüge zu leisten im Bezug Mr. Darwins Fragen, die ich Ihnen nun heute erst beantworte.[3] Ich hoffe dass dieselben dem Zweck entsprechend sind.

Sollten Sie selbst einmal in unsre Nähe hier kommen, würde ich mich gewiss sehr freuen Sie bei uns zu sehen.

Unter besten Grüssen | Ihr ergebenster | F. A. Hagenauer

Answers to Queries about Expression[4]

1. Astonishment is very often expressed by the eyes and mouth being opened wide and the eyebrows raised.

2., I have never seen anything like a blush, but I have seen them looking down to the ground in account of shame.

3., It is seldom that a man in an indignant state frowns or holds the head erect, but may oftener clench his fist.

4. When considering deeply he does frown.

5., Not observed.

6. When in good spirit the eyes sparkle, with the skin round and under them a little wrinkled and with the mouth a little drawn back in the corners.

7. Not observed.

8., A dogged and obstinate expression can clearly be recognised by the mouth being firmly closed and a frown.

9. Not observed.

10. Not observed.

11., fear is expressed in the same manner as by Europeans even still more so, that they would lift up both arms above the head.

12. I have often seen tears coming into their eyes by great laughter.

13. Not observed,

14. Children when sulky do pout—

15. Guilty expressions can be seen by the eyes being generally closed a little; jealousy by a frown.

16. A gentle hiss is uttered as a signal for silence.

17. The head is nodded vertically in affirmation and shaken latterly in negation.

DAR 166: 80

CD ANNOTATION

Top of letter: '10 | Australia' *brown crayon*

[1] For a translation of the German section of this letter, see Appendix I.

² The date is established by an annotation, probably by Mueller, at the top of the letter, which reads: 'date of Poststamp 12/9, received 19 Septb 67'.

³ CD sent queries on expression to Mueller to pass on to contacts who could provide answers pertaining to indigenous Australians (see letter to Ferdinand von Mueller, 28 February [1867]; see the enclosure for the list of queries). It is possible that this letter was sent by Mueller with another letter containing answers to CD's queries (see letter from Ferdinand von Mueller, 8 October 1867).

⁴ CD cited Hagenauer's answers to questions 2 and 15, as well as a later response to question 5, in *Expression*. Hagenauer's later responses were enclosed with the letter from R. B. Smyth, 13 August 1868 (*Correspondence* vol. 16).

From J. D. Hooker [14 September 1867]¹

Kew
Saturday—

Dear Darwin

Very many thanks for the famous collection of plants, which are most acceptable especially the Oxalis.²

I have better news from Norwich, whither I go again today till Monday—but I suppose there is some occult cause for these repeated attacks, which there is little chance of removing even if discovered, whether morbid or proceeding from functional derangement may never be known.³

Ever aff yrs | J D Hooker

My wife had a long letter from Lady Lyell who seems to have been delighted with the Brit. Assn.⁴

DAR 102: 178

¹ The date is established by the references to Maria Hooker's illness and to the British Association for the Advancement of Science meeting at Dundee, which ended on 11 September 1867 (see nn. 3 and 4, below). The Saturday following 11 September 1867 was 14 September.

² CD had received several specimens of different species of *Oxalis*, as well as seeds or specimens of other plants, from Fritz Müller in Brazil (see letter to Fritz Müller, 15 August [1867]). The following plants were recorded as being received from CD at Kew on 7 September 1867:

> Angræcum eburneum
> Cymbidium ensifolium
> Maxillaria
> Stanhopea
> Oxalis rubella. hirta. Spe. spe. *Brazil*
> —   speciosa. multiflora
> —   versicolor. cunaefolia.
> —   bifida. & pectinata
> A quantity of Ophrys insectivora
> 2 Lobelias. 2 [Cordias.]
> A pot of seedling Gesnerias
> —   — —   Borrerias.

(Inwards book, Royal Botanic Gardens, Kew.)

³ Hooker's mother, Maria Hooker, who lived in Norwich, had suffered an attack diagnosed as peritonitis and enteritis (see letter from J. D. Hooker, 31 August 1867).

[4] Hooker refers to Frances Harriet Hooker, Mary Elizabeth Lyell, and the meeting of the British Association at Dundee from 4 to 11 September 1867 (*Report of the thirty-seventh meeting of the British Association for the Advancement of Science, held at Dundee*, p. lxxiii).

To J. V. Carus 16 September 1867

Down. | Bromley. | Kent. S.E.
Sep. 16th. 1867

My dear Sir,

First let me thank you for your very kind and honourable little Biography of myself, which you were so kind as to send me.[1]

With respect to the clean sheets, you may rely on it that they shall be sent the very day I receive them.[2] Three or four sheets were returned ready for the press several months ago, and why they have not been printed off I know not. I have almost finished the first proofs of the *whole work* and on October 1st hope to begin on the second proofs or Revises, so there will be some delay after you receive the next two or three clean sheets, about which I have written to the printers today.

I have difficulty in answering about the Dogs, partly from not knowing the foreign breeds, and partly from the changes which the English breeds have undergone during the last century.

Our Bulldog perhaps is nearest to the Canis molossus, as figured by Brehm in his Illust. Thierleben Vol. I, yet differs *greatly*, and I suspect is peculiar to England.[3] My son has made a tracing from a photograph of a first rate dog.[4] The German badgerhound is figured by Brehm p. 365 and resembles a turnspit in form of body, though the turnspit figured by Brehm is not characteristic. The German badger hound is not known in England and therefore I added this epithet.[5] The fox hound is figured by Brehm p. 376.[6] The deer hound is a term applied to two animals viz: the common deer-hound like a large fox hound; & the *Scotch deer*-hound which is a gigantic rough greyhound.—

The retriever is believed to be a cross of Newfoundland and Spaniel. The Spaniel (a small var.) is figured by Col! Ham. Smith plate 15.[7] The Setter is well figured by Brehm & by Col. Ham. Smith p 379,[8] and is no doubt descended from the Spaniel, but differs greatly in size form and instinct. Please observe that the Canis avicularius according to Brehm p 370 is the pointer. Perhaps I translated Rütimeyer's Jagdhund &c wrongly.[9]

With respect to pigeons what you propose about the frill-back seems correct.[10] The carrier is the Turkische taube, and the barb is the Indische taube But I will send by this post Neumeister's work, which please sometime to return to me.—[11] I have marked with red letters the English names about which I am certain. I never attended much to the Toy pigeons— You will see an archangel figured. The dragon is a subvar. of the carrier. I do not think the Germans have any quasi-generic name for Runts; the French call them "pigeons romains".[12]

It will always give me pleasure to do my very best in answering your questions. Do not be too much discouraged by my first volume for I really think the second is more interesting.

Pray believe me | My dear Sir | Yours very sincerely | Charles Darwin

You are a most conscientious Translator, & I am sure I am grateful for all the trouble, which you have taken.—

[Enclosure]

Bull-Dog
From photograph of "Harold" a two-year old dog weighing 45 pounds

LS(A)
Staatsbibliothek zu Berlin, Sammlung Darmstaedter (Carus 6)

[1] Carus had written a biographical entry on CD for a German encyclopaedia (see *Correspondence* vol. 14, letter from J. V. Carus, 15 November 1866). The encyclopaedia has not been identified and the version of the biography sent to CD has not been found.

[2] Carus was translating *Variation* into German, and had asked CD to have the next set of clean proof-sheets sent to him (see letter from J. V. Carus, 11 September 1867).

[3] CD refers to the first volume of Alfred Edmund Brehm's illustrated guide to animal life (Brehm *et al.* 1864–9). CD had received the first four volumes of the work from his Russian translator, Vladimir

Onufrievich Kovalevsky (see letter to V. O. Kovalevsky, 24 June [1867]). Carus translated 'bulldog' as 'der Bullenbeisser' (Carus trans. 1868, 1: 52).

[4] It is not known which of CD's sons made the enclosed tracing.

[5] See Brehm *et al.* 1864–9, 1: 365. In *Variation* 1: 28, CD referred to 'the turnspit-like German badger-hound', which Carus translated simply as 'deutsche Dachshund' (Carus trans. 1868, 1: 35).

[6] See Brehm *et al.* 1864–9, 1: 376.

[7] CD refers to Charles Hamilton Smith and C. H. Smith 1839–40.

[8] Plate 15 of C. H. Smith 1839–40 has figures of both a spaniel and a setter. The page reference is to Brehm's illustration of a setter (Brehm *et al.* 1864–9, 1: 379).

[9] Brehm *et al.* 1864–9, 1: 370. In *Variation* 1: 19, CD had translated a passage from Ludwig Rütimeyer's study of the fauna of prehistoric Switzerland (Rütimeyer 1861). He translated 'Jagdhund und Wachtel-hund' as 'hounds and setters or spaniels'.

[10] For Carus's suggested translation, see the letter from J. V. Carus, 11 September 1867.

[11] Carus had been unable to find a copy of Gottlob Neumeister's book on pigeon breeding (Neumeister 1837: see letter from J. V. Carus, 11 September 1867). CD's annotated copy of Neumeister 1837 is in the Darwin Library–CUL (see *Marginalia* 1: 640–1).

[12] Neumeister 1837 included fifteen coloured plates showing different pigeon breeds organised into groups. Each plate in CD's copy is heavily annotated by CD and the English names of some breeds added in red pencil. The figure of the archangel pigeon (Gimpel-Taube) is on plate 13. Plate 14, which includes carrier pigeons, does not include a dragon. Carus translated 'runts' as 'Runt-Tauben' (Carus trans. 1868, 1: 175); the word 'Römertauben' is generally used today.

From Andrew Murray 16 September 1867

Duncrivie | By Milnathort
16 Sept 1867

My dear Sir

Your letter has been forwarded to me here, where I am luxuriating in draughts of Native air.—[1]

I am truly obliged by your sending me the Italian Memoirs—the more that it is a token that you mean to take an interest in the New Journal.—[2] I shall look at them when I return to town (next week)—and if they seem suitable will get them reviewed by a Competent Italian Scholar. My Italian goes very much upon Crutches.—

I launched the Journal at the Meeting of the British Association at Dundee—that is—made known the prospectus on the other side and I think the general opinion so far as it reached me was in favour of it.—[3] Of course I am not a very good judge, as I would be the last to hear anything depreciatory.—

I have still room for one or two occasional contributors & if you know any qualified persons who would be content with a remuneration of 8£ per sheet for original articles & £5 for reviews send them to me.—

With many thanks I am | Yours very truly | And^w. Murray

DAR 171: 329

[1] CD's letter has not been found. Murray had attended the annual meeting of the British Association for the Advancement of Science held at Dundee from 4 to 11 September 1867 (*Report of the thirty-seventh*

meeting of the British Association for the Advancement of Science, held at Dundee, p. lxxiii). Duncrievie is a village near Milnathort, a town about twenty-five miles north-east of Stirling (Bartholomew 1943).

[2] Murray had asked CD for advice on the journal he was starting up (see letter from Andrew Murray, 12 August 1867 and n. 1). CD sent the papers he had recently received from Federico Delpino (Delpino 1867a and 1867b; see letter from Federico Delpino, 5 September 1867 and n. 2, and letter from Andrew Murray, 13 December 1867). An article by Murray on Delpino 1867b appeared in the third number of the *Journal of Travel and Natural History* (A. Murray 1868).

[3] Murray wrote his letter on a blank page on the back of the leaflet advertising the *Journal of Travel and Natural History*.

From Asa Gray [after 17 September 1867][1]

Dear Darwin

I have said *yes* to this reminder of an old promise.—[2] But I sh^d like to have time to turn over the pages & ponder upon them, and perhaps to exchange thoughts with you, before I attempt a review of your *opus*.[3] By 1st of Oct, we have an inter-national *book post*, and you could, perhaps, send me sheets somewhat in advance.[4]

A. Gray

Postmark: 'London OC 12 67'
DAR 165: 158

[1] The date is established by the date of the letter to Gray written on the same sheet of paper as this letter (see n. 2, below).

[2] Gray's letter was written on the blank pages of the folded sheet of a letter to him from Edwin Lawrence Godkin, dated 17 September 1867. The letter text follows:

My dear Sir,

Two years ago or more when the *Nation* was first started—you promised me that when Darwin's new book appeared, you would write us a notice of it. I see it is announced now. I write to ask whether convenience & inclination will allow you to take it up for the *Nation*.

I remain, dear Sir, | Very faithfully yours | E L Godkin

Variation had been advertised as forthcoming as early as April 1865 (see *Correspondence* vol. 13, letter to John Murray, 31 March [1865] and n. 2). The *Nation* was a weekly newspaper published in New York, and first appeared on 6 July 1865 (*ANB* s.v. Godkin, Edwin Lawrence).

[3] Gray's review of *Variation* appeared in the 19 March 1868 issue of the *Nation* ([A. Gray] 1868).

[4] No details of the proposed international book post have been found; for more information on the history of US postal rates, see Jane Kennedy 1957.

From John Murray 19 September [1867][1]

50, *Albemarle S^t* | *W.*
Sept 19

My Dear Sir

I have made up the account of the new Edition of Darwin on Species—of wch I have now disposed of about 700 copies—& have much pleasure in enclosing a Cheque for £250" being $\frac{2}{3}$^{ds} of the estimated profits—allowing £50" for advertising of wch £30' have been actually expended—[2]

I enclose also the account of The Orchids,[3] by wch you will perceive that the deficiency has been considerably reduced—

I remain My Dear Sir | Yours very faithfully | John Murray

I hope I have found a suitable Index-maker.[4] I will ask for a sample for your approval

Chas Darwin Esq

[Enclosure]

Darwin on Orchids

Dr
1862

June	To Balance Deficiency[5]	45	5	1
Sept	" Electros[6]	3	6	–
1867				
June	" Binding 100 Copies	3	–	5
	" Advertising 5 Years	20	10	–
		72	1	6

1867

June 30	To Balance Deficiency	22	2	1

Cr
1862

June   By 685 on hand

              4 Presented

520   516 on hand June 1867

165 Sold viz

52 Trade 25 as 24 6/	15	–	–
113 d⁰ " " 6/5[7]	34	19	5

165

1867

June 30	By Balance Deficiency	22	2	1
		72	1	6

June 30   By 516 on hand

DAR 171: 350, 524

CD ANNOTATION[8]
Verso of last page:

250

12

———

238

[1] The year is established by the relationship between this letter and the letter to John Murray, 20 September 1867.

[2] The reference is to the fourth edition of *Origin*, which was published in November 1866 (Freeman 1977). The publisher's ledger records a payment to CD of £250 on 19 September 1867 (see Peckham ed. 1959, p. 776).

[3] Murray refers to *Orchids*.

[4] Murray refers to the index for *Variation*. The book was indexed by William Sweetland Dallas (see letter from John Murray, 1 November [1867]).

[5] This figure was carried over from the last statement Murray had sent for sales of *Orchids* (see *Correspondence* vol. 13, Supplement, letter from John Murray, [1 July – 23 August 1862]).

[6] After receiving the statement for *Orchids*, CD reminded Murray to charge him for the cost of three electrotypes sent to the United States (see *Correspondence* vol. 10, letter to John Murray, 24 August [1862] and n. 4).

[7] Trade purchasers received one free copy in every twenty-five, as was the general practice (Plant 1965, p. 405), and were charged 6s. or 6s. 5d. The retail price of *Orchids* was 9s. See Freeman 1977, p. 113.

[8] CD's annotation refers to his payment of £12 for presentation copies. See letter to John Murray, 20 September 1867 and n. 2.

From J. D. Hooker [20 September 1867][1]

Kew

Friday,

Dear D.

We shall be delighted to see you Sunday or Monday or Tuesday either—& I am sure to be in at 10 am— there will be lunch at 1 pm & you can stay or no as you like.[2]

I left my Mother better at Norwich last Monday but I fear permanently invalided, as the seat of her malady is not clear.[3]

Hoping to see you Sunday Monday or, (or *and*) Tuesday

Ever yr aff | J D Hooker

I felt so above suspicion, that I am not in the least elated at your having found that *I* had not mislaid y[r] Adam Bede.[4]

DAR 102: 179

[1] The date is established by the relationship between this letter and the letter from J. D. Hooker, [14 September 1867], and by the dates of CD's visit to London (see n. 2, below). The Friday following 14 September 1867 was 20 September.

[2] According to CD's 'Journal' (Appendix II), he was in London from 18 to 24 September 1867.

[3] Hooker's mother, Maria Hooker, had suffered an attack diagnosed as peritonitis and enteritis (see letter from J. D. Hooker, 31 August 1867).

[4] Hooker refers to Eliot 1859. CD evidently found his copy, which he had lent to Hooker in 1865, at his brother Erasmus Alvey Darwin's house (see letter from J. D. Hooker, 30 July [1867] and n. 3).

To John Murray 20 September 1867

6 Queen Anne St | W.

Sept 20th 1867

My dear Sir

I acknowledge receipt & thank you for the very pleasant remittance of £250 for the 4th. Edit. of Origin.—[1]

I enclose cheque for Presentation Copies;[2] be so kind as to return me the enclosed receipted.— I am very glad to see that Orchid Book does sell a little.[3] I hope the new Book will be more successful;[4] but Heaven only knows.

I think it will be a very good plan that I may see a portion of Index, before, I presume, it is arranged alphabetically & I will compare it with the early sheets.—[5]

With my ⟨best than⟩ks | yours very sincerely | Ch. Darwin

John Murray Archive

[1] See letter from John Murray, 19 September [1867] and n. 2.

[2] An entry in CD's Account books–banking account (Down House MS) for 20 September 1867 records a payment of £12 for presentation copies.

[3] According to the account Murray sent, 165 copies of *Orchids* had been sold and 516 remained (see enclosure to letter from John Murray, 19 September [1867]).

[4] CD refers to *Variation*; he had almost finished correcting the first proofs of the work and planned to begin working on the second proofs on 1 October 1867 (see letter to J. V. Carus, 16 September 1867).

[5] See letter from John Murray, 19 September [1867] and n. 4.

From John Murray 23 September [1867][1]

Sept 23

Dear Mr Darwin

The note enclosed shows the anxiety of Mr Schweizerbart to get forward with the Translation. I have assured him the sheets will be forwarded as fast as thrown off—[2]

I hope I have found a fit Index-Maker in a Mr Dallas Keeper of the York Museum but I have not yet a reply to my application[3]

I am Dear S | Yours faithfully | John Murray

Ch Darwin Esq

DAR 171: 351

[1] The year is established by the relationship between this letter and the letter from John Murray, 19 September [1867].

[2] The enclosure has not been found. Christian Friedrich Schweizerbart was the director of the German firm, E. Schweizerbart'sche Verlagsbuchhandlung, that published the German translation of *Variation*. Julius Victor Carus, CD's German translator, had also written requesting proof-sheets (see letter from J. V. Carus, 11 September 1867).

[3] William Sweetland Dallas was curator of the Yorkshire Philosophical Society's museum (*Modern English biography*).

From John Scott 24 September 1867

Roy. Botanic Gardens | Calcutta
24th. Sept. 1867

Dear Sir,

I was much pleased to receive your letter by last mail though sorry when I had read it to think that I could do so little to meet your requests.[1] Permit me to say however that I shall always continue proud to do my best: you can always command my services.

With respect to Vandellia, I am unfortunately ignorant of the species you want.— I do my best however, in the circumstances and enclose you fresh seeds of the only two species which I can find in our neighbourhood—[2] Lately I have got a pretty species from Silhet, I expect to find a pod of this also to enclose for you.[3] I may state however that I have seen **no** *closed flowers* on any of them—though there are a profusion of perfect flowers in the two common species. Possibly like Viola, they may produce closed flowers in another season. I shall look to them in the ensuing cold season—[4]

Since you wrote me, I have examined the genus in the Herbarium, and did not find a *single* species without **perfect** *flowers*.[5] As you ask me if ever I have seen Vandellia with perfect flowers, I append enumeration of the species I have examined they may interest you and may possibly thus also hit upon the species which you have in view. I enclose list—all exhibiting perfect flowers—

Perhaps you might like a few seeds of our Viola Roxburghiana, Voigt., which I find produces perfect flowers in the *cold season* **only**; in the *hot* season a sparing number of imperfect, and in the *rains* a profusion of the same, all *perfectly fertile*.[6] I enclose seeds—and I shall later send you seeds of a new species, I had lately from the Sikkim Terai.[7] This also produces some closed flowers only though the collector tells me he saw it in flower in its native habitats in February— the sexual economy is thus probably similar to above—

Leersia has not produced a *single perfect flower* with me: though I have now had it growing freely for upwards of two years. I am sorry I have no seeds of it by me just now—and there are only a very few unripe ones on my plants. I shall send you a few in time for sowing next spring. I have sent seeds of it to Darjeeling for cultivation and return of seeds. By such interchanges I do hope to effect a change in its sexual economy, though thus I have failed.[8]

Now for *Adenanthera*—which indeed is a perplexing case.[9] When you wrote I could not think of ever having seen a bird touch the seed. Fortunately we have plants in the gardens covered with seeds now, & so have specially attended to it— — The seeds assume their rich scarlet hue and become quite hard previous to the dehiscence of pod, this taking place gradually from the apex and accompanied by the tortion of the valves, which internally are of a silky grey colour, and thus exhibit to advantage the pendent scarlet seeds. One can readily imagine what pretty objects large trees of this plant must form when covered (as I am told they frequently are in Molucca)[10] with the dehisced pods, and the full number of seed (12) or more pendent from each— In trees in our Garden, I have only seen at any one time a *few seeds* in *each pod* exposed— The lower in apical seeds having always dropped off ere those in the basal part of pod had been exposed.

I had seen no birds touch the seeds until by the merest accident when passing one of the Small Sulphur-Crested Cockatoos—Cacatua sulphurea—belonging to Dr. Anderson.[11] I showed it a few seeds of Adenanthera when it quickly alighted from its perch and picked a few from my hands, and with little difficulty split their hard testa, and eat with seeming gusto the embryonic parts only rejecting the coloured covering.

I next put the bird on a tree (seed bearing) of Adenanthera: it scrambled quickly up to where the seeds where, grasping the branch with one foot, it with the other caught the pendent pods; retaining this position until it had filled its bill with the *exposed* seeds then re-erecting itself it commenced the splitting process, which it does with facility by grasping a single seed in its foot and holding it in the concave extremity of lower mandible while it crushes it with the strong curved extremity of upper. You will readily understand the adaptability of the Cockatoos bill for such work—and in carrying the co-adaptation further— I may remark that the Cockatoo dislikes the unripe seeds, in general rejecting them. It shows no aptitude either in *opening* the pods, but usually drops them when it has picked out the exposed seeds—were it otherwise Adenanthera might soon be a rare plant in the haunts of Cockatoos; but as it is they are clearly a means for its dissemination. Thus as I have observed it generally takes several seeds into its bill before posing itself for the splitting and I invariably see that fully 50 percent of these are dropped uninjured while it splits the remainder.— — I can't say whether the seeds, if swallowed *accidentally* by the Cockatoo would be passed unimpaired, however I find they are by the Common Fowl (G. domesticus).[12] It naturally rejects the seeds, but I made the experiment by forcing a few thoroughly ripe and hard seeds on one, and a few unripe seeds on another— In the latter they were digested— in the former past uninjured.... . The seeds of Adenanthera unlike those of many species of related genera *e.g.* Prosopis spicigera, Linn., several of the Ingas of which I may notice I. dulcis as one whose seeds are imbedded in a mealy pulp of which the Indian Crow is remarkably fond) are *entirely destitute* of any *pulp*.[13]

In conclusion allow me to refer to the apparent Co-relations which we at times observe between the colours of flowers and seeds, when variations occur in the former—[14] Thus Canavalia gladiata, D.C. has **varieties** with *flowers* and *seeds red* & *fls* & *seeds white*, and also a variety (I must state) with a variation in the flowers (white) while the *seeds* continue *red*. Again, Abrus precatorius has vars. with rose-col. fls. & *red* seeds with a dark eye: *white* flowered var. greyish seeds with eye brown: and a second whitish flowered variety has black seeds with a white eye. Similar cases also occur in Dolichos sinensis; Lablab vulgaris etc—[15] but I need not enlarge as you will no doubt be familiar with many such cases, and have fully considered their bearing on the case in point (Adenanthera)

Reflecting on such variations I had previous to the receipt of your letter, regarded, I believe too lightly the end to which colour in seeds may at times be subservient— The preeminent beauty of the Adenanthera is remarkable— I shall direct my observations more to the subject.

I am always keeping in view your queries about expression—[16] I find it however most difficult to make exact and satisfactory observations— Several of the more simple, I am now prepared to answer you, however, as the time you gave me is yet far from expired I shall continue my observations, and send you them at once as complete as I can—[17]

I am glad to say that I am extremely comfortable in my position here—for which I feel ever thankful to you[18] and remain | Dear Sir | Most respectfully yours | J. Scott

P.S. over I append list of Vandellias

List of Vandellias in which I have observed perfect flowers—

Vandellia	mollis, Benth.—	from my parts of India
—	erecta Benth.—	D^o —
—	nummularafolia, D. Don.	D^o —
—	angustifolia, Benth—	D^o —
—	crustacea,—Benth	D^o —
—	molluginoides, Benth.	
—	multiflora, f. Don.	
—	scabra, Benth—	
—	laxa, Benth[19]	

DAR 157a: 106

CD ANNOTATIONS

1.1 I was ... failed. 5.6] *crossed blue crayon*
3.3 I append] *after opening square bracket, red crayon*
4.1 Perhaps ... above— 4.7] 'Violets' *in margin, red crayon*
4.6 the sexual ... years. 5.2] *scored red crayon*
6.1 Now] *after opening square bracket, blue crayon*
7.1 I had] '(About conspicuous seeds)' *added in margin above, blue crayon; square brackets in original*

[1] CD's letter has not been found, but in a letter to Fritz Müller he mentioned that he had written to India to try to find out how *Adenanthera* seed was disseminated (letter to Fritz Müller, 15 August [1867]; see n. 9, below).

[2] In *Cross and self fertilisation*, p. 90, CD reported having received seeds of *Vandellia nummularifolia* from Scott (see also *Forms of flowers*, pp. 324–5). The other species Scott sent has not been identified. Notes made by CD on 5 and 11 October 1868 recording experiments on *V. nummularifolia* are in DAR 111: 25–6. Notes comparing seed production of open and cleistogamic flowers of *V. nummularifolia*, dated 1 November 1875, are in DAR 111: 28.

[3] Silhet was a city in eastern Bengal (now Sylhet, Bangladesh; *Columbia gazetteer of the world*).

[4] CD was interested in species that produced closed (cleistogamic) flowers, which were invariably self-pollinated. For more on *Viola* in this context, see *Correspondence* vol. 13, letter from B. D. Walsh, 1 March 1865 and n. 12, and *Forms of flowers*, pp. 314–21.

5 The existence of open or 'perfect' (that is, non-cleistogamic) flowers in species that also produced closed flowers lent support to CD's view that no species was perpetually self-fertilised (see *Origin*, pp. 96–101, and *Orchids*, p. 359). CD discussed the relation of cleistogamic to open flowers in *Forms of flowers*, pp. 335–45.

6 CD reported the results of his experiments with *Viola roxburghiana* (a synonym of *V. patrinii*) and cited Scott's observations in *Forms of flowers*, pp. 319–20. CD's notes on *V. roxburghiana* are in DAR 111: 18.

7 The new species referred to was *Viola nana* (a synonym of *V. tricolor*); CD reported receiving the seeds from Scott and discussed his observations on the species in *Forms of flowers*, p. 319. CD's notes on *V. nana* are in DAR 111: 13. The Terai is a region of south Nepal and north India below the southern Himalayan foothills (*Columbia gazetteer of the world*).

8 Scott refers to experiments with the grass *Leersia oryzoides*; he had asked CD for seeds in 1865 (see *Correspondence* vol. 13, letter from John Scott, 21 July 1865). CD had published observations on *Leersia* in 'Three forms of *Lythrum salicaria*', pp. 191–2 n. (*Collected papers* 2: 131). In *Forms of flowers*, p. 335, CD reported sending seeds from cleistogamic flowers to Scott.

9 Scott refers to the fact that *Adenanthera pavonina* produces conspicuous seeds with no apparent nutritive value. See letter to Fritz Müller, 15 August [1867] and n. 11. For more on CD's interest in this species and the dissemination of its seeds, see *Correspondence* vol. 14.

10 The Moluccas (now called Maluku) are an island group of eastern Indonesia between Sulawesi and New Guinea (*Columbia gazetteer of the world*).

11 The lesser sulphur-crested cockatoo (*Cacatua sulphurea*) is native to Sulawesi and many of the surrounding islands, where *Adenanthera pavonina* is also found (*Birds of the world*; Corner 1988, 1: 450). Thomas Anderson was the superintendent of the Calcutta botanic garden.

12 *Gallus domesticus*. For CD's earlier experiment feeding seeds of *A. pavonina* to a fowl, see *Correspondence* vol. 14, letter to J. D. Hooker, 10 December [1866].

13 Like *Adenanthera pavonina*, *Prosopis spicigera*, and *Inga dulcis* (a synonym of *Pithecellobium dulce*) belong to the subfamily Mimosoideae of the family Leguminosae (Allen and Allen 1981). Scott refers to *Corvus splendens*, the house crow (*Birds of the world*).

14 CD discussed correlated variation of flower, leaf, and seed colour in plants in *Variation* 2: 330.

15 Scott refers to *Canavalia gladiata*, *Abrus precatorius*, *Dolichos sinensis* (a synonym of *Vigna sinensis*), and *Lablab vulgaris* (a synonym of *Dolichos lablab*).

16 CD evidently enclosed a copy of his queries about expression (for a printed version, see Appendix IV) in his letter to Scott (see n. 1, above), or in an earlier letter.

17 The printed version of the queries on expression suggested that answers be sent within about a year (see Appendix IV). Scott sent his replies to the queries with his letter of 4 May 1868 (*Correspondence* vol. 16).

18 CD had helped Scott to find employment in India and had given him financial support (see *Correspondence* vol. 13, letter from John Scott, 20 January 1865 and n. 6).

19 Scott refers to *Vandellia mollis*, *V. erecta* (a synonym of *V. multiflora*), *V. nummularafolia*, *V. angustifolia*, *V. crustacea*, *V. molluginoides*, *V. multiflora*, and *V. laxa* (a synonym of *V. scabra*).

From John Lubbock 28 September [1867][1]

High Elms

28 Sep

Dear M�におけるr. Darwin

I return you with many thanks Haliburtons ingenious paper, & send you a paper which gives almost at full length something of my own which is partly in answer to it.[2]

Will you kindly return me the newspaper.

I hope you have been pretty well & that the Book is making satisfactory progress.[3]

We had a capital meeting at Dundee & I have been since in the Orkneys & Shetlands.[4]

Ever | Yours affecy | John Lubbock

C Darwin Esq

DAR 170: 58

[1] The year is established by the reference to the British Association for the Advancement of Science meeting at Dundee (see n. 4, below).

[2] Lubbock refers to Robert Grant Haliburton and a paper discussing superstitions connected with sneezing as a proof of the unity of origin of the human race (Haliburton 1863; CD's copy is in the Darwin Pamphlet Collection–CUL; see also *Correspondence* vol. 11, letter to J. D. Hooker, 23 [June 1863] and n. 8). Lubbock evidently sent CD a copy of a paper, later published in *Transactions of the Ethnological Society*, in which he discussed Haliburton 1863 (Lubbock 1867; CD's annotated copy of an offprint of this article is in the Darwin Pamphlet Collection–CUL). In his paper, Lubbock argued that the existence of similar ideas in distant countries owed its origin not to humans having once lived all in close association with one another, as Haliburton suggested, but rather to 'the original identity of the human mind' (see Lubbock 1867, p. 341).

[3] According to Emma Darwin's diary (DAR 242), CD had recently been unwell with eczema (see also letter from H. B. Jones to Emma Darwin, 1 October [1867]). CD was correcting proof-sheets of *Variation*; he had recently reported that he had almost finished the first proofs and planned to start work on the second proofs at the beginning of October (see letter to J. V. Carus, 16 September 1867).

[4] The British Association held its annual meeting for 1867 at Dundee from 4 to 11 September 1867 (*Report of the thirty-seventh meeting of the British Association for the Advancement of Science, held at Dundee*, p. lxxiii). The Orkneys and Shetlands are groups of islands in the North Sea, off the north coast of Scotland.

From Henry Bence Jones to Emma Darwin   1 October [1867][1]

5 Albion Villas | Folkestone

Oct 1.

Dear Mrs Darwin

This sudden temporary failure of memory by itself and in itself does not appear to me of importance.

Like a temporary affection of the sight which often occurs, it probably is only caused by some irregularity of circulation arising from indigestion.

As long as it only comes by itself and with no other nerve symptoms I do not consider that any concern need be felt about it.

The sudden coming in and out of the Eczema shews that there is some unusual state of the circulation & digestion also.[2]

Probably the increased mental work has determined the appearance of the new symptom in the new place.[3]

Now as to what should be done the best course would be to put mental work aside *altogether* for this month; except on wet days; & to think & do only what is best for the health.

It is a fine month to be out riding walking, driving as much as possible

To be very strict in the diet and to take a teaspoonful or two of compound tincture of Jentian or tincture of Chyretta with a few drops of mineral acid in a wineglass of water half an hour before breakfast & dinner for three weeks[4]

I shall stay on here as long as I can go yachting probably until the 28 of this month when I hope to return to Brook S^{ts}

I live between Boulogne Calais Margate and Dungeness for seven or eight hours daily and I am getting much stronger.[6]

I wish Mr Darwin could have as much air without fatigue as I have; he w^d soon digest better

Pray give him my kind regards I was about to write to him to ask how he was. I shall be very glad of a few words from you at any time about him

Believe me | Yours very truly | H Bence Jones

DAR 168: 78

[1] The year is established by the references to CD's eczema and Jones's improving health (see nn. 2 and 6, below).

[2] Emma Darwin had noted in her diary (DAR 242) for 21 September 1867, 'CD very unwell with eczema all week', and for 27 September, 'C bad'. There is no mention of other symptoms. Jones had been CD's doctor since 1865 (*Correspondence* vol. 13).

[3] CD had been correcting proofs for *Variation*; by mid-September he reported that he had almost finished the first proofs and planned to start on the second proofs at the beginning of October (see letter to J. V. Carus, 16 September 1867).

[4] Jones refers to gentian (*Gentiana lutea*) and chiretta (*Swertia chirata*), both members of the family Gentianaceae. A tincture of either plant was used as a digestive tonic as well as in the treatment of skin diseases. Mineral acid is a generic term for inorganic acids (*OED*). On the use of acids in the treatment of stomach disorders, see Ringer 1869, pp. 80–2.

[5] Jones's London address was 31 Brook Street (*Post Office London directory* 1866).

[6] Jones had been seriously ill with heart disease in late 1866 and early 1867 (see letter to W. D. Fox, 6 February [1867]; see also Kyle 2001).

To J. T. Moggridge 1 October [1867][1]

Down Bromley
Oct. 1.

Dear M^r. Moggridge

I am much obliged to you for telling me of your expected departure and kind offer of assistance, but I have been working all the summer so hard at proof sheets, that I have attended to nothing else, and therefore I have no favour to beg.[2]

I hope you will pass a pleasant winter and that your health will improve.[3] I suppose and hope that you will still attend to ophrys. As there did not seem any probability of the plants which you so kindly gave me undergoing any greater modification, I sent them to Kew where they are much valued.[4]

The plants of Ononis have interested me much.[5] Should you have any opportunity I shall be much obliged if you will make further enquiries about the spontaneous crossing of vars. of common and sweet Peas.[6]

Believe me

Copy
DAR 146: 376

[1] The year is established by the reference to plants sent to Kew (see n. 4, below).

[2] No letter from Moggridge to CD announcing his departure has been found. Moggridge spent winters in France (see n. 3, below), and regularly offered CD assistance with botanical research, often sending seeds and plant specimens (see *Correspondence* vols. 12–14, and this volume, letter from J. T. Moggridge, 22 April [1867]). CD was correcting the proofs of *Variation* and had planned to begin work on second proofs at the beginning of October (see letter to J. V. Carus, 16 September 1867).

[3] Owing to chronic ill health, Moggridge spent most winters at Mentone (now Menton), a town on the French Riviera near the Italian border (R. Desmond 1994).

[4] Moggridge had sent CD specimens of different varieties of *Ophrys* in December 1865 and February 1866 (see *Correspondence* vol. 13, letter from J. T. Moggridge, 27 December [1865], and *Correspondence* vol. 14, letter from J. T. Moggridge, 15 February [1866]). 'A quantity of Ophrys insectivora' (i.e. *Ophrys insectifera*), is recorded as having been received from CD at the Royal Botanic Gardens, Kew, along with several other plant specimens, on 7 September 1867 (Inwards book, Royal Botanic Gardens, Kew); see also letter from J. D. Hooker, [14 September 1867] and n. 2). For more on CD's interest in *Ophrys* varieties, see *Correspondence* vol. 13.

[5] CD was interested in the unopened (later called cleistogamic) flowers of *Ononis*. Moggridge sent CD seeds of *O. columnae* in July and August 1866, and seeds of *O. minutissima* in November 1866 (see *Correspondence* vol. 14, letters from J. T. Moggridge, 5 and 6 July [1866], 3 August [1866], and 9 November [1866]). In a note dated 8 May 1867, CD described the flowers of *O. columnae* and *O. minutissima* raised from seed sent to him by Moggridge (DAR 111: A21). CD discussed his experiments with *O. minutissima* in *Cross and self fertilisation*, pp. 167–8, and his experiments with *O. columnae* and *O. minutissima* in *Forms of flowers*, pp. 325–6.

[6] Moggridge had promised CD that he would try to get evidence of spontaneous crossing in peas (see *Correspondence* vol. 14, letter from J. T. Moggridge, 9 November [1866]; see also *ibid.*, letter to J. T. Moggridge, 13 November [1866]).

From A. R. Wallace   1 October [1867][1]

76½, Westbourne Grove | Bayswater. W.
Oct.ʳ 1ˢᵗ.

Dear Darwin

I am sorry I was not in town when your note came.[2] I took a short trip in Scotland after the Brit. Ass. Meeting; and went up Ben Lawers. It was very cold and wet and I could not find a companion or I should have gone as far as Glen Roy.[3]

My article on *"Creation by Law"* in reply to the Duke of Argyle and the North British Reviewer, is in the present month's Number of the "Quarterly Journal of Science".[4] I cannot send you a copy because they do not allow separate copies to be printed. There is a nice illustration of the *predicted* Madagascar Moth and *Angræcum sesquipedale*.[5]

I shall be glad to know whether I have done it satisfactorily to you, and hope you will not be so very sparing of criticism as you usually are.

I hope you are getting on well with your great book. I hear a rumour that we are to have *one* vol. of it about 'Xmas.[6]

I quite forget whether I told you that I have a little boy, now three months old, and have named him "Herbert Spencer",—(having had a brother Herbert.)[7] I am now staying chiefly in the country at Hurstpierpoint but come up to town once a month at least. You may address simply

> Hurstpierpoint
> *Sussex.*[8]

Hoping your health is tolerable & that all your family are well | Believe me Dear Darwin | Yours very faithfully | Alfred R. Wallace—

Charles Darwin Esq.

DAR 106: B43–4

CD ANNOTATIONS[9]
4.1 I hope ... 'Xmas. 4.2] *scored red crayon*
5.3 Hurstpierpoint] 'Address' *added red crayon*
Top of letter: 'Leach' *red crayon*; 'Crossing' *pencil, circled pencil*
End of letter: 'Victoria Institute' *red crayon*

[1] The year is established by the reference to Wallace's article in the *Quarterly Journal of Science* (A. R. Wallace 1867c).

[2] CD's letter has not been found, but he was in London from 18 to 24 September 1867 (see 'Journal' (Appendix II)).

[3] Wallace refers to the annual meeting of the British Association for the Advancement of Science, which was held in Dundee from 4 to 11 September 1867 (*Report of the thirty-seventh meeting of the British Association for the Advancement of Science held at Dundee*). Ben Lawers is a mountain just north of Loch Tay, Perthshire. Glen Roy is in the district of Lochaber, southern Inverness-shire. CD visited Glen Roy in June 1838 and in 1839 published 'Parallel roads of Glen Roy', in which he presented a theory of the marine origin of the 'roads' (for more on CD and Glen Roy, see *Correspondence* vol. 2 and *Correspondence* vol. 9, Appendix IX; see also Rudwick 1974).

[4] George Douglas Campbell, the eighth duke of Argyll, had published *The reign of law*, a book critical of CD's transmutation theory, in 1867 (G. D. Campbell 1867); the *North British Review* had published an anonymous article by Henry Charles Fleeming Jenkin, also critical of CD's theory, in June 1867 ([Jenkin] 1867). Wallace responded to these criticisms in his article 'Creation by law' in the *Quarterly Journal of Science* (A. R. Wallace 1867c).

[5] The illustration referred to is in A. R. Wallace 1867c, facing p. 471; see also the frontispiece to this volume. CD received a specimen of *Angraecum sesquipedale* in January 1862 and was astounded by the length of its nectary (see *Correspondence* vol. 10, letter to J. D. Hooker, 25 [and 26] January [1862]). In *Orchids*, p. 198, CD concluded: 'in Madagascar there must be moths with probosces capable of extension to a length of between ten and eleven inches!' The illustration in A. R. Wallace 1867c is an artist's rendition of the moth CD had predicted. The predicted moth, *Xanthopan morgani praedicta*, eventually discovered in 1903, was a subspecies of a moth mentioned by Wallace in his article (*Macrosila morganii*, now *Xanthopan morgani*; see A. R. Wallace 1867, p. 477; see also Kritsky 1991).

[6] Wallace refers to *Variation*; both volumes appeared at the end of January 1868 (Freeman 1977).

[7] Wallace's son was born on 22 June 1867 (Raby 2001, p. 194). Wallace's brother, Herbert Edward Wallace, had died of yellow fever in Brazil in 1851 (*ibid.*, p. 76).

[8] While in Hurstpierpoint Wallace stayed at the home of his father-in-law, William Mitten (Raby 2001, p. 194).

[9] CD's annotations are notes for his reply to Wallace (see letter to A. R. Wallace, 12 and 13 October [1867] and nn. 14 and 20).

To Charles Lyell 4 October [1867][1]

> Down. | Bromley. | Kent. S.E.
>
> Oct 4

My dear Lyell

With respect to the points in y^r note I may sometimes have expressed myself with ambiguity.[2] At the end of Chap. 23 where I say "that marked races are not (*often* (you omit "often")) produced by changed conditions", I intended to refer to the *direct* action of such conditions in causing *variation*, & not as leading to the preservation or destruction of certain forms.[3] There is as wide a difference in these two respects, as between voluntary selection by man & the causes which induce variability. I have somewhere in my book referred to the close connection between Nat. Selection & the action of external conditions in the sense which you specify in your note. And in this sense all Natural selection may be said to depend on changed conditions.[4]

In the "Origin" I think I have underrated (& from the cause which you mention) the effects of the *direct* action of external conditions in producing varieties; but I hope in Chap. 23 I have struck as fair a balance as our knowledge permits.[5]

It is wonderful to me that you have patience to read my slips & I cannot but regret it, as they are so imperfect, they must I think give you a wrong impression; & had I sternly refused, you would perhaps have thought better of my book. Every single slip is greatly altered & I hope improved.—

With respect to the human ovule, I cannot find dimensions given, though I have often seen the statement.[6] My impression is that it w^d be just or barely visible if placed on clear piece of glass.— Huxley[7] could answer your question at once.— I have not been well of late, & have made slow progress, but I think my book will be finished by middle of November.

Yours affectionately | C. Darwin

LS(A)
American Philosophical Society (334)

[1] The year is established by the reference to Lyell's comments on the proof-sheets of *Variation* (see n. 3, below).

[2] Lyell's letter has not been found.

[3] CD refers to Lyell's comments on the proofs of *Variation*; Lyell had received the proofs around the end of July (see letter to Charles Lyell, 18 July [1867], and letter from Charles Lyell, 4 August 1867). In the published version, the last sentence of chapter 23 reads, 'Hence, although it must be admitted that new conditions of life do sometimes definitely affect organic beings, it may be doubted whether well-marked races have often been produced by the direct action of changed conditions without the aid of selection either by man or nature' (*Variation* 2: 292).

[4] See *Variation* 2: 278–85.

[5] In *Origin*, p. 132, CD had written, 'How much direct effect difference of climate, food, &c., produces on any being is extremely doubtful. My impression is, that the effect is extremely small in the case of animals, but perhaps rather more in that of plants.' In the third and fourth editions, the words 'perhaps rather' were deleted (see *Origin* 3d ed., p. 149, and 4th ed., p. 158). When the fifth edition was published in 1869, after the publication of *Variation*, the sentences read, 'It is very difficult to decide how far changed conditions, such as of climate, food, &c., have acted in a definite manner.

There is some reason to believe that in the course of time the effects have been greater than can be proved to be the case by any clear evidence' (*Origin* 5th ed., p. 166).

[6] A human egg cell is about a tenth of a millimetre in diameter. In *Variation* 2: 2, CD refers to 'male or female sexual cells, which are so minute as not to be visible to the naked eye'.

[7] Thomas Henry Huxley.

From J. V. Carus 5 October 1867

Leipzig,
Oct 5th. 1867

My dear Sir,

I am sorry I pother you again, but I cannot help it. It is so very difficult to get the technical terms for many breeds which are either unknown in Germany or only described in pamphlets or periodicals unknown or inaccessible to me. You mention a breed of the game-fowl as "Duckwings". Now first of all, game-fowl is the race of the "fighting cocks"; isn't it? I imagine "duckwing" means having similar markings on the wings as the wild duck. Am I right? What then is "Piles"?[1] The Hamburgh breed has "spangled" and "pencilled" forms. Is "spangled" synonymous or pretty nearly so with "shining"? I think "pencilled" is something like "marked with narrow short lines", what we should call "gestrichelt".[2] Does any technical, zoological term exist for *hackles*? I know these feathers, but as you speak of hackles in the loins, I think there ought to be a more general name for this kind of feathers.[3] The 13th. breed of your list is the Sooty fowl.[4] I find in "*Drechsler, die Zucht-Hühner*" a breed called negro-fowl. I rather doubt if these two breeds are the same, as you say, that the hens alone are characterized so as you describe the breed; but I find no other breed in any way agreeing with your description.[5]

For the present I must stop with the translation, as I have no sheets. About a fortnight ago I got the last. Almost at the same time Schweizerbart wrote me, that the first volume was nearly printed off according to an answer he got from Murray.[6] Most likely the rest (pp. 257–420) will come together.

May I keep Neumeister till the sheets of the translation containing the pigeons are corrected?[7]

I beg your pardon that I troubled you anew. Believe me | Yours most truly, | J. Victor Carus

DAR 161: 61

CD ANNOTATIONS
1.5 "fighting cocks"] *underl pencil*
1.6 "Piles"] *underl pencil*
1.7 "spangled"] *underl pencil*
2.3 first ... Murray. 2.4] *scored pencil*
3.1 May ... corrected? 3.2] *scored pencil*

[1] 'Duckwings' and 'piles' are mentioned as game breeds of fowl in *Variation* 1: 227. In Carus trans. 1868, 1: 279, the former is translated as 'entenflüglige' with the original English term in parentheses, while the latter term is left untranslated.

² In Carus trans. 1868, 1: 282, 'spangled' is translated as 'geflitterte' and 'pencilled' as 'gestrichelte'.

³ In Carus trans. 1868, 1: 314, 'hackles' is translated as 'die Schuppenfedern'.

⁴ See *Variation* 1: 230.

⁵ Drechsler and the book on fowl breeding have not been identified. In Carus trans. 1868, 1: 284, 'sooty fowl' is translated as 'schwarzes Huhn'.

⁶ Carus had requested proof-sheets from CD in his letter to CD of 11 September 1867; the German publisher, Christian Friedrich Schweizerbart, had made the same request of CD's publisher, John Murray (see letter from John Murray, 23 September [1867]).

⁷ Carus refers to Gottlob Neumeister's book on pigeon breeding (Neumeister 1837), which CD had recently sent him (see letter to J. V. Carus, 16 September 1867).

From J. V. Carus 7 October 1867

My dear Sir,

I get in this moment the sheets T, U, X, Y, or p 273–336. Unfortunately the sheet S or *p. 257–272* is still wanting.¹ I hasten to tell it to you at once. I apply to you, as the envelop bears the stamp "Bromley" and I therefore conclude that you were sending the sheets.²

Pray send the missing as soon as possible Believe me | Yours most truly | J. Victor Carus

Leipzig, 39, Elsterstrasse
7 Oct 67

DAR 161: 62

CD ANNOTATION
Top of letter: pencil

$$257$$
$$16$$
$$\overline{}$$
$$273$$

¹ Carus refers to proof-sheets of *Variation* (see letter from J. V. Carus, 5 October 1867 and n. 6).

² Post from Down would have been franked in Bromley (*Post Office directory of the six home counties* 1866). See also letter from J. V. Carus, 5 October 1867, n. 6.

To George Warington 7 October [1867]¹

Down. | Bromley. | Kent. S.E.
Oct. 7.

Dear Sir

I hope that you will not think me presumptuous if I cannot resist the pleasure of telling you how much I admire your argument of the origin of species in the Transact. of the Victoria Institute.² The whole case strikes me as placed in the clearest & most spirited light; & I have no where seen so good an abstract. I quite agree with your Chairman that you have put the whole argument better than I have done.[3] But I disagree with you, & it is the only point on which I do disagree, when you say that there is nothing in your article original.[4] As I am writing I will ask you two questions, but if you cannot answer them easily, pray do not take any trouble on the subject; Firstly. Where have you seen an account of inherited baldness & deficient nails; & 2[ndly] of the case of the plane which sent up an evergreen sucker or shoot.[5]

With sincere admiration of your powers of reasoning & illustration I beg leave to remain dear Sir | yours faithfully | Charles Darwin

P.S. I am charmed with M[r] Ince's argument that the ark was much too big to hold only half a dozen primordial types.[6] M[r] Ince w[d] fully appreciate a simple & beautiful theory which Admiral Fitz Roy published some years ago on the extinction of the Ante-diluvial gigantic quadrupeds, namely that the door of the ark was made too small for them to get in[7]

LS
Royal College of Physicians

[1] The year is established by the reference to Warington 1867 (see n. 2, below).

[2] CD refers to Warington's paper 'On the credibility of Darwinism', read on 4 March 1867 and published in the *Journal of the Transactions of the Victoria Institute* (Warington 1867); CD's annotated copy is in the Darwin Pamphlet Collection–CUL. For more on the Victoria Institute, see the letter to A. R. Wallace, 12 and 13 October [1867] and n. 20.

[3] In Warington 1867, p. 62, the chairman of the meeting is reported as having commented, 'I think he [Warington] has done more justice to Darwinism than the book of Darwin himself.'

[4] CD refers to Warington's statement, 'the present paper makes no claim to originality' (Warington 1867, p. 61).

[5] See Warington 1867, pp. 46–7. The descriptions of these cases are scored in CD's copy.

[6] Warington's paper was discussed at three subsequent meetings of the Victoria Institute. At the second of these, William Ince commented on the size of Noah's Ark and the fact that it must have been 'a great deal too large for eight or ten species only' (*Journal of the Transactions of the Victoria Institute* 2 (1867): 95).

[7] CD probably refers to Robert FitzRoy's 'A very few remarks with reference to the Deluge' in *Narrative* 2: 657–82. FitzRoy wrote, 'The small number of enormous animals that have existed since the Deluge, may be a consequence of this shutting out of all but a very few' (*Narrative* 2: 671 n.).

To J. V. Carus 8 October [1867][1]

Down. | Bromley. | Kent. S.E.

Oct 8.

My dear Sir

The game cock is the same as what you call the fighting cock. The "duck-wing" is a sub-breed of the game with wings marked as you suppose. "Piles" is another sub-breed of the game with much white about it, & which I do not think worth describing to you. I would suggest in these cases to use within inverted commas the English names.[2] "Pencilled" fowls have feathers marked with narrow transverse, dark lines: "Spangled" fowls have feathers tipped with a spot of dark colour; these feathers are said to be "laced" when the dark colour runs some way up both margins.[3]

I do not know any proper term for "hackles"; you will see nearly the same elongated feathers on the neck & loins of the cock; those on the loins are often called by Breeders "saddle-feathers".[4] There are two breeds of fowls with black bones; I know of no European name for the breed in which the hens alone are thus characterized.[5] You will I hope have received by this time some additional clean sheets.[6]

Pray keep Newmeister as long as you like.[7]

I beg you always to ask me any questions you like & believe me my dear Sir | yours very faithfully | Ch. Darwin

LS(A)
Staatsbibliothek zu Berlin, Sammlung Darmstaedter (Carus 7)

[1] The year is established by the relationship between this letter and the letter from J. V. Carus, 5 October 1867.
[2] See letter from J. V. Carus, 5 October 1867 and n. 1.
[3] See letter from J. V. Carus, 5 October 1867 and n. 2.
[4] See letter from J. V. Carus, 5 October 1867 and n. 3.
[5] See letter from J. V. Carus, 5 October 1867 and n. 5.
[6] Carus reported receiving additional proof-sheets of *Variation* in his letter to CD of 7 October 1867.
[7] CD refers to Gottlob Neumeister's book on pigeon breeding, which he had sent to Carus (Neumeister 1837; see letter from J. V. Carus, 5 October 1867 and n. 7).

From Ferdinand von Mueller 8 October 1867

To Charles Darwin Esq. with Ferd Muellers best salutation 8/10/67.[1]

[Enclosure][2]

Mansfield
12 Septbr 67

Lieber Herr Doctor.

Sonnabend d 7'ts kam ich hier, an meiner Tour nach Mount Buller,[3] am Sonntag hatte ich die Freude diesen majestätischen Berg mit seiner Schneekappe klar und deutlich zu sehen, seitdem haben wir leider täglich Regen gehabt und sind diese Alpenberge in Wolken eingehüllt, ich beabsichtige in dieser Gegend ein paar Wochen zu verweilen Mt Buller wie Timbertop[4] zu besteigen und Ansichten aufnehmen soweit sich mir diese schöne Gelegenheit zu Gebote steht. Der Zweck dieser Zeilen ist, Ihnen hiermit meinen wärmsten Dank abzustatten für die freundliche Unterstützung und Rath zur Unternehmung dieser Tour; wenn in den Mt Buller Ranges werde ich nicht versäumen mein Augenmerk auch auf Flechten u Schwämme zu lenken und könnte ich sonst während meiner Anwesenheit hier für Ihr Departement[5] hier anderweitig nützlich sein so bitte Sie recht freundlichst es mir mittheilen lassen zu wollen. Auf meine eingesandten Queries des Hn Ch^s Darwin bekam ich von der Missionsstation Framlingham erst vor c^a. 8 Tagen die Antwort, dass die Eingebornen dort zu sehr mit der geringen (*low*) Classe der Europäer in Verbindung wären und er deshalb die Fragen nicht beantworten könne.[6] M^rs. Green (Green) von der Coranderrk Station[7] schreibt mir heute u. A. wie folgt:

"I am verry happy to inform you that we are now free of fever. It was only very recently that it finally disappeared from our station. there is not an individual here but has been attacked with it; Big Lizzie, Jack & Jacky Warren have died in it I was closely confined to bed in it for three weeks just after You were here last.—
— I have not been able to give much attention to your Queries I am only able to answer four of them yet; but I hope now to be able to make special observations on the others and you may depend on me in answering none of them without carefully observing them myself."— —

Answers to Queries about Expressions:

12., I have seen Maggie and Marie laugh until the water from the eyes run down the cheeks.— (NB. Maggie (Tommy Hobrons Lubra & Marie, Simon Wongar Lubra ar both *pure* Aborigines. *Ch^sW*.)[8]

14., the Children when sulky protrude the lips *very much*.

15. Jealous expressions can be observed in the countenance, *the Natives seem not to be able* **to hide** *them.*

17. In Affirmation the head is two or three times nodded vertically with the mouth shut and protruding a little. In Negation it is moved laterally sometimes in silence and sometimes accompanied by "*Eota!*"

The nearest Post Office for any letters intended for me will be
    Mansfield[9]
for the next fourtnight.

Hoping dear Doctor that your health has improved, I remain, | Yours | most obed^t. Servt. | Charles Walter | Landscape Photo. Artist

Ferd: Mueller Esqr M.D. | F.R.S. &c. &c. &c. | Melbourne

DAR 181: 11

CD ANNOTATIONS
Top of letter: 'Australia'; '11' *brown crayon*
End of letter: 'Charles Walter | The answers by Mrs Green.' *pencil*

---

[1] Mueller wrote this in the margin of Charles Walter's letter to him. CD had asked Mueller to forward a handwritten list of his queries about expression to missionaries or colonists who might be in contact with Aborigines (see letter to Ferdinand von Mueller, 28 February [1867]; the queries are transcribed as an enclosure to that letter).

[2] For a translation of the German portion of the letter, see Appendix II.

[3] Mount Buller, Victoria, Australia, is about 150 miles north-east of Melbourne.

[4] Timbertop is a mountain adjacent to Mount Buller.

[5] Mueller was director of the Botanical Gardens, Melbourne.

[6] Walter refers to CD's queries on expression (see n. 1, above). Framlingham, a settlement near Warr-
nambool, Victoria, became a mission in 1865, when the Central Board for the Protection of Aborigines
gave permission to the Church of England Mission to the Aborigines of Victoria to occupy the site
(*Encyclopaedia of Aboriginal Australia*).

[7] Mary Green was the wife of the superintendent of Coranderrk Aboriginal Station near Healesville,
Victoria (Barwick 1972, p. 23; *Encyclopaedia of Aboriginal Australia*). CD acknowledged her contribution
in *Expression*, p. 20.

[8] The native population of Coranderrk was composed of members of the five tribes of the Kulin
confederacy, the Woiwurrung, Jajowrong, Bunurong, Wudthaurung, and Taungerong (see Barwick
1972, pp. 18–25). In 1868, the Kulin population included sixteen half-castes (Barwick 1972, p. 35).
Lubra: an Aboriginal woman (*Chambers*).

[9] Mansfield, Victoria.

From Fritz Müller[1]   [8 October 1867][2]

Itajahy

. . . . . Hätten Sie je gedacht, dass die Familie der Amarantaceen[3] auch auffällige
Samen darbieten möchte, welche die Aufmerksamkeit von Vögeln erregen? Nun,
dies ist der Fall bei einer kletternden Chamissoa unserer Flora. Die schwarzen
Samen sind beinah vollständig von einem weissen Arillus umgeben und bleiben
am Grunde der Kapsel angeheftet, deren obere Hälfte abfällt ("utriculus circumscissus")[4] wie bei Anagallis.[5] Nach Endlicher's Beschreibung scheint es, dass in andern Arten jener Gattung der Arillus viel kleiner ist ("arillo brevi albo, umbilicum
lateraliter cingente"),[6] und so bilden diese Formen ein verbindendes Glied zwischen unserer Art und einer andern Amarantacee (Celosia?), bei der die kleinen,
schwarzen und glänzenden Samen überhaupt keinen Arillus haben und, obwohl
sie nicht sehr ansehnlich sind, dennoch fest angeheftet auf dem Grunde der Kapsel
bleiben, welche nach dem Abfallen ihrer oberen Hälfte einen halbkugeligen Becher
bildet. Die Thatsache, dass ansehnliche Samen an den offnen Hüllen festhaften,
kommt auch bei Monocotyledonen vor. Ich beobachtete sie bei Hedychium (coronarium?—nicht einheimisch hier)[7] und bei einer Marantacee. . . . .

Incomplete
Möller ed. 1915–21, 2: 132

[1] For a translation of this letter, see Appendix I. For an account of the reconstruction of Fritz Müller's
letters to CD, see the letter from Fritz Müller, 17 July 1867, n. 1.

[2] The date is established by the relationship between this letter and the letters to Fritz Müller, 15 August [1867] and 30 January [1868] (*Correspondence* vol. 16). Alfred Möller dated the letter fragment
'September 1867' when he published it in German translation (Möller ed. 1915–21, 2: 132). CD
thanked Müller for information 'about conspicuous seeds' in his letter of 30 January [1868], and also
mentioned receiving Müller's answers on expression written on 5 October. It is likely that Müller
answered the queries on expression on a separate sheet, dated 5 October 1867, which was included
with the letter written on 8 October 1867. Müller had written at the top of CD's letter of 15 August
[1867], 'Received Octobr. 5.— | Answered Octobr. 8.—'

[3] The modern spelling is 'Amaranthaceae'.

[4] 'Arillus': aril; 'utriculus circumscissus': utricle circumscissile (Latin; the characterisation comes from
Endlicher 1836–42, s.v. *Chamissoa*).

[5] The genus *Anagallis* belongs to the family Primulaceae.

[6] Arillo brevi albo umbilicum lateraliter cingente: with a short white aril surrounding the hilum laterally (Latin). Müller refers to Stephan Ladislaus Endlicher and Endlicher 1836–42 (s.v. *Chamissoa*).

[7] *Hedychium coronarium*, though native to tropical Asia, is widely naturalised in tropical America (Mabberley 1997).

To J. V. Carus 10 October [1867][1]

Down. | Bromley. | Kent. S.E.
Oct 10

My dear Sir

I am sorry you have had so much trouble. I cannot imagine how the sheet could have been lost unless opened by the post office, or unless I sent the sheet by mistake to France or Russia in which case I hope it will be returned & it shall be sent to you.[2]

In the meantime I send by this post a *corrected* revise which I trust will serve your purpose—

Dear Sir | yours very sincerely | Ch. Darwin

LS
Staatsbibliothek zu Berlin, Sammlung Darmstaedter (Carus 29)

[1] The year is established by the relationship between this letter and the letter from J. V. Carus, 7 October 1867.

[2] Carus was translating *Variation* into German (Carus trans. 1868). He had reported that one proof-sheet was missing from the sheets that CD had just sent him (see letter from J. V. Carus, 7 October 1867). *Variation* was being translated into French by Jean Jacques Moulinié and into Russian by Vladimir Onufrievich Kovalevsky. See also letter from J. J. Moulinié, 11 October 1867.

From J. J. Moulinié 11 October 1867

Secrétariat | Général | Institut National | Genevois | Genève,
le 11 October *1867*

Dear Sir,

I just receive your last parcel forwarded the 5th October last, and including the—four sheets T, U, X, Y, (273–336); and following the former containing two sheets, Q, R (225–256). I have in-consequence to beg of your kindness to send me the intermediate missing sheet S. (257–272), which I have not received.[1]

I am happy to inform you, Sir, that in a few days, the six first chapters of the French translation will be completely printed,—for I have just sent back to the editor the fourteenth sheet with the last corrections.[2]

I remain Sir yours' truly | J. J. Moulinié

DAR 171: 268

[1] Moulinié refers to proof-sheets of *Variation*, which he was translating into French.

[2] The reference is to Charles-Ferdinand Reinwald, who published the French translation of *Variation* (Moulinié trans. 1868).

To James Samuelson 12 October [1867][1]

Down. | Bromley. | Kent. S.E.
Oct 12

My dear Sir

I am much obliged for your kind present of Vol IV of the Q. Journal of Science, which I received this morning with a note from Messrs Churchill.[2]

I have not hitherto taken in this journal, though I have occasionally procured numbers & read various articles with much interest, because I already take in so many periodicals that I am much in arrear in reading them. I have however just got & read the last No. with Mr Wallace's really admirable article.[3] He is a master of clear argument.

Believe me my dear Sir | yours truly obliged | Charles Darwin

LS
Paul V. Galvin Library, Illinois Institute of Technology

[1] The year is established by the reference to volume 4 of the *Quarterly Journal of Science*, which was published in 1867.

[2] Samuelson was the founder and co-editor of the *Quarterly Journal of Science* (*Men and women of the time* 1899). The journal was published by John Churchill and Sons.

[3] Alfred Russel Wallace's article, 'Creation by law' (A. R. Wallace 1867c), appeared in the October issue of the *Quarterly Journal of Science*. For more on the article, see the letter to A. R. Wallace, 12 and 13 October [1867].

To A. R. Wallace 12 and 13 October [1867][1]

Down. | Bromley. | Kent. S.E.
Oct 12 & 13th

My dear Wallace

I ordered the journal a long time ago, but by some oversight rec^d it only yesterday & read it.[2] You will think my praise not worth having from being so indiscriminate, but if I am to speak the truth, I must say I admire every word.—

You have just touched on the points which I particularly wished to see noticed. I am glad you had the courage to take up Angræcum after the Duke's attack; for I believe the principle in this case may be widely applied.[3] I like the Figure but I wish the artist had drawn a better sphynx.[4]

With respect to Beauty y^r remarks on hideous objects & on flowers not being made beautiful except when of practical use to them strike me as very good.[5]

On this one point of Beauty I can hardly think that the Duke was quite candid. I have used in the concluding paragraph of my present book precisely the same argument as you have, even bringing in the bull dog, with respect to variations not having been specially ordained.[6] Your metaphor of the river is new to me & admirable; but y^r other metaphor in which you compare classification & complex machines does not seem to me quite appropriate, tho' I cannot point out what seems deficient.[7] The point which seems to me strong is that all naturalists admit that there is a *natural* classification, & it is this which descent explains. I wish you had insisted a little more against the N. British on the reviewer assuming that each variation which appears is a strongly marked one; though by implication you have made this *very* plain.[8] Nothing in y^r whole article has struck me more than y^r view with respect to the limit of fleetness in the race horse & other such cases; I shall try & quote you on this head in the proof of my concluding chapter. I quite missed this explanation, tho' in the case of wheat I hit upon something analogous.[9] I am glad you praise the Duke's book for I was much struck with it. The part about flight seemed to me at first very good, but as the wing is articulated by a ball & socket joint, I suspect the Duke w^d find it very difficult to give any reason against the belief that the wing strikes the air more or less obliquely.[10] I have been very glad to see your article & the drawing of the butterfly in "Science Gossip."[11] By the way I cannot but think that you push protection too far in some cases, as with the stripes on the tiger.[12] I have also this m^g read an excellent abstract in Gard. Chron. of y^r paper on nests; I was not by any means fully converted by y^r letter, but I think now I am so; & I hope it will be published somewhere in extenso. It strikes me as a capital generalization, & appears to me even more original than it did at first.[13]

I have had an excellent & cautious letter from Mr Leach of Singapore with some valuable answers on expression which I owe to you.[14]

I heartily congratulate you on the birth of "Herbert Spencer", & may he deserve his name, but I hope he will copy his father's style & not his namesake's.[15] Pray observe, though I fear I am a month too late, when tears are first secreted enough to overflow; & write down dates.[16]

I have finished Vol. 1 of my book & I hope the whole will be out by the end of Nov;[17] if you have the patience to read it through, which is very doubtful, you will find I think a large accumulation of facts which will be of service to you in y^r future papers, & they c^d not be put to better use, for you certainly are a master in the noble art of reasoning.

Have you changed y^r house to Westbourne Grove??[18]

Believe me my dear Wallace | yours very sincerely | Ch. Darwin

This letter is so badly expressed that it is barely intelligible, but I am tired with Proofs[19]

P.S. Mr Warington has lately read an excellent & spirited abstract of the "Origin" before the Victoria Inst. & as this is a most orthodox body he has gained the name of the Devil's Advocate. The discussion which followed during 3 consecutive meetings is very rich from the nonsense talked. If you w^d care to see the number I c^d send it you.[20]

I forgot to remark how capitally you turn the table on the Duke, when you make him create the Angræcum & moth by special creation.—[21]

LS(A)
British Library (Add 46434 f. 96)

¹ The year is established by the reference to Wallace's article 'Creation by law' in the *Quarterly Journal of Science* (A. R. Wallace 1867c); see n. 2 below.

² CD refers to the *Quarterly Journal of Science* and to Wallace's article, 'Creation by law' (A. R. Wallace 1867c; see also letter to James Samuelson, 12 October [1867], and letter from A. R. Wallace, 1 October [1867] and n. 4).

³ In *The reign of law* (G. D. Campbell 1867), pp. 45–6, George Douglas Campbell, the duke of Argyll, had criticised CD's explanation of the development of the long nectary in the orchid *Angraecum sesquipedale*. Wallace countered that the laws of 'multiplication, variation, and survival of the fittest' would '*necessarily* lead to the production of this extraordinary nectary' (A. R. Wallace 1867c, p. 475).

⁴ The illustration (A. R. Wallace 1867c, facing p. 471; see also frontispiece to this volume) was an artist's impression of a hypothetical sphinx moth (family Sphingidae) with a proboscis that could reach the base of the nectary (spur) of *Angraecum sesquipedale* (see letter from A. R. Wallace, 1 October [1867] and n. 5).

⁵ See A. R. Wallace 1867c, p. 482. Campbell had argued that beauty conferred no selective advantage and therefore could only be explained with reference to a 'Creator' (see, for example, G. D. Campbell 1867, pp. 242–8, and A. R. Wallace 1867c, pp. 480–1).

⁶ See A. R. Wallace 1867c, p. 484, and *Variation* 2: 431, where CD wrote, 'Did [the Creator] cause the frame and mental qualities of the dog to vary in order that a breed might be formed of indomitable ferocity, with jaws fitted to pin down the bull for man's brutal sport?'

⁷ See A. R. Wallace 1867c, pp. 477–9, 487.

⁸ The anonymous article in the *North British Review* was by Henry Charles Fleeming Jenkin. See [Jenkin] 1867, pp. 293–4, and Wallace 1867c, pp. 485–6.

⁹ CD quoted Wallace on the limits to fleetness in horses in *Variation* 2: 417. The reference to wheat has not been identified. CD had commented on wheat found in Swiss lake habitations, showing the antiquity of its cultivation, but remarked, 'at the present day new and better varieties occasionally arise' (*Variation* 2: 416).

¹⁰ CD refers to G. D. Campbell 1867, pp. 128–80. Campbell had argued that a bird could only flap its wings in a direction perpendicular to the axis of its body and that no bird could fly backwards (G. D. Campbell 1867, pp. 145–6).

¹¹ CD refers to 'The disguises of insects' (A. R. Wallace 1867e), which included two illustrations of butterflies (figs. 195 and 196).

¹² Wallace had argued in an article in the *Westminster Review* that the tiger's stripes helped to conceal it from its prey (see [A. R. Wallace] 1867a, p. 5). In *Descent* 2: 302, CD cited Wallace on this point, but suggested that sexual selection played a role in determining colour, since males were brighter than females.

¹³ Wallace had presented a paper 'On birds' nests and their plumage' on 9 September 1867 at the British Association meeting at Dundee (*Report of the thirty-seventh meeting of the British Association for the Advancement of Science held at Dundee*, Transactions of the sections, p. 97). An abstract of the paper appeared in the 12 October 1867 issue of *Gardeners' Chronicle* (A. R. Wallace 1867f). A greatly extended version of the paper appeared the next year in the *Journal of Travel and Natural History* (A. R. Wallace 1868–9). CD also refers to the letter from Wallace of 26 April [1867].

¹⁴ Frederick F. Geach had sent answers to CD's queries on expression (see letter from F. F. Geach, June 1867). Wallace had provided CD with Geach's address (see letter from A. R. Wallace, 2 March [1867]).

¹⁵ CD refers to Herbert Spencer Wallace (see letter from A. R. Wallace, 1 October [1867] and n. 7). CD had commented on Herbert Spencer's use of 'awesomely long words' in a letter to J. D. Hooker, 2 October [1866] (*Correspondence* vol. 14).

¹⁶ CD had kept notes on the development of his own children and noted that very early crying was not accompanied by tears (see *Correspondence* vol. 4, Appendix III). In *Expression*, p. 164, CD wrote that infants did not weep until the age of 'from two to three or four months'.

¹⁷ *Variation* was published on 30 January 1868 (Freeman 1977).

[18] CD refers to Wallace's London address, 76½ Westbourne Grove (see letter from A. R. Wallace, 1 October [1867]). The last extant letter from Wallace before this was addressed from 9 St Mark's Crescent.

[19] CD was correcting proof-sheets for *Variation*.

[20] CD refers to George Warington, the Victoria Institute, and Warington 1867 (see also letter to George Warington, 7 October [1867] and n. 2). The Victoria Institute was founded in 1865; its primary objective was to 'investigate fully and impartially the most important questions of Philosophy and Science, but more especially those that bear upon the great truths revealed in Holy Scripture, with the view of defending these truths against the oppositions of Science, falsely so called' (*Journal of the Transactions of the Victoria Institute* 1 (1865–6): vi).

[21] See A. R. Wallace 1867c, pp. 475–7, and n. 4, above.

From J. V. Carus 16 October 1867

39, Elsterstrasse, | Leipzig.
Oct. 16. 1867

My dear Sir,

On page 308 you mention the "mattee" as a plant including coffine. I suppose it is the "arvore do mate", Ilex paraguariensis.[1]

In the table on p 286 you give as the weight of the Call-duck (from Mr. Fox) 717 grains and 713 (below). I suppose the latter figure is a misprint; for this number used for calculating gives rather more than 130 in the last column (1000:130), while 717 gives 128, 5, that is almost 129.[2]

The sheet you kindly sent me I have received in time. Now there are still wanting pp. 337–411. Will I get them soon?[3]

I am, as you may believe me, really anxious to be able to read the second volume, you may trust it doesn't give me much trouble to translate it. I learn so very much by it and I am thoroughly thankful to you for the way in which you here teach us all how to look on, and how to study, nature. May I venture to ask you to send me (if you can in any way spare it) a copy of the work when it is out? I shall always prefer to read the original, besides the treat of having a neat copy in my library, which is rather a pet of mine.

Believe me | My dear Sir | Yours very sincerely | Prof J. Victor Carus

DAR 161: 63

CD ANNOTATION
End of letter: 'Give pages in Vol. 2 *p. 181 [*added ink*] about Hybridism. | =Pangenesis=' *pencil*[4]

[1] Carus was translating *Variation* into German. 'Arvore do maté': maté tree (Portuguese). CD commented on the chemical similarities of maté (*Ilex paraguaiensis*), tea, and coffee in *Variation* 1: 308.

[2] See *Variation* 1: 286, a table giving the weights of the wing bones of various bird species as a proportion of the weight of the whole skeleton. In the first printing of *Variation*, the weight of the call duck's skeleton is given first as 713 grains and later as 717 grains. In the second printing, 717 grains is given in both places (see Freeman 1977 for more on the differences in the two imprints). In the German edition, the weight is given as 717 grains (Carus trans. 1868, 1: 356). William Darwin Fox had sent CD a call-duck skeleton in 1855 (see *Correspondence* vol. 5, letter to W. D. Fox, 31 July [1855]).

[3] Carus refers to proof-sheet 'S' (pp. 257–72) and to the final section of the first volume of *Variation* (see letter to J. V. Carus, 10 October [1867] and n. 2).

[4] CD's annotation is a note for his reply to Carus (see letter to J. V. Carus, 19 October [1867] and nn. 5 and 6).

To Asa Gray 16 October [1867][1]

Down. | *Bromley.* | *Kent. S.E.*
Oct 16

My dear Gray

I send by this post clean sheets of Vol. 1. up to p. 336, & there are only 411 pages in this vol.[2] I am **very** glad to hear that you are going to review my book; but if the Nation is a newspaper I wish it were at the bottom of the sea, for I fear that you will thus be stopped reviewing me in a scientific journal.[3] The first Vol. is all details, & you will not be able to read it; & you must remember that the Chapters on plants are written for naturalists who are not botanists. The last Chap. in Vol. 1 is, however, I think a curious compilation of facts; it is on bud-variation. In Vol. 2 some of the Chaps are more interesting; & I shall be very curious to hear your verdict on the Chap. on close inter-breeding.[4] The Chap. on what I call Pangenesis will be called a mad dream, & I shall be pretty well satisfied if you think it a dream worth publishing; but at the bottom of my own mind I think it contains a great truth.[5] I finish my book with a semi-theological paragraph, in which I quote & differ from you; what you will think of it I know not.[6]

Many thanks for a 2nd note rec^d some time ago on Dionæa.[7]

I have done nothing worth mentioning this summer, as all my time has been consumed in correcting horrid proof sheets. I may mention one little fact which may possibly interest you. A man in Natal sent me a little packet by post of the dung of locusts with the statement that it was believed that locusts brought new plants to the districts which they visited.[8] Six Grasses, belonging to at least two species have germinated out of the dung, & the seeds were fairly enclosed in the little pellets, as I ascertained by dissection. This verifies what I said in the Origin, that many new methods of transport w^d be discovered; for locusts are often blown many 100 miles out to sea.[9]

The rest of the sheets which have all been corrected will be printed off by the middle of Nov^r. & shall be sent to you in 2 or 3 packets. Do not forget to let me have hereafter a copy of the Nation.[10]

My dear Gray | yours most sincerely | Ch. Darwin

LS(A)
Gray Herbarium of Harvard University (95)

[1] The year is established by the relationship between this letter and the letter from Asa Gray, [after 17 September 1867].

[2] CD refers to proof-sheets of *Variation*.

[3] Gray told CD that he planned to write a review of *Variation* for the *Nation*, a weekly newspaper (see

letter from Asa Gray, [after 17 September 1867] and n. 2). The *Nation* was intended to discuss the political and economic issues of the day 'with greater accuracy and moderation' than were found in the daily newspapers, and to print 'sound and impartial criticisms' of books and works of art (*ANB*).

[4] CD refers to chapter 17 of *Variation*, 'On the good effects of crossing, and on the evil effects of close interbreeding' (*Variation* 2: 114–44).

[5] See *Variation* 2: 357–404. For Gray's reaction to pangenesis, see [A. Gray] 1868 and Dupree 1959, pp. 356–7.

[6] For CD's concluding paragraph, see the letter to J. D. Hooker, 8 February [1867], n. 7.

[7] See letter from Asa Gray to J. D. Hooker, [after 6 July 1867]. Gray enclosed a letter from William Marriott Canby describing his experiments with *Dionaea* (Venus fly-trap). CD thanked Gray for Canby's letter in a letter of 8 August [1867]. Evidently, Gray sent further observations, but no additional letter has been found.

[8] James Philip Mansel Weale had sent CD a packet of locust dung with his letter of 7 July 1867. In his reply to Weale, CD had been doubtful of finding any seeds (see letter to J. P. M. Weale, 27 August [1867]).

[9] See *Origin*, p. 363. CD reported the results of his experiments in *Origin* 5th ed., p. 439.

[10] The 19 March 1868 issue of the *Nation*, in which Gray's review appeared, has not been found in the Darwin Archive–CUL.

From J. D. Hooker 18 October 1867

Kew
Oct. 18/67.

Dear old Darwin

I find myself obliged to put off half my visit to you. The Governess has gone away for her holiday, & my wife finds the nurse not to her trust, & what with this & the boys home for the holidays, she wants more help in the house, so I have to go to Bath on Saturday to bring up a friend of her's, who will help her.[1] Moreover she does not like the nurse, with reason I think, & wants to get rid of her, but has not grounds enough for a dismissal; & as I think it not safe to let the nurse have the child after she has warning, I have advised her to take the child to Hastings for change of air, leaving nurse here, & writing up to dismiss her from thence—[2] This is an ugly ruse to play, but it must be done & a prudent friend in the house will be a great help to her. The whole thing is a horrid bore

If the weather is fine Willy & I will run down on Sunday morning to Bromley with our nightshirts in our pockets & walk across Holwood park—[3]

Ever aff | J D Hooker

DAR 102: 180–1

[1] Hooker refers to Frances Harriet Hooker and to his sons William Henslow Hooker, aged 14, Charles Paget Hooker, aged 12, and possibly Brian Harvey Hodgson Hooker, aged 7. The governess, nurse, and friend have not been identified.

[2] The Hookers's youngest child, Reginald Hawthorn Hooker, was ten months old.

[3] Hooker refers to the estate attached to Holwood House, the home of CD's neighbour, Robert Monsey Rolfe, Lord Cranworth. Holwood House was about a mile and a half north-west of Down House (*Post Office directory of the six home counties* 1866 s.v. Keston [Kent]).

From William Houghton 18 October 1867

Preston Rectory | *Wellington Salop*
Oct. 18. '67.

Dear Mr Darwin.

If you can spare 5 minutes, will you kindly inform me whether, if you can remember, your researches on the planarian worms satisfied yourself as to the existence, or absence, of a nervous system in these little beasts?[1] Some German Naturalists ascribe to them a Nervous system.[2] I have lately occupied myself with studying these annelids[3]—i e.—the common fresh water kinds—& can trace no appearance of the most rudimentary nervous system—spite of my utmost endeavours to find some gangliæ.[4]

It is hardly fair to trouble one whose hands are so full of work, & I apologise for this note.

I suppose we shall shortly have your new book.[5] I think it a subject of congratulate that, in these days, any theories if well supported are likely to meet with fair discussion. At any rate we are making progress, despite the women & the parsons!

I hope to be in London next January, & to have the pleasure of once more seeing you at that time.[6]

Eyton goes on as usual. Have you seen his book on the osteology of Birds?[7] Owen thinks Huxley has lately cribbed an idea from him.[8]

I sincerely trust your health improves, & with kindest regards I am | Yours most sincerely | W. Houghton

Charles Darwin Esq | &c &c

PS. | Some months ago I heard, indirectly from a medical man, of a supposed case of fertility in hybrids between the domestic fowl & pheasant—& took a long walk of 12 miles to see into the matter— All I found was that the hens may have occasionally been trod by the cock-pheasants, & that the supposed offspring were simply the result of a connection between the game fowl & the pheasant variety of the domestic cock. (Golden Spangled Hawks'—)

DAR 166: 271

[1] While on the *Beagle* voyage, CD had found several new species of planarians; he later wrote a paper describing twelve terrestrial and five marine species, but did not describe a nervous system ('*Planariae*'; see also *Journal of researches*, pp. 30–1, *Correspondence* vol. 1, and R. D. Keynes 2002, pp. 82–4).

[2] See, for example, Schultze 1851, pp. 21–3.

[3] Houghton mistakenly refers to planarian worms as 'annelids'. Planarian worms belong to the class Turbellaria (free-living acoelomate flatworms). The Annelida were the class (now phylum) of segmented coelomate worms.

[4] In *Journal of researches*, p. 31, CD had commented on the difficulty of preserving planarians for observation, noting, 'immediately the cessation of life allows the ordinary laws of change to act, their entire bodies become soft and fluid, with a rapidity which I have never seen equalled'. In an article on land

planarians (Schultze 1856, p. 33), Max Johann Sigismund Schultze discussed the special preparation of specimens for detailed studies of their internal organs.

[5] Houghton refers to *Variation*, which was published in January 1868.

[6] According to his journal, CD visited London in late November 1867 and in March 1868 (see 'Journal' (Appendix II) and *Correspondence* vol. 16, Appendix II).

[7] The main part of Thomas Campbell Eyton's *Osteologia avium* (Eyton 1867–75) was published in 1867. A supplement was published in 1869 and a second supplement in three parts appeared between 1873 and 1875.

[8] The references are to Richard Owen and Thomas Henry Huxley. Houghton may be referring to Huxley's work on the classification of the bird species *Opisthocomus cristatus* (the hoatzin). In a paper on bird classification read in April 1867, Huxley had briefly discussed the species but admitted that he had had only an incomplete skull to study (T. H. Huxley 1867, p. 435). Eyton later provided Huxley with a complete skeleton of the bird, and Huxley published his conclusions on its classification in a later paper (T. H. Huxley 1868, pp. 304–11; see also *Correspondence* vol. 16, letter from T. C. Eyton, 23 August [1868]).

To J. V. Carus 19 October [1867][1]

Down. | Bromley. | Kent. S.E.
Oct. 19.

My dear Sir

The "Mattee" is the species which you name. I am much obliged for the correction at p. 286 of the figure "713." which is a misprint.[2]

I expect soon to receive the remaining clean sheets of vol 1, & when rec^{d} they shall be sent *without delay*.[3]

I shall have great pleasure in sending you a perfect copy of the book, & you have the best right of any one to claim a copy.[4]

In Vol. 2. p. 181 you will find some pages which you have already translated in the last edit. of the Origin, & which I have here re-inserted; but will you be so kind as to compare them carefully, as I have altered & inserted a few sentences in the hope of making the subject clearer.[5] I do not know what you will think about Chap. 27;[6] I fear it is very wild, but after mature reflection I believe that physiologists will some day be compelled to admit some such doctrine.

With cordial thanks for all the kind interest you take about my book I remain | my dear Sir | yours very sincerely | Ch. Darwin

LS(A)
Staatsbibliothek zu Berlin, Sammlung Darmstaedter

[1] The year is established by the relationship between this letter and the letter from J. V. Carus, 16 October 1867.

[2] See letter from J. V. Carus, 16 October 1867 and nn. 1 and 2.

[3] See letter from J. V. Carus, 16 October 1867 and n. 3.

[4] Carus's name appears on CD's presentation list for *Variation* (see DAR 210.11: 33).

[5] See *Variation* 2: 181–5 and *Origin* 4th ed., pp. 320–4.

[6] CD refers to the chapter on his provisional hypothesis of pangenesis (*Variation* 2: 357–404).

To John Murray   19 October [1867]

Down. | *Bromley.* | *Kent. S.E.*
Oct 19th

My dear Sir

In your advertisement of my Book in the Athenæum there is an absurd error (which I formerly pointed out) viz principles of . . . . . . **Election** instead of Selection. I have no objection to these "Principles" being inserted in the advertisement, but I think that you agreed with me that "The Variation of Animals & Plants under Domestication" was sufficient. for the Title.—[1]

I am getting on well.— Vol. I completed. Vol. 2. *all* sheets corrected the first time, & nearly the half revised. I shall have finished the whole by at latest, if I keep well, by 10th or 12th November.— I hope you have urged Mr Dallas to lose no time about the Index, & that you have ordered the *Revises* to be sent to him: the Revises w^d do perfectly for Index.—[2] You could not, I think, have got a better man than M^r. Dallas.—

I hope to Heaven that the Book will not turn out a failure.—

My dear Sir | Yours very sincerely | Ch. Darwin

Endorsement: '1867. Oct 19 | C Darwin'
John Murray Archive

---

[1] The advertisement for *Variation* appeared in the 19 October 1867 issue of the *Athenæum*, p. 487. The title given was: '*The variation of animals and plants under domestication; or, the principles of inheritance, reversion, crossing, inter-breeding, and election*'. Murray had also made an error in an earlier advertisement of the book under a different title (see *Correspondence* vol. 13, letter to John Murray 2 June [1865] and n. 3).

[2] William Sweetland Dallas had been invited by Murray to prepare the index to *Variation* (see letter from John Murray, 23 September [1867]).

From Joseph Plimsoll   21 October 1867

Exmouth. | 8 Bicton Place
Octr. 21. 1867.

Dear Sir

Feeling, as I do, deeply solicitous for the salvation of your immortal soul, I hope you will pardon the liberty I have thus taken in addressing, by letter, one with whom I have not the honour and the privilege of a personal acquaintance, but for whose great intellectual powers and reputation I entertain a profound respect—and also bear with me, whilst in this brief epistle, I endeavour to bring before your notice those momentous considerations which should move us all to be anxious to become reconciled unto God, and to secure His love and favour, in time, and throughout eternity—and also to state the means which Heaven in its infinite grace, wisdom, and condescension, has instituted for the realization, by fallen man, of those great ends of human existence.

In the first place, then, dear Sir, I beg to submit, respectfully, to your consideration, and serious ponderings—that most welcome and soul-rejoicing declaration of Holy Writ—"Believe in the Lord Jesus Christ, and thou shalt be saved"[1] Will you not at once, dear Sir, without a moment's delay—for who knoweth what a day may bring forth!—avail yourself of this most gracious *promissory* mandate? Will you not hasten to the throne of grace, and entreat the Author and Finisher of our faith (even the Lord Jesus Christ—the Saviour of the world—the Redeemer of fallen humanity—and respecting whose saving power—nay essentialness to salvation—his apostles have affirmed—"there is no other name given under heaven among men whereby we can be saved, but the name of the only begotten Son of God"—)[2] to grant to you this saving faith in Him? True faith is the gift of Heaven. Man can no more beget it in himself—(in his own heart—for it is "with the heart", and not the intellect, merely, "that man believeth unto righteousness"—)[3] than he can create a universe. Mere intellectual faith—or historic faith, as it is sometimes called—that is, the assent of the understanding, only, to the truths of Holy Writ, and a belief in, and avowal (when a man wishes to be considered orthodox in his theology,) of their authenticity and divine origin, may be engendered by the reasoning powers alone. This however, is not the kind of faith that will issue in the procurement of the divine forgiveness, approval, love, and adoption of the subject of such belief into the redeemed and truly filial family of God. Plead earnestly, then, I beseech you, for the bestowment upon you of this holy, heaven-begotten, and heaven-securing faith. Have you not every reason to hope that such a petition will be granted you by an all-merciful and benevolent God? Has not the Saviour himself said—"Whatsoever ye ask the Father in my name he will give it you"?[4] Think, dear Sir, of the stupendous and inestimably-blessed issues of such a boon—as it respects its influence on your eternal destiny—should it be vouchsafed to you, in answer to your earnest and importunate prayers! Is it nothing—or a matter of but small moment—to become the possessor of a divinely implanted principle of spiritual, moral, and intellectual operation in the soul—which will insure for its recipient the assurance—of deliverance from obnoxiousness to the divine wrath—of escape from the condemnation of the eternal judge of quick and dead, and consequent consignment of the condemned to the realms of everlasting woe—even to the gnawings of that worm which dieth not, and to the flames of that fire which never can be quenched?

Is not *this* assurance—the assurance of such an awful doom being averted from us—worth praying for? Then *what is*? "For who"—as the prophet Jeremiah asks—"Who can dwell with everlasting burnings?".[5] And is it *nothing*, to enjoy, throughout this mortal life, the conviction—and a well-founded one—that in the last Great Day, when we shall all be assembled before the "Great White Throne"[6] of the dread Arbiter of our eternal destiny,—before Him, who, on that awful occasion will pronounce our doom (and execute it too) to endless destruction, or our exaltation to the glories and blessedness of the realms of holiness, life, peace, and joy—that the latter adjudication will be our blissful and glorious portion? And will not *this* be worth praying for?... Is it a consciousness of guilt that deters you from entertaining the hope that the Almighty will graciously answer you, should you present yourself as a petitioner at his footstool, and ask Him to impart to you such a transcendent blessing—that consciousness of guilt which all the sons and daughters of Adam must experience, when the soul is in a state of healthful moral sensibility; since all mankind have shared hitherto, and will continue so to do, till the end of time, in that inheritance of sinfulness which has descended to them from our first parents, as the primal transgressors of the divine laws—and whereby they have been actuated, during life, in their thoughts, desires, purposes, aspirations, and conduct—for "there is none", as the Word of God saith, "that doeth good and sinneth not"?[7] If so—if you are the subject of such a consciousness, and that to a great and very distressing degree—then, for the relief of your sin-burdened spirit, I beg still further to quote the language of Inspiration, and for your peace and comfort, to convey to you, through this letter, the all-important and incontrovertible announcements that "the blood of Jesus Christ cleanseth from all sin"[8]—that "God so loved the world, as to give his only-begotten Son, that whosoever believeth in Him might not perish, but have everlasting life"[9]—that "it is a faithful saying, and worthy of all acceptation, that Christ Jesus came into the world to save sinners"[10]—that "whosoever cometh unto Him he will in no wise cast out"[11]—that "He came to seek and to save them that are lost"[12]—that He is "the friend of sinners"[13]—that He came not to call the righteous but sinners to repentance"[14]—that He came not to condemn the world, but that the world through Him might be saved".[15] What more can fallen, rebellious man require more than this to be animated by hope of salvation, if he comply with the requirements of the gospel—or rather requirement—for it is belief alone that is required? Since, dear Sir, it is appointed unto men once to die, and after death the judgment;[16] since we shall all have to appear before the judgment seat of Christ, to answer for the deeds done in the body, whether they be good or whether they be evil; and since it has been affirmed by Him who cannot lie, that to those on the right hand of the Almighty Saviour, will be addressed the invitation—'Come ye blessed of my Father, inherit the Kingdom prepared for you from the foundation of the world'[17]—whilst to those on his left he will say—Depart from me ye cursed into everlasting fire prepared for the devil and his angels"[18]— are you not looking forward with the deepest anxiety to that eventful last Assize, and with the most intense desire that its decisions may be in your favour, and that through faith in the blood and righteousness of Jesus Christ you may on that day win the favourable notice, forgiveness, and commendation of the Lord God Almighty?.

I am, Sir | Yours sincerely— | J. Plimsoll | M.D

DAR 174: 51

[1] Acts 16:31.
[2] See John 3:18.
[3] Rom. 10:10.
[4] John 16:23.
[5] See Isa. 33:14.
[6] Rev. 20:11.
[7] See Eccles. 7:20.

[8] 1 John 1:7.
[9] John 3:16.
[10] 1 Tim. 1:15.
[11] See John 6:37.
[12] See Luke 19:10.
[13] See Matt. 11:19, Luke 7:34.
[14] Luke 5:32.
[15] See John 3:17.
[16] See Heb. 9:27.
[17] Matt. 25:34.
[18] Matt. 25:41.

From A. R. Wallace 22 October [1867][1]

Hurstpierpoint
Octr. 22nd.

Dear Darwin

I am very glad you approve of my article on "Creation by Law" as a whole.[2]

The *"machine metaphor* is *not mine*, but the N.B. reviewers.[3] I merely accept it and show that it is on our side and not against us, but I do not think it at all a *good* metaphor to be used as an *argument* either way. I did not half develope the argument on the *limits of variation*, being myself limited in space; but I feel satisfied that it is the true answer to the *very common* and very strong objection, that "variation has strict limits". The fallacy is the requiring variation in domesticity to go *beyond* the limits of the same *variation* under nature. It *does* do so sometimes however, because the conditions of existence are so different. I do not think a case can be pointed out in which the limits of variation under domestication are not up to or beyond those already marked out in nature, only we generally get in the *species* an amount of change which in nature occurs only in the whole range of the *genus* or *family*[4]

The many cases however in which variation has gone *far beyond nature* and has not yet stopped, are ignored. For instance no wild pomaceous fruit is I believe so large as our *apples*, and no doubt they could be got much larger if flavour &c. were entirely neglected.

I may perhaps push "protection" too far some times for it is my hobby just now,—but as the Lion & the Tiger are I think the *only* two non-arboreal cats, I think the Tiger stripe agreeing so well with its usual habitat is at least a probable case.[5]

I am rewriting my article on Birds' nests for the new "Nat. Hist. Review."[6]

I cannot tell you about the first appearance of *tears*, but it is very early,—the first week or two I think.[7] I can see the Vict. Mag. at the London Library.[8]

I shall read your book *every word*. I hear from Sir C. Lyell that you come out with a *grand [new] theory* at the end, which even the *Cautious!* Huxley is afraid of![9] Sir C. said he could think of nothing else since he read it. I long to see it.

My address is *Hurstpierpoint* during the winter, and when in Town, $76\frac{1}{2}$ *Westbourne Grove*.

I suppose you will now be going on with your book on *Sexual selection* & *Man*, by way of relaxation![10] It is a glorious subject but will require delicate handling.

Yours very faithfully | Alfred R. Wallace—

C. Darwin Esq.

DAR 106: B46–7

CD ANNOTATIONS
1.1 I am ... limits". 2.6] *crossed ink*
2.6 The fallacy ... nature. 2.7] *double scored ink*
4.1 I may ... handling. 9.2] *crossed ink*

[1] The year is established by the relationship between this letter and the letter to A. R. Wallace, 12 and 13 October [1867].

[2] In his letter of 12 and 13 October [1867], CD had praised Wallace's article in the *Quarterly Journal of Science* (A. R. Wallace 1867c) defending CD's theory against criticisms made in G. D. Campbell 1867 and [Jenkin] 1867.

[3] In his anonymous article in the *North British Review*, Henry Charles Fleeming Jenkin made an analogy between the criteria for determining species in the natural world and the criteria for patenting machines (see [Jenkin] 1867, pp. 310–12, and A. R. Wallace 1867c, p. 487).

[4] Jenkin had argued that the limit to variation in any one direction implied that successive variations could not be accumulated over long periods of time ([Jenkin] 1867, pp. 285–6). Wallace countered that the definite limits to variability in certain directions were no objection to the view that all modifications were produced by the gradual accumulation by natural selection of small variations (A. R. Wallace 1867c, pp. 486–7).

[5] See letter to A. R. Wallace, 12 and 13 October [1867] and n. 12.

[6] See letter to A. R. Wallace, 12 and 13 October [1867] and n. 13. The 'new *Natural History Review*' was the *Journal of Travel and Natural History* (see letter from Andrew Murray, 12 August 1867 and n. 1).

[7] See letter to A. R. Wallace, 12 and 13 October [1867] and n. 16.

[8] Wallace refers to the *Journal of the Transactions of the Victoria Institute* and to the London Library, an independent subscription library with premises at 12 St James's Square (*Post Office London directory* 1867; see letter to A. R. Wallace, 12 and 13 October [1867] and n. 20).

[9] Wallace refers to Charles Lyell, Thomas Henry Huxley, and CD's provisional hypothesis of pangenesis (*Variation* 2: 357–404). CD had sent Lyell proof-sheets of *Variation* and was particularly pleased that Lyell had 'noticed Pangenesis' (see letter to Charles Lyell, 22 August [1867]).

[10] The reference is to *Descent*. Earlier, CD told Wallace he had almost resolved to publish a 'little essay on the Origin of Mankind', and that he thought sexual selection had been the main agent in forming the human races (see letter to A. R. Wallace, 26 February [1867]).

From Hermann Müller 23 October 1867

Lippstadt
23th October 1867

Dear Sir!

I am very much obliged to you for your kindness in having sent me the German version of the forth edition of your work on species.[1] I have read several months ago the original of this forth edition and have been as much pleased as surprised at the plenty of new important evidences your theorie has gained since the first

German edition has been published.[2] It has been a great pleasure to me to receive now out of your own hand your work together with your portrait.[3] Being by no means able to return my thanks to you by any however unimportant own work I can only send to you my portrait.[4]

A single observation made by me several weeks ago perhaps could be of some interesting to you. I have never heard or read that Dipterous insects take other than fluid food. Therefore I was surprised by seing that Eristalis tenax, Rhingia rostrata, different species of Syrphus and many other Dipterous insects eat enorm quantities of pollen grains.[5]

With Eristalis tenax I was allowed long time to look at its eating in immediate nearness, and to convince myself that the complicate structure of its underlip is admirably adapted as well to this foot as to the sucking of nectar.

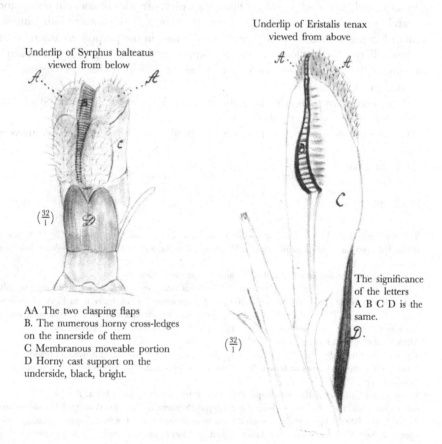

Underlip of Eristalis tenax
viewed from above

Underlip of Syrphus balteatus
viewed from below

$\left(\frac{32}{1}\right)$

AA The two clasping flaps
B. The numerous horny cross-ledges
on the innerside of them
C Membranous moveable portion
D Horny cast support on the
underside, black, bright.

$\left(\frac{32}{1}\right)$

The significance
of the letters
A B C D is the
same.

The two flaps on the free end of the underlip clasp little clusters of pollen grains between them and grind them behind by quick alternate porrections and retractions. The numerous horny cross-ledges on the inner side of the two flaps serve to seize, to hold and to shove behind the single pollengrains The distance of these horny ledges apparently is adapted to the greatness of the pollengrains. For I found the stomach of Eristalis tenax filled for the greatest part with many hundred thousands of the large pollen grains of an Oenothera cultivated in my garden; whilst the stomach of Syrphus balteatus (which species has narrower cross ledges on the underlip) contained only much lesser pollengrains

When sucking the fly put the ends of their bristle shaped mandibles (not represented in my drawing) between the two flaps, thus giving to them a fixed direction and then draws back the lateral portions C.[6]

The cold season has finished to early my just begun observations. It will be, I believe, an interesting proposing to compare accurately the structure of the underlip of stinging and exclusively sucking Dipterons with those who besides eat pollen and to search generally what details of structure the mouth of the flower-visiting insects of all orders has gained in the struggle for existence, in the purpose to attain better the food offered by the flowers. For the complet understanding of the complicate structure of the mouth of Dipterons it is, I believe, indispensable to take notice of its adaption to pollen-food.

If you know any observations published on this object I whould be glad of hearing where they are published.

With the greatest respect and my sincere thanks I remain, dear Sir, | yours very faithfully

Hermann Müller.

DAR 171: 291

---

[1] The reference is to the third German edition of *Origin* (Bronn and Carus trans. 1867), which was translated from the fourth English edition. No presentation list for the third German edition has been found.

[2] Müller refers to the fourth edition of *Origin* and Bronn trans. 1860.

[3] CD may have sent a copy of a photograph of himself taken by Ernest Edwards during his visits to London in November 1865 or April 1866 (see *Correspondence* vol. 13, letter from E. A. Darwin to Emma Darwin, 25 [November 1865] and n. 3), or a copy of one taken by his son William Erasmus Darwin in 1864 (see *Correspondence* vol. 14, letter from W. E. Darwin, 8 May [1866] and n. 10).

[4] Müller's photograph has not been found in the Darwin Archive–CUL.

[5] *Eristalis tenax* (the drone fly), *Rhingia rostrata* (the long-tongued hoverfly), and *Syrphus* all belong to the family Syrphidae (hoverflies or flower flies), dipterous insects that often resemble bees or wasps (F. S. Gilbert 1993).

[6] In modern terminology, the 'two flaps' that Müller refers to are the labella, the 'underlip' is the prementum, and the 'horny cross-ledges' are the pseudotracheae. The 'bristle-shaped mandibles' are probably the labrum-epipharynx (personal communication, Francis Gilbert). *Syrphus balteatus* is now *Episyrphus balteatus*. For more on the structure and function of mouthparts in the Syrphidae, see F. S. Gilbert and Jervis 1998. In a paper on the application of Darwinian theory to flowers and flower-visiting insects, Müller described his observations, made in the summer of 1867, of the behaviour of *Eristalis tenax* visiting flowers of *Oenothera media* (now *O. fruticosa*; H. Müller 1869, p. 59). CD's heavily annotated copy of the paper is in the Darwin Pamphlet Collection–CUL.

From J. V. Carus 30 October 1867

39, Elsterstrasse | Leipzig,
Oct. 30. 1867

My dear Sir,

Mr Schweizerbart wishes to advertise your new book and to give a short prospectus of it.[1] I promised him to write some few lines, but I cannot do it without knowing the exact title of the work. Judging from the inscription on the first page it will be "On the variation of Animals and Plants under Domestication." But I prefer to ask you directly for it. Possibly you had already a proof of the title-page.

When will the remainder of Vol. I come? As the winter term begins now I should be very glad to get it soon.[2]

Pray excuse my hurrying!

Yours very sincerely. | Prof J. Victor Carus

DAR 161: 64

[1] Christian Friedrich Schweizerbart was the director of E. Schweizerbart'sche Verlagsbuchhandlung of Stuttgart, CD's German publisher. Carus was translating *Variation* into German.
[2] Carus had earlier asked for the final proof-sheets of the first volume of *Variation* (letter from J. V. Carus, 16 October 1867). Carus was professor extraordinarius of zoology at Leipzig (*DSB*).

To Charles Lyell 31 October [1867][1]

Down. | *Bromley.* | *Kent. S.E.*
Oct 31st.

My dear Lyell

Emma[2] has got a bad-headach, so I write direct to you— M^r. Weale sent to me from Natal a small packet of dry locust-dung, under $\frac{1}{2}$ oz, with the statement that it is believed that they introduce new plants into a district. This statement, however, must be very doubtful. From this packet, 7 (seven) plants have germinated, belonging to at least two kinds of grasses. There is no error, for I dissected some of the seeds out of the middle of the pellets.[3] It deserves notice that Locusts are sometimes blown far out to sea; I caught one 370 miles from Africa, & I have heard of much greater differences.

You might like to hear the following case as it relates to a migratory bird, belonging to the most wandering of all orders, viz. the Woodcock. The tarsus was firmly coated with mud weighing when dry nine grains, & from this the Juncus bufonius or Toad-rush germinated.[4] By the way the Locust case verifies what I said in the 'Origin' that many possible means of distribution would be hereafter discovered.[5] I quite agree about the extreme difficulty of the distrib: of land Mollusca. You will have seen in the last edit: of 'Origin' that my observations on the effect of sea water have been confirmed.[6] I still suspect that the legs of birds which roost on the ground may be an efficient means; but I was interrupted when going to make trials on this subject & have never resumed it.[7]

We shall be in London in the middle or latter part of Nov.ʳ when I shall much enjoy seeing you. Emma sends her love & many thanks for Lady Lyell's note.[8]

Yours very sincerely | Ch. Darwin

P.S. My brother is at home now— You can of course, use anything in this note.[9]

LS(A)
American Philosophical Society (336)

[1] The year is established by the relationship between this letter and the letter from J. P. M. Weale, 7 July 1867.

[2] Emma Darwin.

[3] James Philip Mansel Weale had enclosed a packet of locust dung with his letter of 7 July 1867. In his Experimental notebook (DAR 157a: 83), CD recorded that 'the dung was put on burnt sand' on 27 August 1867, and by 6 September two grass seeds had germinated. Further germinations were recorded on 7, 9, and 24 September, and on 17 October, CD concluded that seven seeds had germinated altogether.

[4] The woodcock (*Scolopax rusticola*) belonged to the Grallae or Grallatores, an order of birds that included all the waders. It now belongs to the order Charadriiformes, which includes gulls and shorebirds. Herbert George Henry Norman sent CD the foot of a woodcock with some earth attached, from which seeds of *Juncus bufonius* (toadrush) were germinated (see *Correspondence* vol. 14, letter from H. G. H. Norman, 30 November 1866 and n. 2). CD added the information to *Origin* 5th ed., p. 440.

[5] CD had long maintained that seeds might be dispersed over long distances by various means including being transported by birds, and predicted that other means of dispersal would be discovered (see *Origin* 4th ed., pp. 432–3; see also *Correspondence* vol. 14 for some of CD's more recent discussions on dispersal). CD added the information on the seeds germinated from locust dung to *Origin* 5th ed., p. 439.

[6] Lyell's letter has not been found, but CD had puzzled over the means of distribution of land molluscs on oceanic islands for a long time (see *Correspondence* vol. 6, letter to P. H. Gosse, 28 September 1856 and n. 4). In *Origin* 4th ed., pp. 471–2, CD added information on recent experiments confirming his view that hibernating land molluscs could resist immersion in sea-water.

[7] CD had been unable to obtain the eggs of land molluscs to test their resistance to sea-water (see *Correspondence* vol. 6, letter to W. D. Fox, 3 October [1856]).

[8] According to CD's 'Journal' (see Appendix II), CD went to London on 28 November 1867. The reference is to Mary Elizabeth Lyell.

[9] Lyell added the information about seeds in locust dung and in mud on the foot of a woodcock to the tenth edition of *Principles of geology* (C. Lyell 1867–8, 2: 420–1), but did not refer to CD's information in his discussion of the distribution of land shells (*ibid.*, pp. 421–32). Erasmus Alvey Darwin lived at 6 Queen Anne Street, London.

From Henry Napier Bruce Erskine to Frances Julia Wedgwood 1 November 1867

My dear Snow[1]

I dare say you may remember sending me from Down some "queries about expression"[2] I wish I could have sent replies likely to be of service but I fear those I now send do not contain a great deal. Two or three times of late I have tried to watch "a row" in passing through the town—but as soon as I got near the "row" subsided into sulks—the fact is natives are seldom natural before Europeans—(except an exception can be made in the case of Europeanized natives)—they are always either trying to please or determined to be sulky! at all events they are always reserved.[3]

I applied to one or two friends sending them the questions but they all seem to have found much the same difficulties that I experienced— I also tried some native gentlemen but not with much greater success. However it is my all— I can no more!

I suppose your party is all back in London by this time after rusticating in Devonshire— I am sure Hengwrt[4] must have spoilt them for commonplace places! it was a delightful place—and I often look back with pleasure to my visit there. Have Effie and Hope kept up their photography?[5]

I hope this may find all your party well, it is very difficult for me to believe that eight months have not passed since I saw you all— it seems years ago!

I am now at Ahmednuggur a pleasant place and I have a nice house and lovely garden[6] I wish you c^d see it now the "poins settia" (I dont know how it sh^d be spelt!) is looking so lovely covered with its crimson leaves[7]—not little shrubs but good sized bushes 10 or 12 feet in height. I am however just on the point of leaving for the Districts to spend the next seven or eight months in tents a sort of life I do not dislike.

I hope Aunt Rich is pretty well—and Lady Inglis too neither seem as well as one could wish.[8] You have doubtless heard of Claude Turnbulls death at Rajkote[9] I [had] a line from him not long before written in low spirits and complaining of fever, but I did not apprehend danger. He seems to have been a favourite in his Reg^t & they are erecting a monument to his memory

With kind regards to all your party | Ever y^r affect cousin | H N B Erskine

1 November /67

[Enclosure]

Replies to questions about expression received from C. Darwin Esqre.

Quest. 1. Is astonishment expressed by the eyes & mouth being opened wide & by the eye brows being raised?

A. Yes.

Quest. 2. Does shame excite a blush & especially how low down the body does the blush extend?

A. Yes it does, but none of those I have asked have noticed that it extended below the neck, & I have never seen it very decidedly even in the neck.[10]

Quest 3. When a man is indignant or defiant does he frown—hold his body & head erect square his shoulders & clench his fists.

A. A native gentleman who was consulted answers yes to this but I have doubts. M^r. West the Judge in Canara[11] replies "I am not sure that I have ever seen the expression of pure indignation or defiance. It is I think a quivering frown with a slight projection somewhat sideways of the face"

I do not remember ever having seen a native clench his fists—as an Englishman does when angry & excited.

Quest 4. When considering deeply on any subject or trying to understand any puzzle does he frown—or wrinkle the skin beneath the lower eyelids?

A. M^r. West writes. "In trying to comprehend the brows are wrinkled and the mouth closed but not tightly closed. In meditation or the endeavour to recollect the brows are uncontracted, the lips half open the head often a little towards one side."

A native gentleman remarks. "Yes he does so in looking down—but when looking up while in the act of considering or trying to comprehend or recollect the eyes are kept open the brows raised up & the upper part of the forehead wrinkled"[12]

Quest. 5. When in low spirits are the corners of the mouth depressed & the inner corner or angle of the eye brows raised by that muscle which the French call the grief muscle?

Ans. All whom I have asked—Native & European agree in replying "Yes" to this—[13]

Quest 6. When in good spirits do the eyes sparkle with the skin round & under them a little wrinkled & with the mouth a little drawn back at the corners

Ans. Yes.

Quest 7. When a man sneers or snarls at another is the corner of the upper lip over the canine or eye-teeth raised on the side facing the man addressed?

The answer rec^d. from both Europeans & natives to whom I have written is "Yes" but so far as my own experience goes I do not remember to have seen the upper lip raised in the manner described—

Quest 8. Can a dogged or obstinate expression be recognised which is chiefly shown by the mouth being firmly closed a lowering brow & a slight frown?

Yes.

Quest 9. Is contempt expressed by a slight protrusion of the lips & turning up of the nose with a slight expiration?

Yes is the reply I have received from all—but I do not remember ever having heard any expiration when expressing contempt—

Quest 10. Is disgust shown by lower lip being turned down—the upper lip slightly raised with a sudden expiration something like incipient vomiting or spitting out of the mouth?

M^r. West says "I have never seen disgust alone expressed it has always been accompanied by fear or a desire to suppress any sign of emotion" and I must say the same—

A native gentleman answers this in the affirmative—

Quest: 11— Is extreme fear expressed in the same general manner as with Europeans—?

I think so—

Quest: 12. Is laughter ever carried to such an extent as to bring tears into the eyes?

Yes.

Quest— 13. When a man wishes to show that he cannot prevent something being done—or cannot himself do something—does he shrug his shoulders turn inwards his elbows, extend outwards his hands & open the palms?

Ans. M^r. West replies "He shrugs his shoulders and lays his hands uncrossed on his breast—"

I have however frequently seen a native when shrugging his shoulders extend outward his hands & open the palms—but I have never seen the elbows turned inwards in the European way—at least not so markedly.[14]

Quest 14. Do the children when sulky pout or greatly protrude the lips— — — Yes.

Quest 15. Can guilty or shy[15] or jealous expressions be recognised tho' I know not how these can be defined?

M^r. West writes "I have recognised guilty & shy expressions but not jealousy".

A native gentleman writes "When a man feels guilty his countenance is a little darkened & the lips are dried & shrivelled. Shyness is also recognisable"[16]

None of those to whom I referred have been able to state that they have ever observed jealousy expressed—and I am unable at present to give any hint as to how jealousy is expressed.

Quest— 16— As a sign to keep silent is a gentle hiss uttered?

Ans. M^r. West replies "A sound like "ch" is uttered—whence I suppose comes "chhup" (the Hindustani word for silence)

A native gentleman who saw the above reply wrote "Yes but it is adopted from the English— The genuine native sign to keep silent is made by putting the index finger of the right hand upon the nose with eyes closely shut when the sign is made without anger when it is made in anger it is accompanied by a frown."

Quest 17 Is the head nodded vertically in affirmation & shaken laterally in negation?

M^r. West replies "It is slightly nodded vertically in affirmation with a little raising of the outer corners of the eyebrows Negation is sometimes expressed by a lateral shake but more frequently by throwing the head suddenly back & a little on one side with a click of the tongue"

A Native gentleman writes "Negation is sometimes expressed by shaking the head laterally & sometimes by a click of the tongue. Affirmation is sometimes expressed by a vertical nod but more frequently by throwing the head to the left."

Among all the wild hill people the throwing back the head with a click of the tongue is alone used so far as I have seen.[17]

I fear there is not much in the foregoing replies—but as a rule Natives are so much on their good behaviour in the presence of Europeans that the latter have not favourable opportunities for observation— A natives anger or assumed anger bursts out frequently in a torrent of words— the louder he can howl/yell the better he seems pleased— they burst out in this way about the merest trifles—& the squabble seems to end as rapidly as it begun—

31 Octr. 1867 | H N B E

DAR 163: 31–2

CD ANNOTATIONS

1.1 I dare ... sulks— 1.5] *crossed blue crayon*
1.5 the fact ... natives) 1.7] *double scored blue crayon*
2.1 I applied ... difficulties 2.2] *scored blue crayon*
2.2 I also ... more! 2.3] *crossed blue crayon*
3.1 I suppose ... ago! 4.2] *crossed blue crayon*
5.1 Ahmednuggur] *underl blue crayon*
5.1 I am ... memory 6.5] *crossed blue crayon*

Top of letter: 'India | 13' *red crayon*

Enclosure:

Top of enclosure: '1' *blue crayon, circled blue crayon*; 'Henry Erskine | India' *pencil*
Question 3 and answer: *crossed pencil*
Question 17, answer: 'Probably acquired' *added in margin, blue crayon*

---

[1] Snow was Frances Julia Wedgwood's family nickname (B. Wedgwood and Wedgwood 1980).

[2] The letter has not been found. CD had sent out handwritten copies of his queries about expression earlier in the year (see, for example, enclosure to letter from J. P. M. Weale, 27 February [1867]). For a later, printed, version, see Appendix IV.

[3] See *Expression*, p. 21.

[4] Members of the Wedgwood family, including Francis (Snow's uncle) and Katherine Euphemia (Snow's sister), stayed at Lindridge, Teignmouth, Devon, in July 1867 (letter from H. E. Darwin to G. H. Darwin, 23 July [1867] (DAR 245: 280)). Emma Darwin's diary (DAR 242) notes that Francis and Leonard Darwin travelled there on 18 July 1867. Hengwrt, a house at Dolgelly (now Dolgellau) in north-west Wales, had been rented by Snow's family in the summer of 1865 (*Correspondence* vol. 13, letter from E. A. Darwin, 24 August [1865], n. 1).

[5] The references are to Katherine Euphemia Wedgwood and Hope Elizabeth Wedgwood, Snow's sisters.

[6] Ahmadnagar is a city in western central India, east of Bombay (*Columbia gazetteer of the world*). In the Darwin Archive–CUL there is a later letter from Claudius James Erskine to Hope Elizabeth Wedgwood probably answering a query about the answers to CD's questionnaire enclosed with this letter: 'Henry probably wrote from the Ahmednugur District, in the Bombay Presidency— That district in 1867, included several tracts occupied by wild tribes in the neighbourhood of the northern portion of the chain of the Western Ghats—' (DAR 163: 30).

[7] Poinsettia is a shrub native to Mexico (*Euphorbia pulcherrima*).

[8] Snow often stayed with Mary Rich, who was her aunt, and Erskine's (B. Wedgwood and Wedgwood 1980). Mary Rich lived at this time at 7 Bedford Square, London, the home of her friend Mary Inglis (*Post Office London directory* 1866, B. Wedgwood and Wedgwood 1980, p. 287).

[9] The reference is to Claudius James Turnbull. Rajkot is a town in the Sind District of western India (*Columbia gazetteer of the world*).

[10] See *Expression*, p. 316

[11] Raymond West was a judge in Canara (now Kanara), a region in the south part of Bombay Province (*Columbia gazetteer of the world, Imperial gazetteer of India*).

[12] See *Expression*, p. 33.

[13] Citing Erskine, CD noted that this facial expression was familiar among Indians (*Expression*, p. 187).

[14] See *Expression*, p. 268.

[15] On the printed questionnaire (see Appendix IV), the word is 'sly', not 'shy'. See also, however, enclosure to letter to J. P. M. Weale, 27 February [1867].

[16] CD cited Erskine's informant on the expression of shyness among Indians in *Expression*, p. 332.

[17] See *Expression*, p. 276.

From Charles Kingsley 1 November 1867

Eversley Rectory, | Winchfield.
Nov. 1/67.

My dear M[r]. Darwin

I have just found a letter written to you 5 years ago, & never sent.[1] Do me the honour to read it— & even if you do not answer it, think over it

Yours ever attached | C Kingsley

[Enclosure]

Eversley
March 23/62

My dear M[r]. Darwin,

Will you kindly give me your views of an old puzzle of mine? I am told that man is the highest mammal—w[h]. I dont deny. But that is supposed to include the theory of his being the highest *possible* mammal—w[h]. I do deny.

I see two imperfections in man as he is

1. The existence of the mammæ in the male, shewing that the sexes are not yet perfectly separated.

2. The ditrematous condition, w[h]. he has in common with the other mammals.[2]

That the specialty of organs increases as you rise in the scale, is, I suppose an acknowledged law—[3] And therefore, while I see, both in male & female, two diff[t]. secretions passing through the one orifice of the urethra, I cannot but suspect imperfection, & look forward to some higher tri-trematous race.

It is noteworthy, that the fact of the 2 secretions (urinary & sexual) passing through the same orifice) has been in all ages, Brahmin, Buddhist, Monastic, & What not, the physical ground of the contempt of sex, & of all that belongs to sex. No physical fact has played a more important part in the history of religion— W[h]. is, & always will be, the main history of the human mind.

Tell me what you think of this. You I can speak to as I can to no other man—[4]

Yours ever faithfully | C. Kingsley

DAR 169: 36, 30

[1] Kingsley first wrote to CD after receiving a presentation copy of *Origin* (see *Correspondence* vol. 7, letter from Charles Kingsley, 18 November 1859). In early 1862, Kingsley and CD speculated on the 'genealogy of man' (*Correspondence* vol. 10, letter from Charles Kingsley, 31 January 1862, and letter to Charles Kingsley, 6 February [1862]).

[2] Ditrematous: 'having the anal and genital orifices distinct' (*OED*).

[3] CD added passages to the third and later editions of *Origin* on the advancement of organisation and structure by natural selection. CD cited Karl Ernst von Baer's as the best standard of advancement or 'highness'; according to von Baer, advancement or 'highness' was related to the amount of differentiation of different parts of an adult organism and to their degree of specialisation with regard to function (see, for example, *Origin* 3d ed., pp. 133–4, 363–7).

[4] Kingsley's sense of propriety was reflected in his address to the Devonshire Scientific Society in 1871, in which he declined to consider 'physiological and anatomical' aspects of the origin of man (Kingsley ed. 1883, p. 316). For more on Kingsley's attitude to sexuality and his views on the divinity of carnal love, see Barker 2002.

From John Murray 1 November [1867][1]

50, Albemarle S.ᵗ | W.
Nov. 1.

My Dear Sir

Our thoughts about the Index jumped together, as they say, as you will perceive by the enclosed from Mʳ Dallas[2]—to whom I wrote 2 days ago— By this post I send you his Specimen—[3] Will you address him direct at York to save loss of time? I hope he is working to your satisfaction.

Here is a matter wch may interest you—not a "Cock & a Bull" but an Ass & a Bull— I have a MS journey in Asia Minor—the writer of wch a respectable American Missionary[4]—met with the offspring of these two animals & has made a *drawing* of the specimen & states that it is of common occurrence in that country.

I must confess—I do not believe it & if *you* do not—I shall advise the author to suppress the phenomenon

I shᵈ like to know what opinion you entertain, & whether you are acquainted with any such instance of extra-breeding

I am | My Dear Sir | Yours very faithfully | John Murray

Chas Darwin Esq

DAR 171: 352

[1] The year is established by the relationship between this letter and the letter from W. S. Dallas, 4 November 1867.

[2] The enclosure has not been found. Murray had engaged William Sweetland Dallas to prepare the index for *Variation* (see letter from John Murray, 23 September [1867], and letter to John Murray, 19 October [1867]).

[3] In his letter of 19 September [1867], Murray had told CD that he would ask for a sample of the index for CD's approval. The specimen CD received was returned to Dallas and has not been found (see letter to John Murray, 2 November [1867]).

[4] Murray refers to a manuscript later published by him as *Travels in little-known parts of Asia Minor: with illustrations of biblical literature and researches in archaeology* by Henry John Van Lennep (Van Lennep 1870). The published version did not contain the 'ass and bull' story.

From Isaac Anderson-Henry 2 November 1867

Hay Lodge, Trinity, | Edinburgh.
Novʳ. 2/67

My dear Sir

Have you ever seen, or heard of "An Account of the regular Gradation in Man and in different Animals and Vegetables and from the former to the latter illustrated with Engravings adapted to the subject by Charles White Read to the Literary and Philosophical Society of Manchester at different Meetings in the year 1795" published "London 1799"—an extraordinary Book and seems to have suggested to Lamarck if he ever saw it the theory which goes by his name.[1] I have got my hands over the Book in a somewhat singular way. Going in two days ago into a second hand Book Depot I saw a gentleman closely engrossed upon the Book, which after a time he laid down saying he would call for it next *Decemr* After he left I had the curiosity to glance at the Book and the first thing which attracted my attention was a Map or Engraving shewing on a graduated scale all the types among animals up to man, the lowest given being a *Snipe*.[2] As the Book was not to be called for till next month, I asked,—& the Bookseller allowed me to see it; nay more, on my suggesting it, he was agreeable to my sending it on sight to a friend for week. So if you have not seen and should wish to see this Brochure,—a thin Quarto of 200 pages,—I could easily send it to you per post. It is a Clergyman of Dundee who is the buyer at 3/6!

The Author who dedicates to Sir Richd Clayton Bart.[3] was a Medical man in Manchester He sets out in the "Advertisement" that "he has no desire to elevate the Brute Creature to the rank of humanity, nor to reduce the human species to a level with the Brutes"[4]

Please say if you would like to see the book for the limited space allowed

With best wishes | I remain ever | Yours very faithfully | Is: Anderson Henry

DAR 159: 71

[1] Anderson-Henry refers to Charles White and C. White 1799; CD had read the work in May or June 1844, and considered it to be 'a foolish book' (see *Correspondence* vol. 13, Supplement, letter to Henry Denny, [27 July – 10 August 1844] and n. 6). Anderson-Henry also refers to Jean Baptiste de Lamarck.

[2] The reference is to C. White 1799, plate 2, which shows gradations in the perpendicularity of a line drawn between the mouth and the front of the cranium in a series of silhouettes; the line in the silhouette of a European is shown as perpendicular, and in the head of a snipe as horizontal. Silhouettes of humans from other regions, and of an orang-utan, a monkey, several dogs, and a crocodile also appear in the plate to illustrate intermediate angles.

[3] Richard Clayton was the father-in-law of Charles White.

[4] C. White 1799, p. iii.

To Fritz Müller 2 November 1867

Down. | Bromley. | Kent. S.E.
Nov 2. 1867

My dear Sir

I rec^d y^r letter of July 17 two months ago, but did not answer it sooner, partly because you said you were starting for the interior & c^d not attend to science,[1] & partly because I have been working so very hard at my book.[2]

At last it will be finished & published towards the end of this month.[3] It consists of 2 vols 8vo. which shall of course be sent to you by post direct to Itajahy St Caterina Brazil.[4] In your last letter you give me various interesting details on dimorphic plants,[5]—on the orchid with lateral fertile anthers[6] & on sexual differences in crustacea[7]—for all of which I heartily thank you. The Borrerias are growing well[8] & Cocopizum (or some such name) has just germinated.[9] These with the Plumbago

will give me a rich harvest next summer.[10] The Gesnerias are now in flower; there
is much difference according to the age of the flower, & the individual, in the length
of the pistil; but now two plants have appeared with pistils so extraordinarily long
that I begin to think the species must be dimorphic.[11] Escholtzia Californica has
proved with me self-fertile; therefore your experiment w^d be worth trying again.[12]
From what you will see in my book about Passiflora, it w^d be interesting to observe
whether any not annual endemic species is self-sterile like your orchids.[13] You might
like to hear about Adenanthera; I wrote to India on the subject.[14]

I hear from Mr J. Scott that Parrots are eager for the seeds; & wonderful as
the fact is, can split them open with their beaks; They first collect a large number
in their beaks & then settle themselves to split them, & in doing this drop many;[15]
thus I have no doubt they are disseminated, on the same principle that the acorns
of our oaks are most widely disseminated. I hope you will prosper in y^r wild home,
& hereafter find much of interest to observe.

Believe me | my dear Sir | yours very sincerely | Ch. Darwin

P.S. A new part of Hooker & Bentham's Botany has come out & I hope has
been sent to you, but I will before long make enquiries.[16]

LS
British Library (Loan 10: 19)

[1] Müller had moved from Destêrro (now Florianópolis) on Santa Catarina Island to a homestead on the
Itajahy (now Itajaí) river near Blumenau, Brazil (see letter from Fritz Müller, 17 July 1867 and n. 5).
The information on Müller's proposed move and interruption of his scientific work was presumably
in the missing section of the letter from Fritz Müller, 17 July 1867.

[2] *Variation*.

[3] Murray advertised *Variation* as forthcoming in the *Athenaeum*, 19 October 1867, p. 487. It was published
in January 1868 (Freeman 1977).

[4] On Müller's address, see the letter from Fritz Müller, 17 July 1867 and n. 5.

[5] Müller had identified various species of the family Rubiaceae as dimorphic (see letter from Fritz
Müller, 17 July 1867 and n. 3).

[6] The reference is to an unnamed species of *Epidendrum*. See letter from Fritz Müller, 17 July 1867 and
n. 19.

[7] On sexual differences in the amphipod *Brachyscelus diversicolor*, see the letter from Fritz Müller, 17 July
1867 and n. 16. See also letter from Fritz Müller, 1 April 1867.

[8] In his letter of 2 June 1867, Müller noted that the rubiaceous plant that he had sent to CD under
the name *Diodia* was in fact a plant of the closely related dimorphic genus *Borreria*. See also letter
from Fritz Müller, 4 March 1867. In his letter to Müller of 15 August [1867], CD reported that his
Borreria seeds had germinated.

[9] The reference is to *Coccocypselum*, a dimorphic rubiaceous genus (see letter from Fritz Müller, 2 June
1867).

[10] Müller had informed CD of dimorphism in *Plumbago* and sent him seeds (*Correspondence* vol. 14, letter
from Fritz Müller, [2 November 1866], and this volume, letter to J. D. Hooker, 24 [March 1867]).
CD later reported that he had raised three plants (letter to Fritz Müller, 26 May [1867]).

[11] CD reported the germination of hundreds of seeds of *Gesneria*, supplied by Müller, in his letter to
Müller of 26 May [1867]. On CD's earlier doubt about dimorphism in the genus and later publication
of his observations, see *Correspondence* vol. 14, letter to Fritz Müller, [late December 1866 and] 1 January
1867 and n. 8.

[12] Müller had found *Eschscholzia californica* to be self-sterile (*Correspondence* vol. 14, letters from Fritz Müller, 2 August 1866 and [2 November 1866]).

[13] CD refers to *Variation* 2: 137–8, where he recorded that various species of *Passiflora* were self-sterile, although *P. gracilis*, an annual species, was self-fertile. CD had also considered conditions that might lead to reduced fertility (*Variation* 2: 163–71); endemic plants would be less prone to influence by such conditions. Müller had found many Brazilian orchids to be self-sterile (see letters from Fritz Müller, 1 January 1867 and 2 June 1867).

[14] CD had written to John Scott in India, enquiring about the dissemination of seeds of *Adenanthera* (see letter from John Scott, 24 September 1867). On CD's receipt of seed from Müller, and its identification as *A. pavonina*, see *Correspondence* vol. 14, especially the letter to J. D. Hooker, 10 December [1866] and nn. 2 and 3.

[15] See letter from John Scott, 24 September 1867.

[16] CD refers to the third part of the first volume of Joseph Dalton Hooker and George Bentham's *Genera plantarum*, published on 12 October 1867 (Bentham and Hooker 1862–83; see also Stearn 1956, p. 131). CD had previously sent Müller the first two parts of the work (see *Correspondence* vol. 14, letter to Fritz Müller, [late December 1866 and] 1 January 1867 and n. 13, and this volume, letter from Fritz Müller, 4 March 1867).

To John Murray 2 November [1867][1]

Down. | Bromley. | Kent. S.E.
Nov 2.

My dear Sir

Many thanks about the index. I have returned the M.S. with a note to Mr Dallas; it seems very well done but rather too full.[2]

With respect to the story of a hybrid from an ass and a bull (for I presume you refer to such) it is utterly incredible.[3] Many similar stories have been propagated, but on investigation have proved entirely false.

The explanation has generally been a slight monstrosity.

My dear Sir | yours very sincerely | Ch. Darwin

LS
John Murray Archive

[1] The year is established by the relationship between this letter and the letter from W. S. Dallas, 4 November 1867.

[2] CD refers to William Sweetland Dallas's specimen of the index for *Variation* (see letter from John Murray, 1 November [1867]). Neither the enclosure nor CD's note to Dallas has been found.

[3] See letter from John Murray, 1 November [1867] and n. 4.

To J. V. Carus 4 November [1867][1]

Down. | Bromley. | Kent. S.E.
Nov 4.

My dear Sir

I enclose the Title, which as you will observe is corrected.[2] I hardly see how you can write a prospectus until you see at least part of the 2nd volume.[3]

You will have received, since writing, 4 sheets; & 12 more pages completes vol. 1. Vol. 2 consists of 432 pages, & I suppose the clean sheets of the 1st half of this vol. will be soon printed.[4]

My dear Sir | yours sincerely | Ch. Darwin

LS
Staatsbibliothek zu Berlin, Sammlung Darmstaedter

[1] The year is established by the relationship between this letter and the letter from J. V. Carus, 30 October 1867.

[2] The enclosure has not been found. See letter from J. V. Carus, 30 October 1867. For the approved version of the title of *Variation*, see the letter to John Murray, 19 October [1867].

[3] Carus had offered to write a short prospectus for the German edition of *Variation* for its publisher, Christian Friedrich Schweizerbart (see letter from J. V. Carus, 30 October 1867).

[4] Proof-sheets of *Variation* were being sent to Carus for him to translate (see also letter to J. V. Carus, 19 October [1867], and letter from John Murray, 23 September [1867]).

From William Sweetland Dallas 4 November 1867

York
4 Nov. 1867

My dear Sir

I am much obliged by your kind letter of Saturday, & greatly pleased that you approve of the specimen of Index sent to you through Mr Murray.—[1] I notice your objection to the long list of references under the head of *"Dogs"*, & will endeavour to pick out such as may be considered useful *catch-lines*,[2]—or rather I will try to strike out some other principle in the subsequent parts of the book.— The portion relating to the introduction might, I think, remain as it is, unless you positively *object to it*,—the excess is at all events a fault on the right side.—

You would probably notice that I did not Index all the author's names cited in the notes, but only those whose statements are actually quoted,—will you be kind enough to let me know whether this will be satisfactory, or whether you would wish all author's names to appear in the Index, with references to all the places in which they occur.—

I will write at once to Mr Murray to send me the remaining Sheets,—& I think, if nothing unforeseen happens to prevent me, that I ought to be able to finish everything in a fortnight.— I should have made more progress already, but my work on the Zoological Record was interrupted for several weeks by illness in the early part of the summer, & with all my efforts I have been unable to make up the lost time.—[3]

It will be desirable that I should read the first proofs of the Index, & equally so that you should also look over them.—

Believe me, | My dear Sir | Your's very truly | W. S. Dallas.

C. Darwin Esqr. F.R.S.

DAR 162: 2

CD ANNOTATION
Top of letter: 'One *[proof]* & M.S. to you, and other to me | address of Clowes[4] | I s^d read 2^d Revises'

[1] The letter from CD has not been found; in 1867, the Saturday before 4 November was 2 November. See also letter to John Murray, 2 November [1867]. Dallas had been engaged by John Murray to prepare the index to *Variation*, subject to CD's approval of a sample (see letter from John Murray, 1 November [1867] and n. 3).

[2] There are forty-six sub-headings under 'Dogs' in the index of *Variation*.

[3] Between 1864 and 1868, Dallas contributed all the reports and abstracts on arachnids, myriapods, and insects published in the annual volumes of the *Record of Zoological Literature* (*Record of Zoological Literature* 6 (1870): preface). The title *Zoological Record* was adopted in 1870.

[4] CD refers to William Clowes & Sons, the printers of *Variation*.

To Charles Kingsley 6 November [1867][1]

Down. | Bromley. | Kent. S.E.
Nov. 6th.

My dear M^r. Kingsley

The subject to which you refer is quite new to me & very curious. I had no idea that the double function of an excretory passage had ever played a part in the history of religion.[2] I agree with what you say on speciality of organs being the best proof of highness in the scale of beings; nevertheless, when man as a standard of comparison is excluded, as with plants, it seems to be nearly impossible to give a good definition of Highness.[3] I do not feel sure that a passage performing a double function, if performed well, ought to be considered as a sign of lowness. I suppose that the presence of rudiments must be looked at as an imperfection, but it seems very doubtful whether these records of a former, & in most cases lower, state should be viewed as indices of relative lowness in the scale. Some authors, indeed, have used them as proofs of an opposite position.— It is an extraordinary fact that even Man should still bear about his body the plain evidence, as it seems to me, of the former hermaphrodite condition of the parent-form of all the Vertebrata.—[4]

From what you formerly wrote, I had hoped to have seen a review by you on the Reign of Law, but I have not been able to hear of its appearance.[5]

Pray believe me | Yours very sincerely | Charles Darwin

The Charles Darwin Trust: Quentin Keynes' Bequest (DAR 270)

[1] The year is established by the relationship between this letter and the letter from Charles Kingsley, 1 November 1867.

[2] See letter from Charles Kingsley, 1 November 1867. In the enclosure to the letter of 1 November 1867, Kingsley claimed that sex had long been held in contempt by different religions owing to the sharing of urinary and sexual functions by the same organ.

[3] On 'highness', see the enclosure to the letter from Charles Kingsley, 1 November 1867 and n. 3.

[4] Kingsley regarded the genito-urinary anatomy of humans and the presence of mammae in male humans as 'imperfections' (enclosure to letter from Charles Kingsley, 1 November 1867). CD had long

speculated on the hermaphroditic origin of animals (see *Natural selection*, p. 362, *Notebooks*, Notebook D, 158–9, 161–2, 172, 174, Notebook E, 57, 80). For more on CD's earlier views in relation to a hermaphrodite progenitor of the human and other vertebrate species, see *Correspondence* vol. 8, letter to Charles Lyell, 10 January [1860], and *Correspondence* vol. 10, letter to George Maw, 3 July [1862]. Later, CD suggested several possible causes, including former hermaphroditism, to explain rudimentary mammae in male mammals including humans (*Descent* 1: 207–11). See also letter from W. D. Fox, 1 February [1867], n. 11.

[5] The *Reign of law* (G. D. Campbell 1867) was critical of CD's theory; see the letter from Charles Kingsley, 6 June 1867, and the letter to Charles Kingsley, 10 June [1867]. No review of G. D. Campbell 1867 by Kingsley has been found.

To W. S. Dallas 8 November [1867][1]

> *Down. | Bromley. | Kent. S.E.*
> Nov. 8th

My dear Sir

On reflection I fear you will find it endless labour to give all author's names in notes.[2] So use your own discretion; anyhow most ought to be introduced. You might omit the word "note" in the index,[3] as the searcher has only one page to search; & even a word omitted in index is a gain.— When I had strength & made my own index, I used to give every author's name, but added only 2 or 3 words on the subject on which he treated or species or genus treated of by him; but then I never had such an index to make as this ought to be.[4] But as I said before kindly use your discretion & make a good Index.—

Yours very sincerely | C. Darwin

American Philosophical Society (Getz 5671)

[1] The year is established by the relationship between this letter and the letter from W. S. Dallas, 4 November 1867.

[2] CD's earlier letter commenting on the index to *Variation*, which Dallas was preparing, has not been found, but see the letter from W. S. Dallas, 4 November 1867).

[3] Subjects in footnotes and main text are indexed without differentiation in *Variation*.

[4] A series of brief descriptive entries appears after each personal name in the index of *Variation*; sometimes names mentioned in the text are not included in the index.

From Charles Kingsley 8 November 1867

> *Eversley Rectory, | Winchfield.*
> Nov. 8/67.

My dear Mr. Darwin

My thanks for your most interesting letter.[1]

Sex—you will find—plays *the* part in the real ground of all creeds. It is the primæval fact wh. has to be explained, or mis-explained, somehow. I cd write volumes on this. I may write one little one some day—[2]

As you say—the plain fact that man bears the evidence of a former hermaphrodite type are as indisputable—as they are carefully ignored—[3]

The whole question will have to be reconsidered by us—or by some other wiser race—in the next few Centuries.— *& you will be esteemed then as a prophet.*

Yours ever sincerely | C Kingsley

I have found actually a Darwinian Marchioness!!!!![4] So even the Swells of the World are beginning to believe in you. The extreme Radical press is staying off from you, because you may be made a Tory & an Aristocrat of.[5] So goes the foolish ignorant world— It will go, believing & disbelieving not according to facts, but to *convenience.* But do you keep yourself—(as you are) "unspotted from the world" as the good book bids all good men do[6]—& then 500 years hence, men will know what you have done for them.

DAR 169: 37

[1] See letter to Charles Kingsley, 6 November [1867].
[2] See letter from Charles Kingsley, 1 November 1867. Kingsley did not devote a separate work to the subject of sex. The relationship between sex and religion in Kingsley's writing and personal life is explored in Barker 2002.
[3] See letter to Charles Kingsley, 6 November [1867] and n. 4.
[4] The marchioness has not been identified.
[5] Kingsley was a frequent contributor to periodicals and from May to August 1867 ran the progressive *Fraser's Magazine* during the absence of its editor (see letter to Charles Kingsley, 30 April [1867] and n. 2). For an analysis of the Victorian press and its readership, see Shattock and Wolff eds. 1982, Secord 2000, and Cantor and Shuttleworth eds. 2004. For the polarisation of radical and Conservative views in the 1860s and on Kingsley's political views, see Houghton 1957, especially chapters 6 and 7. See also Kingsley ed. 1877.
[6] James 1: 27.

From Edward Wilson 8 November 1867

Hayes | Bromley, Kent.
8 Nov 1867

My dear Mr Darwin

I got your note while in Paris.[1] Before I went, however, I had put your emotional questions in train, sending one list to a friend in Melbourne, one to Queensland, & one to South Australia[2]

My South Australian friend writes me to the effect enclosed, by which you will see that your queries are likely to receive every attention from one of your ardent disciples,[3] & ⟨I⟩ think my Melbourne & Queensland friends will be equally interested in these enquiries[4]

Yours very truly | Ed Wilson

C Darwin Esq

DAR 181: 120

[1] Wilson may have visited the *Exposition universelle* in Paris; the *Exposition* opened on 1 April and closed on 31 October 1867 (see Mainardi 1987).

[2] Wilson had recently become a neighbour of CD's, having retired to Hayes Place, Bromley, in 1867 (*Aust. dict. biog.*). Wilson had evidently received CD's queries about expression (see Appendix IV) and sent them to friends in Australia, where he had lived from 1842. In *Expression*, p. 19, CD acknowledged Wilson's help in securing thirteen sets of answers from Australia.

[3] The enclosure has not been found. Wilson's friend has not been identified.

[4] Wilson's other two friends were probably Robert Brough Smyth of Melbourne and Dyson Lacy of Queensland, both of whom were acknowledged by CD for having provided information for *Expression* (*Expression*, p. 20).

From Francis Elliott Kitchener   9 November 1867

Rugby School, | Rugby.
Nov. 9th. 1867.

Sir,

I hope I may, without impertinence, send you a flower of Aquilegia alpina, from Switzerland from a height of, about 7000, or 8000 ft—

You will notice that each spur is eaten away as if the insects prefered a short cut to the nectary. I found the same had happened in all the full-blown flowers which I gathered— As I gathered the specimens in clouds, I saw no insects about—so that I am unable to say who were the robbers.

The probability, I am afraid, is that they were not Lepidoptera, or the case might have affected the subject treated in p. 51 of your "Orchids":[1] I merely send the fact as possibly of use of you. Please do not trouble yourself to answer this in any way.

Yr obedient Servant | F. E. Kitchener.

Has any one yet investigated the fertilization of the Stapelia, to see, whether the putrid smell may be regarded as a mimetic resemblance to carrion, which benefits the plant by attracting flies, under false pretences of its being a suitable place to lay their eggs?[2]

DAR 169: 39

[1] In *Orchids*, pp. 49–52, CD conjectured that moths (Lepidoptera) sucked fluid from between the inner and outer membranes of the nectaries of *Orchis pyramidalis* (now *Anacamptis pyramidalis*), though he admitted that no case was recorded of moths penetrating a membrane with their probosces.

[2] The pollination of *Stapelia* by flies attracted by the odour of the flowers was considered by Henri Lecoq (Lecoq 1862, pp. 270–1). There is an annotated copy of Lecoq 1862 in the Darwin Library–CUL.

From W. S. Dallas   10 November 1867

York
10 Nov 1867

My dear Sir

I have to thank you for your two notes & will endeavour throughout the Index to select the references to authors as judiciously as possible.—[1] The principle I have followed hitherto is to give those whose positive statements are cited, omitting those only quoted in notes in support of statements in the text,—but as you wish the whole given I think it might be managed without very great difficulty by running through the sheets a second time before completing the Index—[2] I will look to this at the proper time.—

Believe me | Your's very truly | W. S. Dallas.

C. Darwin Esq.[r] F.R.S. | &c &c &c

DAR 162: 3

[1] Dallas refers to the index to *Variation*. See letter to W. S. Dallas, 8 November [1867] and n. 2.
[2] See letter to W. S. Dallas, 8 November [1867] and nn. 3 and 4.

From J. V. Carus 11 November 1867

Leipzig
Nov. 11th. | 1867

My dear Sir,

I should be very sorry if you should think that I intend to write a Prospectus of your work before knowing it.[1] All that Schweizerbart asked me to do was to put down some few lines on the contents and the whole tenor of the first volume, which he could use for his advertisement of the work. And this does not even reach the public but is chiefly if not exclusively sent to booksellers.

Shall I return the proof of title?[2]

On page 325 there are two quotations of "Gardener's Chron. and Agricult Gaz.", notes 73 and 74. The pages are so near each other (729 and 730) that I think I ought to ask you if the years are correct, one gives 1856, the other 1855. I have no access to that periodical here, else I should look myself if all is right.[3]

The first two words on p. 348 made me ask you, what is "gean"; at first I thought "géant" was meant, but the word is on so prominent a place of the page, that it is not likely to be overlooked.[4]

Couldn't you ask Mr Murray to send the last sheet and preface (if there is one) *at once*. Schweizerbart (that is, his successor) is quite out of humour;[5] he saw the original advertised and sends me telegrams and letters to get the end out of me.[6]

I beg to accept my very best wishes for your health and and | Yours very sincerely | J. Victor Carus

DAR 161: 65

[1] See letter to J. V. Carus, 4 November [1867] and n. 3.
[2] See letter to J. V. Carus, 4 November [1867] and n. 2.
[3] Notes 73 and 74 in *Variation* 1: 325 refer respectively to *Gardeners' Chronicle* 1856, p. 729, and *Gardeners' Chronicle* 1855, p. 730.
[4] The first two words in *Variation* 1: 348 are 'Hungarian Gean', a variety of cherry. In his translation (Carus trans. 1868, 1: 439), Carus left the name 'Gean' untranslated.

[5] In autumn 1867, Christian Friedrich Schweizerbart was succeeded by Eduard Friedrich Koch as director of E. Schweizerbart'sche Verlagsbuchhandlung of Stuttgart (*Biographisches Jahrbuch und deutscher Nekrolog* 2 (1898): 227).

[6] Advertisements of the publication of *Variation* appeared in the *Publishers' Circular*, 1 October 1867, p. 566, and the *Athenaeum*, 19 October 1867, p. 487.

From Samuel Wilson to Ferdinand von Mueller[1]   12 November 1867

Longerenong | Wimmera
Nov 12th. 1867.

My Dear Dr. Mueller

I send you answers to a few of the questions sent by Mr Darwin.[2] I could not answer the others with sufficient accuracy. I have endeavoured to get the missionaries in this neighbourhood to answer them but have not yet received their answers—[3] As soon as I do so I shall forward them

And remain | Yours Truly | Saml Wilson

DAR 181: 129

CD ANNOTATIONS

0.1 Longerenong | Wimmera] *underl blue crayon*
0.1 Longerenong … 1867. 0.2] *after opening bracket, blue crayon*
Top of letter: 'Please return' *blue crayon*

[1] Samuel Wilson and Ferdinand von Mueller are both cited in *Expression*.
[2] CD had sent a handwritten copy of his queries about expression with his letter to Mueller of 28 February [1867], asking Mueller to forward it to someone capable of answering the questions from direct observation of Aboriginal peoples. The enclosure has not been found.
[3] The missionaries have not been identified. However, the missionaries Friedrich August Hagenauer and John Bulmer, both, like Wilson, working in Victoria, Australia, are cited in *Expression*, pp. 19–20.

From C. L. Brace   14 November 1867

Hastings-on-Hudson
Nov 14th | 1867

My dear Sir:

While in the mountains—the *Adirondacks* of New York[1]—this summer, I came across a "capital fact" bearing on the distribution of plants—

In that region, when the forest (of beech, birch, sugar-maple & pine) is burnt, there springs up immediately a thick grove of Wild Cherry (P. Pennsylvanica, I think).[2] The foresters' explanation is that these forests for square miles around were some thirty or forty years ago, the favorite roosting-places of the Wild Pigeon who came here by the millions. Their favorite food was the Wild Cherry, & they fed their young on it, so that the soil was thick-sown with the seeds, for miles about—

I am so much rejoiced to hear that you are taking up the *anthropoid* part of your great subject—[3]

With regard to change of type in America, the dentists say it is becoming almost universally the practice to remove some of the back-teeth in children, as the jaw does not seem large enough for the normal number of teeth,[4] & the front are made projecting— Two of my children, though their mother is from the Old World,— —have had to lose sound back-teeth for this reason— The American monkeys I believe have more than the normal number of teeth—[5]

I hope soon to send you an article on the reception of your theory by the Germans—Haeckel, Vogt, Büchner & others—

How little they have contributed to the Science of the subject![6]

Whitney showed me a skull found in California which was as low as the Neanderthal—[7]

We are daily expecting Dr & Mrs Gray—[8]

Believe me dear Sir | Yours with high respect | C. L. Brace

C. R. Darwin Esq

DAR 80: B154–5

CD ANNOTATIONS
1.1 While ... subject.— 3.2] *crossed pencil*
4.3 the normal] 'poor denture of the' *added above and del*
4.3 & the] *after closing square bracket*
4.5 The American ... Gray— 8.1] *crossed pencil*
Top of letter: 'Carl Vogt' *ink*; 'Teeth', *red crayon*

[1] The Adirondack mountains are in northern New York State, between Lake Ontario and Lake Champlain (*Times atlas*).

[2] Brace refers to *Prunus pensylvanica*, the wild red cherry or fire cherry of North America (Bailey and Bailey 1976).

[3] No letter from CD to Brace on this subject has been found. On CD's plans to publish on the origin of humans, see the letter to J. D. Hooker, 8 February [1867].

[4] CD cited this information from Brace in *Descent*, p. 27.

[5] American monkeys have thirty-six teeth and Eurasian monkeys have thirty-two (Nowak 1999, pp. 538, 569). American monkeys were considered to 'stand low in their order' (*Descent*, p. 47).

[6] Brace refers to Ludwig Büchner, Ernst Haeckel, and Carl Vogt. Brace's review of works by Büchner, Haeckel, Fritz Müller, and Vogt, 'Darwinism in Germany', was published in 1870 (Brace 1870). In the review, Brace reaffirmed his belief that German science had contributed little to CD's transmutation theory and criticised some German scientists for subscribing to the theory because of its congruence with their political and religious views. There is a presentation copy of Brace 1870 in the Darwin Pamphlet Collection–CUL. For more on the reception of CD's theory in Germany, see, for example, *Correspondence* vol. 11, letter from Friedrich Rolle, 26 January 1863 and nn. 6–14. See also Corsi and Weindling 1985, Montgomery 1988, Junker 1989, Engels ed. 1995, and Nyhart 1995.

[7] Brace refers to Josiah Dwight Whitney and to the 'Calaveras skull' found in February 1866 in a mine in Calaveras County, California. The skull was 130 feet below the surface, beneath a layer of lava. Whitney, the state geologist of California, acquired the skull and announced its discovery in July 1866 at a meeting of the California Academy of Sciences (Whitney 1866). He claimed that it was evidence of the existence of Pliocene humans in North America. For more on the discovery and

the ensuing controversy about the authenticity of the skull, see Dexter 1986. On the skulls found in the Neanderthal valley in Prussia, and Thomas Henry Huxley's view of their significance in relation to the antiquity of the human species, see *Correspondence* vol. 11, letter from J. D. Hooker, [15 March 1863], n. 19. For discussion of the Calaveras skull and human antiquity in relation to *Origin* and *Descent*, see Grayson 1983, pp. 210–13.

8 Asa and Jane Loring Gray. Brace was Jane Loring Gray's nephew (Dupree 1959, p. 192).

To J. V. Carus 14 November [1867][1]

Down. | *Bromley.* | *Kent. S.E.*
Nov 14—

My dear Sir

As you w^d probably like to finish Vol. 1 I send by this post a corrected revise which will be intelligible to you.[2] You can return it hereafter with Newmeister,[3] for I often find old revises come in useful. Never mind about the title-page. Herr Sweizerbart is unreasonable, & if I were in your place I w^d silence him.[4]

As all the 2nd revises of Vol 2 except of the 4 last sheets are finally corrected, the printers before long must print off; but on account of the index which is not yet finished I suppose the book will not be published till early in Dec^r.[5]

My publisher always sells by auction his books in advance & to my great surprize he sold at the auction 1200 copies of my book.[6]

I thank you sincerely for your intended corrections; but Gean is correct, & is the name of a whole group of cherries; but in what language I know not.[7] The 2 references to Gardener's Chronicle are correct, but when I saw the odd coincidence, I thought I had made a mistake.[8]

Rely on me there shall be no delay in sending clean sheets.

Believe me my dear Sir | yours sincerely | Ch. Darwin

LS(A)
Staatsbibliothek zu Berlin, Sammlung Darmstaedter

[1] The year is established by the relationship between this letter and the letter from J. V. Carus, 11 November 1867.

[2] CD refers to the first volume of *Variation*.

[3] CD had sent Carus a copy of a book by Gottlob Neumeister on pigeon breeding (Neumeister 1837; see letter to J. V. Carus, 16 September 1867).

[4] CD refers to the new director of the publishing firm E. Schweizerbart'sche Verlagsbuchhandlung, whose name was in fact Eduard Friedrich Koch. See letter from J. V. Carus, 11 November 1867 and n. 5.

[5] The index was being compiled by William Sweetland Dallas (see, for example, letter from W. S. Dallas, 10 November 1867). *Variation* was published in January 1868 (Freeman 1977).

[6] On publishers' customary November 'sale dinners', see J. Murray 1908–9, p. 540. Freeman 1977 records 1250 copies of *Variation* sold at John Murray's autumn sale in 1867.

[7] See letter from J. V. Carus, 11 November 1867 and n. 4. The word 'gean' is of Romance origin (*OED*).

[8] See letter from J. V. Carus, 11 November 1867 and n. 3.

To J. D. Hooker 17 November [1867]

Down. | Bromley. | Kent. S.E.
Nov. 17th

My dear Hooker

Congratulate me, for I have finished last revise of last sheet of my Book. It has been an awful job,—7½ months correcting the press— the book from much small type does not look big, but is really very big.[1] I have had hard work to keep up to the mark, but during last week only few revises came, so that I have rested & feel more myself. Hence, after our long mutual silence, I enjoy myself by writing a note to you, for the sake of exhaling & hearing from you.— On account of Index, I do not suppose that you will receive your copy till middle of next month.—[2] I shall be intensely curious to hear what you think about pangenesis; though I can see how fearfully imperfect even in mere conjectural conclusions it is, yet it has been an infinite satisfaction to me somehow to connect the various large groups of facts, which I have long considered, by an intelligible thread.[3] I shall not be at all surprised if you attack it & me with unparalleled ferocity.— It will be my endeavour to do as little as possible for some time, but shall soon prepare a paper or two for Linn. Soc.—[4] In a short time we shall go to London for 10 days, but the time is not yet fixed.[5] Now I have told you a deal about myself; & do let me hear a good deal about your own past & future doings. Can you pay us a visit?[6] Early in December Woolner is coming here to make a bust of me for my Brother;[7] & a most horrid bore it is; but he being here w^d not interfere with your visit if you could come. I have seen no one for an age & heard no news. I enclose some Himalayan Balsam seed, which my wife collected for you, as you said you wanted it—but the seed does not appear very good.[8]

Ever my dear Hooker | Yours most truly | C. Darwin

About my book, I will give you a bit of advice, skip the *whole* of Vol I, except last Chapt. (& that need only be skimmed) & skip largely in 2^d vol., & then you will say it is very good book.—[9]

Endorsement: '/67'
DAR 94: 35–6

[1] The reference is to *Variation*. In his 'Journal', CD recorded receiving the first proof on 1 March and finishing revisions on 15 November 1867 (see Appendix II). On the size of the book and the use of two different sizes of type, see the letter to John Murray, 8 January [1867].

[2] The index of *Variation* was being prepared by William Sweetland Dallas, who had expected to complete it by about 18 November 1867 (letter from W. S. Dallas, 4 November 1867). However, the task progressed more slowly than Dallas had predicted (see letter from W. S. Dallas, 20 November 1867). *Variation* was published on 30 January 1868 (see 'Journal' (Appendix II)); Hooker mentioned in his letter of 1 February 1868 (*Correspondence* vol. 16) that he had received his copy.

[3] Chapter 27 of *Variation* was headed 'Provisional hypothesis of pangenesis'. On pangenesis, see also the letter to Charles Lyell, 22 August [1867] and n. 4.

[4] CD's two papers, 'Illegitimate offspring of dimorphic and trimorphic plants' and 'Specific difference in *Primula*', were read at the meetings of the Linnean Society on 20 February and 19 March 1868, respectively, and subsequently appeared in the *Journal of the Linnean Society*.

[5] CD visited London from 28 November to 7 December 1867 ('Journal' (Appendix II)).

[6] Hooker next visited Down on 21 December 1867 (Emma Darwin's diary (DAR 242)).

[7] The references are to Thomas Woolner and Erasmus Alvey Darwin.

[8] Numerous species of balsam (*Impatiens*) are endemic to the Himalayas; for a contemporary list, see J. D. Hooker and Thomson 1859, pp. 117–18. Himalayan balsam is now *I. glandulifera* (A. Huxley *et al.* eds. 1992), but was formerly *I. roylei*, which was described as endemic and very common in the western Himalayas (J. D. Hooker and Thomson 1859, pp. 117, 127–8). Receipt of these seeds at Kew was recorded on 27 November 1867 with the note: 'The lady who gathered it mixed seeds of white Balsam with it. *Dead*' (Royal Botanic Gardens, Kew, Inwards book). No letter from Hooker requesting balsam seed has been found.

[9] The last chapter of the first volume of *Variation* is headed 'On bud-variation, and on certain anomalous modes of reproduction and variation' (*Variation* 1: 373–411).

From Isabella Elinor Aylmer   18 November 1867

Havelock House | Shanklin | Isle of Wight
18 Nov. 1867—

Sir,

As I am about to write for once a week a paper upon Chillingham and the "Wild Cattle"—for a picture of the latter, drawn by M⟨r⟩ Bradley, I should feel very much obliged ⟨if⟩ you would have ⟨the⟩ great kindness to g⟨ive⟩ me the results ⟨of your⟩ examination into ⟨the⟩ character and original breed of these Cattle—[1] Lord & Lady Tankerville[2] mentioned you as an authority upon the subject and I should have written direct to her Ladyship, had I ⟨n⟩ot wished to ask your permission to ⟨in⟩sert your observations ⟨into⟩ my article—

Trusting you will kindly favour me with as early an answer as you can—I am Sir | yours faithfully | I. E. Aylmer

DAR 159: 135

[1] The article 'The Chillingham Cattle' appeared anonymously in *Once a Week*, 1 February 1868, pp. 98–101, with an illustration by Basil Bradley. The text included an extensive quotation of information provided by CD, which is very similar to the information on the Chillingham cattle in *Variation* 1: 81, 83–4, and 2: 119.

[2] Charles Augustus Bennet, sixth earl of Tankerville, and Olivia Bennet.

From Asa Gray   18 November 1867

Cambridge Mass,
Nov. 18, 1867

My Dear Darwin

I was rejoiced by your favor of Oct. 16, and the sheets of the new book up to p. 336 (sign Y)[1] Thus far I have read only *dogs* & *cats*[2]—but expect soon to have some hours in a rail-way carriage— then, if not before, I shall read it up. I do not care to read it in driblets, bit by bit. But I shall be all ready for the treat you promise me in vol. 2ⁿᵈ,—"semi-theology" and all.[3]

Thanks for the facts about locusts' dung,—which I read to Wyman.[4] We hope you mean to print it.[5]

No time to write more now.—

Ever Your | A. Gray

DAR 165: 159

[1] CD had sent proof-sheets of most of the first volume of *Variation* to Gray with his letter of 16 October [1867]. Signature Y was pages 321 to 336 of the first volume.

[2] The first chapter of *Variation* is 'Domestic dogs and cats' (*Variation* 1: 15–48).

[3] In his letter to Gray, 16 October [1867], CD wrote that *Variation* ended with a 'semi-theological paragraph' in which he quoted and disagreed with Gray. See letter to J. D. Hooker, 8 February [1867], n. 7.

[4] See letter to Asa Gray, 16 October [1867] and n. 8. Gray refers to Jeffries Wyman.

[5] The information about the germination of grass seeds taken from the locust dung was added to *Origin* 5th ed., p. 439.

From E. A. Darwin to Emma Darwin [before 19 November 1867][1]

Dear Emma

Many thanks for having settled my affairs. I do hope I have not put you to inconvenience but I had no idea that Cumb^d was going down to you so soon.[2] It seems very doubtful whether Jos[3] will be able to come as Caroline[4] says he is very much pulled down. Lyell was talking yesterday with great admiration of Pangenesis.[5] What great news about Woolmer— I hope it wont be very hideous which is the most that I expect.[6]

yours affec, | E D

Tithes[7]

DAR 105: B121

[1] The date is established by the references to Charles Lyell's opinion of pangenesis, Thomas Woolner's bust of CD, and the tithe payment (see nn. 5, 6, and 7, below).

[2] The reference is to the family of Hensleigh and Frances Emma Elizabeth Wedgwood, who lived at 1 Cumberland Place, Regent's Park, London (*Post Office London directory* 1867). Frances and her daughter Katherine stayed at Down from 21 to 23 November 1867 (Emma Darwin's diary (DAR 242)).

[3] Josiah Wedgwood III.

[4] Caroline Sarah Wedgwood was Josiah Wedgwood's wife and CD and Erasmus's sister.

[5] Lyell had evidently seen and commented on CD's hypothesis of pangenesis in August 1867, after reading the proof-sheets of *Variation* (see letter to Charles Lyell, 22 August [1867] and n. 4).

[6] CD wrote to Joseph Dalton Hooker that Woolner was coming to Down to make a bust of him for Erasmus in early December (letter to J. D. Hooker, 17 November [1867]).

[7] Erasmus usually sent payments for tithes to CD in November or early December. No enclosure to this letter has been found. CD recorded receipt of cheques for tithes from Josiah Wedgwood III and Erasmus on 19 November 1867 (CD's Account books–banking account (Down House MS)). For more on CD's and Emma's incomes, see A. Desmond and Moore 1991, p. 396.

From J. D. Hooker 19 November 1867

Royal Gardens Kew
Nov 19/67.

My dear Darwin

I do not indeed congratulate—myself—on your book being done & the truce to our taciturnity—[1] Knowing from Lyell[2] that you were sore pressed, I did not like to bother you. I shall not be inclined to challenge Pangenesis,[3] I am 'umbled by your victory over my continental hypothesis.[4] (I won't give up Greenland though—I will have a "rag of Protection")[5]

As for me I have been & am Sic vos non vobissing rather too much even for my liking—& I really do like that sort of dilettanteing for my neighbours—[6] I have just concluded Bootts Carices, & am at the distribution of the copies (as much bother as any thing)—[7] I am printing Harvey's Genera of Cape Plants[8]—& revising the English Edition of DeCandolles "Laws of Botanical Nomenclature, which will be a good thick pamphlet.[9]

In the Garden I am very busy laying out grounds & planting all over, & doing a vast deal for better or for worse. Also I have induced the Board to put the whole Heating apparatus, (which has been messed & jobbed till Curator & Foremen are driven wild,) into my hands instead of the Surveyor of Works, & I have elaborated a plan for rearranging the whole in 25 Houses & 3 Museums, & have put out all for estimates from 3 Tradesmen.[10] I shall effect an enormous saving, & have all properly heated too. Also I am planning one new range of Houses to supersede 7 old ones, & which will not only save 6 fires, but save Smith & myself a deal of labor.[11]

Smith has been very bad since July, has considerable heart disease & functions all out of order— he was away a month in Cornwall, & is now gone for a month to Brighton— this is a severe blow to me.— The whole Garden system is however in such good order that I can conduct the out of door duties in his absence with pleasure. I can trust all my 7 foremen[12]—& Oliver reigns supreme in the Herbarium, & takes some of the correspondence—[13] he has taken to mineralogy as an amusement & collected some beautiful things in Skye & elsewhere.

I shall be delighted to come in December & will hold myself free whenever most convenient to you. & be glad to meet Woolner. I suppose there is a chance of *my* getting your bust now—which you seem to have forgotten all about.[14]

I have met Huxley several times lately, he has two children ill with S. Fever,—the first, my Godson, had it mild, I hope the second, a girl, will be equally favorable.[15]

I have just heard that the Endemic Umbelliferous plant of S.t Helena, which is a species of a Cape genus, takes exactly the same abnormal form & Palm like habit as one of the Endemic Madeiran species of the European genus "Ferula".— this is a good case of conditions.[16]

I expect the first installment of Seychelles Island plants very soon.[17]

Thanks for the Balsam seed—also for the advice about your book,[18] but the chances are that I shall not find time to read it. at all till I have forgotten the advice.

Have you read Sintram, I never did before, what a *grewsome* story it is.—[19]

Gen. Plant. jogs on, I am at a family Rubiaceæ, that takes an immense deal of dissection & gets on proportionately slowly.[20]

We are all well— | Ever aff Yrs | Jos D Hooker

Do you know that most Brambles have an odd habit of actually thrusting the ends of their surculi down into the ground when a sort of callus forms at the tip & makes root & new plant— This is quite different from a Strawberry runner, I think, that buds at the side, like an ordinary surculus.

DAR 102: 182–4, DAR 47: 191

CD ANNOTATIONS[21]

3.1 In the ... Labor. 3.9] 'Your work stuns one' *added pencil*
4.1 Smith ... July,] *scored red crayon*
4.5 Oliver ... favorable. 6.2] *crossed ink*
4.5 Oliver ... advice. 9.3] 'Nov. 19ᵗʰ 1867— Laws of Variation' *added ink*
5.1 I shall ... December] *scored red crayon*
5.2 I suppose ... now 5.3] *scored red crayon*
7.1 I have ... conditions. 7.4] *scored red crayon*; 'like N. Amer *[illeg]*' *pencil*; 'Due to similar constitution? When growing to larger size, for suppose they do grow larger??' *ink*
9.1 Thanks ... advice. 9.3] *crossed ink*
13.1 Do you know ... habit] *scored red crayon*
Top of letter: 'Keep 2ᵈ sheet | Law of Variation' *pencil*

[1] See letter to J. D. Hooker, 17 November [1867].
[2] Charles Lyell.
[3] CD had asked Hooker's opinion of his hypothesis of pangenesis (see letter to J. D. Hooker, 17 November [1867] and n. 3).
[4] The humble Uriah Heep in Charles Dickens's *David Copperfield* did not pronounce the 'h' in 'humble'. In letters to CD in 1866, Hooker argued that the distribution of plants to islands could be explained by continents having formerly been greater in extent (*Correspondence* vol. 14). Hooker and CD had long disagreed over continental extension (see, for example, *Correspondence* vol. 13, letter to J. D. Hooker, 22 and 28 [October 1865] and nn. 11–13). They also disagreed on the influence of occasional transport (or trans-oceanic migration), as postulated by CD in *Origin*, chapter 12 (see, for example, *Correspondence* vol. 14). In his lecture on insular floras at the meeting of the British Association for the Advancement of Science in August 1866 (J. D. Hooker 1866a), Hooker in large part retracted his earlier objections to CD's theory. For CD's relief at persuading Hooker of his argument, see *Correspondence* vol. 14, letter to J. D. Hooker, 30 August [1866] and n. 5. For more on CD's long-standing criticism of continental extension theories, see, for example, *Correspondence* vol. 9, letter to Daniel Oliver, 30 November [1861] and nn. 7 and 8.
[5] On Hooker's hypothesis that Greenland represented the western boundary of the European flora, see *Correspondence* vol. 14, letter from J. D. Hooker, 9 August 1866 and n. 17. Hooker apparently alludes to the 'rag of protection' to which Charles William Wentworth Fitzwilliam had objected in the bill for the repeal of the Corn Laws (*Annual register* 1846, 1: 78); previously, Richard Cobden had referred to agricultural protection as an 'old, tattered and torn flag' (Bright and Rogers eds. 1870, 1: 282).

[6] *Sic vos non vobis*: a Latin phrase attributed to Virgil and used of those from whose work others reap the reward (Brewer 1898, p. 1183).

[7] Hooker supervised the engravings and production of the fourth volume of Francis Boott's *Illustrations of the genus* Carex (Boott 1858–67), adding occasional notes (*ibid.*, vol. 4, preface). Boott died in 1863. See also *Correspondence* vol. 12, letter from J. D. Hooker, [23 November 1864] and n. 13.

[8] Hooker had edited W. H. Harvey 1868, a new, enlarged edition of William Henry Harvey's *Genera of South African plants* (W. H. Harvey 1838), in accordance with the late author's wishes (W. H. Harvey 1868, preface).

[9] Alphonse de Candolle's laws of nomenclature had been adopted at the international botanical congress held in Paris in August 1867 (Candolle 1867). The English translation, *Laws of botanical nomenclature adopted by the international botanical congress held at Paris in August, 1867*, was published in 1868 by L. Reeve & Co. of London.

[10] Hooker refers to the Board of Works and Public Buildings and their surveyor (R. Desmond 1995, pp. 180, 241), and to the curator of the Royal Botanic Gardens, Kew, John Smith. For Hooker's programme of planting at Kew at this period, see R. Desmond 1995, pp. 226–8, 371. For more on improvements to the glasshouses of Kew and their heating systems, see R. Desmond 1995, pp. 230–1, 371. A museum was established at Kew in 1848 and a second opened in 1857 (R. Desmond 1995, pp. 193, 368–9). Hooker's plans for a museum for the timber specimens were not realised until after his retirement (R. Desmond 1995, p. 288; see also Allan 1967, p. 208).

[11] Hooker refers to the curator, John Smith. A new range of greenhouses was built in 1868 and 1869 (R. Desmond 1995, pp. 231, 371).

[12] Since becoming director of the Royal Botanic Gardens, Kew, in 1865, Hooker had taken measures to increase the efficiency of his staff, including replacing a number of foremen (Allan 1967, pp. 212–3).

[13] Daniel Oliver was keeper of the herbarium.

[14] On Hooker's earlier request for a bust of CD by the sculptor, Thomas Woolner, see *Correspondence* vol. 11, letter from Emma Darwin to J. D. Hooker, 26 December [1863] and n. 1. Hooker visited Down on 21 December 1867 (Emma Darwin's diary (DAR 242)).

[15] Leonard Huxley was Hooker and CD's godson (A. Desmond 1994–7, 1: 290–1). It is not known which of Thomas Henry Huxley's daughters, Jessie, Marian, Rachel, Nettie, or Ethel, had scarlet fever.

[16] On the 'law of the conditions of existence', see *Origin*, p. 206. Hooker's source was probably George Bentham, who was working on the Umbelliferae for *Genera plantarum* (Bentham and Hooker 1862–83; see letter from J. D. Hooker, 4 February 1867). The umbellifer endemic to St Helena is *Sium helenianum* (Hemsley 1885, 2: 68–9); *Sium* is represented in South Africa by a single species, *S. thunbergii* (W. H. Harvey 1868, p. 140; see also n. 8, above). The Madeiran species referred to is probably either a *Melanoselinum* or *Monizia edulis*; Hooker had noted the unusual palm-like habit of these plants and their occurrence on Madeira in J. D. Hooker 1866a, p. 7. The species of *Melanoselinum* endemic in Madeira is *M. decipiens* (Lowe 1868). The only European species of *Ferula*, *F. communis*, does not occur in Madeira (see Lowe 1868). For previous discussion of the effect of external conditions on morphology in the Umbelliferae (now Apiaceae), see *Correspondence* vol. 13, letter from Charles and Emma Darwin to J. D. Hooker, [10 July 1865] and n. 8.

[17] Edward Perceval Wright was collecting plants in the Seychelles (see letter from J. D. Hooker, 31 March 1867). Wright later described several Seychelles plants and found the general character of the islands' flora to conform with the 'general rules' given in Hooker's lecture on insular floras (E. P. Wright 1868; see also J. D. Hooker 1866a).

[18] See letter to J. D. Hooker, 17 November [1867] and nn. 8 and 9.

[19] Friedrich Heinrich Karl Fouqué, Baron de la Motte's *Sintram and his companions* (Fouqué 1815) is an allegorical novel concerned with death and sin. A new English translation had recently been published (Fouqué 1867).

[20] Hooker refers to the *Genera plantarum* (Bentham and Hooker 1862–83). The joint and separate contributions of the authors are documented in Bentham 1883.

[21] CD's annotations are for his letter to Hooker of 25 November [1867].

From Edward Cresy 20 November 1867

Metropolitan Board of Works | Spring Gardens
20 Nov 67.

My dear Sir,

The titles of the books I spoke to you about are—

Lebrun Charles— Conferences sur l'expression des differents caracteres des passions— Paris 4to. 1667.[1]

Lebrun Charles— Traitè de la physionomie aux rappat de la physionomie humaine avec celle des animaux 1 vol fol 56 plates. (no date)[2]

Le Grand & Baltard— Dissertation sur le traitè de Lebrun. Atlas fol 37 plates Paris 1827. published at £2.2[3]

Both the Passions & the Dissertation were in Bohn's catalogue 15 yrs ago but I do not see them in Quaritch's current catalogue, but they are not very rare & Murray could soon get them for you—[4]

My copy of Sir C Bells Anatomy of Expression is 4to London 1806. & that is given in the catalogue as the first edition—[5]

I have not forgotten the academic enquiries & have just got a batch of proof or rather revise from Paris in returning which I will propound the matter to my friend—[6]

I hope you continue as flourishing as when we saw you this sharp morning touches up my back terribly. twas better on Wednesday—

With kindest remembrances to all your circle—

Yours very truly | E Cresy

C Darwin Esq—

DAR 161: 249

CD ANNOTATION
Top of letter: 'Lebrun'; 'Not in London L or Athenæum | Brit. Mus.' *pencil*

[1] The reference is erroneous; Charles Le Brun delivered his lecture on the expression of the passions in spring 1668 (Montagu 1994, p. 142).

[2] Square brackets in original. For a detailed history of the numerous illustrated editions of Le Brun's work on human expression, see Montagu 1994, pp. 175–87. CD referred to Le Brun in *Expression*, quoting from *Conférence tenue en l'académie royale de peinture et de sculpture par Charles Le Brun, sur l'expression générale et particulière*, which was published in the ninth volume of Lavater 1820, pp. 257–302; there is an annotated copy of Lavater 1820 in the Darwin Library–CUL (see *Marginalia* 1: 484–5).

[3] Cresy refers to Jacques-Guillaume Legrand, Louis-Pierre Baltard, and Legrand and Baltard 1827.

[4] Cresy refers to Henry George Bohn, Bernard Quaritch, and John Murray.

[5] Cresy refers to Charles Bell and to Bell 1806. There is a heavily annotated copy of the third edition of Bell's *Anatomy of expression* (Bell 1844) in the Darwin Library–CUL; the inscription indicates that CD acquired it from Erasmus Alvey Darwin in November 1866 (see *Marginalia* 1: 47–9). CD referred frequently to the third edition in *Expression*. See also letter to A. R. Wallace, [12–17] March [1867] and n. 8.

[6] Cresy had spent time in Paris while training as an architect (*DNB*).

From W. S. Dallas 20 November 1867

Yorkshire Philosophical Society | *York*
20 Nov. 1867

My dear Sir

Except in the two or three earlier chapters of your book I have adopted your plan of giving every author's name,—the others I must fill in before completion.—[1] It makes the labour very great, however, & I cannot get on so quickly as I could wish.—[2]

Instead of 6 sheets *per noctem*[3] upon which I calculated, I can only get through 3 or at most 4— Still I hope to finish Vol. I. tomorrow night, & I have now up to page 192 of Vol. 2 in rough impressions.— I shall press on as fast as I can, but the work cannot be done quickly & well, as its very nature requires that much consideration should be given to almost every paragraph.—

Believe me | Your's very truly | W. S. Dallas.

The proofs can be managed in any way most convenient to you.—

Cha⁵. Darwin Esq.

DAR 162: 4

[1] On the inclusion of authors' names in the index of *Variation*, see the letter to W. S. Dallas, 8 November [1867], and the letter from W. S. Dallas, 10 November 1867.

[2] In his letter of 4 November 1867, Dallas had expressed the hope that he would finish the index 'in a fortnight'.

[3] *Per noctem*: per night (Latin). Dallas was curator of the Yorkshire Philosophical Society's museum (*Modern English biography*); he also worked for the *Zoological Record* (letter from W. S. Dallas, 4 November 1867 and n. 3).

To W. E. Darwin 20 November [1867][1]

Down
Nov. 20th

My dear William

I enclose cheque for £200, & & please acknowledge receipt.—[2]

I have thought of a better & *safer* plan than lending it. & that is giving it. Anyhow I w^d. not have been such an old Jew as to take 6 per cent.— I believe I did say something about helping you, so now I pay about half of Overend & your duty.—[3] I am sorry the Bank will be bad this Xmas; but I supposed it c^d. not be otherwise. You are quite right to pay off whole Duty & get done with it.

I am so glad that you will pay us & Eras.⁴ a little visit.— We do not yet know when we go up.— Woolner comes here, the Devil take him, early in December.⁵

My dearest old fellow | Your affect. Father | C. Darwin

DAR 210.6: 123

[1] The year is established by the date on which CD paid the cheque to William (see n. 2, below).

[2] CD recorded a payment of £200 to William on 20 November 1867 (CD's Account books–banking account (Down House MS)).

[3] Against his record of the payment, CD noted that the cheque was in respect of Overend and succession duty (CD's Account books–banking account (Down House MS)). For more on the failure of the financial firm, Overend, Gurney & Co., see *Correspondence* vol. 14, letter from J. D. Hooker, 13 May 1866, n. 15. The succession duty probably related to William's inheritance from his aunt, Susan Elizabeth Darwin, who died in October 1866 (*Correspondence* vol. 14, letter to W. E. Darwin, 8 November [1866]).

[4] Erasmus Alvey Darwin.

[5] Thomas Woolner planned to visit Down in December to make a bust of CD (see letter to J. D. Hooker, 17 November [1867]).

From R. M. Rolfe 20 November 1867

40 Upper Brook St
20 Nov 1867

My dear M[r] Darwin

I send you my Cheque for the Downe Charities of which you are so good as to be the Guardian—[1]

I suppose this is the proper time for the payment, inasmuch as I have just received the enclosed unsigned receipt but you will know how to deal with the matter properly[2]

Yours very truly | Cranworth

DAR 161: 236

[1] Rolfe was a regular subscriber to the Down Friendly Club and the Down Coal and Clothing Club, of which CD was treasurer (Freeman 1978; see also *Correspondence* vol. 14, letter from Harriet Lubbock, [April? 1866], n. 4). CD recorded a contribution of £2 2s. from Rolfe in 1867 in his Down Coal and Clothing Club account book (Down House MS).

[2] The enclosure has not been found.

From Samuel James Augustus Salter 20 November 1867

17 New Broad St. | E.C.
November 20. 1867.

My dear Sir,

I recollect some time ago seeing, in the Field Newspaper, some questions put by you, asking for information relative to the colouring & marking of horses.[1] If I am not mistaken your queries had reference to dun-coloured horses approximating asses in their characteristic markings. Would you be so very kind as to tell me if you obtained any information on this head and if it is published.[2]

I have been making observations on this subject for years; and I have a good many notes and some sketches, which I mean to put together & bring before the Zoological Society.[3]

I find that the three stripe-markings, supposed to be characteristic of Asinus are all found in Equus.

1st. The stripe down the back, common in *duns*, occasional in *chestnuts*.

2nd. Stripes on the legs rare and only seen in *duns*.

3rd. Cross-stripe over the withers **very** rare indeed; and when present generally associated with *yellow dun-colour*.

This latter marking is so extremely uncommon that I find many persons, associated with horses all their lives, deny that it ever occurs. I have however seen some twenty examples in the last ten years.

I feel sure you will excuse my troubling you about this question.

Believe me, | My dear Sir, | Yours sincerely, | James Salter.

DAR 177: 15

[1] CD had requested information about the colour of horses in letters published in the *Field* on 27 April, 25 May, and 15 June 1861 (*Correspondence* vol. 9, letters to the *Field*, [before 27 April 1861], [before 25 May 1861], and [before 15 June 1861]).

[2] CD summarised his findings on this topic in *Variation* 1: 55–61, acknowledging information from several sources, including respondents to his queries in the *Field* in 1861 (see n. 1, above).

[3] Salter did not publish on the subject, and CD did not add his observations to the section on the markings of horses in later editions of *Origin*.

From John Edward Gray 21 November 1867

Brit. Mus
21 Nov 1867

My Dear Darwin

Thanks for sending the skins & skeletons of the Domestic & other animals they shall be well cared for[1]

I do not know if your take in the Proceedings Zool Soc I have been making a zoological arrangement of Sponges & would send you a copy of the paper if you do not take in the Proceedings hithertoo the arrangement of them has been entirely empirical—[2]

I will take care that the Boxes shall be returned to you & will you direct Mr G Snow to call for them any day next week[3]

You were quite right not to pay the carriage

With kindest Regards & Congratulations at your Book being so nearly out as to allow you to send the specimens[4]

Ever yours faithfully | J. E Gray

The Boxes have arrived but unfortunately the man left without our giving him the direction to call & they were left in the Hall & the Porter too idle to let us know the man was there

The Boxes shall be Ready for him on Monday or any day after

DAR 165: 212

[1] CD provided the British Museum with specimens of ducks, pigeons, and *Gallus bankiva*, ancestor of the domestic fowl; these were exhibited at the Natural History Museum in 1910, and their relevance

to CD's work on *Variation* was noted in the guide to the exhibition (British Museum (Natural History) 1910, especially pp. 273–4).

[2] CD subscribed to the *Proceedings of the Zoological Society of London*. Gray's paper (J. E. Gray 1867) is a review of the classification of sponges with proposals for new taxonomic divisions; there is a copy in the collection of unbound journals, Darwin Library–CUL.

[3] George Snow, the coal dealer at Down, operated a carrier service between the village and London on Thursdays (*Post Office directory of the six home counties* 1866).

[4] Gray refers to *Variation*. See also n. 1, above.

From Joseph Plimsoll[1]   21 November 1867

Exmouth | 8 Bicton Place
Nov.ʳ 21. 1867

Jesus said unto her, I am the resurrection and the life: he that believeth in me, though he were dead, yet shall he live. And whosoever liveth and believeth in me, shall never die. Believest thou this? John xi.25.26.

Pause my soul, over those divine, and glorious words of the Almighty Redeemer! What man, what prophet, what servant of the Lord; what angel, but He that is the Angel of the Covenant, One with the Father over all, God blessed for ever, could assume such language, and vindicate that assumption as Jesus did, both by his own Resurrection, and that of Lazarus?

And mark, my soul, the many precious things contained in this sweet scripture. Observe the blessing itself, even resurrection and life. Observe the source, the author, the fountain of it, Jesus, thy Saviour. Observe *for* whom this stupendous mercy is designed, and to whom conveyed, namely, the dead in trespasses and sins, and for the dying, languishing frames of believers. And lastly, observe how absolute the thing itself is; they *shall* live. O! precious words of a most precious Saviour! And may I not say to thee, my soul, as Jesus did to Mary, after proclaiming himself under this glorious distinction of character; 'Believest thou this?' Canst thou answer as she did, 'Yea, Lord, I believe that thou art the Christ the Son of God, which should come into the world'? This is a blessed confession to witness before God. For if I believe that Jesus be indeed the Christ of God, every other difficulty is removed to the firm belief that, as the Father hath life in himself, even so hath the Son life in himself, and whom he will he quickeneth. Witness then for me, every looker on, angels and men, that my soul heartily, cordially, fully subscribes to the same precious truth; and in the same language as Mary. *Yea, Lord*, I would say to every word of thine concerning thy sovereignty, grace, and love; as thou hast said it, so I accept it: in the very words of thine I take it, and cry out, yea, Lord, even so be it unto me according to thy word. And now, my soul, under all remaining seasons of deadness, coldness, backslidings, wanderings, and the like, never henceforth forget from whom all revivals can only come. Never look within for them: for there is no power of resurrection in thyself. 'Can these dry bones live?'[2] Yes, if Jesus quickens! And is Jesus less to quicken thee than thy connection with Adam to have killed thee? Oh! how plain is it that the very wants of the soul correspond to the very fulness of Jesus to answer them. And therefore, when the Lord Jesus saith, I am the Resurrection and the Life, he comes to seek employment in this glorious character, to quicken the dead and revive the living. Oh! Lord! give me to hear thy blessed voice this day, and my soul shall live, and live to praise thee.

The above is an extract from a work entitled "Morning and Evening Portion"— by the late D[r] Hawker. Vicar of Charles—Plymouth.[3]

DAR 174: 52

[1] The correspondent is identified from the handwriting and from his address (see letter from Joseph Plimsoll, 21 October 1867).

[2] Ezek. 37:3–4.

[3] The reference is to a collection of bible commentaries for every day of the year (Hawker 1829), by Robert Hawker. The extract is the morning reading for 8 May. In 1868, Plimsoll published *The vicar of Charles. A poem in commemoration of Plymouth's great preacher (Hawker) in a preceding age* (Plymouth: W. Cann).

From Karl von Scherzer 21 November 1867

8, Postgasse. Vienna
November 21st. 1867

Dear Sir,

I took the liberty to send you by the interference of our Embassy in London a volume containing "*the results of measurements taken on numerous individuals of different races during the campaign of the Austrian frigat Novara round the world in 1857–59,*" and beg to accept this work as a token of my highest esteem and admiration.[1] The measurements were taken by myself and the late D[r] Schwarz (physician on board the frigat); the results were gained by my indefatigable friend D[r] A. Weisbach and tend to support and to strengthen that wonderful law of which you are the venerated originator![2]

I have also given orders that a copy of the linguistic portions of the Novara-publications may be forwarded to you, and so will be the ethnographical part, which I am just now preparing for print.[3]

There is some hope, that another Expedition will be sent by our government to China and Japan in order to make treaties of commerce with the respective governments.[4] It is very likely, that I shall accompany that Expedition as Imperial Commissioner, and if this be the case, you will permit me to ask you for instructions with respect to *Scientific Desiderata.*

I do not want to add, how much I should be delighted to become of any service to you and your scientific pursuits. Allow me yet to remark, that D[r] Karl Vogt, who is now lecturing on the primitive history of man at Cologne and Aix-la Chapelle, has got a sight of the proof-sheets of the results of my measurements and was so much interested with them, that he will make them an object of his lectures.[5] Should the work be happy enough to enjoy and gain also *your* approval, this will considerably strenghten our efforts and endeavours and contribute to let us persevere in these most difficult pursuits. I have the honor to remain with the highest consideration Sir, | your most obedient | Dr Karl v. Scherzer.

DAR 177: 49

[1] Scherzer refers to part 2 (body measurement) of volume 2 (anthropology) of *Novara* expedition 1861–75; it was published in 1867. There is a lightly annotated copy in the Darwin Library–CUL (*Marginalia* 1: 854, s.v. Weisbach)). For CD's comments following the announcement of the proposed expedition, see *Correspondence* vol. 6, letter to Charles Lyell, 11 February [1857].

[2] In *Descent* 1: 216, 275, and 2: 320, CD cited Scherzer, Eduard Schwarz, and Albin Weisbach on anatomical differences between human races, and on anatomical features being more variable in men than in women. Scherzer refers to CD's theory of the transmutation of species by means of natural selection.

[3] Volume 5 (linguistics) of *Novara* Expedition 1861–75 was compiled by Friedrich Müller and covered Australian, Austronesian, Indian, and African languages; it has not been found in the Darwin Library–CUL or Darwin Library–Down. Part 3 (ethnography) of volume 2 of *Novara* Expedition 1861–75, published in 1868, was also compiled by Müller. There is a lightly annotated copy in the Darwin Library–CUL (*Marginalia* 1: 608–9, s.v. Müller, F.). For a detailed account of the *Novara* expedition, see Scherzer 1861–3.

[4] For an account of the expedition to Siam (now Thailand), China, and Japan, see Scherzer 1872.

[5] CD was aware of Carl Vogt's work on the origin of humans (see letter to Edward Blyth, 23 February [1867] and n. 3). For more on Vogt's controversial lectures on the subject, given in various European countries between 1867 and 1869, see W. Vogt 1896, pp. 177–84.

To J. J. Moulinié   22 November [1867][1]

Down. | *Bromley.* | *Kent. S.E.*
Nov 22

My dear Sir

I send by this day's the 2nd & 3rd sheets of vol. 2.[2] I hope you are making good progress in your dull & laborious undertaking.[3] My object in writing now is to ask you whether vol. 1 will be published separately before vol. 2? In this case, I sh^d be much obliged if you w^d give me the name & address of the publisher in Paris as I have mislaid it.

Pray forgive me for troubling you & believe me | with much respect | yours very faithfully | Charles Darwin

LS
Bibliothèque Publique et Universitaire de Genève, Ms suppl. 66, f. 1–2

[1] The year is established by the relationship between this letter and the letter from J. J. Moulinié, 25 November 1867.

[2] CD refers to proof-sheets of *Variation*. The sheets were probably for *Variation* 2: 1–32.

[3] Moulinié was translating *Variation* into French (see letter from J. J. Moulinié, 3 May 1867).

To E. Schweizerbart'sche Verlagsbuchhandlung 22 November [1867][1]

Down. | Bromley. | Kent. S.E.

Nov. 22

Dear Sir

As you have the right of translation of my present & previous works without payment I do not hesitate to ask you to let me have nine copies of the present translation as by the enclosed list.[2]

When the 1st vol is ready (and of course the 2nd vol likewise) will you be so good as to despatch the volumes according to the addresses given in the enclosed paper, & be so kind as to paste in "From the author" (as enclosed) in each copy excepting that for myself

Believe me dear Sir | yours faithfully | Charles Darwin

[Enclosure]

  to myself, Down, Bromley Kent[3]

  (2) Prof. Rütimeyer,[4] Bâle, Switzerland

  (3) Prof. Oswald Heer, Zurich, do.

  (4) Dr. Hildebrand[5] Bonn

  (5) Prof. Caspary Kœnigsburg[6] Prussia

  (6) Prof. Ernst Häckel[7] Jena Saxe Weimar

  (7) Dr. Rolle, Homburg vor der Hoehe, Frankfurt am Main[8]

  (8) Prof. Oscar Schmidt,[9] Hochschule Graz

  (9.)Hermann von Nathusius of Hundisburg, to care of Wigandt und Hempel of Berlin[10]

  Charles Darwin

LS(A)
Württembergische Landesbibliothek, Stuttgart (Cod. hist. 4° 333a, 77, 1)

[1] The year is established by the relationship between this letter and the letter from Eduard Koch, 11 December 1867.

[2] E. Schweizerbart'sche Verlagsbuchhandlung had published German translations of *Orchids* and the first three editions of *Origin* (Bronn trans. 1860, 1862, 1863, and Bronn and Carus trans. 1867). See also *Correspondence* vols. 8, 10, and 14. CD had renounced his right to profit from the German translation of *Origin* (*Correspondence* vol. 8, letter to H. G. Bronn, 4 February [1860] and n. 6). See also letter to E. Schweizerbart'sche Verlagsbuchhandlung, [19 March 1867].

[3] CD's copy of the German translation of *Variation* (Carus trans. 1868) is in the Darwin Library–Down.

[4] Ludwig Rütimeyer.

[5] Friedrich Hildebrand.

[6] Robert Caspary of Königsberg.

[7] Ernst Haeckel.

[8] Friedrich Rolle of Bad Homburg near Frankfurt am Main.

[9] Oskar Schmidt.

[10] CD refers to Hermann Engelhard von Nathusius and to the firm of Wiegandt und Hempel, Nathusius's publisher.

To J. D. Hooker 25 November [1867]

Down. | Bromley. | Kent. S.E.
Nov. 25th

My dear Hooker

I was heartily glad to get your letter, & to hear of your doings, which are so multifarious as to stun one: so many jobs on hand would fairly distract me.—[1]

We go to London to Erasmus'es (6. Q. Anne St) on the 28th & return home on Decr 7th.—[2] Woolner comes here, I believe, on Dec. 9th & I suppose & fear will be above a week about his work; so, if you possibly can, do pray come here any time after the 9th.—[3] I thought that you had given up all idea about my bust—[4] —pray excuse plain language, but you cannot be such an ass as to think of a marble bust.— I shall be proud & glad to give you a cast, & surely that will do.[5] The bust is making for Erasmus; & we are fighting here, for Emma votes for a marble copy & I maintain it is absurd, & plaister of paris just as good, or any good enough.—

I am very sorry to hear about poor Smith's health; for I took a great fancy for him, the day we walked round the gardens: it must be a fearful evil for you.—[6]

That is a very curious fact which you mention about the St. Helena Umbellifer; but can the "palm-like" growth be due to similar conditions? ought it not rather to be said that there is something in the constitution of the whole order, which leads them to take this form of growth, when the conditions favour their growth to a great size; for I presume they do grow very big?—[7]

I knew about the Brambles; & they excited in me a few years ago much just indignation; for when I found that stems, placed obliquely in a glass of water, bent upwards in absolute darkness, apparently guided in opposition to the force of gravity, I felt convinced that the ends of Bramble-shoots would bend downwards; but they bent neither up nor down; their flexibility & weight apparently guiding them to the ground.[8] I have often seen grey roots protruding before the end of the shoots had reached the ground.—

My dear old friend | Yours affectionately | C. Darwin

Endorsement: '/67'
DAR 94: 37–8

[1] See letter from J. D. Hooker, 19 November 1867.

[2] See also CD's 'Journal' (Appendix II). Erasmus Alvey Darwin, CD's brother, lived at 6 Queen Anne Street, London.

[3] Hooker had expressed his wish to visit Down in December 1867 while Thomas Woolner was making CD's bust (see letter from J. D. Hooker, 19 November 1867).

[4] See letter from J. D. Hooker, 19 November 1867 and n. 14.

[5] Cast: i.e. copy.

[6] John Smith was curator at the Royal Botanic Gardens, Kew. He had been suffering from heart problems since July (see letter from J. D. Hooker, 19 November 1867).

[7] See letter from J. D. Hooker, 19 November 1867 and n. 16. *Sium helenianum* attains between eight and twelve feet in height (Hemsley 1885, 2: 68–9). The herbaceous family Umbelliferae (now Apiaceae) includes a number of other tall species (Mabberley 1997).

[8] See letter from J. D. Hooker, 19 November 1867. In his undated notes on movement in bramble

stems in DAR 157.2: 87, 88, CD recorded that oblique bramble stems in water turned upwards in darkness, while strawberry stolons turned downwards. In a note dated 21 February 1864 (DAR 157.1: 59), CD concluded that bramble stems did not have the power of movement possessed by some climbing plants. CD's conclusions about the influences of contact, light, water vapour, and gravity on the movement of stems were later published in *Movement in plants*, but with no reference to brambles (*Rubus* sp.).

From J. J. Moulinié 25 November 1867

Geneva
25th Nov. 1867.

Dear Sir,

I am happy to inform you that I have last week forwarded to my Editor in Paris, the end of the copy of the first volume.[1] I am actually expecting the first proofs of the Eigth chapter, and conjecture, for I have not yet received them, that the six first chapters must now be printed and finished, for I have long ago, given the signature for press for them.

As for what concerns the publishing of the first volume separately, I must leave that to your own decision, being a far more competent judge than I can be, of what may be most convenient.[2] Unless there be peculiar and important motives to do otherwise, I think that in a general manner, it is perhaps better, when it can be done, to publish a work complete, and lay at once the whole subject before the reader, However, what you will decide on the matter, will be done.

I remain, my dear Sir your's | very respectfully | J J Moulinié

P.S. The french publisher at Paris is: M^r. C. Reinwald, Libraire-Editeur, 15, rue des Saints-Pères.[3]

DAR 171: 269

[1] Moulinié refers to the first volume of *Variation*, which he was translating into French, and to Charles-Ferdinand Reinwald (see n. 3, below).
[2] See letter to J. J. Moulinié, 22 November [1867].
[3] See letter to J. J. Moulinié, 22 November [1867]. Reinwald published the French editions of many of CD's books (see Freeman 1977).

To Anton Dohrn 26 November [1867]

Down. | Bromley. | Kent. S.E.
Nov. 26th

Dear Sir

I thank you sincerely for having sent me your paper on the morphology of the Arthropoda, which I was very curious to read.[1] It is a most deeply interesting subject, & if you finally succeed in showing how far the head-organs are homologous in the various classes, you will indeed have achieved a triumph in science.—[2] Allow me to thank you cordially for the generous & much too honourable manner in which you refer to my work.—[3]

If you could get fresh specimens of Scalpellum (Cirripedia), you would probably succeed in finding just-hatched larvæ in the sack; & these larvæ are so large that they are excellent for observation.[4]

I cannot yet quite persuade myself that the view which I have taken in my vol. on Balanidæ p. 105, of the homologies of the appendages is wrong. I still believe (though Fritz Müller writes to me that he thinks I am mistaken) that I saw in the anterior-lateral horns of the Carapace, the prehensile antennæ in process of development.[5] I sincerely wish you success in your studies, than which nothing can be more interesting.

I received some time ago a valuable memoir from you on certain ancient fossil insects, for which I am much obliged.[6]

When you next see Prof. E. Häckel, pray give him my kindest remembrances.[7]

With the most sincere respect, I remain | Dear Sir | Yours very faithfully | Ch. Darwin

Postmark: NO 27 67
Bayerische Staatsbibliothek (Ana 525, Ba 694)

[1] CD refers to a paper read by Dohrn on 5 September 1867 at the annual meeting of the British Association for the Advancement of Science (Dohrn 1867; see also *Report of the thirty-seventh meeting of the British Association for the Advancement of Science held at Dundee*, Transactions of the sections, p. 82, for a brief summary of the paper). CD's lightly annotated copy of Dohrn 1867 is in the Darwin Pamphlet Collection–CUL.

[2] Dohrn had claimed to have found that the carapace and head-plate in the larval forms of the crustacean genus *Cuma* and the insect genus *Phryganea* (caddis-flies) were homologous structures that became the branchial apparatus in adult *Cuma* and the antennae in adult *Phryganea* (Dohrn 1867, p. 86). Earlier, Fritz Müller had argued that larval stages in insects were acquired rather than inherited, but did not completely discount the possibility of a common ancestry for Crustacea and Insecta (F. Müller 1864, pp. 80–1; Dallas trans. 1869, pp. 118–21; Müller expanded his argument considerably for the English translation).

[3] Dohrn wrote that the application of Darwinian principles would solve difficulties that had beset studies of the morphology of arthropods for more than fifty years (Dohrn 1867, p. 86).

[4] CD had used *Scalpellum vulgare* and *Balanus balanoides* to illustrate the earliest larval stage of cirripedes in *Living Cirripedia* (1854), pp. 670–1, and plate 29, figs. 8 and 9.

[5] In *Living Cirripedia* (1854), pp. 104–5, CD described seeing an 'articulated organ' within both the minute curved horns (frontal filaments) and great lateral horns (frontolateral or anterior lateral horns) of the larva of *Scalpellum vulgare*; he argued that these became the first and second antennae, respectively, and that the second antennae developed into the prehensile antennae of the final larval stage (usually called 'pupa' by CD, but 'cypris' became the accepted term; see *Living Cirripedia* (1851), pp. 9–10). For CD's view of the importance of homological relations for classification, see *Correspondence* vol. 4, Appendix II, pp. 391–9. CD's interpretation led to a long-standing debate with James Dwight Dana on the homologies of larval and adult organs (see *Correspondence* vols. 4 and 5 and Newman 1993, pp. 372–4). Müller's letter on the subject has not been found, but in 1865, CD had asked Müller to investigate observations made by August David Krohn that were at odds with CD's interpretation of cirripede morphology (see *Correspondence* vol. 13, letter to Fritz Müller, 10 August [1865]). However, Müller, like Krohn, found the prehensile antennae were formed from the foremost limbs (in CD's terminology, the uniramous natatory legs; see *Living Cirripedia* (1854), pp. 670–1, and plate 29, figs. 8–10), not the anterior lateral horns, as CD believed (see Krohn 1860, p. 427, and F. Müller 1864, pp. 61–2 (Dallas trans. 1869, pp. 90–2)). For a comparison of CD's homologies with modern ones, see Newman 1993, p. 373 and fig. 7.

[6] The reference is to a paper on a newly discovered fossil insect that, according to Dohrn, demonstrated a link between Hemiptera and Neuroptera (Dohrn 1866, p. 334). CD's heavily annotated copy is in the Darwin Pamphlet Collection–CUL.

[7] Dohrn was Ernst Haeckel's student and assistant at Jena (*Correspondence* vol. 14, letter from Ernst Haeckel, 28 January 1866 and n. 7).

From Anton Dohrn 30 November 1867

Jena.

30. Novemb. 1867.

Dear Sir!

I need not tell You, how happy Your letter has made me.[1] If there was anything exciting my study, it was to change the manner of practical Zoology into that shape, which it must take after Your Origin of Species. There is a peculiarity in human spirit, that enables one to unite two quite different opinions. In *theory* people is of opinion, Natural Selection with all its Consequences is right,—and meanwhile in *praxi*[2] they follow the old rule and feel not the inconsistency of such doing.

It was my endeavour to try, whether my favourite animals the Arthropoda would not allow a reformation by applying Your Principles. I have worked since two years nothing but Embryology,—and I see now, this was the right way.[3]

The little Paper, I sent You, will scarcely be able to give a true idea of what I stated, to anybody; it is too short and the matter too complicated for such brief communication.[4] I am happy to say, that the first Volume of my larger work on the Morphology of the Arthropoda is almost ready to be printed. It contains the general foundations, and the special application to the Crustacea.[5] It is wonderful with how great a surety the genealogical tribe is to be stated and how simply the morphological specialities are to be understood as soon as they are brought under the principle of Natural Selection.

I can tell You perhaps by some few words one of the most striking facts in Cirripeds. I cannot enter into the proves, but only give You the result and ask confidence for my investigations.

The rudiment of the dorsal Spine of Zoëa is to be found in all classes of the Crustacea,—except in the Copepoda, one of the freshest and youngest. Perhaps there it is overlooked; I myself never treated them in special. This proves that *all* Crustacea have passed through the Zoëa, and this is necessary for the declaration of a quantity of facts, which without such view never could be declared.[6] You find the Rudiment even in Lepas, in the Cypris state.[7] It is situated above the mandibles on the back, between the prehensile Antennae.

The same Rudiment You meet in Evadne Nordmanni;[8] but when I first saw this remarkable creature, I was struck by the shape of it, for instead of seeing a rudimentary, functionless organ, I found an organ so well shaped, as ever an organ that has a distinct function to undergo. But I did not see any function. You can believe my joy, when I found in a small Paper of Professor Leuckart in the Archiv für Naturgeschichte, that this little organ, called by him "saugnapfartiges

Haftorgan" was used by the little animal to fix its body to the glass, wherein it was observed.[9] I now understood the not-rudimentary character.

This same thing You meet in Lepas. But soon there it becomes elongated and as long as the prehensile antennae, near which it is fastened to the body, on which the animal afterwards is to be found. It grows and grows, its muscular character is more and more developed, the ovaries are placed into it and finally it represents—the petiolum[10] of the Cirripedes! Is'nt that most striking? Such a change of function?

What belongs to the Extremities of the Cirripeds, their close affinity to the Cladocera and Phyllop⟨o⟩da[11] enables us to follow another interpretation. The observations of Krohn, Mecznikow, and Pagenstecher state, that the first Antennae bear the Cement Apparatus,—and I think, they are right.[12] What is to be observed in the prehensile antennae might be the Schalendrüse,—the homologous organ of the Grüne Drüse and similar organs.[13] The second Antennae and the Mandibles, the remaining Nauplius-Extremities are lost. Now we have a labrum and three pairs of maxillae; in Your nomenclature mandible, inner maxilla and outer maxilla.[14] What You describe as Palpus is, I think, the underlip or tongue (Savigny) that little bifid organ, which we meet in every Arthropodous animal opposite to the Labrum as the hinder wall of the mouth.[15]

Thus all is in agreement. And surely, I would never have found but by applying genealogical ideas, and I cannot tell how strongly even at every step I am indebted to Your leading ideas, that bring a splendid light into the Arthropoda-Confusion.

I'll not enter into other chapters. I only will promise, that the first copy of my book will be forwarded to You; and surely, it will be my greatest pride, if *You* could say me, that there is something valuable in it and if You acknowledge my leading ideas as those Principles that Your book has impregnated on every free young spirit.

I know, dear Sir, You don't like Compliments—and I dare say I cannot make them, remembering that word of Shakespeare: "And what they call compliments is like the encounter of two dog-apes".[16] So I am sure You'll make a difference between Compliments and deep Veneration, which flows out of the heart of a young ardent champion for truth and Liberty. Pardon my words, but I feel it a duty to tell You them. Enthusiasm is one of the most beautiful privileges of Youth and it is very often the orgin of good and lasting things. You may therefore imagine, that it was a great desire of mine to see and to speak You, when I was last summer in England. But I dared not trouble You, knowing that Your health is not so strong as we all might wish.[17] The more I owe You my greatest thanks, that You have sent me that letter, which will give me, as I might call it my scientific knighthood. I thank You, dear Sir!

Yours ever truly devoted | Anton Dohrn

Professor Haeckel sends his most sincere regards, and expresses with me the best wishes for Your health.

DAR 162: 203

[1] Letter to Anton Dohrn, 26 November [1867].

[2] *In praxi*: in practice (German; *Brockhaus-Wahrig*).

[3] Dohrn's doctoral work had been on insect anatomy. Early in 1866, he became a student of Ernst Haeckel at Jena and concentrated on crustacean morphology and embryology (Heuss 1991, pp. 46, 48).

[4] Dohrn refers to Dohrn 1867; see letter to Anton Dohrn, 26 November [1867] and n. 3.

[5] Dohrn never published the larger work he envisioned. He became aware that some of the observations on which he based his conclusions about insect–crustacean homologies were wrong (letter from Anton Dohrn to Adolf Stahr and Fanny Lewald, 29 May 1868 (Archive, Stazione Zoologica 'Anton Dohrn' Bc.60.85); see also Kühn 1950, p. 30, and Groeben 1982, p. 89 n. 1). He eventually published a series of papers on the structure and development of the Arthropoda, ten of which were collected as a book (Dohrn 1870; see also *Jenaische Zeitschrift für Medizin und Naturwissenschaft* 5 (1870) and 6 (1871) and *Zeitschrift für wissenschaftliche Zoologie* 20 (1870) and 21 (1871) for the ten papers in Dohrn 1870, and two further papers that appeared in 1871).

[6] In *Living Cirripedia* (1854), pp. 108–9, CD noted the similarity of structure of a larval form of *Chthamalus stellatus* and 'the so-called Zoea, or larva of certain Podophthalmia' (Podophthalmia is an older name for stalk-eyed crustaceans), but did not mention the dorsal spine. For an elaboration of Dohrn's views on the remnants of the zoea stage in the embryonic development of different members of the Crustacea, see Dohrn 1870, pp. 142–63.

[7] 'Cypris state': the last (post-naupliar) stage in the development of cirripedes, so called because the cyprid larva resembles the bivalve appearance of the ostracod genus *Cypris* (R. C. Moore and McCormick 1969). CD referred to the final larval stage as the 'pupa'.

[8] *Evadne nordmanni* is a marine crustacean in the order Cladocera (see also n. 11, below).

[9] Dohrn refers to Rudolf Leuckart and to Leuckart 1859. 'Saugnapfartiges Haftorgan': suctorial disc-like prehensile appendage (German). Leuckart actually wrote 'Haftapparat' or 'prehensile apparatus' (Leuckart 1859).

[10] Dohrn evidently refers to the peduncle of *Lepas*, but probably erred in translating the German word 'Stiel', which can also be translated as 'petiole'. In Dohrn 1870, pp. 154–6, Dohrn argued that the peduncle (Stiel) was derived from an organ homologous with the suctorial disc of *Daphnia* and the Phyllopoda, which he had interpreted as the rudiment of the zoea spine (Zoëastachel).

[11] The Cladocera and Phyllopoda were formerly suborders of the order Branchiopoda (Ziegler ed. 1909); in modern classification, Cladocera is a suborder and Phyllopoda a subclass of the class Branchiopoda (see J. W. Martin and Davis 2001).

[12] The references are to August David Krohn, Ilya Ilyich Mechnikov, and Heinrich Alexander Pagenstecher. Krohn's observations were published in Krohn 1860. For CD's view of the homologies of these organs, see the letter to Anton Dohrn, 26 November [1867] and n. 5. Dohrn later revised his position on the homologies and agreed with CD's interpretation (see Dohrn 1870, pp. 156, 172).

[13] 'Schalendrüse': shell gland (maxillary gland; see R. C. Moore and McCormick 1969). 'Grüne Drüse': green gland (antennal gland; see R. C. Moore and McCormick 1969). Dohrn's view that the shell gland in cirripedes and green gland in decapods were homologous organs was evidently derived from Fritz Müller (see F. Müller 1864, p. 61, and Dallas trans. 1869, pp. 90–1).

[14] Dohrn refers to CD's nomenclature in *Living Cirripedia* (1851), but in *Living Cirripedia* (1854), p. 107, CD wrote that he considered his first impression that the limbs were mandibles and two pairs of maxillae as untenable (see Newman 1993, pp. 374–5, for a discussion of CD's interpretation of the homologies).

[15] Palpus: oval setose mandibular endopod of cirripedes, attached directly to the mandible in the Acrothoracica or to the lateral margin of the labrum in the Thoracica (R. C. Moore and McCormick 1969; see also *Living Cirripedia* (1854), pp. 664–6 and plate 26). Dohrn also refers to Marie Jules César Lelorgne de Savigny.

[16] *As you like it*, 2.5.26–7: 'but that they call compliment is like the encounter of two dog-apes'.

[17] Dohrn had attended the annual meeting of the British Association held at Dundee from 4 to 11 September 1867 (*Report of the thirty-seventh meeting of the British Association for the Advancement of Science held at Dundee*, p. lxxiii). For more on Dohrn's activities in England and Scotland in the summer

of 1867, see Heuss 1991, pp. 60–5. Dohrn had probably been told about CD's health by Haeckel (see n. 3, above). Haeckel had visited CD in October 1866 (see *Correspondence* vol. 14, letter to Ernst Haeckel, [20 October 1866]).

To J. J. Moulinié 30 November [1867][1]

Down. | Bromley. | Kent. S.E. [6 Queen Anne Street, London]
Nov. 30th

Dear Sir

I am very much obliged for your note.—[2] I am particularly glad that M. Reinwald does not intend to publish the 1st. vol. separately.[3] I quite agree with you that in all cases it is desirable that the whole work shd. appear at the same time, and in the present case it is especially desirable, as I hope & think that the second volume is a little more interesting than the first; & if it had been possible, ought to have formed the first volume.—[4]

The remaining sheets of Vol. 2 will soon be sent to you.

Whenever the work is complete & nearly ready for publication, perhaps you be so kind as to inform me.

You will rejoice when your labour is over, as I do myself.— How far the book is worth publication, I from the first, as you may remember, doubted. But it has had by anticipation a large sale, my publisher having disposed of 1263 copies at a sort of auction which he holds at the commencement of our publishing season.—[5]

With much respect & thanks for the honour which you have done me in under taking the Translation I remain Dear Sir | Yours very faithfully | Charles Darwin

Bibliothèque Publique et Universitaire de Genève, Ms suppl. 66, f. 3–4

[1] The year is established by the relationship between this letter and the letter from J. J. Moulinié, 25 November 1867.

[2] Letter from J. J. Moulinié, 25 November 1867.

[3] CD refers to Charles-Ferdinand Reinwald and to the French translation of *Variation* (Moulinié trans. 1868).

[4] See letter from J. J. Moulinié, 25 November 1867.

[5] See also letter to J. V. Carus, 14 November [1867] and n. 6.

From Henry James Slack 30 November 1867

London | 34 Camden Sqr NW
30 Nov. 67

My dear Sir

I have long wished that the *Intellectual Observer*, which I have edited from its beginning, should publish some of your important papers on natural history subjects.[1]

I have not ventured to ask you before before, because I feared that from health & occupation you might not be able to accede to the request, but let me assure you that if at any time, it would be convenient to publish portions of your researches in a scientific magazine, I should be much obliged by your letting me have them for the I. O[2]

Permit me also to say that as Hon Sec of the Royal Microscopical Society, that any paper from you on points of minute structure would be highly appreciated by that body.[3]

I remain | My dear Sir | Yours faithfully | Henry. J. Slack

Charles Darwin Esq FRS | &c &c

DAR 177: 180

[1] The *Intellectual Observer*, established in 1862, was one of the most popular science periodicals in Britain during the 1860s, appealing to a young and well-educated readership (Ellegård 1957, p. 29). Its contributors included William Bernhard Tegetmeier and Alfred Russel Wallace (see *Correspondence* vols. 11 and 12, and this volume, letter to A. R. Wallace, 6 July [1867]).

[2] CD did not contribute an article to the *Intellectual Observer*.

[3] The Royal Microscopical Society published the *Quarterly Journal of Microscopical Science*. CD never contributed an article.

To *Hardwicke's Science-Gossip* [before 1 December 1867][1]

As in the August and September numbers, you have published an account of hedgehogs apparently carrying away pears and crabs sticking on their spines, you may think the following statement worth insertion as a further corroboration.[2] I have received this account in a letter dated August 5, 1867, from Mr. Swinhoe at Amoy:[3]—"Mr. Gisbert, the Spanish Consul at Amoy, informs me that when he was an engineer on the roads in Spain some years ago, he was fond of shooting and roaming about the country. He states that in the Sierra Morena, a strawberry-tree (Arbutus unedo?) was very abundant, and bore large quantities of red, fruit-like, fine, large, red strawberries. These gave quite a glow to the woods. The district in the mountain chain he refers to, is on the divisional line between the provinces of Seville and Badajos. Under these trees hedgehogs occurred innumerable, and fed on the fruit, which the Spaniards call Madrône. Mr. Gisbert has often seen an Erizo (hedgehog) trotting along with at least a dozen of these strawberries sticking on its spines. He supposes that the hedgehogs were carrying the fruit to their holes to eat in quiet and security, and that to procure them they must have rolled themselves on the fruit which was scattered in great abundance all over the ground beneath the trees."—

Charles Darwin.

Hardwicke's Science-Gossip 3 (1867): 280

[1] The date is established by the date of publication of the letter, 1 December 1867 (*Hardwicke's Science-Gossip* 3: 280). See also *Collected papers* 2: 137.

[2] Two short reports of hedgehogs collecting fruit on their spines appeared in the August and September issues (*Hardwicke's Science-Gossip* 3 (1867): 184–5, 213).

[3] The quotation is from the letter from Robert Swinhoe, 5 August 1867, with slight modifications. Mr Gisbert has not been further identified.

From Daniel MacKintosh 1 December 1867

20 Sussex Street, | Winchester,
1st Dec. 1867.

Sir,

Having some idea of the extent to which your time must be occupied, I should not have troubled you with acknowledging the receipt of your very kind and considerate letter,[1] were it not that for some time past I have been longing for an opportunity of consulting you on one or two points connected with Denudation.[2]

You may, perhaps, have noticed a number of articles by me in the Geological Magazine which have given rise to a rather warm controversy on the origin of escarpments, valleys, and plains.[3] With the exception of a little assistance from Mr Hull (of the Ord. Survey) and Mr Kinnahan (Irish Survey) I have been left to fight the battle with the Subaërial school singlehanded.[4] In endeavouring to answer opponents, I have been gradually led not to place *too much* reliance on sea-coast action, and after allowing a certain amount of influence to ice, I have been driven to oceanic currents, periodically increased in intensity by sudden upheavals or depressions of the earth's crust, as the main excavators of valleys.[5]

My object in writing is to take the liberty of asking if you have published any thing, or know of any thing that has been published, on the *excavating power* of currents, and whether you think that their action on the chalk of the south of England (with or without ice) would be sufficient to explain the hollowing, rounding scoring, *escarpmenting*, and terracing, which form so striking a feature of the chalk downs.[6] For upwards of a year I have been wandering among these downs with the view of generalizing all the facts connected with the terracing and scoring of their slopes. When I ventured a short time ago (too inconsiderately) to assert that there were raised beaches among the chalk downs, Mr. Poulett Scrope ridiculed the idea in the Geological Mag., and referred all the terraces to the action of the plough.[7] I think I can now demonstrate that, however much the terraces may have been either enhanced or defaced by cultivation, there are thousands which are of natural origin. The most puzzling characteristic is their very frequent *want* of horizontality and parallelism, which at first might suggest the idea of currents rather than sea-coast action. But I have noticed a similar absence of horizontal parallelism among the smaller terraces of the North of Scotland & elsewhere. Would you kindly inform me if this be a characteristic of any of the terraces you have discovered in S. America, and whether unequal upheaval, or irregular formation during oscillations of the land, would offer an explanation.[8] The finest series I have seen is near Stockbridge, on the side of the Andover and Romsey Railway.[9] They are parallel, but gently inclined longitudinally. They are covered with fractured flints, mixed with thoroughly rounded pebbles.

I enclose the very rough & imperfect sketch I took on the spot.[10]

Hoping you will kindly excuse the liberty I take in asking for a hint or two on these subjects when you happen to have a little leisure, | I am, Sir, | Your very obliged & humble Ser.ᵗ, | D. MacKintosh

This will be my address for more than a week to come—afterwards Chichester

P.S. I shall gladly embrace the first opportunity of seeing the work you refer to, and shall call attention to the fact probably in the Geological Magazine.[11]

DAR 171: 7

[1] CD's letter has not been found. See letter from Daniel MacKintosh, 8 December [1867] and n. 2.

[2] CD had written about denudation in two periodicals in the 1840s (see *Correspondence* vol. 2, letter to Charles Maclaren, [15 November – December 1842] (also published in *Collected papers* 1: 171–4), and 'On the transportal of erratic boulders from a lower to a higher level' (*Collected papers* 1: 218–27)). CD also considered denudation in *Origin*, pp. 285–7, 308, estimating the rate of its action in the formation of the Weald (see *Correspondence* vol. 12, letter from A. C. Ramsay, 10 July 1864 and n. 3).

[3] See MacKintosh 1865a, 1865b, 1866a–d, and 1867a–c, in which MacKintosh argued that greater recognition should be given to the effect of marine influences on the surface geology of Britain. In response to MacKintosh's papers, various writers, notably Joseph Beete Jukes, George Maw, George Poulett Scrope, and William Whitaker, published articles in the *Geological Magazine* during 1866 and 1867. These authors emphasised the importance of subaerial denudation, that is, erosion resulting from fluvial and glacial action rather than marine action (Challinor 1978). For a more detailed contemporary account of subaerial denudation, see Greenwood 1866. See also *Correspondence* vol. 13, letter to J. D. Hooker, [29 July 1865], n. 11.

[4] MacKintosh refers to Edward Hull and George Henry Kinahan and to the Ordnance Survey and the Geological Survey of Ireland. For Hull's views on denudation and their relevance to the controversy surrounding MacKintosh's papers, see E. Hull 1867. Kinahan emphasised the similar appearances of geological features formed by marine and subaerial agencies, including ice (Kinahan 1867).

[5] See, for example, MacKintosh 1867a, pp. 137–8.

[6] CD considered the action of sea currents in excavating coastal land forms, though not specifically chalk, in *South America*, chapters 1–3. CD wrote on the denudation of Wealden chalk by coastal erosion in *Origin*, pp. 285–7; the Weald is a district bounded by the North and South Downs in Hampshire, Surrey, Sussex, and Kent. In 1865, CD wrote to Charles Lyell, approving of the account of the denudation of the Weald in the sixth edition of Lyell's *Elements of geology* (C. Lyell 1865), which included an argument for marine erosion as the main agent in the formation of the Weald (see *Correspondence* vol. 13, letter to Charles Lyell, 21 February [1865], n. 5).

[7] MacKintosh had claimed that thousands of raised beaches were to be found in the chalk downlands of southern England (see MacKintosh 1866a, p. 69, and MacKintosh 1866b, p. 155). Scrope challenged MacKintosh's view of the marine origin of terraces on the chalk downs in Scrope 1866.

[8] CD had interpreted inclined terraces at Coquimbo, Chile, as raised beaches on ground that had subsequently undergone unequal elevation (*South America*, pp. 41–5).

[9] The terraces, on the south-west side of the village of Stockbridge, between the towns of Andover and Romsey, Hampshire, are described in MacKintosh 1869, pp. 86–8.

[10] The enclosure has not been found; the terraces at Stockbridge are illustrated in MacKintosh 1869, p. 87.

[11] The work CD refers to has not been identified. MacKintosh acknowledged CD in the preface to his book on the scenery of England and Wales, and claimed CD as a supporter of his views on marine weathering (MacKintosh 1869). For more on MacKintosh's views in the context of nineteenth-century geology in England, see Oldroyd 1999.

From Joseph Plimsoll 3 December 1867

Exmouth, Devon. | 8 Bicton Place
Decr 3rd. 1867

Dear Sir

"For to-day the Lord hath wrought salvation in Israel"
1 Samuel XI. 13.^{1}

This salvation, the result of the merit of Him who is above the law, is the provision of sovereign, undeserved, unmerited grace. Salvation was not deserved by us, nor won by us; but done for us, and is offered to us from the throne every day that we hear the gospel preached or taught. There was no obligation on God's part thus to interpose; there was no right on our part thus to obtain it. The Lawgiver, above the law, interposed in his sovereignty; why he passed by the fallen fiends that are in hell, and lighted, in his love and mercy, upon us, is one of those deep facts that ought to make us very thankful, very humble, and to feel more profoundly than we have ever felt before, what a magnificent salvation is that which is by grace, and not of works, lest any man should boast. Grace originated it all, and love without parallel or precedent executed it, and wisdom incrutable devised it. Christ Jesus, without constraint from above or claim below, interposed and died for us: and therefore as salvation by grace, and to the praise of grace, is the happy fact below, salvation to our God and to the Lamb is the never-ceasing, joyful song of the redeemed above. This salvation, thus free and sovereign, thus based on the strongest, surest foundations, is offered to all. Whatever it may be in its application to any, it is freely, *bona fide*, offered to every human being that hears it. It is not true that it is to be preached only to the elect; it is to be preached to sinners as such, without admitting the element of elect or non-elect, predestinated or unpredestinated; it is unto all and upon all that will take it; it recognizes no distinctions; it overflows all the sand-ridges of social division; it rises to, and reaches, and gives pardon to the greatest sin; (even the sin of infidelity—nay of atheism itself—and that in its most flagrant, rampant, revolting, God-incensing forms—as exhibited in what is termed the ^{x}development theory—of which, Sir, you are the chief exponent in this country—if not the author. Think of that Mr Darwin—to your soul's present and everlasting comfort and joy!—and for so much grace displayed in your behalf, cast yourself at the feet of your great deliverer, saviour, and benefactor—exclaiming—"Oh! the depths of the grace, and mercy, and love of God in Christ Jesus! in pardoning such a wretch as I am, who have so outraged reason, and thy divine attributes—and not only pardoning, but receiving me into thy favour and thy redeemed family—not even *mentioning* my enormities of transgression—but blotting them all out of the book of thy remembrance;^{2} calling on all the holy angels to rejoice at my restoration—saying—"for this my son was dead, but is alive again, was lost and is found."— therefore bring forth the *best* robe and put it on him; let the fatted calf be killed; let there be a sumptuous banquet on the occasion of his return to his home, and his father's heart; and great rejoicing amongst all the inmates of my celestial palace, in commemoration of his repentance, towards me and faith in my beloved Son".[3] Will you not then, dear Sir, be moved to say, mentally, if not orally, "Love so amazing, so divine, demands my soul my life, my all".[4]

But to return from this long parenthetical digression, to which I have been impelled by my deep solicitude for your soul's salvation—);[5] it follows, and pursues, and lays hold on the oldest and the worst of sinners; so that if any man perish, it is not because God will not save him, it is not because salvation cannot reach and overwhelm his sins, as the ocean would bury them in its depths; but solely and wholly, because one goes to his farm, another to his merchandise, another to his home, and anywhere and everywhere; giving to things that perish an importance that he denies to the salvation of his immortal soul. Your right to hear the gospel—oh! wondrous grace—is just your own self-inflicted ruin. If you be not sinners there is no salvation for you; if you be sinners—the oldest, the chiefest, the wickedest, the worst (even the notorious exponents and champions of atheism and infidelity, in all its various phases, and malignant manifestations and results—subtle and disguised, or open and avowed—such as are so rife—so rampant in the present day—this day, predicted in the Book of the Revelations, when "the "doctrines of devils"[6]—will be extensively preached.)[7] there is for you this very day pardon for the greatest sin, cancelling of the longest life of transgression; and God, instead of being unwilling to receive you, the instant he sees you in the far distant horizon he gives notice to choirs of cherubim and seraphim, and they will join in the glorious anthem that ever sounds and is ever sweet: "Let us rejoice; this my son was lost, and is found; was dead, and is now alive".[8] Our very disfranchisement from heaven is our franchise to Christ; that which keeps us out of heaven is that which makes us welcome to the sacrifice of Christ Jesus. I wish we could look upon the gospel less as law; I wish you could look upon all these truths as some dead things that lie far remote from us, or transcendantly above us. It is now—and what a thought!— it is now true that Jesus died and suffered, and God loved and planned, and prophets wrote, and psalmists sung, and evangelists have written, and apostles have preached, as truly for thee, my brother, as if there never had been, is not now, and will not be, another individual but thyself in the world. This salvation is received by faith alone. It is by grace—that is, it is undeserved; it is offered to all; it is received by faith alone. Do not think of faith as of some abstraction, some grand thing for theologians to talk about, but not meant for the ordinary level of mankind. Faith in Christ is so far identical with faith in other things. On Saturday night—if you are a merchant—you deposit all the stores of the week in some banking house. Now, notwithstanding an occasional breach of trust here and there, you have confidence in your banker that he will safely keep what you put in his hands. That is an act of faith. The only difference here is that what you deposit in the hands of the Lord of glory never can meet with disaster; for you can say when you have done it, "I know in whom I have believed; and that he is able to keep what I have committed to him against that day". When you take a five-pound or a ten-pound Bank of England note, you give your goods for it, and you get in exchange that bank note. What is that? Your faith in those copper-plate words, and in that water-mark, and in that piece of paper, and in the institution from which it comes. If you had not confidence in it, of course you would not accept it. What is faith in Christ? Just taking God at his word; believing what he says is true, and acting upon it; that is to say, carrying it into personal and practical action. Faith is confidence in God's word, and in Christ our sacrifice. Oh! what a tremendous thought at a judgment-seat, if God should tell us, "What! you could not believe! Did you put faith in a Bank of England note? Did you put faith in Coutts— in Drummond,[9] and did you never dare to put faith in me? you could not take my word! What an awful! It is true faith is the gift of God and the inspiration of the Holy Spirit. But this does not modify what I have said. We injure the gospel by mixing it up with incomprehensible abstractions. The simplest thing in the world is salvation; the simplest thing in the world is the way of being saved; and what you are called upon to do is to take God at his word. If when offered this Bank of England note for your goods, instead of taking it for what it is worth, you were to begin to try what sort of paper it was made of, whether the ink was indelible, and to copy the pictures on it, and to admire the exquisite mechanism, if I may so call it, of the bank-note, you would waste time and show want of confidence. I want you less to criticise this and that in the Bible, and oftener to open it, and take God at his word. When he says, "Jesus Christ is come into the world to save sinners, of whom I am chief"[10]—do not ask, "Is it for me?" ask rather, Why not for me? Your qualification is sin; your fitness is sin; and if you be a sinner, the chiefest of sinners, why, you are just the very person that Jesus Christ came into the world to save. Faith is in all its simplicity flying to the city of refuge, washing, that you may be clean. It is neither doing, nor buying, nor waiting, nor hesitating; it is a feast made for you, you have only to sit down and eat it; it is a wedding garment spun, woven for you, you have only to put it on. It is an ark sailing on the sea, and you are floating on a shattered wreck: you have only to get in, and be wafted to the haven of everlasting rest and peace. This is the gospel. I should only spoil the simplicity of this magnificent thing, if I were to add more; except to pray that the spirit of God may so apply it to your heart that you may be able to say "This day the Lord hath wrought salvation in my heart",[11] to his glory, and my present and eternal good.

The above, which I have been at considerable pains to transcribe for your perusal—is an extract from the writings of an eminent Clergyman of the Church of Scotland.[12] That it may—through the grace of God—the love of Christ—and the omnipotent and resistless might of God the Holy Ghost, be made the means of your being turned from darkness to light, from spiritual death to spiritual life—and translated from the Kingdom of Satan into the Kingdom of God's dear Son—from obnoxiousness to everlasting destruction and woe, to a certainty of resurrection to endless life, glory, and bliss—is the prayer | of dear Sir—Yours faithfully. | J. Plimsoll M.D

^x"The development theory"¹³—it may well be called indeed! This is a very appropriate and expressive designation for it—inasmuch as it involves a full development of the truth—"the wisdom of the wise is foolishness"¹⁴—and especially of that declaration of the prophet Jeremiah—"The heart is deceitful above all things and desperately wicked".¹⁵— "the world by wisdom knew not God"¹⁶—as the Apostle Paul affirms—is likewise herein exemplified—*atheism* being the logical result of the development theory.

DAR 174: 53

¹ The letter text, apart from the sections in parentheses (see nn. 5 and 7, below), the final paragraph and the postscript, is a transcription of a sermon by an unidentified preacher.

² Ps. 51:1–3.

³ The writer alludes to the parable of the prodigal son (Luke 15:11–32).

⁴ The quotation is from the hymn 'When I survey the wondrous cross' by Isaac Watts (*Church hymnal*, no. 247).

⁵ The section in parentheses is apparently Plimsoll's interpolation into the quoted sermon. See n. 1, above.

⁶ 1 Tim. 4:1.

⁷ See n. 5, above.

⁸ Luke 15.

⁹ The writer refers to the banking firms Coutts & Co. and Messrs Drummond.

¹⁰ See 1 Tim. 1:15.

¹¹ See 1 Sam. 11:13.

¹² The author has not been further identified.

¹³ Development theory or hypothesis: the doctrine of evolution, applied especially to that form of the doctrine taught by Lamarck (*OED*). A contemporary reviewer had referred to *Origin* as the latest form of the development theory (see also *Correspondence* vol. 8, letter to Charles Lyell, 6 June [1860] and n. 10); George Henry Lewes, Herbert Spencer, and Hewett Cottrell Watson were amongst CD's contemporaries who had written about development theory (Lewes 1853, Spencer 1858–74, 1: 389–95, and Watson 1845).

¹⁴ 'For it is written, I will destroy the wisdom of the wise, and will bring to nothing the understanding of the prudent' (1 Cor. 1:19).

¹⁵ Jer. 17:9.

¹⁶ 'For after that in the wisdom of God the world by wisdom knew not God, it pleased God by the foolishness of preaching to save them that believe' (1 Cor. 1:21).

From Julius von Haast 4 December 1867

Glückauf Christchurch
Decb 4. 1867

My dear M^r. Darwin

One of the Gentlemen to whom I submitted your Queries about expression,¹ the Rev^d J Stack, Maori Missionary Kaipoi has sent me his answers, but they were written on such thick foolscap & so wide apart, that it would have made quite a thick letter.² I therefore took the liberty to have them copied & beg to enclose them, trusting that they will be of interest to you. I have no doubt that the other Gentlemen in the North Island have forwarded their answers directly to you.³

I heard with great pleasure from our mutual friend D[r] Hooker,[4] that you enjoy now much better health & trust that this will continue, both to your comfort and in the interest of Science. Some few months ago, I had the pleasure to forward to you one of my reports on the headwaters of the Rakaia, which will make you acquainted with some of our glacier period phenomena.—[5] I have been busy lately with articulating 6 skeletons of Dinornis for our Museum & they form a really curious group & as I believe unique.[6] I enclose you a small photograph of them.[7] I have numbered the different species for your guidance.

No 1. Dinornis	giganteus—	9′ 10″
" 2 "	Robustus	8′ 5″
" 3 "	Elephantopus	5′ 3″
" 4 "	Crassus—	4′ 4″
" 5 "	casuarinus	5′ 2″
" 6 "	didiformis	4′ 3″

I have introduced the skeletons of a Kiwi (Apterix Owenii) & a human figure for comparison.—[8]

Wishing you further restoration to health believe me my dear M[r] Darwin | Yours most faithfully | Julius Haast

Ch[s]. Darwin Esq[re] FRS.
Down, Bromley | Kent—

[Enclosure]

D[r]. Darwins Queries about expression.

1. "*Is astonishment expressed by the eyes and mouth being opened wide and by the eyebrows being raised?*"

It is, but the action is more observable in some individuals than in others. The habitual endeavour of the maori is to conceal the workings of internal feeling and it is only when an individual is off his guard or the force of passion breaks through the habitual restraint put upon the feelings that any outward manifestations are noticeable either in the expression of the face or the attitude of the person.[9]

2. "*Does shame excite a blush when the color of the skin allows it to be visible?*"

It does. Hundreds of instances have come under my observation An amusing one occurred only a few days ago. An old man partly tatooed and rather darker than the average of his people, let a section of land (his all) to an Englishman for a term of years for a small rental just sufficient to keep him in clothes.—For some time there has been quite a mania among the Kaiapoi maoris for the possession of gigs and dog carts, and this seized my old friend who came to me to know whether he could not draw the rent for four years in one lump sum, to enable him to buy a gig that he had set his heart upon. The idea of this poor old clumsey ragged fellow driving in his private carriage was so absurd, that I burst out laughing; the old man blushed to the roots of his hair.[10]

3. "*When a man is indignant or defiant, does he frown hold his body and head erect, square his shoulders, and clench his fists?*"

He does. But immediately he begins to speak he loses the stiff set and tries to express his rage by the violent action of every part of his body.

I watched a man and woman quarrelling a few days ago and set down in my note-book the following particulars.—[11]

—Eyes dilated.— body swayed violently backwards and forwards. Head inclined forward toward the antagonist— fists clenched, now thrown behind the body, now brought forward and directed toward eachothers faces.

6. *"When in good spirits do the eyes sparkle with the skin round and under them a little wrinkled and with the mouth a little drawn back?"*

I took the following notes whilst watching a group of maoris who were much amused by something that was being told them.

—Eyes sparkling, half closed. Teeth exposed. Flesh drawn into a round lump on the cheeks.— Corners of the eyes all gathered up.[12]

14. *"Do children when sulky pout, or greatly protrude the lips?"*

Yes. I have seen them constantly do so.— Man sometimes and woman very frequently.[13]

I could have answered more of D^r. Darwin's questions but as he so particularly requests that memory may not be trusted to in doing so, I forbear till I can do so from actual observation of persons not likely to have copied their mode of expressing their feelings from Europeans.[14]

signed | James Stack

DAR 166: 12; DAR 177: 243

CD ANNOTATIONS
Enclosure:
11.1 —Eyes ... up. 11.2] *scored red crayon; scored pencil;* 'I ought to quote this' *pencil*
Top of enclosure: '*New Zealand*' ink; '32 | N. Zealand' *red crayon*

[1] CD sent Haast queries about expression with his letter to Haast of 27 February [1867]. For the names of those who received copies of CD's queries on expression from Haast, see the letter from Julius von Haast, 12 May – 2 June 1867 and nn. 2–4.

[2] Haast refers to James West Stack and to Kaiapoi, near Christchurch, on South Island, New Zealand (*Columbia gazetteer of the world*). For more on Stack and his view of CD's transmutation theory, see H. F. von Haast 1948, especially pp. 514–15, and Reed ed. 1935, pp. 61–7.

[3] No replies from the other recipients have been found in the Darwin Archive–CUL. See also letter from Julius von Haast, 12 May – 2 June 1867, n. 5.

[4] Joseph Dalton Hooker.

[5] Haast refers to J. F. J. von Haast 1866. See also letter from Julius von Haast, 12 May – 2 June 1867 and n. 8.

[6] Haast refers to *Dinornis*, the extinct genus of moas of New Zealand, and to the Canterbury Museum, Christchurch, New Zealand, which he founded (H. F. von Haast 1948, pp. 113, 123). The museum had opened to the public on 3 December 1867 (*ibid.*, p. 504). Haast first described the bones in J. F. J. von Haast 1868. Haast was assisted in articulating the skeletons by Frederick Richardson Fuller (H. F. von Haast 1948, p. 482). For more on the discovery of the moa bones in December 1866, see H. F. von Haast 1948, pp. 481–6.

[7] The enclosed photograph has not been found in the Darwin Archive–CUL. The Canterbury Museum has two photographs of the articulated *Dinornis* skeletons, taken by Daniel Louis Mundy. One, which

may be the same as that sent to CD, is reproduced facing p. 118. An engraving after one of the photographs appeared in the *Illustrated London News*, 8 February 1868, p. 144; the accompanying text named the species and gave their heights in feet and inches as in this letter.

[8] The male figure in the photograph reproduced facing p. 118 is probably Haast; the smallest skeleton could be that of the kiwi, *Apterix owenii*, a bird related to the moas. For an overview of subsequent taxonomic changes affecting species assigned by Haast to *Dinornis*, see, for example, Anderson 1989, pp. 17–38.

[9] See *Expression*, p. 279.

[10] See *Expression*, pp. 317–18.

[11] See *Expression*, p. 248.

[12] CD concluded that all human races expressed good spirits similarly, referring to observations made in New Zealand, among other places, but without naming Stack (*Expression*, p. 213).

[13] See *Expression*, p. 233.

[14] CD wrote at the end of the questionnaire, 'Memory is so deceptive on subjects like these that I hope it may not be trusted to' (enclosure to letter to Julius von Haast, 27 February [1867]). Only the five answers given here are recorded under Stack's name in CD's records of responses to his queries on expression (DAR 186: 1–29).

From I. E. Aylmer 5 December 1867

Havelock House | Shanklin | Isle of Wight
5th. Dec 67.

Dear Sir

I return your proofs, with hearty thanks for your courtesy and kindness.[1]

I have made extracts from them in my paper on Chillingham, and if you will permit me, I shall let you see the proof in order, that you may be perfectly satisfied with the use I have made of your labour.[2]

Lord and Lady Tankerville[3] or old friends of mine— And I was very anxious to do justice to their beautiful place and fine old Cattle, partly as a return for much personal kindness—and partly because being a Northumbrian myself, I like our county to have some honour paid it in the current literature of the day.

Believe me yours | gratefully | I E Aylmer

DAR 159: 136

[1] CD had sent Aylmer proof-sheets of *Variation* containing information on the wild white cattle of Chillingham Castle, Northumberland. See letter from I. E. Aylmer, 18 November 1867 and n. 1.

[2] The proof has not been found in the Darwin Archive–CUL. However, Aylmer's anonymous article included a substantial quotation from CD (see letter from I. E. Aylmer, 18 November 1867, n. 1).

[3] Charles Augustus Bennet, sixth earl of Tankerville, and Olivia Bennet.

From Thomas Woolner 6 December 1867

29, Welbeck Street, | W
Dec 6th 67

Dear M[r]. Darwin

I was afraid that on account of a severe cold I caught last Monday upon another just going away that I should have been too knocked up to begin work next

Monday,[1] but today felt so much better that I had had the clay packed and Railway people sent to to fetch it, when I learned that my young brother had to start for Monte Video on the 14th. of this month instead of Jan: as I expected; and the consequence is that I have so much to arrange for him that I cannot possibly get away just yet.[2] I hope this will not in any way cause you inconvenience; but to me it is very provoking as I thought that at last I saw my way to a clear beginning of the bust.[3] So soon as I get clear of an entanglement of things suddenly come upon me beside this primary one (which I trust will take only several days beyond the 14th.) I will write and ask if I may come then.—

Very truly yours | Thos: Woolner

Best thanks for your note anent trains.[4]

Please give the enclosed to M[rs]. Darwin— it is from my Mon: to M[rs]. A. Peel now in Wrexham Church.[5]

DAR 181: 160

[1] Woolner had been engaged to make a bust of CD, who expected him to start work at Down on Monday 9 December 1867 (letter to J. D. Hooker, 25 November [1867]). See also letter from J. D. Hooker to Emma Darwin, 11 December 1867. On 11 December 1867, Hooker wrote to Woolner inviting him to Kew to recover from influenza (Woolner 1917, pp. 280–1).

[2] The reference is probably to Henry Woolner (see Woolner 1917, p. 70).

[3] CD finally sat for Woolner in November 1868 (*Correspondence* vol. 16, letter to J. D. Hooker, 26 November [1868]). Woolner finished his marble bust of CD in 1870 (Woolner 1917, p. 340). In June 1869, Woolner completed a medallion of CD, which he sent to William Erasmus Darwin; it was later manufactured by Josiah Wedgwood & Sons (Woolner 1917, pp. 283, 340). See also Freeman 1978, pp. 94, 306.

[4] The note has not been found in the Darwin Archive–CUL.

[5] Woolner made the monument 'Heavenly welcome' in memory of Mary Ellen Peel and her eldest son; it was erected in Wrexham church in 1867 (Woolner 1917, pp. 242, 339). The enclosure has not been found in the Darwin Archive–CUL, but may have been a collotype of the work (see *ibid.*, facing p. 126).

To Albert Charles Lewis Gotthilf Günther 7 December [1867][1]

Down. | Bromley. | Kent. S.E.

Dec 7

My dear Sir

I enclose, as you desired, my photograph. Allow me to thank you cordially for your great kindness in giving me so much information, which is of real value to me.[2] Should any cases occur to you of well-marked sexual differences in snakes, batrachians, or lizards (about which I forgot to ask) will you have the kindness to make a memorandum on the subject.[3] I think I remember that the males of certain lizards in S. America had a scarlet throat; but this may have been an error.[4] In the Zoolog. gardens the keeper shewed me the males & females of the rattle-snake, & they differed considerably in colour;—the males being much more yellow.[5] He gave me positive evidence of the sexes, & said all that he had seen presented the same difference. I have told my publisher to send a copy of my book on "Variation &c" which will be published in 2 or 3 weeks, to you at the Brit. Museum; but I much fear it will contain very little that can interest you[6]

Believe me | my dear Sir | yours very sincerely | Charles Darwin

LS
Shrewsbury School

[1] The year is established by the relationship between this letter and the letter from A. C. L. G. Günther, 19 December 1867.

[2] No letter from Günther requesting CD's photograph has been found in the Darwin Archive–CUL. CD probably saw Günther during his visit to London from 28 November until 7 December 1867 (CD's 'Journal' (Appendix II)). The information provided by Günther probably included details of sexual differences in fish as reported extensively in *Descent*.

[3] Günther is cited on sexual differences in species of batrachians and reptiles in *Descent*, chapter 12. See also letter from A. C. L. G. Günther, 19 December 1867. CD's notes of information from Günther for *Descent* about snakes, batrachians, and fish are in DAR 82.

[4] The lizard species has not been identified. However, CD cited the blue, green, and coppery-red throat pouch of the Chilean lizard *Proctotretus tenuis* (now *Liolaemus tenuis*) as an example of males being more brightly coloured than females in *Descent* 2: 37. For descriptions of the lizards collected in South America by CD, including information on sexual dimorphism, see *Reptiles*. For CD's notes on the lizards he collected in South America, see R. D. Keynes ed. 2000.

[5] CD refers to the gardens of the Zoological Society of London in Regent's Park, London. The information on the rattlesnake is given in *Descent* 2: 29. For names of some contemporary keepers, see Scherren 1905, pp. 127, 143. CD had visited London from 28 November until 7 December 1867 (see n. 2, above).

[6] CD refers to John Murray and *Variation*, which was published on 30 January 1868 (CD's 'Journal' (Appendix II)).

To Charles Lyell 7 December [1867][1]

Down. | Bromley. | Kent. S.E.
Dec 7

My dear Lyell

I send by this post the Article in Vict. Institute[2] With respect to frog's spawn, if you remember in yr boyhood having ever tried to take a small portion out of the water you will remember that it is most difficult. I believe all the birds in the world might alight every day on the spawn of batrachians & never transport a single ovum.[3] With respect to the young of molluscs, undoubtedly if the bird to which they were attached alighted on the sea, they w^d be instantly killed; but a land bird w^d I sh^d think never alight except under dire necessity from fatigue. This however has been observed near Heligoland, & land-birds after resting for a time on the tranquil sea, have been seen to rise & continue their flight. I cannot give you the reference about Heligoland without much searching.[4]

This alighting on the sea may aid you in your unexpected difficulty of the too easy diffusion of land-molluscs by the agency of birds.[5]

I much enjoyed my mornings talk with you[6] & believe me | my dear Lyell | ever yours | Ch. Darwin

LS(A)
American Philosophical Society (337)

[1] The year is established by the reference to the *Journal of the Transactions of the Victoria Institute*, which was first published in 1867, and by the information provided to Lyell for the second volume of the tenth edition of *Principles of geology* (C. Lyell 1867–8; see n. 3, below), which was published in 1868.

[2] CD probably refers to Warington 1867 (see letter to George Warington, 7 October [1867] and n. 2, and letter to A. R. Wallace, 12 and 13 October [1867] and n. 20).

[3] Lyell quoted CD's observation that frog's spawn does not adhere to the feet of birds in C. Lyell 1867–8, 2: 413. CD had confirmed an earlier observation that batrachians do not occur on oceanic islands in *Origin*, p. 393. See also *Correspondence* vol. 5, letter to J. D. Hooker, 10 June [1855].

[4] Heligoland (now Helgoland) is an island off the coast of Denmark (*Columbia gazetteer of the world*). The migration of birds by way of Heligoland is discussed in Baird 1866, but without reference to land-birds resting on the sea; there is a copy of Baird 1866 in the Darwin Pamphlet Collection–CUL. Heinrich Gätke, who contributed information to Baird 1866, later wrote that he had observed three instances of small land-birds resting on the sea half a mile from Heligoland (Gätke 1895, p. 71).

[5] Lyell considered the range of distribution of terrestrial molluscs, including the possibility of their transport by birds, especially waterfowl, in *Principles of geology* (see C. Lyell 1867–8, 2: 372–7, 399). However, there is no reference in C. Lyell 1867–8 to CD's example from Heligoland.

[6] CD evidently saw Lyell during his visit to London from 28 November until 7 December 1867 (CD's 'Journal' (Appendix II)).

From Edward Wilson 7 December 1867

Hayes. | Bromley, Kent.
7th. Dec[r]. 1867

My dear M[r]. Darwin

We are taking steps to get out the Humble Bee to Australia,[1] & are anxious to avail ourselves of the ice house in which a shipment of Salmon Ova is being forwarded within the next few weeks to Otago.[2]

Mr. Woodbury, the great Apiarian, tells us, that the only way to get it out is to send forward Queen-bees during their condition of hibernation.[3]

The question is, where to find them, and my neighbour M[r]. Reed[4] tells me, that some of y[r]. sons have a special genius in that way— Would you oblige me by giving them a hint of what we want & inducing them if possible, to put us in the way of finding a few specimens.[5]

As the Salmon Ova will be sent away as soon as it can be obtained from the fish there is not much time to lose, but I should think it will be the end of the month before anything will be finally done.

I am my dear M[r]. Darwin | Y[rs]. very sincerely | Edw Wilson

LS
DAR 181: 121

CD note:
Boys given up & not at home— Very difficult to find *They are very *[different]* to *[illeg]* *[interl]* Of all men in Eng Sir J. L. most likely, because he investigated a curious parasite[6] according I went to him this mng, asking him, if he c^d find any to send them direct to you.—

[1] Humble-bees (*Bombus* spp.) are not indigenous to Australia; although a species of *Bombus* was introduced at some time before 1912, it did not persist (Franklin 1912, pp. 186, 202–3). Wilson was influential in introducing many animals and plants to Australia. For more on Wilson's activities in this area, see Gillbank 1986.

[2] Salmon do not occur naturally in New Zealand (Doak 1972). A technique in which ova were transported on moss inside 'ice houses' on board ship was used to introduce salmon and trout from England to Australia and New Zealand; the first successful shipment arrived in Tasmania in 1864 (Nicols 1882, pp. 9–31). Wilson refers to the shipment of salmon ova to the province of Otago, New Zealand, on board the *Celestial Queen*, which sailed from England in January 1868 (Nicols 1882, p. 237). For more on the introduction of salmon and trout to Australia and New Zealand, as pioneered by James Arndell Youl in association with Wilson, see Nicols 1882. The work of the Acclimatization Society of New Zealand, including its policy of introducing species of economic benefit, such as salmon, is described in H. F. von Haast 1948, pp. 278–93.

[3] Thomas White Woodbury introduced Ligurian honey bees to Australia; at Wilson's request, Woodbury had despatched four stocks of these bees from London to Melbourne in September 1862 (R. H. Brown 1975, p. 30).

[4] George Varenne Reed.

[5] On the assistance CD received from his children in his observations of male *Bombus*, see, for example, Freeman 1968.

[6] CD refers to his neighbour John Lubbock, whose research on parasites of *Bombus terrestris* and other *Bombus* species had involved the collection of humble-bees during the winter months (see Lubbock 1864).

From W. S. Dallas 8 December 1867

York
8 Decr. 1867

My dear Sir

On Page 104 of Vol. I. I think notes 4 & 5 are transposed.— Will you look at them & send me word?.—[1] On page 132, are acknowledgments of indebtedness with regard to Pigeons,—these I have taken no notice of.— On Page 275 you refer to *Phasianus Amherstii*, which I think should be *Amherstiæ*,—it is *Lady* Amherst's Pheasant.—[2] On p. 282 the Egyptian goose is said to be *Tadorna ægyptiaca*,— Is this intentional?— *Tadorna* is the genus of the *Sheldrakes* according to general acceptation, & the Egyptian Goose does not seem to me to be a Sheldrake.—[3] These are questions or remarks that I am obliged to make.—

In speaking of Pigeons you use the term Dragon for one of the breeds,—this was originally Dragoon, but I have adopted the Dragon.—[4]

I venture also to mention that the expression "In the name of God the compassionate, the merciful" referred to at p. 205, is the common adjuration at the opening of all works among the Arabs,— you will find it given or noticed by Lane in his translation of the Arabian nights,[5]—together with a concluding phrase "God is all knowing" which is peculiarly appropriate to the contents of many of those tales—[6]

I am pushing on with the second volume, having completed the first notes & all,— you were in the right in one of your letters to speak of the names in notes as making the work almost interminable.—[7] I fear I shall be causing delay in the publication, but the nature of the work is such that it cannot be hurried over.—[8] I hope the Index when done may be regarded as worthy of the work, which will be the highest praise that could be given to it,— the further I advance the more I am astonished at the wonderful array of facts brought together & at the manner in which you bring them to bear.—

Believe me, yours's very truly | W. S. Dallas.

DAR 162: 5

CD ANNOTATIONS
1.1 On ... transposed.—] 'yes' *pencil, circled pencil*
1.3 On ... Pheasant.— 1.5] '7 from bottom' *pencil*
1.5 On ... Sheldrake.— 1.7] '14 from top' *pencil*

[1] Dallas refers to *Variation*; he was using the proof-sheets to compile the index. No reply to this letter has been found. Dallas's correction was incorporated in the published text (see also letter to J. V. Carus, 10 December [1867]).

[2] The correction was incorporated in the published text (see also letter to J. V. Carus, 10 December [1867]).

[3] The name *Tadorna aegyptiaca* for the Egyptian goose is used in *Variation* 1: 282 and 2: 68; however, *Tadorna* is not used in the index, both references instead being indexed under *Anser aegyptiaca*. In the second printing of *Variation*, CD changed *Tadorna* to *Anser* in the text. See also letter to J. V. Carus, 10 December [1867].

[4] 'Dragon' is the term used in the index (*Variation* 2: 448).

[5] In his notes on the introduction to the *Arabian Nights*, Edward William Lane stated that it was customary for Muslims to begin books with this phrase (Lane trans. 1839–41, 1: 16).

[6] According to Lane, Arab writers used the phrase 'God is all knowing' when giving information of uncertain veracity (Lane trans. 1839–41, 1: 24).

[7] See letter to W. S. Dallas, 8 November [1867].

[8] For Dallas's account of the reasons for delay in his indexing of *Variation*, see *Correspondence* vol. 16, letter from W. S. Dallas, 8 January 1868.

From Daniel MacKintosh 8 December [1867][1]

20 Sussex Street, | Winchester
8th Dec.

Sir

I am truly obliged by your kind letter received a few days ago, and am looking forward with great pleasure to reading your work at the Geol. Society.[2]

I am glad to find there are "inclined terraces" in South America.[3] The inclination of the terraces of many parts of the Chalk downs, is one of the principal objections to their having been made for agricultural purposes.[4]

I can never give in to the escarpments of Sussex & Kent having been formed by Rain & Streams. The *plains* at their bases, in many places cutting equally through the gault & chalk, are inexplicable by an agent which can only act by *deepening*; and making V-shaped valleys[5]

Kent, near Maidstone[6]

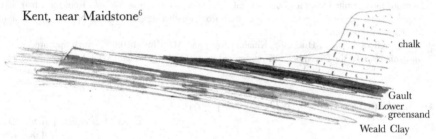

Your definition of the distinctive offices of the sea (widening) & freshwater (deepening) seems to me to furnish the Key to the whole subject[7]

Near Royston & elsewhere in the E. Central Counties the valleys at the base of the chalk escarpments have been cut down, through chalk, to the Gault, *not* excavated along the Gault (Mr. Searles Wood.)[8]

In the number of the Geological Magazine for the present month there are three letters on Denudation in answer to Mr. Whitaker, which shew the present state of the Controversy.[9]

With many thanks for your kindness, I beg to remain, | Sir, Your very faithful & | obliged Ser.ᵗ, | D. MacKintosh.

DAR 171: 6

[1] The year is established by the relationship between this letter and the letter from Daniel MacKintosh, 1 December 1867.

[2] The letter has not been found in the Darwin Archive–CUL. MacKintosh probably refers to *South America* (see letter from Daniel MacKintosh, 1 December 1867 and n. 8). CD's first published comments on terraces in South America were in 'Observations of proofs of recent elevation on the coast of Chili', p. 49 (*Proceedings of the Geological Society of London* 2 (1838): 446–9; see also *Collected papers* 1: 43).

[3] See also letter from Daniel MacKintosh, 1 December 1867 and n. 8.

[4] See letter from Daniel MacKintosh, 1 December 1867 and n. 9.

[5] Mackintosh's belief in the sea as an agent of erosion had been challenged by others who believed in subaerial denudation, that is, erosion resulting from fluvial and glacial action rather than marine action (Challinor 1978). See also letter from Daniel MacKintosh, 1 December 1867 and n. 3. In his book *Rain and rivers*, George Greenwood argued that these agents had determined the physical character of Wealden Sussex and Kent (Greenwood 1866, pp. 57–61). See also letter from Daniel MacKintosh, 1 December 1867, n. 6. In earlier letters to CD, Joseph Beete Jukes had offered a similar interpretation (*Correspondence* vol. 10, letters from J. B. Jukes, 25 May 1862 and 30 May 1862).

[6] The sketch is derived from figure 2 in Foster and Topley 1865 (see MacKintosh 1869, p. 102).

[7] For CD's early view of the physical characteristics of valleys formed by rivers, see *Correspondence* vol. 5, letter to Charles Lyell, 18 February [1854] and n. 3.

[8] The reference is to Searles Valentine Wood's description of a geological section through parts of Cambridgeshire and Hertfordshire (Wood 1867). According to Wood, a chalk scarp near Royston, Hertfordshire, resulted from a trough several miles wide 'cut down *on both sides* from the Chalk and Upper Glacial clay into the Gault' rather than from weathering (Wood 1867, pp. 401–2). See also MacKintosh 1869, pp. 102–3.

[9] The references are to E. Hull 1867, Kinahan 1867, and MacKintosh 1867c, which appeared in the *Geological Magazine* for December 1867, prompted by an article by William Whitaker (Whitaker 1867).

To Abraham Dee Bartlett 9 December [1867][1]

Down. | Bromley. | Kent. S.E.
Dec. 9th

My dear Sir

Would you have the kindness to send me on a slip of paper the name of the 3 or 4 Tringa-like Birds in the Aquarium, which never, except once, assumed the proper summer plumage.[2] Please just add whether you have known this with more than the 3 or 4 individuals, which you showed me.[3]

I much wish I could persuade you to try with differently coloured worsted or rags, whether the Bower-bird prefers gay colours.—[4]

I thank you most sincerely for all the interesting information which you so often give me.—[5]

My dear Sir | Yours very faithfully | Ch. Darwin

Archives of The New York Botanical Garden (Charles Finney Cox Collection) (Darwin 31)

[1] The year is established by the relationship between this letter and the letter from A. D. Bartlett, 9 December 1867.

[2] CD refers to the aquarium in the gardens of the Zoological Society of London (Scherren 1905, pp. 107–8) and to the genus *Tringa* (now *Calidris*) (*Birds of the world* 3: 519–20). In *Descent* 2: 82, CD noted that captivity in the Zoological Gardens affected the reproductive system as well as the development of ornamental summer plumage in *Tringa*.

[3] CD had visited London from 28 November until 7 December 1867 (CD's 'Journal' (Appendix II)).

[4] CD refers to birds of the family Ptilonorhynchidae; their preference for brightly coloured objects is described in *Descent* 2: 69–71, 112.

[5] Bartlett was cited in *Variation*, the sixth edition of *Origin*, *Descent*, and *Expression*. See also *Correspondence* vols. 8–14.

From A. D. Bartlett 9 December 1867

Zoological Society's Gardens, | Regent's Park, | London, N.W.
Decr 9th *1867*.

My Dear Sir

In reply to your note[1] I beg to say the name of the birds in question is the *Knot* (Calidris canutus) I have kept eight or nine of them and out of this number only one put on any of the breeding or summer dress,[2] while in the same cage *two Turnstones* (Strepsilas interpres)[3] *both assumed their breeding plumage in the Summer*, all the birds appear equally well in health and condition, the *Turnstones* have now moulted into the winter garb, the coloured worsted rags &c shall be tried immediately.[4]

believe me to be | Yours faithfully | A. D. Bartlett.

Chas Darwin Esq

DAR 84.1: 38–9

CD ANNOTATION

Top of letter: 'If the summer Plumage of the Knot, is the same in 2 sexes, we learn that the summer plumage is nuptial & common to the 2 sexes—as occurs with a multitude of Birds.—'[5] *ink*

[1] Letter to A. D. Bartlett, 9 December [1867].
[2] CD referred to the bird by its older name, *Tringa canutus*, in *Descent* 2: 82, citing Bartlett on specimens in the Zoological Gardens in London failing to develop their summer plumage.
[3] *Strepsilas interpres* is now *Arenaria interpres* (see *Birds of the world* 3: 515–16).
[4] See letter to A. D. Bartlett, 9 December [1867].
[5] These points are developed more fully in *Descent* 2: 81–2.

To the Linnean Society 9 December [1867][1]

Down. | *Bromley.* | *Kent. S.E.*
Dec 9.

My dear Sir

I have read the paper with care, & it seems to me better than I expected, though badly arranged. As far as I can judge the whole of the first part, with the exception of a few introductory sentences, (which I have struck out) must be published. No one I think without specimens c^{d} make a good abstract. With respect to the latter part of the paper you will be a much better judge than I: at p. 11., where I have put a pencil cross, the subject changes, & again at p. 12.; whether either of these discussions ought to be retained, I really cannot decide.[2]

I have rec^{d} other accounts, from the same author & others, of the curious imitation of plants in S. Africa.[3]

The climbing of the convolvulus is also a curious point with reference to the same plant when grown in Ireland; but I must beg you to decide whether these extraneous passages ought to remain.[4]

With respect to the plates, it is obvious that all cannot & do not deserve to be engraved; I w^{d} suggest fig 6, 7, 2 & 3 in Pl. 1. to be engraved on the same block & inserted at p. 7 of the M.S.[5]

I enclose a title for the wood block in case you approve of my suggestion.

My dear Sir | yours very faithfully | Ch. Darwin

[Enclosure]

All from Plate I for a woodcut

Fig 1.	(fig 6 of m.s.)	Fig 3.	(Fig 2. of m.s.)
– 2.	(Fig 7. of m.s.)	– 4.	(Fig 3 of m.s.)

(Beneath the 4 cuts insert in small type)

Fig. 1.   Under surface of Labellum of Bonatea Darwinii (magnified)

Fig. 2.   Pollinium of do in natural position (magnified)

Fig 3   Under surface of Labellum of Bonatea species (from Mʳ. Trimen)

Fig 4.   Pollinium of do (from do.)[6]

LS(A)

Linnean Society of London, SP1249, 1253

[1] The year is established by the relationship between this letter and the letter to J. P. M. Weale, 9 December [1867].

[2] The reference is to a manuscript of Weale 1867, a paper on *Bonatea* that was read at the Linnean Society on 7 March 1867 but not published until 1869. See letter from J. P. M. Weale, 9 January 1867, and letter to the Linnean Society, 22 February [1867]. The manuscript, with CD's pencil markings, is in the Linnean Society archives (SP1249). CD crossed through in pencil three sentences from the first part of Weale's manuscript concerning the difficulty he had in acquiring specimens of the plant. He put a pencil cross on page 11 of the manuscript at the head of a paragraph where Weale suggested that species in South Africa were related more as subspecies than as 'distinctive specific forms', and another on page 12, where Weale discussed *Ipomoea argyraeoides*. Both passages were omitted from the version published in the *Journal of the Linnean Society*.

[3] The flower of *Bonatea darwinii* was described as resembling a white butterfly in Weale 1867, p. 470.

[4] In the manuscript of Weale 1867 (see n. 2, above), Weale included a discussion of *Ipomoea argyraeoides*, which he said showed in his observation no tendency to climb, even though growing in comparatively moist situations. He wrote that he had found one plant, growing in the shelter of the Bedford mountain range, that showed a slight tendency to twine in its upper branches. In 'Climbing plants', pp. 24–5, CD had suggested that *Ipomoea argyraeoides* was one of a number of plants that rarely showed a tendency to climb in South Africa, but that twined when grown in Ireland, in moister conditions. See also *Correspondence* vol. 14, letters from W. H. Harvey, 8 November [1864] and 10 November 1864. Weale's discussion was omitted from the version of the paper published in the *Journal of the Linnean Society*.

[5] See letter to the Linnean Society, 22 February [1867] and n. 3.

[6] CD's captions were used in the published paper: Weale had supplied several more diagrams.

From John Lubbock 9 December [1867][1]

15, Lombard Street. E.C.

9 Dec.

My dear Mʳ. Darwin

You will find the passage in P. 6 of Prehistoric Times. It is a very remarkable statement, & something like his prophetic allusion to the dark rays of light which Tyndall has quoted.[2]

As to the Humble Bees they would if once landed in Australia probably destroy some native insect & I wish M.ʳ Wilson would leave the Australian fauna alone.[3]

I will not forget the clubs.[4]

Believe me | Yours most sincerely | John Lubbock

DAR 170: 60

[1] The year is established by the relationship between this letter and the letter from Edward Wilson, 7 December 1867.

[2] Lubbock refers to a quotation from Lucretius about the successive use of stone and wood, then bronze and iron, as weapons, given in his book *Pre-historic times* (Lubbock 1865, p. 6). He also refers to Lucretius's speculation that the sun produced invisible radiation, which was quoted by John Tyndall in 'On calorescence' (Tyndall 1865, p. 1).

[3] See letter from Edward Wilson, 7 December 1867 and nn. 1 and 3.

[4] Lubbock refers to the Down Friendly Club and the Down Coal and Clothing Club, of which CD was treasurer (Freeman 1978). The Down Friendly Club was a local savings and insurance society; the Coal and Clothing Club supplied parishioners with inexpensive coal and clothes in exchange for regular savings. In his Down Coal and Clothing Club account book (Down House MS), CD recorded a contribution of £2 from John and Ellen Frances Lubbock in December 1867. On CD's involvement in local charities, see Browne 2002, pp. 452–4.

To St George Jackson Mivart 9 December [1867][1]

Down. | Bromley. | Kent. S.E.

Dec 9ᵗʰ

[thanking him for his "Memoir on the Append. skeleton of the Primates" which he says he is glad to receive at present "as I am now attending to some point in the natural history of man".[2]] I hope that we may some time soon meet in London, for I am anxious to make the acquaintance of one who has done such good work in comparative anatomy.[3]

With my best thanks, | pray believe me | Dear Sir | Yours faithfully | Ch. Darwin

Incomplete
Sotheby's London (21 and 22 July 1988, pp. 212–13)

[1] The year is established by the reference to Mivart 1867 (see n. 2, below), and by CD's expression of his desire to make Mivart's acquaintance; the two men met before 6 April 1868 (see *Correspondence* vol. 16, letter from St G. J. Mivart, 6 April 1868).

[2] CD refers to Mivart's paper 'On the appendicular skeleton of the primates' (Mivart 1867). No offprint of Mivart 1867 has been found in the Darwin Archive–CUL; however, CD's copy of the journal in which it appeared is in his collection of unbound journals in the Darwin Library–CUL. Mivart's paper is partly uncut and lightly annotated. CD cited Mivart 1867 in *Descent* 1: 196–7.

[3] Mivart's papers on the comparative anatomy of primates were published between 1864 and 1873 (J. W. Gruber 1960, pp. 31–2); for a bibliography of Mivart's publications in this and other fields, see J. W. Gruber 1960. For an account of Mivart's researches in the field of comparative anatomy, see Blum 1996. CD referred to Mivart's findings in this field in the sixth edition of *Origin*. Mivart was made an FRS in 1869 in recognition of his work on comparative anatomy (Royal Society of London, certificates of election and candidature).

To Philip Lutley Sclater 9 December [1867][1]

Down. | Bromley. | Kent. S.E.
Dec.[r] 9th

Dear Sclater

I have no index of Proc. Zoolog. Soc.—at least no separate one, for I have not looked through each volume,—so I sh.[d] be very glad of both indices, & they can be sent by same address as the Journal &c.—[2]

My list of Fellows is dated 1858; can you send me a more recent one?[3]

Some one has sent me the Intellectual Observer for December but it does not contain any article by you; hence I suppose that on Barbets is in the November No.[r][4] If I do **not** hear from you in the course of the next week, I will order the Nov.[r] number.

A talk with you always does me good, & I enjoy it much.—[5]

Yours very sincerely | Ch. Darwin

American Philosophical Society (338)

[1] The year is established by the reference to the article in the *Intellectual Observer* (see n. 4, below).

[2] CD refers to two cumulative indexes to the *Proceedings of the Zoological Society of London*. The index for 1830 to 1847 was published in 1866, and that for 1848 to 1860 in 1863; there is a copy of each in the collection of unbound journals in the Darwin Library–CUL.

[3] No list of fellows of the Zoological Society of London has been found in the Darwin Library–CUL.

[4] CD may have received the December issue of the *Intellectual Observer* with the letter of 30 November 1867 from its editor, Henry James Slack. Sclater's article on barbets was published in the November issue (Sclater 1867); there is a lightly annotated copy in the Darwin Pamphlet Collection–CUL.

[5] CD frequently visited Sclater while in London (Lightman ed. 2004, s.v. Sclater, Philip Lutley); CD was in London from 28 November until 7 December 1867 (CD's 'Journal' (Appendix II)).

To Herbert Spencer 9 December [1867][1]

Down | Bromley | Kent. S.E.
Dec 9th.

My dear M.[r] Spencer

I thank you very sincerely for your kind present of your First Principles.[2] I earnestly hope that before long I may have strength to study the work as it ought to be studied, for I am certain to find or refind much that is deeply interesting. In many parts of your Principles of Biology I was fairly astonished at the prodigality of your original views.[3] Most of the chapters furnished suggestions for whole volumes of future researches. As I have heard that you have changed your residence I am forced to address this to Messrs. Williams & Norgate;[4] and for the same reason I gave some time ago the same address to M.[r] Murray for a copy of my book on Variation etc which is now finished but delayed by the index-maker.[5]

Pray believe me, with sincere thanks | Yours very truly | Ch. Darwin

Copy
DAR 147: 485a

[1] The year is established by the reference to work on the index of *Variation*.

[2] CD refers to the second edition of *First principles* (Spencer 1867); there is a copy in the Darwin Library–Down. There is a copy of the first edition (Spencer 1860–2) in the Darwin Library–CUL.

[3] For a bibliographical account of Spencer's *Principles of biology* (Spencer 1864–7) and its relationship to *First principles* (Spencer 1860–2), see *Correspondence* vol. 12, letter from A. R. Wallace, 2 January 1864, n. 20. For more on CD's opinion of Spencer 1864–7, see, for example, *Correspondence* vol. 14, letter to J. D. Hooker, 30 June [1866] and n. 8. CD's annotated instalments of Spencer 1864–7 are in the Darwin Library–CUL as a bound volume (see *Marginalia* 1: 769–73).

[4] Williams & Norgate were Spencer's publishers. In September 1866, Spencer moved to a boarding house in Queen's Gardens, Bayswater, where he lived for more than twenty years. However, he was absent from London in the summer of 1867 owing to the death of his mother, and made a tour of England in the autumn of 1867 (Spencer 1904, 2: 145–62).

[5] CD refers to John Murray and to William Sweetland Dallas's delay in completing the index of *Variation*. See also letter from W. S. Dallas, 8 December 1867.

From Francis Walker 9 December 1867

Elm Hall | Wanstead
Dec^ber. 9th. 1867

My dear Sir,

I have looked at the Homopterous Insects armed with spines or with horns, & I believe that there is no difference in this respect between male & female. In the Homoptera there is no contrast between the sexes, but only the general difference between one sex & the other.[1] This rule excludes the Aphididæ & the Coccidæ, the latter being most remarkable in the want of resemblance between the sexes. I have, as yet not published on variation in Insects except a few remarks on the question of species in my list of the Diptera of the Eastern Isles in the Journ Linn. Soc—[2] I am about to publish some notes on the Aphididæ—a translation of Passerini's synopsis of the species & remarks on the nomenclature, & I have introduced here & there some words on the mutability of species.[3] I will write to you all that may appear from me on the latter subject. I have made only one experiment on Aphides. The Aphis of the holly & the Aphis of the ivy seem to be exactly alike tho' they have been described as two species. I took some shoots of ivy infested by Aphides & tied them with thread to shoots of holly, but tho' the Aphides wandered to the holly they did not feed on it & soon perished. Probably the Aphis requires the same kind of food from its birth for tho' some species migrate to a great variety of plants they hardly feed after they have migrated & pass out of existence when they have brought forth a new generation[4] I believe that the migrations initiate yearly many species (so called) which have no continuity, but pass away before winter. I am just now making out a list of the Blattidæ or Cockroaches— this tribe is remarkable in four particulars— 1st.—the great difference in some species between male & female 2nd the incomplete metamorphose of some species or their perpetuity in the apparently larva-state 3rd. the facility with which some species accomodate themselves to artificial circumstances & thus become cosmopolitan—& 4th. the disguises which they put on or their mimicry of the Coleoptera, Myriapoda

& terrestrial Crustacea with which they are associated in their natural habitations—
They thus seem to epitomize some tribes that are inferior to them in structure & to
indicate others of higher organization—[5] Some insects are said to photograph the
colours near them— those that are disguised may be said to morphograph other
tribes with which they dwell or to be examples of morphography.[6]

I enclose my Carte de visite— may I ask for the favor of yours?—[7]

With my best wishes, | Yours truly | F. Walker

P.S. Your Carte de Visite has just arrived— many thanks for it.[8]

DAR 82: A48–9

CD ANNOTATIONS

1.1 I have ... female. 1.2] *scored red crayon*
1.2 In the ... the sexes. 1.5] 'Homoptera' *added and underl blue crayon*
1.5 I have, ... winter. 1.20] *crossed ink*
1.21 1st. ... larva-state 1.23] *scored red crayon*
1.21 difference.... & female 1.22] 'I suppose (not in ornament)' *added ink*
1.23 3rd. ... best wishes, 3.1] *crossed ink*
Top of letter: 'Perhaps serve as defences like Bates' Ants'[9] *ink*

[1] Walker made an extensive study of the Homoptera collected by Alfred Russel Wallace in the Malay
Archipelago; his catalogue of the collection had been read at the Linnean Society in February 1867
(Walker 1867).
[2] Walker refers to his synopsis of the Diptera discovered by Wallace in the Malay Archipelago (Walker
1865); in it, he commented on similarities between species and on geographical distribution, ques-
tioning whether some species of *Laphria*, for example, should be considered as varieties (Walker 1865,
pp. 1–6). In *Descent* 1: 348–9, CD cited Walker on differences between the sexes in the dipterous genus
Bibio.
[3] Giovanni Passerini published two works on aphids (Passerini 1860 and Passerini 1862–3); Walker
refers to the English publication of Passerini's systematic arrangement of the aphids, supplemented
by Walker's notes (Walker 1868–70). For more on Walker's work on the taxonomy of aphids, see
Doncaster 1961.
[4] Walker described *Aphis ilicis*, the holly aphid, and *A. hederae*, the ivy aphid, and the experiment of
changing the food source, in Walker 1868–70, p. 1998.
[5] Walker's *Catalogue of the specimens of Blattariae in the collection of the British Museum* included representatives
of the family Blattidae (Walker 1868). Walker reiterated his third and fourth points in Walker 1868,
p. 233.
[6] Walker's neologism has not persisted in the sense in which he uses it.
[7] Walker's photograph has not been found in the Darwin Archive–CUL. The photographic 'carte' or
'carte de visite', smaller than a conventional portrait, became popular in the 1860s and several were
made of CD between 1864 and the end of his life (see Browne 1998, pp. 253–80).
[8] It is not clear which 'carte de visite' CD sent. It may have derived from an 1864 photograph of
CD by his son William Erasmus Darwin (see *Correspondence* vol. 12), or from photographs taken by
Ernest Edwards, probably in November 1865 or April 1866 (see *Correspondence* vol. 13, letter from E. A.
Darwin to Emma Darwin, 25 [November 1865], n. 3). See also letter from E. Schweizerbart'sche
Verlagsbuchhandlung, 22 March 1867, n. 8.
[9] CD may refer to Henry Walter Bates's account of differences in the sizes of mandibles in a number
of Brazilian ant species; different types of worker ants of the same species exhibited these differences
(Bates and Smith 1855). There is an annotated copy of Bates and Smith 1855 in the collection of
unbound journals in the Darwin Library–CUL.

To J. P. M. Weale 9 December [1867][1]

Down. | *Bromley.* | *Kent. S.E.*
Dec 9.

My dear Sir

Had I not been lately much over worked I sh[d] have written before to tell you how much I & others have been interested in the result of my trial of the locust-dung.[2] No less than 7 grasses have germinated, & are now growing well; they belong to at least 2 kinds, & I hope they will flower.[3] The seeds were in the middle of the pellets. Sir C. Lyell was so much interested in the case that he is going to cite it in a new edit. of the Principles.[4]

You said that some persons believe that weeds are thus introduced, & I sh[d] be much obliged for any information on this head.[5] The Linn. Soc. has been so much pressed with papers that the Council is only now sending your paper on Bonatea to press. They want it shortened but I hardly see how this can be done; it w[d] be a pity if the concluding extraneous remarks sh[d] be left out; but perhaps this will be done; I do not know as yet.[6]

The Council c[d] not possibly afford to engrave all the figures but will give one wood-block.[7] I am sorry to tell you that 2 authors viz. D[r] Hildebrand of Bonn & Delpino in Italy have been attending with great care & have published on the fertilization of the Asclepiadæ.[8] I wrote before to thank you & say how valuable to me y[r] information on expression has been,[9] & I sh[d] be very thankful for any further information.

Believe me my dear Sir | with sincere thanks | yours very faithfully | Ch. Darwin

I have had inserted your 2 or 3 corrections into your paper.—[10]

LS(A)
University of Virginia Library, Special Collections, Darwin Evolution Collection (3314)

[1] The year is established by the relationship between this letter and the letter from J. P. M. Weale, 7 July 1867, and by the reference to C. Lyell 1867–8 (see n. 4 below).

[2] A small packet of locust dung was enclosed with the letter from J. P. M. Weale, 7 July 1867. See also letter to Asa Gray, 16 October [1867].

[3] This information was included in *Origin* 5th ed., p. 439.

[4] CD had communicated the results of his experiment in his letter to Charles Lyell, 31 October [1867]; they were reported in C. Lyell 1867–8, 2: 420–1.

[5] See letter from J. P. M. Weale, 7 July 1867.

[6] See letter to the Linnean Society, 9 December [1867] and nn. 2 and 4.

[7] For CD's recommendations about the figures, see the letter to the Linnean Society, 9 December [1867].

[8] Weale had remarked on similarities in the floral adaptations of asclepiads and orchids (see letter from J. P. M. Weale, 7 July 1867 and n. 14). In his book on the separation of the sexes in plants, Friedrich Hildebrand referred to Federico Delpino's first scientific paper, published in 1865, on fertilisation in the Asclepiadaceae (Hildebrand 1867a, p. 58, n. 3; see also Pancaldi 1991, p. 123 and n. 58). There is an annotated copy of Hildebrand 1867a in the Darwin Library–CUL. Hildebrand also described the fertilisation of *Asclepias cornuti* (now *A. syriaca*) in Hildebrand 1866c, and reported Delpino's researches on pollination in the Asclepiadaceae in Hildebrand 1867b. There are annotated copies of Hildebrand 1866c and Hildebrand 1867b in the Darwin Pamphlet Collection–CUL.

⁹ See letter to J. P. M. Weale, 27 August [1867].
¹⁰ Weale communicated the corrections in his letter of 7 July 1867.

To J. V. Carus 10 December [1867]¹

Down. | Bromley. | Kent. S.E.
Dec. 10th

My dear Sir

If not too late for 1st vol, then append to 2d vol. the following errata.² The last one is an extraordinary & quite unintelligible blunder on my part.

In haste believe me yours very sincerely | C. Darwin

Vol I. p. 104. Transpose foot-notes 4 & 5.
Vol I. p. 275 (7 lines from bottom) for Phasianus amherstii
        read Phas. *amherstiæ.*
Vol I. p. 282, (14 lines from bottom) for Tadorna Ægyptiaca
        read *Anser Ægyptiacus.*—³

You will have received all sheets of Vol. 2, except last two & index & title pages.— The book is delayed by the index-maker.—⁴

Staatsbibliothek zu Berlin, Sammlung Darmstaedter

¹ The year is established by the relationship between this letter and the letter from W. S. Dallas, 8 December 1867.
² The errors listed by CD were in the proof-sheets of *Variation* and were identified by William Sweetland Dallas, who was indexing the work. See letter from W. S. Dallas, 8 December 1867 and nn. 1–3.
³ The corrections were not incorporated into the first German edition (Carus trans. 1868, 1: 130, 342, 351), but were incorporated into the second (Carus trans. 1873, 1: 114–15, 306, 314); '*Anser Ægyptiacus*' was printed as '*Anser aegyptica*'.
⁴ See n. 2, above.

From W. S. Dallas 10 December 1867

Yorkshire Philosophical Society | York
10 Dec.ʳ 1867

My dear Sir

Your letter received this morning was written I suppose before you received a letter from me, which I thought would have reached you yesterday morning.—¹

I am pushing forward as fast as I can with the second volume & hope to finish it this week, which I hope will obviate all chance of injurious delay—² The work is, however, fearfully heavy, & my only hope is that when the Index is finished you will think it worth the labour bestowed upon it.—

Believe me | Your's very truly | W. S. Dallas

Chaˢ Darwin Esq

DAR 162: 6

[1] The letter from CD has not been found. Dallas's letter was that of 8 December 1867.
[2] Dallas was indexing *Variation*.

From J. P. M. Weale [10 December 1867][1]

On the mountain top here I have discovered Pelargonium Bowkeri Harv: hitherto only known in the Transkeian country.[2] This & various other observations lead me to believe that these inland mountain spurs detached mountain tops & valleys are islands of a former coast flora & fauna, but I will say more of this towards the conclusion.

I have received some answers from M^r J H Bowker to your questions.[3]

A Kafir[4] on being asked whether he thought East London was as fine a place as King Williamstown made a distinct "Achg!" like incipient vomiting, far more distinctly than any European could.[5]

Kafirs when in grief place their hands especially the women on their heads. They likewise do so to express surprise. They also place one hand on the chin & mouth in grief.

While in the magistrate's court a short while ago I observed Christian Gaika listening attentively to a Kafir case. His chin was stretched forward & his forehead wrinkled up with the eyes rather prominent & directed forward & upward towards the witness box.

I had occasion whilst in Kafir land[6] to chastise a servant, who had care of my horse for allowing the headstall to be stolen. Although not guilty himself I am convinced he knew who had taken it.

The women give way to grief very violently but not for a very long period. I noticed this the case of a young woman, Goondoo, who died from catching cold after child birth from gastric fever. Her sister Umfazi wept copiously; but as she, Goondoo, was very delirious before death none of the women would nurse her, but placed food in the hut.

You speak sanguinely about the civilization of the natives, & the fact that Christian Gaika can write.[7] This appears to me an error into which most people in England fall, & I trust you will not think it unnecessary if I make some comments thereon.

Although by no means desirous of running down Missionary work, I must own that in my opinion their teaching is to little effect.

The principles on which they work—excepting always the Moravians & some others[8]—is almost purely an appeal to the emotions, & the longer a Kafir has been on a Mission Station the worse servant he is.

There is little or no attempt at inculcating by precept & habit either industry or economy. I have had occasion to visit some of the stations—one conducted by a very kind hearted & well meaning man, as far as I could see the Revd Mr Philip of Hankey⁹—& I

AL incomplete
DAR 181: 42

CD ANNOTATIONS
1.1 On ... conclusion. 1.5] *crossed blue crayon*
2.1 I have] *after opening square bracket red crayon*
7.1 The women ... & I 11.4] *crossed blue crayon*
First page: '19' *added brown crayon*
End of letter: 'J. P. Mansel Weale 10th of December 1867' *ink*

[1] The date is established by CD's annotation.
[2] Weale lived in Bedford, Cape Colony, South Africa (letter from J. P. M. Weale, 7 July 1867). The Transkeian territories were also in Cape Colony; they bordered the Indian Ocean north-east of the Kei river (*Columbia gazetteer of the world*).
[3] In his letter of 7 July 1867, Weale informed CD that he had sent CD's queries about expression to James Henry Bowker and others. Bowker's answers have not been found in the Darwin Archive–CUL. He is not cited in *Expression*.
[4] For the nineteenth-century use of the term 'kafir', see the letter to J. P. M. Weale, 27 February [1867], n. 3.
[5] East London is a city in the Eastern Cape province of South Africa. King William's Town, thirty miles west-north-west of East London, was capital of the province of British Kaffraria until 1865. (*Columbia gazetteer of the world*.)
[6] Kafir land: i.e. British Kaffraria, now part of the Eastern Cape province of South Africa (*Columbia gazetteer of the world*).
[7] See letter to J. P. M. Weale, 27 August [1867] and n. 5. For Christian Gaika's answers to CD's queries about expression, see the enclosure to the letter from J. P. M. Weale, 7 July 1867.
[8] On the high regard in which the South African Moravian missionaries were held in the nineteenth century, see, for example, Groves 1948–58, 2: 130, 251.
[9] Thomas Durant Philip. Hankey is in the Eastern Cape Region of South Africa near Port Elizabeth (*Times atlas*).

From W. S. Dallas 11 December 1867

Yorkshire Philosophical Society | York
11th. Decr 1867

My dear Sir

I write to apologise to you if the remarks in my note of Sunday last seem to you so much in the nature of criticism as I fear they do.—[1] Nothing was further from my intention.— The questions as to the use of the name *Tadorna*, the name of the Pheasant & the transposition of the notes were necessary ones,— the use of Dragon instead of Dragoon is of not the least consequence, but as I was more familiar with the latter name I noted it in passing,—& the Arab matter I mentioned, not as pointing out an error, but because I thought you might feel interested in the explanation of the point.—[2]

I last night timed myself & found that in 5 hours hard work I got through 3 sheets less 2 pages, & this is more than I have been able to do sometimes since working the whole of the references together.—[3] The subjects often seem to elude the Index-maker, as the noticing every point taken up would often be to transpose whole pages in an altered form to the Index,—& on the other hand it is difficult, or sometimes impossible to hit upon any middle course between this elaborate Analysis & the simple notice of the general subject.—

Of course this note requires no answer, but you may depend on my pushing forward at my best speed.—

Believe me | Your's very truly | W. S. Dallas

C. Darwin Esq[r].

DAR 162: 7

[1] See letter from W. S. Dallas, 8 December 1867.
[2] See letter from W. S. Dallas, 8 December 1867 and nn. 3–6.
[3] Dallas was compiling the index of *Variation*. Each proof-sheet comprised sixteen pages.

From J. D. Hooker to Emma Darwin 11 December 1867

Royal Gardens Kew
Dec 11/67.

Dear Mrs Darwin

I should like to go to Down on Friday or Saturday *week* 20th. or 21st. & read a novel—[1] I have told Mr Woolner of my proposal—[2] he is very unwell with Influenza

My wifes votes are unfortunately promised or she would have been delighted to have helped your candidate.[3]

Very sincerely Yrs | Jos D Hooker

DAR 102: 185

[1] Hooker visited Down on 21 December 1867 (Emma Darwin's diary (DAR 242)).
[2] Hooker had earlier expressed his wish to visit Down during early December 1867 (see letter from J. D. Hooker, 19 November 1867). See also letter from Thomas Woolner, 6 December 1867 and n. 1.
[3] Hooker refers to Frances Harriet Hooker; neither the candidate nor the election has been identified.

From Charles Kingsley 11 December 1867

Trinity Lodge, | *Cambridge.*
Dec[r] 11/67

My dear M[r]. Darwin

I have been here 3 or 4 days;[1] & have been accidentally drawn, again & again, into what the world calls Darwinism, & you & I & some others fact & science— I have been drawn thereinto, simply because I find everyone talking about it to anyone who is supposed to know (or mis-know) anything about it: all shewing how men's minds are stired.

I find the best & strongest men coming over. I find one or 2 of them like Adams (& Cayley) fighting desperately.[2]

1. Because, being really great men, they know so much already w^h. they cannot coordinate with your theories (at least as yet) & say (as they have a right) "I will stand by what I do know from mathematics, before I give in to what I *dont* know from ——

That last dash is the key of the position. They dont know. The dear good fellows have been asking me questions.—e.g. "You dont say that there are links between a cat & a dog? If so, what are they?— To w^h. I have been forced to answer—my dear fellow, you must read & find out for yourself— I am not bound to answer such a question as that. I am not bound to teach you the *alphabet*, while you are solemnly disputing about my translation of the *language*.

That is what it comes to, my dear & honoured *master*, for so I call you openly where I can, among "great swells', as well as here in Cambridge— Why men dont agree with you, is because they dont know facts: & what I do is—simply to say to every one, as I have been doing for 3 days past "Will you kindly ascertain a few facts—or at least ascertain what facts there are, to be known or disproved, before you talk on this matter at all?"—& I find, in Cambridge, that the younger M.A's. are not only willing, but greedy, to hear what you have to say; & that the elder, (who have of course more old notions to overcome) are facing the whole question in a quite different tone from what they did 3 years ago. I wont mention names for fear of "compromising" men who are in an honest, but "funky" stage of conversion: but I have been surprised, coming back for 3 or 4 days, at the change since last winter.

I trust you will find the good old university (w^h. has always held to physical science & free thought—& allows—as she always has done—anybody to believe anything reasonable, *provided he dont quarrel with his neighbours*) to be your finest standing ground in these isles.

I say this—especially now—because you will get, I suppose, an attack on you by an anonymous "Graduate of Cambridge"[3]—w^h. I found in the hands of at least one very wise & liberal man—who admired it very much—but knew nothing of *The Facts*: he shewed it me, & in the first 3 pages I opened at hazard, I pointed him out 2 or 3 capital cases of ignorance or omission, on w^h. I declined to read any more of the book, as coming from a man who knew—or did not choose to know—anything about *The Facts*. He was astonished, when I told him that the man was an ignoramus, or worse, & could be proved such. & I think I have done him good. & so it will be with many more—

Excuse the bad writing— I have a pen w^h. if natural selection influenced pens, w^d have been cast into the fire long ago: but the disturbing moral element makes me too lazy to cast it thereinto—& to find a new one.

I have—as usual—a thousand questions to ask you—& no time, nor brain, to ask them now.

But ever I am— | Your affte pupil | C Kingsley

Dont trouble yourself to answer me. But if you write to me, I return to Eversley tomorrow.[4]—*& give my love to Lubbock.*[5]

DAR 169: 38

[1] Kingsley went to Cambridge twice a year to deliver his professorial lectures (Kingsley ed. 1877, 2: 153).

[2] The Cambridge mathematicians John Couch Adams and Arthur Cayley were friends (*DNB*, s.v. Cayley, Arthur).

[3] Kingsley refers to Robert Mackenzie Beverley's *The Darwinian theory of the transmutation of species examined by a graduate of the University of Cambridge* ([Beverley] 1867).

[4] Eversley, Hampshire.

[5] John Lubbock.

From Eduard Koch[1] 11 December 1867

Geehrter Herr!

Im Besitz Ihres sehr geehrten Briefes vom 22 Novem. habe ich heute das Vergnügen Ihnen den I Band in *zwei* Exemplaren per Post zugehen zu lassen.[2]

Die in Auftrag gegebenen Freiexemplare werden wir bestens besorgen und seiner Zeit auch den zweiten Band nachsenden.[3]

Für die Zusicherung des Uebersetzungs Rechts auch Ihrer spaeter erscheinenden Werke sage ich Ihnen meinen besten Dank und werde ich stets bestrebt sein durch eine dem Gegenstand würdige Ausstattung Ihre volle Zufriedenheit zu erwerben.[4]

Mit der Uebersetzung des zweiten Bandes hat H Prof. Carus bereits begonnen und hoffe ich Ihnen bis Maerz oder April *1868* den fertigen Band vorlegen zu koennen.[5]

Ich bitte mir umgehend gefälligst mittheilen zu wollen bis wann das Werk in England ausgegeben wird;[6] bei den eigenthümlichen deutschen Verhaeltnissen sollte ich nothwendig das Werk noch dieses Jahr ausgeben,[7] da bei Ausgabe nach Neujahr eine Uebersicht des Absatzes erst 1869 erfolgen koennte. Sie würden mich also durch baldigste Mittheilung sehr erfreuen.

Empfangen Sie zum Schluss die Versicherung meine vorzüglichsten Hochachtung und Verehrung | E. Schweizerbart'sche Verlagshdg | Eduard Koch

Stuttgart den 11 Dez. 1867

DAR 169: 42

[1] For a translation of this letter, see Appendix I.

[2] In his letter to E. Schweizerbart'sche Verlagsbuchhandlung of 22 November [1867], CD had requested only one copy of the German edition of *Variation* (Carus trans. 1868) for himself.

[3] CD had enclosed a list of recipients of presentation copies of Carus trans. 1868 in his letter to E. Schweizerbart'sche Verlagsbuchhandlung, 22 November [1867]. This list was supplemented with the names of Carl Gegenbauer, Hermann Müller, and Rudolf Carl Virchow in the version retained by CD (DAR 210.11: 33).

[4] Koch evidently misunderstood CD's letter of 22 November [1867]. CD had stated that Koch's firm had translation rights for his 'present & previous works'.

[5] Julius Victor Carus was translating *Variation* into German; the translation was published in two volumes.

[6] *Variation* was published on 30 January 1868 (Freeman 1967).

[7] The year 1868 appears on the title pages of both volumes of Carus trans. 1868.

To Charles Kingsley 13 December [1867][1]

Down. | Bromley. | Kent. S.E.
Dec 13th.

My dear M[r]. Kingsley

Although you are so kind as to tell me not to write, I must send a few lines to thank you for your letter.[2] It is very interesting & surprising to me that you find at Cambridge after so short an interval a greater willingness to accept the views which we both admit. I do not doubt that this is largely owing to a man so eminent as yourself venturing to speak out. The mass of educated men will always sooner or later follow those, whose knowledge they recognize on any especial study; & this being the case I feel no doubt that views closely akin to those which I have advocated will ultimately be universally admitted. The younger working naturalists are almost all coming round: recently one of the paleontologists in Jermyn St[3] told me that he did not know a single rising man who did not largely adopt my views and I hear that this is the case likewise in Germany.[4]

I literally did not find, nor did Sir C. Lyell,[5] one single new idea in the Graduates' of Cambridge book.[6] My work on Variation Under Domestication is delayed by the index but will appear about the close of the year; and I have told Murray to send a copy to Eversley.[7] You will find the greater part quite unreadable—a mere encyclopedia of facts—but certain portions may, & I heartily hope will, interest you.

With hearty thanks for all your kindness | My dear M[r]. Kingsley, yours very sincerely | Charles Darwin

LS
Quaritch (dealers)

[1] The year is established by the relationship between this letter and the letter from Charles Kingsley, 11 December 1867.

[2] Letter from Charles Kingsley, 11 December 1867.

[3] The Royal School of Mines, the Geological Survey of Great Britain, and the Museum of Practical Geology occupied the same building in Jermyn Street, London (Reeks 1920, p. vii).

[4] In his letter of 12 May 1867, Ernst Haeckel expressed the view that CD's work would be increasingly accepted by the younger scientists in Germany. On the reception of CD's theory in Germany, see also the letter from C. L. Brace, 14 November 1867, n. 6.

[5] Charles Lyell.

[6] [Beverley] 1867; there is a copy in the Darwin Library–Down.

[7] Kingsley resided in Eversley, Hampshire. His name appears on the presentation list for *Variation* (DAR 210.11: 33).

To Eduard Koch 13 December [1867]

Down. | Bromley. | Kent. S.E.
Dec 13th.

Dear Sir

I am much obliged for your note. I shall probably receive the translation in a day or two.[1] The English edition is delayed only by the index, but will appear at the close of this year or very soon afterwards. Hence you may publish the German edit. on the last day of Dec^r.[2]

If you will refer to my former note you will find that I spoke of having given my previous works for translation to your firm without payment, but I said nothing about any future work.[3] Will you be so good as to strike out of the list previously sent the name of "Prof. Ernst Häckel of Jena" & insert in its place the name of "Prof. Gegenbauer of Jena".[4]

I hope the translation of my book may be successful in Germany, and I remain | Dear Sir | yours faithfully | Charles Darwin

I intend sending an English copy to Prof. E. Häckel

Endorsement: '13/12 67.'
LS(A)
Württembergische Landesbibliothek, Stuttgart, Cod. hist. 4° 333a, 77, 2

[1] Koch had informed CD that he was sending him two copies of the first volume of the German translation of *Variation* (Carus trans. 1868; see letter from Eduard Koch, 11 December 1867).
[2] See letter from Eduard Koch, 11 December 1867 and nn. 6 and 7.
[3] See letter to E. Schweizerbart'sche Verlagsbuchhandlung, 22 November [1867] and n. 2.
[4] CD refers to Ernst Haeckel and Carl Gegenbaur.

From Andrew Murray 13 December 1867

67 Bedford G^{ns} | Kensington
13 Decr 1867

My dear Sir

You may be thinking that I sh^d. have returned S^r. Delpino's two pamphlets ere now.—[1] I have written a review or notice of both, but keep the pamphlets themselves until I see my MSS in type in case of further reference being needed And I find that I will not have space for them in my first number— I shall get them into the second—so unless I hear that you wish the pamphlets sooner I shall retain them until they have served their purpose—[2]

Believe me | Yours sincerely | And^w. Murray

DAR 171: 330

[1] CD had sent two papers by Federico Delpino (Delpino 1867a and 1867b) for review in the *Journal of Travel and Natural History*, which had been launched by Murray (see letter from Andrew Murray, 16 September 1867 and n. 2).
[2] The review was published in the third number of the journal (Murray 1868).

From J. D. Hooker 17 December 1867

Royal Gardens Kew

Dear Darwin

I hope to get away by the afternoon (4 pm or so) train from Victoria on Saturday.[1]
Woolner[2] tells me he goes down on Sunday Morning.
Ever yr affec | Jos D Hooker
Get a good novel ready for me.

Dec 17/67

DAR 102: 186

[1] Hooker had mentioned his intention of visiting Down on Friday 20 or Saturday 21 December 1867 in his letter to Emma Darwin, 11 December 1867.
[2] Thomas Woolner.

From Joseph Plimsoll 17 December 1867

Exmouth. | 8 Bicton Place
Dec.r 17. 1867

Dear Sir

I hope you will pardon my importunity—in pressing on your notice the claims of Eternity and Salvation[1]—prompted as that importunity is by a deep solicitude for your soul's deliverance from a state of unbelief—(most perilous to its everlasting well-being; security, and happiness)—in the revelations contained in the Word of God, respecting the world's, and man's creation—the being and attributes of Jehovah—and His purposes of mercy and grace to our race, through the sacrifice and death of His beloved Son—our Lord and Saviour Jesus Christ. I am anxious to know, dear Sir, if you have yet surrendered your heart to the demands of infinite love and grace—as manifested in the incarnation, sufferings, and sin-atoning blood, of the world's Redeemer, and as verbally expressed in the command of the Apostles of Him, who is "the brightness of the Father's glory, and the express image of His person"[2]—" Believe in the Lord Jesus Christ, and thou shalt be saved".[3] Have you yet obeyed the injunction—"Kiss the Son, lest He be angry, and ye perish from the way, when His wrath is kindled, but a little,— blessed are all they who put their trust in Him"—[4] "Man's chief end is to glorify God, and enjoy Him for ever"—[5] "Thou shalt love the Lord thy God, with all thine heart, and mind, and soul, and strength"[6]

Are you thus fulfilling the great design of your creation, and of your being endowed with the attributes of an intelligent and immortal nature? Has not the thought of the Saviour dying on the cross of Calvary to atone for your sins, and to reconcile you to God, produced any softening and penitence-engendering influence on your mind and heart? Cannot you yet say, whilst contemplating that self-immolation of a spotless, innocent, and ineffably-holy, as well as a compassionate and most philanthropic being—thus dying that you might live—enduring the penalty due to the violation of God's laws, by man, in order that you and others may be delivered from the curse denounced against human transgression, and the hell of torment to which it will inevitably conduct you—(and which will be eternal in its duration),—unless you believe the gospel of Christ—exercise repentance towards God, and faith in the atonement made by the blood-shedding and death of His Son;— I ask—cannot you yet say, after contemplating the terrible price which was thus paid for your ransom from everlasting death and woe—and in virtue of your belief and trust in which,—as the means of your obtaining the divine pardon, acceptance, and favour—the offer of eternal salvation is so freely and graciously made to you—"the love of Christ constraineth me"?[7] If not—then, hard, and unbelieving, indeed, must your heart be! May the almighty Spirit of God—in his infinite mercy and love to poor, perishing man—break that hard-heartedness, and destroy the enmity of your carnal mind against God,—through His omnipotent and resistless might. This is His blessed province;—His sole prerogative. May He exercise it in your behalf—and thus magnify the riches of His goodness, and the glory of His almightiness, by making you willing in the day of His power—even as He did, by the awakening—sin-convincing—and soul-converting power He displayed, in making Saul of Tarsus[8]—the previously fierce, and blood-thirsty persecutor of the infant church—a new creature in Christ Jesus. Yes! blessed thought! He can convert your soul, as well as that of Saul's—and make you—as He did that illustrious man—a sincere, zealous, and devoted servant of the Lord Jesus Christ; can make you a powerful champion, and a worthy exponent of that faith, which by your "development theory"[9]—you have been endeavouring to sap the foundations of—; can effect in your heart, and life, and prospects, aims, and influence on your fellow men, as marvellous a transformation as that wrought in the disciple of Gamaliel,[10] when from being a cruel persecutor of the Church, a bitter enemy of the doctrines of the Cross, and to Him, whose death gives to the Cross all its glory, and a labourer to stifle, at its birth, the religion of the Man of Nazareth—he became a chosen vessel of mercy to a lost and ruined world—the great Apostle of the Gentiles—the tender, and loving foster-parent of the infant Church—and one of the holiest, and most devoted servants of the living God.

That you, in like manner, may undergo the spiritual change, in which true conversion to God consists—and show forth the praises of Him who has thus called you out of nature's darkness into God's marvellous light, and made you a new creature by faith in Christ Jesus, and caused you to rejoice with joy unspeakable and full of glory—is—dear Sir, the sincere wish and fervent prayer of Yours faithfully | J. Plimsoll | M.D. R.N.

DAR 174: 54

[1] See also letters from Joseph Plimsoll, 21 October 1867, 21 November 1867, and 3 December 1867.
[2] See Heb. 1:3.
[3] Acts 16:31.
[4] Ps. 2:12.

[5] The quotation is from 'Man's chief end', the first chapter of *A body of practical divinity* (London, 1692) by Thomas Watson.

[6] See Mark 12:30 and Luke 10:27.

[7] See 2 Cor. 5:14.

[8] Saul was the name used by the apostle Paul before his conversion to Christianity (see Acts 13:9); Tarsus, now in southern Turkey, was his birthplace (*Columbia gazetteer of the world*).

[9] On the development theory, see the letter from Joseph Plimsoll, 3 December 1867 and n. 13.

[10] The apostle Paul had been educated by the teacher Gamaliel prior to his conversion on the road to Damascus (Acts 22).

From A. C. L. G. Günther   19 December 1867

British Museum
19.12.67.

My dear Sir

Many thanks for your photograph.[1]

The keeper in the Zoolog. Gardens is quite right with regard to the difference in color in the sexes of Rattlesnakes.[2] This case is perfectly analogous to that of the common viper, which I mentioned to you the other day. I do not know of any sexual structural (external) difference in snakes, except that the tail of males is generally longer & more slender.

Male snakes are always smaller, & generally more brightly coloured, & with the markings more distinctly defined than females.[3]

As regards other Reptiles, I will mention some remarkable cases in Indian species, & thinking that you are a subscriber to the Ray Society, & have at hand my book on Indian Reptiles, I may refer you to the pages.[4] If you have not the book, let me know it, & I will supply you with further details.

Page 130 & 131. Pl. 13. figs F & G. *Ceratophora stoddartii* & *aspera*. The rostral appendage is perfectly analogous to the comb of a cock, being but little developed in the female & young.[5]

Page 132. *Cophotis* The dorsal crest is much more developed in the male than in the female.[6] This is one of numerous similar instances, all of which show that in Lizards with a dorsal crest, this is more developed in the male. *Iguana* is such an example.

p. 135. pl. 14. fig. A. *Sitana* is a lizard with a very large gular pouch, entirely absent in the female. There are other Lizards in which both sexes are equally provided with a gular pouch.[7]

p. 143 | *Calotes nigrilabris* offers a remarkable instance of sexual difference in coloration of the head & body.[8]

Sexual differences in coloration of snakes I have mentioned particularly in *Tragops dispar* p. 304, *Dipsas cynodon* p. 308, *Trimeresurus erythrurus* p. 386.[9]

As regards Frogs, the case of *Megalophrys montana* p. 413–414 is very striking; the male has a rostral appendage (skinny), absent in the female; it has a pointed appendage above the eye, much less developed in the female; & some tubercles on the back, also absent in the female.[10]

So much for today. If I come across other cases, I shall note them. I think, to appreciate fully the instances mentioned, you ought to look at the specimens when you come to the Museum.[11]

Many thanks for the promised book, on Variation; you underrate me when you say, that it will contain very little which can interest me.[12]

Yours sincerely | A Günther

DAR 82: B72–4

CD ANNOTATIONS

1.1 Many ... females. 3.2] *crossed ink*
2.4 except ... slender. 2.5] *double scored blue crayon*
3.1 Male ... females. 3.2] *scored blue crayon*
4.1 As ... body. 8.2] *crossed pencil*
5.1 Page ... example. 6.4] 'Reptiles' *added above blue crayon*
6.1 Cophotis] 'ceylanica' *added pencil*
6.2 This ... male. 6.3] *scored red crayon*; 'White band plainer— the *stripes more [clearly] defined*' *added* pencil[13]
7.1 p. 135.] '3' *added at top of page, blue crayon, circled blue crayon*
8.1 p. 143 ... body. 8.2] 'very variable in above | see Cophotis' *added pencil*
9.1 Sexual ... p. 304, 9.2] *scored blue crayon*; 'Snakes' *added blue crayon*
9.1 Sexual ... p. 386. 9.2] *crossed ink*
9.2 dispar] 'Tragops' *added pencil*; '4' *added at top of page, blue crayon, circled blue crayon*
10.1 As ... me. 12.2] *crossed pencil*; 'Batrachia' *added above blue crayon*

[1] See letter to A. C. L. G. Günther, 7 December [1867].
[2] See letter to A. C. L. G. Günther, 7 December [1867] and n. 5.
[3] CD and Günther probably met during CD's visit to London from 28 November until 7 December 1867 (CD's 'Journal' (Appendix II)). CD cited Günther on the sexual differences of snakes in *Descent* 2: 29–32.
[4] Günther refers to Günther 1864; there is an annotated copy in the Darwin Library–CUL (see *Marginalia* 1: 352).
[5] In his copy of Günther 1864, CD marked passages on the pages indicated by Günther (see *Marginalia* 1: 352). In *Ceratophora stoddartii* (the rhino-horned lizard) and *C. aspera* (the rough-horned lizard), pronounced appendages develop on the heads of males alone (Günther 1864, pp. 130–1; *Descent* 2: 34). Figures F and F′ of plate 13 in Günther 1864 were reproduced in *Descent* 2: 34 to indicate the difference between male and female *C. stoddartii*.
[6] The reference is to *Cophotis ceylanica* (the pygmy lizard; see Günther 1864, p. 132, and *Descent* 2: 32).
[7] Figure A of plate 14 in Günther 1864 shows the expanded gular pouch of the male *Sitana minor* (the fan-throat lizard; now *S. ponticeriana*) and is reproduced in *Descent* 2: 33.
[8] See Günther 1864, p. 143. CD noted the different coloration of body and lips in *Calotes nigrilabris* (the black-cheek lizard) in *Descent* 2: 36. See also plate facing p. 151.
[9] CD included brief information on sexual differences in *Dipsas cynodon* (the dog-toothed cat snake; now *Boiga cynodon*) and *Tragops dispar* (Günther's vine snake; now *Ahaetulla dispar*) in *Descent* 2: 29–30.
[10] CD cited this information from Günther 1864 in *Descent* 2: 26. *Megalophrys montana* (the horned frog) is now *Megophrys montana*.

[11] Günther refers to the British Museum.

[12] Günther refers to *Variation*. See letter to A. C. L. G. Günther, 7 December [1867].

[13] The dorsal scales, and the white band between mouth and shoulder, are more distinct in males of *Cophotis ceylanica* (Günther 1864, p. 132). CD quoted Günther on this subject in *Descent* 2: 32, 36.

From Frederick Du Cane Godman 21 December [1867][1]

Park Hatch | Godalming.

Dear Sir

You will perhaps remember my writing to you some time ago saying that I was going to the Azores and asking you if you could direct my attention to any special objects of interest that were likely to be found there.[2]

Shortly after my return I wrote a short paper on the birds of those islands, in the "Ibis" but did not think it contained any thing of sufficient interest to be worth your notice, so did not send you a copy as I would otherwise have done.[3]

A short time ago I met M[r]. Bates[4] who told me that he thought you might like to see it & I therefore now send you a copy and I trust you will excuse the liberty—

M[r]. Crotch has worked out a small collection of coleoptera which I also brought with me and has published a paper in the Zool: Proc: upon them. This I have no doubt you will have seen, but in case you have not, I also send a copy—[5] M[r]. W. C. Watson has already written on the flora[6] and the land shells also are well worked out by Mess[rs]. Morelet & Drouet[7]—and with all this material I think of writing a collected account of the fauna & flora of the Archipelago with a few remarks of my own which suggest themselves.[8]

Believe me | Yrs truly | F. Du Cane Godman

Dec[r]. 21st.

DAR 165: 59

[1] The year is established by the reference to Crotch 1867 (see n. 5, below).

[2] The letter has not been found. Godman went to the Azores in 1865 (Godman 1870).

[3] Godman's paper on the birds of the Azores appeared in *Ibis* in 1866 (Godman 1866); there is a lightly annotated copy in the Darwin Pamphlet Collection–CUL.

[4] Henry Walter Bates.

[5] Godman refers to George Robert Crotch and Crotch 1867; CD's annotated copy of the article is in his collection of unbound journals in the Darwin Library–CUL. For a bibliography of Crotch's work, see Smart and Wager 1977.

[6] Godman refers to papers on the flora of the Azores by Hewett Cottrell Watson (Watson 1843–7). CD had formerly corresponded with Watson about the flora of the Azores (see *Correspondence* vol. 5, letters from H. C. Watson, 20 November [1854] and 11 July [1855]).

[7] Godman refers to Arthur Morelet and Henri Drouet. See Drouet 1858 and 1861, and Morelet 1860.

[8] Godman's *Natural history of the Azores* was published in 1870 (Godman 1870); there is a lightly annotated copy in the Darwin Library–CUL (see *Marginalia* 1: 330).

To A. C. L. G. Günther 21 December [1867][1]

Down. | *Bromley.* | *Kent. S.E.*
Dec 21st

My dear Sir

Very many thanks for your note. You give me exactly the information which I wanted. I have looked at your Plates & Text & the cases are capital & quite new to me.— It was odd & stupid in me not to think of looking for this object to your work.—[2]

With sincere thanks | Yours very truly | Ch. Darwin

American Philosophical Society (339)

[1] The year is established by the relationship between this letter and the letter from A. C. L. G. Günther, 19 December 1867.
[2] In his letter of 19 December 1867, Günther referred CD to plates and passages in his book on Indian reptiles (Günther 1864) for information about sexual differences.

From Henry Holland 21 December [1867][1]

Brook Street
Decr 21

My dear Charles

The pages I enclose will probably interest you. They are taken from the proceedings of the Manchester Phil. Society, which I received this morning;[2] It is now perhaps the best of the Provincial Scientific bodies. I have been a Member of it, for very nearly 60 years.[3]

I received a few days ago a copy of the attack upon you from the *Author*; This gives me no clue whatever to his name or *nature*.[4] The nearest approach I make, is that of being very intimate with a friend of his, who fully admits knowing him, but under strict injunction not to disclose him.

I find that very few people have even heard of the volume.

Ever yours aff^y. | H Holland

Charles Darwin Esqr

DAR 166: 247

[1] The year is established by the relationship between this letter and the letter from Henry Holland, 27 January [1868] (*Correspondence* vol. 16).
[2] The enclosure was an article on species of plants that had colonised land cleared for planting at Tatton Park, Cheshire (Hurst and Carter 1867); there is a copy in the Darwin Pamphlet Collection–CUL.
[3] The Manchester Literary and Philosophical Society was founded in 1781; for more on its history, see Hayhurst 1967. In the *Memoirs of the Literary and Philosophical Society of Manchester*, Holland is listed as a corresponding member, elected in 1812 (see, for example, *ibid.* 3d ser. 2 (1865), Council of the Literary and Philosophical Society of Manchester (1864)).
[4] The reference is to *The Darwinian theory of the transmutation of species examined by a graduate of the University of Cambridge* by Robert Mackenzie Beverley ([Beverley] 1867; see letter to Charles Kingsley, 13 December [1867] and n. 6).

To F. Du C. Godman 23 December [1867]¹

Down. | Bromley. | Kent. S.E.
Dec 23

Dear Sir

I am much obliged to you for sending me your paper which I shall probably receive by tomorrow's post.² I should have been very sorry to have missed seeing it, which might have happened as I do not take in the Ibis.³ I had already seen Mr. Crotch's paper.⁴ I am particularly glad to hear that you intend to work up all the materials into a single paper & draw your own conclusions, for that certainly is the way to gain valuable results.⁵

Believe me dear Sir | yours very faithfully | Ch. Darwin

LS
Courtesy of Joseph Sakmyster (dealer)

¹ The year is established by the relationship between this letter and the letter from F. Du C. Godman, 21 December [1867].
² CD refers to Godman 1866. See letter from F. Du C. Godman, 21 December [1867] and n. 3.
³ Godman 1866 was published in *Ibis*, the journal of the British Ornithologists' Union, of which Godman was a founder member.
⁴ Crotch 1867. See letter from F. Du C. Godman, 21 December [1867] and n. 5.
⁵ See letter from F. Du C. Godman, 21 December [1867] and n. 8.

From J. D. Hooker [23 December 1867?]¹

Royal Gardens Kew
Monday

Dear Darwin

I am disgusted with myself for leaving the tin box in the Dog Cart pray send it by Willy—² the plants will keep well enough—³

Ever Yrs | J D Hooker

Also I left a shirt!

DAR 102: 197

¹ The date is conjectured on the basis of the letter's position in the Darwin Archive–CUL and its references to a visit by Hooker to Down. Hooker visited Down on Saturday 21 December 1867 (Emma Darwin's diary (DAR 242)).
² Emma Darwin's diary (DAR 242) for 21 December 1867 records, 'Dʳ Hooker Wᵐ'; 'Wᵐ' could refer to William Erasmus Darwin or to Hooker's son William Henslow Hooker, known as 'Willy' (see, for example, letter from J. D. Hooker, 13 April 1867). Hooker also referred to W. E. Darwin as 'Willy' (see, for example, letter from J. D. Hooker, 26 [and 27] March 1867).
³ CD sometimes had plant specimens sent packed in tin boxes lined with damp moss or paper for preservation during transport (see, for example, *Correspondence* vol. 9, letter to J. D. Hooker, 17 [July 1861], and letter to D. F. Nevill, 12 November [1861]).

To Roland Trimen 24 December [1867][1]

Down. | *Bromley.* | *Kent. S.E.*
Dec. 24th

My dear Sir

If you are not engaged, will you give me the great pleasure of your company here next Saturday, & stay the Sunday with us. We dine at 7 oclock.—[2]

You would have to come by Train to Bromley, but I am sorry to say this place is six miles from the Station.

I am bound to tell you that my health is very uncertain & I am continually liable to bad days, & even on my best days I cannot talk long with anyone; but if you will put up with the best will to see as much of you as I can, I hope that you will come.—

Pray believe me, | my dear Sir | Yours very sincerely | Ch. Darwin

Royal Entomological Society (Darwin letters and papers, box 21: 61)

[1] The year is established by the fact that Trimen visited CD at Down at the end of 1867 (see Poulton 1909, p. 28 n. 2). See also n. 2, below.
[2] Trimen, who lived in South Africa, was in London in late 1867 to work on a paper about mimicry in African butterflies (Trimen 1868; see Poulton 1920, p. xxiv). For CD's letters to Trimen between 1863 and 1871, and a commentary on their relationship, see Poulton 1909, pp. 213–46.

From W. S. Dallas 26 December 1867

York
26 Dec 1867

My dear Sir

I am vexed to the heart that you should have occasion to write to me again about this Index.—[1] I cannot plead illness, although as I am suffering again from the same pain in the back that laid me up in the summer, no doubt this has impeded my work, but the real cause of delay lies in the nature of the work itself.[2] I have, as I told M^r. Murray[3] in a note written today, consumed more than half a ream of paper in writing out the analysis of the book.— This part of the business is now completed, however, & the slips are in process of being cut up & sorted,—I shall have a considerable portion in the printer's Hands on Monday morning.—[4]

It is now many weeks since I have left off work until at least between 1 & 2 oclock in the morning, but very frequently, either through being utterly worn out, or from some other cause, I have not got through more than a sheet or sheet & a half of your book in 3 or 4 hours.—[5] This, I hope is the last letter of this kind I shall have to write—[6]

Believe me | Your's very truly | W. S. Dallas.

C. Darwin Esqr

DAR 162: 8

[1] Dallas was compiling the index for *Variation*. The letter from CD to Dallas has not been found.

[2] In his letter of 4 November 1867, Dallas mentioned that his work had been interrupted by illness during several weeks in the summer of 1867.

[3] John Murray.

[4] *Variation* was printed in London by William Clowes & Sons.

[5] Dallas had previously written that he was completing nearly three sheets (each consisting of sixteen pages) in five hours (letter from W. S. Dallas, 11 December 1867).

[6] Dallas had written about the difficulties of compiling the index, and about the consequent delay to publication, in his letters of 8, 10, and 11 December 1867.

To John Smith 26 December [1867][1]

Down. | Bromley. | Kent. S.E.
Dec. 26th

Dear Sir

I am very much obliged to you for your note on crossing the Victoria regia, received a few days ago from D[r]. Hooker.—[2]

I very much hope that you feel sufficient interest in the subject to try some more experiments; for if you obtained anything at all like the same result, the case would be extremely curious.— There is, I feel sure, very much to be made out about the fertility of plants when fertilised by their own pollen.—[3]

With many thanks, for your interesting communication, I remain | Dear Sir | Yours faithfully | Charles Darwin

American Philosophical Society (Getz 5740)

[1] The year is established by the relationship between this letter, the letter from J. D. Hooker, 31 August 1867, and the letter to J. D. Hooker, 3 April [1868] (*Correspondence* vol. 16).

[2] The note about *Victoria regia* (now *V. amazonica*) has not been found. Joseph Dalton Hooker visited Down on 21 December 1867 (Emma Darwin's diary (DAR 242)).

[3] CD was interested in *Victoria regia* (originally *Euryale amazonica*, now *V. amazonica*) because of Robert Caspary's claim that it bred by repeated self-fertilisation (see Caspary 1865, pp. 19–20, and *Correspondence* vol. 14, letter from Robert Caspary, 25 February 1866 and n. 13). CD discussed pollination and the setting of seed in *E. amazonica* and *E. ferox* in *Cross and self fertilisation*, pp. 358, 365.

From Clair James Grece 29 December 1867

Dear Sir,

There has lately appeared a learned and elaborate English grammar, very far beyond aught of its kind which has been published before. In fact there is no other grammar of the English language with which it can be at all compared. It is a work of prodigious research and erudition, the author having traced in the historical manner, the descent of the finest shades of usage from the teutonic and romance sources whose confluence formed the English tongue.

It is marvellous how a foreigner could ever come to discriminate shades of usage whose subtlety almost baffles the apprehension of a native. He has brought to his work a thorough acquaintance with the science of language, and of thought, its counterpart, and an intimate knowledge of eight tongues. It can be compared only with Kuehner's Greek or Grimm's German grammar;[1] and, like these, it is the production of a German, Professor Maetzner of Berlin, and it is written in the German language.[2]

This work, which can scarcely fail to take rank, sooner or later, as a leading authority on our tongue, and as an indispensable complement to Webster's, Wedgwood's or Richardson's dictionary,[3] in every well-stocked library, ought not, I think, to be confined to such Englishmen and Americans as can read German, wherefore I have conceived the project of devoting my unoccupied time to translating it. But, as this would entail a very considerable expenditure of effort, the work consisting of three octavo volumes, each of about 500 pages, it is important, on commencing it, to ensure that I shall be able to unite myself with a publisher, as else much time and labour might be consumed in vain. I have been recommended to negotiate with M.ʳ Murray, and, it was hinted that if I were sustained by an introduction from one of his authors, his attention would be more likely to be fairly bestowed upon the project than if I approached him abruptly.[4]

It at once occurred to me that you would not object to give me a letter to M.ʳ Murray in which you might refer to me as the person mentioned in the note to your "Origin of Species".[5] Of my competence to render the ideas from the High Dutch into the English idiom so far as they are capable of being transfused, you can, of course, say nothing.

I am, Dear Sir, | yours very truly, | Clair J. Grece

Redhill, Surrey;
29.ᵗʰ December, 1867

DAR 165: 221

[1] The references are to Raphael Kühner and Jakob Ludwig Karl Grimm, and to Kühner 1834–5 and Grimm 1819–37.
[2] Grece refers to Eduard Mätzner and his *Englische Grammatik* (Mätzner 1860–5).
[3] On Noah Webster's and Charles Richardson's English dictionaries, see, for example, Micklethwait 2000. Hensleigh Wedgwood compiled a dictionary of English etymology (H. Wedgwood 1859–65).
[4] Grece refers to John Murray. See letter to John Murray, 31 December [1867] and n. 4.
[5] Grece had previously pointed out to CD a passage in Aristotle that indicated, according to Grece, some understanding on the part of 'the ancients' of the principles of natural selection (see *Correspondence* vol. 14, letter from C. J. Grece, 12 November 1866 and n. 3); CD added this information to the historical sketch in *Origin* 4th ed., p. xiii n., acknowledging Grece as its source.

To John Murray 31 December [1867][1]

Down. | Bromley. | Kent. S.E.
Dec 31.ˢᵗ

My dear Sir

M.ʳ Clair Grece has asked me to give him an introduction to you. He wishes to publish & translate from the German a large & profound work on the principles of english grammar, & the believes the work to possess the highest value.[2] I have not the pleasure of a personal acquaintance with M.[r] Grece, but he was so kind as to send me translations of several passages in Greek authors bearing on Natural Selection;[3] and I believe him to be endowed with much learning. I do not doubt that he would bring out an excellent translation of the work which interests him so much.[4]

I remain, my dear Sir | yours sincerely | Charles Darwin

To | John Murray Esq.[re] | 50 Albemarle St

LS(A)
John Murray Archive

[1] The year is established by the relationship between this letter and the letter from C. J. Grece, 29 December 1867.
[2] See letter from C. J. Grece, 29 December 1867 and n. 2.
[3] See letter from C. J. Grece, 29 December 1867 and n. 5.
[4] John Murray later published Grece's translation of Eduard Mätzner's English grammar (Grece trans. 1874).

From Roland Trimen 31 December 1867

71, Guildford Street, | Russell Square | London, W.C.
31st. December, 1867.

My dear Mr Darwin,

In a wilderness of *ocelli*,[1] I have stupidly forgotten the exact nature of the information which you indicated as requisite. Am I right in thinking that you wanted one or two instances of gradation in the developement of ocellated spots in the same species? or, if that cannot be found, of gradations from a simple spot to a many-ringed ocellus in closely-allied species?[2]

With reference to the relative abundance of the sexes, which is apparently very generally in favour of the ♂ in butterflies, I find that Wallace (Trans. Linn. Soc. XXV "On the Papilionidæ of the Malayan Region") notes that the ♀ s of *Ornithoptera Cræsus* (peculiar to the island of Batchian) were "more plentiful" than the ♂ s.[3] This is a case in which there is a very wide difference between the sexes, the ♂ being splendidly coloured with shifting green and orange on a black ground, while the ♀ is (like those of all the nearly-allied species) dull-brown with whitish markings.

Mr Waterhouse Jun.[r][4] has kindly given me the name of the Cape Beetle which I mentioned to you. It is *Peritrichia cinerea*, a small, long-legged Lamellicorn belonging to the characteristic S. African *Pachycnemidæ*. The ♂ is slaty-grey clothed with whitish hairs, and the ♀ rufous-brown with pale-yellowish hairs. Though the ♀ has been described as a distinct species (P. *proboscidea*), I found that the sexes were placed together as one species in the British Museum collection, but without the knowledge that they *were sexes* of the same insect.[5]

I have so often taken the two in copulâ that you may rely on the correctness of this observation.

I will not forget to send you any points of importance which appear to bear upon your present special subject of research, if I am fortunate enough to chance upon them.

With my respects to M^{rs}. Darwin, I remain | Very faithfully yours, | Roland Trimen.

C. Darwin Esqre.

DAR 82: A30–1

CD ANNOTATIONS
1.1 In a ... closely-allied species? 1.5] *crossed blue crayon*
2.1 With ... markings. 2.7] 'Lepidoptera' *added blue crayon*
2.6 shifting ... ground,] *underl red crayon*
3.1 M^r. Waterhouse ... observation. 4.2] 'Coleoptera' *added blue crayon; crossed ink*
5.1 I will ... them. 5.3] *crossed blue crayon*

[1] Trimen refers to the wing markings of butterflies; he was working on a paper about mimicry in butterflies (see Poulton 1920, p. xxiv).
[2] No letter from CD requesting this information has been found, but CD may have asked for it when Trimen visited Down (see letter to Roland Trimen, 24 December [1867]). In *Descent* 2: 133, CD cited Trimen for information on the gradation of the ocelli of the South African butterfly *Cyllo leda* (now *Melanitis leda*).
[3] Trimen refers to Alfred Russel Wallace's paper 'On the phenomena of variation and geographical distribution as illustrated by the *Papilionidæ* of the Malayan region' (A. R. Wallace 1864a). CD cited A. R. Wallace 1864a on the numerical predominance of females in *Ornithoptera croesus* in *Descent* 1: 310. Batchian (now Bacan) is an island in Indonesia (*Columbia gazetteer of the world*).
[4] Frederick Herschel Waterhouse.
[5] With acknowledgment to Trimen and Waterhouse, CD cited *Peritrichia* as a rare example of a beetle species in which one sex differed from the other in colour (*Descent* 1: 367–8 and n.). The family Pachycnemidae is no longer recognised; *Peritrichia* is now classified within the subfamily Hopliini of Scarabideae (F. Krell, Natural History Museum, personal communication).

From A. C. L. G. Günther [late December 1867 or early January 1868][1]

My dear Sir

There is a whole family of fishes (Cyprinodontidæ) exhibiting sexual differences as remarkable as any in Reptiles or Birds.[2] It was very stupid in me not to think of it when you were in the Museum.[3] I have given the specimens of two of the most remarkable forms to my artist, who will make drawings of them.[4]

As soon as they are finished, I will send you proofs of the two plates.[5]

With the compliments of the season | Yours sincerely | A Günther

DAR 82: B75

CD ANNOTATIONS
Top of letter: 'A mere crest or **uncoloured** ornament c^d not cause more danger to female than male.— | Fresh Water?'[6] *pencil*; 'Musical Fish S. Brazil'[7] *ink*

Bottom of 1st page: 'How about *numbers* of males in other fishes in which males differ'[8] *pencil, del ink*
End of letter: 'Colours?? | Female lays **first**, & never attends to eggs or young— Are males greatly in excess?? | Polygamy' *ink*; '— Stickle-backs[9] | Get names of other Fish'[10] *pencil*

[1] The date is established by the seasonal greeting and the relationship between this letter and the letter from A. C. L. G. Günther, 13 May 1868 (*Correspondence* vol. 16).

[2] In his letter of 19 December 1867, Günther had provided information on sexual differences in reptiles. CD cited Günther on sexual differences in size and ornamentation in the Cyprinodontidae in *Descent* 2: 7, 9–10. There are many contemporary notes by CD on sexual differences in animals, including fish, reptiles, and birds, in DAR 82.

[3] Günther was employed at the British Museum. CD had visited London from 28 November until 7 December 1867 (CD's 'Journal' (Appendix II)).

[4] The reference is to George Henry Ford. The two forms Günther refers to are probably *Mollienesia petenensis* and *Xiphophorus hellerii*: CD cited Günther 1859–70 and Günther 1864–6 on the differences between the sexes of these two Cyprinodont species in *Descent* 2: 9–10 and figure 29.

[5] Chapter 12 of *Descent* includes illustrations of four examples of sexual differences in fish. Drawings for the plates were executed by Ford, who also drew reptiles and amphibians for chapter 12 under Günther's supervision (*Descent* 2: 4–11). See also *Correspondence* vol. 16, letter from A. C. L. G. Günther, 13 May 1868.

[6] In *Descent* 2: 18–23, CD concluded that ornamentation in fish occurs more frequently in males, and considered this in relation to their vulnerability to predation. In *Descent* 2: 9, CD noted that fish of the Cyprinodontidae inhabit fresh water.

[7] CD was aware of fish in Brazil that made melodious sounds (see letter from Fritz Müller, 17 July 1867 and n. 18). CD mentioned the topic in *Descent* 2: 23, 331.

[8] In *Descent* 1: 307–9, CD discussed the numerical proportions of the sexes in fish; he discussed numerical proportions of the sexes in relation to sexual selection in *ibid.*, pp. 263–5. CD's notes on the numerical proportions of the sexes in animals and humans are in DAR 85 and 86; a note on the proportion of Cyprinodont sexes is in DAR 86: A49.

[9] In *Descent* 2: 19, CD noted that the fertilised ova of most fish are left unprotected by both sexes after oviposition. On numerical proportions of the sexes, see n. 8, above. In *Descent* 1: 271 and 2: 2, CD referred to the polygamous habit of the common stickleback, *Gasterosteus leiurus* (now *G. aculeatus aculeatus*).

[10] In *Descent* 2: 14, CD referred to sticklebacks, together with tench, roach, perch, salmon, and pike, as examples of fish in which striking colour develops during the breeding season in males only.

APPENDIX I
Translations of letters

From E. Schweizerbart'sche Verlagsbuchhandlung[1] 24 January 1867

Stuttgart
24 January 1867

Most honoured Sir!

I was glad to see from your kind lines of 30 October that having Prof. Carus as translator of your book is agreeable to you.[2]

He has already sent me a part of the manuscript and this has been sent to the printers. I will arrange for the printing to be done as quickly as possible and I shall be pleased to send you the proofs as and when they become ready.

Today I have another request for you, most honoured Sir; I asked Williams & Norgate about the possible availability of a picture of you, should one exist that is better than the one I had made from a photograph for the 2^d German edition, but these gentlemen have as yet left me without an answer.[3] I had in mind, you see, to get a steel engraved portrait for the 3^d edition if I could find a good original. Now, could I ask you to tell me whether the photograph in the 2^d edition is a good one or whether you could send me a copy of a better one.[4]

In the hope that this year's very cold winter has not impaired your health, I am with the highest esteem | your most devoted | E Schweizerbart[5]

a few days ago we had 25° frost, today 9° warmth (Réaumur)[6]

DAR 177: 74

[1] For a transcription of this letter in its original German, see pp. 48–9. It was written by Christian Friedrich Schweizerbart, head of the Stuttgart publishing firm E. Schweizerbart'sche Verlagsbuchhandlung (see n. 5, below).

[2] CD's letter to E. Schweizerbart'sche Verlagsbuchhandlung of 30 October 1866 has not been found, but see the letter from E. Schweizerbart'sche Verlagsbuchhandlung, 26 October 1866 (*Correspondence* vol. 14). Julius Victor Carus was translating the fourth English edition of *Origin* for a third German edition. Beginning with Carus's letter of 7 November 1866, he and CD exchanged four letters in November 1866 discussing the new edition; see also this volume, letter from J. V. Carus, 18 January 1867, and letter to J. V. Carus, 22 January [1867].

[3] The publishers and booksellers Williams & Norgate specialised in foreign scientific literature. The photograph of CD in Bronn trans. 1863 is the same as the photograph, taken *circa* 1857, that appears as the frontispiece to *Correspondence* vol. 8.

[4] CD's reply to this letter has not been found, but see the letter from E. Schweizerbart'sche Verlagsbuchhandlung, 22 March 1867 and n. 8.

[5] In his business communications, C. F. Schweizerbart continued to use the signature of his uncle, Wilhelm Emanuel Schweizerbart, from whom he had purchased the publishing firm in 1841 (*Jubiläums-Katalog*, pp. x–xi).

[6] Schweizerbart gives the temperatures based on the Réaumur thermometer scale, which has the freezing-point of water at 0°, and the boiling-point at 80°. The approximate equivalents in Celsius would be −31° centigrade and 11° centigrade.

From E. Schweizerbart'sche Verlagsbuchhandlung[1]   22 March 1867

E. Schweizerbart, Commissions- und Verlagsbuchhandlung. Buchdruckerei
Stuttgart
22 March 1867

Most honoured Sir!

Your friendly letter of the 19th prompts me to answer immediately.[2] Please accept my sincere thanks for the new demonstration of your favour in most unselfishly offering me clean sheets of your larger work now in press, titled: *The variation of animals & plants under domestication*", for translation into German.[3]

I am most pleased to accept your offer and I shall write to Prof. Carus[4] today to ask him to deal with the German edition; I do not doubt that he will be happy to undertake this commission and therefore will contact you himself.

Concerning the woodcuts, I will contact Mr. Murray shortly myself and need only ask you to tell him, when the occasion arises, to kindly let me have **good** stereotypes of the woodcuts for your work.[5]

The printing of the new edition of the German Origin is progressing rapidly, and the first delivery (12 sheets) can be dispatched in a couple of days. When it is all ready you will receive copies as you wish.[6]

I must also thank you for kindly sending me your portraits. I was not a little surprised to receive instead of the familiar fine figure of the earlier picture, that of a more elderly gentleman with grey beard.[7]

I sent Herr Carus a copy of the large picture, made here, whereupon he wrote to me that the smaller one that he received from you (presumably the one with the chair-back) seems rather better to him, it is the *full profile*.

It strikes us that the earlier picture may appeal more to some; so I take the liberty of asking whether it perhaps matters to you that a newer picture is used in which the change in your appearance is evident. *I will use the picture you indicate to me*, please just let me know your decision.[8]

With the highest esteem and admiration | your most devoted | E Schweizerbart

DAR 177: 75

[1] For a transcription of this letter in its original German, see pp. 157–8. It was written by Christian Friedrich Schweizerbart, the head of E. Schweizerbart'sche Verlagsbuchhandlung, who used the signature E. Schweizerbart in business communications.

[2] See letter to E. Schweizerbart'sche Verlagsbuchhandlung, [19 March 1867].

³ Schweizerbart refers to *Variation* (see letter to E. Schweizerbart'sche Verlagsbuchhandlung, [19 March 1867]).

⁴ Julius Victor Carus.

⁵ John Murray, CD's publisher, provided Schweizerbart with electrotypes of the illustrations for *Variation*; Murray's ledger records a payment of £10 from Schweizerbart in September 1867 (John Murray Archive).

⁶ Carus had recently finished translating the fourth edition of *Origin* (Bronn and Carus trans. 1867; see letter to J. V. Carus, 17 February [1867]). No presentation list for Bronn and Carus trans. 1867 has been found. CD's copy is in the Darwin Library–Down.

⁷ The image used in the first two German editions of *Origin* (Bronn trans. 1860 and Bronn trans. 1863) was a reproduction of a photograph taken around 1857 (DAR 225: 175; reproduced as the frontispiece to *Correspondence* vol. 8).

⁸ CD had sent Carus a photograph of himself with his letter of 10 November 1866 (*Correspondence* vol. 14). The photograph Carus received was one made by CD's son William Erasmus Darwin in 1864, reproduced as the frontispiece to *Correspondence* vol. 12 (Carus photograph album, Smithsonian Institution Archives, Record Unit 7315). An engraving made from this photograph was used for the frontispiece to Bronn and Carus trans. 1867. CD may have sent Schweizerbart a later photograph, taken by Ernest Edwards in 1865 or 1866 (see *Correspondence* vol. 13, letter from E. A. Darwin to Emma Darwin, 25 [November 1865] and n. 3).

From J. D. Hooker 23 March 1867

[Enclosure]¹

Paris,
9 March 1867—

Dear Mr Hooker,

I cannot tell you how happy your letter has made me. I thank you for it, as well as the charming portrait which you were so courteous to enclose and which will receive pride of place in my collection of contemporary celebrities. I also send you and Mrs Hooker my congratulations on the addition to your family.²

I have thought a lot about what you told me in your letter of the structure of different cucurbits. There are a number of things that are still obscure about this family, but that will become clearer, I hope, when studied in live specimens.

You will receive, at the same time as this letter, a small box containing seeds of something that I have provisionally named *Microphœnix decipiens*.³ These seeds, or rather these fruits, look quite like little dates, and perhaps not without reason. Their story will, I believe, be a source of some astonishment to you.

These apparent dates are quite simply the fruits of *Chamærops humilis*, var. *arborescens*, that have been *fertilised by the pollen of the date palm* (Phœnix). The experiment was made in Hyères, by M. Denis,⁴ a *grand amateur* of horticulture, who has a number of adult palms in his garden. Two years ago, he had the idea of shaking bunches of flowering male date palms on bunches of female chæmerops, also in flower, and strange to say, the fruits of chæmerops thus fertilised grew *twice as large* as in the case of normal fertilisation, and noticably more elongated, in a word, more finger-shaped. I made sure these seeds were sown at Hyères, last year (1866), and the resulting young plants show very clear signs of hybridity in the shape of their leaves, which are pinnate and no longer palmate, like those of the Chæmerops, their female parent.

I advised M. Denis to repeat his experiment in 1866; he did, and the same phenomenon occurred as in 1865. Last December, while at Hyères, I saw on a tree fruits fertilised by the date palm— I took a few of them and I am happy to send them to you, thinking that you would like to observe a new kind of hybrid, that is to say, one from two plants of very different genera and even classified in two distinct sections by Martius.—[5] This fact will change the theory of hybridisation as we know it today.

You interest me greatly, dear Mr Hooker, by what you tell me on the effects of cold on certain plants. I knew of the very hardy Jubæa, though I would not have believed it capable of such an amazing feat as that of which you told me. At Montpellier, it tolerates anything, even frosts of -12° centigrade, under cover. I consider it hardy to the same degree as *Cham. humilis*, which does perfectly well throughout the whole olive-growing region, but can't tolerate living outside it. *Cham. Fortunei*[6] is, in my opinion, the hardiest of all; it is the only one that can successfully be kept *without cover* at Bordeaux and in the Landes.

Regarding frozen plants, *In my opinion*, there is a great reform needed in English horticultural meteorology, and truly, dear sir, you would do the world a good service by taking the initiative here. This reform would consist in *Substituting the centigrade thermometer* for the Fahrenheit thermometer. The latter is really so badly designed that even those who are used to using it do not always succeed in making it comprehensible. I find proof of this all the time in *Gardener's Chron.* when I read things like: Effects of the late frosts—The late Winter &c.[7] Since Horticulturalists don't have a common formula to indicate the degree of cold they are talking about, the result is an arbitrary manner of expressing themselves, which means that very often they don't understand each other. All these uncertainties would disappear if these gentlemen would adopt Centigrade, in which the starting point (zero) coincides with a natural phenomenon, the temperature at which ice forms, the temperature zero, which is neither hot nor cold. Nothing is easier than to make oneself understood, when one says: -6; -2; +3 &c. Moreover this thermometer graduates just as well in milligrade as in Centigrade, and even tenths of a milligrade if one wants.

I will be really happy to see you again, dear sir;[8] we are now talking about an experimental garden that is being planned. Meanwhile, please accept the expression of my most sincere and affectionate regard

    your devoted colleague   Ch Naudin

Royal Botanic Gardens, Kew, DC 143: 643

[1] For a transcription of this enclosure in its original French, see pp. 161–2.
[2] Naudin refers to Frances Harriet Hooker and Reginald Hawthorn Hooker.
[3] The name was later published in *Revue Horticole* (1885): 513–14.
[4] Alphonse Amaranthe Dugomier Denis. See also letter from J. D. Hooker, 14 March 1867 and n. 7.

[5] Carl Friedrich Phillip von Martius.

[6] *Chamaerops fortunei* is a synonym of *Trachycarpus fortunei*.

[7] The winter of 1866 to 1867 was unusually cold. In an editorial in *Gardeners' Chronicle* for 26 January 1867 (p. 73), readers were asked to record 'the results of the recent severe weather'. In the issue for 2 March 1867 (pp. 210–11), several replies were published. Some writers referred to degrees below freezing, others to degrees below zero, and one gave both the temperature and its relation to freezing point ('10°, or 22° below freezing'; *ibid.*, p. 211).

[8] Hooker was going to Paris to attend the International Horticultural Exhibition (see letter from J. D. Hooker, 14 March 1867 and n. 6).

From Fritz Müller 1 April 1867

[Enclosure][1]

Claus, Free-living copepods. Leipzig, 1863; p. 35;[2]

Finally, we can discuss the remarkable iridescence on the external integument of some *Sapphirina* males, with which earlier researchers were already familiar. On *Cook*'s last voyage, *Anderson* had already observed the phenomenon in his *Oniscus fulgens*, a form which is obviously identical with *Bank*'s *Carcinium opalinum* and *Thompson*'s *Sapphirina* . . . . *Meyen*, who observed *Sapphirina* in the Azores, described the same feature as extremely flexible scales with continual colour changes.[3] Sunk in the depths the animal appears the most brilliant violet-red surrounding a purple centre, but when caught it loses the colours, which were only induced by the interrruption of light rays playing on the surface of its body. On the latter, however, he found on the underside 4-, 5- or 6-sided scales, whose flat edges lined up against each other displayed a brilliant surface; through these, Meyen imagined, as through a row of prisms, the light was refracted, and thus with each movement of the animal a colour change was produced. An extemely detailed description of the colour phenomena just mentioned is given by *Gegenbauer* (Müller's Archiv 1858 p. 67).[4] This researcher also locates its source in these polygonal . . areas, which he traces back to the connected layer of flat cells under the cuticle, *to the matrix of the chitonous integument.* "With females," says our researcher, "the cell content is bright overall throughout its life, males on the contrary display in life almost the same phenomenon in each cell under the microscope as that which can be observed in the free living animal. With both the light that is transmitted and the incident light, the shifting play of colours from cell to cell can be observed and while, in the latter case, only a metallic sheen sparkles, in the former, a dioptric colour-play is visible in addition to this phenomenon. Often, one cell is set off from its neighbour with great clarity by the colour or metallic shimmer, appearing yellow, red or blue with the most varied nuances of one colour merging into another, but without all the intermediate colours, without green, violet or orange. Both main colours, on the other hand, occur with the catoptric phenomenon, in which blue plays the main part. If one considers the appearance in a single cell, one finds, thereby, that there is a transition from blue to red without the intermediate colour, that in one part of the cell, perhaps a corner of it, the blue turns pale, becomes almost gray and then suddenly at this spot a red fringe appears that broadens over the cell, expanding in size as the blue fades so that presently the whole cell appears blue. The same applies to yellow. The colour of any one cell is completely independent of neighbouring cells. Thus yellow appears in the middle of red, red in the middle of blue. However, the phenomenon can extend to a neighbouring cell; blue from the edge of a blue cell goes over into the adjacent cell, which is still red, and thus one colour occasionally expands over a large section. Sometimes a colourless spot suddenly appears in one and the same cell, in the middle or on the margin, larger or smaller, while the remaining part is still resplendent in full colour. If one now changes the light from transmitted to incidental, then the spot glows with a full metallic gleam, while the remaining part, coloured before and after, is dark. The periods of time during which this phenomenon takes place are variable, often the colour changes three times in a second, often one colour lasts many seconds. With the death of the little animal, at which time the fine-grained contents crowd together towards the centre, the whole phenomenon is extinguished.'

I have quoted *Gegenbaur*'s complete description, because my own observations to confirm the nature of the colour-change phenomenon have not agreed with the details of the sequence of colour change discovered by Gegenbaur. Nevertheless, I cannot concur in all points with Gegenbaur. First, according to my observations, the polygonal areas certainly form a layer under the cuticle, but they are: 1, not cells; 2, they lie only under the dorsal surface; 3, they cannot be seen in the female sex. That they are not simple cells of the epithelial matrix can be seen not only from their size, which in *Sapph. auronitens* is 0.08 mm, in *Saph. fulgens* 0.1 mm in diameter, so they cannot be compared in any way with the small matrix cells of other copepods, but especially because of the retention of the boundary. The polygonal plates are not enclosed by a firm membrane, but show a very fine serrated outline. We have thin plates of a fine-grained substance, plates that are bordered by a suture-like interlocking edge, and (in *Sapph. nitens*) the edges frequently display thick and fine borders similar to certain butterfly scales. Nuclei, depicted by *Gegenbaur* as one of three areas distinguished by him, I have never observed clearly and consistently, and for this reason I cannot consider these areas to be cells. It is far more likely that this whole complex corresponds to amalgamated and altered cells of the matrix, for which I could find no proof of a second, deeper layer of epithelium. Further, I have stressed that the iridescence is in no way extinguished with the death of the animal, which only terminates the wonderful change of colours, the transformation of the same part from blue to red etc. I find the gold-green metallic gleam (*S. auronitens*) as well as the greenish violet play of colour (*S. fulgens*) to have been splendidly retained in some specimens preserved for years in a dilute glycerine solution. In *Sapphirinella mediterranea*, the same polygonal, finely bordered areas are found under the cuticle and show a light violet under incident light, a pale yellow shimmer by transmitted light.

I will not hazard a detailed explanation of the colour phenomena referred to. There can, of course, be no question of a comparison of the polygonal laminae with a series of glass prisms, that divide light into the colours of the spectrum; rather it is to do with interference phenomena, found in the fine-grained, sometimes fractured and fissured structure of the laminae. We remain completely in the dark on the highly remarkable colour variation in life in individual polygonal areas, which, although not controlled by the animal's will, seem to be dependent on the processes of metabolism, which is also where the cause of the illuminating power of *Sapphirina* has to be looked for.

l.c. pg. 152. The male of *Sapphirina nigromaculata Claus* has no iridescence.

DAR 81: 167

[1] For a transcription of this enclosure in its original German, and the letter with which it was enclosed, see pp. 196–201.

[2] In *Descent* 1: 335–6, CD cited Claus's description of colour change in *Sapphirina*.

[3] William Anderson's identification of *Oniscus fulgens* is discussed in the account of Cook's third voyage to the Pacific Ocean from 1776 to 1780, during which Cook was killed (Cook and King 1784, 2: 257). John Vaughan Thompson's account of *Sapphirina* is in Thompson [1828–34]; Franz Julius Ferdinand Meyen's is in Meyen 1834, pp. 153–5. Claus also refers to Joseph Banks.

[4] Claus refers to the *Archiv für Anatomie, Physiologie und wissenschaftliche Medizin*, which was edited by Johannes Peter Müller between 1834 and 1858. The lengthy quotation given by Claus is from Gegenbaur 1858, pp. 66–8, with some brief omissions.

From Carl Vogt[1] 8 April 1867

Geneva

8 April 1867

Sir and highly honoured Master,

I have just received from my publisher and friend, Mr. J. Ricker, bookseller in Giessen (Hesse) a letter in which he informs me that you are going to publish a work in two volumes entitled "Domesticated animals and cultivated plants etc.[2] Mr. Ricker inquires whether I would be inclined to translate this book into German, and if it is the case that you have not yet promised the rights of translation to someone else whether you would be willing to entrust a German edition to him and to me as publisher and translator? Mr Ricker adds "Would you ask Mr Darwin whether he requests the rights of the author for the German translation and what sum he would require and in addition whether his publisher would allow the use of the plates for the illustrations that will ornament the work and at what cost?"[3]

At the time of my last visit to England some years ago, Mr. Wallace[4] and others told me that you were too poorly and that, consequently, I must give up my plan to go to see you. I rejoice now to see that you are improved enough to be able to consider the publication of a major work, for you, Sir, have opened new pathways for science— we proudly call ourselves your disciples.

I would not have addressed these lines to you if your former translator, Mr Bronn, had not unhappily succumbed to his prolonged illness.[5] Undoubtedly you know my name as a man of science— I also have some recognition as a translator

of "Vestiges of the natural history of creation", of the Lectures of Mr. Huxley, etc.[6]
Accept, dear Sir and master, my devoted regards, | C Vogt

Prof. C. Vogt | Pleinpalais | *Geneva* (Switzerland)

DAR 180: 10

[1] For a transcription of this letter in its original French, see pp. 214–15.

[2] Vogt was born in Giessen, studied there until 1835, and taught zoology there between 1846 and 1848, when he went to Geneva (Judel 2004). Franz Anton Ricker, Vogt's friend and publisher, took over his brother Joseph's publishing and bookselling business in Giessen in 1835 after his brother's death, keeping the name J. Ricker'sche Buchhandlung (*Hessische Biographien* 3: 301). Vogt also refers to *Variation*.

[3] CD was in the process of arranging for the German translation of *Variation* to be translated by Julius Victor Carus and published by E. Schweizerbart'sche Buchhandlung; see letter to John Murray, 4 April [1867].

[4] Alfred Russel Wallace.

[5] Heinrich Georg Bronn, who translated the first and second German editions of *Origin* (Bronn trans. 1860 and 1863), and *Orchids* (Bronn trans. 1862), died in 1862.

[6] Vogt was professor of geology at Geneva, and had applied CD's theory of descent to humans in his *Vorlesungen über den Menschen* (C. Vogt 1863). His translation of *Vestiges of the natural history of creation* ([Chambers] 1844) was published in 1851 (C. Vogt trans. 1851); his translation of Thomas Henry Huxley's *On our knowledge of the causes of the phenomena of organic nature* (T. H. Huxley 1863b) was published in 1865 (C. Vogt trans. 1865).

From Carl Vogt[1] 17 April 1867

Geneva
17 April | 1867

Dear Sir and Master,

Enclosed is a photograph as you wished.[2] I could almost say with Goethe:

Da hast Du, weil Du's willst, mein garstig Gesicht—
Aber meine Liebe, die siehst Du nicht!
[There you have, as you wished, my ugly face—
But you cannot see my love upon it!][3]

Your photograph reminded me to an astonishing degree of my late friend, Theodore Parker of Boston.[4]

I sent on your letter to Mr. Ricker— he will try to console himself.[5]

Shortly I will have the pleasure of sending you a memoir on the Microcephalics or man-apes, the fruit of the studies for which the war gave me leisure in the past year.[6] I come to the conclusion that this abnormal conformation is an atavism that leads back to the point of departure of the two stocks, man and apes—but that this point of departure is no longer represented in modern creation. I think that I have elucidated the question in a satisfactory manner—as far as the facts at our disposal permit.

As for your new work regarding domesticated animals, you will no doubt be aware of Mr. Nathasius's fine paper on the domestic pig and its metamorphoses.

This is extremely important since he demonstrates in a decisive manner the changes that have occurred in the conformation of the head and especially in the snout following domestication.[7]

I remain sincerely | Your devoted | C. Vogt

Mr. Ch. Darwin

DAR 180: 11

[1] For a transcription of this letter in its original French, see p. 227.

[2] CD asked for Vogt's photograph in his letter to Vogt of 12 April [1867]; it has not been found in the Darwin Archive–CUL.

[3] Vogt slightly modified a quotation from a poem by Johann Wolfgang von Goethe, 'Das garstige Gesicht':

's ist ungefähr das garst'ge Gesicht—
Aber meine Liebe siehst du nicht.
[This is nearly the most ugly face—
But you cannot see my love upon it.]

See Goethe 1988, p. 413.

[4] CD sent his photograph with his letter to Vogt of 12 April [1867]. Parker, an American Unitarian theologian, social reformer, and writer, died in 1860.

[5] Franz Anton Ricker was Vogt's friend and publisher; he had hoped to publish a translation of *Variation* by Vogt. In fact, *Variation* was translated by Julius Victor Carus and published by Christian Friedrich Schweizerbart (Carus trans. 1868).

[6] Vogt refers to his *Mémoire sur les microcéphales* (C. Vogt 1867); there is an inscribed, annotated copy in the Darwin Library–CUL (see *Marginalia* 1: 824–6). CD thanked Vogt for it in his letter to Vogt of 7 August [1867]. Vogt assembled a collection of microcephalic skulls at the meeting of the natural history section of the Institut Genevois in June 1866 (C. Vogt 1867, p. 4), near the time of the Seven Weeks War between Prussia, Austria, and allied German states in June and July 1866 (*EB*).

[7] Vogt refers to *Variation*, and to Hermann Engelhard von Nathusius and Nathusius 1864. There is a heavily annotated copy of Nathusius 1864 in the Darwin Library–CUL (see *Marginalia* 1: 630–5); CD frequently cited this and other works by Nathusius in *Variation*. On changes in the shape of the skull in domestic pigs, see *Variation* 1: 71–3.

From Carl Vogt[1]   23 April 1867

Geneva
23 April | 1867.

Dear Sir and honoured master!

A few days after receiving your letter, which I have shown to Mr. Ricker, I met Colonel Moulinié, whose name you doubtless know.[2] He is one of my oldest pupils, I might add— he has done excellent work on trematode larvae and moreover he has carefully translated my "Lectures on Man" into French.[3] I spoke to him about your new book.[4] "Goodness," he said to me, "that would be something for me! My mother is, as you know, English,[5] I speak English as my mother tongue and I have nothing to do at the moment, so that I could devote myself body and soul to a translation of Mr. Darwin's book that would be better, I hope, than the one Mlle. Royer did for his book on species."[6] Mr. Moulinié has therefore written to our publisher, Mr. Reinwald,[7] 15 rue des S^ts. Pères in Paris and I have the pleasure of enclosing a copy of the letter from my friend Reinwald.

It goes without saying that it would be a real pleasure to me to be able to review Mr. Moulinié's work and to add a preface.[8] If then, you and your publisher have not yet disposed of the rights for the French translation and if the conditions that Mr. Reinwald mentions at N°. 3 are not too onerous, I would beg you to favour my two friends. With regard to questions 1 and 2 put by Mr. Reinwald, I have passed on to him the information contained in your last letter.[9]

As to the conditions relating to the rights of translation, I believe that Mr. R. is right, as I know from my own experience. France is, on the whole, the worst market for scientific books that one can imagine— novels and textbooks for schools and colleges are what sell— scientific books are the worst speculations for French booksellers.

Please be assured of my deep devotion, my dear Sir and master | C Vogt

[Enclosure]

Paris
⟨ ⟩ April 1867

Colonel Moulinié Geneva, 15 Rue du Montblanc
Sir

I had the honour of receiving your letter of the 22nd in which you make me a proposition that suits me perfectly.

Like you, I believe that Mr Darwin's new book, "*Domesticated Animals and Cultivated Plants*",[10] could be usefully translated into French. The book in my ⟨opinion⟩ will meet with success, if it isn't too expensive. We should first establish:

1st. What is the length of this publication

2nd. Whether it will be accompanied by ⟨wood⟩cuts or steel-plate engravings

3^d. Under what conditions the author or publisher ⟨ ⟩ would want to grant you the ⟨French⟩ translation right with the exclusion of all other translations in France ⟨ ⟩ and in Switzerland.

Naturally I will approach ⟨ ⟩ only when you permit me. I even think ⟨ ⟩ better that this request should be made directly ⟨ ⟩ Darwin rather than to Mr Murray[11] (⟨ ⟩ English ⟨ ⟩ often impose conditions that are too onerous ⟨ ⟩ the continent) and also that this demand ⟨ ⟩ by a man of letters than by a bookseller.

If Mr. Vogt were to be in correspondence with the English author, and if he were to allow us to involve him in the negotiation, for example by promising to give our translation a preface, or by some other kind of collaboration, we will almost certainly succeed with the author.

But above all, we must know to what we are committing ourselves. Therefore I await your clarification on the subject and have no doubt that once the preliminaries are settled, we will all agree perfectly about the rest.[12]

Please accept my thanks again for your kind communication and believe that I will be delighted if we ⟨　⟩ renew our relations and make them fruitful. | I have the honour of being, Sir, your humble servant| C Reinwald

DAR 180: 12, DAR 176: 90

[1] For a transcription of this letter in its original French, see pp. 233–4.

[2] Vogt refers to CD's letter to him of 12 April [1867], to his friend the publisher Franz Anton Ricker, and to Jean Jacques Moulinié.

[3] Vogt refers to Moulinié 1856 and Moulinié trans. 1865. Moulinié trans. 1865 was a translation of C. Vogt 1863.

[4] *Variation.*

[5] Moulinié's mother has not been further identified.

[6] Clémence Auguste Royer translated the third English edition of *Origin* (Royer trans. 1862); she published a second French edition, also based on the third English edition, in 1866 (Royer trans. 1866). For CD's opinions of these translations, see *Correspondence* vols. 10 and 13. See also J. Harvey 1997.

[7] Charles-Ferdinand Reinwald.

[8] Vogt did write the preface to Moulinié's translation of *Variation* (Moulinié trans. 1868).

[9] See letter to Carl Vogt, 12 April [1867].

[10] *Variation.*

[11] Reinwald refers to John Murray, CD's publisher.

[12] CD's response has not been found, but see the letter from C.-F. Reinwald, [May 1867].

From Ernst Haeckel[1] 12 May 1867

Jena
12. May 1867

My dear, most admired friend!

Having only been back from my travels a few days, I found your kind letter of 12. April along with the new edition of your epoch-making work.[2] For both I express my most heartfelt thanks. Unfortunately, I can see from your letter that you have not received a long letter that I wrote to you several months ago, and that contained an enclosure for our friend Huxley. Given the great disorder that *dominates* the sloppy Spanish post on the Canary Islands, I certainly cannot say I am surprised.[3] On the other hand, I did receive your first letter, sent to Madeira, although 2 months later, on Lanzarote.[4]

Now, let me first of all dear friend, repeat my *most heartfelt* thanks for the exceedingly friendly reception that I enjoyed in your dear home.[5] That day, which I was privileged to spend with you and to which I had looked forward for so long, will always be unforgettable to me.

I cannot tell you how exceptionally happy you made me by allowing me to visit you, and what immense satisfaction I got from becoming personally acquainted with the naturalist who, as reformer of the theory of descent and discoverer of natural selection, has had a greater influence on the direction of my studies and my life's work than anyone else. Once more I send you and your dear family my most heartfelt and sincere thanks.

I hope your health is improving now, and you can again devote your great knowledge and your thoughtful examination of nature to the progress of our sciences. When on my travels I got to know the wonderful climate of Madeira and the Canary Islands, I often wished that you were with us in order permanently to strengthen your weakened health. I myself returned to my old strength there and recovered completely from the strains inflicted on me by work in the last few years.

It gives me very great satisfaction that you liked my general morphology overall, and *your praise* is the *highest reward* for my efforts. I am especially pleased, however, about your frank, *critical* remarks on the shortcomings of the work, since this shows that you consider me worthy of the *most honest friendship*. Your reproach about the excessive sharpness of my critical attacks and the bitterness of my polemics is certainly justified.[6] My best friends here, especially Prof. Gegenbaur,[7] have also reproached me strongly for this. In my defence, I can only say that when I wrote the book in dismal solitude last summer and winter, I was in an extremely bitter mood and suffering from irritation of the nerves. I also felt too angry about your opponents' impertinent and stupid attacks to let them get away with it unpunished. In any case, my attitude has been very damaging to me personally and I have been attacked no less vehemently from many sides. But I am *completely indifferent* to this, since my personal reputation and the esteem of contemporaries matter little to me. Long may my many enemies attack my work strongly; it just contributes to its spread. Whether they rebuke and malign me is unimportant to me.

But I should be very sad if my too vehement attacks were to damage the good cause for which we both fight. My friend Gegenbaur, whom I love like a brother, is of this opinion and he is very indignant about my literary excesses. You too, most honoured sir, seem to share this fear.

I admit that in this respect I am not quite of your opinion and that I find you unduly worried. Your apprehension would be justified if one could expect an independent, impartial attitude and a fair judgement of both sides from the majority of today's naturalists. But unfortunately this is emphatically *not* the case. Rather, we see that the large majority do not make an independent judgement, do not have a distinct and clear conviction of the claims of truth. And so, I think, it is necessary to speak the truth loud and clear and not spare the opponents' weaknesses.

It seems to me, we are now dealing with a *radical reform* of the whole of science, a reform that you, most honoured Sir, have initiated with your grounding of the theory of descent in mechanical causation. Such a reformation, which has everywhere to fight immense obstacles and prejudices, can never be won with soft words and benevolent persuasion. Rather, energetic attacks and merciless blows are everywhere necessary in demolishing the old edifice of persistent errors. As with all struggles, here too, the bold attacker has great advantages and thus I thought it wiser to attack mercilessly myself than to be assaulted by my malevolent opponents.

When I see how unfairly and wrongly your great work is judged in various ways, how even you personally are maligned, then all my respect for the great audience of naturalists vanishes. Your extraordinary *humility* is seen as *weakness* and your admirable *self-criticism* is interpreted as lack of firm conviction. Of course with good, understanding and thinking men you have only gained by this. But unfortunately these are in the minority. If instead I have approached the issue much more trenchantly and polemically, then I really can hardly believe I have damaged the cause in the long term. In any case, I hope I have contributed something to the diffusion of your great ideas and your pioneering reform, and to have formulated firmly some important and relevant questions and brought them into the public domain, where open discussion can only promote the topic and reveal the truth. Like you, esteemed sir, I trust, first and foremost, the receptive and unprejudiced nature of *younger* scientists, and I have no doubt that we will surely encounter much greater and keener interest among the now flourishing generation. Already in the few days since I have been back, I have noticed this again. I immediately started giving my lectures on the theory of descent once more,[8] and I feel with lively enjoyment the warm support, even more spirited than before, that the ⟨   ⟩ academic youth gives our cause.

By the way, in future I shall heed your benevolent, paternal advice, for which I sincerely thank you, and I shall rein in my all too keen and passionate pen. You would grant me the most honest friendship, dear Sir, if in future you continue to draw my attention so openly to the mistakes in my work.

I will soon send you a printed report on my travels.[9] I was only briefly in Madeira, just 14 days on Tenerife (where I climbed the Pic),[10] but 3 months on *Lanzarote*, where I primarily studied the development of the glorious *Siphonophores* and Atlantic *Radiolaria*.[11] In March I spent 14 days in Morocco and 14 days in Gibraltar. In the first half of April I travelled through Seville, Cordova, *Granada* (!) Madrid; the second half I spent studying the remarkable Exposition universelle in Paris.[12] My health was splendid all the time and I am very satisfied with the result of the journey.

I may, perhaps, come to England some time this year (in the autumn), and I would be very pleased if I could see you again in better health. In the meantime, dear Sir, please accept again my warmest thanks for your sincere friendship, and the request that it may continue in future.

With sincerest regards to M^rs and Miss Darwin,[13] with all my heart, your obedient servant | Ernst Haeckel

DAR 166: 44

[1] For a transcription of this letter in its original German, see pp. 254–7.

[2] Haeckel had spent from November 1866 to March 1867 travelling and doing research on Tenerife and Lanzarote; he also visited Morocco, Portugal, and the Spanish mainland (see Haeckel 1867). Haeckel also refers to the letter from CD of 12 April [1867], and the fourth edition of *Origin*. Haeckel's name appears on the presentation list for this book (see *Correspondence* vol. 14, Appendix IV).

[3] Neither the letter to CD nor the letter to Thomas Henry Huxley has been found. For transcripts of the extant correspondence between Haeckel and Huxley, see Uschmann and Jahn 1959–60. Haeckel wrote to Huxley on 12 May 1867 (Uschmann and Jahn 1959–60, pp. 11–12; the letter is at the Imperial College of Science and Technology), mentioning the enclosure to the letter to CD that went astray. The Canary Islands, which include Tenerife and Lanzarote, are governed by Spain (*EB*).

[4] Haeckel presumably refers to the letter from CD of 8 January 1867.

[5] Haeckel had visited CD at Down House in October 1866 (see *Correspondence* vol. 14). On their meeting, see also A. Desmond and Moore 1991, pp. 539–40.

[6] See letter to Ernst Haeckel, 12 April [1867] and n. 6.

[7] Carl Gegenbaur.

[8] Haeckel gave lectures on CD's theory of descent both to students at Jena University and to the general public, but these seem to have been given in the winter (see Uschmann 1959, p. 45; see also *Correspondence* vol. 14, letter from Ernst Haeckel, 11 January 1866 and n. 14). His lectures given in the winter of 1867 to 1868 were published as Haeckel 1868 (see Uschmann 1959, p. 45), which was translated as Haeckel 1876.

[9] Haeckel sent a preliminary report on his travels to CD with his letter of 28 June 1867; this report has not been found in the Darwin Archive–CUL. A report was also published on 12 September 1867 in the *Jenaische Zeitschrift für Medicin und Naturwissenschaft* (Haeckel 1867).

[10] The highest point of the Pico de Teide on Tenerife is 3770 m (*Times atlas*). For Haeckel's account of the climb, see Uschmann 1984, pp. 92–6.

[11] Haeckel published his research on siphonophores in 1869 (Haeckel 1869).

[12] See also letter from J. D. Hooker, 13 April 1867.

[13] Emma and Henrietta Emma Darwin.

From Camille Dareste[1] 19 May 1867

Lille;
19 May 1867.

Sir

Perhaps you have not forgotten that you received, some five years ago, a paper from a Professor of natural history in Lille on the artificial creation of monstrosities. The letter that you wrote to him, which he preserves carefully, showed him that you really wished to understand his work and read it with some interest.[2] This letter was for him, or rather for me since I refer to myself, the greatest of all encouragements; for I regret to say that my work is at the moment very little appreciated in France. Only one man understood it well and that was Is. Geoffroy Saint-Hilaire. Since his death there has been alas, only a very faint interest in it.[3]

Since I last wrote to you, I have continued my work with great perseverance, and I have been able to observe the embryological development of almost every stage of monstrosity. This work will, I hope, be completed in a year, and I will then be able to have it published.[4] Up to now, I have had to satisfy myself with making the major results known in the proceedings of the Académie des Sciences in Paris;[5] and since these notes are not published as offprints, I have not been able to send them to you. When this preliminary work is completed, I will be able, I hope, to discover the conditions under which monsters develop, and thus produce this or that anomaly at will. I already possess a certain number of preliminary indications on this subject; and I have every reason to believe that I can follow them up. I will then keep you informed about everything I succeed with in a subject that is too close to your own studies not to interest you a little.[6]

Meanwhile, I am sending you a little work that is directly related to a question that you yourself have observed. This concerns the *niata* cattle of South America.[7]

I had occasion to study one of these animals which was born spontaneously in Lille from an ordinary Flemish cow and which exactly reproduced the characteristics of the *niata*. Had this animal survived to the age of puberty, it would have passed on to its descendants its distinguishing characteristics. This fact accords perfectly with your work on species; I therefore decided to send it to you.[8]

And speaking of this, I must tell you that the conclusions that I wished to draw relative to the formation of the *niata* race have been rather vigorously opposed in France. There are some people so blinded by false theories that they do not want to admit the existence of the *niãta* race and reject your testimony completely.[9] I would be happy if you could give it once again in the clearest possible way.[10]

Equally, I had occasion to observe the production of two individuals of the Du Saxe poultry breed that possessed cephalic and cerebral characters similar to Polish hens. A hernia of the brain that later becomes enclosed in a bony envelope results from the ossification of the fontanelle.[11]

Please excuse me, Sir, for troubling you by speaking of my work: but I believe that it is linked to yours by very close bonds. I am always hoping that my research will lead me to some new discovery regarding the formation of species and that my work might one day take its place in Science, after your book, although at a rather great distance (*longo sed proximus intervallo*).[12]

That is one of the thoughts that sustains me in the often thankless career that I have undertaken.

Please accept the expression of my admiration and respect. | Camille Dareste

Professor at the Faculté des Sciences of Lille

DAR 162: 43

[1] For a transcription of this letter in its original French, see pp. 266–8.

[2] Dareste wrote to CD on 8 February 1863, enclosing a copy of Dareste 1863; CD replied in a letter dated 16 February [1863] (see *Correspondence* vol. 11). There is a lightly annotated copy of Dareste 1863, in which Dareste provides an account of monstrosities artificially induced in chicken eggs, in the Darwin Pamphlet Collection–CUL.

[3] Isidore Geoffroy Saint-Hilaire died in 1861; he had worked on teratology, like his father, Etienne Geoffroy Saint-Hilaire, whose work influenced Dareste (Tort 1996).

[4] Dareste published a book summarising his research in teratology in 1877 (Dareste 1877).

[5] Dareste published articles on teratology regularly in the *Comptes rendus hebdomadaires des séances de l'Académie des sciences*.

[6] CD cited Dareste a number of times on teratology in the second volume of *Variation*, and commended his work in *Descent* 2: 338.

[7] Niata: 'An abnormally small variety of cattle, found in South America' (*OED*). CD discussed niata cattle in *Journal of researches* 2d ed., pp. 145–6.

[8] There is a copy of Dareste 1867, *Rapport sur un veau monstreaux*, in the Darwin Pamphlet Collection–CUL. Dareste suggested that monstrous forms, under the right conditions, could originate new races.

[9] Dareste had been involved in an exchange of papers with André Sanson in the *Comptes rendus hebdomadaires des séances de l'Académie des sciences*. Sanson denied the existence of a race of niata cattle in South America; Dareste had cited CD as one of his authorities. See *Comptes rendus* 64 (1867): 423–6, 669–70, 743–5, 822–4, 1101–3. See also *Correspondence* vol. 11, letters from Armand de Quatrefages, [28 March –] 11 April 1863 and n. 4, and 19 May [1863] and n. 4.

[10] CD discussed niata cattle in *Variation* 1: 89–91; see letter to Camille Dareste, 23 May 1867.

[11] See Dareste's first paper in *Comptes rendus* (see n. 9, above): 'Mémoire sur le mode de production de certaines races d'animaux domestiques'. CD discussed the skulls of Polish and other crested breeds of fowl in *Variation* 1: 262–5.

[12] *Longo sed proximus intervallo*: 'The next, but after a long interval. A poor second' (H. P. Jones ed. 1900).

From Albert Gaudry[1]   22 May 1867

Sir,

Permit me to send you a work on an ancient reptile that seems intermediate between true Triassic reptiles and Devonian fish.[2] I take advantage of this opportunity to thank you for the letter you were kind enough to send me a few months ago.[3] I was very honoured that a naturalist as eminent as you should attach any interest to my research on the transitions of fossil animals. It is not without reason that I am regarded as one of your admirers, since, although I do not share all your views on the explanation of the transformations of living things, at least these transformations seem to me every day more probable, and your book on the origin of species contributed mightily to throw light on them.[4] I do not know of a finer study than that of the descent of living beings that occurs during ages of indefinite extent, and I will endeavour to give evidence for this descent drawn from palaeontology. As far as explanations of the transformations are concerned, I avoid addressing them, because such a difficult subject can be approached only by a naturalist such as you, who has thorough experience and extensive knowledge. I will also admit that by studying embryogeny and by seeing that there are indeed causes of which God alone knows the secret, I think that both in the evolution of species and the embryogeny of individuals, there are facts that the greatest geniuses will not be able to discover.

Your letter found me on Mt Léberon, near Cucuron (Vaucluse) in a very rich fossil bed in which there is found the same fauna as in Pikermi. I have not yet cleaned up the bones that I have brought back, but what I can already see makes me think that the animals of Mt Léberon present such small differences from those of Greece that it is difficult not to admit some links of a real kinship, although they are not of identical parentage.[5]

I would be really pleased, dear Sir, if your health and your great works allowed you the opportunity to visit Paris; I would show you in the Jardin des Plantes some fossils which seem to me curiously intermediate.

Please accept the expression of my most respectful sentiments. | Albert Gaudry

12 Rue Taranne. Paris.
22 May 1867.

DAR 165: 15

[1] For a transcription of this letter in its original French, see p. 274.

[2] Gaudry refers to his 'Mémoire sur le reptile (*Pleuracanthus Frossardi*) découvert par M. Frossard à Muse (Saône-et-Loire)' (Gaudry 1867a). There is an inscribed and scored offprint of this article in the Darwin Pamphlet Collection–CUL. Gaudry also enclosed a second paper; see letter to Albert Gaudry, 27 May [1867] and n. 2. The name *Pleuracanthus* had already been assigned to a genus of beetles: most members of *Pleuracanthus* (freshwater sharks) have been reassigned to the genus *Xenacanthus*.

[3] See *Correspondence* vol. 14, letter to Albert Gaudry, 17 September [1866].

[4] After excavations at Pikermi in Attica in 1855 and 1860, Gaudry had reconstructed several skeletons of new species that appeared to be intermediate between species already known. He believed that such intermediates were evidence for transmutation, but differed from CD in believing that these transmutations were the result of continuing creation by God. See Gaudry 1862–7, 1: 365–70. Gaudry had sent CD an extract from this work in 1866 (see *Correspondence* vol. 14, letter to Albert Gaudry, 17 September [1866]).

[5] Gaudry published on the fossils of Mont Léberon (now Montagne de Lubéron) in the south of France in Gaudry *et al.* 1873. He made comparisons with findings at Pikermi and other sites on pp. 75–98.

From Ernst Haeckel[1] 28 June 1867

Jena
28. June 1867.

My dear, highly honoured friend!

I begin my letter this time with some personal news that I know will interest you very much, given your friendly feelings towards me. I have become engaged again and I hope to have found in my bride a true and good feminine heart that will rebuild the happiness of my life. My bride is *Agnes Huschke*, the youngest daughter of the late professor of anatomy here.[2] She is a simple and natural, very sensible and cheerful girl, who I hope will, in many respects, make up for the absence of my late altogether excellent and unforgettable wife.[3] We will marry in mid-August and then travel together to the Swiss Alps and Northern Italy. Thus, I will not return to England in the autumn, as I had hoped, and must delay the joy of seeing you until next year.

I enclose a preliminary report of my trip to the Canaries. I am busy writing a more detailed description of the trip.[4] Then I will work on the siphonophores that I examined on Lanzarote.[5] At present I am occasionally also again observing the strange *Protamoeba primitiva* that I described on p. 133 in vol. I of my Morphology (in the footnote).[6] These simplest of all organisms, homogeneous, simple, contractile, living lumps of protein, lacking any differentiation of matter, without any organs or definite form, are indeed extremely odd, and I believe that they facilitate the hypothesis of autogeny considerably.[7] Now such very simple, still completely un-differentiated organisms, in which matter has never yet constituted a definite form, must be regarded as the primitive, autogenously arising, original form of all other organisms. The original Monera[8] did perhaps differ, but since the differences in their chemical constitution could only have been very slight, I think the question, apparently so important, of whether to postulate one or several original forms of the different organic phyla loses much of its fundamental meaning. In fact it seems to me basically a matter of indifference.

With this account of the journey you will also receive an honorary diploma from the K. K. Zoologisch-botanischen Gesellschaft of Vienna.[9] Its secretary sent it to me with the request to forward it to you, and at the same time he asks you to notify the Zool. bot. Gesellschaft in Vienna directly to formally confirm your receipt of the diploma. The very same researcher (cand. jur. Aug. Kanitz,[10] Blutgasse 3, Vienna), who has worked very zealously for our cause in *Vienna* and especially in *Hungary*, sends you through me his photograph[11] and humbly requests that you send him *your photograph* with your own signature. You could send it either direct or through me.

I hope, highly honoured friend, your next letter will inform me that your health is improving with the arrival of the beautiful summer. I imagine that your country estate, whose charm I remember with pleasure, constitutes a quite healthy and refreshing summer retreat, especially on account of its tranquility and peacefulness in contrast with noisy London.

Please give my most humble regards to your honoured wife. I remain most respectfully | yours sincerely | Ernst Haeckel

DAR 166: 45

[1] For a transcription of this letter in its original German, see pp. 308–10.
[2] The reference is to Emil Huschke, who died in 1858.
[3] Haeckel's first wife, Anna Sethe, died in 1864.
[4] The enclosure has not been found. Haeckel published a report on the trip as well as an account of climbing Mount Teide in Tenerife (Haeckel 1867 and 1870).
[5] Haeckel's study of the developmental history of the Siphonophorae attempted to show the colonial nature of these animals (Haeckel 1869). Lanzarote is one of the Canary Islands.
[6] See Haeckel 1866, 1: 133–4 n. Haeckel had written earlier about a similar primitive organism (see *Correspondence* vol. 13, letter from Ernst Haeckel, 11 November 1865 and nn. 11 and 12).
[7] On Haeckel's hypothesis of autogeny, or spontaneous generation, see Haeckel 1866, 1: 179–90. For more on Victorian views on spontaneous generation, see Strick 2000.
[8] For Haeckel's definition of 'Monera' (moner), see Haeckel 1866, 1: 135. According to Haeckel, they were the primordial, anucleate ancestors of all animals (see S. J. Gould 1977, p. 170).
[9] See Appendix III.
[10] The reference is to August Kanitz. 'Cand. jur.': Candidatus juris (Latin); a law degree.
[11] Kanitz's photograph has not been found.

From Fritz Müller[1] 17 July 1867

Itajahy, Santa Catharina,
17. July 1867.

. . . . . I have to apologise for not thanking you earlier for kindly sending me Dr. Hildebrand's book on fertilisation. I have read it with great interest. I have also received Dr. Hooker's extremely interesting publication "Lecture on insular floras".[2]

The family Rubiaceae seems to be very rich in dimorphic plants; although now in winter many fewer plants are flowering than in any other season, I still observed yet another three dimorphic species: 1) a delicate Hedyotis somewhat similar to Erica, 2) a second species of Suteria with brown calyx and yellow corolla, which is said to be the favourite food of tapirs, 3) the Manettia bicolor, which is very common here.[3] I know another two species of Manettia with red flowers, which I suspect are also dimorphic, although all the plants that I have observed so far were long-styled.

I have also found a trimorphic white Oxalis here; I have seen various short-styled and medium-styled but no long-styled yet; a single plant had the stamens of the long-styled one in conjunction with the styles of the medium-styled ones, so that the style was as long as the longest stamens.[4]

Before leaving Desterro,[5] I examined most of the capsules that resulted from my crossing experiments with orchids; most had not yet matured.[6] Almost all seeds were apparently good in the capsules of:

> Cattleya intermedia ♀ and Cattl. elegans[7] ♂
> Epidendrum vesicatum ♀ and Ep. raniferum ♂
>   "   "   ♀ and Ep. glumaceum[8] ♂
>   "   raniferum ♀ and Ep. cinnabarinum ♂
>   "   "   ♀ and Brassavola fragans Leon[9] ♂!
> Cirrhaea saccata (?) ♀ and Cirrhaea dependens (?)[10] ♂
> Oncidium micropogon ♀ and Gomeza[11] sp. ♂
> Gomeza sp. ♀ and Oncidium micropogon ♂

There were many more good than bad seeds in the capsules of:

> Cattleya elatior var. Leopoldi ♀ and Laelia purpurata ♂
>   "   intermedia ♀ and Cattl. elatior var. Russeliana ♂.[12]

Approximately half the seeds were apparently good in the capsule of:

> Cattleya elatior ♀ fertilised with Cattl. intermedia ♂.

At least $\frac{9}{10}$ of the seeds were bad in the capsules of:

> Cattleya elatior ♀ crossed with Cattl. elegans ♂ and
> Oncidium (from Theresopolis)[13] ♀ crossed with Sigmatostalix ♂.

An extraordinarily small number of apparently good seeds were found in the capsules of:

> Cattleya elatior var. Russeliana ♀ and Epidendrum cinnabarinum ♂
> Notylia sp. ♀ and Ornithocephalus sp. ♂ (!)[14]

All seeds were bad in the capsules of:

> Cattleya elatior ♀ and Epidendron[15] Schomburgkii (?) ♂
> Epidendron vesicatum ♀ and Epid. Schomburgkii (?) ♂
>   "   "   ♀ and Epid. cinnabarinum ♂
> Sigmatostalix sp. ♀ and Ornithocephalus sp. ♂.

I found a self-fertilised capsule of Ornithocephalus was much smaller and contained many more bad seeds than another capsule fertilised with the pollen from a different plant. . . . .

With regard to your query about sexual differences in lower animals, one ought perhaps to mention one of our amphipods, Brachyscelus diversicolor F.M.[16] The male of this species, which lives on some of our larger acalephs (Rhizostoma cruciatum Less. = Rhacopilus cruciatus and cyanolobatus Agass. and Chrysaora Blossevillei Less. = Lobrococis Blossevillei Agass.),[17] differs not only in the antennae, the first pair of which is very thick and covered in olfactory hair whereas the second is extraordinarily long (this second pair is absent on the ♀ and young ♂)—but also in colour. The female is usually milky white or dull yellowish, the male is dark reddish-brown or blackish. The species has unusually large eyes, as indeed most Hyperina have, and so it is not unlikely that the female is attracted by the colour of the male.

In this connection I would like to add that there is a fish in the sea by Santa Catharina that produces very melodious sounds that may also serve to attract the opposite sex.[18] The sounds are like the ringing of distant church bells. I have only heard them on quiet evenings when these maritime musicians swam by a rock close to the coast, but I never saw the fish.

To come back once more to the Epidendron with fertile lateral anthers, in which the pollen tubes are emitted even before the opening of the flower: it seems that the great diversity of the lateral anthers observed in one case is favourable to the view that they have only been acquired recently as the result of a chance reversion to a long lost character.[19]

With the greatest respect, believe me, dear Sir, that I am sincerely yours | Fritz Müller.

Möller ed. 1915–21, 2: 130–1

[1] For a transcription of this letter in the German of its published source, see p. 322–4. All Fritz Müller's letters to CD were written in English (see Möller ed. 1915–21, 2: 72 n.); most of them have not been found. Many of the letters were later sent by Francis Darwin to Möller, who translated them into German for his *Fritz Müller: Werke, Briefe und Leben* (Möller ed. 1915–21). Möller also found drafts of some Müller letters among Fritz Müller's papers and included these in their original English form (*ibid.*, 2: 72 n.). Where the original English versions are missing, the published versions, usually appearing in German translation, have been used.

[2] The references are to Hildebrand 1867a and J. D. Hooker 1866a (see letters to Fritz Müller, 25 March [1867] and 22 April [1867]).

[3] In *Forms of flowers*, pp. 131–3, 135, CD reported Müller's information on heterostyly in *Hedyotis*, *Suteria*, and *Manettia bicolor*. *Suteria* is now subsumed within the genus *Psychotria*; *M. bicolor* is a synonym of *M. luteo-rubra*.

[4] CD mentioned Müller's observation of a white-flowered trimorphic *Oxalis* in *Forms of flowers*, p. 180.

[5] Müller had lived in Destêrro (now Florianópolis) from 1856, but had just returned to live on the Itajahy (now Itajaí) river near Blumenau, where he bought a new property in September 1867. Müller typically gave his address simply as 'Itajahy' until the mid-1870s, when he began to add 'Blumenau', but only consistently gave his address as 'Blumenau' from 1878 (Möller ed. 1915–21, 3: 94; see also West 2003, pp. 149–50).

[6] Müller had begun a series of crossing experiments with orchids earlier in the year (see letter from Fritz Müller, 1 April 1867).

[7] *Cattleya elegans* is a synonym of *Laelia elegans*.

[8] *Epidendrum glumaceum* is a synonym of *Prosthechea glumacea* (see Higgins 1997, p. 378).

[9] *Brassavola fragans* (named by Charles Antoine Lemaire; Möller mistranscribed 'Lem' as 'Leon') is a synonym of *B. tuberculata*.

[10] *Cirrhaea dependens* is a synonym of *C. viridipurpurea*.

[11] *Gomeza* is a synonym of *Gomesia*.

[12] *Cattleya elatior* is a synonym of *C. guttata*; the varieties mentioned by Müller are now *C. guttata* var. *leopoldii* and *C. guttata* var. *russelliana*.

[13] Now Queçaba (West 2003).

[14] Müller probably found it surprising that a cross between two fairly widely separated species was possible. In orchids, natural hybrids are usually confined to intergeneric crosses within a subtribe, making a cross at this level rare (see Dressler 1981, pp. 147–8). *Notylia* is now a member of the orchid tribe Cymbidieae, while *Ornithocephalus* belongs to the tribe Maxillariae.

[15] 'Epidendron': a mistranscription of *Epidendrum*.

[16] For CD's query, see the letter to Fritz Müller, 22 February [1867]. The name *Brachyscelus diversicolor* was never published by Müller. For a description of all currently recognised species within the genus, see Vinogradov *et al.* 1996, pp. 488–96. Müller had discussed sexual differences in the antennae of *Brachyscelus* in F. Müller 1864, p. 53 (see also Dallas trans. 1869, pp. 78–9; CD annotated this section in his copy; see *Marginalia* 1: 609).

[17] *Rhizostoma cruciata* and *Rhacophilus cruciatus* are both synonyms of *Catostylus cruciatus* (Kramp 1961, p. 370). *Chrysaora blossevillei* is now considered to be a doubtful species (Kramp 1961, p. 324). Two species of *Chrysaora* have been recorded in Brazil, *C. lactea* and *C. plocamia* (Migotto *et al.* 2002, p. 23). It is likely, because of its size, that the species Müller refers to is *C. plocamia*.

[18] The fish has not been identified, but of thirty-four possible sound-producing species in the families Batrachoididae and Sciaenidae in Brazil, the Bocon toadfish, *Amphichthys cryptocentrus*, is the only reef-associated species known to occur at depths of a metre or less (Joseph Luczkovich, personal communication). CD mentioned musical sounds made by fish in *Descent* 2: 23, 331 (see also Pauly 2004, pp. 193–4).

[19] Müller first mentioned this orchid in a letter of 1 June 1866 that has not been found (see *Correspondence* vol. 14, letter to Fritz Müller, 23 August 1866). He gave an account of it in letters to Max Johann Sigismund Schultze (2 June 1866) and Hermann Müller (1 July 1866; the letters are reproduced in Möller ed. 1915–21, 2: 83–4, 86–9). Müller described it as a species of the orchid tribe Epidendreae with three fertile anthers, two in the outer whorl and one in the inner whorl (see Möller ed. 1915–21, 2: 87–8). He later published his observations in F. Müller 1868a, 1869, and 1870; CD added a reference to Müller's work in *Orchids* 2d ed., p. 148.

From Edouard Bornet[1] [before 20 August 1867][2]

Antibes | Alpes Maritimes

Dear Sir,

We are now at the time of year during which M. Jordan indicated that *Draba verna* seeds should be sown. I take the liberty of sending you ten different kinds that we cultivated last winter in Antibes and that are interesting to observe.[3] Their cultivation is very easy and requires virtually no care.

M. Jordan recommends sowing them "towards the end of August or the beginning of September, in pots that are planted almost level with the soil and that are kept rather distant from each other, until the seeds are well grown. They will at first develop into beautiful rosettes during the autumn and will then produce ⟨ ⟩ flowers, from the first fine days of spring."

The seedlings of *Papaver* have been very adversely affected ⟨   ⟩ the bad weather and have done very badly. Only one of the crosses that I made has come out well, that is the hybrid of *Pap. dubium* (*modestum, Jord.*) L. fertilised by the pollen of *Pap. Rhœas* (*cruciatum, Jord.*).[4] The plant was extraordinarily vigorous. It had the appearance and the fruits of *P. dubium* but the flowers were much larger, redder, and often marked with a large black cross. The stigmatic rays were violet and strongly papillate as in *P. Rhœas.* This hybrid was absolutely sterile. The anthers did not contain any pollen. The ovules, which I examined with care, appeared to me to be well formed.

Last year I received the memoir that you had the kindness to send me and I beg you to accept my most sincere thanks.[5]

Please accept, Sir, the assurance of my respectful regards | Ed. Bornet

DAR 160: 256

[1] For a transcription of this letter in its original French, see p. 352.

[2] The year is established by Bornet's reference to receiving the memoir sent by CD (see n. 5, below). The date is established by the relationship between this letter and the letter to Edouard Bornet, 20 August [1867].

[3] Alexis Jordan described twenty forms of *Draba verna* (now *Erophila verna*), classifying these as separate species with the genus name *Erophila*, in the first part of *Icones ad floram Europæ novo fundamento instaurandam spectantes* (Jordan and Fourreau 1866–1903). Bornet's experimental work was done in the garden of Gustave Alphonse Thuret at Antibes in the south of France.

[4] *Papaver modestum* is described in Jordan 1851, p. 215; *P. cruciatum* is described in Jordan 1860, pp. 465–6. Jordan maintained that *P. rhoeas* as described by other authors corresponded to approximately twenty separate species by his account (Jordan 1860, p. 467).

[5] CD had sent a copy of 'Climbing plants' to Bornet by the same post as his letter to Bornet of 1 December 1866 (see *Correspondence* vol. 14).

From Federico Delpino[1]   5 September 1867

Distinguished Sir

Full of admiration for the great talent of your most distinguished honour, I venture to offer you two of my publications as a token of my respect.[2]

If it pleases you to glance at them, you will notice at once how much influence your admirable works on the variability of species and on the fertilisation of orchids had on me.[3]

Concerning the first work, while I accept entirely *the model of variation* that has been so ingeniously uncovered and expounded by your honour, it seems to me that it needs to be interpreted in a spiritualistic manner, which (unless I am mistaken) would dispose of all the objections that have been raised against the selfsame model.[4]

Concerning the second work, you will see that I have had the good luck of finding that the law of mixed marriages mediated by insects occurs in equal proportions in the asclepiads and other plant families as in orchids.[5]

I beg your pardon for the liberty I have taken, and have the honour of professing my deep respect

Your true admirer and servant | Delpino Federico

Italy | Genoa | Chiavari 5 September 1867

DAR 162: 142

[1] For a transcription of this letter in its original Italian, see p. 365.

[2] Delpino sent his paper on Friedrich Hildebrand's essay on the distribution of the sexes in plants and the prevention of self-fertilisation (Delpino 1867a) and another paper on the mechanisms to ensure fertilisation in seed plants (Delpino 1867b). CD's annotated copies (together with three additional pages of notes affixed to Delpino 1867a) are in the Darwin Pamphlet Collection–CUL.

[3] Delpino refers to *Origin* and *Orchids*.

[4] Delpino refers to Delpino 1867a. On Delpino's philosophical position and his attempt to reconcile Darwinian transmutation theory with his own teleological approach, based on Kantian nature philosophy, see Pancaldi 1991, pp. 107–36.

[5] See Delpino 1867b. Delpino alludes to CD's statement: 'it is apparently a universal law of nature that organic beings require an occasional cross with another individual' (*Orchids*, p. 1; see also *Origin*, p. 97). Delpino had published an earlier article on the mechanism to ensure fertilisation in asclepiads, with some considerations on final causes and CD's theory of the origin of species (Delpino 1865).

From F. A. Hagenauer to Ferdinand von Mueller[1]   [12 September 1867][2]

Aborigena Mission Station | Lake Wellington, Gippsland

My dear Sir

I apologise that I have not been able to satisfy your request with regard to Mr. Darwin's questions sooner; I am sending the replies today.[3] I hope that they are suitable.

Should you ever be in the vicinity, I would certainly be very glad to see you.

With best wishes | Yours sincerely | F. A. Hagenauer

DAR 166: 80

[1] For a transcription of this portion of the letter in its original German, as well as the English part, which includes Hagenauer's responses to CD's queries about expression, see p. 369.

[2] The date is established by an annotation, probably by Mueller, at the top of the letter, which reads: 'date of Poststamp 12/9, received 19 Septb 67'.

[3] CD sent queries on expression to Mueller to pass on to contacts who could provide answers pertaining to indigenous Australians (see letter to Ferdinand von Mueller, 28 February [1867]; see the enclosure for the list of queries). It is possible that this letter was sent by Mueller with another letter containing answers to CD's queries (see letter from Ferdinand von Mueller, 8 October 1867).

From Ferdinand von Mueller 8 October 1867

[Enclosure][1]

Mansfield

12 September 67

Dear Doctor.

Saturday the 7th I arrived here on my tour to Mount Buller;[2] on Sunday I had the pleasure of seeing clearly and distinctly this majestic mountain covered with snow; we have unfortunately had rain every day since and these alpine mountains have been enveloped in clouds. I plan to stay in this area for a couple of weeks, to climb Mt Buller and Timbertop[3] and to photograph some views, should a good opportunity arise. The reason for these lines is to thank you warmly for your kind support and advice in the planning of this tour. When in the Mount Buller Range, I will not fail to direct my attention to the lichens and fungi, and if I can be of service to your department[4] in any other way during my stay here, please let me know. In response to Mr Ch[s] Darwin's queries that I forwarded to the missionary station in Framlingham, just 8 days ago I received the answer that the indigenous peoples there are too much in contact with the poorer (*low*) class of Europeans, and he therefore cannot answer the queries.[5] Mrs Green (Green) of Coranderrk Station[6] has written the following to me today, as per your instructions:

DAR 181: 11

[1] For the original German version of this part of Charles Walter's letter to Mueller, along with the rest of the letter, written in English, see pp. 390–1.
[2] Mount Buller, Victoria, Australia, is about 150 miles north-east of Melbourne.
[3] Timbertop is a mountain adjacent to Mount Buller.
[4] Mueller was director of the Botanical Gardens, Melbourne.
[5] Walter refers to CD's queries on expression (see n. 1, above). Framlingham, a settlement near Warr-nambool, Victoria, became a mission in 1865, when the Central Board for the Protection of Aborigines gave permission to the Church of England Mission to the Aborigines of Victoria to occupy the site (*Encyclopaedia of Aboriginal Australia*).
[6] Mary Green was the wife of the superintendent of Coranderrk Aboriginal Station near Healesville, Victoria (Barwick 1972, p. 23; *Encyclopaedia of Aboriginal Australia*). CD acknowledged her contribution in *Expression*, p. 20.

From Fritz Müller[1] [8 October 1867][2]

Itajahy

. . . . . Would you ever have thought that the Amarantaceae[3] family could also present conspicuous seeds that attract the attention of birds? Well, this is the case with a climbing Chamissoa in our flora. The black seeds are almost completely surrounded by a white arillus and remain attached to the base of the capsule, the upper half of which falls off ("utriculus circumscissus")[4] as with Anagallis.[5] According to Endlicher's description it seems that in other species of that genus the arillus is much smaller ("arillo brevi albo, umbilicum lateraliter cingente"),[6] and so these forms constitute a connecting link between our species and another Amarantaceae (Celosia?), in which the small, black, and shiny seeds have no arillus at all, and although not very sizeable, remain firmly attached to the base of the capsule, which forms a hemispherical cup when its upper half has fallen off. The fact that large seeds stick to the open hull also occurs in monocotyledons. I observed this in a Hedychium (coronarium?—not indigenous here)[7] and in a Marantaceae. . . . .

Incomplete

Möller ed. 1915–21, 2: 132

[1] For a transcription of this letter in the German of its published source, see p. 392. For an account of the reconstruction of Fritz Müller's letters to CD, see the letter from Fritz Müller, 17 July 1867, n. 1.

[2] The date is established by the relationship between this letter and the letters to Fritz Müller, 15 August [1867] and 30 January [1868] (*Correspondence* vol. 16). Alfred Möller dated the letter fragment 'September 1867' when he published it in German translation (Möller ed. 1915–21, 2: 132). CD thanked Müller for information 'about conspicuous seeds' in his letter of 30 January [1868], and also mentioned receiving Müller's answers on expression written on 5 October. It is likely that Müller answered the queries on expression on a separate sheet, dated 5 October 1867, which was included with the letter written on 8 October 1867. Müller had written at the top of CD's letter of 15 August [1867], 'Received Octobr. 5.— | Answered Octobr. 8.—'

[3] The modern spelling is 'Amaranthaceae'.

[4] 'Arillus': aril; 'utriculus circumscissus': utricle circumscissile (Latin; the characterisation comes from Endlicher 1836–42, s.v. *Chamissoa*).

[5] The genus *Anagallis* belongs to the family Primulaceae.

[6] Arillo brevi albo umbilicum lateraliter cingente: with a short white aril surrounding the hilum laterally (Latin). Müller refers to Stephan Ladislaus Endlicher and Endlicher 1836–42 (s.v. *Chamissoa*).

[7] *Hedychium coronarium*, though native to tropical Asia, is widely naturalised in tropical America (Mabberley 1997).

From Eduard Koch[1] 11 December 1867

Honoured Sir!

Having your very honoured letter of 22 Nov. to hand, I am pleased to have *two* copies of the first volume sent to you today.[2]

We will take care of distributing presentation copies as directed and will also send out the second volume in due course.[3]

You have my sincere thanks for the assurance of translation rights for your later works, and I will always strive to earn your total satisfaction by producing the work to a high standard.[4]

Prof. Carus has already started on the translation of the second volume and I hope to be able to send the completed volume to you in March or April *1868*.[5]

Please inform me at your earliest convenience when the work is to be published in England;[6] due to peculiar German circumstances, I must necessarily publish it this year,[7] since publication after New Year would mean that a general idea of sales would only be possible in 1869. Your earliest reply would therefore please me very much.

In closing, please accept the assurance of my highest respect and admiration |
E. Schweizerbart'sche Verlagshdg | Eduard Koch

Stuttgart 11 Dec. 1867

DAR 169: 42

[1] For a transcription of this letter in its original German, see p. 479.

[2] In his letter to E. Schweizerbart'sche Verlagsbuchhandlung of 22 November [1867], CD had requested only one copy of the German edition of *Variation* (Carus trans. 1868) for himself.

[3] CD had enclosed a list of recipients of presentation copies of Carus trans. 1868 in his letter to E. Schweizerbart'sche Verlagsbuchhandlung, 22 November [1867]. This list was supplemented with the names of Carl Gegenbauer, Hermann Müller, and Rudolf Carl Virchow in the version retained by CD (DAR 210.11: 33).

[4] Koch evidently misunderstood CD's letter of 22 November [1867]. CD had stated that Koch's firm had translation rights for his 'present & previous works'.

[5] Julius Victor Carus was translating *Variation* into German; the translation was published in two volumes.

[6] *Variation* was published on 30 January 1868 (Freeman 1967).

[7] The year 1868 appears on the title pages of both volumes of Carus trans. 1868.

APPENDIX II
Chronology 1867

This appendix contains a transcription of Darwin's 'Journal' for the year 1867. Darwin commenced his 'Journal' in August 1838 and continued to maintain it until December 1881. In this small notebook, measuring 3 inches by $4\frac{1}{2}$ inches, Darwin recorded the periods he was away from home, the progress and publication of his work, and important events in his family life.

The version published by Sir Gavin de Beer as 'Darwin's Journal' (de Beer ed. 1959) was edited before the original 'Journal' had been found and relied upon a transcription made by an unknown copyist. The original, now in the Darwin Archive in Cambridge University Library (DAR 158), reveals that the copyist did not clearly distinguish between the various types of entries it contains and that the transcription made was incomplete.

From 1845 onward, Darwin recorded all that pertained to his work (including his illnesses, since these accounted for time lost from work) on the left-hand pages of the 'Journal', while the periods he was away from home, and family events, were noted on the right-hand pages. In order to show clearly Darwin's deliberate separation of the types of entries he made in his 'Journal', the transcription has the left- and right-hand pages labelled.

All alterations, interlineations, additions, and the use of a different ink or pencil have been noted. In addition, the editors have inserted additional information relevant to Darwin's correspondence throughout this transcription of the 'Journal' for 1867. These interpolations are enclosed in square brackets to distinguish them from Darwin's own entries, the source of the information being given in the footnotes.

[Left]

1867

Last chapt of "Variation under Domestication" & begining of Man Essay—[1] First proof arrived March 1st.— Revises finished Nov. 15th.

I began this Book beginning of 1860 (& then had some M.S), but owing to interruptions from my[2] illness, & illness of children: from various editions of Origin & papers, especially Orchis Book & Tendrils, I have spent 4 years & 2 months over it.—[3] 1260 copies sold at Murray's Sale—[4]

The book not actually published till Jan. 30 1868.—

[Right]

1867

[16 or 18 January. CD and Charles Kingsley may have met.][5]

Feb. 13 to 21 to Erasmus.—[6]

[18 February. CD called on Henry Walter Bates.][7]

June 17 to 24 do. do.

[3 August. Joseph Dalton Hooker and William Henslow or Charles Paget Hooker visited.][8]

[*Circa* 21 August. Vladimir Onufrievich Kovalevsky visited.][9]

Sept 18 to 24 do. poorly all time.

[20 October. J. D. Hooker and W. H. Hooker may have visited.][10]

Nov. 28 to Dec. 7 do.— very well.[11]

[CD met Charles Lyell and Philip Lutley Sclater in London, and visited the gardens of the Zoological Society of London with Abraham Dee Bartlett.][12]

[21 December. J. D. Hooker visited.][13]

[22 December. Thomas Woolner may have visited.][14]

[28–9 December. Roland Trimen visited.][15]

---

[1] '"Variation under Domestication"' interl after del 'Dom A'. CD refers to *Variation*. See letter to John Murray, 18 March [1867]. On the 'Man Essay', see the letter to J. D. Hooker, 8 February [1867] and n. 16.

[2] 'my': interl.

[3] For CD's and his children's illnesses since 1860, see *Correspondence* vols. 8–14. Since 1860, CD had published the second, third, and fourth editions of *Origin*, *Orchids*, 'Dimorphic condition in *Primula*', 'Three sexual forms of *Catasetum tridentatum*', 'Two forms in species of *Linum*', 'Three forms of *Lythrum salicaria*', 'Climbing plants', and a number of shorter notes and papers (see *Collected papers*).

[4] '1260' over '1200'. Freeman 1977 reports 1250 copies of the first printing of *Variation* were sold at John Murray's autumn sale.

[5] Letter from R. M. Rolfe, 14 January 1867. See also Hutchinson 1914, 1: 91–2.

[6] CD's brother, Erasmus Alvey Darwin, lived at 6 Queen Anne Street, London.

[7] Letter to A. R. Wallace, 23 February 1867.

[8] Letter from J. D. Hooker, 30 July [1867].

[9] DAR 84.2: 193; see also letter to Cassell, Petter, & Galpin, 24 August [1867].

[10] Letter from J. D. Hooker, 18 October 1867.

[11] This visit ended after 5 December (Emma Darwin's diary (DAR 242)).

[12] Letter to Charles Lyell, 7 December [1867], letter to A. D. Bartlett, 9 December [1867], letter to P. L. Sclater, 9 December [1867].

[13] Emma Darwin's diary (DAR 242).

[14] Letter from J. D. Hooker, 17 December 1867.

[15] See letter to Roland Trimen, 24 December [1867] and n. 1.

APPENDIX III
Diplomas presented to Charles Darwin

In 1867, Darwin received the following diplomas. Although not letters in the conventional sense, they represent significant communication between Darwin and scientific organisations, and the citations in such diplomas often provide valuable indications of those aspects of Darwin's work that were considered worthy of honour. In view of this, they have been included here, together with a translation.

From the Zoological and Botanical Society of Vienna 23 May 1867

Die
kaiserlich königliche
zoologisch botanische Gesellschaft
in
Wien
ernennt
Seiner Hochwohlgeboren Herrn
Sir Charles Darwin
als
Mitglied
Wien, am 23. Mai *1867*

Colloredo-Mansfeld[1]
Präsident

Prof. D^r. A. E. Reuss[2]   Georg Ritter von Frauenfeld[3]
Vice Präsident   *Secretär*

[Translation]

The Imperial Royal Zoological Botanical Academy of Vienna elects the honourable gentleman Sir Charles Darwin *as member*

Vienna, 23. May *1867*

Colloredo-Mansfeld | *President*

Prof. D^r. A. E. Reuss | *Vice-president*

Georg Ritter von Frauenfeld | *Secretary*

[1] Joseph Franz Hieronymus, prince of Colloredo-Mansfeld.
[2] August Emanuel von Reuss.
[3] Georg von Frauenfeld.

From the Imperial Academy of Sciences of St Petersburg 29 December 1867

Imperialis Academia Scientiarum Petropolitana virum illustrissimum

Carolum Darwin Londinensem

sagacissimum rerum naturalium perscrutatorem socium ab epistolarum commercio in sectione biologica rite elegit electumque literis his publicis renunciavit die XXIX mensis decembris anni MDCCCLXVII.

Imperialis Academiae Scientiarum Petropolitanae

Praeses Comes Fr. Lütke[1]
Praesidis vices gerens V. Bouniakowsky.[2]
Secretarius perpetuus C. Vesselofski[3]

N⁰. 701.

[Translation]

The Imperial Academy of Sciences of St Petersburg has duly elected as a corresponding member in the biological section Charles Darwin of London, a most learned investigator of natural history, and has announced his election in these letters patent on 29 December 1867.

Imperial Academy of Sciences of St Petersburg
President Count Fr. Lütke
Vice-president V. Bouniakowsky.
Perpetual secretary C. Vesselofski
N⁰. 701.

DAR 229: 18

[1] Feodor Petrovich Lütke.
[2] Victor Iakovlevich Bouniakowsky.
[3] Constantin Stepanovich Vesselofski.

APPENDIX IV
Darwin's *Queries about expression*

In 1867, Darwin sent out a number of handwritten copies of a list of queries on human expression (see, for example, enclosure to letter to Fritz Müller, 22 February [1867]). In the Darwin Archive–CUL, three printed copies of these queries have been found (DAR 53: B2, DAR 181: 41, DAR 186: 1). There is also what is apparently a proof copy of this printed version (DAR 96: 1) with handwritten corrections on it, some by Darwin, that are incorporated into the three other copies. It is likely that this printed version was not made until late 1867 or early 1868 (see Freeman and Gautrey 1972 and 1975), and that it represents a near final version of the queries; in particular, questions 2, 5, 6, 10, and 13 contain additions that rarely appear in extant early versions of the queries, but that do appear in the version that CD reproduced in the introduction to *Expression*. The final version of the printed queries in the Darwin Archive–CUL is transcribed below for its own intrinsic interest and as a convenience to readers wishing to see the printed questions when answers to a now unavailable list of queries appear in the *Correspondence*. At least two additional copies of Darwin's printed list of queries survive in other archives (see Freeman and Gautrey 1975).

QUERIES ABOUT EXPRESSION.

(1.) Is astonishment expressed by the eyes and mouth being opened wide, and by the eyebrows being raised?

(2.) Does shame excite a blush when the colour of the skin allows it to be visible? and especially how low down the body does the blush extend?

(3.) When a man is indignant or defiant does he frown, hold his body and head erect, square his shoulders and clench his fists?

(4.) When considering deeply on any subject, or trying to understand any puzzle, does he frown, or wrinkle the skin beneath the lower eyelids.

(5.) When in low spirits, are the corners of the mouth depressed, and the inner corner of the eyebrows raised by that muscle which the French call the "Grief muscle?" The eyebrow in this state becomes slightly oblique, with a little swelling at the inner end; and the forehead is transversely wrinkled in the middle part; but not across the whole breadth, as when the eyebrows are raised in surprise.

(6.) When in good spirits do the eyes sparkle, with the skin a little wrinkled round and under them, and with the mouth a little drawn back at the corners?

(7.) When a man sneers or snarls at another, is the corner of the upper lip over the canine or eye tooth raised on the side facing the man whom he addresses?

(8.) Can a dogged or obstinate expression be recognized, which is chiefly shewn by the mouth being firmly closed, a lowering brow and a slight frown?

(9.) Is contempt expressed by a slight protrusion of the lips and by turning up the nose, with a slight expiration?

(10.) Is disgust shewn by the lower lip being turned down, the upper lip slightly raised, with a sudden expiration, something like incipient vomiting, or like something spat out of the mouth?

(11.) Is extreme fear expressed in the same general manner as with Europeans?

(12.) Is laughter ever carried to such an extreme as to bring tears into the eyes?

(13.) When a man wishes to shew that he cannot prevent something being done, does he shrug his shoulders, turn inwards his elbows, extend outward his hands and open the palms; with the eyebrows raised?

(14.) Do children when sulky, pout or greatly protrude the lips?

(15.) Can guilty, or sly, or jealous expressions be recognized? though I know not how these can be defined.

(16.) As a sign to keep silent, is a gentle hiss uttered?

(17.) Is the head nodded vertically in affirmation, and shaken laterally in negation?

Observations on natives who have had little communication with Europeans would be of course most valuable, though those made on any natives would be of much interest to me.

General remarks on expression are of comparatively little value; and memory is so deceptive that I earnestly beg it may not be trusted.

A definite description of the countenance under any emotion or frame of mind, with a statement of the circumstances under which it occurred, would possess much value. An answer within six or eight months, or even a year, to any *single* one of the foregoing questions would be gratefully accepted. In sending answers, the questions need not be copied, but reference may be made to the numbers of each query.

CHARLES DARWIN,

DOWN, BROMLEY, KENT,
    1867.

MANUSCRIPT ALTERATIONS AND COMMENTS

The alteration notes and comments are keyed to the letter texts by paragraph and line numbers. The precise section of the letter text to which the note applies precedes the square bracket. The changes recorded are those made to the manuscript by CD; changes of hand in letters written partly by CD and partly by amanuenses are also recorded. Readers should consult the Note on editorial policy in the front matter for details of editorial practice and intent. The following terms are used in the notes as here defined:

del	deleted
illeg	illegible
interl	interlined, i.e., inserted between existing text lines
omitted	omitted by the editors to clarify the transcription
over	written over, i.e., superimposed

To John Murray 3 January [1867]
2.2 as cancelled] 'as' *interl*
2.3 & on Dogs & on Plants 2.4] *altered from* '& Dogs Chapter &'
2.4 chapters] *interl*
2.8 the whole] *interl*
4.1 Man] 'M' *over* 'm'

To William Bernhard Tegetmeier 6 January [1867]
1.3 for your] *after del* 'you'

To Thomas Henry Huxley 7 January [1867]
1.2 book] *after del* 'tra'
1.6 all people] *above del* 'anyone'
1.6 that] *interl*
1.6 would be] *interl*

To Ernst Haeckel 8 January 1867
3.12 again. . . . Darwin 4.2] *in CD's hand*

To John Murray 8 January [1867]
2.6 of Book] *after del* 'or ¼'
2.7 could] *after del* 'wo'
3.1 a red] 'a' *interl*
3.2 printed] *interl*
4.4 the M.S.] 'the' *interl*

To Joseph Dalton Hooker 9 January [1867]
1.3 &] *after del* 'which will str'
1.3 no] *interl*
1.3 by zones, 1.4] *interl*
1.5 Webb] *altered from* 'Webb's'
1.5 zones on] 'on' *over* 'in'
1.6 viewed the] 'the' *over illeg*
1.8 have] 'h' *over illeg*
3.3 countries] *interl*
3.6 (Tyndall)] *interl*
5.3 The volumes] *above del* 'They'
5.4 printed] *interl*

To J. D. Hooker 15 January [1867]
1.6 than the others 1.7] *interl in CD's hand*

To William Turner 15 January [1867]
2.1 good] *interl*
2.2 which I copied] *interl*
3.1 the] *interl*
4.1 (from . . . &c)] *interl*
4.2 coccygeal] *interl*
4.2 spinal] *after del* 't'
4.3 an] *interl*
7.1 traces of] *interl*
7.1 supra-] *inserted*
7.2 humerus of] *interl*

To Alfred Newton 19 January [1867]
3.3 ie ... species? 3.4] *added in CD's hand*
3.4 or in] 'in' *interl*

To J. D. Hooker 21 January [1867]
1.1 Four ... bottom] *above del* 'At near base'
2.1 instead] *after del* 'Ins'

To J. D. Hooker 29 January [1867]
2.1 far] *interl*
2.3 much] *interl*
2.5 geographical] *interl*
2.7 when there] *interl*
2.8 fumarole] *above del* 'hot-springs &c'
2.9 each] *interl after del* 'the'
2.9 has] *interl*
2.14 oceanic islands] *interl*

To John Murray 29 January [1867]
1.3 hear immediately] 'immediately' *interl*

To William Turner 1 February [1867]
2.3 in the notion] *interl*
2.5 I do] *after del* 'Now it seems'
2.6 both] *after del illeg*
2.10 hairy] *interl*
2.11 the front] 'the' *interl*
2.12 On the] 'the' *interl*
2.13 and as] 'and' *above del* 'not'
2.13 not hairy] 'not' *interl*
5.2 alone] *interl*

To J. D. Hooker 8 February [1867]
1.3 would] *interl*
1.5 & taste] *interl*
1.7 thought that] 'that' *after del* 'when'
1.7 of the articles] *interl*
1.8 not very profound.] *interl*
1.9 the Saturday] 'the' *interl*
1.11 this counter-] *interl*
1.15 material] *above del* 'less'
1.16 & of] *over comma*
1.16 the structure of 1.17] *interl*
1.23 staggered] *altered from* 'stagger'
2.2 in ... permits] *interl*
2.5 has played] *altered from* 'plays'
2.7 funny] *above del* 'funny'
2.7 from a lady] *interl*
2.9 expected] *above del* 'thought'
3.1 on,] *interl*
3.3 has ... been a 3.4] *interl above del* 'is a'
3.4 is a] *added*

4.1 those of a] *interl*
4.3 but] *interl*
5.1 -sheet] *added*
5.1 in a note] *interl*
5.1 ideal] *interl*
5.1 he] *after del* 'it'
7.4 the interval] 'the' *interl*

To Edward Blyth [18 February 1867]
5.2 go] *after del illeg*

To Fritz Müller 22 February [1867]
1.6 of yours] *interl in CD's hand*
2.5 and] *interl in CD's hand*
2.12. they] *in CD's hand above del* 'two males'
3.2 to which] *interl in CD's hand*
3.2 unfortunately I] 'I' *interl in CD's hand*
3.6 though ... Negros.] *interl in CD's hand*
3.7 even] *interl in CD's hand*
Enclosure:
Question 9: 'nose with a'] 'with a' *interl in CD's hand*

To Edward Blyth 23 February [1867]
3.5 common ... sexes] *interl in CD's hand*
3.5 confined to] *interl in CD's hand above del* 'of'
4.3 on ... Antelopes] *interl in CD's hand*

To Alfred Russel Wallace 26 February [1867]
1.9 respect to colour in] *interl in CD's hand*

To Julius von Haast 27 February [1867]
Enclosure:
Question 6: drawn back] *in CD's hand; above del* 'extended'

To James Philip Mansel Weale 27 February [1867]
Enclosure: in hand of amanuensis
Question 2: Especially ... extend?] *added in CD's hand*
Question 5: & contracted] *interl in CD's hand*

To Ferdinand von Mueller 28 February [1867]
1.4 soon] *interl in CD's hand*
3.1 Feb 28th] *in CD's hand*
Enclosure: in hand of amanuensis

To Alfred Newton 4 March [1867]
1.5 (Zoolg. ... Birds)] *interl in CD's hand*
1.8 Casuarinus] 'C' *over* 'K' *in CD's hand*
3.1 P.S. ... selected.— 4.3] *in CD's hand*

To W. B. Tegetmeier 5 March [1867]
1.4 male] *interl*
2.2 & saddle feather] *interl*
2.4 drakes] *after del* 'mallards'
2.9 mode] *after del illeg*
2.9 the trial] 'the' *interl*

To A. R. Wallace [12–17] March [1867]
1.2 add] *above del* 'give'
1.4 (or . . . published) 1.5] *interl*
1.5 way] *interl*
1.9 (not *quite* truly)] *interl*
1.10 shall] *above del* 'would the sub should'
1.11 for . . . up] *interl*
1.13 interested] *after del interl* 'gr'
2.2 with you] *inserted*
2.2 that] *interl*
2.4 some] *after del illeg*
2.11 have] *interl*
3.2 get] *interl*

To J. D. Hooker 17 March [1867]
3.4 with] *after del comma*
3.5 species of] *interl*
4.1 are] *interl*
4.5 , but . . . sentence of,] *interl*
4.5 second-hand] *above del* '2^d hand'
5.4 flourish] *after del illeg*
5.4 & he] 'he' *interl*
5.5 , spreading by seed,] *interl*
6.7 believe it,] *interl*

To John Murray 18 March [1867]
1.4 you] *after del* 'ha'
2.1 can] *after del illeg*
3.2 & . . . Russia] *interl*

To E. Schweizerbart'sche Verlagsbuchhandlung [19 March 1867]
1.1 have . . . my new] *above del* 'I am beginning to print my'
1.2 under] *before omitted point*
1.4 G.] *interl*
1.8 I] *after del* 'and'
1.9 been] *after del* 'appeared.'
1.9 on] *after del* 'I make my offer contingent'
1.10 You] *after del* '& I w^d wish for an an answer after, say 10'
1.10 negotiate with] *above del* 'negotiate first'
1.11 before . . . so that 1.12] *above del* 'after, say if you have'
1.12 might] *above del* 'would'

1.12 early] *above del* 'the'
1.12 the] *above del* 'my'

To J. D. Hooker 24 [March 1867]
2.1 I] *interl*
2.1 in] *above del* 'at'
3.1 Müller . . . tree!] *added in margin*
3.1 counted] *above del* 'couted'
5.1 dimorphic] *interl*
5.2 he has] 'has' *interl*
5.2 kind] *interl*
5.4 dimorphic] *interl*
5.6 a piece of good] *interl*
5.7 of R. Brown] 'of' *interl*

To Vladimir Onufrievich Kovalevsky 26 March [1867]
4.2 you] *interl*

To John Lubbock 26 March [1867]
3.3 new] *above del* 'next'
4.2 Mahomet] 'M' *over* 'm'

To William Erasmus Darwin 27 [March 1867]
5.1 has] *interl*

To Maxwell Tylden Masters [28 March – 5 April 1867]
1.2 to me] *interl*
1.2 little] *after del* 'no'
1.3 taken] *interl*
1.3 now] *after del* 'know'
1.3 know of] *interl*
1.5 however] *added pencil*
1.5 to be . . . effective)] *above del* 'to be good *as in the sense [interl and del] by fertilising other individuals or other species)'
1.5 as itself] *after del* 'effective'
1.5 easily] *interl*
1.5 pollen taken from 1.6] *interl*
1.6 of the same species, or from] *interl*
1.6 quite] *interl after del pencil* 'other &'
1.8 with plants] *interl pencil*
1.8 regular or] *interl pencil after del pencil* 'a process or' *above del illeg*'
1.8 ⌈union⌉] *pencil*
1.8 between] *after del pencil* 'crosses' *altered from* 'cross' *pencil*
2.1 If] *after del* '(You can if you think fit make any use of these remarks)'
2.1 If . . . use them] *pencil*
2.1 value] *above del* 'use'

2.1 with ... communication] *interl pencil*
3.1 to] *interl*
3.1 copy of his 3.2] *interl*
4.1 no] *after omitted point*
4.1 carefully] *pencil above del ink* 'dissect'
4.1 interesting specimen 4.2] *pencil above del pencil*
'ovaria'
4.2 remarkable] *after del* 'very'
4.3 do not] *interl after interl and del* 'first'
4.4 until] *interl*
4.4 even] *interl*
4.5 ascertain] *after del* 'distinguish *by dissection
the female [*interl*] *or male [*interl pencil*] the
sexes of Orchids.—'
4.5 whether ... [whether] 4.6] *interl pencil*
4.6 has started] *pencil below del pencil* '*these plants
[*above del* 'orchids']'
4.7 pollen.] *after del* 'male organs or'

To Henry Walter Bates 30 March [1867]
2.1 late] *interl*
2.3 what] *after del* 'that'

To W. B. Tegetmeier 30 March [1867]
1.7 soon] *after omitted point*
2.3 how] *after del* 'whether'
2.4 the] *interl*

To Stephen Paul Engleheart? [April 1867?]
1.1 (Wednesday)] *interl*

To J. D. Hooker 4 April [1867]
3.1 The] 'T' *over* 't'

To Thomas Blunt 5 April [1867]
3.2 longer any] *interl*

To J. D. Hooker 5 April [1867]
Enclosure 1
1.5 Do] *after del* 'Do you [*illeg*]'
2.1 Copied from ancient notes of mine.—] *added
in left margin*
Enclosure 2
1.1 (I possess ... grow in P. Santo.—) 1.3]
square brackets in MS
1.2 (& ... Santo)] *interl*
2.1 (You remember ... England.—) 2.2] *square
brackets in MS*

To Carl Wilhelm von Nägeli [after 8 April 1867]
2.1 2 or 3 days ago 2.2] *interl*
2.2 Photograph] *after del* 'your [*illeg*]'

2.2 & your several memoirs] *added*
2.3 Delay ... letter] *interl after del* 'your memoirs'

To Julius Victor Carus 11 April [1867]
1.2 my book] *above del* 'it'
2.1 large] *interl*
2.3 clean] *interl*

To Asa Gray 15 April [1867]
1.3 few] *after del* 'cor'
3.1 old] *after del illeg*
4.6 is almost] 'is' *interl*
4.6 compilation] *first* 'i' *interl*
4.7 5] *over* '4'
5.5 growth] *after del* 'those'
5.6 by seed] *interl*
5.7 but] *over* '&'
5.7 have been] *after del* 'get'

To J. D. Hooker 15 [April 1867]
2.2 are failures] *interl*
6.1 P.S. ... Monday.— 6.4] *added pencil*
6.1 on Saturday] *interl pencil*
6.2 with pony-carriage 6.3] *interl pencil*
6.3 you] *interl pencil*
6.3 near to Station.] *interl pencil*

To J. V. Carus 18 April [1867]
1.6 heard of] 'of' *interl*

To Fritz Müller 22 April [1867]
1.4 all] *interl in CD's hand*
2.9 by insects] *interl in CD's hand*
2.11 both produced by] *in CD's hand above del* 'borne
on'
3.1 I will ... C. Darwin 4.1] *in CD's hand*

To A. R. Wallace 29 April [1867]
1.2 very briefly] *interl in CD's hand*
2.6 her] *in CD's hand above del* 'its'
7.1 It is curious ... note.— 8.3] *in CD's hand*
Enclosure: in CD's hand
1.3 & beauty] *interl*
1.3 viz] *added*
1.5 ever] *interl*
1.6 extended ... your] *above del* 'ventured on th'

To J. D. Hooker [12] May [1867]
3.6 (9 ... oz.)] *square brackets in MS*

To V. O. Kovalevsky 16 May [1867]
0.2 May 16th ... faithfully 2.1] *in hand of amanuensis*

To Charles Lyell 1 June [1867]
1.6 Birds is] 'is' *above del* 'are'
1.8 to Humming Birds.] *interl*
1.10 being] *after del illeg*
2.3 he] *after del* 'there'

To Daniel Oliver 1 June [1867]
2.4 to some] 'to' *interl*

To Charles Lyell 9 June [1867]
1.3 almost] *interl*

To Charles Kingsley 10 June [1867]
1.2 on the "Origin"] *interl*
2.4 interesting ... &] *interl*
2.12 diversity of structure] *above del* 'this head'
2.15 letter by Wallace] *interl*
3.1 looks at] *above del* 'views'
3.12 remained] *above del* 'are'
4.4 of years. 4.5] *interl*
5.6 of Nature] *interl*
5.6 more energetic.] *above del* 'greater'
5.6 extremely] *above del* 'more nearly'
5.8 not] *above* 'no where'
5.8 stated] *before del caret*
5.8 is probably] *after del* 'may be'
5.10 each] *interl*
5.11 in comparison ... of a 5.12] *interl above del* 'of'
5.12 the genealogical] 'the' *added pencil*
5.12 explains] *after del pencil* 'view'
5.13 of Review] *after* 'I think in' *interl & del*
5.13 I think] *interl*
5.17 those which first] *interl above del* 'under'
5.17 caused the] *interl after del illeg*
5.17 to vary] *above del* 'which first deviated'
5.18 the sudden] 'the' *above del illeg*
5.18 of structure] *interl*
5.18 such] *interl*
6.1 When ... long beak] *above del* 'I sh^d like to speak'
6.1 for instance] *interl*
6.2 saying, ... done] *interl above del* 'sometimes speaking'
6.2 incautiously] *after del* 'done'
6.3 that of] *interl*
6.4 born,] *after del* 'each'
6.4 a beak] 'a' *interl*
6.5 conditions] *after del illeg*
6.5 or habits of life] *interl*
6.5 longer] *after del* 'the use of a'
6.5 all] *above del illeg*
7.1 The] *after del* 'After'

7.1 augmented 7.2] *interl*
7.2 have given] *after del* 'do'
7.3 not] *interl*
7.3 The Reviewer ... *[*variation*/]* *added in margin, pencil*

To T. H. Huxley 12 June [1867]
5.2 & term] *above del* '& it'
5.2 refer to the] *interl*
6.2 this word] 'this' *interl*

To J. D. Hooker [16 June 1867]
1.3 to see us.] *interl*

To A. R. Wallace [24 June 1867]
2.2 i.e. ornamental] *interl in CD's hand*
5.1 P.S ... day.— 5.3] *in CD's hand*

To Ernst Haeckel 4 July [1867]
1.1 you] *interl*
4.1 day] *interl*
4.4 the sooner] 'the' *interl*

To Charles Lyell 18 July [1867]
1.7 completed] *interl*
1.8 & others] *interl*
3.6 so frequently 3.7] *interl*
3.7 areas,] *interl*
3.9 in ... archipelagoes] *interl*
3.10 mechanical ... chemical 3.11] *interl*
3.11 ie] *above del* 'with'
3.11 & cracking. 3.12] *interl*
3.12 remarkable absence,] 'absence' *interl*
3.13 the absence 3.14] 'absence' *above del* 'presence'
3.14 on this side.] *interl*
3.14 long] *interl*
3.19 coast-lines] *above del* 'their'

To Henrietta Emma Darwin 26 July [1867]
2.2 either] *first* 'e' *over* 'k'
2.3 chapter] *after del* 'til'

To William Bowman 30 July [1867]
1.9 have] *interl*
2.6 merely] *interl*

To Fritz Müller 31 July [1867]
1.18 Erica-] *interl above* 'heath-'
1.25 together] *interl*

To John Murray 4 August [1867]
1.6 & Plates (uncoloured)] *interl in CD's hand*

To John Murray 6 August [1867]
1.2 in ... box.—] *added in CD's hand*

To William Bowman 7 August [1867]
1.3 entirely] *after del* 'from'

To Benjamin Dann Walsh 9 August [1867]
2.6 for ... reference] *interl in CD's hand*
2.10 it is, ... Ch. Darwin 3.1] *in CD's hand*

To Fritz Müller 15 August [1867]
1.1 for a necklace] *interl*
1.4 highly] *after del* 'f'
2.1 that] *interl*
2.4 a little] *interl*
2.4 or] *over* '&'
2.5 & close] *interl*
3.1 &] *over* '—'
5.1 & Gesneria] *interl*
5.2 seeds] *interl*
5.3 After ... Kew.— 5.4] *added*

To Charles Lyell 22 August [1867]
2.5 (which ... old!) *interl in CD's hand*
2.6 it be] *interl in CD's hand*
2.6 as a] 'a' *interl in CD's hand*
2.6 hypothesis,] *interl in CD's hand*
5.1 You ... Origin. 5.4] *in CD's hand*
5.2 who ... Russian,] *interl*
5.3 in Russia] *interl*

To J. P. M. Weale 27 August [1867]
1.2 the case is 1.3] *interl in CD's hand*
1.3 curious] *interl in CD's hand*

To Alfred Wrigley [September 1867]
1.1 experience] *after del* 'it'
1.1 an] *interl pencil*
1.1 of the ... passed, 1.2] *interl pencil*
1.3 exertion] *pencil, above del* 'work'
1.3 for] *pencil, above del* 'at'
1.4 small] *interl pencil above del* 'no'
2.2 Xmas] 'X' *over* 'C'
2.2 hope] *after del* 'beg'
2.5 myself—] *pencil, altered from* 'me.'
4.2 respect] *interl pencil*

To J. D. Hooker 2 September [1867]
4.1 (but ... yourself) 4.2] *added in CD's hand*
4.3 sold] *in CD's hand above del* 'sowed'
4.3 vars. of] *interl in CD's hand*

To J. V. Carus 16 September 1867
2.5 or Revises,] *interl in CD's hand*
4.7 a term] 'a' *interl in CD's hand*
4.8 common deer-hound] 'deer-hound' *interl in CD's hand*
5.3 & ... Smith] *interl in CD's hand*
9.1 You ... taken.— 9.2] *in CD's hand*
Enclosure: 45] *in CD's hand below del* '50' *below del illeg*

To Charles Lyell 4 October [1867]
1.3 (you ... "often")] *interl in CD's hand*
1.5 in these two respects 1.6] *altered from* 'in this respect' *in CD's hand*
1.8 which] *interl in CD's hand*
1.9 And ... conditions. 1.10] *added in CD's hand*
2.1 (& ... mention)] *interl in CD's hand*
3.3 & had ... Darwin 5.1] *in CD's hand*
3.3 perhaps] *interl*
4.2 or barely] *interl*

To J. V. Carus 8 October [1867]
1.4 cases] *interl*
1.4 within] *above del* 'in'
1.6 spot of] *interl*
2.3 by Breeders] *interl*

To A. R. Wallace 12 and 13 October [1867]
2.3 may] *after del* 'alluded to'
2.3 Figure] *above del* 'illustration'
4.1 of Beauty] *interl*
4.4 specially] *interl*
4.10 which appears] *interl*
4.10 a strongly] 'a' *interl*
4.10 one] *interl*
4.12 & other such cases;] *interl*
4.13 the proof of] *interl*
5.2 on expression] *interl*
7.1 the whole] *above del* 'it'
10.1 This ... Proofs 10.2] *added in CD's hand*
12.1 I ... creation.— 12.2] *added in CD's hand*

To Asa Gray 16 October [1867]
4.1 which ... corrected] *interl in CD's hand*

To J. V. Carus 19 October [1867]
3.2 of any one] *interl*
4.6 some day] *interl*

To John Murray 19 October [1867]
1.5 for] *over* '.—'
2.1 sheets] *interl*

2.2 the whole] *interl*
2.3 or] *after del* 'of'

To Charles Lyell 31 October [1867]
2.2 firmly] *interl in CD's hand*
5.1 P.S. . . . note.] *in CD's hand*

To Charles Kingsley 6 November [1867]
1.3 being] *above del* 'as'

To William Sweetland Dallas 8 November [1867]
1.3 the word] 'the' *added*
1.6 subject on] 'on' *before del caret*
1.6 or species . . . him] *interl*
1.8 &] *over* '.—'

To J. V. Carus 14 November [1867]
1.3 old] *interl in CD's hand*
2.2 which . . . finished 2.3] *interl in CD's hand*

To J. D. Hooker 17 November [1867]
1.1 of my Book.] *interl*
1.12 will be] *above del* 'is'
1.17 of] *above del* 'for'
1.18 & . . . it is;] *interl*
1.18 but he] 'he' *altered from* 'his'
3.1 About] *above del* 'About'

To William Erasmus Darwin 20 November [1867]
2.1 *safer*] *interl*

To E. Schweizerbart'sche Verlagsbuchhandlung
22 November [1867]
Enclosure: *in CD's hand*
8.1 Graz] 'z' *after del* 't'

To J. D. Hooker 25 November [1867]
2.1 London to] 'to' *interl*
2.3 pray] *interl*

4.3 whole] *interl*

To Anton Dohrn 26 November [1867]
3.2 homologies of] 'of' *over* 'is w'
3.3 (though] *parenthesis over comma*
3.4 the prehensile] *interl*

To Charles Lyell 7 December [1867]
1.1 in Vict. Institute] *interl in CD's hand*
1.9 have . . . to] *interl in CD's hand*

To Abraham Dee Bartlett 9 December [1867]
1.2 Tringa-like] *interl*

To the Linnean Society 9 December [1867]
1.3 (which . . . out)] *interl in CD's hand*
4.2 engraved] 'engraved' *in CD's hand above del* 'cut'
Enclosure:
2.1 (fig 6 . . .)] *square brackets in original*
2.1 (Fig 2. . . .)] *square brackets in original*
3.1 (Fig 7. . . .)] *square brackets in original*
3.1 (Fig 3 . . .)] *square brackets in original*
4.1 (Beneath . . . type)] *square brackets in original*
5.1 Darwinii] *after del* 'speciosa'
6.1 position (magnified)] '(magnified)' *interl*

To J. P. M. Weale 9 December [1867]
5.1 I . . . paper.—] *added in CD's hand*

To Eduard Koch 13 December [1867]
4.1 I . . . Häckel] *added in CD's hand*

To Albert Charles Lewis Gotthilf Günther
21 December [1867]
1.1 which I] 'I' *over* 'you'

To John Murray 31 December [1867]
3.1 To . . . St] *added in CD's hand*

BIOGRAPHICAL REGISTER

This list includes all correspondents and all persons mentioned in the letters and notes that the editors have been able to identify. Dates of letters to and from correspondents are given in chronological order. Letters to correspondents are listed in roman type; letters from correspondents in italic type; third-party letters are listed in italic type with the name of the recipient given in parentheses.

Adams, John Couch (1819–92). Astronomer and mathematician. Co-discoverer, by mathematical calculation, of the planet Neptune. Fellow and tutor, St John's College, Cambridge, 1843–52; Pembroke College from 1853. Lowndean professor of astronomy and geometry, Cambridge University, 1859–92; director of the Cambridge Observatory, 1861–92. FRS 1849. (*DSB, ODNB.*)

Adams Express Company. American express delivery company. Incorporated in 1854. (*DAB* s.v. Adams, Alvin).

Agassiz, Elizabeth Cabot Cary (1822–1907). Educator. A founder of the educational establishment for women that later became Radcliffe College, Cambridge, Massachusetts. President of Radcliffe College, 1894–9; honorary president, 1900–3. Married Louis Agassiz in 1850. (*ANB.*)

Agassiz, Jean Louis Rodolphe (Louis) (1807–73). Swiss-born zoologist and geologist. Professor of natural history, Neuchâtel, 1832–46. Emigrated to the United States in 1846. Professor of zoology and geology, Harvard University, 1847–73. Established the Museum of Comparative Zoology at Harvard in 1859. Foreign member, Royal Society of London, 1838. (*ANB, DAB, DSB, Record of the Royal Society of London.*)

Anderson, John Anderson (*c.* 1811–91). Scottish nurseryman. Ran a nursery in Perth, in partnership with his uncle, Archibald Turnbull. Sold the business in 1883. (R. Desmond 1994; letter from Isaac Anderson-Henry, 20 May 1867.)

Anderson, Thomas (1832–70). Scottish physician and botanist. Entered the Bengal medical service in 1854. Superintendent, Calcutta botanic garden, 1861–8; conservator of forests, Bengal, 1864–6; retired because of ill health. Instituted experiments that led to the successful cultivation of *Cinchona* in India. (Lightman ed. 2004.)

Anderson, William (b. *c.* 1748 d. 1778). Surgeon and naturalist. Accompanied Captain James Cook on two of his voyages. (*ODNB.*)

Anderson-Henry, Isaac (1800–84). Scottish lawyer and horticulturalist. Practised as a solicitor in Edinburgh from 1834 or earlier. Retired from practice in 1861, when his wife succeeded to the estates of Woodend, Perthshire, and, as a condition of the entail, took the additional name Henry. Established a garden in Edinburgh known for its many rare plants. Collected plants from the Andes, north-western Himalayas, and New Zealand. Interested in plant hybridisation and acclimatisation. Member of various scientific societies in London and Edinburgh. President of the Botanical Society of Edinburgh, 1867–8. (*County families* 1878; *Curtis's botanical magazine dedications*, pp. 175–6; *Gardeners' Chronicle*, 22 March 1873, p. 399; *Post Office Edinburgh directory* 1834–84 (1861–84 s.v. Henry, Isaac Anderson); *Proceedings of the Linnean Society of London* (1886–7): 42–4.)

3 April 1867, 20 May 1867, 22 May [1867], 24 May 1867, 25 May 1867, 2 November 1867

Anon.

22 [March? 1867]

D. Appleton & Co. New York publishing house. Founded by Daniel Appleton (1785–1849) in 1831. His son William Henry Appleton (1814–99) was taken into partnership in 1838. American publishers of works by CD and Herbert Spencer. (*ANB.*)

1 February 1867 (Asa Gray)

Argyll, 8th duke of. *See* Campbell, George Douglas.

Aristotle (384–322 B.C.). Greek philosopher. Author of many works, including *Historia animalium.* (*EB.*)

Arran, 4th earl of. *See* Gore, Philip Yorke.

Athenæum.

1 January 1867

Avebury, 1st Baron. *See* Lubbock, John.

Aylmer, Isabella Elinor (d. 1908). Author. Of Northumbrian descent. (Aylmer 1931, p. 235; *CDEL* supplement, s.v. Aylmer, Mrs Fenton; letter from I. E. Aylmer, 5 December 1867.)

18 November 1867, 5 December 1867

Ayres, Philip Burnard (1813–63). Surgeon and botanist. Pupil of John Lindley; studied cryptogams. Practised medicine in Thame, Oxfordshire, for ten years. Lecturer on chemistry at Charing Cross Hospital; physician to the Islington dispensary, 1851. Edited the *Pharmaceutical Times*. Superintendent of quarantine on Mauritius, 1856–63. (R. Desmond 1994, *Modern English biography*.)

Babington, Charles Cardale (1808–95). Botanist, entomologist, and archaeologist. Involved in natural history activities at Cambridge for more than forty years; an expert on plant taxonomy. A founding member of the Cambridge Entomological Society and the Cambridge Antiquarian Society. Editor of *Annals and Magazine of Natural History* from 1842. Chairman, Cambrian Archaeological Association, 1855–85. Professor of botany, Cambridge University, 1861–95. FRS 1851. (*DNB, DSB.*)

Backhouse, James (1825–90). Botanist. Collected plants in Norway in 1851; Ireland in 1854; and Scotland in 1859. Also collected plants around Teesdale,

Backhouse, James, cont.
England. Son of James Backhouse (1794–1869), with whom he ran a nursery in York from 1845. (R. Desmond 1994, *ODNB* s.v. Backhouse family.)

Bacon, Francis, Baron Verulam and Viscount St Alban (1561–1626). Lawyer, statesman, and philosopher. Lord Chancellor, 1618–21. Created Baron Verulam, 1618; Viscount St Alban, 1621. (*DSB, ODNB.*)

Baer, Karl Ernst von (1792–1876). Estonian zoologist and embryologist. Professor of anatomy at Königsberg University, 1819; professor of zoology, 1826–34. Professor of zoology at the Academy of Sciences, St Petersburg, 1834–67. Demonstrated the existence of the mammalian egg, 1826. Propounded the influential view that embryological development proceeds from the general to the specific. Foreign member, Royal Society of London, 1854. (*DSB, NDB, Record of the Royal Society of London.*)

Baird, Spencer Fullerton (1823–87). American zoologist and scientific administrator. Professor of natural history, Dickinson College, 1846. Assistant secretary and curator of the Smithsonian Institution, 1850; secretary, 1878–87. Science editor for *Harper's Weekly*, 1871–9. (*ANB.*)

Baker, John Gilbert (1834–1920). Botanist. Draper in Thirsk, Yorkshire, 1847–64. Active in the Thirsk Natural Historical Society. Assistant in the herbarium at the Royal Botanic Gardens, Kew, 1866–90; keeper of the herbarium and library, 1890–9. Lectured on botany at the London Hospital Medical School, 1869–81, and at the Chelsea Physic Garden, 1882–96. Contributed to a wide range of colonial floras. Authority on ferns. FRS 1878. (*DSB, ODNB.*)

Baker, Samuel White (1821–93). Traveller, sportsman, and author. Established an English colony in Ceylon (Sri Lanka), where he lived for nine years. Explored the tributaries of the Nile, 1861–4. Served as governor-general, under the Egyptian government, of the equatorial Nile basin, 1869–73. Knighted, 1866. FRS 1869. (*ODNB.*)

Bakewell, Robert (1725–95). Stockbreeder and farmer. Farmed at Dishley, Leicestershire. Improved breeds of sheep and cattle. Produced the new Leicestershire breed of sheep, the Leicestershire longhorn, and a breed of black horses. Improved grassland by watering. (*ODNB.*)

Baltard, Louis-Pierre (1764–1846). French architect and artist. Studied in Paris and Rome. Professor of architecture, Ecole polytechnique, from 1794; Ecole des beaux-arts, from 1819. Strongly interested in the fine arts and engraving. Illustrated and wrote a number of works on travel, architecture, history, and portraiture. (*DBF.*)

Banks, Joseph (1743–1820). Botanist. Travelled around the world on HMS *Endeavour* with Captain James Cook, 1768–71. President of the Royal Society, 1778–1820. Created baronet, 1781; privy councillor, 1797. FRS 1766. (*ODNB.*)

Barber, Mary Elizabeth (1818–99). British-born naturalist, artist, and writer in South Africa. Sister of James Henry Bowker. Emigrated to South Africa with her family in 1820. Married Frederick William Barber, a chemist, in 1845. Studied birds, moths, reptiles, and plants, and corresponded with a number of leading scientists, providing them with specimens and drawings. Published a number of scientific papers. (*DSAB, ODNB.*)

[after February 1867]

Barkly, Henry (1815–98). Colonial administrator. Served as governor of British Guiana, 1849–53; Jamaica, 1853–6; Victoria, 1856–63; Mauritius, 1863–70; and Cape Colony, 1870–77. Sent plants to the Royal Botanic Gardens, Kew. Investigated the ferns of Jamaica and Mauritius and its dependencies. Knighted, 1853. FRS 1864. (Gunn and Codd 1981, *ODNB, Proceedings of the Royal Society of London* 75 (1905): 23–5.)

Barrow, John (1764–1848). Promoter of exploration and travel writer. A founder of the Royal Geographical Society. Civil servant in South Africa, 1797–1803. Second secretary of the Admiralty, 1804–6, 1807–45. Promoted imperial expansion, the exploration of Australia and the course of the Niger river, and the search for a north-west passage. Created baronet, 1835. FRS 1805. (*ODNB.*)

Bartlett, Abraham Dee (1812–97). Taxidermist and zoo superintendent. Taxidermist, *circa* 1834–52. Superintendent of the natural history department, Crystal Palace, 1852–9; of the Zoological Society's gardens, Regent's Park, 1859–97. (*Modern English biography, ODNB.*)

9 December [1867], *9 December 1867*

Bates, Henry Walter (1825–92). Entomologist. Undertook a joint expedition to the Amazon with Alfred Russel Wallace, 1848–9; continued to explore the area, after Wallace returned to England, until 1859. Provided the first comprehensive scientific explanation of the phenomenon subsequently known as Batesian mimicry. Published an account of his travels, *The naturalist on the River Amazons*, in 1863. Assistant secretary, Royal Geographical Society of London, 1864–92. President, Entomological Society of London, 1868, 1869, and 1878. FRS 1881. (*DSB, ODNB.*)

11 March 1867, 29 March 1867, 30 March [1867]

Beaton, Donald (1802–63). Scottish gardener. An expert on bedding schemes and hybridisation. Trained in the gardens at Beaufort Castle, Inverness-shire. Gardener to William Gordon-Cumming at his estate in Altyre, at the Dickson & Turnbull nursery in Perth, and at the Caledonian Horticultural Society in Edinburgh. Head gardener and estate manager to William Gordon of Haffield House, Herefordshire, 1829–37; gardener to Thomas Harris of Kingsbury, Kilburn, London, then head gardener to William Fowle Middleton of Shrubland Park, Suffolk, from 1840. A regular contributor to the *Gardener's Magazine* and the *Cottage Gardener*. (*Cottage Gardener*, 28 November 1854, pp. 153–8; R. Desmond 1994; Hadfield *et al.* 1980.)

Beck, John (1801–73). Banker. Baptised by the Reverend George Augustus Case at the Unitarian Church, High Street, Shrewsbury, Shropshire, 20 July 1801. Attended Case's day-school in Shrewsbury with Erasmus Alvey Darwin, then Shrewsbury School, leaving in 1816. Married Susanna Margaret Badger in the

Beck, John, cont.

parish church of Ellesmere, Shropshire, in 1824; their three children were also baptised in the Church of England. A partner in the Shrewsbury and Welshpool Bank, Shrewsbury; retired between 1829 and 1831. By 1841 he was living apart from his family. Left an estate valued at under £100. (Parish records, Shropshire Records and Research Centre (Shrewsbury High St. baptism register, 20 July 1801; Ellesmere marriage register, 30 March 1824; Shrewsbury St Julian baptism register, 12 March 1825; Meole Brace baptism register, 30 September 1829, 1 April 1831; Will of John Beck, 1873/257); Death certificate, district of St Chad, Shrewsbury, 15 February 1873; Census returns 1841 and 1851 (Public Record Office HO 107/925/65 and HO 107/1992/749); letter from John Beck, 6 October 1864; *Commercial directory for Shropshire*; *Shrewsbury School register*.)

Becker, Lydia Ernestine (1827–90). Leading member of the women's suffrage movement, botanist, and astronomer. Published *Botany for novices* (1864); awarded a Horticultural Society Gold Medal, 1865. Founder and president of the Manchester Ladies' Literary Society, 1867. Secretary to the Manchester National Society for Women's Suffrage from 1867. Member of the Manchester School Board, 1870. Editor of and regular contributor to the *Women's Suffrage Journal* from 1870. Secretary to the London central committee of the National Society for Women's Suffrage from 1881. (Blackburn 1902, *Journal of Botany* 3 (1865): 164, *Macmillan dictionary of women's biography*, *ODNB*.)

6 February 1867

Bell, Charles (1774–1842). Anatomist and surgeon. Best known for his investigations of the nervous system and the expression of emotions in humans. Illustrated his own works. Co-owner of and principal lecturer at the Great Windmill Street School of Anatomy, London, 1814–25. Surgeon at the Middlesex Hospital, 1812–36. Professor of surgery, Edinburgh University, 1836. Knighted, 1831. FRS 1826. (*DSB*, *ODNB*, *Record of the Royal Society of London*.)

Bell, Marion (1787–1876). Daughter of Charles Shaw of Ayr. Married Charles Bell in 1811. Following the death of her husband in 1842, lived with her brother, Alexander Shaw; their house became a centre of literary and scientific society. Published her husband's letters in 1870. (*DNB* s.v. Bell, Charles (1774–1842), and Shaw, Alexander (1804–90); *Modern English Biography*.)

Belt, Thomas (1832–78). Geologist, naturalist, and mining engineer. Member of Tyneside Naturalist's Field Club. Joined the Australian gold rush in 1852, and studied geology. Returned to England in 1862 and established himself as a consultant mining engineer; worked in Nova Scotia, Wales, Nicaragua (1868–72), Siberia and southern Russia, and the United States. Fellow of the Geological Society of London. (Lightman ed. 2004, *ODNB*.)

12 January 1867, 15 January [1867]

Bennet, Charles Augustus, 6th earl of Tankerville (1810–99). Styled Lord Ossulston, 1822–59. MP for North Northumberland, 1832–59. Became sixth earl in 1859. Queen Victoria's lord steward of the household, 1867–8. (*Burke's peerage* 1999, *Modern English biography*.)

Bennet, Olivia, countess of Tankerville (1830/1–1922). Daughter of the sixth duke of Manchester. Married Charles Augustus Bennet in 1850. (*Burke's peerage* 1999, *The Times*, 17 February 1922, p. 13.)

Bentham, George (1800–84). Botanist. Moved his botanical library and collections to the Royal Botanic Gardens, Kew, in 1854, and was provided with facilities there for his research from 1861. President of the Linnean Society of London, 1861–74. Published *Genera plantarum* (1862–83) with Joseph Dalton Hooker. FRS 1862. (*DSB, ODNB*.)

Bentley, Richard (1794–1871). Publisher. Had premises from 1829 at New Burlington Street. Started *Bentley's Miscellany* in 1836. Published Charles Dickens, Benjamin Disraeli, George Cruikshank, and others. (*DNB*.)

Berthelot, Sabin (1794–1880). French naturalist and traveller. Studied in the Canary Islands, 1820–30; directed the botanic garden at La Orotava, Tenerife. Agent in Santa Cruz, Tenerife, 1847; consul, 1867. (*DBF*.)

Beverley, Robert Mackenzie (1797/8–1868). Poet and controversialist. Matriculated at Trinity College, Cambridge, 1816; LL.B., 1821. Published works critical of the church, Cambridge University, and, anonymously, Darwinian theory. (*Alum. Cantab., Modern English biography*.)

Billroth, Christian Albert Theodor (Theodor) (1829–94). German surgeon. MD, Berlin, 1852; habilitated in surgery and pathological anatomy, 1856. Appointed professor of surgery, Zurich, 1860; Vienna, 1867. His textbook on surgical pathology and therapy ran to sixteen editions and was translated into ten languages. (*DBE*.)

Birkbeck, Robert (1836–1920). Married Mary Harriet Lubbock in 1857. Resided at Kinlock Hourn, Inverness-shire. (*Burke's landed gentry* 1898.)

Blasius, Johann Heinrich (1809–70). German ornithologist and zoologist. Professor of natural history, Collegium Carolinum, Brunswick, 1836–70. Founder and director of the natural history museum in Brunswick from 1859 or later. (*ADB, DBE, NDB*.)

Blunt, Thomas (1803–74). Chemist. In business at Wyle Cop in Shrewsbury with Joseph Birch Salter, and later with his son, Thomas Porter Blunt. (Deirdre Haxton, personal communication.)

    5 April [1867]

Blunt, Thomas Porter (1842–1929). Pharmaceutical chemist. BA, Oxford, 1864. Registered as a chemist and druggist in 1868, having been in business before that date. Registered as a pharmaceutical chemist in 1878, having taken the major examination of the Pharmaceutical Society in 1877. In business with his father, Thomas Blunt, at Wyle Cop, Shrewsbury. Public analyst for Shropshire, Montgomery, and Merioneth; responsible for the botanical collection at Shropshire Museum. (*Alum. Oxon.*; Census returns 1861 (PRO RG9/1873 St Chad's

Blunt, Thomas Porter, cont.

district); Deirdre Haxton, personal communication; *Post Office directory of Gloucestershire, with Bath, Bristol, Herefordshire, and Shropshire* 1870; Royal Pharmaceutical Society; *Shrewsbury Chronicle*, 15 February 1929.)

Blunt and Salter. Pharmaceutical and analytical chemists and druggists, and soda water manufacturers, at Wyle Cop, Shrewsbury. (*Post Office directory of Gloucestershire, with Bath, Bristol, Herefordshire, and Shropshire* 1863.)

Blyth, Edward (1810–73). Zoologist. Druggist in Tooting, London, *circa* 1832–7. Wrote and edited zoological works under the pseudonym Zoophilus. Curator of the museum of the Asiatic Society of Bengal, Calcutta, India, 1841–62. Provided CD with information on the plants and animals of India in correspondence between 1855 and 1858. Returned to Britain in 1863, and continued to write on zoology and on the question of the origin of species. (*Correspondence* vols. 5–7, *DSB, ODNB*.)

[18 February 1867], [19 February 1867], *19 February 1867*, 23 February [1867], *24 February 1867, [2–30 March 1867]*

Boardman, Alexander F. (1819–76). American dry goods merchant. Born in Haiti. Educated in New Hampshire schools; attended Bowdoin College, but left because of weak eyes. Established a large and successful dry goods business in Brunswick, Maine, in 1840. (Wheeler and Wheeler 1878.)

26 January 1867

Bohn, Henry George (1796–1884). London bookseller and publisher. (*ODNB*.)

Bonnet, Charles (1720–93). Swiss naturalist and philosopher. Although educated for the law, his main interest was natural history. Discovered parthenogenesis in aphids in 1746. Studied invertebrate regeneration, entomology, and plant physiology. Advocate of preformation. Wrote influential works on theoretical biology. (*DSB*.)

Boole, Mary Everest (1832–1916). Writer and educator. Educated privately in France. In 1852, studied calculus with George Boole, whom she married in 1855. Librarian, Queen's College, London, 1865–73. Secretary to the doctor and philosopher James Hinton from 1873. Popular writer on mathematics and philosophy. Originator of Boole's Sewing Cards, an aid in teaching geometry. Many of her works focused on the psychology of learning. (*BDWS, ODNB*.)

Booth, Edward Thomas (1842–90). Ornithologist. Educated at Harrow and Trinity College, Cambridge. Collected specimens of birds of Great Britain, and stuffed them himself. The collection was eventually housed in the Dyke Road Museum, Brighton, and bequeathed to the town on Booth's death. (*Alum. Cantab.*, *Ibis* 6th ser. 2 (1890): 271–2.)

Boott, Francis (1792–1863). American-born physician and botanist. Resident in England from 1816. MD, Edinburgh, 1824. Lecturer on botany and materia medica, Webb Street School of Medicine, London, 1825. Conducted a successful medical practice in London, 1825–32. Secretary, Linnean Society of London, 1832–9; treasurer, 1856–61; vice-president, 1861–3. (R. Desmond 1994, *ODNB*.)

Bornet, Jean-Baptiste-Edouard (Edouard) (1828–1911). French botanist. Collaborated with Gustave Adolphe Thuret at Thuret's garden in Antibes. Worked especially on algae and lichens. (*DBF.*)

[before 20 August 1867], 20 August [1867]

Bouniakowsky, Victor Iakovlevich (1804–89). Russian mathematician. Vice-president, St Petersburg Academy of Sciences, 1864–89. (*GSE* s.v. Buniakovskii, Victor Iakovlevich.)

Bowker, James Henry (1822–1900). South African farmer, soldier, civil servant, and naturalist. Brother of Mary Elizabeth Barber. Inspector in the Frontier Armed and Mounted Police, Cape Colony, 1855; commandant, 1858; commanding officer, 1870. High commissioner's agent in Basutoland, 1868. Chief commissioner at the diamond fields of Griqualand West, after leading the expedition that secured their annexation to the Cape Colony in 1871. Major contributor of specimens to the South African museum's collection of Lepidoptera. Collaborated with Roland Trimen to publish *South-African butterflies: a monograph of the extra-tropical species* (Trimen 1887–9). (*DSAB*, Trimen 1887–9.)

Bowman, William (1816–92). Ophthalmic surgeon. Author of works on physiology. Assistant surgeon, King's College Hospital, London, 1839–56; surgeon, 1856–62; elected joint professor of physiology and of general and morbid anatomy, King's College, London, 1848; assistant surgeon at the Royal London Ophthalmic Hospital, Moorfields, 1846–51; surgeon, 1851–76. Created baronet, 1884. FRS 1841. (*DSB, ODNB.*)

30 July [1867], *5 August 1867,* 7 August [1867]

Brace, Charles Loring (1826–90). American philanthropist and social reformer. After studying theology at Yale University and the Union Theological Seminary, New York City, toured Europe, 1850–1. One of the founders of the New York City Children's Aid Society, 1853; secretary from 1853. (*ANB.*)

29 April 1867, 14 November 1867

Bradley, Basil (1842–1904). British landscape and genre painter. (*Dictionary of British artists.*)

Bradley, Charles (1816/17–83). Clergyman. Private tutor in Southgate, Middlesex. Fellow of Worcester College, Oxford. (*Alum. Oxon., Crockford's clerical directory,* letter from J. B. Innes, 1 September [1867].)

Brehm, Alfred Edmund (1829–84). German zoologist and traveller. Travelled in Egypt and the Sudan, Spain, Norway and Lapland, Siberia, and Turkestan. Studied natural history in Jena and Vienna, 1853–6. Director of the zoological gardens, Hamburg, 1862–7. Founded the aquarium in Berlin, 1867. Published books and articles on zoology. (*ADB, BHGW.*)

Brent, Bernard Peirce (1822–67). Bird-fancier and author. Studied pigeon breeding in France and Germany. (*CDEL*; family information.)

Bridges, Thomas (b. *c.* 1842 d. 1898). Dictionary compiler, missionary, and sheep farmer. Missionary on Keppel Island, West Falklands, 1856–68; established a mission at Ushuaia in the Beagle Channel, 1867. Settled permanently in Ushuaia

Bridges, Thomas, cont.

with his wife, Mary Ann Varder, and daughter in 1871. Abandoned missionary work in 1884 and became a sheep farmer near Ushuaia from 1887. Author of a Yámana dictionary. (E. L. Bridges 1948, *ODNB*.)

Bridgewater, 8th earl of. *See* Egerton, Francis Henry.

Briggs, Mark. Coachman. Darwin family coachman at The Mount, Shrewsbury, until 1866; thereafter lived in a cottage near The Mount. (*Emma Darwin* 1904, 2: 13.)

Bronn, Heinrich Georg (1800–62). German palaeontologist. Professor of natural science at Heidelberg University, 1833. Translated and superintended the first German editions of *Origin* (1860) and *Orchids* (1862). (*DSB*, *NDB*.)

Brooke, Charles Anthoni Johnson (1829–1917). Naval officer, colonial officer, second raja of Sarawak. Served in the Royal Navy, 1842–52. Joined his uncle James Brooke, raja of Sarawak, in 1852. Named as heir, 1867; became second raja on his uncle's death in 1868. In later life, spent part of each year in England. Knighted, 1888. (*ODNB*.)

Brooke, James (1803–68). Soldier and colonial governor. East India Company military service, 1819–30; served in the Burma War. At the invitation of Mudda Hassim, uncle of the Malay sultan of Brunei and nominal ruler of Borneo, assumed government of Borneo as raja of Sarawak, 1841. Commissioner and consul-general to the sultan and the independent chiefs in Borneo, 1847–55, and governor of the island of Labuan and its dependencies, 1847–52. Knighted, 1848. (*DNB*; *Foreign Office list* 1856.)

Brooks, J. H. Colonial civil servant. Medical officer, Seychelles, from 1858 until his retirement in 1883. (*Colonial Office list.*)

Brown, John Croumbie (1808–95). Scottish clergyman and botanist. LL.D., University of Aberdeen, 1858. Served the London Missionary Society in St Petersburg, Russia, and then in Cape Town, 1844–8. Minister to a United Presbyterian Church in Aberdeen, 1849. Lecturer in botany, Joint Medical School, Aberdeen, 1853. Colonial botanist, 1863–6, attending particularly to agricultural problems; also chair of botany at the South Africa College, Cape Town. Returned to his Aberdeen ministry in 1866. Fellow of the Linnean Society of London, 1867. (A. C. Brown 1977, p. 465, R. Desmond 1994, Gunn and Codd 1981.)

Brown, Robert (*c.* 1767–1845). Scottish nurseryman. Ran a nursery in Perth. Discovered *Menziesia caerulea* and other rare natives of the Scottish Highlands. Visited North America with James McNab, 1834. (*Gardeners' Chronicle* (1867): 755–6, R. Desmond 1994.)

Brown, Robert (1773–1858). Scottish botanist. Naturalist to the expedition surveying the coast of Australia, 1801–5; published descriptions of the plants he collected. Librarian to the Linnean Society of London, 1805–22; to Sir Joseph Banks, 1810–20. Continued as curator of Banks's collections after his death in 1820 and negotiated their transfer to the British Museum in 1827. Keeper of the botanical collections, British Museum, 1827–58. Discoverer of 'Brownian motion'. FRS 1811. (*DSB, ODNB.*)

Brown, William. Antique furniture and curiosity dealer. Had premises at 1 Spur Street, Leicester Square, and 14 Wardour Street, Soho, London. (*Post Office London directory* 1866.)

Brownlee, Charles Pacalt (1821–90). South African civil servant. Assistant commissioner for the Ngqika (Gaikas), 1848; commissioner, 1849–52, and from 1853. On annexation of British Kaffraria to Cape Colony in 1866, became civil commissioner of the newly proclaimed district of Stutterheim. Resident magistrate of Somerset East, 1868–72. Secretary for native affairs, Cape Colony, 1872–8; chief commissioner for native affairs, 1878. Chief magistrate at Kokstad, 1878–85. (*DSAB.*)

Buccleuch, 5th duke of. *See* Scott, Walter Francis.

Büchner, Friedrich Karl Christian Ludwig (Ludwig) (1824–99). German materialist philosopher and physician. Lecturer in medicine, especially forensic medicine, at Tübingen University, 1854–5. Following the publication of his first work, *Kraft und Stoff* (1855), he was debarred from academic teaching and returned to general medicine. (*DBE, NDB.*)

Buckland, Francis Trevelyan (1826–80). Naturalist, popular science-writer, and surgeon. Son of William Buckland. Trained and practised medicine at St George's hospital, London, 1848–53. Assistant surgeon in the second Life Guards, 1854–63. Staff writer for the *Field*, 1856–65. In 1865, appointed scientific referee to the South Kensington Museum, where he established an exhibit on pisciculture. Launched a weekly journal, *Land and Water*, in 1866. Inspector of salmon fisheries from 1867. (Bompas 1885, *ODNB.*)

   9 March 1867

Buckland, William (1784–1856). Geologist and clergyman. Professor of mineralogy, Oxford University, 1813; reader in geology, 1819–49. President of the Geological Society of London, 1824–5 and 1840–1. Dean of Westminster from 1845. FRS 1818. (*DNB, DSB.*)

Buckle, Henry Thomas (1821–62). Historian. Heir to a London shipowner. Travelled widely and published a number of volumes on the history of English and European civilisation. (*DNB.*)

Buffon, comte de. *See* Leclerc, Georges Louis.

Buller, Walter Lawry (1838–1906). New Zealand interpreter, lawyer, and ornithologist. Native interpreter in the Magistrate's Court, Wellington, 1855. Resident magistrate in Manawatu, 1862; transferred to Wanganui, 1865. Published *Essay on the ornithology of New Zealand* (1865). Went to London, 1871–4, to publish the *History of the birds of New Zealand* (1873). Practised as a barrister in New Zealand, 1874–86. Returned to London as commissioner for the Colonial and Indian Exhibition in 1886. Knighted, 1886. FRS 1879. (*DNZB, ODNB.*)

Bulmer, John (1833–1913). Missionary. Emigrated to Australia, arriving in Melbourne in 1853. Worked as a carpenter and goldminer in Victoria until 1855. Volunteered to help establish a mission at Yelta, on the Murray River, in 1855.

Bulmer, John, cont.

Helped establish the Lake Tyers Mission in Gippsland in 1860, and worked there until his death. (*Encyclopaedia of Aboriginal Australia.*)

Bunbury, Charles James Fox, 8th baronet (1809–86). Botanist. Collected plants in South America, 1833–4; in South Africa, 1838–9. Accompanied Charles Lyell to Madeira in 1853. Justice of the peace and deputy lieutenant for Suffolk; high sheriff, 1868. Succeeded to the baronetcy, 1860. FRS 1851. (*County families* 1875, R. Desmond 1994, Sarjeant 1980–96.)

Bunsen, Robert Wilhelm Eberhard (1811–99). German chemist. Professor of chemistry, University of Marburg, 1842; University of Heidelberg, 1852–89. Carried out important work in spectroscopy in the 1860s. Foreign member, Royal Society of London, 1858. (*DSB, NDB.*)

Burchell, William John (1781–1863). Explorer and naturalist. Schoolmaster and acting botanist to the East India Company, St Helena, South Africa, 1805–10. Collected plants in southern Africa, 1811–15, and Brazil, 1825–9. (R. Desmond 1994, *DNB.*)

Butler, Samuel (1774–1839). Educationalist and clergyman. Headmaster of Shrewsbury School, 1798–1836. Bishop of Lichfield and Coventry, 1836; of Lichfield, 1836–9. (*DNB.*)

Campbell, George Douglas, 8th duke of Argyll (1823–1900). Scottish statesman and author of works on science, religion, and politics. A defender of the concept of design in nature. Chancellor of St Andrews University, 1851–1900. President of the Royal Society of Edinburgh, 1860–4. Privy seal, 1852–5, 1859–60, 1860–6, and 1880–1; postmaster-general, 1855–8 and 1860; secretary of state for India, 1868–74. Succeeded to the dukedom in 1847. FRS 1851. (*ODNB.*)

Canby, William Marriott (1831–1904). American botanist, businessman, and philanthropist. Lived in Wilmington, Delaware. Published several articles on insectivorous plants. Amassed a substantial herbarium, which was sold to the College of Pharmacy, City of New York. (Harshberger 1899.)

Candolle, Alphonse de (1806–93). Swiss botanist, lawyer, and politician. Active in the administration of the city of Geneva until 1860. Responsible for the introduction of postage stamps to Switzerland. Professor of botany and director of the botanic gardens, Geneva, from 1835. Concentrated on his own research after 1850. Foreign member, Royal Society of London, 1869. (*DSB, Record of the Royal Society of London.*)

Carpenter, Louisa (1812–87). Daughter of Joseph Powell, a merchant in Exeter. Married William Benjamin Carpenter in 1840. (*ODNB* s.v. Carpenter, William Benjamin.)

Carpenter, William Benjamin (1813–85). Naturalist. Fullerian Professor of physiology at the Royal Institution of Great Britain, 1844–48; physiology lecturer, London Hospital, 1845–56; professor of forensic medicine, University College, London, 1849–59. Registrar of the University of London, 1856–79. President of the British Association for the Advancement of Science, 1872. Founding mem-

ber of the Marine Biological Association. FRS 1844. (*DNB*, *DSB*, *Modern English biography*, Royal Institution of Great Britain.)
[13–16 February 1867]

Carus, Julius Victor (1823–1903). German comparative anatomist. Conservator of the Museum of Comparative Anatomy, Oxford University, 1849–51. Professor extraordinarius of comparative anatomy and director of the zoological museum, University of Leipzig, 1853. Translated the third German edition of *Origin* (1867) and, subsequently, twelve other works by CD. (*DSB*, *NDB*.)
18 January 1867, 22 January [1867], *11 February 1867*, 17 February [1867], *5 April 1867*, 11 April [1867], *15 April 1867*, 18 April [1867], *11 September 1867*, 16 September 1867, *5 October 1867*, *7 October 1867*, 8 October [1867], 10 October [1867], *16 October 1867*, 19 October [1867], *30 October 1867*, 4 November [1867], *11 November 1867*, 14 November [1867], 10 December [1867]

Caspary, Johann Xaver Robert (Robert) (1818–87). German botanist. Director, Bonn herbarium, 1856. Professor of botany and director of the botanic gardens at the University of Königsberg from 1858. Specialised in aquatic plants. (*ADB*.)

Cassell, Petter, & Galpin. Publishing and printing firm with premises at La Belle Sauvage printing works, Ludgate Hill, London. (*Post Office London directory* 1866.)
24 August [1867]

Cayley, Arthur (1821–95). Mathematician and conveyancer. Fellow of Trinity College, Cambridge, 1842–6. Studied law at Lincoln's Inn and was called to the bar in 1849. Practised law until 1863. Sadlerian Professor of pure mathematics at Cambridge, 1863–95. FRS 1852. (*DSB*, *ODNB*.)

Cheadle, Walter Butler (1835–1910). Physician. Accompanied William Wentworth Fitzwilliam, Viscount Milton, on his travels in north-west America and across the Rocky Mountains, 1862–4. Elected physician to the Western General Dispensary, London, 1865; assistant physician to St Mary's Hospital, London, 1867; assistant physician to the Hospital for Sick Children, London, 1869–92. (*ODNB*.)

John Churchill & Sons. Medical publishing firm with premises at 11 New Burlington Street, London. (*Post Office London directory* 1866.)

Clark, George. Naturalist. Resident of Mahébourg, Mauritius. Corresponding member of the Zoological Society of London. (G. Clark 1866, *Proceedings of the Zoological Society of London* (1869).)

Clarke, Benjamin (1813–90). Systematic botanist. Trained as a medical practitioner, but rarely practised. Devised his own system of classification. Fellow of the Linnean Society of London, 1845. (R. Desmond 1994, *Journal of Botany* 28 (1890): 84–6.)
12 March 1867, 25 March 1867

Claus, Carl Friedrich (1835–99). German zoologist. Studied medicine, mathematics, and zoology at Marburg and Giessen, 1854–7. Professor of zoology, Marburg, 1863; Göttingen, 1870. Professor of zoology and head of the institute

Claus, Carl Friedrich, cont.
of zoological and comparative anatomy at Vienna, 1873. Founder and first director of the zoological research station at Trieste, 1873. Did major research in environmental influences on variability, especially in Crustacea. A strong supporter of Darwin in both his writing and lecturing. His zoology textbook was a standard work in the last three decades of the nineteenth century. (*DBE, NDB, OBL.*)

Clayton, Richard, 1st baronet (1745–1828). Translator. Called to the bar, 1771. Recorder of Wigan, 1815–28; constable of Lancaster Castle; British consul, Nantes, 1825–8. Married Ann, daughter of Charles White, in 1780. Published several translations of works in French, Latin, and Italian. Created baronet, 1774. (*ODNB.*)

Clowes, William (1807–83). London printer. Together with his brothers Winchester (1808–62) and George (1814–86), took over the printing firm established by his father, William Clowes (1779–1847). (*ODNB.*)

Clowes, William & Sons. Printers. William Clowes (1807–83), eldest son of William Clowes (1779–1847), joined his father's printing business in 1823; the name of the firm was changed to William Clowes & Sons in 1839. Printed the official catalogue of the Great Exhibition of 1851. Introduced improvements in type-music printing. Printers to John Murray. (*ODNB.*)

Cobden, Richard (1804–65). Manufacturer and politician. Leading member of the Anti-Corn Law League, 1838–46. MP for Stockport, 1841–7; for West Riding, Yorkshire, 1847–57; for Rochdale, 1859–65. Vocal advocate for free trade and for non-intervention. (*ODNB.*)

Colenso, William (1811–99). English-born New Zealand botanist and missionary. Worked with the London firm of Richard Watts, printer to the Church Missionary Society, 1833. Left to run a printing press for the CMS in Paihia in the Bay of Islands, New Zealand, in 1834. Published Christian religious works in the Maori language and official publications in English and Maori. Ordained deacon, 1844, and went with his family to a new mission station in Hawke's Bay. Had a son by a Maori woman in 1851, and was suspended as deacon and dismissed from the mission, 1852. Elected to the Provincial Council, 1859, and the General Assembly, 1861–6. Published scientific papers and was commissioned by the General Assembly to produce a Maori dictionary, which he worked on until his death. FRS 1886. (*DNZB, Record of the Royal Society of London.*)

Colling, Charles (1751–1836). Stockbreeder. One of the earliest and most successful improvers of shorthorn cattle. Had a farm at Ketton, near Darlington, Durham. (*ODNB.*)

Colling, Robert (1749–1820). Stockbreeder. Sold to his brother, Charles Colling, the bull from which the improved stock of shorthorns was bred. Farmed at Barmpton, Durham. (*ODNB.*)

Colloredo-Mannsfeld, Joseph Franz Hieronymus Fürst von (1813–95). Fifth prince of Colloredo-Mansfeld. (*Genealogisches Handbuch des Adels.*)

Columella, Lucius Junius Moderatus (*fl.* A.D. 50). Agricultural writer of Gades (Cádiz). Author of *De re rustica* and *De arboribus*. (*Oxford classical dictionary* 3d ed. rev.)

Colvile, James William (1810–80). Judge in India. Appointed advocate-general of Bengal, 1845; chief justice of Bengal, 1855–9. President of the Asiatic Society of Bengal. Knighted, 1848. FRS 1875. (*ODNB*.)

Comte, Isidore Auguste Marie François Xavier (Auguste) (1798–1857). French philosopher. Private secretary to Claude Henri de Rouvroy, comte de Saint-Simon, 1817–23. Founded the Association Polytechnique, a group devoted to education of the working classes, in 1830; and the Société Positiviste, devoted to the promulgation of the 'Cult of Humanity', in 1848. Adopted the term 'positivism' for his philosophy. (*DSB*.)

Cook, James (1728–79). Naval officer, navigator, explorer, and marine surveyor. Commander of several voyages of discovery. Circumnavigated the world, 1768–71 and 1772–5. FRS 1776. (*DSB, ODNB, Record of the Royal Society of London*.)

Cooper, James Fenimore (1789–1851). American novelist. (*ANB*.)

Corbet, Athelstan (1837–1926). Clergyman. BA, Cambridge, 1860; MA, 1866. Ordained deacon, 1861; priest, 1862. Rector of Adderly, 1863–1900. (*Alum. Cantab.*, *Crockford's clerical directory*.)

Cranworth, 1st Baron Cranworth of. *See* Rolfe, Robert Monsey.

Cresy, Edward (1824–70). Surveyor and civil engineer. Son of Edward Cresy (1792–1858), the architect and civil engineer who advised CD about the purchase and improvement of Down House. Worked as an architectural draftsman in his father's office as a young man. Assisted his father in preparing his *Cyclopædia of civil engineering* in 1845. Assistant surveyor under the commissioners of sewers, 1849; afterwards engineer. Principal assistant clerk at the Metropolitan Board of Works, 1859. Architect to the fire brigade, 1866. Founder member of the Geologists' Association, 1858; president, 1864–5; vice-president, 1865–70. (*Annual Report of the Geologists' Association* 1859–70; Census returns 1861 (Public Record Office RG9/422: 118a); *DNB* s.v. Cresy, Edward (1792–1858); *Engineer* 30 (1870): 409.)
   6 June 1867, 20 November 1867

Cresy, Mary Louis (b. 1820/1). Wife of Edward Cresy (1824–70). (Census returns 1861 (Public Record Office, RG9/422: 118a), family communication.)

Crotch, George Robert (1842–74). Entomologist. Specialised in the Coleoptera, especially Coccinellidae and Erotylidae. BA, St John's, Cambridge, 1846. Second assistant librarian, Cambridge University Library, from 1867. Rearranged the insect collection at the Museum of Zoology and Comparative Anatomy, Cambridge, in 1871. Embarked for North America to collect specimens in 1872. Worked at the Museum of Comparative Zoology, Harvard, 1873–4. (Smart and Wager 1977.)

Crüger, Hermann (1818–64). German pharmacist and botanist. Apothecary in Trinidad in the West Indies from 1841; government botanist and director of the botanic garden, Trinidad, from 1857. Collected plants in Jamaica, Trinidad, and

Crüger, Hermann, cont.
Venezuela. (R. Desmond 1994, S[chlechtenda]l 1864.)
Cunningham, Robert Oliver (1841–1918). Naturalist. MD Edinburgh. Collected plants in South America. Naturalist to HMS *Nassau* on an expedition to the Strait of Magellan and the west coast of Patagonia, 1866–9. Professor of natural history, Belfast, 1871–1902. (Cunningham 1871, R. Desmond 1994, Newman 1993.)
Cuvier, Jean Léopold Nicolas Frédéric (Georges) (1769–1832). French systematist, comparative anatomist, palaeontologist, and administrator. Professor of natural history, Collège de France, 1800–32; professor of comparative anatomy, Muséum d'Histoire Naturelle, 1802–32. Permanent secretary to the Académie des Sciences from 1803. Foreign member, Royal Society of London, 1806. (*DBF*, *DSB*.)
Dallas, William Sweetland (1824–90). Entomologist, author, and translator. Prepared lists of insects for the British Museum, 1847–58. Curator of the Yorkshire Philosophical Society's museum, 1858–68. Assistant secretary to the Geological Society of London, 1868–90. Translated Fritz Müller, *Für Darwin* (1869); prepared the index for *Variation* and the glossary for *Origin* 6th ed. Editor, *Annals and Magazine of Natural History*, 1868–90, *Popular Science Review*, 1877–80. (Freeman 1978, *Geological Magazine* n.s. decade 3, vol. 7 (1890): 333–6, *Modern English biography*, Sarjeant 1980–96.)
    4 November 1867, 8 November [1867], *10 November 1867*, *20 November 1867*, *8 December 1867*, *10 December 1867*, *11 December 1867*, *26 December 1867*
Dana, James Dwight (1813–95). American geologist and zoologist. Geologist and mineralogist with Charles Wilkes's expedition to the South Seas, 1838–42; wrote reports on the geology, zoophytes, and Crustacea. An associate editor of the *American Journal of Science and Arts* from 1846. Professor of natural history, Yale University, 1855–64; professor of geology and mineralogy, 1864–90. Foreign member, Royal Society of London, 1884. (*ANB*, *DSB*, *Record of the Royal Society of London*.)
Dareste, Gabriel-Madeleine-Camille (Camille) (1822–99). French zoologist. A specialist in experimental embryology. Doctor of medicine, 1847. Doctor of science, 1851. Taught natural history at various provincial institutions. Professor of zoology, University of Lille, 1864–72. Professor of ichthyology and herpetology, Muséum d'Histoire Naturelle, Paris, 1872. Director of the laboratory of teratology, later attached to the Ecole des Hautes-Etudes, 1875. Awarded the grand prize in physiology by the Académie des Sciences for *Recherches sur la production artificielle de monstruosités* (1877). (*DBF*, *Dictionnaire universel des contemporains*.)
    19 May 1867, 23 May 1867,
Darwin, Elizabeth (Bessy) (1847–1926). CD's daughter. (*Darwin pedigree*, Freeman 1978.)
Darwin, Emma (1808–96). Youngest daughter of Josiah Wedgwood II. Married CD, her cousin, in 1839. (*Emma Darwin* (1904) and (1915).)
Darwin, Erasmus Alvey (1804–81). CD's brother. Attended Shrewsbury School, 1815–22. Matriculated at Christ's College, Cambridge, 1822; Edinburgh Univer-

sity, 1825–6. Qualified in medicine but never practised. Lived in London from 1829. (*Alum. Cantab.*, Freeman 1978.)

[before 3 February 1867?] (Emma Darwin), *3 March 1867, 22 [March 1867], 17 [July 1867], [before 19 November 1867]* (Emma Darwin)

Darwin, Francis (1848–1925). CD's son. Botanist. BA, Trinity College, Cambridge, 1870. Qualified as a physician but did not practise. Collaborated with CD on several botanical projects, 1875–82. Lecturer in botany, Cambridge University, 1884; reader, 1888–1904. Published *LL* and *ML*. President of the British Association for the Advancement of Science, 1908. Knighted, 1913. FRS 1882. (*DSB, ODNB.*)

Darwin, George Howard (1845–1912). CD's son. Mathematician. BA, Trinity College, Cambridge, 1868; fellow, 1868–78. Studied law in London, 1869–72; called to the bar in 1872 but did not practise. Plumian Professor of astronomy and experimental philosophy, Cambridge University, 1883–1912. President of the British Association for the Advancement of Science, 1905. Knighted, 1905. FRS 1879. (*DSB, Men-at-the-bar, ODNB.*)

27 May [1867], *[3 June 1867]*

Darwin, Henrietta Emma (1843–1927). CD's daughter. Married Richard Buckley Litchfield (*Alum. Cantab.*) in 1871. Assisted CD with some of his work. Edited *Emma Darwin* (1904) and (1915). (*Burke's landed gentry* 1952, *Correspondence* vol. 11, Freeman 1978.)

26 July [1867]

Darwin, Horace (1851–1928). CD's son. Civil engineer. BA, Trinity College, Cambridge, 1874. Apprenticed to an engineering firm in Kent; returned to Cambridge in 1877 to design and make scientific instruments. Founder and director of the Cambridge Scientific Instrument Company. Mayor of Cambridge, 1896–7. Knighted, 1918. FRS 1903. (*Alum. Cantab., ODNB.*)

Darwin, Leonard (1850–1943). CD's son. Military engineer. Attended the Royal Military Academy, Woolwich. Commissioned in the Royal Engineers, 1870; major, 1889. Served on several scientific expeditions, including those for the observation of the transit of Venus in 1874 and 1882. Instructor in chemistry and photography, School of Military Engineering, Chatham, 1877–82. Intelligence service, War Office, 1885–90. Liberal Unionist MP, Lichfield division of Staffordshire, 1892–5. President, Royal Geographical Society of London, 1908–11; Eugenics Education Society, 1911–28. Chairman, Bedford College, London University, 1913–20. (M. Keynes 1943, Sarjeant 1980–96, *WWW.*)

Darwin, Robert Waring (1766–1848). CD's father. Physician. Had a large practice in Shrewsbury and resided at The Mount. Son of Erasmus Darwin (*ODNB*) and his first wife, Mary Howard. Married Susannah, daughter of Josiah Wedgwood I (*ODNB*), in 1796. FRS 1788. (Freeman 1978.)

Darwin, Susan Elizabeth (1803–66). CD's sister. Lived at The Mount, Shrewsbury, the family home, until her death. (*Darwin pedigree*, Freeman 1978.)

Darwin, William Erasmus (1839–1914). CD's eldest son. Banker. BA, Christ's

Darwin, William Erasmus, cont.

College, Cambridge, 1862. Partner in the Southampton and Hampshire Bank, Southampton, 1861. Chairman of the Southampton Water Company. Amateur photographer. (*Alum. Cantab.*, Darwin 1914.)

   27 [March 1867], 20 November [1867]

Davidson, Thomas (1817–85). Artist and palaeontologist. Fellow of the Geological Society of London. Expert on fossil brachiopods. FRS 1857. (*ODNB*, Sarjeant 1980–96.)

Dawes, Richard (1793–1867). Clergyman. Mathematical tutor and bursar of Downing College, Cambridge, 1818. Rector of King's Somborne, Hampshire, 1836–50; dean of Hereford, 1850; master of St Catherine's Hospital, Ledbury, 1861. Founded schools for the children of the poor and took an interest in Ledbury's national schools. (*DNB*.)

Dawkins, William Boyd (1837–1929). Geologist and palaeontologist. Member of the Geological Survey of Great Britain, 1861–9. Curator of natural history, Manchester Museum, 1869. Professor of geology, Owens College, Manchester, 1874–1908. Specialised in fossil mammals. FRS 1867. (*ODNB*.)

   22 August 1867, 26 August [1867], *27 August 1867, 7 September 1867*

Dawson, John William (1820–99). Canadian geologist and educationalist. Investigated the geology of the maritime provinces with Charles Lyell in 1842 and 1852. Superintendent of education for common schools in Nova Scotia, 1850. Appointed principal and professor of geology at McGill University, 1855. Knighted, 1884. FRS 1862. (*DNB, DSB*.)

Delpino, Federico (1833–1905). Italian botanist. Travelled extensively for botanical purposes as a youth and in 1873. Civil servant, ministry of finances, Turin, 1852–6; assistant in botanic garden and museum, Florence, 1867; lecturer, Vallombrosa school of forestry, 1871; professor of botany and director of botanic garden, Genoa, 1875–84; professor, University of Bologna, 1884; professor of botany and head of botanic garden, Naples, 1894. (*DBI*, Mayerhöfer 1959–70, Penzig 1905.)

   5 September 1867

De Morgan, Augustus (1806–71). Mathematician and historian. BA, Trinity College, Cambridge, 1827. Professor of mathematics, London University, 1828–31 and 1836–66. Elected a fellow of the Royal Astronomical Society, 1828; served at various times as secretary and vice-president. Member of the Society for the Diffusion of Useful Knowledge, which published many of his works. (*ODNB*.)

De Morgan, Sophia Elizabeth (1809–92). Author and spiritualist. Daughter of William Frend. Married Augustus De Morgan in 1837. Author of *From matter to spirit; the result of ten year's experience in spirit manifestations* (1863). Active in the anti-slavery movement, the reform of prisons and workhouses, the founding of Bedford College for Women, and the women's suffrage movement. (*ODNB* s.v. De Morgan, Augustus, and De Morgan, William Frend.)

Denis, Alphonse Amaranthe Dugomier (1794–1876). French local politician.

Mayor of Hyères, France, 1830–46. Introduced many exotic plants to the city and supported a garden of acclimatisation. (*DBF.*)

Derby, earl of. *See* Stanley, Edward George Geoffrey Smith.

Dickens, Charles (1812–70). Novelist. (*ODNB.*)

Disraeli, Benjamin (1804–81). Author and Conservative politician. Prime minister, 1868, 1874–6. Created earl of Beaconsfield, 1876. (*ODNB.*)

Dohrn, Felix Anton (Anton) (1840–1909). German zoologist. Studied medicine and zoology at various German universities. PhD, Breslau, 1865. Studied with Ernst Haeckel and became Haeckel's first assistant at Jena, where he habilitated in 1868. Founded the Zoological Station at Naples, built between 1872 and 1874. The station was the first marine laboratory, and served as a model for other similar institutions throughout the world. (*DBE, DSB,* Heuss 1991.)
26 November [1867], *30 November 1867*

Dominy, John (1816–91). Gardener. Worked at the nurseries of James and James Veitch in Exeter (1834–41) and Chelsea (*c.* 1846–80). Grew the first known artificially produced orchid hybrid, *Calanthe dominii*. Hybridised nepenthes and fuchsias. (R. Desmond 1994.)

Douglas, Dunbar James, 6th earl of Selkirk (1809–85). Became sixth earl of Selkirk in 1820. FRS 1831. (*Burke's peerage.*)

Drouet, Henri (1829–1900). French administrator and zoologist. Went on a scientific voyage to the Azores, French Guyana, and Angola, in 1854. Became principal private secretary to the prefect of Vienna in 1861. Adviser to the prefectures of Ardennes, Vaucluse, and the Côte-d'Or, 1863–4. Sometime inspector of prison services in Algeria. (*DBF.*)

Duchenne, Guillaume Benjamin Amand (1806–75). French physician. Studied medicine at Paris, then practised in his native town of Boulogne-sur-Mer. Returned to Paris in 1855 and carried out experiments on the therapeutic use of electricity. One of the founders of neurology. (*DBF.*)

Duncan, Peter Martin (1821–91). Physician, zoologist, and geologist. Physician, Essex and Colchester Hospital, 1848–59; consultant physician, county asylum and Oldham Club. Practised at Blackheath from 1860. Professor of geology, King's College, London, 1870. Secretary of the Geological Society of London, 1864–70; president, 1876–8. Specialist on living and fossil corals and Mesozoic echinoids. FRS 1868. (*DNB, Medical directory* 1849–76.)

Edwards, Ernest (1837–1903). Photographer. Ran a photographic studio in London, specialising in portraits, 1864–9. Moved to Boston, and later New York, where he opened commercial firms in landscape photography and colour printing. (Johnson 1990, *Post Office London directory,* Pritchard 1994.)

Egerton, Francis Henry, 8th earl of Bridgewater (1756–1829). Scholar and patron of learning. His literary works, many on his notable ancestors, were mostly privately published. Bequeathed £8000 in his will to commission works illustrating the 'power, wisdom, and goodness of God as manifested in the Creation'; the money to be divided among eight persons. The resulting essays have

Egerton, Francis Henry, cont.
   become known as the Bridgewater treatises. FRS 1781. (*ODNB*, *Record of the Royal Society of London*, Topham 1993.)
Elers, David (b. 1656?). Dutch-born silversmith and potter. Came to London with his brother, John Philip, in the 1680s. They had a pottery at Vauxhall producing slipcast salt-glaze stoneware, and from about 1690 until 1700 a pottery at Bradwell Wood in Staffordshire producing red stoneware. The brothers were famous for their innovative techniques. (G. Elliott 1998.)
Elers, John Philip (b. 1664?). Dutch-born silversmith and potter. Came to London with his brother, Paul, in the 1680s. John went to Dublin in 1700 and was set up in business by Lady Barrington. (G. Elliott 1998.)
Eliot, George. *See* Evans, Marian (or Mary Anne).
Elliot, Charles (1801–75). Naval officer and colonial governor. Entered the navy in 1815, lieutenant, 1822, commander of a hospital ship, 1826, post rank, 1828. After 1828, served in the foreign or colonial office, including posts in Guiana, China during the opium war, and Texas. Served as governor of Bermuda, 1846–54, of Trinidad, 1854–6, and of St Helena, 1863–9. In retirement received honorary naval promotions, including admiral in 1865. Knighted, 1856. (*ODNB*.)
Elliot, Walter (1803–87). East India Company servant and archaeologist. Commissioner for the administration of the Northern Circars, 1845–54. Member of the council of the governor of Madras, 1854–60. Wrote articles on Indian natural history and culture. Knighted, 1866. FRS 1878. (*ODNB*, *Record of the Royal Society of London.*)
Endlicher, Stephan Ladislaus (1804–49). German botanist. (*NDB.*)
Engelmann, Georg (George) (1809–84). German-born physician and botanist. MD, University of Würzburg, 1831. Went to the US in 1832. A founder of the St Louis Academy of Science, Missouri, 1856, and member of numerous scientific societies. Played a principal role in the establishment of the botanical garden at St Louis. Made fundamental contributions to the classification and taxonomy of many plant families, especially grapes, cacti, and yuccas. Discovered disease-resistant grape species and the role of the pronuba moth in pollination of yuccas. Provided thousands of plant specimens to collections in Berlin and St Petersburg. (*ANB*, *DAB.*)
Wilhelm Engelmann. German publishing firm. Founded in 1833, based in Leipzig. Publisher of *Zeitschrift für wissenschaftliche Zoologie* from 1848. (*NDB*, *Zeitschrift für wissenschaftliche Zoologie.*)
Engleheart, Stephen Paul (1831/2–85). Surgeon. Member of the Royal College of Surgeons of England, 1859; licentiate of the Royal College of Surgeons of Edinburgh, 1860. Surgeon in Down, Kent, 1861–70. Medical officer, Second District, Bromley Union, 1863–70; divisional surgeon of police, 1863–70. Resident in Shelton, Norfolk, 1870–81; in Old Calabar, Nigeria, 1882–5. (*Medical directory* 1861–86, *Post Office directory of the six home counties* 1862.)
   [April 1867?]

Erskine, Claudius James (1821–93). Civil servant in India. Private secretary to the governor of Bombay, 1843; secretary, judicial department of Bombay, 1854; director of public instruction, Western India, 1855–9; high court judge, Bombay, 1862–3; member of the Bombay Council, 1865. Retired because of ill health, 1867. Son of Maitland and William Erskine. (*DNB* and *ODNB* s.v. William Erskine (1773–1852), *Modern English biography*.)

Erskine, Henry Napier Bruce (1831–93). Civil servant in India. Arrived in Bombay in 1853; sub-collector, Ahmadnagar and Shollapoor, 1868; collector and magistrate, Nasik, 1869; commissioner, northern division of Bombay, 1877–9; Sind, 1879–87. Son of Maitland and William Erskine. (*ODNB* s.v. William Erskine (1773–1852), *India list, Modern English biography*.)
    1 November 1867 (Frances Julia Wedgwood)

Evans, Marian (or Mary Anne) (1819–80). Novelist. Published under the name George Eliot. (*ODNB*.)

Eyton, Thomas Campbell (1809–80). Shropshire naturalist. Friend and Cambridge contemporary of CD. Author of several works on natural history. On coming into possession of the family estate at Eyton, Shropshire, in 1855, he built a museum for which he formed a collection of skins and skeletons of European birds. An active magistrate throughout his adult life. (*ODNB*.)

Farrar, Frederic William (1831–1903). Anglican clergyman and headmaster. Master at Harrow school, 1855–70. Canon of Westminster, 1876–95. Dean of Canterbury, 1895–1903. Promoted scientific education. Noted for his school stories, writings on language, and biographies of Christian figures. FRS 1866. (*DNB*.)
    5 March 1867, 7 March [1867]

Fenton, Francis Dart (1820–5? – 1898). English-born New Zealand public administrator and musician. Solicitor in Huddersfield; emigrated to New Zealand for reasons of health, 1850. Taught music at a mission school in Auckland; became clerk in the Registry of Deeds in 1851. Held a succession of judicial and administrative positions in the public service; chief judge of the Native Land Court, 1865–82. Published *Observations on the state of the aboriginal inhabitants of New Zealand* (1859), the result of the first census of the Maori people. (*DNZB*.)

FitzRoy, Robert (1805–65). Naval officer, hydrographer, and meteorologist. Commander of HMS *Beagle*, 1828–36. Tory MP for Durham, 1841–3. Governor of New Zealand, 1843–5. Superintendent of the dockyard at Woolwich, 1848–50. Chief of the meteorological department at the Board of Trade, 1854; chief of the Meteorological Office from 1855. Rear-admiral, 1857; vice-admiral, 1863. FRS 1851. (*DNB, DSB*.)

Fitzwilliam, Charles William Wentworth, 3d and 5th Earl Fitzwilliam (1786–1857). Politician. Whig MP for Malton, 1806; Yorkshire, 1807–30; Peterborough, 1830; Northamptonshire, 1831–3. President of the Yorkshire Philosophical Society, 1831–57; president of the British Association for the Advancement of Science, 1831. President of the Statistical Society of London three times between 1838 and 1855. Succeeded to the earldom in 1833. FRS 1811. (*ODNB*.)

Fitzwilliam, William Wentworth, Viscount Milton (1839–77). Politician. Entered Trinity College, Cambridge, in 1859. MP for South West Yorkshire, 1865–72. (*Alum. Cantab.*, *Complete peerage*, vol. 5, *Modern English biography*.)

Ford, George Henry (1809–76). Artist, lithographer, and illustrator. Assistant to Andrew Smith, 1821–4. Artist, Cape Town Museum, 1825; expedition for exploring Central Africa, 1834–6. Moved to England in 1837. Artist, British Museum, 1837 until *circa* 1875. Friend of Albert Charles Lewis Günther; illustrator of several of Günther's works; provided illustrations for the second volume of *Descent*. (Gunther 1972, Gunther 1975.)

Forster, Johann Georg Adam (Georg) (1754–94). German naturalist, writer, and ethnologist. Son of Johann Reinhold Forster. Translator and naturalist on James Cook's second voyage round the world, 1772–5. Professor of natural history, Kassel, 1779–84; Vilneus, 1784–7. Librarian, Mainz, 1788–92. Representative of German national convention in Paris, 1792. (*DBE*, *ODNB*.)

Forster, Johann Reinhold (1729–98). German traveller, naturalist, and geographer. Accompanied James Cook on his circumnavigation of the globe, 1772–5. Foreign member, Royal Society, 1772. (*DSB*, *NDB*.)

Fouqué, Friedrich Heinrich Karl, Baron de la Motte (1777–1843). German writer. Served in the Prussian army in campaigns against the French in 1794 and 1813. Best known for novels with themes derived from medieval chivalry and northern mythology. (*DBE*, *EB*.)

Fox, Charles Woodd (1847–1908). Barrister. Son of Ellen Sophia and William Darwin Fox. (*Alum. Oxon.*, *Repton School register*.)

Fox, Frederick William (b. 1855). Son of Ellen Sophia and William Darwin Fox. (*Darwin pedigree*.)

Fox, Robert Gerard (b. 1849). Justice of the peace, Hampshire. Son of Ellen Sophia and William Darwin Fox. (*Darwin pedigree*.)

Fox, William Darwin (1805–80). Clergyman. CD's second cousin. A friend of CD's at Cambridge; introduced CD to entomology. Maintained an active interest in natural history throughout his life and provided CD with much information. Rector of Delamere, Cheshire, 1838–73. Spent the last years of his life at Sandown, Isle of Wight. (*Alum. Cantab.*, *Autobiography*, *Correspondence*.)

1 February [1867], 6 February [1867]

Frankland, Edward (1825–99). Chemist. Professor of chemistry, Putney College for Civil Engineering, 1850, and Owens College, Manchester, 1851–7. Lecturer in chemistry, St Bartholomew's Hospital, London, 1857–64. Professor of chemistry, Royal Institution, 1863–8, and Royal College of Chemistry, 1865. President of the Chemical Society, 1871–3; of the Institute of Chemistry, 1877–80. Knighted, 1897. FRS 1853. (*DNB*, *DSB*.)

Frauenfeld, Georg von (1807–73). Austrian naturalist. Founding member of the zoological-botanical society, Vienna; secretary, 1851–73. Took part in the *Novara* expedition, 1859. (*OBL*.)

Froude, James Anthony (1818–94). Historian and writer. Disciple of Thomas

Carlyle; published *History of England* (1856–70), *The English in Ireland in the eighteenth century* (1872–4), and *The English in the West Indies* (1888). Editor of *Fraser's Magazine* 1860–74. Regius professor of modern history, Oxford University, 1892–4. (*DNB.*)

Fuller, Frederick Richardson (b. *c.* 1830 d. 1876). Taxidermist. Emigrated from England to Australia, 1849. Settled in Otago, New Zealand, *circa* 1862 and worked on ornithological collections. Assisted Julius von Haast from 1865. Taxidermist at the Canterbury Museum, Christchurch, New Zealand. Committed suicide after being dismissed from his position. (Greenaway 2000.)

Gaika, Christian (1831–1905). Xhosa interpreter and constable. Educated by missionaries. Interpreter and constable in the division of Bedford, Cape Colony, from 1867 or before until 1887 or later. Son of Ngqika (Gaika), a Xhosa chief. Brother of Chief Sandili. (*Cape of Good Hope general directory*, letter from J. P. M. Weale, 7 July 1867, Shanafelt 2003, pp. 822–4, 842.)

Gaskell, Elizabeth Cleghorn (1810–65). Novelist. (*ODNB.*)

Gärtner, Karl Friedrich von (1772–1850). German physician and botanist. Practised medicine in Calw, Germany, from 1796, but left medical practice in 1800 to pursue a career in botany. Travelled in England and Holland in 1802. Studied plant hybridisation from *circa* 1824. Elected a member of the Deutsche Akademie der Naturforscher Leopoldina, 1826. Ennobled, 1846. (*ADB, DBE, DSB.*)

Gätke, Heinrich (1814–97). German ornithologist. Worked as a merchant and a painter, then moved to Helgoland where he became a government secretary. A self-taught ornithologist, his publications drew attention to Helgoland (in the North Sea) as a centre of bird migration. The Prussian government purchased his bird collection in 1891 for the North Sea museum of the Biological Institute. (*DBE.*)

Gaudry, Albert-Jean (Albert) (1827–1908). French palaeontologist. Assistant to his brother-in-law, Alcide d'Orbigny, professor of palaeontology at the Muséum d'Histoire Naturelle. Carried out excavations at Pikermi, Attica, in 1855 and 1860, and published *Animaux fossiles et géologie de l'Attique* (1862–7). Studied the fossils of small reptiles and batrachians, 1866–92. Taught a course in palaeontology at the Sorbonne, 1868–71; appointed professor of palaeontology at the Muséum, 1872. (*DBF, DSB.*)

22 May 1867, 27 May [1867]

Geach, Frederick F. Mining engineer in Malaya. Worked for the Portuguese government in Timor, where he met Alfred Russel Wallace in 1861, and was working in Malacca (Melaka) in 1867. (Marchant ed. 1916; letter from A. R. Wallace, 2 March [1867].)

June 1867

Gegenbaur, Carl (or Karl) (1826–1903). German anatomist and zoologist. A supporter of CD; emphasised the importance of comparative anatomy in evolutionary reconstruction. Professor extraordinarius of zoology, Jena, 1855–8; professor of anatomy and zoology, 1858–62; of anatomy, 1862–73. Professor of anatomy

Gegenbaur, Carl (or Karl), cont.
and comparative anatomy, Heidelberg, 1873–1901. Elected to the Deutsche Akademie der Naturforscher Leopoldina, 1857. (*DBE, DSB, NDB.*)

Geoffroy Saint-Hilaire, Etienne (1772–1844). French zoologist. Professor of zoology, Muséum d'Histoire Naturelle, 1793. Devoted much attention to embryology and teratology. (*DBF, DSB.*)

Geoffroy Saint-Hilaire, Isidore (1805–61). French zoologist. Succeeded his father, Etienne Geoffroy Saint-Hilaire, as professor at the Muséum d'Histoire Naturelle in 1841. Continued his father's work in teratology. Became professor of zoology at the Sorbonne in 1850. (*DBF, DSB.*)

Gerstaecker, Carl Edouard Adolph (Adolph) (1828–95). German zoologist. Studied medicine at the University of Berlin and passed the state medical examination in 1852 but never practised. Habilitated in zoology, University of Berlin, 1856; became keeper of the entomological collection at the Berlin zoological museum, 1856. Docent in zoology at the agricultural institute, Berlin, 1864; professor extraordinarius, 1874. Professor of zoology, University of Greifswald, 1876. Co-editor with Julius Victor Carus of the *Handbuch der Zoologie* (1863), for which he wrote the chapter on arthropods. (*ADB, NDB.*)

Gervais, Paul (1816–79). French zoologist and palaeontologist. Assistant, Muséum d'Histoire Naturelle, Paris, 1835–45. Professor of zoology and comparative anatomy, Faculté des sciences de Montpellier, 1845; head of the faculty, 1856. Professor of comparative anatomy, Muséum d'Histoire Naturelle, 1868. (Tort 1996.)

Gibbons, Helen Gordon (1772–1855). Daughter of the reverend Edmund Dana, vicar of Wroxeter, who married the third daughter of Lord Kinnaird. Married John Gibbons and had three children, including William Henry Kinnaird Gibbons. (Deirdre Haxton, personal communication.)

Gibbons, John (1769–1858). Clergyman. Admitted as pensioner to Clare College, Cambridge, 1786; achieved fifth place in mathematics; MA 1794; fellow of the college, 1791–1800. Rector of Harley, Shropshire, 1805–58. (*Alum. Cantab.*)

Gibbons, William Henry Kinnaird (b. 1802). Son of John and Helen Gordon Gibbons. Attended Oswestry School and Christ's College, Cambridge. Resided in Harley, Shropshire, then East Grinstead, Sussex, and London. (*Alum. Cantab.*; Deirdre Haxton, personal communication; letter from W. H. K. Gibbons, 7 February 1867.)

7 February 1867

Gibbs, George (1815–73). American ethnologist and geologist. Graduated from Harvard University with a law degree, 1838; practised in New York City. Left New York in 1849; settled in Astoria, Oregon. Compiled vocabulary lists of indigenous languages and worked on maps of the area west of the Sacramento Valley in 1851. Worked as geologist and ethnologist surveying a railroad route to the Pacific in 1853. Moved in 1854 to Washington Territory, where he continued his ethnological and surveying work; worked for the North-west Boundary Survey, 1857–8. Returned east in 1860, and settled in Washington, D.C., in 1861.

Wrote on Native American linguistics; the Smithsonian Institution published several of his works. Clerk to a commission investigating British claims in the American North-West, 1865–9. (*ANB*.)

31 March 1867

Giebel, Christoph Gottfried Andreas (1820–81). German palaeontologist and zoologist. PhD, Halle, 1845; professor extraordinarius, 1858; professor, 1861. (*NDB*.)

Gmelin, Johann Georg (1709–55). German botanist, naturalist, and explorer. Professor of chemistry and natural history at the Academy of Sciences, St Petersburg, 1731–47. Professor of botany and chemistry at Tübingen University, 1749. (*DSB, NDB*.)

Godkin, Edwin Lawrence (1831–1902). Irish-born newspaper and magazine editor. BA, Queen's College, Belfast, 1851. War correspondent for the London *Daily News*, 1854–6. Emigrated to the United States, 1856. Studied law and practised briefly in New York, 1857–9. Editor-in-chief, the *Nation*, 1865–81. Co-editor, New York *Evening Post*, 1881; editor 1883–1900. (*ANB*.)

Godman, Frederick Du Cane (1834–1919). Ornithologist and entomologist. Educated at Eton, and Trinity College, Cambridge. Travelled in Central America and the Azores. Undertook, with Osbert Salvin, a comprehensive serial publication on the botany and zoology of Central America. FRS 1882. (*Ibis* 9th ser. 2 (1908): 81–92, *Record of the Royal Society of London*.)

21 December [1867], 23 December [1867]

Goethe, Johann Wolfgang von (1749–1832). German poet and naturalist. (*DSB, NDB*.)

Goodsir, John (1814–67). Scottish surgeon and anatomist. Surgeon in Anstruther, Fifeshire, 1835–40. Conservator, museum of the Royal College of Surgeons, Edinburgh, 1841–3. Conservator, human and comparative anatomy museum, Edinburgh University, 1840; curator, anatomy and pathology museum, 1843; demonstrator in anatomy, 1844; professor of anatomy, 1846. FRS 1846. (*DNB, DSB*.)

Gore, Philip Yorke, 4th earl of Arran (1801–84). Diplomat. Chargé d'affaires in Buenos Aires, 1832–4. Succeeded to the earldom on the death of his uncle in 1837. (*Burke's peerage, Complete peerage, Modern English biography*.)

Gould, John (1804–81). Ornithologist and artist. Taxidermist to the Zoological Society of London, 1828–81. Described the birds collected by CD on the *Beagle* expedition. FRS 1843. (*ODNB*.)

Gratiolet, Louis Pierre (1815–65). French anatomist and anthropologist. Laboratory assistant, Muséum d'Histoire Naturelle, Paris, 1842–53; lectured on anatomy, 1844–50; director of anatomical studies, 1853–62. Deputy to the professor of zoology, Faculty of Science, Paris, 1862–3; professor, 1863–5. (*DSB*.)

Gray, Asa (1810–88). American botanist. Fisher Professor of natural history, Harvard University, 1842–88. Wrote numerous botanical textbooks and works on North American flora. President of the American Academy of Arts and Sciences,

Gray, Asa, cont.

1863–73; of the American Association for the Advancement of Science, 1872; a regent of the Smithsonian Institution, 1874–88. Foreign member, Royal Society of London, 1873. (*DAB, DSB, Record of the Royal Society of London.*)

26 March 1867, 15 April [1867], *[after 6 July 1867]* (J. D. Hooker), 8 August [1867], *[after 17 September 1867]*, 16 October [1867], *18 November 1867*

Gray, Jane Loring (1821–1909). Daughter of Charles Greely Loring, Boston lawyer and politician, and Anna Pierce Brace. Married Asa Gray in 1848. Edited the *Letters of Asa Gray* (1893). (Barnhart comp. 1965; Dupree 1959, pp. 177–84.)

Gray, John Edward (1800–75). Botanist and zoologist. Assistant keeper of the zoological collections at the British Museum, 1824; keeper, 1840–74. President, Botanical Society of London, 1836–56. FRS 1832. (R. Desmond 1994, *DNB.*)

21 November 1867

Grece, Clair James (d. 1905/6?). Solicitor and philologist. Town clerk of Reigate and clerk to the Reigate Local Board. Member of the Philological Society, 1862–1905. Wrote works on various legal topics and translated a German study of English grammar. (*Post Office directory of the six home counties* 1866, *Transactions of the Philological Society.*)

29 December 1867

Green, Mary (d. 1919). Wife of the superintendent of Coranderrk Aboriginal Station, John Green. Emigrated from Scotland in 1858. Supervised the care of orphans at Coranderrk, 1863–74. Supplied answers to CD's queries on expression. (Barwick 1972.)

Gregory, William (1803–58). Scottish chemist. MD, Edinburgh, 1828. Studied chemistry on the continent. Professor of medicine and chemistry, King's College, Aberdeen, 1839. Chair of chemistry, Edinburgh, 1844. Interested in animal magnetism and mesmerism. (*DNB.*)

Greenwood, George (1799–1875). Soldier, arboriculturalist, and geomorphologist. Studied rain and rivers as agents of erosion. (Sarjeant 1980–96.)

Griffith, George (1834/5–1902). Schoolmaster. BA, Oxford, 1856; MA, 1859. Lecturer in natural science, 1857. Assistant master, natural science, Harrow, 1867–93. (*Alum. Oxon., Harrow school register.*)

Grimm, Jakob Ludwig Karl (1785–1863). German philologist. After studying law in Marburg, became a librarian in the Kingdom of Westphalia, 1807. Civil servant in the Electorate of Hesse and participant in the Congress of Vienna, 1815. Librarian, Kassel, 1816–29; Göttingen, 1829–37. Worked on the *Deutsches Wörterbuch* from 1838. Published on history of religion, mythology, and law, as well as linguistics. (*DBE.*)

Grimmer, W. (*fl.* mid 1860s). South African surgeon. District surgeon in the division of Colesberg, Cape Colony. (*Cape of Good Hope general directory.*)

Grove, William Robert (1811–96). Lawyer, judge, and natural philosopher. Professor of experimental philosophy, London Institution, 1847. An active member of the Royal Society of London; treasurer and chairman of the executive committee of the Philosophical Club, 1847; Royal Medallist, 1847. Member of the Royal Commission on the Law of Patents, 1864. Appointed to the bench, Court of Common Pleas, 1871. FRS 1840. (*DNB, DSB*.)

Günther, Albrecht Carl Ludwig Gotthilf (Albert Charles Lewis Gotthilf) (1830–1914). German-born zoologist. Began his association with the British Museum in 1857; made catalogues of the museum's specimens of amphibia, reptiles, and fish; officially joined the staff in 1862. Assistant keeper of the zoological department, 1872–5; keeper, 1875–95. Edited the *Record of Zoological Literature*, 1864–9. FRS 1867. (*ODNB, NDB*.)

7 December [1867], *19 December 1867*, 21 December [1867], *[late December 1867 or early January 1868]*

Gurney, John Henry (1819–90). Financier and ornithologist. MP, King's Lynn, Norfolk, 1854–65. (*Modern English biography*, Stenton 1976.)

Gurney, Samuel (1816–82). Banker. Partner in the firm of Overend, Gurney and Co., money-lenders. MP for Penryn and Falmouth, 1857–65. Member of the committee of the Anti-slavery Society; president of the Aborigines' Protection Society. Director of several telegraph companies. Fellow of the Linnean Society of London. (*Friends' biographical catalogue, Modern English biography*, Stenton 1976.)

Haast, John Francis Julius (Julius) von (1822–87). German-born explorer and geologist. Travelled to New Zealand in 1858 to report on the prospects for German emigration. Explored the western districts of Nelson province at the request of the provincial government in 1859. Appointed geologist to Canterbury province, 1861. Conducted the first geological survey of Canterbury province, 1861–8. Became a British national in 1861. Founded the Philosophical Institute of Canterbury in 1862, and the Canterbury Museum in 1863. Professor of geology, Canterbury College, 1876–87. Member of the senate of the University of New Zealand, 1879–87. Knighted, 1886. FRS 1867. (*DNZB, DSB*, H. F. von Haast 1948, *ODNB*.)

27 February [1867], *12 May – 2 June 1867, 4 December 1867*

Haeckel, Agnes (1842–1915). Daughter of Emil Huschke. Second wife of Ernst Haeckel, whom she married in 1867. (Krauße 1987.)

Haeckel, Anna (1835–64). Née Sethe. Cousin of Ernst Haeckel, whom she married in 1862. (*DSB* s.v. Haeckel, Ernst; Uschmann 1984, p. 317.)

Haeckel, Ernst Philipp August (Ernst) (1834–1919). German zoologist. MD, Berlin, 1857. Lecturer in comparative anatomy, University of Jena, 1861–2; professor extraordinarius of zoology, 1862–5; professor of zoology and director of the Zoological Institute, 1865–1909. Specialist in marine invertebrates. Leading populariser of evolutionary theory. His *Generelle Morphologie der Organismen* (1866) linked morphology to the study of the phylogenetic evolution of organisms. (*DSB, NDB*.)

8 January 1867, 12 April [1867], *12 May 1867*, 21 May [1867], *28 June 1867*, 4 July [1867]

Hagenauer, Friedrich August (1829–1909). Moravian missionary. Founded the

Hagenauer, Friedrich August, cont.
mission school in Wimmera, Victoria, Australia, 1859. Set up Ramahyuck mission for Aborigines at Lake Wellington, Gippsland, Victoria, 1863. (*Aust. dict. biog.*) *[12 September 1867]* (Ferdinand von Mueller)

Haliburton, Robert Grant (1831–1901). Canadian lawyer and ethnologist. Active in many Nova Scotian societies; founder (1862) and vice-president of the Nova Scotia Institute of the Natural Sciences. Secretary to the Nova Scotia commissioners for the International Exhibition in London, 1862. Advocated protectionism; a founding member of the Canada First Movement. Published anthropological and ethnological studies. (*DCB.*)

Hanbury, Daniel (1825–75). Pharmacologist. Qualified as a pharmaceutical chemist in 1857. Partner in the London firm Allen & Hanbury of Lombard Street, London, 1868–70. Member of a number of scientific societies in London, including the Royal Society, the Linnean Society, the Chemical Society, and the Microscopical Society. Member of the board of examiners of the Pharmaceutical Society, 1860–72. FRS 1867. (*ODNB.*)

Hanstein, Johannes Ludwig Emil Robert (Johannes) von (1822–80). German botanist. Studied horticulture in Berlin and Potsdam, 1838–43; PhD, botany, Berlin, 1848. Curator, botanical museum, Berlin, 1861. Professor of botany and director of botanical gardens, Bonn, 1865. (*NDB.*)

Hardwicke, Robert (1822–75). Publisher. Printer and publisher at 192 Piccadilly, London, from 1852. One of the founders of the Quekett Microscopical Club, 1865. Published *Hardwicke's Science Gossip*, 1865–75. Fellow of the Linnean Society of London, 1863. (*Modern English biography, ODNB.*)

Hardwicke's Science-Gossip
[before 1 December 1867]

Hart, A. E. Trader and coffee-grower near Delli (Dili), east Timor. Owned and lived on a plantation two miles from town. (Raby 2001.)

Harvey, William Henry (1811–66). Irish botanist. Colonial treasurer in Cape Town; collected plants in South Africa, 1836–42. Keeper of the herbarium, Trinity College, Dublin, from 1844; professor of botany, Royal Dublin Society, 1848–66; professor of botany, Trinity College, Dublin, 1856–66. Published works on South African plants, including *Flora Capensis* (1859–65) with Otto Wilhelm Sonder. Specialist in marine algae. FRS 1858. (R. Desmond 1994, *DNB, DSB.*)

Haughton, Samuel (1821–97). Irish clergyman, mathematician, geologist, and palaeontologist. Professor of geology, Dublin University, 1851–81. Became registrar of the medical school after graduating in medicine in 1862. Co-editor of the *Natural History Review*, 1854–60. President of the Royal Irish Academy, 1887. FRS 1858. (R. Desmond 1994, *DNB*, Sarjeant 1980–96.)

Hawker, Robert (1753–1827). Clergyman and author of religious works. Curate of the parish of Charles, near Plymouth, Cornwall, from 1779; vicar, 1784–1827. Regarded as an ultra-Calvinist; author of *Sermons on the divinity of Christ* (1792). (*ODNB.*)

Hector, James (1834–1907). Scottish geologist. Surgeon and geologist on the government expedition to the western parts of British North America, 1857–60. Geologist to the provincial government of Otago, New Zealand, 1861–5. Director of the Geological Survey of New Zealand, 1865. Director of the meteorological department of the New Zealand Institute, of the Colonial Museum, and of the botanical garden, Wellington, 1866–1903. Knighted, 1887. FRS 1866. (*DNZB.*)

Heer, Oswald (1809–83). Swiss biogeographer, palaeontologist, and botanist. An expert on Tertiary flora. Lecturer in botany, University of Zurich, 1834–5; director of the botanic garden, 1834; associate professor, 1835–52; professor of botany and entomology, 1852–83. (*DSB, NDB.*)

Henslow, George (1835–1925). Clergyman, teacher, and botanist. BA, Christ's College, Cambridge, 1858. Curate of Steyning, 1859–61; of St John's Wood Chapel, 1868–70; of St James's Marylebone, 1870–87. Headmaster at Hampton Lucy Grammar School, Warwick, 1861–4; at the Grammar School, Store Street, London, 1865–72. Lecturer in botany at St Bartholomew's Hospital, 1866–80. Honorary professor of botany at the Royal Horticultural Society, 1880–1918. Author of a number of religious books, including *Plants of the Bible* (1907), and of children's books on natural history. Younger son of John Stevens Henslow. (*Alum. Cantab., Crockford's clerical directory*, R. Desmond 1994, Lightman ed. 2004, *ODNB* s.v. Henslow, John Stevens.)
  [*c. August 1867?*], *15 August 1867*

Henslow, John Stevens (1796–1861). Clergyman, botanist, and mineralogist. CD's teacher and friend. Professor of mineralogy, Cambridge University, 1822–7; professor of botany, 1825–61. Extended and remodelled the Cambridge botanic garden. Curate of Little St Mary's Church, Cambridge, 1824–32; vicar of Cholsey-cum-Moulsford, Berkshire, 1832–7; rector of Hitcham, Suffolk, 1837–61. (*DSB, Historical register of the University of Cambridge, ODNB.*)

Herbert, John Maurice (1808–82). Lawyer. BA, St John's College, Cambridge, 1830; fellow, 1832–40. Barrister, 1835. County court judge, South Wales, 1847–82. Friend of CD's. (*Alum. Cantab., Correspondence* vol. 1, *Modern English biography.*)
  3 May 1867, 7 May [1867]

Herbert, Mary Anne. Poet. Wife of John Maurice Herbert. (Herbert ed. 1877.)

Herbert, William (1778–1847). Naturalist, classical scholar, linguist, and clergyman. Noted for his work on plant hybridisation. Rector of Spofforth, Yorkshire, 1814–40. Dean of Manchester, 1840–7. (*DSB, ODNB.*)

Heron, Robert, 2d baronet (1765–1854). Whig politician. MP for Grimsby, 1812–18; Peterborough, 1819–47. Succeeded to a baronetcy upon his father's death in 1805. (*ODNB.*)

Herschel, John Frederick William, 1st baronet (1792–1871). Astronomer, mathematician, chemist, and philosopher. Member of many learned societies. Carried out astronomical observations at the Cape of Good Hope, 1834–8. Master of the Royal Mint, 1850–5. Created baronet, 1838. FRS 1813. (*DNB, DSB.*)

Hicks, Henry (1837–99). Geologist and medical practitioner. Practised medicine at St David's, 1862–71, then at Hendon, Middlesex; proprietor of a women's asylum at Hendon. President, Geological Society of London, 1896–8. Wrote more than sixty papers on geological subjects. FRS 1885. (*Modern English biography, ODNB.*)

Hildebrand, Friedrich Hermann Gustav (Friedrich) (1835–1915). German botanist. After studying mineralogy, geology, and agriculture at Berlin, he took up botany, studying at Bonn, then from 1855 to 1858 at Berlin, where he received his doctorate. Habilitated at Bonn, becoming privat-dozent there, in 1859. Professor of botany, Freiburg im Breisgau, 1868–1907. Worked mainly on hybridity, dimorphism, and generation. (Correns 1916, Junker 1989, Tort 1996.)
  18 March 1867, 20 March [1867]

Hirst, Thomas Archer (1830–92). Mathematician. Lecturer in mathematics and natural philosophy, Queenwood College, Hampshire, 1853–6. Appointed mathematical master of University College School, London. Professor of physics, University College, London, 1865; of mathematics, 1866–70. Assistant registrar of the University of London, 1870–83. General secretary of the British Association for the Advancement of Science, 1866–70. Director of naval studies, Royal Naval College, Greenwich, 1873–83. From 1861 his research specialised in pure geometry. Took a prominent part in the founding of the London Mathematical Society in 1865; president, 1872–4. FRS 1861. (*DNB.*)

Hodgson, Brian Houghton (1801?–94). Diplomat and Nepalese scholar. In the service of the East India Company from 1816; assistant resident in Nepal, 1820; resident, 1833–43. Acquired an important collection of Sanskrit and Tibetan manuscripts. Wrote extensively on the geography, ethnography, and natural history of India and the Himalayas. Lived in Darjeeling, 1845–58; thereafter settled in England. FRS 1877. (*ODNB.*)

Holland, Henry, 1st baronet (1788–1873). Physician. Related to Josiah Wedgwood I. Physician in ordinary to Prince Albert, 1840; to Queen Victoria, 1852. President of the Royal Institution of Great Britain, 1865–73. Created baronet, 1853. FRS 1815. (Caroe 1985, *DNB*, *Emma Darwin* (1904), *Record of the Royal Society of London.*)
  21 December [1867]

Hooker, Brian Harvey Hodgson (1860–1932). Fifth child of Frances Harriet and Joseph Dalton Hooker. (Allan 1967 s.v. 'Hooker pedigree'.)

Hooker, Charles Paget (1855–1933). Physician and surgeon. Third child of Frances Harriet and Joseph Dalton Hooker. Trained at St Bartholomew's Hospital, London; made a licentiate of the Royal College of Physicians of London and the Royal College of Surgeons, Edinburgh, 1879, before being appointed to the staff of the Hertfordshire General Infirmary. Cottishall Cottage Hospital, Norfolk, 1880–5; Cirencester Cottage Hospital, Gloucestershire, 1885–1912. (Allan 1967, *Medical directory* 1881–1933, *Medical who's who* 1914.)

Hooker, Frances Harriet (1825–74). Daughter of John Stevens Henslow. Married Joseph Dalton Hooker in 1851. Assisted her husband significantly in his published work. Translated *A general system of botany, descriptive and analytical*, by Emmanuel Le Maout and Joseph Decaisne (1873). (Allan 1967, Lightman ed. 2004.)

[6 April 1867]

Hooker, Joseph Dalton (1817–1911). Botanist. Worked chiefly on taxonomy and plant geography. Son of William Jackson Hooker. Friend and confidant of CD. Accompanied James Clark Ross on his Antarctic expedition, 1839–43, and published the botanical results of the voyage. Appointed palaeobotanist to the Geological Survey of Great Britain, 1846. Travelled in the Himalayas, 1847–9. Assistant director, Royal Botanic Gardens, Kew, 1855–65; director, 1865–85. Knighted, 1877. FRS 1847. (*DSB, ODNB*.)

9 January [1867], *[12 January 1867]*, 15 January [1867], *20 January 1867*, 21 January [1867], 29 January [1867], *4 February 1867*, 8 February [1867], *12 February 1867, 14 March 1867*, 17 March [1867], *20 March 1867*, 21 March [1867], *23 March 1867*, 24 [March 1867], *26 [and 27] March 1867*, 29 [March 1867], *31 March 1867, 3 April 1867*, 4 April [1867], 5 April [1867], *13 April 1867, [13 April? 1867]* ([W. E. Darwin?]), 15 [April 1867], 25 [April 1867], [12] May [1867], *17 May 1867*, [21 May 1867], *24 May 1867*, 26 [May 1867], [16 June 1867], *18 June 1867*, [23 June 1867], *4 July 1867, [27 July 1867]*, 29 July [1867], *30 July [1867]*, *17 August 1867, 31 August 1867*, 2 September [1867], *[14 September 1867], [20 September 1867], 18 October 1867*, 17 November [1867], *19 November 1867*, 25 November [1867], *11 December 1867* (Emma Darwin), *17 December 1867, [23 December 1867?]*

Hooker, Maria (1797–1872). Eldest daughter of Dawson Turner. Married William Jackson Hooker in 1815; acted as her husband's secretary. Mother of Joseph Dalton Hooker. (Allan 1967 s.v. 'Turner pedigree', R. Desmond 1994.)

Hooker, Reginald Hawthorn (1867–1944). Sixth child of Joseph Dalton and Frances Harriet Hooker. Took B-ès-Sc. in Paris, and studied mathematics at Trinity College, Cambridge, 1886–9. Assistant to the director of the Intelligence Department of the Board of Agriculture, and subsequently head of the statistical branch until 1927. Secretary, Royal Statistical Society; president, Royal Meteorological Society, 1920–1. Married Olive Marion Rücker in 1911. (Allan 1967, *Alum. Cantab.*, Royal Meteorological Society.)

Hooker, William Henslow (1853–1942). Eldest child of Frances Harriet and Joseph Dalton Hooker. Civil servant, India Office, 1877–1904. Encouraged imperial ties between metropolitan institutions (particularly the Royal Botanic Gardens, Kew) and British East Africa, *circa* 1896–1906. (Allan 1967; *India list* 1904–5; *Zanzibar Gazette*, 5 February 1896, p. 6, and 28 November 1900, p. 5.)

Hooker, William Jackson (1785–1865). Botanist. Father of Joseph Dalton Hooker. Regius professor of botany, Glasgow University, 1820–41. Appointed first director of the Royal Botanic Gardens, Kew, 1841. Knighted, 1836. FRS 1812. (*DSB, ODNB*.)

Hopkins, William (1793–1866). Mathematician and geologist. Tutor in mathematics at Cambridge University. President of the Geological Society of London,

Hopkins, William, cont.

1851–3. Specialised in quantitative studies of geological and geophysical questions. FRS 1837. (*DNB, DSB, Record of the Royal Society of London.*)

Horsman, Samuel James O'Hara (d. 1887?). Clergyman. Educated at Trinity College, Dublin. Curate of All Saints, Northamptonshire, 1858; of St Matthew's, Rugby, 1860. Ordained priest, 1860. Assistant minister and acting chaplain to the forces, Stirling Castle, 1862; curate of St Philip's, Liverpool, 1864; of Acton Trussell, Staffordshire, 1865; curate of Down, 1867–8; of St Luke's, Marylebone, London, 1868; of St George the Martyr, Southward, London, 1880; of St Mark's, Regent's Park, London, 1883; rector of Condicote, Gloucestershire, 1884. (*Correspondence* vol. 16, letter from S. J. O'H. Horsman, 2 June [1868], *Crockford's clerical directory,* J. R. Moore 1985, pp. 470, 477.)

Houghton, William (1828–95). Clergyman and naturalist. BA, Oxford, 1850; MA 1853. Wrote numerous articles on natural history for popular scientific and literary journals. (*Crockford's clerical directory, Wellesley index.*)
     18 October 1867

Hugo, Victor (1802–85). French poet, dramatist, and romance-writer. (*EB.*)

Hull, Edward (1829–1917). Irish geologist. Entered the Geological Survey of Great Britain in 1850. Worked in north Wales with Joseph Beete Jukes, and in various English counties. District surveyor, Geological Survey of Scotland, 1867–8; director of the Geological Survey of Ireland, 1869–90. Professor of geology at the Royal College of Science for Ireland, 1869–90; president of the Royal Geological Society of Ireland, 1873–5. Led an expedition to the Wadi Araba on behalf of the Palestine Exploration Fund, 1883–4. FRS 1867. (*ODNB.*)

Humboldt, Friedrich Wilhelm Heinrich Alexander (Alexander) von (1769–1859). Prussian naturalist, geographer, and traveller. Official in the Prussian mining service, 1792–6. Explored northern South America, Cuba, Mexico, and the United States, 1799–1804. Travelled in Siberia in 1829. Foreign member, Royal Society of London, 1815. (*DBE, DSB, NDB.*)

Huschke, Agnes. *See* Haeckel, Agnes.

Huschke, Emil (1797–1858). German physiologist and anatomist. MD, Jena, 1820; professor extraordinarius of anatomy and physiology, 1823; professor, 1838. (*DBE.*)

Hutton, Frederick Wollaston (1836–1905). Geologist and army officer. Served with the Royal Welsh Fusiliers in the Crimea and India, 1855–8. Captain, 1862. Left the army in 1865, and emigrated to New Zealand in 1866. Employed by Auckland provincial government to survey coal deposits. Assistant geologist to the geological survey of New Zealand, 1871–3. Provincial geologist of Otago, 1873–6. Professor of natural science, University of Otago, 1877–80. Professor of biology, Canterbury College, 1880–93. Curator of Canterbury Museum from 1893. FRS 1892. (*DNZB,* Stenhouse 1990, G. M. Thomson 1884–5.)

Huxley, Ethel Gladys (1866–1941). Daughter and youngest child of Henrietta Anne and Thomas Henry Huxley. Married artist–author John Collier in 1889.

Because of the British law against marrying a deceased wife's sister they married in Norway. (R. W. Clark 1968, p. 110, A. Desmond 1994–7, 1: 348.)

Huxley, Jessie Oriana (1858–1927). Daughter of Henrietta Anne and Thomas Henry Huxley. Married Fred Waller, architect, in 1877. Shared her mother's interest in Moravian principles of education, and published an article, 'Mental and physical training of children', in 1889. (Bibby 1959, R. W. Clark 1968, A. Desmond 1994–7, Waller 1889.)

Huxley, Leonard (1860–1933). Biographer, editor, and poet. Son of Henrietta Anne and Thomas Henry Huxley. Assistant master at Charterhouse, 1884–1901. Assistant editor, *Cornhill Magazine*, 1901–16; editor from 1916. Married Julia Frances Arnold, niece of Matthew Arnold, in 1885. (R. W. Clark 1968, *ODNB*.)

Huxley, Marian (1859–87). Artist. Daughter of Henrietta Anne and Thomas Henry Huxley. Studied art at the Slade School, London. Exhibited at the Royal Academy, 1880–4. Married artist–author John Collier, 1879. Her sketch of CD, made when she was 18, hangs in the National Portrait Gallery. (Bibby 1959; *Bryan's dictionary of painters and engravers*; R. W. Clark 1968, p. 97 and *passim*; A. Desmond 1994–7; Petteys 1985.)

Huxley, Nettie (1863–1940). Singer and illustrator. Daughter of Henrietta Anne and Thomas Henry Huxley. Married Harold Roller, joint owner of a firm of picture restorers, in 1889, but spent most of her time travelling in Europe with her daughter, supporting herself as a singer. (Bibby 1959, pp. 15, 275, 283; R. W. Clark 1968, pp. 111, 252, and *passim*.)

Huxley, Rachel (1862–1934). Daughter of Henrietta Anne and Thomas Henry Huxley. Married Alfred Eckersley, a civil engineer, in 1884 and lived in various countries until his death in 1895 in San Salvador. Returned to London, where she ran a laundry business until her marriage to Harold Shawcross, when she moved to Lancashire. (R. W. Clark 1968, pp. 98, 109, 129, 165, and *passim*.)

Huxley, Thomas Henry (1825–95). Zoologist. Assistant-surgeon on HMS *Rattlesnake*, 1846–50, during which time he investigated Hydrozoa and other marine invertebrates. Lecturer in natural history, Royal School of Mines, 1854; professor, 1857. Appointed naturalist to the Geological Survey of Great Britain, 1854. Hunterian Professor, Royal College of Surgeons of England, 1862–9. Fullerian Professor of physiology, Royal Institution of Great Britain, 1855–8, 1866–9. President of the Royal Society of London, 1883–5. FRS 1851. (R. W. Clark 1968, A. Desmond 1994–7, *DSB*, *ODNB*.)

[*before 7 January 1867*], 7 January [1867], 12 June [1867]

Ince, William (1825–1910). Theologian and classicist. Ordained deacon, 1850; priest, 1852. Tutor, Exeter College, Oxford, 1850. Junior proctor, 1856–7; select preacher before the university, 1859, 1870, and 1875; Oxford preacher at the Chapel Royal, Whitehall, 1860–2. Appointed regius professor of divinity, Oxford, and canon of Christ Church, 1878. (*ODNB*.)

Inglis, Mary (1787–1872). Eldest daughter of Joseph Seymour Biscoe of Pendhill Court, Bletchingley, Surrey. Married the politician Sir Robert Harry Inglis in

Inglis, Mary, cont.

1807. With her husband, became guardian to nine orphaned children. (*ODNB* s.v. Inglis, Robert Harry.)

Innes, John Brodie (1817–94). Clergyman. Perpetual curate of Down, 1846–68; vicar, 1868–9. Left Down in 1862 after inheriting an entailed estate at Milton Brodie, near Forres, Scotland; changed his name to Brodie Innes in 1861 as required by the entail. Priest in charge of Milton Brodie Mission and general licentiate of the diocese of Moray, 1861. Chaplain to the Bishop of Moray, 1861–80 and 1886–94. (*Clergy list, County families 1864, Crockford's clerical directory,* Freeman 1978, J. R. Moore 1985.)

1 September [1867]

Jameson, William (1796–1873). Scottish botanist. Licentiate of the Royal College of Surgeons, Edinburgh. Emigrated to South America in 1820. Practised medicine in Guayaquil, Ecuador, 1822–6. Professor of chemistry, University of Quito, 1827–35, 1846–58; professor of botany, 1827–35, 1846–60. Appointed assayer of the Mint in Quito, Ecuador, 1832; director, 1861. Prepared a synopsis of the flora of Ecuador. (*ODNB, Transactions and Proceedings of the Botanical Society [of Edinburgh]* 12: 19–28.)

Jardine, William, 7th baronet (1800–74). Naturalist. A founder of the *Magazine of Zoology and Botany,* 1836. Commissioner for the salmon fisheries of England and Wales, 1860. President, Dumfriesshire and Galloway Natural History and Antiquarian Society, 1862–74. FRS 1860. (Gladstone 1913, *ODNB, Record of the Royal Society of London.*)

Jenkin, Henry Charles Fleeming (1833–85). Engineer and university teacher. Studied natural philosophy at Genoa. Apprenticed at Fairbairn's works, Manchester, 1851. Worked as a draughtsman and marine telegraph engineer in London from 1855. Carried out important experiments on the resistance and insulation of electrical cables with William Thomson (later Lord Kelvin). Largely occupied in the fitting out of submarine telegraph cables, 1858–73. Appointed professor of civil engineering, University College, London, 1866; professor of engineering, Edinburgh University, 1868. Promoted the formation of a sanitary association, Edinburgh, 1877–8. Wrote miscellaneous papers on literature, science, and political economy. FRS 1865. (*ODNB.*)

Jenner, Edward (1749–1823). Surgeon. Pioneer of smallpox vaccination. (*DSB, ODNB.*)

Jerdon, Thomas Claverhill (1811–72). Zoologist. Joined the East India Company as assistant surgeon in the Madras service in 1835; later civil surgeon at Telicherry and major. Resigned from government service in 1868 and returned to England in 1872. Author of works on the birds and mammals of India. (Lightman ed. 2004, *ODNB.*)

Jones, Henry Bence (1814–73). Physician and chemist. Physician to St George's Hospital, 1846–62. Secretary of the Royal Institution, 1860–72. In his chemical studies, devoted himself especially to the application of chemistry to pathology and medicine. FRS 1846. (*ODNB*.)

   1 October [1867] (Emma Darwin)

Jones, Thomas Rupert (1819–1911). Geologist and palaeontologist. Medical assistant, 1842–50. Appointed assistant secretary of the Geological Society of London, 1855; editor of the *Quarterly Journal* of the Geological Society, 1850–65. Lecturer on geology, Royal Military College, Sandhurst, 1858–62; professor of geology, 1862–80. An authority on Entomostraca and Foraminifera. FRS 1872. (*ODNB*.)

Jordan, Claude Thomas Alexis (Alexis) (1814–97). French botanist. Conducted field research, 1836–46, to complete and correct existing French floras. Assembled an important private herbarium. After giving up his botanical expeditions, worked in his own experimental gardens, trying to demonstrate the stability of species. A strong opponent of transmutation theory. (*DSB*, Tort 1996.)

Jukes, Joseph Beete (1811–69). Geologist. Geological surveyor of Newfoundland, 1839–40. Naturalist aboard HMS *Fly* in the survey of the north-east coast of Australia, 1842–6. Geologist with the Geological Survey of Great Britain working in North Wales and the English midlands, 1846–50. Local director of the Geological Survey of Ireland, 1850–67; director, 1867–9. Lecturer on geology at the Royal College of Science, Dublin, from 1854. President of the Geological Society of Dublin, 1853–4. FRS 1853. (*ODNB*.)

Kanitz, August (1843–96). Austro-Hungarian botanist. Professor of botany and director of the botanic garden at Klausenburg, 1872. (*BLKO*, *OBL*, *Taxonomic literature*.)

Kant, Immanuel (1724–1804). German philosopher. (*DSB*, *NDB*.)

Kelvin, Lord. *See* Thomson, William.

Kempson, Jessie (1867–1939). Daughter of Louisa Frances and William John Kempson.

Kempson, Louisa Frances (1834–1903). Daughter of Jessie and Henry Allen Wedgwood. Emma Darwin's niece. Married William John Kempson in 1864. (Freeman 1978.)

   20 June 1867 (Emma Darwin)

Kinahan, George Henry (1829–1908). Irish geologist. Joined the Geological Survey of Ireland in 1854; senior geologist, 1861; district surveyor, 1869–90. President of the Royal Geological Society of Ireland, 1879–81. (*ODNB*.)

King's College, London

   22 February [1867]

Kingsley, Charles (1819–75). Author and clergyman. Appointed professor of English, Queen's College for Women, London, 1848. Regius professor of modern history, Cambridge University, 1860–9. Rector of Eversley, Hampshire, 1844–75. Chaplain to the queen, 1859–75. (*ODNB*.)

   30 April [1867], *6 June 1867*, *10 June [1867]*, *1 November 1867*, 6 November [1867], *8 November 1867*, *11 December 1867*, 13 December [1867]

Kingsley, Frances Eliza (1814–91). Née Grenfell. Married Charles Kingsley in 1844. Acted as his amanuensis and edited his memoirs, and compilations of his writings, after his death. (*ODNB.*)

Kirchhoff, Gustav Robert (1824–87). German physicist. Professor extraordinarius, Breslau, 1850–4; Heidelberg, 1854–75. Professor of theoretical physics, Berlin, 1875–86. Made major contributions to the study of electromagnetic radiation and electrical currents. A close friend of Robert Wilhelm Eberhard Bunsen; the two collaborated in laying the foundation of the method of spectral analysis, 1857–63. (*DSB, NDB.*)

Kitchener, Francis Elliott (1838–1915). Schoolteacher and botanist. President, Rugby School Natural History Society, 1867–74. Fellow of the Linnean Society of London, 1867. (R. Desmond 1994.)
   9 November 1867

Knight, William Angus (1836–1916). Clergyman and writer. Minister of St Enoch's Church, Dundee, 1866–76. Professor of moral philosophy, St Andrews, 1876–1903. LLD, Glasgow, 1879. (*Fasti ecclesiæ Scoticanæ.*)
   20 August 1867

Koch, Eduard Friedrich (Eduard) (1838–97). German publisher. Took over E. Schweizerbart'sche Verlagsbuchhandlung in 1867, after which the firm published mostly scientific titles. Published a multi-volume edition of CD's works, translated by Victor Carus. (*Biographisches Jahrbuch und deutscher Nekrolog* 2 (1898): 227.)
   11 December 1867, 13 December [1867]

Kölliker, Rudolf Albert von (1817–1905). Swiss anatomist and physiologist. Professor of physiology and comparative anatomy, University of Würzburg, 1847–64; professor of anatomy, 1849–97. Foreign member, Royal Society of London, 1860. (*DSB.*)

Kovalevsky, Alexander Onufrievich (1840–1901). Russian embryologist. Brother of Vladimir Onufrievich Kovalevsky. Held academic posts at various Russian universities; professor of histology, St Petersburg, 1890–4. His studies of ascidian embryology revealed that tunicates were chordates and gave strong support to Darwinian transmutation theory. Foreign member of the Royal Society of London, 1885. (*DSB, GSE* s.v. Tunicata.)

Kovalevsky, Vladimir Onufrievich (1842–83). Russian palaeontologist. Graduated from the School of Jurisprudence in 1861. Thereafter published, translated, and edited works by CD, Charles Lyell, Louis Agassiz, and others. Studied natural science and palaeontology, travelling throughout Europe, 1869–74. Submitted doctoral thesis on the palaeontology of horses at the university of Jena in 1872. Associate professor, Moscow University, 1880–3. (*DSB.*)
   15 March 1867, 26 March [1867], 2 April 1867, 24 April [1867], 2 May [1867], 14 May 1867, 15 May 1867, 16 May [1867], 20 May [1867], *[after 24? May 1867]*, 3 June [1867], 24 June [1867]

Krohn, August David (1803–91). Russian-born zoologist, anatomist, and embryologist. Worked in Bonn; travelled extensively. (Blyakher 1982.)

Krohn, John (*fl.* 1860s). British wine merchant of Russian descent. In 1858, together with his brother Nicholas, founded the wine producing and exporting firm of Krohn Brothers & Co. in Madeira. (Cossart 1984.)

Krohn, Nicholas (*fl.* 1860s). British wine merchant of Russian descent. In 1858, together with his brother John, founded the wine producing and exporting firm of Krohn Brothers & Co. in Madeira. (Cossart 1984.)

Kuhn, Friedrich Adelbert Maximilian (Max) (1842–94). German botanist. Student of Alexander Braun at Berlin. Completed his dissertation in 1867. Assistant at the royal herbarium, Berlin, 1866–8. Teacher at Königstadt Realschule, 1870; professor, 1889. Wrote several major taxonomic works on ferns. (*ADB.*)

Kühner, Raphael (1822–78). German philologist. Studied at Göttingen, then taught at the Lyceum in Hannover. Published a popular elementary Latin grammar in 1841, and later a Greek grammar. (*ADB.*)

Lacy, Dyson. Supplied information to CD on expression of the emotions among indigenous Australians. (*Correspondence* vol. 16, letter from Dyson Lacy, [before 18 September 1868], *Expression.*)

Lamarck, Jean Baptiste Pierre Antoine de Monet (Jean Baptiste) de (1744–1829). French naturalist. Held various botanical positions at the Jardin du Roi, 1788–93. Appointed professor of zoology, Muséum d'Histoire Naturelle, 1793. Believed in spontaneous generation and the progressive development of animal types; propounded a theory of transmutation. (*DSB.*)

Lane, Edward William (1801–76). Scholar of Arabic language and literature. Writer on Egypt and translator of the *Thousand and one nights.* Edited selections from the Koran. Compiled a standard Arabic–English lexicon. (*ODNB.*)

Langton, Emily Catherine (Catherine) (1810–66). CD's sister. Married Charles Langton in 1863. (*Darwin pedigree.*)

Latham, John (1740–1837). Physician and ornithologist. A founder of the Linnean Society, 1788. FRS 1775. (*ODNB.*)

Layard, Edgar Leopold (1824–1900). Civil servant and naturalist. Served in the Ceylon civil service, 1846–54; in the Cape of Good Hope civil service, 1854–70. Studied the birds and shells of Ceylon (Sri Lanka). Founded the South African museum; curator, 1855–72. Published on the birds of South Africa (1867). (*DSAB, Modern English biography.*)

Leadbeater, John. Bird dealer and taxidermist at 19 Brewer Street, Golden Square West, London. Ornithologist to Queen Victoria. (*Post Office London directory* 1858.)

Le Brun, Charles (1619–90). French designer and artist. Worked in the court of King Louis XIV. From 1663, director of Les Gobelins tapestry works. (*Larousse du XXe siècle.*)

Leclerc, Georges Louis, comte de Buffon (1707–88). French naturalist, philosopher, and mathematician. Keeper, Jardin du Roi, 1739–88. His theory of transmutation is outlined in *Histoire naturelle* (1749–1804). FRS 1739. (*DBF, DSB, Record of the Royal Society of London.*)

Lecoq, Henri (1802–71). French naturalist and vulcanologist. Professor of natural history, University of Clermont-Ferrand, and director of the city's botanic garden, 1826–54. Taught at the Preparatory School of Medicine and Pharmacy, Clermont-Ferrand, from 1840; professor, science faculty, from 1854. Published widely on botany, agriculture, and meteorology. (*Grande encyclopédie*, Sarjeant 1980–96.)

Legrand, Jacques-Guillaume (1743–1807). French architect. Author of works on fine art, architecture, and topography. (*Dictionnaire général des artistes de l'école française.*)

Leidy, Joseph (1823–91). American anatomist, microscopist, and palaeontologist. Appointed dissecting assistant to William Horner, professor of anatomy at the University of Pennsylvania, 1844. Professor of anatomy, University of Pennsylvania, 1853–91. President of the Academy of Natural Sciences of Philadelphia, 1881–91. Published extensively on North American fossil vertebrates, and also studied freshwater Protozoa. (*ANB, DSB.*)

Lemaire, Antoine Charles (Charles) (1800–71). French botanist. Studied at Paris and became professor of classical literature there before turning to botany. Edited horticultural and botanical journals from 1835, notably *L'Illustration Horticole*, 1854–68. In Belgium, *circa* 1854–69. Published works on cacti and other succulent plants. (*Gardeners' Chronicle* 1871, p. 870, *Taxonomic literature.*)

Lesson, René Primevère (1794–1849). French naval surgeon and naturalist. Medical officer on board the *Coquille*, 1822–5. Published the results of his zoological investigations made during the voyage. Taught botany at the naval medical school, Rochefort, 1829–31; appointed professor of pharmacy, 1831. Became the top-ranking navy pharmacist in 1835. (*DSB, NBU.*)

Leuckart, Karl Georg Friedrich Rudolf (Rudolf) (1822–98). German zoologist. Lecturer in zoology, University of Göttingen, 1847–50. Professor extraordinarius of zoology, University of Giessen, 1850–5; professor, 1855–69. Appointed professor of zoology in Leipzig in 1869, dean of the philosophical faculty in 1873, and rector from 1877 to 1878. Carried out important work in morphology and parasitology. (*DBE, DSB, NDB.*)

Lewes, George Henry (1817–78). Writer. Author of a biography of Goethe (1855). Contributed articles on literary and philosophical subjects to numerous journals. Editor, *Fortnightly Review*, 1865–6. Published on physiology and on the nervous system in the 1860s and 1870s. Lived with Mary Ann Evans (George Eliot) from 1854. (Ashton 1991, *ODNB.*)

Lichtenstein, Martin Karl Heinrich (1780–1857). German naturalist and traveller. MD, Jena, 1802. Tutor to the son of the governor of Cape Colony, and house doctor, 1804–6, during which time he studied the natural history of the Cape. Surgeon-major in the Dutch army before returning to Germany, where he wrote up the results of his travels. Appointed professor of zoology, Berlin, 1811; director of the zoological museum, 1813–57. Travelled and collected in South America. (*ADB.*)

Lindley, John (1799–1865). Botanist and horticulturalist. Assistant in Joseph Banks's library and herbarium, 1819. Garden assistant secretary, Horticultural Society of London, 1822–6; assistant secretary, 1826–41; vice-secretary, 1841–58; honorary secretary, 1858–62. Lecturer on botany, Apothecaries' Company, from 1836. Professor of botany, London University (later University College, London), 1829–60. Horticultural editor of the *Gardeners' Chronicle* from 1841. FRS 1828. (R. Desmond 1994, *DSB*, *ODNB*.)

Linnaeus, Carolus. *See* Linné, Carl von.

Linné, Carl von (Carolus Linnaeus) (1707–78). Swedish botanist and zoologist. Professor of practical medicine, University of Uppsala, 1741; professor of botany, diatetics, and materia medica, 1742; court physician, 1747. Proposed a system for the classification of the natural world, and reformed scientific nomenclature. FRS 1753. (*DSB*, *Record of the Royal Society of London*.)

Linnean Society of London
  22 February [1867], 9 December [1867]

Lister, Joseph (1827–1912). Surgeon. Assistant surgeon to the Royal Infirmary, Edinburgh, 1856–60; professor of surgery, Glasgow University, 1860–9; of clinical surgery, 1869–77. Announced his system of antiseptic surgery in the *Lancet* in 1867. Professor of surgery, King's College, London, 1877–92. FRS 1860. (*DSB*, *ODNB*.)

Lord, John Keast (1818–72). Naturalist and traveller. Served as veterinary surgeon in the Crimea, 1855–6. Naturalist to the Boundary Commission sent to British Columbia, 1858. Joined the staff of the journal *Land and Water* in 1866 before being engaged by the viceroy in archaeological and scientific research in Egypt. Appointed manager of the newly established Brighton Aquarium four months before his death. (*ODNB*.)

Loudon, John Claudius (1783–1843). Landscape gardener and horticultural writer. Travelled in northern Europe, 1813–15; in France and Italy, 1819–20. Urban and rural landscape designer. A prolific author; founded and edited the *Gardener's Magazine*, 1826–43, and the *Magazine of Natural History*, 1828–36. (R. Desmond 1994, *DNB*.)

Lowe, Edward Henry (d. 1874). Wharf keeper at Mardol Quay in Shrewsbury. Purchased The Mount, Shrewsbury, in 1867, and rented it out. (Harris 2004.)

Lowe, Richard Thomas (1802–74). Clergyman and botanist. English chaplain in Madeira, *circa* 1832–52. Rector of Lea, Lincolnshire, 1852–74. Published a flora of Madeira (1868–72). Friend of Thomas Vernon Wollaston. (R. Desmond 1994, *ODNB*.)

Lubbock, Ellen Frances (1834/5–79). Daughter of Peter Hordern, clergyman, of Chorlton-cum-Hardy, Lancashire. Married John Lubbock in 1856. (*Burke's peerage* 1970, Census returns 1861 (Public Record Office RG9/462: 75).)

Lubbock, John, 4th baronet and 1st Baron Avebury (1834–1913). Banker, politician, and naturalist. Son of John William Lubbock and a neighbour of CD's in Down. Studied entomology and anthropology. Worked at the family

Lubbock, John, cont.
 bank from 1849; head of the bank from 1865. Liberal MP for Maidstone, Kent,
 1870–80; for London University, 1880–1900. Succeeded to the baronetcy in 1865.
 Created Baron Avebury, 1900. FRS 1858. (*DSB*, Hutchinson 1914, *ODNB*, *Record
 of the Royal Society of London.*)
   16 January 1867, 17 January [1867], *12 February 1867, 25 March 1867,*
   26 March [1867], *29 March 1867, 28 September [1867], 9 December [1867]*
Lubbock, Mary Harriet (d. 1910). John Lubbock's elder sister. Married Robert
 Birkbeck in 1857. (*Burke's landed gentry* 1898.)
Lucretius (Titus Lucretius Carus) (*c.* 94–55 or 51 B.C.?). Roman philosophical
 poet. (*Oxford Classical dictionary* 3d ed. rev.)
Lütke, Feodor Petrovich (1797–1882). Russian navigator, geographer, and arctic
 explorer. Admiral, 1855; count, 1866. Founding member, Russian Geographical
 Society, 1845; director, 1845–50, 1857–72. President, St Petersburg Academy of
 Sciences, 1864–82. (*GSE* s.v. Litke, Fedor Petrovich.)
Lyell, Charles, 1st baronet (1797–1875). Scottish geologist. Uniformitarian geologist whose *Principles of geology* (1830–3), *Elements of geology* (1838), and *Antiquity of man*
 (1863) appeared in many editions. Professor of geology, King's College, London,
 1831. President of the Geological Society of London, 1835–7 and 1849–51; of the
 British Association for the Advancement of Science, 1864. Travelled widely and
 published accounts of his trips to the United States. CD's scientific mentor and
 friend. Knighted, 1848; created baronet, 1864. FRS 1826. (*DSB*, *ODNB*.)
   1 June [1867], 9 June [1867], 12 June 1867, 18 July [1867], *4 August 1867,*
   22 August [1867], 4 October [1867], 31 October [1867], 7 December [1867]
Lyell, Mary Elizabeth (1808–73). Eldest child of Leonard Horner. Married
 Charles Lyell in 1832. Fluent in French and German; helped Charles Lyell
 with translation. (Freeman 1978, *ODNB* s.v. Charles Lyell.)
McCarthy, J. (*fl.* mid 1860s). South African surgeon. Surgeon at Katberg Convict
 Station, Cape Colony. (*Cape of Good Hope general directory.*)
MacKintosh, Daniel (1815–91). Scottish geomorphologist and Quaternary geolo-
 gist. Author of *The scenery of England and Wales* (1869), and of papers on the drift
 and erratics of North Wales. Fellow of the Geological Society of London, 1871.
 Lecturer in physical geography, Liverpool College. (Letter from Daniel Mac-
 Kintosh, 14 October 1879 (*Calendar* no. 12257), *Geological Magazine* (1891) 28: 432,
 Sarjeant 1980–96.)
   1 December 1867, 8 December [1867]
McLennan, John Ferguson (1827–81). Scottish lawyer and social anthropologist.
 MA, King's College, Aberdeen, 1849. Studied mathematics at Trinity College,
 Cambridge. Began practising law in Edinburgh, 1857. Published on kinship, mar-
 riage, and the law. Moved to London, 1870. Appointed parliamentary draughts-
 man for Scotland, 1871. Regarded as one of the founders of modern British
 social anthropology. (*ODNB*.)
MacOwan, Peter (1830–1909). Botanist and educator. BA, chemistry, London

University, 1857. Principal, Shaw College, Grahamstown, South Africa, 1862. Science master, Gill College, Somerset East, 1869–81. professor of botany at South African College, 1881–9, director of Cape Town Botanic Garden, 1881–92, and curator of Cape Government Herbarium, 1881–1905. Government botanist, 1892–1905. Founder and secretary of the South African Botanical Exchange Society. (*DSAB*, Gunn and Codd 1981.)

Maillet, Benoît de (1656–1738). French diplomat and traveller. General consul of the king of France in Cairo, 1692–1708; consul in Livorno, 1708–14; inspector of French establishments in the Levant and Barbary States, 1715–20; retired to Paris, 1720–1, then Marseilles. Published *Description de l'Egypte* (1735). His *Telliamed*, based on geological observations made during his travels in Egypt and other Mediterranean countries, was published posthumously (1748). (Carozzi trans. and ed. 1968, *DSB*, *NBU*.)

Mantell, Walter Baldock Durrant (1820–95). English-born New Zealand public servant and naturalist. Emigrated to New Zealand, arriving in 1840. Commissioner for extinguishing native titles, Middle Island (South Island), 1848–51; commission of Crown lands for the Southern District of the Province of New Munster, 1851–6. Visited England, 1855–9; returned to New Zealand and was elected to the House of Representatives for Wallace, 1861–6. Attempted unsuccessfully to secure the fulfilment of promises made to the Ngai Tahu at the time of the original land purchases. Retired from the House and accepted a seat on the Legislative Council, 1866. Worked with Charles Lyell and Richard Owen. (*DNZB*.)

Marsh-Caldwell, Anne (1791–1873). Novelist. Family friend of the Wedgwoods. (*ODNB*.)

Maskelyne, Mervyn Herbert Nevil Story- (Nevil) (1823–1911). Mineralogist. Lectured on mineralogy, Oxford University, 1850–7; professor of mineralogy, 1856–95. Keeper of the mineral department, British Museum, 1857–80. MP for Cricklade, 1880–5; North Wiltshire, 1885–92. FRS 1870. (*Alum. Oxon.*, V. Morton 1987, *ODNB*.)

Masters, Maxwell Tylden (1833–1907). Botanist, journal editor, and general medical practitioner. Sub-curator, Fielding Herbarium, University of Oxford, *circa* 1853–7. GP at Peckham from 1856. Lecturer on botany at St George's Hospital medical school, 1855–68. Editor of the *Gardeners' Chronicle*, 1865–1907. Active in the Royal Horticultural Society, succeeding Joseph Dalton Hooker as the chairman of the scientific committee; secretary of the International Horticultural Congress, 1866. FRS 1870. (Clokie 1964, pp. 106, 208; *ODNB*.)
28 March 1867, [28 March – 5 April 1867]

Mätzner, Eduard (b. 1805). German teacher and poet. Teacher, Bromberg Gymnasium. Director, Luisenschule, Berlin, 1838. (Brümmer 1875–6.)

Maw, George (1832–1912). Tile manufacturer, geologist, botanist, and antiquarian. Partner with his younger brother Arthur in the encaustic tile company, Maw & Co., of Brosley, Shropshire. Established a well-known garden at his

Maw, George, cont.

residence at Benthall Hall, Shropshire; an expert on crocuses. Wrote on the geology of western England and North Wales. Travelled to Morocco and Algeria with Joseph Dalton Hooker in 1871 and independently in 1873, writing on the geology of these countries. (Benthall 1980; R. Desmond 1994; *Gardeners' Chronicle*, 12 February 1881, pp. 205–6, 208, 209; Sarjeant 1980–96.)

Mechnikov, Ilya Ilyich (1845–1916). Russian zoologist, microbiologist, and pathologist. Worked on comparative physiology in Germany with Rudolf Leuckart and others in the mid-1860s. Professor of zoology and comparative anatomy, University of Odessa, 1872–82. Worked with Louis Pasteur in Paris, from 1888; at the Pasteur Institute, 1888–1916. Founder of comparative pathology. Discovered phagocytes. Awarded the Copley Medal, 1906; Nobel Prize, 1908, for his immunological work (*DSB*, Williams ed. 1994.)

Meigs, James Aitken (1829–79). American physician, teacher, and anthropologist. Graduated from Jefferson Medical College in 1851. Professor of climatology and physiology at the Franklin Institute, Philadelphia, 1854–62. Librarian, Academy of Natural Sciences of Philadelphia, 1856–9. Held several teaching posts at medical colleges in Philadelphia as well as numerous hospital appointments. Became professor of the Institutes of Medicine and Medical Jurisprudence, Jefferson Medical College, 1868. (*DAB*.)

Meller, Charles James (b. *c.* 1835 d. 1869). Surgeon and botanist. Surgeon and naturalist on David Livingstone's African expedition, 1860–3. Superintendent, Botanic Garden, Mauritius, 1865. Collected plants on the Zambesi, and in Madagascar and Mozambique. (R. Desmond 1994.)

Meyen, Franz Julius Ferdinand (1804–40). Prussian botanist and physician. Naturalist on board the *Prinzess Louise*, 1830–2. Published the results of the voyage, 1834–43. Carried out microscopic investigations of plants and animals. (*ADB*, *DSB*.)

Meyer, Hermann Julius (1826–1909). German publisher. Head of the publishing firm Bibliographisches Institut, 1856–84. (*NDB*.)

    30 July 1867

Middleton, Charles Henry (1828–1915). Clergyman. Undergraduate at Christ's College, Cambridge, 1847–52. Ordained priest, 1852; curate of Wath-on-Dearne, Yorkshire, 1855–9; of St Peter's, Leeds, 1861–3; chaplain of Leeds Gaol, 1863–7; vicar of Lingen, Herefordshire, 1870–4; assistant chaplain, Savoy Chapel, 1881–5. Fellow of the Linnean Society of London, 1867. (*Alum. Cantab.*)

    20 [1867?]

Mill, John Stuart (1806–73). Philosopher and political economist. (*DSB*, *ODNB*.)

Milton, John (1820–80). Accountant. Attended King's College, London. Clerk at the War Office from 1840; assistant accountant general, 1860; accountant general of the Army, 1871. Commissioner of income tax, 1871–5. Nephew of Fanny Trollope (*ODNB*), and reader of manuscripts for the John Murray publishing house. Knighted, 1878. (A. Fraser 1996, p. 16.)

Milton, Viscount. *See* Fitzwilliam, William Wentworth.

Miquel, Friedrich Anton Wilhelm (1811–71). Dutch botanist. Described the flora of the Dutch East Indies. Director, Rotterdam botanic garden, 1835–46. Professor of botany, Amsterdam, 1846–59; Utrecht, 1859–71. (*DSB.*)

Mitten, William (1819–1906). Pharmaceutical chemist and bryologist. Authority on mosses and liverworts. Corresponded with William Jackson Hooker. Resided at Hurstpierpont, Sussex. Described the liverworts for Joseph Dalton Hooker's *Flora Novae Zelandiae* (1853–5) and *Flora Tasmaniae* (1860). Father-in-law of Alfred Russel Wallace. (*Journal of Botany* 44 (1906): 329–32.)

Mivart, St George Jackson (1827–1900). Comparative anatomist. Converted to Catholicism, 1844. Called to the bar, 1851, but never practised. Established his reputation as an anatomist by his studies on primates. Lecturer in comparative anatomy, St Mary's Hospital Medical School, London, 1862–84. Secretary, Linnean Society of London, 1874–80; vice-president, 1892. Professor of the philosophy of natural history, University of Louvain, 1890–3. Excommunicated, 1900. Vigorous critic of Darwinism. Attempted to reconcile evolutionary theory and Catholicism. FRS 1869. (*DNB.*)
    9 December [1867]

Moggridge, John Traherne (1842–74). Entomologist and botanist. Wintered in Mentone, France, and studied the flora of the area. (R. Desmond 1994, *Gardeners' Chronicle* n.s. 2 (1874): 723.)
    6 March [1867], 15 March [1867], 22 April [1867], 29 May [1867], 1 October [1867]

Moore, Thomas (1821–87). Gardener. Gardener, Botanic Garden at Regent's Park, London, 1844–7. Curator, Physic Garden, Chelsea, London, 1848–87. Published major works on ferns. Assistant editor, *Gardeners' Chronicle*, 1841–66; co-editor, 1866–81. Co-editor, *Gardener's Magazine of Botany*, 1850–1; *Garden Companion and Florist's Guide*, 1852; *Orchid Album*, 1881–7. Editor, *Florist and Pomologist*, 1868–74. (*Annals of Botany* 1888: 409–10, R. Desmond 1994, *ODNB*, *Proceedings of the Linnean Society of London* (1886–7): 41–2.)

Morelet, Arthur (b. 1809). French naturalist. Travelled in Algeria, Portugal, the Azores, Cuba and central America. President, Dijon Academy of the sciences, arts and letters. (*Catalogue général de la librairie française.*)

Morton, Samuel George (1799–1851). American physician and ethnologist. Known for his research on human skulls. Recording secretary of the Academy of Natural Sciences of Philadelphia, 1825–9; corresponding secretary, 1831; vice-president, 1840, president, 1849. Professor of anatomy in the medical department, Pennsylvania College, 1839–43. (*ANB, DSB.*)

Moulinié, Jean Jacques (1830–73). Swiss zoologist and militia inspector. Translated *Variation, Descent*, and the sixth edition of *Origin* into French. (Tort 1996.)
    3 May 1867, 16 May [1867], *2 June 1867, 11 October 1867*, 22 November [1867], *25 November 1867*, 30 November [1867]

Mudie's Select Library. Commercial subscription library, founded in 1842 by

Mudie's Select Library, cont.

Charles Edward Mudie (1818–90), with main premises at New Oxford Street, London, from 1852. (*EB* s.v. Mudie, Charles Edward.)

Mueller, Ferdinand Jakob Heinrich (Ferdinand) von (1825–96). German-born explorer and botanist. Emigrated to Australia in 1847. Government botanist, Victoria, 1852. Botanist to the North West Australia Expedition, 1855–7. Director of the Botanical Gardens, Melbourne, 1857–73. President of the Australian Association for the Advancement of Science, 1890. FRS 1861. (*Aust. dict. biog.*, R. Desmond 1994.)

    28 February [1867], *8 October 1867*

Müller, Friedrich (1834–98). Austrian linguist and Sanskrit scholar. PhD, Tübingen, 1858. Privat-dozent, Vienna 1861; professor extraordinarius, 1866; professor of Sanskrit and comparative language, 1869. Wrote the linguistic and ethnographic portions of the report of the *Novara* scientific expedition (1857–9). (*OBL*.)

Müller, Heinrich Ludwig Hermann (Hermann) (1829–83). German botanist and entomologist. Brother of Johann Friedrich Theodor (Fritz) Müller. School-teacher in Schwerin, 1854–5. Studied blind cave insects in Krain, 1855. Teacher of natural sciences at the Realschule in Lippstadt, 1855–83; became director of the school. After settling in Lippstadt, studied the local flora, in particular the mosses. CD's *Orchids* directed Müller's attention to the pollination and fertilisation of flowers, on which he published several papers and books. (P. Gilbert 1977, *Science* 2 (1883): 487–8.)

    23 March 1867, 29 March [1867], *1 April [1867]*, 16 August [1867], *23 October 1867*

Müller, Johann Friedrich Theodor (Fritz) (1822–97). German naturalist. Emigrated to the German colony in Blumenau, Brazil, in 1852. Taught science at the Lyceum in Destêrro (now Florianópolis), 1856–67. Appointed *Naturalista Viajante* of the National Museum, Rio de Janeiro, 1876–92. His anatomical studies on invertebrates and work on mimicry provided important support for CD's theories. (*ADB, DBE*, Möller ed. 1915–21, *NDB*.)

    1 January 1867, 2 February 1867, 7 February [1867], 22 February [1867],
    4 March 1867, 25 March [1867], *1 April 1867*, 22 April [1867], 26 May [1867],
    2 June 1867, 17 July 1867, 31 July [1867], 15 August [1867], *[8 October 1867]*,
    2 November 1867

Müller, Johannes Peter (1801–58). German comparative anatomist, physiologist, and zoologist. Became professor of anatomy and physiology at Berlin University in 1833. Foreign member, Royal Society, 1840. (*ADB, DSB*.)

Mundy, Daniel Louis (1826/7–81). English landscape, portrait, and natural history photographer in New Zealand and Australia. Arrived in New Zealand in 1864, and established a photographic business; Julius von Haast and James Hector were among his clients. Moved to Australia in 1875. (*DNZB*.)

Murchison, Roderick Impey, 1st baronet (1792–1871). Geologist and army officer. Served in the British army, 1807–15. Noted for his work on the Silurian system. President of the Geological Society of London, 1831–3 and 1841–3; of the British Association for the Advancement of Science, 1846; of the Royal Geographical Society of London, 1843–4, 1851, 1857–8, 1862–70. Director-general of the Geological Survey of Great Britain, 1855. Knighted, 1863; created baronet, 1866. FRS 1826. (*DSB, Journal of the Royal Geographical Society, ODNB.*)

Murie, James (1832–1925). Physician and naturalist. MD, Glasgow, 1857; appointed pathologist to the Glasgow Royal Infirmary, 1857. Naturalist and medical officer on John Petherick's expedition to the upper White Nile, 1861–3. Prosector to the Zoological Society, 1865–70. Secretary, Linnean Society of London, 1876–80; librarian, 1880–8. (R. Desmond 1994; *Proceedings of the Linnean Society of London* (1925–6): 92–4.)

Murray, Andrew Dickson (Andrew) (1812–78). Lawyer, entomologist, and botanist. Practised law in Edinburgh; later moved to London. Assistant secretary to the Royal Horticultural Society, 1861; member of the council from 1865; member of the scientific committee from 1868. An expert on insects harmful to crops. In entomology, specialised in the Coleoptera; in botany, in the Coniferae. (*ODNB.*)
12 August 1867, 16 September 1867, 13 December 1867

Murray, John (1808–92). Publisher, and author of guide-books. CD's publisher from 1845. (Freeman 1978, *ODNB* s.v. Murray family, publishers.)
2 January [1867], 3 January [1867], 8 January [1867], *9 January [1867]*,
10 January [1867], 15 January [1867], 27 January [1867], *28 January [1867]*,
29 January [1867], *30 January [1867]*, 31 January [1867], 18 March [1867],
19 March [1867], 20 March [1867], 4 April [1867], 10 April [1867], *17 April [1867]*, 4 August [1867], *5 August [1867]*, 6 August [1867], *19 September [1867]*,
20 September 1867, *23 September [1867]*, 19 October [1867], *1 November [1867]*,
2 November [1867], 31 December [1867]

Nägeli, Carl Wilhelm von (1817–91). Swiss botanist. Maintained a teleological view of evolution. Originally studied medicine, but transferred to botany under Alphonse de Candolle at Geneva. Worked for eighteen months with Matthais Jacob Schleiden at the University of Jena, then worked in Zurich, where he collaborated with Carl Cramer, 1845–52. Professor of botany, University of Freiburg, 1852; professor of botany, University of Munich, 1857. (*DSB* s.v. Naegeli, Carl Wilhelm von.)
[after 8 April 1867]

Nathusius, Hermann Engelhard von (1809–79). German livestock breeder. Studied zoology at the University of Berlin, 1828–30. Turned to agriculture, specialising in cattle and horse breeding, from 1830. Director, state economic board of Saxony, and advisor to the ministry of agriculture, 1869. Chairman, agricultural institute, Berlin, and lecturer in animal husbandry from 1870. (*DBE, NDB.*)

Naudin, Charles Victor (1815–99). French botanist. Joined the herbarium staff at the Muséum d'Histoire Naturelle and became professor of zoology at the Collège

Naudin, Charles Victor, cont.

Chaptal, Paris, in 1846. Resigned his professorship almost immediately owing to a severe nervous disorder. Appointed aide-naturaliste at the Muséum d'Histoire Naturelle, 1854. Established a private experimental garden at Collioure in 1869, earning his living by selling seeds and specimens. First director of the state-run experimental garden at Antibes, 1878. Experimented widely on plants, particularly on acclimatisation and hybridity. Published a theory of transmutation based on hybridisation. (*DSB, Taxonomic literature.*)

Neill, Patrick (1776–1851). Printer. Vice-president of the Botanical Society of Edinburgh, 1836. Founder member of the Caledonian Horticultural Society and secretary for forty years. Collected Scottish plants and had a garden at Canonmills. Promoted the establishment of the Edinburgh Zoological Garden. (R. Desmond 1994, *Gardeners' Chronicle* (1851): 663–4.)

Neumeister, Gottlob. German pigeon fancier and author. (*NUC.*)

Nevill, Geoffrey (1843–85). Conchologist. Worked for his father's mercantile business until he was ordered to go abroad because of ill health. Formed a large collection of shells from the Cape, Mauritius, Bourbon, and the Seychelles in the 1860s, before proceeding to Calcutta, where he took up an appointment as assistant superintendent at the Indian Museum. Published books and articles on land-shells. Corresponding member of the Zoological Society of London. (*Nature* 31 (1885): 435, *Royal Society catalogue of scientific papers.*)

Newton, Alfred (1829–1907). Zoologist and ornithologist. Travelled extensively throughout northern Europe and North America on ornithological expeditions, 1854–63. Editor of *Ibis*, the journal of the British Ornithologists' Union, 1865–70. Professor of zoology and comparative anatomy, Cambridge University, 1866–1907. FRS 1870. (*DNB.*)

19 January [1867], *21 January 1867*, 23 January [1867], *1 March 1867*, 4 March [1867]

Newton, Edward (1832–97). Ornithologist and colonial official. Entered the colonial service, where he held a succession of appointments, including colonial secretary of Mauritius, 1868–78, and lieutenant-governor of Jamaica, 1878–83. Co-authored papers with his elder brother, Alfred Newton. Knighted, 1887. (*Modern English biography*, A. F. R. Wollaston 1921.)

Ngqika, Christian. *See* Gaika, Christian.

Norman, George Warde (1793–1882). Writer on finance. Merchant in the Baltic timber trade, 1810–30. A director of the Bank of England, 1821–72. A founder member of the Political Economy Club, 1821. Public works commissioner, 1831–76. A director of the Sun Insurance Office, 1830–64. Succeeded to his father's estate at the Rookery, Bromley Common, Kent, in 1830. A family friend of the Darwins. (*Burke's landed gentry* 1965, *ODNB, Post Office directory of the six home counties* 1859.)

Norman, Herbert George Henry (b. 1838). Barrister. Son of Henry Norman of Oakley, Kent; nephew of George Warde Norman. BA, Christ Church, Oxford,

1859. Entered Lincoln's Inn, London, 1860. Called to the bar, 1863. (*Alum. Oxon.*, Filmer 1977, *Men-at-the-bar*, *Post Office directory of the six home counties* 1862.)

Nunn, John (b. 1803). Sailor. Sailed on the *Royal Sovereign* to Kerguelen's Land, 1825. He was stranded, with other survivors, until 1829 after the *Favorite* was wrecked while leaving the island. (Nunn 1850.)

Ogle, William (1827–1912). Physician and naturalist. MD, 1861. Lecturer on physiology at the medical school, and assistant physician, St George's Hospital, 1869–72. Superintendent of statistics, General Register Office, 1880–1903. Translated Aristotle's *De partibus animalium* into English in 1882. Published on flower structure and mechanisms for pollination. (*Alum. Oxon.*, *The Times*, 15 April 1912, p. 9.)

29 March [1867]

Oldenbourg, Rudolf (1811–1903). German publisher. Founder of the Munich-based firm R. Oldenbourg, 1858. Business manager of the Munich branch of J. G. Cotta'schen Buchhandlung, 1836–69. (Bosl ed. 1983, *NDB*.)

Oliver, Daniel (1830–1916). Botanist. Assistant in the herbarium of the Royal Botanic Gardens, Kew, 1858; librarian, 1860–90; keeper, 1864–90. Professor of botany, University College, London, 1861–88. FRS 1863. (R. Desmond 1994, *List of the Linnean Society of London*, 1859–91.)

8 April 1867, 1 June [1867]

Owen, Mary Susan (b. 1836). CD's niece. Daughter of Henry Parker and CD's sister Marianne. Married Edward Mostyn Owen in 1866. (*Darwin pedigree*.)

Owen, Richard (1804–92). Comparative anatomist. Assistant conservator of the Hunterian Museum, Royal College of Surgeons of England, 1827; Hunterian Professor of comparative anatomy and physiology, 1836–56. Superintendent of the natural history departments, British Museum, 1856–84; prime mover in establishing the Natural History Museum, South Kensington, in 1881. President of the British Association for the Advancement of Science, 1858. Described the *Beagle* fossil mammal specimens. Knighted, 1884. FRS 1834. (*DSB, ODNB*.)

Pagenstecher, Heinrich Alexander (1825–89). German zoologist and surgeon. Succeeded Heinrich Georg Bronn as professor of zoology and palaeontology, Heidelberg University, 1863. Director of the Zoological Institute, Heidelberg. Director of the Natural History Museum, Hamburg, 1882. Author of *Allgemeine Zoologie* (1875–81). (*NDB*.)

Paget, James, 1st baronet (1814–99). Surgeon. Assistant surgeon at St Bartholomew's Hospital, London, 1847; surgeon, 1861–71. Arris and Gale Professor of anatomy and surgery at the Royal College of Surgeons of England, 1847–52. Lectured on physiology in the medical school, St Bartholomew's, 1859–61; on surgery, 1865–9. Appointed surgeon-extraordinary to Queen Victoria, 1858; serjeant-surgeon, 1877. Created baronet, 1871. FRS 1851. (*ODNB*.)

9 July 1867

Paget, John (1808–92). Agriculturalist and writer. Studied medicine but never practised. Travelled on the continent, 1833–7; married a Hungarian noblewoman

Paget, John, cont.
in 1836 and settled in Hungary, farming an estate at Gyéres, Transylvania. Became a Hungarian citizen in 1847. A zealous member of the Unitarian church of Transylvania. (*ODNB.*)

Parker, Charles (1831–1905). Clergyman. CD's nephew. Fourth son of Henry Parker and CD's sister Marianne. BA, University College, Oxford, 1850; MA, 1857. Vicar of Ford parish, 1863–70, followed by three additional appointments from 1870 until 1883. (*Alum. Oxon.*, *Crockford's clerical directory* 1886, *Darwin pedigree*, *Shrewsbury Chronicle*, 24 November 1905, p. 5.)

Parker, Francis (Frank) (1829–71). Solicitor in Chester. CD's nephew. Third son of Henry and Marianne Parker. (*Darwin pedigree.*)
   22 April 1867

Parker, Henry (1788–1856). Physician. MD, Edinburgh, 1814. Listed as physician to the Shropshire Infirmary, 1847–50. Married Marianne Darwin, CD's sister, in 1824. (*Darwin pedigree, Medical directory.*)

Parker, Henry (1827/8–92). Fine art specialist. Scholar, Oriel College, Oxford, 1846–51; fellow, 1851–85. Son of CD's sister, Marianne Parker. (*Alum. Oxon.*, *CDEL*, *Darwin pedigree.*)

Parker, Marianne (1798–1858). CD's eldest sister. Married Henry Parker (1788–1856) in 1824. (*Darwin pedigree.*)

Parker, Mary Susan. *See* Owen, Mary Susan.

Parker, Robert (b. 1825). CD's nephew. Eldest son of Henry Parker and CD's sister Marianne. (*Darwin pedigree.*)

Parker, Theodore (1810–60). American Unitarian clergyman. Campaigned against slavery and was actively involved in attempts to protect fugitive slaves. (*ANB.*)

Passerini, Giovanni (1816–93). Italian botanist. Professor of botany and director of the botanical garden at Parma, 1844–49 and 1853–93. (*Taxonomic literature.*)

Pastrana, Julia (1834–60). Mexican performer. An indigenous Mexican from the western slope of the Sierra Madre. Suffered from excessive hairiness and physical deformities of the head, resulting in what was thought to be an ape-like visage; her disorders are now thought to have been severe gingival hyperplasia (overdevelopment of the gum) and congenital hypertrichosis with terminal hair. Danced in performances, and was exhibited as a human curiosity in North America, Europe, and Russia; was mummified after death and exhibited further. (Bondeson 1997, pp. 216–44.)

Paterson Simons & Company. British merchant house in nineteenth-century Singapore involved in the tin and gutta-percha trades. William Paterson and Henry Minchin Simons became partners with William Wemys Ker, of Ker, Rawson & Company, in 1853. When Ker retired in 1859, the company became Paterson & Simons. (Turnbull 1972, Wong 1965.)

Paxton, Joseph (1803–65). Gardener and architect. Gardener at the Horticultural Society gardens in Chiswick, 1823; foreman, 1824–6. Head gardener at

Chatsworth, Derbyshire, 1826–50. Designed the layout of and constructions for the Great Exhibition in 1851. Superintended the re-erection of the Crystal Palace at Sydenham, 1853–4; director of the gardens there, 1854–65. Liberal MP for Coventry, 1854–65. Knighted, 1851. (R. Desmond 1994, *Modern English biography*, *ODNB*.)

Peel, Mary Ellen (1830–63). Wife of Archibald Peel, JP in Denbighshire and friend of Thomas Woolner. (*County families* 1880, Woolner 1917, facing p. 126.)

Peel, Robert, 2d baronet (1788–1850). Politician. Tory prime minister, 1834–5 and 1841–6. (*ODNB*.)

Philip, Thomas Durant (1819–90). Missionary. Teacher for the London Missionary Society at Hankey, South Africa, 1847–76. (Hampden-Cook 1926.)

Phillips, John (1800–74). Geologist. Keeper of the Yorkshire Museum, 1825–40. Assistant secretary, British Association for the Advancement of Science, 1832–62. Professor of geology, King's College, London, 1834–9. Palaeontologist to the Geological Survey of Great Britain, 1841–4. Deputy reader in geology, Oxford University, 1853; professor, 1860–74. FRS 1834. (*DSB*, *ODNB*.)

Plimsoll, Joseph (1806/7–85). Naval surgeon, poet, and writer on religious topics. (*BLC*, Census returns 1881 (Devon Record Office), *Navy list* 1855–85.)
    21 October 1867, 21 November 1867, 3 December 1867, 17 December 1867

Pringsheim, Nathanael (1823–94). German botanist. PhD, Berlin, 1848; privatdozent, 1851; professor extraordinarius, 1864. Professor and director of the botanic garden, Jena, 1864–9; director, phytophysiological institute, 1865–9. Founding editor, *Jahrbücher für wissenschaftliche Botanik*, 1857–94. A founder and president of the German botanical society, 1882–94. Helped establish the biological station at Helgoland. Worked on algal sexuality and later on the function of chlorophyll. (*NDB*, *DSB*.)

Pritchard, Charles (1808–93). Clergyman and astronomer. Headmaster, 1834–62, of Clapham Grammar School, where he established an observatory. Hulsean Lecturer, Cambridge University, 1867. Savilian Professor of astronomy, Oxford University, 1870–93. President of the Royal Astronomical Society, 1866–8. FRS 1840. (*Alum. Cantab.*, *DSB*, *ODNB*.)

Quaritch, Bernard Alexander Christian (Bernard) (1819–99). German-born book-dealer. Worked for Henry George Bohn (1796–1884), 1842–4 and 1845–7; established an independent business near Leicester Square, London, in 1847; moved to 15 Piccadilly in 1860. (*ODNB*.)

Quatrefages de Bréau, Jean Louis Armand de (Armand de Quatrefages) (1810–92). French zoologist and anthropologist. Doctorate in the physical sciences, University of Strasbourg, 1830; doctorate in medicine, 1832. Founded *Journal de médicine et de chirurgie de Toulouse*, 1836. Moved to Paris and took a doctorate in the natural sciences, 1840. Professor of natural history at the Lycée Henri IV, Paris, 1850; professor of anthropology, Muséum d'Histoire Naturelle, 1855. Foreign member, Royal Society of London, 1879. (*DSB*, *Record of the Royal Society of London*.)

Ramsay, Marmaduke (d. 1831). BA, Jesus College, Cambridge, 1818; fellow and tutor, 1819–31. (*Alum. Cantab.*)

Reed, George Varenne (1816–86). Clergyman. Curate of Hayes, Kent, 1837–9; of Tingewick, Buckinghamshire, 1839–54. Rector of Hayes, 1854–86. Tutor to George Howard, Francis, Leonard, and Horace Darwin. (*Alum. Cantab.*, Freeman 1978, J. R. Moore 1977.)

Reichenbach, Heinrich Gustav (1824–89). German botanist. Appointed professor extraordinarius of botany, Leipzig, in 1855. Director of the botanic gardens in Hamburg, 1863–89. Wrote extensively on orchids. (*ADB.*)

Reinwald, Charles-Ferdinand (1812–91). German-born bookseller and editor. Founded a business exporting French books in Paris, 1849. Editor in particular of foreign scientific works, and of the *Dictionnaire universel de la langue française*, by M. P. Poitevin. Published the *Catalogue annuel de la librairie française*. (*Dictionnaire universel des contemporains.*)
     [May 1867]

Rengger, Johann Rudolph (1795–1832). German physician, explorer, and naturalist. Travelled in South America (especially Paraguay), 1818–26. (*ADB*, *Neuer Nekrolog der Deutschen* 10 (1832): 699–700, Schumann 1888.)

Reuss, August Emanuel von (1811–73). Czech-born palaeontologist, geologist, and mineralogist. Born in Bohemia; studied philosophy and medicine in Prague; MD, 1833. Professor of mineralogy, Prague, 1849; Vienna, 1863–73. (*DSB*, *OBL.*)

Rich, Mary (1789–1876). Daughter of Sir James Mackintosh. Half-sister of Frances Emma Elizabeth Wedgwood. Married Claudius James Rich (*ODNB*) in 1808. (B. Wedgwood and Wedgwood 1980.)

Richard, Adolph (1822–72). French doctor. Surgeon in Paris hospitals. (Tort 1996.)

Richardson, Charles (1775–1865). Lexicographer. Published his *New English dictionary* between 1835 and 1837. (*ODNB.*)

Ricker, Franz Anton (1816–92). German bookseller and publisher. Took over his brother Joseph's publishing business, J. Ricker'sche Buchhandlung, in Giessen, Hesse, in 1835. (*Hessische Biographien* 3: 301.)

Ricker, Joseph (1804–34). German bookseller and publisher. Set up a publishing business in Giessen, Hesse, in 1832. (*Hessische Biographien* 3: 301.)

Rivers, Thomas (1798–1877). Nurseryman. Succeeded to the family business in Sawbridgeworth, Hertfordshire, in 1827. Specialised in the cultivation of roses and fruit. Author of works on rose and fruit culture; contributed extensively to gardening journals. A founder of the British Pomological Society, 1854. (*ODNB.*)
     26 April 1867, 9 September 1867

Rivers, Thomas Francis (1831–99). Nurseryman. Son of Thomas Rivers (1798–1877). Known for his experimental work on prolonging the season of various fruit trees. (R. Desmond 1994, *Journal of Horticulture* 39 (1899): 161–2.)

Robertson, Peter S. (1818–79). Nurseryman. Gardener, Royal Botanic Garden, Edinburgh, 1837–43. Manager, Peter Lawson and Son, Edinburgh. (R. Desmond 1994.)
     30 April 1867

Rohrbach, Paul (1846–71). German botanist. (*ADB*, Barnhart comp. 1965.)
23 February [1867]

Rolfe, Laura (1807–1868). Daughter of Thomas Carr of Frognal, Hampstead, Middlesex, and of Esholt Heugh, Northumberland. Married Robert Monsey Rolfe in 1845. (*ODNB* s.v. Rolfe, Robert Monsey.)

Rolfe, Robert Monsey, 1st Baron Cranworth of Cranworth (1790–1868). Statesman and jurist. Whig MP for Penryn and Falmouth, 1832–9. Solicitor-general, 1834 and 1835–9. Created Baron Cranworth of Cranworth, 1850. Lord justice of appeal, 1851–2. Lord chancellor, 1852–8 and 1865–6. Holwood Park, his country residence, was a mile and a half north of Down House, the Darwin's residence. (*Dod's parliamentary companion*, Freeman 1978, *ODNB*.)
14 January 1867, 20 November 1867

Rolle, Friedrich (1827–87). German geologist, palaeontologist, and natural history dealer. Assistant at the Kaiserlich-königliche Hofmineralien-Cabinett, Vienna, 1857–9; associate, 1859–62. Returned to Bad Homburg in 1862. Author of *Ch. Darwin's Lehre von der Entstehung der Arten und ihre Anwendung auf die Schöpfungsgeschichte* (1863). (*ADB, BLKO*, G. P. R. Martin and Uschmann 1969, Sarjeant 1980–96.)

Ross, James Clark (1800–62). Naval officer and polar explorer. Joined the navy in 1812. Discovered the northern magnetic pole in 1831. Employed on the magnetic survey of the United Kingdom, 1835–8. Commander of an expedition to the Antarctic, 1839–43; and of a search expedition for John Franklin (*ODNB*), 1848–9. Knighted, 1843. FRS 1828. (*ODNB*.)

Rossmässler, Emil Adolph (1806–67). German natural historian and writer. After studying theology in Leipzig, became a schoolteacher in Weida, 1827–30. Professor of zoology, Forest and Agricultural Academy, Tharundt, 1830. From 1850, worked as a popular writer on natural history after being forced to retire for political and religious reasons. (*ADB, DBE*).

Rothrock, Joseph Trimble (1839–1922). American physician, botanist, and forester. In 1860, entered the Lawrence Scientific School, Harvard University, where he was a student of, and an assistant to, Asa Gray. Enlisted in the 131st Pennsylvania Infantry in 1862; commissioned captain in the 20th Pennsylvania Cavalry in 1863; honourably discharged, 1864. Professor of botany, Pennsylvania State Agricultural College, 1867. Botanist and surgeon to the government survey in Colorado, New Mexico, and California, 1873. Professor of botany, University of Pennsylvania, 1877–1904. (*DAB*.)
31 March 1867 (Asa Gray), *22 August 1867* (Asa Gray)

Royer, Clémence Auguste (1830–1902). French author and economist. Studied natural science and philosophy in Switzerland. In Lausanne in 1859, founded a course on logic for women. Translated *Origin* into French in 1862. (*Dictionnaire universel des contemporains*, J. Harvey 1997.)

Rüppell, Wilhelm Peter Eduard Simon (Eduard) (1794–1884). German naturalist, traveller, and invertebrate palaeontologist. (*ADB*, Sarjeant 1980–96.)

Rütimeyer, Karl Ludwig (Ludwig) (1825–95). Swiss palaeozoologist and geographer. Professor of zoology and comparative anatomy, University of Basel, 1855;

Rütimeyer, Karl Ludwig, cont.

rector, 1865; professor in the medical and philosophical faculties, 1874–93. Made important contributions to the natural history and evolutionary palaeontology of ungulate mammals. (*DSB.*)

   4 May [1867]

Salt, George Moultrie (1825–1907). County attorney. Eldest son of Thomas Salt. Admitted to practice, 1845. Partner in the law firm Salt & Sons, Shrewsbury, 1848–1907. (Harris 2004, *Law list* 1848–1907.)

   16 March [1867]

Salt, William (1831–81). County attorney. Third son of Thomas Salt. Admitted to practice, 1854. Partner in the law firm Salt & Sons, Shrewsbury, 1855–81. (Harris 2004, *Law list* 1855–81.)

Salt & Sons. Shrewsbury law firm.

   17 July 1867

Salter, John William (1820–69). Geologist and palaeontologist. Apprenticed to James de Carle Sowerby in 1835. Assistant to Adam Sedgwick, 1842–6. Assistant to Edward Forbes on the Geological Survey of Great Britain, 1846; palaeontologist to the survey, 1854–63. (*ODNB.*)

   4 January [1867], 14 May 1867, 18 June 1867

Salter, Samuel James Augustus (1825–97). Surgeon. House surgeon and physician, King's College Hospital; dental surgeon to Guy's Hospital, London. Fellow of the Linnean Society of London, 1853. FRS 1863. (*Modern English biography*, R. Desmond 1994.)

   20 November 1867

Salter, Sarah (Sally) (1819–92). Daughter of James de Carle Sowerby. Married John William Salter in 1846. (Secord 1985.)

Samuelson, James (1829–1918). Businessman, science teacher, and writer on scientific and social issues. Inherited his father's business of seed crushing and oil extraction. Worked as science teacher in Liverpool. Founder, with Robert Hardwicke, and editor of the *Popular Science Review*, 1862–4. Helped launch the *Quarterly Journal of Science* and occasionally acted as editor until 1870. Called to the bar, 1870, and then campaigned as social reformer and writer. (Lightman ed. 2004.)

   12 October [1867]

Sandile (1820–78). Xhosa chief. Paramount chief of the Rarabe. Son of Ngqika (Gaika), head of part of the Xhosa tribe, and Sutu, a Tshatshu (or Gcina) princess. Assumed the duties of paramount chief *circa* 1840. Fought in border wars, 1846–7, 1850–2, and 1878; was forced to cede increasing amounts of territory to the British in South Africa, his remaining territory being annexed to Cape Colony in 1866. (*Dictionary of African biography* vol. 3, *DSAB.*)

Sanford, William Ayshford (1818–1902). Palaeontologist. Resided at Nynehead Court, Somerset. Matriculated, Trinity College, Cambridge, 1837. President of the Somerset Archaeological Society. Co-authored papers with William Boyd Dawkins on mammalian palaeontology. (*Alum. Cantab.*, *Burke's landed gentry* 1906.)

Sanson, André (1826–1902). French veterinarian and animal breeder. Author of a number of books and articles on animal physiology, breeding, and veterinary medicine. Teacher at the Ecole Veterinaire de Toulouse, 1856; professor of zoology and zootechny at the Ecole d'Agriculture, Grignon, 1872. Chief editor and contributor to *Culture*, 1859–70. (Glaeser 1878, *Larousse du XX^{e}siècle*.)

Savigny, Marie Jules César Lelorgne de (1777–1851). French naturalist. Accompanied Napoleon's expedition to Egypt as an invertebrate zoologist, 1798–1802. (*DSB*.)

Scherzer, Karl von (1821–1903). Austrian scientific traveller and diplomat. Principal scientist of the *Novara* expedition. Austrian consul in London, 1875–8. (*BLKO*.)
21 November 1867

Schmidt, Eduard Oskar (Oskar) 1823–86. German zoologist. Professor extraordinarius of zoology, Jena, 1849; director of the zoological museum, 1851. Professor of zoology and comparative anatomy, Graz, 1857; director of the agricultural and zoological museum, 1863. Professor of zoology and zootomy, Strasbourg, 1872. His major research interest was the anatomy of sponges. His inaugural lecture supporting Darwinism, made in 1865 at the University of Graz, led to conflict with the Catholic Church in Austria and sparked a wider debate between Catholic and German nationalist circles at the university. (*ADB, OBL*.)

Schultze, Max Johann Sigismund (1825–74). German anatomist. MD, Greifswald, 1849; Privat-dozent, 1850–4. Assistant professor of anatomy, Halle, 1854–9. Professor of anatomy and director of the anatomical institute, Bonn, 1859. Founder and editor of *Archiv für mikroskopische Anatomie*, 1865–74. Noted for his work in microscopy, the reform of cell theory, and descriptive and taxonomic studies of rhizopods and sponges. (*DSB, DBE*.)

Schwarz, Eduard (1831–62). Austrian physician and ethnologist. MD, Vienna 1856. Ship's doctor on the *Novara* scientific expedition, 1857–9. Took systematic anthropometric measurements (with instruments he invented) of different races encountered on the voyage. (*OBL*.)

Schweizerbart, Christian Friedrich (1805–79). German publisher. Director of E. Schweizerbart'sche Verlagsbuchhandlung of Stuttgart, 1841–67. Publisher of the German translations of *Origin* (1860, 1863, 1867) and *Orchids* (1862). (*Jubiläums-Katalog*.)

Schweizerbart, Wilhelm Emanuel (1785–1870). German publisher. Founded E. Schweizerbart'sche Verlagsbuchhandlung in Stuttgart in 1830. Publisher of many German scientific works. Retired in 1841. (*Jubiläums-Katalog*.)

E. Schweizerbart'sche Verlagsbuchhandlung. German publishing company in Stuttgart; founded by Wilhelm Emanuel Schweizerbart in 1830; conducted by his nephew Christian Friedrich Schweizbart from 1841. (*Jubiläums-Katalog*.)
24 January 1867, [19 March 1867], *22 March 1867*, 22 November [1867]

Sclater, Philip Lutley (1829–1913). Lawyer and ornithologist. One of the founders of *Ibis*, the journal of the British Ornithologists' Union, 1858; editor, 1858–65 and 1878–1912. Secretary of the Zoological Society of London, 1860–1903. FRS 1861. (*DSB*, Scherren 1905.)
9 December [1867]

Scott, John (1836–80). Scottish botanist. Gardener at several different country estates, before becoming foreman of the propagating department at the Royal Botanic Garden, Edinburgh, in 1859. Through CD's patronage emigrated to India in 1864, and worked briefly on a *Cinchona* plantation before taking a position as curator of the Calcutta botanic garden in 1865. Seconded to the opium department, 1872–8. Carried out numerous botanical experiments and observations on CD's behalf. Fellow of the Linnean Society of London, 1873. (Lightman 2004, *ODNB*.)

    22 January 1867, 24 September 1867

Scott, Walter Francis, 5th duke of Buccleuch and 7th duke of Queensbury (1806–84). Politician. MA, St Johns, Cambridge, 1827. Lord privy seal, 1842–6. President of the Highland Agricultural Society, 1831–5 and 1866–9. President of the British Association for the Advancement of Science, 1867. Knighted, 1835. FRS 1833. (*ODNB, Record of the Royal of Society.*)

Scrope, George Poulett (1797–1876). Geologist and political economist. Liberal MP for Stroud, Gloucestershire, 1833–67. Studied the volcanic districts of Italy, Sicily, France, and Germany. FRS 1826. (*DSB, ODNB.*)

Sedgwick, Adam (1785–1873). Geologist and clergyman. Woodwardian Professor of geology, Cambridge University, 1818–73. Prebendary of Norwich Cathedral, 1834–73. President, Geological Society of London, 1829–31; British Association for the Advancement of Science, 1833. FRS 1821. (*DSB, ODNB, Record of the Royal Society of London.*)

Selkirk, 6th earl of. *See* Douglas, Dunbar James.

Sethe, Anna. *See* Haeckel, Anna.

Shakespeare, William (1564–1616). Poet and dramatist. (*ODNB.*)

Shaw, Alexander (1804–90). Scottish surgeon. MA, University of Glasgow, 1822. Entered Middlesex Hospital, London, as a pupil, 1822; assistant surgeon, 1836, surgeon 1842. Also entered the Great Windmill Street School of Anatomy, 1827. Member of Royal College of Surgeons, 1828; fellow 1843; served on its college council, 1858–65. Published on and edited works by Sir Charles Bell, his brother-in-law. (*ODNB.*)

    22 March 1867 (E. A. Darwin)

Shaw, James (1826–96). Scottish writer and schoolmaster. Apprenticed as a pattern designer. Worked in the calico printing trade. Began training as a schoolmaster in 1855. After a succession of posts, became schoolmaster in Tynron, Dumfriesshire, 1862–96. Member of the Dumfriesshire and Galloway Natural History and Antiquarian Society. (R. Wallace ed. 1899.)

Sibson, Francis (1814–76). Physician. Apprenticed to a surgeon in Edinburgh, 1828; studied at Guy's and St Thomas's Hospitals, London; resident surgeon and apothecary, Nottingham General Hospital, 1835–48. Set up a consulting practice in London in 1848; physician to Joseph Dalton Hooker and his family. Physician at St Mary's Hospital, 1851–71. Lectured on medicine. Published on the physiology and pathology of respiration. FRS 1849. (*ODNB, Physicians.*)

Thomas Sims & Co. Photographers with premises at 76 Westbourne Grove, Bayswater, London. (*Post Office London directory* 1866.)

Slack, Henry James (1818–96). Author and journalist. Worked on provincial journals, 1846–52. Proprietor and editor of the *Atlas* from 1852; editor, from 1862, of the *Intellectual Observer*, continued as the *Student*, 1868–71. Wrote under the name Little John for the *Weekly Times*. Author of popular science papers. Secretary and president, 1878, of the Royal Microscopical Society. (*ODNB*.)
    30 November 1867

Smith, Andrew (1797–1872). Army surgeon. Stationed in South Africa, 1820–36. Principal medical officer at Fort Pitt, Chatham, 1836; deputy inspector-general, 1845. Director-general, army and ordnance medical departments, 1853–8. Blamed in the press for the scandalous state of medical care of British troops in the Crimean campaign, but was exonerated of dereliction of duty by parliamentary committees. An authority on South African zoology. Knighted, 1858. FRS 1857. (*ODNB*.)
    26 March 1867

Smith, Charles Hamilton (1776–1859). Flemish-born army officer and writer on natural history. Wrote on historical costume and war, and produced many natural history illustrations. Knighted, 1834. FRS 1824. (*ODNB*.)

Smith, James Edward (1759–1828). Botanist. Purchased the library and collections of Carl von Linné in 1784. Founded the Linnean Society of London in 1788 and served as president until his death. Knighted, 1814. FRS 1785. (*DSB, ODNB*.)

Smith, John (1821–88). Scottish gardener. Gardener to the duke of Roxburgh; to the duke of Northumberland at Syon House, Middlesex, 1859–64. Curator, Royal Botanic Gardens, Kew, 1864–86. (R. Desmond 1994.)
    26 December [1867]

Smith, Mary (b. 1826/7). Born in Northumberland. Wife of John Smith (1821–88). (Kew Census returns 1871 (Public Record Office RG10/869/144 v.).)

Smyth, Robert Brough (1830–89) Mining engineer. Emigrated to Victoria, Australia, 1852. Secretary, Board of Mines, 1860; honorary secretary, Board for the Protection of Aborigines, 1860; director of the geological survey of Victoria, 1871–6. Fellow of the Linnean Society of London, 1874. (*Aust. dict. biog., ODNB*.)

Snow, George (1820/1–85). Coal-dealer, Down, Kent. Operated a weekly carrier service between Down and London. (Census returns 1861 (Public Record Office RG9/462: 72); gravestone inscription, Down churchyard; *Post Office directory of the six home counties* 1862.)

Soga, Tiyo (1829–71). Xhosa missionary. Educated at church schools in South Africa; trained as a missionary in Glasgow with the United Presbyterian Church. Qualified as a minister in Edinburgh in 1856. Opened a new mission among the Ngqika (Gaikas), on the Mgwali river, British Kaffraria, thirty miles from King William's Town, in 1857. (*DSAB*.)

Solander, Daniel (1733–82). Swedish botanist. Pupil of Linnaeus. Employed cataloguing natural history specimens at the British Museum, 1763. Engaged by Joseph Banks to accompany him on Cook's voyage in the *Endeavour*, 1768–71, and to Iceland, 1772. Keeper of the natural history department, British Museum, 1773. FRS 1764. (*DSB, ODNB*.)

Sowerby, George Brettingham Jr (1812–84). Conchologist and illustrator. Assisted his father, George Brettingham Sowerby (1788–1854), in a business selling natural history specimens; succeeded to the business in 1854. Illustrated numerous works on shells. Fellow of the Linnean Society of London, 1844. (*DNB* s.v. Sowerby, George Brettingham the elder.)

Sowerby, James de Carle (1787–1871). Naturalist and scientific illustrator. An expert on fossil conchology. Founding member of the Royal Botanic Society and Gardens, Regent's Park, London, 1838; secretary, 1839–69. Brother of George Brettingham Sowerby (1788–1854). (R. Desmond 1994, *DNB*.)

Spencer, Herbert (1820–1903). Philosopher. Apprenticed as a civil engineer on the railways, 1837–41. Became sub-editor of the *Pilot*, a newspaper devoted to the suffrage movement, in 1844. Sub-editor of the *Economist*, 1848–53. From 1852, author of books and papers on transmutation theory, philosophy, and the social sciences. (*DSB, ODNB*.)

9 December [1867]

Spottiswoode, William (1825–83). Mathematician and physicist. Succeeded his father as queen's printer in 1846. Throughout his life pursued mathematical studies in which he supplied new proofs of known theorems and also did important original work; produced a series of memoirs on the contact of curves and surfaces. President of the mathematical section of the British Association for the Advancement of Science, 1865; of the Royal Society of London, 1878–83. FRS 1853. (*DNB*.)

Sprengel, Christian Konrad (1750–1816). German botanist. Rector of the Great Lutheran Town School, Spandau, where he taught languages and natural science, 1780–94. Moved to Berlin, where he worked as a private tutor. Published his major work on botany in 1793. (*ADB, DSB*.)

Stack, James West (1835–1919). New Zealand missionary, clergyman, and interpreter. Travelled to London with his family in 1848; trained as a teacher at the Church Missionary Society Training College, Islington. Returned to New Zealand in 1852. Taught at CMS industrial schools at Maraetai, at Waikato Heads, and Te Kohanga. Worked for the Maori mission in Christchurch, 1859–88. Ordained priest, 1862. Worked as a government interpreter, 1860–80, and later as inspector of native schools on South Island. Left New Zealand, 1898. Wrote on Maori subjects. (*DNZB*.)

Stainton, Henry Tibbats (1822–92). Entomologist. Founder of the *Entomologist's Annual*, 1855–74, and of the *Entomologist's Weekly Intelligencer*, 1856–61. Secretary to the Ray Society, 1861–72; to the biology section of the British Association for the Advancement of Science, 1864 and 1867–72. Co-founder, *Entomologist's Monthly Magazine*, 1864. FRS 1867. (*ODNB*.)

Stanley, Edward George Geoffrey Smith, 14th earl of Derby (1799–1869). Scholar, orator, and Conservative prime minister. MP for North Lancashire, 1832–44. Chief secretary for Ireland, 1830–3. Colonial secretary, 1833–4 and 1841–5. Prime minister, 1852, 1858–9, 1866–8. Appointed chancellor of Oxford University, 1852. Supported gradual parliamentary reform. (*ODNB.*)

Stirling, Waite Hockin (1829–1923). Clergyman and missionary. Secretary, Patagonian Mission Society, 1857–62. Superintendent missionary of Tierra del Fuego, 1862–9. Bishop of the Falkland Islands, 1869–1900. Assistant bishop of Bath and Wells, 1901–11; canon of Wells, 1901–20; precentor, 1903–20. (*Alum. Oxon., Crockford's clerical directory* 1923, Hazlewood 2000, Macdonald 1929.)

Story-Maskelyne, Mervyn Herbert Nevil. *See* Maskelyne, Mervyn Herbert Nevil Story-.

Sulivan, Bartholomew James (1810–90). Naval officer and hydrographer. Lieutenant on HMS *Beagle*, 1831–6. Surveyed the Falkland Islands in HMS *Arrow*, 1838–9. Commander of HMS *Philomel*, 1842–6. Resided in the Falkland Islands, 1848–51. Commanded HMS *Lightning* in the Baltic, 1854–5. Naval officer in the marine department of the Board of Trade, 1856–65. Admiral, 1877. Knighted, 1869. (*ODNB.*)

    11 January 1867, 15 January [1867], *16 August [1867]*

Sutton, Seth. A keeper at the zoological gardens of London. (Archives of the Zoological Society of London, 'Daily occurrences', September 1860; Freeman 1978; letter from Seth Sutton, 8 August 1867.)

    8 August 1867

Swettenham, Richard Paul Agar (1845–99). School inspector. Contemporary of George Howard Darwin's at Trinity College, Cambridge. Studied mathematics. HM inspector of schools, Northern Division. (*Alum. Cantab.*)

Swinhoe, Robert (1836–77). Diplomat and ornithologist. Attached to the British consulate in Hong Kong, 1854; in Amoy (Xiamen), China, 1855. British vice-consul, Formosa (Taiwan), 1860–5; consul, 1865–73. Acting consul, Amoy, 1865–71; Ning-po (Ningbo), 1871–3. Consul, Ning-po, 1873–5. Collected plants and animals in Eastern Asia. FRS 1876. (*Foreign Office list* 1877, Hall 1987, *ODNB.*)

    [27 February 1867], *5 August 1867*

Sykes, William Henry (1790–1872). Naturalist and military officer in the East India Company. Statistical reporter to the Bombay government, 1824–31. Royal commissioner on lunacy, 1835–45. Elected to the board of directors of the East India Company, 1840; chairman, 1856. FRS 1834. (R. Desmond 1994, *ODNB.*)

Tait, Peter Guthrie (1831–1901). Scottish mathematician and physicist. Studied mathematics at Edinburgh University and Peterhouse, Cambridge. Fellow at Peterhouse, 1852–4. Professor of mathematics, Queen's College, Belfast, 1854–60; of natural philosophy, Edinburgh, 1860–1901. Published widely on mathematics and physics; published *Natural philosophy* (1867) with William Thomson (later Lord Kelvin). (*ODNB.*)

Tankerville, countess of. *See* Bennet, Olivia.

Tankerville, 6th earl of. *See* Bennet, Charles Augustus.

Tegetmeier, William Bernhard (1816–1912). Editor, journalist, lecturer, and naturalist. Pigeon-fancier and expert on poultry. Pigeon and poultry editor of the *Field*, 1864–1907. Secretary of the Apiarian Society of London. (*Field*, 23 November 1912, p. 1070; *ODNB*; Richardson 1916.)

    6 January [1867], 5 March [1867], *29 March 1867*, 30 March [1867], 20 April [1867], 4 August [1867]

Tendler & Co. Nineteenth-century publishing company in Vienna. (*BLC*.)

    29 May 1867 (John Murray)

Theile, Friedrich Wilhelm (1801–79). German physician and anatomist. MD, Jena, 1825; director, pharmaceutical institute, 1828; professor extraordinarius of anatomy, 1831. Professor of anatomy, Bern, 1834. Medical advisor and member of the medical commission, Weimar, 1853. (*DBE*.)

Thompson, John Vaughan (1779–1847). Zoologist and army surgeon. Surgeon with the British army, 1799–1816, including postings in the West Indies and Madagascar. Surgeon to the forces in Cork, Ireland, and later deputy inspector-general of hospitals, 1816–35. Medical officer in charge of the convict settlements of New South Wales, Australia, 1835–47. Studied Polyzoa, Cirripedia, and Crustacea. Fellow of the Linnean Society of London, 1810. (*DSB, ODNB*.)

Thomson, Thomas (1817–78). Naturalist. MD, Glasgow, 1839. Travelled to India as assistant surgeon to the East India Company. Curator of the Asiatic Society's museum, Calcutta, 1840. Travelled and was taken prisoner in Afghanistan, 1840–2. Accompanied Joseph Dalton Hooker to the Himalayas, 1850–1, and collaborated with him at Kew on various botanical publications. Superintendent of the Calcutta botanic garden and professor of botany at the Calcutta Medical College from 1855 until his return to England in 1861. FRS 1855. (R. Desmond 1994, *ODNB*.)

Thomson, William, Lord Kelvin (1824–1907). Scientist and inventor. Professor of natural philosophy, Glasgow, 1846–99. Formulated laws of equivalence and transformation in thermodynamics and a doctrine of available energy. Pioneered telegraphic systems and assisted in the laying of the first transatlantic cable. Wrote on the age and cooling of the earth. Proposed a hydroelectric scheme for Niagara. Created Baron Kelvin of Largs, 1892. FRS 1851. Awarded the Copley Medal, 1883. (*DSB, ODNB*.)

Thurber, George (1821–90). American botanist and chemist. Trained as a pharmacist; lecturer in chemistry, Franklin Society of Providence, Rhode Island; botanist, quartermaster, and commissary on the United States–Mexico boundary survey, 1850–3, during which time he collected many new species of plants. MA, Brown University, 1853; with the United States Assay Office, New York, 1853–6. Lecturer in botany and materia medica, College of Pharmacy, New York, 1856–60, 1865–6; professor of botany and horticulture, Michigan State Agricultural College, 1859–63. Editor of the *American Agriculturist*, 1863–85. Specialised in grasses. (*ANB, DAB*.)

Thuret, Gustave Adolphe (1817–75). French botanist. Lawyer and attaché in Constantinople, 1840. Gave up his diplomatic career to devote himself to botanical research. Published a number of papers on the seaweed genus *Fucus*. Resided in Cherbourg, 1852–6, and later at Cap d'Antibes on the Mediterranean in order to carry out research on living algae. (*DSB.*)

Thwaites, George Henry Kendrick (1811–82). Botanist and entomologist. Superintendent of the Peradeniya botanic gardens, Ceylon (Sri Lanka), 1849; director, 1857–80. FRS 1865. (R. Desmond 1994, *ODNB.*)

Tollet, George (1767–1855). Landowner and agricultural reformer. Changed his name from Embury in 1796, on succeeding to the estate of Betley Hall, Staffordshire. Justice of the peace and deputy lieutenant of Staffordshire. A close friend of Josiah Wedgwood II. (*Burke's landed gentry* 1879 s.v. Wicksted of Betley Hall, Freeman 1978.)

Tollet, Georgina. Daughter of George Tollet. A close friend of the Wedgwoods and Darwins. Edited the manuscript of *Origin*. (Freeman 1978.)

Trail, Robert (1796–1873). Surveyor of taxes in Aberlady, Scotland. Member of the Botanical Society of Edinburgh. (James Denwoodie, Scots Ancestry Researcher, Edinburgh; *Transactions of the Botanical Society of Edinburgh.*)
    5 April 1867

Trimen, Roland (1840–1916). Zoologist and entomologist. Emigrated to South Africa in 1858. Arranged the Lepidoptera at the South African Museum. Held civil-service positions in the Commission of Land and Public Works, the governor's office, and the colonial secretary's office. Became part-time curator of the South African Museum in 1873; full-time curator, 1876. FRS 1883. (*DSAB.*)
    24 December [1867], *31 December 1867*

Trübner, Johann Nicolaus (Nicholas) (1817–84). German-born publisher and philologist. Foreign corresponding clerk for Longmans (publishers), 1843–51. In partnership with Thomas Delf and David Nutt, established a successful London publishing house specialising in Asian literature and translations of works in philology, philosophy, and religion. (*ADB, ODNB.*)
    7 May 1867

Trübner & Co. Booksellers at Paternoster Row, London, specialising in Asian, American, and European publications (*Post Office London directory*). Founded by Nicholas Trübner (*ODNB*).
    26 February 1867

Turnbull, Archibald (*c.* 1790–1875). Scottish nurseryman. Partner with his uncle, William Dickson (died 1835), in a nursery in Perth. Founder member of the Perthshire Horticultural Society. (R. Desmond 1994.)

Turnbull, Claudius James (d. 1867). Army officer. British army captain in India. Died on active service. (*India list*, January 1868, p. 541.)

Turner, William (1832–1916). Anatomist and administrator. Senior demonstrator to Sir John Goodsir, professor of anatomy, University of Edinburgh, 1854–67; professor of anatomy, 1867–1916; principal, 1903–16. Published papers on anthropology and comparative anatomy from 1854. Knighted, 1886. FRS 1877. (*DSB, ODNB.*)
    15 January [1867], 1 February [1867], *8 February 1867*, 11 February [1867]

Tylor, Edward Burnett (1832–1917). Anthropologist. Educated at the School of the Society of Friends, Grove House, Tottenham. Author of *Primitive culture* (1871). Keeper of Oxford University Museum, 1883; reader in anthropology, Oxford University, 1883; professor, 1896; professor emeritus, 1909. President of the Anthropological Society, 1879–80, 1891–2. Helped to establish anthropology as a legitimate field of scientific enquiry. Knighted, 1912. FRS 1871. (*Men and women of the time* 1899, *ODNB*.)

Tyndall, John (1820–93). Irish physicist, lecturer, and populariser of science. Studied in Marburg and Berlin, 1848–51. Professor of natural philosophy, Royal Institution of Great Britain, 1853–87; professor of natural philosophy, Royal School of Mines, 1859–68; superintendent of the Royal Institution, 1867–87. Scientific adviser to Trinity House and the Board of Trade, 1866–83. FRS 1852. (*ODNB*, *DSB*.)

Van Lennep, Henry John (1815–89). Turkish-born missionary and educator. Educated in the United States, 1830–9. Acted as a missionary in Turkey for the American Board of Commissions for Foreign Missions, 1839–69. Settled in the United States in 1869. (*DAB*.)

Velie, Jacob W. (b. 1829). Curator. Moved from New York to Rock Island, Illinois, where he had a dental practice, 1856. Assistant, Chicago Academy of Sciences, 1871; secretary and curator, 1872. Made several field trips throughout America to replenish the natural historical collection lost in the fire of 1871. (Andreas 1884, p. 536.)

Vesselofski, Constantin Stepanovich (1819–90). Russian economist and statistician. Permanent secretary, St Petersburg Academy of Sciences, 1857–90. (*GSE* s.v. Veselovskii, Konstantin Stepanovich.)

Virgil (Publius Vergilius Maro) (70–19 B.C.). Roman poet. (*Oxford classical dictionary* 3d ed. rev.)

Virchow, Rudolf Carl (1821–1902). German physician, pathologist, medical reformer, and politician. Professor of pathological anatomy at the University of Würzburg, 1849–56. Professor of pathological anatomy and director of the Pathological Institute, University of Berlin, from 1856. Foreign member, Royal Society of London, 1884; awarded the Copley Medal, 1892. (*BLA*, *DBE*, *DSB*, *Record of the Royal Society of London*, Wrede and Reinfels eds. 1897.)

Vogt, Carl (1817–95). German naturalist. After receiving a doctorate in Giessen, 1839, moved to Switzerland and worked with Louis Agassiz on a treatise on fossil and freshwater fish until 1846. Professor of zoology, Giessen, 1846. Forced to leave the German Federation for political reasons in 1849; settled in Geneva. Professor of geology, Geneva, 1852; director of the Institute of Zoology, 1872. (*ADB*, *DSB*, Judel 2004.)
  8 April 1867, 12 April [*1867*], *17 April 1867*, *23 April 1867*, 7 August [*1867*]

Walker, Francis (1809–74). Entomologist. Specialised in chalcids, small parasitic Hymenoptera. Described the Chalcididae CD collected on the *Beagle* voyage. Catalogued a number of the insect collections in the British Museum. (*Entomologist*

7 (1874): 260–4, *Entomologist's Monthly Magazine* 11 (1874): 140–1.)

    9 December 1867

Wallace, Alfred Russel (1823–1913). Naturalist. Collector in the Amazon, 1848–52; in the Malay Archipelago, 1854–62. Independently formulated a theory of evolution by natural selection in 1858. Lecturer and author of works on protective coloration, mimicry, and zoogeography. President of the Land Nationalisation Society, 1881–1913. Wrote on socialism, spiritualism, and vaccination. FRS 1893. (*DSB, ODNB*.)

    23 February 1867, *24 February [1867]*, 26 February [1867], *2 March [1867]*,
    7 March [1867], *11 March [1867]*, [12–17] March [1867], *26 April [1867]*,
    29 April [1867], *1 May 1867*, 5 May [1867], *[19 June 1867]*, [24 June 1867],
    6 July [1867], *1 October [1867]*, 12 and 13 October [1867], *22 October [1867]*

Wallace, Herbert Edward (1829–51). Brother of Alfred Russel Wallace. Joined his brother in Brazil in 1849 and spent a year on the Amazon as a collector. Contracted yellow fever and died in Para (now Belém), Brazil. (Raby 2001, A. R. Wallace 1905.)

Wallace, Herbert Spencer (1867–74). Son of Alfred Russel Wallace. (Raby 2001.)

Walsh, Benjamin Dann (1808–69). Entomologist, farmer, and timber merchant. Student at Trinity College, Cambridge, 1827–31; fellow, 1833. Emigrated to the United States, where he farmed in Henry County, Illinois, 1838–51; lumber merchant, Rock Island, Illinois, 1851–8. Retired from commerce in about 1858 and concentrated on entomology, making contributions to agricultural entomology. Suggested the use of natural enemies to control insect pests. Author of several papers in agricultural journals. Associate editor of the *Practical Entomologist*, 1865. Acting state entomologist, Illinois, 1867. (*Alum. Cantab., DAB*.)

    [25 February 1867], 23 March [1867], 9 August [1867]

Walsh, Rebecca. Née Finn. Wife of Benjamin Dann Walsh; moved with him to the United States in 1838. (*ANB* s.v. Walsh, Benjamin Dann.)

Walter, Charles (1831–1907). Australian photographer, botanist, and journalist of German descent. Emigrated to Victoria, Australia, in 1856; employed as a plant collector by Ferdinand von Mueller. First Australian photojournalist; much of his work recorded local Aboriginal people. (*Dictionary of Australian artists.*)

Ward, Swinburne. Colonial civil servant. Entered the Admiralty, 1847; private secretary to the secretary to the Admiralty, 1849–55; private secretary to the governor of Ceylon (Sri Lanka); appointed private secretary to the governor of Madras, 1860; civil commissioner of the Seychelles, 1862–8; auditor-general of Mauritius, 1868–76. (*Colonial Office list.*)

Warington, George (1840–74). Author. BA, Cambridge, 1872. Master at Charterhouse school for a short time. Wrote mostly on religious topics. (*Modern English biography.*)

    7 October [1867]

Waterhouse, Frederick Herschel (1845–1919). Entomologist. Son of George Robert Waterhouse. Librarian at the Zoological Society of London, 1872–1912.

Waterhouse, Frederick Herschel, cont.
Described, in 1879, some of the beetles collected by CD on the *Beagle* voyage. Specialised in the Coleoptera. (*Entomologist's Monthly Magazine* 56 (1920): 17.)

Waterhouse, George Robert (1810–88). Naturalist. A founder of the Entomological Society of London, 1833. Curator of the Zoological Society of London, 1836–43. Assistant in the mineralogical branch of the natural history department of the British Museum, 1843–50; keeper, mineralogical and geological branch, 1851–6; keeper, geology department, 1857–80. Described CD's mammalian and entomological specimens from the *Beagle* voyage. (*DNB.*)
    5 March [1867?]

Watson, Hewett Cottrell (1804–81). Botanist, phytogeographer, and phrenologist. Edited the *Phrenological Journal*, 1837–40. Collected plants in the Azores in 1842. Wrote extensively on the geographical distribution of plants. (*DSB, ODNB.*)

Watson, Thomas (d. 1686). Puritan minister. BA, Cambridge, 1639. Chaplain to Lady Vere. Ejected from the rectory of St Stephen Walbrook, London, 1662. Licensed to preach at his house in Dowgate. Prosecuted for holding conventicles, 1681. Kept a school at Bethnal Green. Author of *A body of practical divinity* and other religious works. (*ODNB.*)

Watton, Mary Anne (b. 1827). Daughter of John and Anne Whitehurst of Shrewsbury. Married George Baskerville Watton of Dorrington Grove near Shrewsbury in 1866. Rented the Mount, Shrewsbury, 1866–7. Listed as a widow, living at her brother's house in Shrewsbury, 1871. (Census returns 1871 (Public Record Office RG9/2777: 108); *International Genealogical Index* 1992; *Shrewsbury Chronicle*, 6 July 1866.)

Weale, James Philip Mansel (1838 – after 1899). Naturalist, farmer, and writer. BA, Oxford, 1860. Resided in Cape Colony, South Africa, from about 1860 to about 1878. Supported the formation of a museum in Port Elizabeth. Farmed property (Brooklyn) near King William's Town. CD communicated his papers on orchid pollination to the Linnean Society of London; his papers on South African insects appeared in the *Transactions of the Entomological Society*. (Gunn and Codd 1981; *Royal Society catalogue of scientific papers*; *The Times*, 26 September 1899, p. 9; Weale 1891.)
    9 January 1867, 22 February [1867], 27 February [1867], *7 July 1867*, 27 August [1867], 9 December [1867], *[10 December 1867]*

Webb, Philip Barker (1793–1854). Botanist and traveller. Visited Madeira and the Canary Islands, 1828–30, and, with Sabin Berthelot, was the co-author of *Histoire naturelle des Iles Canaries* (1836–50). FRS 1824. (R. Desmond 1994, *ODNB.*)

Webster, Noah (1758–1843). American lexicographer and journalist. Author of grammars, spelling books, political and scientific essays, and a medical history. Published newspapers and a magazine. Compiled several dictionaries including *An American dictionary of the English language* (1828). (*DAB.*)

Wedgwood, Caroline Sarah (1800–88). CD's sister. Married Josiah Wedgwood III, her cousin, in 1837. (*Darwin pedigree.*)

Wedgwood, Frances Emma Elizabeth (Fanny) (1800–89). Second child of James Mackintosh and Catherine Allen. Married Hensleigh Wedgwood in 1832. (Freeman 1978, O'Leary 1989.)

Wedgwood, Frances Julia (Snow) (1833–1913). Novelist, biographer, historian, and literary critic. Daughter of Hensleigh and Frances Emma Elizabeth Wedgwood. Published two novels in her mid-twenties, under the pseudonym Frances Dawson. Wrote book reviews and an article on the theological significance of *Origin*. Conducted an intense friendship with Robert Browning between 1863 and 1870. Published a study of John Wesley (1870), and helped CD with translations of Linnaeus in the 1870s. Published *The moral ideal: a historical study* (1888). Active in the anti-vivisection movement from the 1860s. (B. Wedgwood and Wedgwood 1980, *ODNB*.)

Wedgwood, Francis (Frank) (1800–88). Master-potter. Partner in the Wedgwood pottery works at Etruria, Staffordshire, until 1876. Emma Darwin's brother. Married Frances Mosley in 1832. (*Alum. Cantab.*)

Wedgwood, Hensleigh (1803–91). Philologist. Emma Darwin's brother. Qualified as a barrister in 1828, but never practised. Fellow, Christ's College, Cambridge, 1829–30. Police magistrate at Lambeth, 1831–7; registrar of metropolitan carriages, 1838–49. An original member of the Philological Society, 1842. Published *A dictionary of English etymology* (1859–65). Married Frances Emma Elizabeth Mackintosh in 1832. (Freeman 1978, *ODNB*.)

Wedgwood, Hope Elizabeth (Dot) (1844–1934). Daughter of Hensleigh and Frances Emma Elizabeth Wedgwood. Second wife of Godfrey Wedgwood. (Freeman 1978.)

Wedgwood, Josiah I (1730–95). Master-potter. Founded the Wedgwood pottery works at Etruria, Staffordshire. Grandfather of CD and Emma Darwin. Greatly interested in experimental chemistry. Contributed several papers on the measurement of high temperatures to the Royal Society of London's *Philosophical Transactions*. Actively associated with scientists and scientific societies. FRS 1783. (*DSB*, *ODNB*.)

Wedgwood, Josiah III (1795–1880). Master-potter. Partner in the Wedgwood pottery works at Etruria, Staffordshire, 1841–4; moved to Leith Hill Place, Surrey, in 1844. Emma Darwin's brother. Married CD's sister Caroline, his cousin, in 1837. (Freeman 1978.)

Wedgwood, Katherine Elizabeth Sophy (Sophy) (1842–1911). Daughter of Caroline Wedgwood and Josiah Wedgwood III. CD's niece. (*Darwin pedigree*, Freeman 1978.)

Wedgwood, Katherine Euphemia (Effie) (1839–1931). Daughter of Hensleigh and Frances Emma Elizabeth Wedgwood. Married Thomas Henry Farrer in 1873. (*Burke's peerage* 1980, Freeman 1978).

Wedgwood, Lucy Caroline (1846–1919). Daughter of Caroline Wedgwood and Josiah Wedgwood III. CD's niece. Married Matthew James Harrison in 1874. (*Darwin pedigree*, Freeman 1978, B. Wedgwood and Wedgwood 1980.)

Wedgwood, Margaret Susan (1843–1937). Daughter of Caroline Wedgwood and Josiah Wedgwood III. CD's niece. Married Arthur Charles Vaughan Williams in 1869. Mother of Ralph Vaughan Williams (*DNB*). (*Emma Darwin* (1915), Freeman 1978.)

Wedgwood, Sarah Elizabeth (Elizabeth) (1793–1880). Emma Darwin's sister. Resided at Maer Hall, Staffordshire, until 1847, then at The Ridge, Hartfield, Sussex, until 1862. Moved to London before settling in Down in 1868. (*Emma Darwin* (1915), Freeman 1978.)

Weir, John Jenner (1822–94). Naturalist and accountant. Worked in HM customs as an accountant, 1839–85. Studied entomology, especially microlepidoptera; conducted experiments on the relations between insects and insectivorous birds and published papers in 1869 and 1870. Member of the Entomological Society of London from 1845, serving many times on the council. Fellow of the Linnean Society of London, 1865; Zoological Society, 1876. (*Science Gossip* n.s. 1 (1894): 49–50.)

Weisbach, Albin (1833–1901). German mineralogist and crystallographer. Described more than a dozen new mineral species, including argyrodite. (Sarjeant 1980–96.)

Wells, William Charles (1757–1817). American-born physician. Born in South Carolina, of Scottish parents, and educated in Scotland, 1768–71. Elected assistant physician to St Thomas's Hospital, London, 1795; physician, 1800–17. Awarded the Rumford Medal of the Royal Society for his 'Essay on dew' (1814). Wrote *An account of a female of the white race of mankind, part of whose skin resembles that of a negro* (1818), which has been thought to anticipate CD's theory of natural selection. FRS 1793. (*ODNB*.)

West, Raymond (1832–1912). Civil servant and judge in India. Appointed registrar of the Bombay high court, 1863, district judge, Canara, 1866; judicial commissioner, Sind, 1868; high court judge, Bombay, 1873–86. University teacher in Indian law, Cambridge, 1895–1907; reader in Indian and Islamic law, 1901–7. Knighted, 1888. (*ODNB*.)

Westwood, John Obadiah (1805–93). Entomologist and palaeographer. Founding member of the Entomological Society of London, 1833; honorary president, 1883. Hope Professor of zoology, Oxford University, 1861–93. Entomological referee for the *Gardeners' Chronicle*. Royal Society Royal Medallist, 1855. (*ODNB*, *Transactions of the Entomological Society of London* 1 (1833–6): xxxiv.)

Whitaker, William (1836–1925). Geologist. Employed by the Geological Survey, 1857–96. Considered a pioneer of English hydrogeology. President, Geological Society of London, 1898–1900. FRS 1887. (*ODNB*.)

White, Charles (1728–1813). Surgeon. Co-founder of the Manchester Infirmary, 1752, the Manchester Lying-in Hospital, 1790, and the Manchester Literary and Philosophical Society, 1781. Published *An account of the regular gradation in man, animals and vegetables* (1799). FRS 1762. (*ODNB*.)

Whitney, Josiah Dwight (1819–96). American geologist. Professor of chemistry, University of Iowa, 1855–8. California state geologist, 1860–74. Professor of geology, Harvard, 1865; dean of the School of Mining and Practical Geology, 1868. (*ANB*.)

Wilberforce, Samuel (1805–73). Clergyman. Rector of Brighstone, Isle of Wight, 1830–40; of Alverstoke, Hampshire, 1840–3. Chaplain to Prince Albert, 1841. Dean of Westminster, 1845. Bishop of Oxford, 1845–69. Bishop of Winchester, 1869–73. FRS 1845. (*ODNB, Record of the Royal Society of London.*)

Williams & Norgate. Booksellers and publishers specialising in foreign scientific literature, with premises at Covent Garden, London, and South Frederick Street, Edinburgh. A partnership between Edmund Sydney Williams (1817–91) and Frederick Norgate. Publishers of the *Natural History Review*. (*Modern English biography* s.v. Williams, Edmund Sidney.)

Wilson, Edward (1813–78). Australian newspaper proprietor. Left London for Australia in 1841; proprietor of the *Argus* newspaper from 1848. Founded the Acclimatisation Society of Victoria in 1861. Moved back to England in 1864; moved to Hayes Place, Bromley, Kent, in 1867. Continued to be active in Australian affairs and promoting assisted emigration. A founder of the Colonial Society, 1868, and member of the council from 1868. (*Aust. dict. biog., ODNB.*)

8 November 1867, 7 December 1867

Wilson, James Maurice (1836–1931). Schoolmaster and clergyman. BA, Cambridge, 1859. Assistant master, Rugby school, 1859–79. Ordained deacon and priest, 1879. Headmaster, Clifton College, 1879–90. Archdeacon of Manchester, 1890–1905. Canon of Worcester, 1905–26. (*Alum. Cantab., ODNB.*)

Wilson, Samuel (1832–95). Australian landowner and politician. From 1869, owned and managed extensive agricultural lands, including Longerenong station, Wimmera, Victoria, where he lived. Developed Australia's largest sheep flock. Knighted, 1875. Retired to England, 1881. (*Aust. dict. biog.*)

12 November 1867 (Ferdinand von Mueller)

Withecombe, John Rees (1816–1904). Surgeon. Studied medicine at Guy's Hospital, London; MD, 1842. Entered the Indian Army (Bengal) in 1842; promoted surgeon, 1856; retired in 1859. Continued medical practice in Richmond, Surrey. (*Medical directory*, Plarr 1930.)

Wolf, Josef (1820–99). German-born painter and illustrator. Specialised in animals. Apprenticed as a lithographer in Koblenz before studying art at the Antwerp Academy. Arrived in London in 1848. Established a studio with Johann Baptist Zwecker at 59 Berners Street, London, *circa* 1860. Illustrated many natural history publications including a number of the plates for Henry Walter Bates's *The naturalist on the river Amazons* (1863). (Bates 1863, p. vi; *ODNB*; *Post Office London directory* 1859–63.)

Wollaston, Thomas Vernon (1822–78). Entomologist and conchologist. Passed many winters in Madeira, where he collected insects and shells. Wrote a series of works on the Coleoptera. (*ODNB*.)

Wood, Searles Valentine (1798–1880). Geologist. Served in the East India Company's mercantile fleet, 1811–26. Became a partner, with his father, in a bank at Hasketon, near Woodbridge, until 1839. Appointed curator of the museum of

Wood, Searles Valentine, cont.
the Geological Society of London, 1838; elected fellow of the Geological Society, 1839. Studied the fossils of the East Anglian crag pits, and of the Hampshire Tertiaries, and French Eocene fossil Mollusca. (*ODNB.*)

Woodbury, Thomas White (1818–71). Journalist and beekeeper. From 1850 devoted himself exclusively to beekeeping. Introduced Ligurian bees to Britain in 1859 and developed the 'Woodbury hive', marketed by the London apiarian specialists George Neighbour & Sons. As 'A Devonshire Beekeeper', a regular contributor to the *Cottage Gardener*, the *Journal of Horticulture*, the *Gardeners' Chronicle*, and *The Times*. (R. H. Brown 1975, Ron Brown 1994, Dodd 1983, H. M. Fraser 1958, Neighbour 1865.)

Woodward, Henry (1832–1921). Geologist and palaeontologist. Joined his brother, Samuel Pickworth Woodward, as an assistant in the geological department, British Museum, in 1858; keeper of geology, 1880–1901. Co-founder of the *Geological Magazine*, 1864; editor, 1865–1918. President of many scientific organisations, including the Geological Society of London, Palaeontographical Society, Royal Microscopical Society, Malacological Society, Geologists' Association, and Museums Association. FRS 1873. (*Geological Magazine* (1921) 58: 481–4, *WWW.*)

Woolner, Henry. Brother of Thomas Woolner. (Woolner 1917, pp. 60, 70.)

Woolner, Thomas (1825–92). Sculptor and poet. Member of the Pre-Raphaelite brotherhood. Established his reputation in the 1850s with medallion portrait sculptures of Robert Browning, Thomas Carlyle, and William Wordsworth. Went on to make acclaimed busts of CD, Charles Dickens, Thomas Henry Huxley, Adam Sedgwick, and Alfred Tennyson, and life-size studies of Francis Bacon, John Stuart Mill, and William Whewell. (*DNB.*)
6 *December 1867*

Wright, Charles (1811–85). American botanical collector. Botanist to the United States North Pacific Exploring and Surveying Expedition, 1853–6. Investigated the botany of Cuba, 1856–67. (*DAB.*)

Wright, Edward Perceval (1834–1910). Irish naturalist. One of the editors of the *Natural History Review*, 1854–65. Curator of the university museum in Dublin, 1857. Lecturer in zoology, Trinity College, Dublin, 1858–68; professor of botany and keeper of the herbarium, 1869. Resigned the professorship in 1904 owing to ill health, but continued to superintend the herbarium. His principal research was in marine zoology. (*ODNB.*)

Wright, Thomas (1809–84). Scottish physician and palaeontologist. Medical officer for Cheltenham and the surrounding districts, and surgeon to Cheltenham General Hospital. Formed an extensive collection of Jurassic fossils. FRS 1879. (*ODNB.*)
24 *August 1867*

Wrigley, Alfred (1817–98). Mathematician and educator. Professor of mathematics at the Royal Military College, Addiscombe, Surrey, 1841–61. Headmaster, Clapham Grammar School, 1862–82. (*Alum. Cantab.*, *Modern English biography.*)
[September 1867]

Wyman, Jeffries (1814–74). American comparative anatomist and ethnologist. Curator of the Lowell Institute, Boston, 1839–42. Travelled in Europe, 1841–2. Professor of anatomy and physiology, Hampden-Sydney College, Virginia, 1843–8. Hersey Professor of anatomy, Harvard College, 1847–74. Curator, Peabody Museum of Archaeology and Ethnology, Harvard, 1866–74. (*ANB*, *DSB*.)

Wynne, Ambrose (b. 1853/4). Printer's compositor. Son of George and Margaret Wynne. (Census returns 1871 (Public Record Office RG10/2777).)

Wynne, Arthur (b. 1850/1). Carpenter. Son of George and Margaret Wynne. (Census returns 1871 (Public Record Office RG10/2777).)

Wynne, George (1813–74). Gardener. Susan Elizabeth Darwin's gardener at The Mount, Shrewsbury, probably from about 1858 until her death in 1866. Continued to live in Shrewsbury until his death. (Census returns 1871 (Public Record Office RG10/2777); Deidre Haxton, personal communication; letter from Salt & Sons, 17 July 1867; Susan Elizabeth Darwin's will, Probate Registry, York.)

Wynne, John (b. 1842/3). Carpenter. Son of George and Margaret Wynne. (Census returns 1871 (Public Record Office RG10/2777).)

Wynne, Margaret (b. 1813/14). Wife of George Wynne. (Census returns 1871 (Public Record Office RG10/2777).)

Yarrell, William (1784–1856). Zoologist. Newspaper agent and bookseller in London. Author of standard works on British birds and fishes. Fellow of the Linnean Society of London, 1825. Member of the Zoological Society of London from its formation in 1826. (*ODNB*.)

Youl, James Arndell (1811–1904). Australian agriculturalist and magistrate. Farmed in Van Diemen's Land (Tasmania), 1827–54. Lived in London from 1854; honorary agent for Tasmania, 1861–3, and honorary secretary and treasurer of the Australia Association. Co-founder of the Colonial Society, 1868. Discovered a means of transporting live fish ova from Britain to Tasmania, Australia, and New Zealand. Knighted, 1891. (*ODNB*.)

BIBLIOGRAPHY

The following bibliography contains all the books and papers referred to in this volume by author–date reference or by short title. Short titles are used for some standard reference works (e.g., *ODNB*, *OED*), for CD's books and papers, and for editions of his letters and manuscripts (e.g., *Descent*, *LL*, *Notebooks*). Works referred to by short titles are listed in alphabetical order according to the title; those given author–date references occur in alphabetical order according to the author's surname. Notes on manuscript sources are given at the end of the bibliography.

ADB: *Allgemeine deutsche Biographie*. Under the auspices of the Historical Commission of the Royal Academy of Sciences. 56 vols. Leipzig: Duncker & Humblot. 1875–1912.

Agassiz, Elizabeth Cary. 1885. *Louis Agassiz: his life and correspondence*. 2 vols. London: Macmillan and Company.

Agassiz, Jean Louis Rodolphe. 1866. Physical history of the valley of the Amazons. *Atlantic Monthly* 18: 49–60, 159–69.

——. 1866–76. *Geological sketches*. 2 vols. Boston: Ticknor & Fields; J. R. Osgood.

——. 1867. The geological formation of the valley of the Amazon. The river, its basin and tributories. The ancient glaciers in the tropics. The aquatic animals of the Amazon. The land animals of South America. The monkeys and native inhabitants. [Six lectures read at the Cooper Institute, New York, 5, 11, 12, 18, 20, and 26 February 1867.] *New York Herald Tribune*, 6 February 1867, p. 8, 12 February 1867, p. 5, 13 February 1867, p. 5, 19 February 1867, p. 8, 21 February 1867, p. 5, 27 February 1867, p. 8.

Agassiz, Jean Louis Rodolphe and Agassiz, Elizabeth. 1868. *A journey in Brazil*. Boston: Ticknor and Fields.

Allaby, Michael, ed. 1999. *A dictionary of zoology*. 2d edition. Oxford and New York: Oxford University Press.

Allan, Mea. 1967. *The Hookers of Kew, 1785–1911*. London: Michael Joseph.

Allen, Oscar Nelson and Allen, Ethel K. 1981. *The leguminosae: a source book of characteristics, uses and nodulation*. London: Macmillan.

Alum. Cantab.: *Alumni Cantabrigienses. A biographical list of all known students, graduates and holders of office at the University of Cambridge, from the earliest times to 1900*. Compiled by John Venn and J. A. Venn. 10 vols. Cambridge: Cambridge University Press. 1922–54.

Alum. Oxon.: *Alumni Oxonienses: the members of the University of Oxford, 1500–1886: . . . with a record of their degrees. Being the matriculation register of the university.* Alphabetically arranged, revised, and annotated by Joseph Foster. 8 vols. London and Oxford: Parker & Co. 1887–91.

ANB: *American national biography.* Edited by John A. Garraty and Mark C. Carnes. 24 vols. and supplement. New York and Oxford: Oxford University Press. 1999–2002.

Anderson, Atholl. 1989. *Prodigious birds: moas and moa-hunting in prehistoric New Zealand.* Cambridge: Cambridge University Press.

Anderson-Henry, Isaac. 1863. Crossing strawberries. *Journal of Horticulture* n.s. 4: 45–6.

——. 1867a. On the hybridization or crossing of plants. [Read 14 March 1867.] *Transactions of the Botanical Society [of Edinburgh]* 9: 101–15.

——. 1867b. On pure hybridisation; or, crossing distinct species of plants. [Read 14 November 1867.] *Transactions of the Botanical Society [of Edinburgh]* 9: 206–31.

Andreas, Alfred Theodore. 1884. *History of Cook County Illinois. From the earliest period to the present time.* Chicago: A. T. Andreas.

Annual register: *The annual register. A view of the history and politics of the year.* 1838–62. *The annual register. A review of public events at home and abroad.* N.s. 1863–1946. London: Longman & Co. [and others].

Anon. 1867. Popularizing science. *Nation* 5: 32–4.

Ashton, Rosemary. 1991. *G. H. Lewes: a life.* Oxford: Clarendon Press.

Audubon, John James. 1827–38. *The birds of America, from original drawings.* 4 vols. London: the author.

Aust. dict. biog.: *Australian dictionary of biography.* Edited by Douglas Pike *et al.* 14 vols. [Melbourne]: Melbourne University Press. London and New York: Cambridge University Press. 1966–96.

Autobiography: *The autobiography of Charles Darwin 1809–1882. With original omissions restored.* Edited with appendix and notes by Nora Barlow. London: Collins. 1958.

Aylmer, Fenton John. 1931. *The Aylmers of Ireland.* London: Mitchell Hughes and Clarke.

Backhouse, James. 1856. *A monograph of the British Hieracia.* York: William Simpson.

Bailey, Liberty Hyde and Bailey, Ethel Zoe. 1976. *Hortus third: a concise dictionary of plants cultivated in the United States and Canada.* Revised and expanded by the staff of the Liberty Hyde Bailey Hortorium. New York: Macmillan. London: Collier Macmillan.

Baird, Spencer Fullerton. 1866. The distribution and migrations of North American birds. *American Journal of Science and Arts* 41: 78–90, 184–92, 337–47.

Bajema, Carl Jay, ed. 1984. *Evolution by sexual selection theory prior to 1900.* New York: Van Nostrand Reinhold Co.

Baker, John Gilbert. 1863. *North Yorkshire; studies of its botany, geology, climate and physical geography.* London: Longman, Green, Longman, Roberts, & Green.

Baker, John Gilbert. 1866. Descriptions of six new species of simple-fronded Hymenophyllaceæ. [Read 15 March 1866.] *Journal of the Linnean Society (Botany)* 9 (1867): 335–40.

Baker, John Gilbert and Nowell, John. 1854. *A supplement to Baines' Flora of Yorkshire.* London: William Pamplin.

Baring-Gould, Sabine and Bampfylde, C. A. 1989. *A history of Sarawak under its two white rajahs, 1839–1908.* Singapore: Oxford University Press. [Reprint of 1909 edition by Henry Sotheran & Co., London.]

Barker, Charles. 2002. Erotic martyrdom: Kingsley's sexuality beyond sex. *Victorian Studies* 44: 465–88.

Barnhart, John Hendley, comp. 1965. *Biographical notes upon botanists . . . maintained in the New York Botanical Garden Library.* 3 vols. Boston, Mass.: G. K. Hall.

Barrett, Paul H. 1980. *Metaphysics, materialism, and the evolution of mind. Early writings of Charles Darwin.* With a commentary by Howard E. Gruber. Chicago: University of Chicago Press.

Barrow, John. 1801–4. *An account of travels into the interior of southern Africa, in the years 1797 and 1798.* 2 vols. London: T. Cadell and W. Davies.

Barrow, Logie. 1986. *Independent spirits: spiritualism and English plebeians 1850–1910.* London: Routledge & Kegan Paul.

Bartholomew, John. 1943. *The survey gazetteer of the British Isles: including summary of 1931 census and reference atlas.* 9th edition. Edinburgh: John Bartholomew & Son.

Barton, Ruth. 1998. 'Huxley, Lubbock, and Half a Dozen Others': professionals and gentlemen in the formation of the X Club, 1851–1864. *Isis* 89: 410–44.

Barwick, Diane. 1972. Coranderrk and Cumeroogunga: pioneers and policy. In *Opportunity and response: case studies in economic development*, edited by Trude Scarlett Epstein and David H. Penny. London: C. Hurst & Company.

Bates, Henry Walter. 1861. Contributions to an insect fauna of the Amazon valley. *Lepidoptera: Heliconidæ.* [Read 21 November 1861.] *Transactions of the Linnean Society of London* 23 (1860–2): 495–566.

———. 1863. *The naturalist on the River Amazons. A record of adventures, habits of animals, sketches of Brazilian and Indian life, and aspects of nature under the equator, during eleven years of travel.* 2 vols. London: John Murray.

———. 1886–90. *Pectinicornia and Lamellicornia.* Vol. 2, pt 2 of *Biologia Centrali-Americana. Insecta. Coleoptera*, edited by Frederick Du Cane Godman and Osbert Salvin. London: R. H. Porter; Dulau & Co.

Bates, Henry Walter and Smith, Frederick. 1855. Descriptions of some species of Brazilian ants belonging to the genera *Pseudomyrma*, *Eciton* and *Myrmica* (with observations on their economy). [Read 1 January 1855.] *Transactions of the Entomological Society* n.s. 3 (1854–6): 156–69.

BDWS: The biographical dictionary of women in science: pioneering lives from ancient times to the mid-20th century. Edited by Marilyn Ogilvie and Joy Harvey. 2 vols. New York and London: Routledge. 2000.

Bechstein, Johann Matthäus. 1789–95. *Gemeinnützige Naturgeschichte Deutschlands nach allen drey Reichen. Ein Handbuch zur deutlichern und vollständigern Selbstbelehrung beson-ders für Forstmänner, Jugendlehrer und Oekonomen.* 4 vols. Leipzig: Siegfried Lebrecht Crusius.

Bell, Charles. 1806. *Essays on the anatomy of expression in painting.* London: Longman, Hurst, Rees, and Orme.

———. 1833. *The hand: its mechanism and vital endowments as evincing design.* Fourth Bridge-water treatise. London: William Pickering.

———. 1836. *The nervous system of the human body: as explained in a series of papers read before the Royal Society of London.* 3d edition. Edinburgh: Adam and Charles Black. London: Longman, Rees, Orme, Brown, Green, and Longman.

———. 1844. *The anatomy and philosophy of expression as connected with the fine arts.* Preface by George Bell, and an appendix on the nervous system by Alexander Shaw. 3d edition, enlarged. London: John Murray.

Bellon, Richard. 2001. Joseph Hooker's ideals for a professional man of science. *Journal of the History of Biology* 34: 51–82.

Belt, Thomas. 1888. *The naturalist in Nicaragua. A narrative of a residence at the gold mines of Chontales; journeys in the savannahs and forests; with observations on animals and plants in reference to the theory of evolution of living forms.* 2d edition, revised and corrected. London: Edward Bumpus.

Benthall, Paul. 1980. George Maw: a versatile Victorian. *National Trust Studies* (1980): 11–20.

Bentham, George. 1883. On the joint and separate work of the authors of Bentham and Hooker's 'Genera plantarum'. [Read 19 April 1883.] *Journal of the Linnean Society (Botany)* 20 (1884): 305–8.

Bentham, George and Hooker, Joseph Dalton. 1862–83. *Genera plantarum. Ad exem-plaria imprimis in herbariis Kewensibus servata definita.* 3 vols. in 7. London: A. Black [and others].

[Beverley, Robert Mackenzie.] 1867. *The Darwinian theory of the transmutation of species examined by a graduate of the University of Cambridge.* London: James Nisbet & Co.

BHGW: Biographisch-literarisches Handwörterbuch zur Geschichte der exacten Wissenschaften enthaltend Nachweisung über Lebensverhältnisse und Leistungen von Mathematikern, As-tronomen, Physikern, Chemikern, Mineralogen, Geologen usw. By Johann Christian Poggendorff. 5 vols. Leipzig: Johann Ambrosius Barth; Verlag Chemie. 1863–1926.

Bibby, Cyril. 1959. *T. H. Huxley. Scientist, humanist and educator.* London: Watts.

Billroth, Christian Albert Theodor. 1863. *Die allgemeine chirurgische Pathologie und Ther-apie in fünfzig Vorlesungen: ein Handbuch für Studirende und Aerzte.* Berlin: G. Reimer.

Birds of the world: Handbook of the birds of the world. By Josep del Hoyo et al. 8 vols. to date. Barcelona: Lynx editions. 1991–.

BLA: Biographisches Lexikon der hervorragenden Aerzte aller Zeiten und Völker. Edited by A. Wernich *et al.* 6 vols. Vienna and Leipzig: Urban and Schwarzenberg. 1884–8.

Blackburn, Helen. 1902. *Women's suffrage: a record of the Women's Suffrage Movement in the British Isles with biographical sketches of Miss Becker.* London: Williams & Norgate.

Blasius, Johann Heinrich. 1857. *Naturgeschichte der Säugethiere Deutschlands und der angrenzenden Länder von Mitteleuropa.* Brunswick: Vieweg und Sohn.

BLC: The British Library general catalogues of printed books to 1975. 360 vols. and supplement (6 vols.). London: Clive Bingley; K. G. Saur. 1979–88.

BLKO: Biographisches Lexikon des Kaiserthums Oesterreich, enthaltend die Lebensskizzen der denkwürdigen Personen, welche seit 1750 in den österreichischen Kronländern geboren wurden oder darin gelebt und gewirkt haben. By Constant von Wurzbach. 60 vols. Vienna: L. C. Zamarski. 1856–91.

Blum, Christopher Olaf. 1996. St. George Mivart: Catholic natural philosopher. PhD thesis: University of Notre Dame, Indiana.

Blyakher, L. Y. 1982. *History of embryology in Russia from the middle of the eighteenth century to the middle of the nineteenth century.* Translated from the Russian by H. I. Youssef and B. A. Malek, with an introduction by Jane Maienschein. Washington, D.C.: Al Ahram Center for Scientific Translations for the Smithsonian Institution and the National Science Foundation.

Blyth, Edward. 1837. On the reconciliation of certain apparent discrepancies observable in the mode in which the seasonal and progressive changes of colour are effected in the fur of mammalians and feathers of birds; with various observations on moulting. *Magazine of Natural History* n.s. 1: 259–63, 300–11.

——. 1841. An amended list of the species of the genus *Ovis. Annals and Magazine of Natural History* 7: 195–201, 248–61.

——. 1842–3. A monograph of the Indian and Malayan species of Cuculidæ, or birds of the cuckoo family. *Journal of the Asiatic Society of Bengal* n.s. 11: 897–928, 1095–112; 12: 240–7.

——. 1866–7. The ornithology of India. A commentary on Dr. Jerdon's *Birds of India. Ibis* n.s. 2 (1866): 225–58, 336–76; 3 (1867): 1–48, 147–85.

Boase, Charles William. 1894. *An alphabetical register of the commoners of Exeter College, Oxford.* Oxford: printed at Baxter's Press.

Bócsa, Iván and Karus, Michael. 1998. *The cultivation of hemp: botany, varieties, cultivation and harvesting.* Translated by Chris Filben. Sebastapol, Calif.: Hemptech.

Bompas, George C. 1885. Life of Frank Buckland. London: Smith, Elder & Co.

Bondeson, Jan. 1997. *A cabinet of medical curiosities.* Ithaca, N.Y.: Cornell University Press.

Boott, Francis. 1858–67. *Illustrations of the genus* Carex. 4 pts. London: William Pamplin (pts 1, 2, and 3), L. Reeve & Co. (pt 4).

Bosl, Karl, ed. 1983. *Bosl's Bayerische Biographie.* Regensburg: Friedrich Pustet.

Braby, Michael F. 2005. Provisional checklist of genera of the Pieridae (Lepidoptera: Papilionoidea). *Zootaxa* 832: 1–16.

Brace, Charles Loring. 1870. Darwinism in Germany. *North American Review* 110: 284–99.

Bramwell, David. 1976. The endemic flora of the Canary Islands; distribution, relationships and phytogeography. In *Biogeography and ecology in the Canary Islands*, edited by Günther Kunkel. The Hague: Junk.

Branca, Gaetano and Travella, Stefano, trans. 1869–74. *La vita degli animali.* By

Alfred Edmund Brehm. (Translation of Brehm *et al.* 1864–9.) Revised by Michele Lessona and Tommaso Salvadori. 6 vols. Turin: Unione tip.

Brandon-Jones, Christine. 1995. Long gone and forgotten: reassessing the life and career of Edward Blyth, zoologist. *Archives of Natural History* 22: 91–5.

Brehm, Alfred Edmund, *et al.* 1864–9. *Illustrirtes Thierleben. Eine allgemeine Kunde des Thierreichs, etc.* 6 vols. Hildburghausen: Bibliographisches Institut.

Brewer, Ebenezer Cobham. 1898. *The reader's handbook of famous names in fiction, allusions, references, proverbs, plots, stories, and poems.* New and enlarged edition. London: Chatto & Windus.

Bridges, Esteban Lucas. 1948. *Uttermost part of the earth.* London: Hodder and Stoughton.

Bright, John and Rogers, James E. Thorold, eds. 1870. *Speeches on questions of public policy by Richard Cobden, M.P.* 2 vols. London: Macmillan and Co.

British Museum (Natural History). 1910. *Memorials of Charles Darwin: a collection of manuscripts, portraits, medals, books and natural history specimens to commemorate the centenary of his birth and the fiftieth anniversary of the publication of* The origin of species. Special guide no. 4. 2d edition. London: British Museum. [Facsimile reprint. *Bulletin of the British Museum (Natural History) Historical Series* 14 (1988): 235–98.]

Brock, William Hodson. 1981. Advancing science: the British Association and the professional practice of science. In *The parliament of science; the British Association for the Advancement of science, 1831–1981,* edited by Roy MacLeod and Peter Collins. Northwood, Middlesex: Science Reviews.

Brockhaus-Wahrig: Brockhaus-Wahrig: deutsches Wörterbuch. Edited by Gerhard Wahrig *et al.* 6 vols. Wiesbaden: Brockhaus. Stuttgart: Deutsche Verlags-Anstalt. 1980–4.

Bronn, Heinrich Georg, trans. 1860. *Charles Darwin, über die Entstehung der Arten im Thier- und Pflanzen-Reich durch natürliche Züchtung, oder Erhaltung der vervollkommneten Rassen im Kampfe um's Daseyn.* Stuttgart: E. Schweizerbart.

——, trans. 1862. *Charles Darwin, über die Einrichtungen zur Befruchtung Britischer und ausländischer Orchideen durch Insekten und über die günstigen Erfolge der Wechselbefruchtung.* Stuttgart: E. Schweizerbart.

——, trans. 1863. *Über die Entstehung der Arten im Thier- und Pflanzen-Reich durch natürliche Züchtung; oder, Erhaltung der vervollkommneten Rassen im Kampfe um's Daseyn.* By Charles Darwin. 2d edition. Stuttgart: E. Schweizerbart.

Bronn, Heinrich Georg and Carus, Julius Victor, trans. 1867. *Über die Entstehung der Arten durch natürliche Zuchtwahl oder die Erhaltung der begünstigten Rassen im Kampfe um's Dasein.* 3d edition. Translated by Heinrich Georg Bronn. Revised and corrected from the fourth English edition by Julius Victor Carus. Stuttgart: E. Schweizerbart'sche Verlagshandlung und Druckerei.

Brooke, Charles Anthoni Johnson. 1866. *Ten years in Saráwak.* 2 vols. London: Tinsley Brothers.

Brown, Alesander Claude. 1977. The amateur scientist. In *A history of scientific endeavour in South Africa,* edited by A. C. Brown. Cape Town: Royal Society of South Africa and Rustica Press.

Brown, R. H. 1975. *One thousand years of Devon beekeeping*. Devon: Devon Beekeepers Association.

Brown, Robert. 1831. On the organs and mode of fecundation in Orchideæ and Asclepiadeæ. [Read 1 and 15 November 1831.] *Transactions of the Linnean Society of London* 16 (1833): 685–745.

Brown, Ron. 1994. *Great masters of beekeeping*. Borrowbridge, Somerset: Bee Books New & Old.

Browne, Janet. 1983. *The secular ark. Studies in the history of biogeography*. New Haven, Conn., and London: Yale University Press.

——. 1985. Darwin and the expression of the emotions. In *The Darwinian heritage*, edited by David Kohn. Princeton: Princeton University Press in association with Nova Pacifica.

——. 1995. *Charles Darwin. Voyaging. Volume I of a biography*. New York: Alfred A. Knopf.

——. 1998. I could have retched all night. Darwin and his body. In *Science incarnate. Historical embodiments of natural knowledge*, edited by Christopher Lawrence and Steven Shapin. Chicago and London: University of Chicago Press.

——. 2002. *Charles Darwin. The power of place. Volume II of a biography*. London: Pimlico.

Browne, Janet and Messenger, Sharon. 2003. Victorian spectacle: Julia Pastrana, the bearded and hairy female. *Endeavour* 27 (4): 155–9.

Brümmer, Franz. 1875–6. *Deutsches Dichter-Lexikon: biographische und bibliographische Mittheilungen über deutsche Dichter aller Zeiten*. 2 vols. Eichstätt: Krüll'sche Buchhandlung.

Bryan's dictionary of painters and engravers. New edition revised and enlarged under the supervision of George C. Williamson. 5 vols. London: George Bell & Sons. 1903–5.

Buckland, William. 1836. *Geology and mineralogy considered with reference to natural theology*. Sixth Bridgewater treatise. 2 vols. London: William Pickering.

Buckle, Henry Thomas. 1857–61. *History of civilization in England*. 2 vols. London: John W. Parker & Son.

Burchell, William John. 1822–4. *Travels in the interior of Southern Africa*. 2 vols. London: Longman, Hurst, Rees, Orme, and Brown.

Burchfield, Joe D. 1990. *Lord Kelvin and the age of the earth. With a new afterword*. Chicago and London: University of Chicago Press.

Burke's landed gentry: A genealogical and heraldic history of the commoners of Great Britain and Ireland enjoying territorial possessions or high official rank but unvisited with heritable honours. Burke's genealogical and heraldic history of the landed gentry. By John Burke, *et al.* 1st–18th edition. London: Henry Colburn [and others]. 1833–1969.

Burke's peerage: A genealogical and heraldic dictionary of the peerage and baronetage of the United Kingdom. Burke's peerage and baronetage. 1st– edition. London: Henry Colburn [and others]. 1826–.

Burkhardt, Frederick H. 1988. England and Scotland: the learned societies. In *The comparative reception of Darwinism*, edited by T. F. Glick. Chicago and London: University of Chicago Press.

Burkill, Isaac Henry. [1965.] *Chapters on the history of botany in India.* [Calcutta: Botanical Survey of India.]

Calendar: A calendar of the correspondence of Charles Darwin, 1821–1882. With supplement. 2d edition. Edited by Frederick Burkhardt *et al.* Cambridge: Cambridge University Press. 1994.

Camardi, Giovanni. 2001. Richard Owen, morphology and evolution. *Journal of the History of Biology* 34: 481–515.

Campbell, George Douglas. 1855. Address of the president. *Report of the twenty-fifth meeting of the British Association for the Advancement of Science held at Glasgow*, pp. lxxiii–lxxxvi.

[——.] 1862. [Review of *Orchids* and other works.] *Edinburgh Review* 116: 378–97.

——. 1864. Opening address, 1864–5 session. [Read 5 December 1864.] *Proceedings of the Royal Society of Edinburgh* 5 (1862–6): 264–92.

——. 1865. The reign of law. *Good Words* (1865): 52–8, 126–33, 227–32, 269–74.

——. 1867. *The reign of law.* London: Alexander Strahan.

Campbell, Ina Erskine, ed. 1906. *George Douglas, eighth duke of Argyll . . . (1823–1900): autobiography and memoirs.* 2 vols. London: John Murray.

Candolle, Alphonse de. 1867. *Lois de la nomenclature botanique adoptées par le congrès international de botanique tenu à Paris en août 1867, suivies d'une deuxième édition de l'introduction historique et du commentaire qui accompagnaient la rédaction préparatoire présentée au congrès.* Geneva and Basle: H. Georg. Paris: J.-B. Baillière et fils.

——. 1883. *Nouvelles remarques sur la nomenclature botanique.* Geneva, Bale-Lyon: H. Georg.

Cantor, Geoffrey N. and Shuttleworth, Sally, eds. 2004. *Science serialized: representations of the sciences in nineteenth-century periodicals.* Cambridge, Mass., and London: MIT Press.

Cape of Good Hope general directory: The Cape of Good Hope commercial directory and general business guide. General directory and guide-book to the Cape of Good Hope and its dependencies. Cape Town, South Africa: Saul Solomon & Co. 1868–87.

Caroe, Gwendolen Mary. 1985. *The Royal Institution: an informal history.* With a final chapter by Alban Caroe. London: John Murray.

Carozzi, Albert Victor, trans. and ed. 1968. *Telliamed or conversations between an Indian philosopher and a French missionary on the diminution of the sea.* By Benoît de Maillet. Urbana: University of Illinois Press.

Carpenter, William Benjamin. 1864. Additional note on the structure and affinities of Eozoön Canadense. [Read 23 November 1864.] *Quarterly Journal of the Geological Society of London* 21 (1865): 59–66.

——. 1866. Supplemental notes on the structure and affinities of Eozoon Canadense. [Read 10 January 1866.] *Quarterly Journal of the Geological Society of London* 22: 219–28.

Carus, Julius Victor, trans. 1863. *Zeugnisse für die Stellung des Menschen in der Natur.* By Thomas Henry Huxley. Brunswick: Vieweg und Sohn.

——, trans. 1868. *Das Variiren der Thiere und Pflanzen im Zustande der Domestication.*

By Charles Darwin. 2 vols. Stuttgart: E. Schweizerbart'sche Verlagshandlung (E. Koch).

Carus, Julius Victor. 1872. *Geschichte der Zoologie bis auf Joh. Müller und Charl. Darwin.* Munich: R. Oldenbourg.

——, trans. 1873. *Das Variiren der Thiere und Pflanzen im Zustande der Domestication.* By Charles Darwin. 2d edition. 2 vols. Stuttgart: E. Schweizerbart'sche Verlagshandlung (E. Koch).

Carus, Julius Victor and Engelmann, Wilhelm. 1861. *Bibliotheca zoologica.* 2 vols. Leipzig: Wilhelm Engelmann.

Carus, Julius Victor and Gerstaecker, Carl Edouard Adolph. 1863–75. *Handbuch der Zoologie.* 2 vols. Leipzig: Wilhelm Engelmann.

Caspary, Johann Xaver Robert. 1865. Über botanische Untersuchungen, welche in Bezug auf Darwin's Hypothese, dass kein Hermaphrodit sich durch eine Ewigkeit von Generationen befruchten könne, gemacht sind. *Schriften der k. Physikalisch-ökonomischen Gesellschaft zu Koenigsberg* 6: 11–21.

Catalogue général de la librairie française: Catalogue général de la librairie française pendant 25 ans (1840–1865). Edited by Otto Lorenz. 4 vols. Paris: O. Lorenz. 1868.

CDEL: A critical dictionary of English literature, and British and American authors, living and deceased, from the earliest accounts to the middle of the nineteenth century . . . with forty indexes of subjects. By S. Austin Allibone. 3 vols. London: Trübner. Philadelphia: Childs & Peterson; J. B. Lippincott. 1859–71. *A supplement to Allibone's critical dictionary of English literature and British and American authors. Containing over thirty-seven thousand articles (authors), and enumerating over ninety-three thousand titles.* By John Foster Kirk. 2 vols. Philadelphia and London: J. B. Lippincott. 1891.

Challinor, John. 1978. *A dictionary of geology.* 5th edition. Cardiff: University of Wales Press.

Chambers: The Chambers dictionary. Edinburgh: Chambers Harrap Publishers. 1998.

[Chambers, Robert.] 1844. *Vestiges of the natural history of creation.* London: John Churchill.

Church hymnal. 5th edition. Melody Edition. Oxford: Oxford University Press. 2000.

Clark, George. 1866. Account of the late discovery of dodo's remains in the island of Mauritius. *Ibis* n.s. 2: 141–6.

Clark, Ronald W. 1968. *The Huxleys.* London: Heinemann.

Clarke, Benjamin. 1866. *A new arrangement of phanerogamous plants, with especial reference to relative position, including their relations with the cryptogams.* London: n.p.

——. 1870. *On systematic botany and zoology, including a new arrangement of phanerogamous plants, with especial reference to relative position, and their relations with the cryptogamous; and a new arrangement of the classes of zoology.* London: n.p.

Claus, Carl Friedrich. 1863. *Die frei lebenden Copepoden mit besonderer Berücksichtigung der Fauna Deutschlands, der Nordsee und des Mittelmeeres.* Leipzig: Wilhelm Engelmann.

Clergy list: The clergy list . . . containing an alphabetical list of the clergy. London: C. Cox [and others]. 1841–89.

'Climbing plants': On the movements and habits of climbing plants. By Charles

Darwin. [Read 2 February 1865.] *Journal of the Linnean Society (Botany)* 9 (1867): 1–118.

Clokie, Hermia Newman. 1964. *An account of the herbaria of the department of botany in the University of Oxford.* Oxford: Oxford University Press.

Collected papers: *The collected papers of Charles Darwin.* Edited by Paul H. Barrett. 2 vols. Chicago and London: University of Chicago Press. 1977.

Colloms, Brenda. 1975. *Charles Kingsley: the lion of Eversley.* London: Constable. New York: Barnes & Noble.

Colonial Office list: *The Colonial Office list . . . or, general register of the colonial dependencies of Great Britain.* London: Edward Stanford; Harrison & Sons. 1862–99.

Columbia gazetteer of the world: *The Columbia gazetteer of the world.* Edited by Saul B. Cohen. 3 vols. New York: Columbia University Press. 1998.

Commercial directory for Shropshire: *Pigot & Co.'s national commercial directory; 1828–9; comprising a directory and classification of the merchants, bankers, professional gentlemen, manufacturers and traders, in all the cities, towns, sea-ports, and principal villages in the following counties, viz. Cheshire Derbyshire Nottinghamshire Shropshire.* London and Manchester: J. Pigot & Co. 1828–9.

Complete peerage: *The complete peerage of England, Scotland, Ireland, Great Britain, and the United Kingdom, extant, extinct, or dormant.* By George Edward Cokayne. Revised edition. Edited by Vicary Gibbs, *et al.* 12 vols. London: St Catherine Press. 1910–59.

Cook, James and King, James. 1784. *A voyage to the Pacific Ocean. . . . In His Majesty's ships the Resolution and Discovery. In the years 1776, 1777, 1778, 1779, and 1780.* 3 vols. London: G. Nicol and T. Caddell.

Coral reefs: *The structure and distribution of coral reefs. Being the first part of the geology of the voyage of the* Beagle, *under the command of Capt. FitzRoy RN, during the years 1832 to 1836.* By Charles Darwin. London: Smith, Elder & Co. 1842.

Corner, Edred John Henry. 1988. *Wayside trees of Malaya.* 3d edition. 2 vols. Kuala Lumpur: Malayan Nature Society.

Correns, C. 1916. Friedrich Hildebrand. *Berichte der deutschen botanischen Gesellschaft* 34 (pt 2): 28–49.

Correspondence: *The correspondence of Charles Darwin.* Edited by Frederick Burkhardt *et al.* 14 vols to date. Cambridge: Cambridge University Press. 1985–.

Corsi, Pietro and Weindling, Paul J. 1985. Darwinism in Germany, France and Italy. In *The Darwinian heritage*, edited by David Kohn. Princeton, N.J.: Princeton University Press in association with Nova Pacifica.

Cossart, Noël. 1984. *Madeira: the island vineyard.* London: Christie's Wine Publications.

County families: *The county families of the United Kingdom; or, royal manual of the titled & untitled aristocracy of Great Britain & Ireland.* By Edward Walford. London: Robert Hardwicke; Chatto & Windus. 1860–93. *Walford's county families of the United Kingdom or royal manual of the titled and untitled aristocracy of England, Wales, Scotland, and Ireland.* London: Chatto & Windus; Spottiswoode & Co. 1894–1920.

Crane, Jocelyn. 1975. *Fiddler crabs of the world (Ocypodidae: genus Uca).* Princeton, N.J.: Princeton University Press.

Crockford's clerical directory: *The clerical directory, a biographical and statistical book of reference for facts relating to the clergy and the church. Crockford's clerical directory etc.* London: John Crockford [and others]. 1858–1900.

Cross and self fertilisation: *The effects of cross and self fertilisation in the vegetable kingdom.* By Charles Darwin. London: John Murray. 1876.

Crotch, George Robert. 1867. On the Coleoptera of the Azores. [Read 28 March 1867.] *Proceedings of the Zoological Society of London* (1867): 359–91.

Crüger, Hermann. 1864. A few notes on the fecundation of orchids and their morphology. [Read 3 March 1864.] *Journal of the Linnean Society (Botany)* 8 (1865): 127–35.

Cunningham, Robert Oliver. 1871. *Notes on the natural history of the Strait of Magellan and west coast of Patagonia, made during the voyage of H.M.S. 'Nassau' in the years 1866, 67, 68, & 69.* Edinburgh: Edmonston and Douglas.

Curtis's botanical magazine dedications: *Curtis's botanical magazine dedications 1827–1927.* Portraits and biographical notes compiled by Ernest Nelmes and William Cuthbertson. London: Bernard Quaritch for the Royal Horticultural Society. [1931.]

Cuvier, Georges. 1817. Faites sur le cadavre d'une femme connue à Paris et à Londres sous le nom de Vénus Hottentotte. *Mémoires du Musée Nationale d'Histoire Naturelle* 3: 259–74.

——. 1840. *Cuvier's animal kingdom: arranged according to its organisation.* Mammalia, birds, and reptiles by Edward Blyth. The fishes and radiata by Robert Mudie. The molluscous animals by George Johnston. The articulated animals by J. O. Westwood. London: Wm. S. Orr and Co.

——. 1849. *The animal kingdom, arranged after its organisation . . . by the late Baron Georges Cuvier.* Translated and adapted to the present state of science. Mammalia, birds, and reptiles by Edward Blyth. The fishes and radiata by Robert Mudie. The molluscous animals by George Johnston. The articulated animals by J. O. Westwood. A new edition with additions by W. B. Carpenter and J. O. Westwood. London: W. S. Orr and Co.

——. 1859. *The animal kingdom, arranged after its organisation . . . by the late Baron Georges Cuvier.* Translated and adapted to the present state of science. Mammalia, birds, and reptiles by Edward Blyth. The fishes and radiata by Robert Mudie. The molluscous animals by George Johnston. The articulated animals by J. O. Westwood. A new edition with additions by W. B. Carpenter and J. O. Westwood. London and Edinburgh: A. Fullerton & Co. [Reprint of Cuvier 1849, with added plates.]

DAB: *Dictionary of American biography.* Under the auspices of the American Council of Learned Societies. 20 vols., index, and 10 supplements. New York: Charles Scribner's Sons; Simon & Shuster Macmillan. London: Oxford University Press; Humphrey Milford. 1928–95.

Dallas, William Sweetland, trans. 1869. *Facts and arguments for Darwin.* By Fritz Müller. London: John Murray.

Dareste, Camille. 1863. Recherches sur les conditions de la vie et de la mort chez les monstres ectroméliens, célosomiens et exencéphaliens, produits artificiellement dans l'espèce de la poule. [Read 23 January 1863.] *Mémoires de la Société Impériale des Sciences de l'Agriculture et des Arts de Lille* 10: 39–82. [Reprinted in *Annales des Sciences Naturelles (Zoologie)* 4th ser. 20: 59–99.]

———. 1867. *Rapport sur un veau monstreux.* (Extracted from the *Archives du Comice Agricole de l'Arrondisement Lille.*) Lille: Blocquel-Castiaux.

———. 1877. *Recherches sur la production artificielle des monstruosités, ou essai de tératogénie expérimentale.* Paris: C. Reinwald.

Darwin, Francis. 1914. William Erasmus Darwin. *Christ's College Magazine* 29: 16–23.

Darwin pedigree: Pedigree of the family of Darwin. Compiled by H. Farnham Burke. N.p.: privately printed. 1888. [Reprinted in facsimile in *Darwin pedigrees*, by Richard Broke Freeman. London: printed for the author. 1984.]

Davidson, Thomas. 1867. On *Waldheimia venosa*, Solander, *sp. Annals and Magazine of Natural History* 3d ser. 20: 81–3.

Davies, N. B. 2000. *Cuckoos, cowbirds and other cheats.* London: T & A D Poyser.

Davitashvili, Leo Shiovich. 1951. *V. O. Kovalevsky.* 2d edition. Moscow: Academy of Science of the USSR.

Dawkins, William Boyd. 1866. On the fossil British oxen. Part I. *Bos urus*, Cæsar. [Read 21 March 1866.] *Quarterly Journal of the Geological Society of London* 22: 391–401.

———. 1867. On the fossil British oxen. Part II. *Bos longifrons*, Owen. [Read 20 February 1867.] *Quarterly Journal of the Geological Society of London* 23: 176–84.

———. 1868a. On the dentition of *Rhinoceros Etruscus*, Falc. [Read 8 January 1868.] *Quarterly Journal of the Geological Society of London* 24: 207–18.

———. 1868b. On the prehistoric Mammalia of Great Britain. *Transactions of the International Congress of Prehistoric Archaeology* 3d session (1868): 269–90.

Dawkins, William Boyd and Sanford, William Ayshford. 1866–72. *British pleistocene felidae.* 4 pts. Vol. 1 of *A monograph of the British Pleistocene Mammalia.* London: Palaeontographical Society.

Dawson, John William. 1864. On the structure of certain organic remains in the Laurentian limestones of Canada. [Read 23 November 1864.] *Quarterly Journal of the Geological Society of London* 21 (1865): 51–9.

DBE: Deutsche biographische Enzyklopädie. Edited by Walter Killy *et al.* 12 vols. in 14. Munich: K. G. Saur. 1995–2000.

DBF: Dictionnaire de biographie Française. Under the direction of J. Balteau *et al.* 20 vols. (A–Leblois) to date. Paris: Librairie Letouzey & Ané. 1933–.

DBI: Dizionario biografico degli Italiani. Edited by Alberto M. Ghisalberti *et al.* 62 vols. (A–Labriola) to date. Rome: Istituto della Enciclopedia Italiana. 1960–.

DCB: Dictionary of Canadian biography. Edited by George W. Brown *et al.* 13 vols. and index to first 12 vols. Toronto, Ontario: University of Toronto Press. 1966–94.

De Beer, Gavin, ed. 1959. Darwin's journal. *Bulletin of the British Museum (Natural History). Historical Series* 2 (1959–63): 3–21.

Decaisne, Joseph. 1836. Remarques sur les affinités du genre *Helwingia*, et établissement de la famille des Helwingiacées. *Annales des Sciences Naturelles (Botanique)* 2d ser. 6: 65–76.

Delpino, Federico. 1865. Relazione sull'apparechio della fecondazione nelle asclepiadee. Aggiuntevi alcune considerazioni sulle cause finali e sulla teoria di Carlo Darwin intorno all'origine delle specie. *Gazzetta Medica di Torino* 2d ser. 15 (1865): 372–4, 382–4, 390–1, 398–400.

———. 1867a. Sull'opera 'La distribuzione dei sessi nelle piante e la legge che osta alla perennità della fecondazione consanguinea' del prof. Hildebrand. Con note critiche. *Atti della Società Italiana delle Scienze Naturali in Milano* 10: 3–34.

———. 1867b. *Sugli apparecchi della fecondazione nelle piante antocarpe (fanerogame): sommario di osservazioni fatte negli anni 1865–1866*. Florence: Cellini.

[De Morgan, Sophia Elizabeth.] 1863. *From matter to spirit. The result of ten years' experience in spirit manifestations*. By C.D. with a preface by A.B. London: Longman, Green, Longman, Roberts & Green.

Descent: *The descent of man, and selection in relation to sex*. By Charles Darwin. 2 vols. London: John Murray. 1871.

Desmond, Adrian. 1982. *Archetypes and ancestors: palaeontology in Victorian London, 1850–1875*. London: Blond & Briggs.

———. 1994–7. *Huxley*. 2 vols. London: Michael Joseph.

Desmond, Adrian and Moore, James. 1991. *Darwin*. London: Michael Joseph.

Desmond, Ray. 1992. *The European discovery of the Indian flora*. Oxford: Oxford University Press [in association with the] Royal Botanic Gardens.

———. 1994. *Dictionary of British and Irish botanists and horticulturists including plant collectors, flower painters and garden designers*. New edition, revised with the assistance of Christine Ellwood. London: Taylor & Francis and the Natural History Museum. Bristol, Pa.: Taylor & Francis.

———. 1995. *Kew: the history of the Royal Botanic Gardens*. London: Harvill Press with the Royal Botanic Gardens, Kew.

———. 1999. *Sir Joseph Dalton Hooker, traveller and plant collector*. Woodbridge, Suffolk: Antique Collectors' Club with the Royal Botanic Gardens, Kew.

Dexter, Ralph W. 1986. Historical aspects of the Calaveras skull controversy. *American Antiquity* 51: 365–9.

Dictionary of African biography. Editor-in-chief, L. H. Ofusu-Appiah. 3 vols. to date. New York: Reference Publications. 1977–.

Dictionary of Australian artists: *The dictionary of Australian artists. Painters, sketchers, photographers and engravers to 1870*. Edited by Joan Kerr. Melbourne: Oxford University Press. 1992.

Dictionary of British artists: *The dictionary of British artists, 1880–1940*. Compiled by J. Johnson and A. Greutzner. Woodbridge, Suffolk: Antique Collectors' Club. 1976.

Dictionary of New Zealand English: *Dictionary of New Zealand English. A dictionary of New Zealandisms on historical principles*. Edited by H. W. Orsman. Auckland, New Zealand: Oxford University Press. 1997.

Dictionnaire général des artistes de l'école française: *Dictionnaire général des artistes de l'école française, depuis l'origine des arts du dessin jusqu'à nos jours: architects, peintres, sculpteurs, graveurs et lithographes.* 2 vols. and supplement. Paris: Librairie Renouard. 1882–6.

Dictionnaire universel des contemporains: *Dictionnaire universel des contemporains contenant toutes les personnes notables de la France et des pays étrangers . . . Ouvrage rédigé et continuellement tenu à jour avec le concours d'écrivains et des savants de tous les pays.* Edited by Louis Gustave Vapereau. Paris: Libraire Hachette. 1858. 3d edition, 1865. 4th edition, 1870. 5th edition, 1880. 6th edition, 1893.

Di Gregorio, Mario A. 1984. *T. H. Huxley's place in natural science.* New Haven and London: Yale University Press.

'Dimorphic condition in *Primula*': On the two forms, or dimorphic condition, in the species of *Primula*, and on their remarkable sexual relations. By Charles Darwin. [Read 21 November 1861.] *Journal of the Proceedings of the Linnean Society (Botany)* 6 (1862): 77–96. [*Collected papers* 2: 45–63.]

Disraeli, Benjamin. 1858. *Lord George Bentinck: a political biography.* New edition. London: G. Routledge.

DNB: Dictionary of national biography. Edited by Leslie Stephen and Sidney Lee. 63 vols. and 2 supplements (6 vols.). London: Smith, Elder & Co. 1885–1912. *Dictionary of national biography 1912–90.* Edited by H. W. C. Davis *et al.* 9 vols. London: Oxford University Press. 1927–96.

DNZB: A dictionary of New Zealand biography. Edited by G. H. Scholefield. 2 vols. Wellington, New Zealand: Department of Internal Affairs. 1940. *The dictionary of New Zealand biography.* Edited by W. H. Oliver *et al.* 5 vols. Auckland and Wellington, New Zealand: Department of Internal Affairs [and others]. 1990–2000.

Doak, Wade. 1972. *Fishes of the New Zealand region.* Auckland, New Zealand: Hodder and Stoughton.

Dodd, Victor. 1983. *Beemasters of the past.* Hebden Bridge, Yorkshire: Northern Bee Books.

Dod's parliamentary companion: The parliamentary pocket companion . . . compiled from official documents, and from the personal communications of members of both houses. Dod's parliamentary companion. London: Whittaker, Treacher, & Arnot; Whittaker & Co. 1833–1914.

Doescher, Rex A. 1981. *Living and fossil brachiopod genera 1775–1979: lists and bibliography.* Smithsonian contributions to paleobiology no. 42. Washington D.C.: Smithsonian Institution Press.

Dohrn, Anton. 1866. *Eugereon Boeckingi*, eine neue Insectenform aus dem Todtliegenden. *Palaeontographica* 13: 333–9.

——. 1867. On the morphology of the Arthropoda. [Read before the British Association, 5 September 1867.] *Journal of Anatomy and Physiology* 2 (1868): 80–6.

——. 1870. *Untersuchungen über Bau und Entwicklung der Arthropoden.* 2 vols. Leipzig: W. Engelmann.

Doncaster, John Priestman. 1961. *Francis Walker's aphids*. London: British Museum (Natural History).

Downie, N. M. and Arnett, R. H., Jr. 1996. *The beetles of northeastern North America*. 2 vols. Gainsville, Fla.: Sandhill Crane Press.

Dressler, Robert L. 1981. *The orchids: natural history and classification*. Cambridge, Mass., and London: Harvard University Press.

———. 1993. *Phylogeny and classification of the orchid family*. Cambridge: Cambridge University Press.

Drouet, Henri. 1858. *Mollusques marins des îles açores*. Paris: Baillière.

———. 1861. *Eléments de la faune açoréenne*. Paris: J. B. Baillière & Fils; J. Rothschild.

DSAB: *Dictionary of South African biography*. Edited by W. J. de Kock *et al*. 4 vols. Pretoria and Cape Town: Nasionale Boekhandel Beperk [and others]. 1968–81.

DSB: *Dictionary of scientific biography*. Edited by Charles Coulston Gillispie and Frederic L. Holmes. 18 vols. including index and supplements. New York: Charles Scribner's Sons. 1970–90.

Dubow, Saul. 1995. *Scientific racism in modern South Africa*. Cambridge: Cambridge University Press.

Duchenne, Guillaume Benjamin Amand. 1862. *Mécanisme de la physionomie humaine, ou analyse électro-physiologique de l'expression des passions*. Paris: Ve Jules Renouard, Libraire.

Dupree, Anderson Hunter. 1959. *Asa Gray, 1810–1888*. Cambridge, Mass.: Belknap Press of Harvard University.

EB: *The Encyclopædia Britannica. A dictionary of arts, sciences, literature and general information*. 11th edition. 29 vols. Cambridge: Cambridge University Press. 1910–11.

EB 8th ed.: *The Encyclopaedia Britannica: or, dictionary of arts, sciences, and general literature*. 8th edition. 22 vols. Edinburgh: Adam and Charles Black. 1853–60.

EB 9th ed.: *The Encyclopaedia Britannica : a dictionary of arts, sciences, and general literature*. 9th edition. 24 vols. and index. Edinburgh: A. and C. Black. 1875–89.

Eiseley, Loren. 1979. *Darwin and the mysterious Mr. X: new light on the evolutionists*. New York: E. P. Dutton.

Ekman, Paul. 1998. Introduction, afterword, and commentary to the third edition of *The expression of the emotions in man and animals*, by Charles Darwin. London: HarperCollins Publishers.

Eliot, George. 1859. *Adam Bede*. 3 vols. Edinburgh: William Blackwood.

Ellegård, Alvar. 1957. The readership of the periodical press in mid-Victorian Britain. *Göteborgs Universitets Årsskrift* 63 (3): 1–41.

Elliott, Brent. 2004. *The Royal Horticultural Society: a history 1804–2004*. Chichester: Phillimore & Co.

Elliott, Gordon. 1998. *John and David Elers and their contemporaries*. London: Jonathan Horne Publications.

Emma Darwin (1904): *Emma Darwin, wife of Charles Darwin. A century of family letters*. Edited by Henrietta Litchfield. 2 vols. Cambridge: privately printed by Cambridge University Press. 1904.

Emma Darwin (1915): *Emma Darwin: a century of family letters, 1792–1896.* Edited by Henrietta Litchfield. 2 vols. London: John Murray. 1915.

Encyclopaedia of Aboriginal Australia: *The encyclopaedia of Aboriginal Australia: Aboriginal and Torres Strait Islander history, society and culture.* General editor: David Horton. 2 vols. Canberra: published by Aboriginal Studies Press for the Australian Institute of Aboriginal and Torres Strait Islander Studies. 1994.

Endersby, James John. 2002. Putting plants in their place: Joseph Hooker's philosophical botany, 1838–65. PhD dissertation, Cambridge University.

Endlicher, Stephan Ladislaus. 1836–42. *Genera plantarum secundum ordines naturales disposita.* With 4 supplements; in 2 vols. Vienna: Friedrich Beck.

Engels, Eve-Marie, ed. 1995. *Die Rezeption von Evolutionstheorien im neunzehnten Jahrhundert.* Suhrkamp Taschenbuch Wissenschaft, Nr. 1229. Frankfurt am Main: Suhrkamp.

Eyton, Thomas Campbell. 1838. *A monograph on the Anatidae, or duck tribe.* London: Longman, Orme, Brown, Green & Longman.

——. 1867–75. *Osteologia avium; or, a sketch of the osteology of birds.* 1 vol. and 2 supplements. Wellington, Shropshire: R. Hobson.

Expression: The expression of the emotions in man and animals. By Charles Darwin. London: John Murray. 1872.

Farrar, Frederick William. 1867. On some defects in public school education. [Read 8 February 1867.] *Proceedings of the Royal Institution of Great Britain* 5 (1866–9): 26–44.

Fasti ecclesiæ Scoticanæ: Fasti ecclesiæ Scoticanæ. The succession of ministers in the Church of Scotland from the Reformation. By Hew Scott. 7 vols. Edinburgh: Oliver and Boyd. 1915–28.

Fausto-Sterling, Anne. 1995. Gender, race, and nation: the comparative anatomy of 'Hottentot' women in Europe, 1815–1817. In *Deviant bodies: critical perspectives on difference in science and popular culture*, edited by Jennifer Terry and Jacqueline Urla. Bloomington and Indianapolis: University of Indiana Press.

'Fertilization of orchids': Notes on the fertilization of orchids. By Charles Darwin. *Annals and Magazine of Natural History* 4th ser. 4 (1869): 141–59. [*Collected papers* 2: 138–56.]

Fichman, Martin. 2004. *An elusive Victorian; the evolution of Alfred Russel Wallace.* Chicago and London: University of Chicago Press.

Filmer, J. L. 1977. The Norman family of Bromley Common. *Bromley Local History* 2: 16–24.

Fitzwilliam, William Wentworth (Viscount Milton) and Walter Butler Cheadle. [1865.] *The north-west passage by land, being the narrative of an expedition from the Atlantic to the Pacific.* London: Cassell, Petter, and Galpin.

Fleagle, John G. 1999. *Primate adaptation and evolution.* 2d edition. San Diego: Academic Press.

Foreign Office list: The Foreign Office list. London: Harrison & Sons. 1852–1965.

Forms of flowers: The different forms of flowers on plants of the same species. By Charles Darwin. London: John Murray. 1877.

Foster, Clement Le Neve and Topley, William. 1865. On the superficial deposits of the valley of the Medway, with remarks on the denudation of the Weald. [Read 24 May 1865.] *Quarterly Journal of the Geological Society of London* 21: 443–74.

Fouqué, Friedrich Heinrich Karl, Baron de la Motte. 1815. *Sintram und seine Gefährten: eine nordische Erzählung nach Albrecht Dürer.* Vienna: Haas.

——. 1867. Sintram and his companions. A northern romance after Albert Dürer. In *Undine and other tales,* translated by F. E. Bunnett. Leipzig: Tauchniz. London: Sampson, Low, Marsten, Low and Searle.

Franklin, Henry James. 1912. The Bombidae of the New World. *Transactions of the American Entomological Society* 38: 177–486.

Fraser, Angus. 1996. A publishing house and its readers, 1841–1880: the Murrays and the Miltons. *Papers of the Bibliographical Society of America* 90: 4–47.

Fraser, H. Malcolm. 1958. *History of beekeeping in Britain.* London: Bee Research Association.

Freeman, Richard Broke. 1968. Charles Darwin on the routes of male humble bees. *Bulletin of the British Museum (Natural History) Historical Series* 3 (1962–9): 177–89.

——. 1977. *The works of Charles Darwin: an annotated bibliographical handlist.* 2d edition. Folkestone, Kent: William Dawson & Sons. Hamden, Conn.: Archon Books, Shoe String Press.

——. 1978. *Charles Darwin: a companion.* Folkestone, Kent: William Dawson & Sons. Hamden, Conn.: Archon Books, Shoe String Press.

Freeman, Richard Broke and Gautrey, Peter Jack. 1972. Charles Darwin's *Queries about expression. Bulletin of the British Museum (Natural History) Historical Series* 4 (1970–5): 205–19.

——. 1975. Charles Darwin's *Queries about expression. Journal of the Society for the Bibliography of Natural History* 7 (1974–6): 259–63.

Friends' biographical catalogue: Biographical catalogue: being an account of the lives of Friends and others whose portraits are in the London Friends' institute. London: Friends' institute. 1888

Frodin, David G. 2001. *Guide to standard floras of the world: an annotated, geographically arranged systematic bibliography of the principal floras, enumerations, checklists and chorological atlases of different areas.* 2d edition. Cambridge: Cambridge University Press.

Fryer, Gavin and Akerman, Clive, eds. 2000. *The reform of the Post Office in the Victorian era and its impact on economic and social activity.* 2 vols. London: Royal Philatelic Society.

Gaskell, Elizabeth Cleghorn. 1848. *Mary Barton.* 2 vols. London: Chapman and Hall.

——. 1855. *North and south.* 2 vols. London: Chapman and Hall.

Gätke, Heinrich. 1895. *Heligoland as an ornithological observatory: the result of fifty years' experience.* Translated by Rudolph Rosenstock. Edinburgh: David Douglas.

Gaudry, Albert. 1862–7. *Animaux fossiles et géologie de l'Attique, d'après les recherches faites en 1855–56 et en 1860 sous les auspices de l'Académie des Sciences.* 1 vol. and atlas. Paris: Libraire de la Société Géologique de France.

——. 1867a. Mémoire sur le reptile (*Pleuracanthus Frossardi*) découvert par M. Frossard à Muse (Saône-et-Loire). *Nouvelles Archives du Muséum d'Histoire Naturelle* 3: 21–40.

———. 1867b. Sur le reptile découvert par M. Ch. Frossard à Muse, près d'Autun. [Read 18 February 1867.] *Bulletin de la Société Géologique de France* 2d ser. 24 (1866–7): 397–401.

———, *et al.* 1873. *Animaux fossiles du Mont Léberon (Vaucluse). Etude sur les vertébrés par Albert Gaudry. Etude sur les invertébrés par P. Fischer et R. Tournouër.* Paris: Libraire de la Société Géologique de France.

Gayon, Jean. 1998. *Darwinism's struggle for survival: heredity and the hypothesis of natural selection.* Translated by Matthew Cobb. Cambridge: Cambridge University Press.

Gegenbaur, Carl. 1858. Mittheilungen über die Organisation von *Phyllosoma* und *Sapphirina. Archiv für Anatomie, Physiologie und wissenschaftliche Medicin* (1858): 43–81.

———. 1870. *Grundzüge der vergleichenden Anatomie.* 2d edition. Leipzig: Wilhelm Engelmann.

Geison, Gerald L. 1969. Darwin and heredity: the evolution of his hypothesis of pangenesis. *Journal of the History of Medicine* 24: 375–411.

Genealogisches Handbuch des Adels. Vol. 1 of *Genealogisches Handbuch der Fürstlichen Häuser.* Director: Hans Friedrich v. Ehrenkrook. Glücksburg/Ostsee: C. A. Starke. 1951.

Geoffroy Saint-Hilaire, Isidore. 1832–7. *Histoire générale et particulière des anomalies de l'organisation chez l'homme et les animaux, ouvrage comprenant des recherches sur les charactères, la classification, l'influence physiologique et pathologique, les rapports généraux, les lois et les causes des monstruosites, des variétés et des vices de conformation, ou traité de tératologie.* 3 vols. and atlas. Paris: J. B. Baillière.

Gerbe, Z., trans. 1869–73. *La vie des animaux illustrée, ou description populaire du règne animal.* By Alfred Edmund Brehm. (Translation of parts 1 and 2 of Brehm *et al.* 1864–9.) 4 vols. Paris: J. B. Baillière et fils.

Gerstaecker, Carl Edouard Adolph. 1863. Arthropoden. In vol. 2 of *Handbuch der Zoologie,* by Carl Edouard Adolph Gerstaecker and Julius Victor Carus. Leipzig: Wilhelm Engelmann.

Gervais, Paul. 1854–5. *Histoire naturelle des mammifères, avec l'indication de leurs moeurs, et de leurs rapports avec les arts, le commerce et l'agriculture.* 2 vols. Paris: L. Curmer.

Gibbs, George. 1877. Tribes of western Washington and northwestern Oregon, part 2. In vol. 1 of *Contributions to North American ethnology.* Washington, D.C.: US Government Printing Office.

Giebel, Christoph Gottfried Andreas. 1866. Eine antidarwinistische Vergleichung des Menschen- und der Orangschädel. *Zeitschrift für die gesammten Naturwissenschaften* 27: 401–19.

Gilbert, Francis S. 1993. *Hoverflies.* Revised edition. Slough: Richmond Publishing.

Gilbert, Francis S. and Jervis, Mark. 1998. Functional, evolutionary and ecological aspects of feeding-related mouthpart specializations in parasitoid flies. *Biological Journal of the Linnean Society* 63: 495–535.

Gilbert, Pamela. 1977. *A compendium of the biographical literature on deceased entomologists.* London: British Museum (Natural History).

Gillbank, Linden. 1986. The origins of the Acclimatisation Society of Victoria: practical science in the wake of the gold rush. *Historical Records of Australian Science* 6: 359–74.

Gladstone, Hugh S. 1913. _The history of the Dumfriesshire and Galloway Natural History and Antiquarian Society._ Dumfries: Dumfriesshire and Galloway Natural History and Antiquarian Society.

Glaeser, Ernest. 1878. _Biographie nationale des contemporains, rédigée par une société de gens de lettres, sous la direction de M. Ernest Glaeser._ Paris: Glaeser & Cie.

Godman, Frederick Du Cane. 1866. Notes on the birds of the Azores. _Ibis_ n.s. 2: 88–109.

——. 1870. _Natural history of the Azores or Western Islands._ London: Van Voorst.

Goethe, Johann Wolfgang von. 1988. _Gedichte 1800–32._ Edited by Karl Eibl. Frankfurt am Main: Deutscher Klassiker Verlag.

Gould, James L. and Gould, Carol Grant. 1997. _Sexual selection: mate choice and courtship in nature._ New York: Scientific American Library.

Gould, John. 1873. _Birds of Great Britain._ 5 vols. London: the author.

Gould, Stephen Jay. 1977. _Ontogeny and phylogeny._ Cambridge, Mass., and London: Belknap Press of Harvard University Press,

——. 1990. _Wonderful life. The Burgess Shale and the nature of history._ London: Hutchinson Radius.

——. 1997. _The mismeasure of man._ Revised and expanded edition. London: Penguin Books.

Grande encyclopédie: La grande encyclopédie inventaire raisonné des sciences, des lettres et des arts. Edited by F. Camille Dreyfus _et al._ 31 vols. Paris: H. Lamirault; Société Anonyme de la Grande Encyclopédie. [1886–1902.]

Gratiolet, Louis Pierre. [1854.] _Mémoire sur les plis cérébraux de l'homme et des primatès._ Paris: Arthus Bertrand.

Graves, Joseph L., Jr. 2002. _The emperor's new clothes; biological theories of race at the millennium._ New Brunswick, N.J., and London: Rutgers University Press.

Gray, Asa. 1862. Fertilization of orchids through the agency of insects. _American Journal of Science and Arts_ 2d ser. 34: 420–9.

——. 1867. _Manual of the botany of the northern United States: including the district east of the Mississippi and north of North Carolina and Tennessee, arranged according to the natural system._ 5th edition. New York: Ivison, Phinney, Blakeman & Co.

[——.] 1868. [Review of _Variation._] _Nation_ 6: 234–6.

Gray, Jane Loring, ed. 1893. _Letters of Asa Gray._ 2 vols. London: Macmillan and Co.

Gray, John Edward. 1843. _List of the specimens of Mammalia in the collection of the British Museum._ With a list of genera and their synonyma. London: by order of the Trustees.

——. 1852. _Catalogue of the specimens of Mammalia in the collection of the British Museum. Part III. Ungulata Furcipeda._ London: British Museum.

——. 1867. Notes on the arrangement of sponges, with the description of some new genera. [Read 9 May 1867.] _Proceedings of the Zoological Society of London_ (1867): 492–558.

Grayson, Donald K. 1983. _The establishment of human antiquity._ New York: Academic Press.

Grece, Clair James, trans. 1874. *An English grammar: methodical, analytical and historical.* 3 vols. By Eduard Mätzner. London: John Murray.

Greenaway, Richard L. N. 2000. *Rich man, poor man, environmentalist, thief. Biographies of Canterbury personalities written for the millennium and for the 150th anniversary of the Canterbury settlement.* Christchurch, New Zealand: Christchurch City Libraries.

Greenwood, George. 1866. *Rain and rivers; or, Hutton and Playfair against Lyell and all comers.* 2d edition. London: Longmans, Green, & Co.

Gregory, Frederick. 1977. *Scientific materialism in nineteenth century Germany.* Dordrecht, Netherlands, and Boston, Mass.: D. Reidel Publishing Company.

Gregory, William. 1851. *Letters to a candid inquirer on animal magnetism.* London: Taylor, Walton, & Maberly.

Grene, Marjorie Glickman and Depew, David J. 2004. *The philosophy of biology: an episodic history.* Cambridge: Cambridge University Press.

Grimm, Jakob Ludwig Karl. 1819–37. *Deutsche Grammatik.* 4 vols. Göttingen: bei Dieterich.

Groeben, Christiane. 1982. *Charles Darwin 1809–1882, Anton Dohrn 1840–1909: correspondence.* Naples: Macchiaroli.

Groves, C. P. 1948–58. *The planting of Christianity in Africa.* 4 vols. London: Lutterworth Press.

Gruber, Howard Ernest. 1981. *Darwin on man. A psychological study of scientific creativity.* 2d edition. Chicago: University of Chicago Press.

Gruber, Jacob W. 1960. *A conscience in conflict. The life of St. George Jackson Mivart.* New York: Columbia University Press for Temple University Publications.

GSE: Great Soviet encyclopedia. Edited by Jean Paradise *et al.* 31 vols. and index. New York: Macmillan. London: Collier Macmillan. 1973–83. [Translation of the 3d edition of *Bol'shaia Sovetskaia entsiklopediia,* edited by A. M. Prokhorov.]

Gunn, Mary and Codd, L. E. 1981. *Botanical exploration of Southern Africa.* Cape Town: A. A. Balkema.

Günther, Albert Charles Lewis Gotthilf. 1859–70. *Catalogue of acanthopterygian fishes in the collection of the British Museum.* 8 vols. London: by order of the Trustees.

——. 1864. *The reptiles of British India.* London: Ray Society.

——. 1864–6. An account of the fishes of the states of Central America, based on collections made by Capt. J. M. Dow, F. Godman, Esq., and O. Salvin, Esq. [Read 22 March 1864 and 13 December 1866.] *Transactions of the Zoological Society of London* 6 (1863–7): 377–494.

Gunther, Albert E. 1972. The original drawings of George Henry Ford. *Journal of the Society for the Bibliography of Natural History* 6: 139–42.

——. 1975. *A century of zoology at the British Museum through the lives of two keepers, 1815–1914.* London: Dawsons of Pall Mall.

Haast, Heinrich Ferdinand von. 1948. *The life and times of Sir Julius von Haast, explorer, geologist, museum builder.* Wellington, New Zealand: privately published.

Haast, John Francis Julius von. 1862. Observations on the birds of the western districts of the Province of Nelson, New Zealand. *Ibis* 4: 103–6.

Haast, John Francis Julius von. 1866. *Report on the head-waters of the Rakaia, with twenty illustrations and two appendices.* Christchurch: government of Canterbury province.

———. 1868. On the measurements of *Dinornis* bones, obtained from excavations in a swamp, situated at Glenmark, on the property of Messrs. Kermode and Co., up to February 15, 1868. [Read before the Wellington Philosophical Society, 28 July 1868.] *Transactions of the New Zealand Institute* 1: 80–9.

Hadfield, Miles, *et al.* 1980. *British gardeners. A biographical dictionary.* London: A. Zwemmer in association with The Condé Nast Publications.

Haeckel, Ernst. 1866. *Generelle Morphologie der Organismen. Allgemeine Grundzüge der organischen Formen-Wissenschaft, mechanisch begründet durch die von Charles Darwin reformirte Descendenz-Theorie.* 2 vols. Berlin: Georg Reimer.

———. 1867. Eine zoologischen Excursion nach den Canarischen Inseln. *Jenaische Zeitschrift für Medicin und Naturwissenschaft* 3: 313–28.

———. 1868. *Natürliche Schöpfungsgeschichte. Gemeinverständliche wissenschaftliche Vorträge über die Entwickelungslehre im Allgemeinen und diejenige von Darwin, Goethe und Lamarck im Besonderen, über die Anwendung derselben auf den Ursprung des Menschen und andere damit zusammenhängende Grundfragen der Naturwissenschaft.* Berlin: Georg Reimer.

———. 1869. *Zur Entwickelungsgeschichte der Siphonophoren. Beobachtungen über die Entwicklungsgeschichte der Genera Physophora, Crystallodes, Athorybia, und Reflexionen über die Entwicklungsgeschichte der Siphonophoren im allgemeinen.* Utrecht: C. van der Post, jr.

———. 1870. Eine Besteigung des Pik von Teneriffe. *Zeitschrift der Gesellschaft für Erdkunde zu Berlin* 5: 1–28.

———. 1876. *The history of creation: or the development of the earth and its inhabitants by the action of natural causes. A popular exposition of the doctrine of evolution in general, and that of Darwin, Goethe, and Lamarck in particular.* Translation of Haeckel's *Natürliche Schöpfungsgeschichte.* Translation revised by E. Ray Lankester. 2 vols. London: Henry S. King & Co.

Haliburton, Robert Grant. 1863. *New materials for the history of man, derived from a comparison of the customs and superstitions of nations.* Halifax, Nova Scotia: n.p.

Hall, Philip B. 1987. Robert Swinhoe (1836–1877), FRS, FZS, FRGS: a Victorian naturalist in Treaty Port China. *Geographical Journal* 153: 37–47.

Hampden-Cook, Ernest. 1926. *The register of Mill Hill school, 1807–1926.* Privately printed.

Harris, Donald F. 2004. Notes on the real-estate and financial affairs of Dr. Robert Waring Darwin of Shrewsbury. Typescript deposited at Cambridge University Library.

Harrow School register: The Harrow School register, 1800–1911. 3d edition. Edited by M. G. Dauglish and P. K. Stephenson. London: Longmans, Green, and Co. 1911.

Harshberger, John W. 1899. *The botanists of Philadelphia and their work.* Philadelphia: T. C. Davis & Son.

Hartley, Lucy. 2001. *Physiognomy and the meaning of expression in nineteenth-century culture.* Cambridge: Cambridge University Press.

Harvey, Joy. 1997. *'Almost a man of genius': Clémence Royer, feminism, and nineteenth-century science.* New Brunswick, N.J., and London: Rutgers University Press.

Harvey, William Henry. 1838. *The genera of South African plants, arranged according to the natural system.* Cape Town, South Africa: A. S. Robertson.

———. 1868. *The genera of South African plants, arranged according to the natural system.* 2d edition. Edited by Joseph Dalton Hooker. Cape Town, South Africa: J. C. Juta. London: Longman, Green, Reader, and Dyer.

Harvey, William Henry and Sonder, Otto Wilhelm. 1859–65. *Flora Capensis: being a scientific description of the plants of the Cape Colony, Caffraria, & Port Natal.* 3 vols. Dublin: Hodges, Smith & Co.

Havergal, Francis Tebbs. 1881. *Monumental inscriptions in the cathedral church of Hereford.* London: Simpkin, Marshall & Co. Walsall: W. H. Robinson. Hereford: Jakeman & Carver.

Hawker, Robert. 1829. *The poor man's morning and evening portions; being a selection of a verse of scripture, with short observations for every day in the year.* London: E. Palmer.

Hayhurst, H. 1967. 'The Lit. and Phil.': its past and its future. *Memoirs and Proceedings, Manchester Literary and Philosophical Society* 109: 5–17.

Hazlewood, Nick. 2000. *Savage. The life and times of Jemmy Button.* London: Hodder and Stoughton.

Hemsley, William Botting. 1885. *Report on the botany of the Bermudas and various other islands of the Atlantic and southern oceans.* 2 vols. Part of *The report of the scientific results of the voyage of H.M.S. Challenger during the years 1873–76.* London: HMSO.

Henslow, George. 1865. Note on the structure of *Medicago sativa*, as apparently affording facilities for the intercrossing of distinct flowers. [Read 16 November 1865.] *Journal of the Linnean Society (Botany)* 9 (1867): 327–9.

———. 1866. Note on the structure of *Indigofera*, as apparently offering facilities for the intercrossing of distinct flowers. [Read 19 April 1866.] *Journal of the Linnean Society (Botany)* 9 (1867): 355–8.

———. 1868. Note on the structure of *Genista tinctoria*, as apparently affording facilities for the intercrossing of distinct flowers. [Read 16 April 1868.] *Journal of the Linnean Society (Botany)* 10 (1869): 468.

Herbert, John Maurice, ed. 1877. *Poems by the late Mary Anne Herbert.* London: Harrison and Son.

Heron, Robert. 1835. Notes on the habits of the pea-fowl. [Read 14 April 1835.] *Proceedings of the Zoological Society of London* (1833–5) pt 3: 54.

Herschel, John Frederick William. 1866. *Familiar lectures on scientific subjects.* London and New York: Alexander Strahan.

Hessische Biographien. In association with Karl Esselborn and Georg Lehnert; edited by Herman Haupt. 3 vols. Darmstadt: Hessischer Staatsverlag. 1918–34.

Heuss, Theodor. 1991. *Anton Dohrn: a life for science.* Translated from the German by Liselotte Dieckmann. Berlin: Springer Verlag.

Higgins, Wesley E. 1997. A reconsideration of the genus *Prosthechea* (Orchidaceae). *Phytologia* 82: 370–83.

Hildebrand, Friedrich Hermann Gustav. 1863. Die Fruchtbildung der Orchideen, ein Beweis für die doppelte Wirkung des Pollen. *Botanische Zeitung* 21: 329–33, 337–45.

——. 1865. Bastardirungsversuche an Orchideen. *Botanische Zeitung* 23: 245–9.

——. 1866a. On the necessity for insect agency in the fertilisation of *Corydalis cava*. *International Horticultural Exhibition* 1866, pp. 157–8.

——. 1866b. Ueber die Vorrichtungen an einigen Blüthen zur Befruchtung durch Insektenhülfe. *Botanische Zeitung* 24: 73–8.

——. 1866c. Ueber die Befruchtung von *Asclepias Cornuti*. *Botanische Zeitung* 24: 376–8.

——. 1866–7a. Ueber die Befruchtung von Aristolochia Clematitis und einiger anderer Aristolochia-Arten. *Jahrbücher für wissenschaftliche Botanik* 5: 343–58.

——. 1866–7b. Ueber die Nothwendigkeit der Insektenhülfe bei der Befruchtung von Corydalis cava. *Jarhbücher für wissenschaftliche Botanik* 5: 359–63.

——. 1867a. *Die Geschlechter-Vertheilung bei den Pflanzen und das Gesetz der vermiedenen und unvortheilhaften stetigen Selbstbefruchtung*. Leipzig: Wilhelm Engelmann.

——. 1867b. Federigo Delpino's Beobachtungen über die Bestäubungsvorrichtungen bei der Phanerogamen. *Botanische Zeitung* 25: 265–70, 273–8, 281–7.

Hill, George Birkbeck. 1880. *The life of Sir Rowland Hill and the history of penny postage*. 2 vols. London: Thomas de la Rue & Co.

Historical register of the University of Cambridge: The historical register of the University of Cambridge, being a supplement to the Calendar *with a record of university offices, honours, and distinctions to the year 1910*. Edited by J. R. Tanner. Cambridge: Cambridge University Press. 1917.

Hodge, Frederick Webb, ed. 1910. *Handbook of American Indians north of Mexico*. 2 vols. Washington: Government Printing Office.

Hodgson, Brian Houghton. 1847. On the tame sheep and goats of the sub-Himálayas and of Tibet. *Journal of the Asiatic Society of Bengal* 16, pt 2: 1003–26.

Hooker, Joseph Dalton. 1844–7. *Flora Antarctica*. 1 vol. and 1 vol. of plates. Pt 1 of *The botany of the Antarctic voyage of HM discovery ships* Erebus and Terror *in the years 1839–1843, under the command of Captain Sir James Clark Ross*. London: Reeve Brothers.

——. 1845. An enumeration of the plants of the Galapagos Archipelago; with descriptions of those which are new. [Read 4 March, 6 May, and 16 December 1845.] *Transactions of the Linnean Society of London* 20 (1846–51): 163–233.

——. 1846. On the vegetation of the Galapagos Archipelago, as compared with that of some other tropical islands and of the continent of America. [Read 1 and 15 December 1846.] *Transactions of the Linnean Society of London* 20 (1846–51): 235–62.

——. 1854. *Himalayan journals; or, notes of a naturalist in Bengal, the Sikkim and Nepal Himalayas, the Khasia Mountains, &c.* 2 vols. London: John Murray.

——. 1864–7. *Handbook of the New Zealand flora: a systematic description of the native plants of New Zealand and the Chatham, Kermadec's, Lord Auckland's, Campbell's, and MacQuarrie's Islands*. 2 vols. London: Lovell Reeve & Co.

———. 1866a. Insular floras. [Read 27 August 1866.] *Gardeners' Chronicle* (1867): 6–7, 27, 50–1, 75–6.

———. 1866b. Abstract of Dr. Hooker's lecture on insular floras. In *The British Association for the Advancement of Science. Nottingham meeting, August, 1866. Report of the papers, discussions, and general proceedings*, edited by William Tindal Robertson. Nottingham: Thomas Forman. London: Robert Hardwicke.

———. 1866c. Considérations sur les flores insulaires. *Annales des Sciences Naturelles (Botanique)* 5th ser. 6: 267–99.

———. 1867. On insular floras: a lecture. *Journal of Botany* 5 (1867): 23–31.

Hooker, Joseph Dalton and Thomson, Thomas. 1859. Præcursores ad Floram Indicam.— Balsamineæ. [Read 16 June 1859.] *Journal of the Proceedings of the Linnean Society of London (Botany)* 4 (1860): 106–57.

Hooker, William Jackson and Baker, John Gilbert. 1868. *Synopsis filicum; or, a synopsis of all known ferns including the Osmundaceæ, Schizæaceæ, Marattiaceæ, and Ophioglossaceæ (chiefly derived from the Kew Herbarium)*. London: Robert Hardwicke.

Hooker, William Jackson, Sowerby, James de Carle [*et al.*]. 1831–63. *Supplement to the English Botany of the late Sir J. E. Smith and Mr. Sowerby*. 5 vols. London: n.p.

Houghton, Walter Edwards. 1957. *The Victorian frame of mind: 1830–1870*. New Haven: Yale University Press.

House, Kay Seymour. 1965. *Cooper's Americans*. [Columbus, Ohio]: Ohio State University Press.

Hull, David L. 1973. *Darwin and his critics: the reception of Darwin's theory of evolution by the scientific community*. Cambridge, Mass.: Harvard University Press.

Hull, Edward. 1867. Mr Whitaker on 'subaerial denudation'. *Geological Magazine* 4: 567–9.

Humboldt, Alexander von. 1814–29. *Personal narrative of travels to the equinoctial regions of the New Continent, during the years 1799–1804. By Alexander de Humboldt and Aimé Bonpland*. Translated into English by Helen Maria Williams. 7 vols. London: Longman, Hurst, Rees, Orme, & Brown; J. Murray; H. Colburn.

Hurst, Henry Alexander and Carter, George. 1867. On plants appearing in successive years on land prepared for plantations, in Cheshire. [Read 2 December 1867.] *Proceedings of the Literary and Philosophical Society of Manchester* 7 (1867–8): 62–6.

Hutchinson, Horace Gordon. 1914. *Life of Sir John Lubbock, Lord Avebury*. 2 vols. London: Macmillan.

Hutton, Frederick Wollaston. 1861. Some remarks on Mr Darwin's theory. *Geologist* 4: 132–6, 183–8.

Huxley, Anthony, *et al.*, eds. 1992. *The new Royal Horticultural Society dictionary of gardening*. 4 vols. London: Macmillan.

Huxley, Leonard, ed. 1900. *Life and letters of Thomas Henry Huxley*. 2 vols. London: Macmillan.

———, ed. 1918. *Life and letters of Sir Joseph Dalton Hooker, OM, GCSI*. Based on materials collected and arranged by Lady Hooker. 2 vols. London: John Murray.

[Huxley, Thomas Henry.] 1860. Darwin on the origin of species. *Westminster Review* n.s. 17: 541–70.

Huxley, Thomas Henry. 1863a. *Evidence as to man's place in nature.* London: Williams & Norgate.

———. 1863b. *On our knowledge of the causes of the phenomena of organic nature. Being six lectures to working men, delivered at the Museum of Practical Geology.* London: Robert Hardwicke.

———. 1866. *Lessons in elementary physiology.* London: Macmillan.

———. 1867. On the classification of birds; and on the taxonomic value of the modifications of certain of the cranial bones observable in that class. [Read 11 April 1867.] *Proceedings of the Zoological Society of London* (1867): 415–72.

———. 1868. On the classification and distribution of the *Alectoromorphæ* and *Heteromorphæ*. [Read 14 May 1868.] *Proceedings of the Zoological Society of London* (1868): 294–319.

'Illegitimate offspring of dimorphic and trimorphic plants': On the character and hybrid-like nature of the offspring from the illegitimate unions of dimorphic and trimorphic plants. By Charles Darwin. [Read 20 February 1868.] *Journal of the Linnean Society of London (Botany)* 10 (1869): 393–437.

Imperial gazetteer of India. New edition. 26 vols. Oxford: Clarendon Press. 1907–9.

Ince, Joseph, ed. 1876. *Science papers, chiefly pharmacological and botanical. By Daniel Hanbury.* London: Macmillan and Co.

India list: The East-India register and directory. 1803–44. *The East-India register and army list.* 1845–60. *The Indian Army and civil service list.* 1861–76. *The India list, civil and military.* 1877–95. *The India list and India Office list.* 1896–1917. London: Wm. H. Allen [and others].

Insectivorous plants. By Charles Darwin. London: John Murray. 1875.

International Horticultural Exhibition 1866: *International Horticultural Exhibition and Botanical Congress, held in London, from May 22nd to May 31st, 1866. Report of Proceedings.* London: Truscott, Son, & Simmons.

Jann, Rosemary. 1996. Darwin and the anthropologists: sexual selection and its discontents. In *Sexualities in Victorian Britain*, edited by Andrew H. Miller and James Eli Adams. Bloomington and Indianapolis: Indiana University Press.

[Jenkin, Henry Charles Fleeming.] 1867. The origin of species. *North British Review* 46: 277–318.

Jenner, Edward. 1788. Observations on the natural history of the cuckoo. *Philosophical Transactions of the Royal Society of London* 78: 219–37.

Jerdon, Thomas Claverhill. 1862–4. *The birds of India; being a natural history of all the birds known to inhabit continental India, with descriptions of the species, genera, families, tribes, and orders, and a brief notice of such families as are not found in India, making it a manual of ornithology specially adapted for India.* 2 vols. in 3. Calcutta: the author.

Johnson, William S. 1990. *Nineteenth-century photography: an annotated bibliography, 1839–1879.* Boston: G. K. Hall & Co.

Jones, Hugh Percy, ed. 1900. *A new dictionary of foreign phrases and classical quotations*

. . . *with English translations or equivalents*. London: Charles William Deacan.

Jones, Thomas Rymer, trans. [1869–73.] *Cassell's book of birds*. By Alfred Edmund Brehm. (Translation of part 2 of Brehm *et al*. 1864–9.) 4 vols. London: Cassell & Co.

Jordan, Alexis. 1851. Pugillus plantarum novarum. *Mémoires de l'Académie Nationale des Sciences, Belles-Lettres, et Arts de Lyon. Classe des Sciences* 2d ser. 1: 212–358.

——. 1860. Diagnoses d'espèces nouvelles ou méconnues, pour servir de matériaux à une flore de France réformée. *Annales de la Société Linnéenne de Lyon* 7: 373–518

Jordan, Alexis and Fourreau, Jules. 1866–1903. *Icones ad floram Europæ novo fundamento instaurandam spectantes*. Paris: Savy.

Journal and remarks: *Journal and remarks. 1832–1836*. By Charles Darwin. Vol. 3 of *Narrative of the surveying voyages of His Majesty's ships Adventure and Beagle between the years 1826 and 1836, describing their examination of the southern shores of South America, and the Beagle's circumnavigation of the globe*. London: Henry Colburn. 1839. [Separately published as *Journal of researches*.]

Journal of researches: *Journal of researches into the geology and natural history of the various countries visited by HMS Beagle, under the command of Captain FitzRoy, RN, from 1832 to 1836*. By Charles Darwin. London: Henry Colburn. 1839.

Journal of researches 2d ed.: *Journal of researches into the natural history and geology of the countries visited during the voyage of HMS Beagle round the world, under the command of Capt. FitzRoy RN*. 2d edition, corrected, with additions. By Charles Darwin. London: John Murray. 1845.

Jubiläums-Katalog: *Jubiläums-Katalog der E. Schweizerbart'schen Verlagsbuchhandlung (Erwin Nägele) G.m.b.H., Stuttgart, 1826–1926*. Stuttgart: E. Schweizerbart. 1926.

Judel, Claus Günther. 2004. Der Liebigschüler Carl Vogt als Wissenschaftlicher, Philosoph und Politiker. *Giessener Universitätsblätter* 37: 51–6.

Junker, Thomas. 1989. *Darwinismus und Botanik. Rezeption, Kritik und theoretische Alternativen im Deutschland des 19. Jahrhunderts*. Stuttgart: Deutscher Apotheker Verlag.

Kennedy, Jane. 1957. Development of postal rates: 1845–1955. *Land Economics* 33: 93–112.

Kennedy, Joseph. 1962. *A history of Malaya A.D. 1400–1959*. London: Macmillan & Co.

Keynes, Margaret. 1943. *Leonard Darwin, 1850–1943*. Cambridge: privately printed at Cambridge University Press.

Keynes, Richard Darwin, ed. 1988. *Charles Darwin's* Beagle *diary*. Cambridge: Cambridge University Press.

——, ed. 2000. *Charles Darwin's zoology notes & specimen lists from H.M.S. Beagle*. Cambridge: Cambridge University Press.

——. 2002. *Fossils, finches and Fuegians: Charles Darwin's adventures and discoveries on the Beagle, 1832–1836*. London: HarperCollins Publishers.

Kinahan, George Henry. 1867. On cliffs and escarpments. *Geological Magazine* 4: 569–70.

King, William and Rowney, T. H. 1866. On the so-called 'Eozoonal Rock'. *Quarterly Journal of the Geological Society of London* 22: 185–218.

Kingsley, Frances, ed. 1877. *Charles Kingsley: his letters and memories of his life.* 2 vols. London: Henry S. King & Co.

——, ed. 1883. *Charles Kingsley: his letters and memories of his life.* London: Kegan Paul, Trench & Co.

Kölliker, Rudolf Albert von. 1850–4. *Microskopische Anatomie oder Gewebelehre des Menschen.* Vol. 2: *Specielle Gewebelehre* (2 parts). Vol. 1 not published. Leipzig: Wilhelm Engelmann.

——. 1852. *Handbuch der Gewebelehre des Menschen für Aerzte und Studirende.* Leipzig: Wilhelm Engelmann.

Kottler, Malcolm Jay. 1980. Darwin, Wallace, and the origin of sexual dimorphism. *Proceedings of the American Philosophical Society* 124: 203–26.

——. 1985. Charles Darwin and Alfred Russel Wallace: two decades of debate over natural selection. In *The Darwinian heritage*, edited by David Kohn. Princeton: Princeton University Press in association with Nova Pacifica.

Kovalevsky, Alexander Onufrievich. 1865a. Beiträge zur Anatomie und Entwickelungsgeschichte des *Loxosoma Neapolitanum* sp. n. [Read 30 November 1865.] *Mémoires de l'académie Impériale des Sciences de St.-Pétersbourg* 7th ser. 10.2 (1866): 1–10.

——. 1865b. Entwickelungsgeschichte der Rippenquallen. [Read 30 November 1865.] *Mémoires de l'académie Impériale des Sciences de St.-Pétersbourg* 7th ser. 10.4 (1866): i–viii, 1–28.

——. 1866a. Anatomie des *Balanoglossus* delle Chiaje. [Read 11 January 1866.] *Mémoires de l'académie Impériale des Sciences de St.-Pétersbourg* 7th ser. 10.3: 1–18.

——. 1866b. Entwickelungsgeschichte der einfachen Ascidien. [Read 1 November 1866.] *Mémoires de l'académie Impériale des Sciences de St.-Pétersbourg* 7th ser. 10.15: 1–19.

——. 1866c. Beiträge zur Entwickelungsgeschichte der Holothurien. [Read 1 November 1866.] *Mémoires de l'académie Impériale des Sciences de St.-Pétersbourg* 7th ser. 11.6 (1867): 1–8.

——. 1866d. Entwickelungsgeschichte des *Amphioxus lanceolatus.* [Read 20 December 1866.] *Mémoires de l'académie Impériale des Sciences de St.-Pétersbourg* 7th ser. 11.4 (1867): 1–17.

[Kovalevsky, Vladimir Onufrievich], trans. 1866–70. *Illyustrirovannaya zhin zhivotnykh; vseobshchaya istoriya zhivotnago tsarstva.* By Alfred Edmund Brehm. (Russian translation of Brehm *et al.* 1864–9.) 4 vols. in 5. St Petersburg: Kukol-Yasnopolskii.

Kovalevsky, Vladimir Onufrievich, trans. 1868–9. *Proiskhozhdenie vidov. Otdel 1. Izmeneniya zhivotnyk i rastenii vsledstvie pirucheniya. Piruchenniya zhivotniya i vozdelanniya rasteniya.* By Charles Darwin. (Russian translation of *Variation.*) Edited by Ivan Mikhailovich Sechenov, the botanical parts by Alexander Heard. 2 vols. St Petersburg: F. S. Suchchinskii.

Kramp, Paul Lassenius. 1961. *Synopsis of the medusae of the world.* Special volume of *Journal of the Marine Biological Association* 40.

Krauße, Erika. 1987. *Ernst Haeckel.* 2d edition. Leipzig: B. G. Teubner.

Kritsky, Gene. 1991. Darwin's Madagascan hawk moth prediction. *American Entomologist* 37: 206–9.

Krohn, August David. 1860. Beobachtungen über die Entwickelung der Cirripedien. *Archiv für Naturgeschichte* 26: 1–8. [Reprinted in *Annals and Magazine of Natural History* 3d ser. 6 (1860): 423–8.]

Kühn, Alfred. 1950. *Anton Dohrn und die Zoologie seiner Zeit.* Naples: Stazione Zoologica di Napoli.

Kühner, Raphael. 1834–5. *Ausfürliche Grammatik der griechischen Sprache.* 2 vols. Hannover.

Kyle, Robert A. 2001. Henry Bence Jones – physician, chemist, scientist and biographer: a man for all seasons. *British Journal of Haematology* 115: 13–18.

Lamarck, Jean Baptiste Pierre Antoine de Monet de. 1809. *Philosophie zoologique; ou exposition des considérations relatives à l'histoire naturelle des animaux; à la diversité de leur organisation . . . et les autres l'intelligence de ceux qui en sont doués.* 2 vols. Paris: Dentu; the author.

Lamas, Gerardo. 2004. Pieridae. In *Checklist: Part 4A. Hesperioidea–Papilionoidea,* edited by Gerardo Lamas. Gainesville: Association for Tropical Lepidoptera/Scientific Publishers.

Landau, Sidney I., ed. 1986. *International dictionary of medicine and biology.* 3 vols. New York: John Wiley & Sons.

Lane, Edward William, trans. 1839–41. *The thousand and one nights, commonly called, in England, the Arabian nights' entertainments.* 3 vols. London: Charles Knight and Co.

Larousse du XX^{e}siècle. Edited by Paul Augé. 6 vols. Paris: Librairie Larousse. 1928–33.

Larson, James L. 1971. *Reason and experience. The representation of natural order in the work of Carl von Linné.* Berkeley, Los Angeles, and London: University of California Press.

Lavater, Jean-Gaspard. 1820. *L'art de connaître les hommes par la physionomie.* New edition by M. Moreau de la Sarthe. 10 vols. Paris: Depélafol.

Law list: Clarkes' new law list; being a list of the judges and officers of the different courts of justice . . . and a variety of other useful matter. London: W. Clarke. 1816–40. *The law list; comprising the judges and officers of the different courts of justice: counsel, special pleaders, draftsmen, conveyancers, attorneys, notaries, &c., in England and Wales . . . and a variety of other useful matter.* London: V. & R. Stevens & G. S. Norton [and others]. 1841–1969.

Layard, Edgar Leopold. 1862. Notes on the sea-birds observed during a voyage in the Antarctic Ocean. *Ibis* 4: 97–100.

Lecoq, Henri. 1862. *De la fécondation naturelle et artificielle des végétaux et de l'hybridation. Considérée dans ses rapports avec l'horticulture, l'agriculture et la sylviculture. Contenant les moyens pratiques d'opérer l'hybridation et de créer facilement des variétés nouvelles.* 2d edition. Paris: Libraire agricole de la Maison rustique.

Legrand, Jacques-Guillaume and Baltard, Louis Pierre. 1827. *Dissertation sur un traité* de C. le Brun: concernant le rapport de la physionomie humaine avec celle des animaux. London: J. Carpenter.

Leuckart, Rudolf. 1852. *Vesicula prostatica*. In vol. 4 of *The cyclopaedia of anatomy and physiology*, edited by Robert Bentley Todd. London: Longman, Brown, Green, and Longmans.

———. 1859. Über das Vorkommen eines saugnapfartigen Haftapparates bei den Daphniaden und verwandten Krebsen. *Archiv für Naturgeschichte* 25: 262–5.

Lewes, George Henry. 1853. The development hypothesis of the 'Vestiges'. *Leader* (1853): 784–5, 812–14, 832–4, 883–4.

Liddell, Henry George and Scott, Robert, comps. 1996. *A Greek–English lexicon*. Revised and augmented by Henry Stuart Jones with the assistance of Roderick McKenzie. Oxford: Clarendon Press.

Lightman, Bernard, ed. 2004. *Dictionary of nineteenth-century British scientists*. 4 vols. Bristol: Thoemmes Press.

Lindley, John. 1853. *The vegetable kingdom; or, the structure, classification, and uses of plants, illustrated upon the natural system*. 3d edition with corrections and additional genera. London: Bradbury & Evans.

Linnaeus, Carolus. 1788–93. *Systema naturæ*. Edited by Johann Friedrich Gmelin. 13th edition. 3 vols. in 10. Leipzig: Georg Emanuel Beer.

List of the Geological Society of London. London: [Geological Society of London]. 1864–1934.

List of the Linnean Society of London. London: [Linnean Society of London]. 1805–1939.

Lister, Joseph. 1853. Observations on the muscular tissue of the skin. *Quarterly Journal of Microscopical Science* 1: 262–8.

———. 1856. On the minute structure of involuntary muscular fibre. [Read 1 December 1856 before the Royal Society of Edinburgh.] *Quarterly Journal of Microscopical Science* 6 (1858): 5–14.

Living Cirripedia (1851): *A monograph of the sub-class Cirripedia, with figures of all the species. The Lepadidæ; or, pedunculated cirripedes*. By Charles Darwin. London: Ray Society. 1851.

Living Cirripedia (1854): *A monograph of the sub-class Cirripedia, with figures of all the species. The Balanidæ (or sessile cirripedes); the Verrucidæ, etc*. By Charles Darwin. London: Ray Society. 1854.

Lord, John Keast. 1866. *The naturalist in Vancouver Island and British Columbia*. 2 vols. London: Richard Bentley.

Loudon, John Claudius. 1841. *An encyclopædia of plants*. London: Longman, Orme, Brown, Green, and Longmans.

Lowe, Richard Thomas. 1868. *A manual flora of Madeira and the adjacent islands of Porto Santo and the Desertas*. Vol. 1, *Dichlamydeæ*. London: John van Voorst.

Lubbock, John. 1864. Notes on *Sphaerularia Bombi*. *Natural History Review* n.s. 4: 265–70.

———. 1865. *Pre-historic times, as illustrated by ancient remains, and the manners and customs of modern savages*. London and Edinburgh: Williams & Norgate.

———. 1866. On *Pauropus*, a new type of centipede. [Read 6 December 1866.] *Journal of the Linnean Society (Zoology)* 9 (1867): 179–80.

———. 1867. On the origin of civilisation and the primitive condition of man. [Read 26 November 1867.] *Transactions of the Ethnological Society of London* n.s. 6: 328–41.

———. 1870. *The origin of civilisation and the primitive condition of man.* London: Longmans, Green, and Co.

Lurie, Edward. 1960. *Louis Agassiz: a life in science.* Chicago: University of Chicago Press.

Lydekker, Richard. 1898. *Wild oxen, sheep and goats of all lands, living and extinct.* London: Rowland Ward.

Lyell, Charles. 1830–3. *Principles of geology, being an attempt to explain the former changes of the earth's surface, by reference to causes now in operation.* 3 vols. London: John Murray.

———. 1851. *A manual of elementary geology; or, the ancient changes of the earth and its inhabitants as illustrated by geological monuments.* 3d edition, revised. London: J. Murray.

———. 1853. *Principles of geology; or, the modern changes of the earth and its inhabitants considered as illustrative of geology.* 9th edition, entirely revised. London: John Murray.

———. 1858. On the structure of lavas which have consolidated on steep slopes; with remarks on the mode of origin of Mount Etna, and on the theory of 'craters of elevation'. [Read 10 June 1858.] *Philosophical Transactions of the Royal Society of London* 148: 703–86.

———. 1863. *The geological evidences of the antiquity of man with remarks on theories of the origin of species by variation.* London: John Murray.

———. 1865. *Elements of geology, or the ancient changes of the earth and its inhabitants as illustrated by geological monuments.* 6th edition, revised. London: John Murray.

———. 1867–8. *Principles of geology, or the modern changes of the earth and its inhabitants considered as illustrative of geology.* 10th edition. 2 vols. London: John Murray.

Lyell, Katherine Murray, ed. 1881. *Life, letters and journals of Sir Charles Lyell, Bart.* 2 vols. London: John Murray.

Mabberley, David J. 1997. *The plant-book. A portable dictionary of the vascular plants.* 2d edition. Cambridge: Cambridge University Press.

McCracken, Donal P. 1997. *Gardens of empire: botanical institutions of the Victorian British empire.* London: Leicester University Press.

Macdonald, Frederick C. 1929. *Bishop Stirling of the Falklands. The adventurous life of a soldier of the cross whose humility hid the daring spirit of a hero & an inflexible will to face great risks.* London: Seely, Service & Co.

MacKintosh, Daniel. 1865a. Marine denudation illustrated by the Brimham rocks. *Geological Magazine* 2: 154–8.

———. 1865b. A tourist's notes on the surface-geology of the Lake-district. *Geological Magazine* 2: 299–306.

———. 1866a. The sea against rain and frost; or the origin of escarpments. *Geological Magazine* 3: 63–70.

———. 1866b. The sea against rivers: on the origin of valleys. *Geological Magazine* 3: 155–60.

MacKintosh, Daniel. 1866c. Denudation.— Reply to Mr G. Poulett Scrope and Mr J. B. Jukes. *Geological Magazine* 3: 280–2.

———. 1866d. Results of observations on the cliffs, gorges, and valleys of Wales. *Geological Magazine* 3: 387–98.

———. 1867a. Mr Maw, Professor Jukes, and others on denudation. *Geological Magazine* 4: 136–9.

———. 1867b. Pholas-borings, denudation, and deposition in S.E. Devon. *Geological Magazine* 4: 295–9.

———. 1867c. Lyell, Jukes, and Whitaker on surface-geology. *Geological Magazine* 4: 571–5.

———. 1869. *The scenery of England and Wales: its character and origin*. London: Longmans, Green, and Co.

McLennan, John Ferguson. 1865. *Primitive marriage: an inquiry into the origin of the form of capture in marriage ceremonies*. Edinburgh: Adam and Charles Black.

———. 1970. *Primitive marriage: an inquiry into the origin of the form of capture in marriage ceremonies*. Edited and with an introduction by Peter Rivière. Reprint of the first edition. Chicago: University of Chicago Press.

Macmillan dictionary of women's biography: *The Macmillan dictionary of women's biography*. 2d edition. Compiled and edited by Jennifer Uglow, assisted (1st edition) by Frances Hinton. London and Basingstoke, Hampshire: Macmillan Press. 1989.

Maillet, Benoît de. 1750. *Telliamed: or, discourses between an Indian philosopher and a French missionary, on the diminution of the sea, the formation of the earth, the origin of men and animals, and other curious subjects, relating to natural history and philosophy*. Translated from the French. London: T. Osborne.

Mainardi, Patricia. 1987. *Art and politics of the Second Empire: the universal expositions of 1855 and 1867*. New Haven, Conn., and London: Yale University Press.

Mangelsdorf, Paul C. 1974. *Corn. Its origin, evolution, and improvement*. Cambridge, Massachusetts: Belknap Press of Harvard University Press.

Marchant, James, ed. 1916. *Alfred Russel Wallace. Letters and reminiscences*. 2 vols. London: Cassell and Company.

Marginalia: Charles Darwin's marginalia. Edited by Mario A. Di Gregorio with the assistance of Nicholas W. Gill. Vol. 1. New York and London: Garland Publishing. 1990.

Mariager, P., trans. 1871. *Fuglenes liv, populairt fremstillet*. By Alfred Edmund Brehm. Copenhagen.

———, trans. 1873. *Pattedyrenes liv, populairt fremstillet*. By Alfred Edmund Brehm. Copenhagen: P. G. Philipsen.

Mariager, P. and Feddersen, Arthur, trans. 1879. *Krybdyrenes og paddernes, fiskenes samt de lavere dyrs liv*. By Alfred Edmund Brehm. Copenhagen: P. G. Philipsen.

[Marsh-Caldwell, Anne.] 1845. *Mount Sorel; or the heiress of the De Veres*. 2 vols. London: Chapman and Hall.

Martin, Gerald P. R. and Uschmann, Georg. 1969. *Friedrich Rolle 1827–1887, ein Vorkämpfer neuen biologischen Denkens in Deutschland*. Leipzig: Johann Ambrosius Barth.

Martin, Joel W. and Davis, George E. 2001. *An updated classification of the recent Crustacea*. Science series 39. Los Angeles: Natural History Museum of Los Angeles County.

Mathew, William M. 1981. *The house of Gibbs and the Peruvian guano monopoly*. London: Royal Historical Society.

Mätzner, Eduard. 1860–5. *Englische Grammatik*. 3 vols. Berlin: Weidmannsche Buchhandlung.

Mayerhöfer, Josef. 1959–70. *Lexikon der Geschichte der Naturwissenschaften: Biographien, Sachwörter und Bibliographien*. Vol. 1. Vienna: Verlag Brüder Hollinek.

Mayr, Ernst. 1972. Sexual selection and natural selection. In *Sexual selection and the descent of man, 1871–1971*, edited by Bernard Campbell. London: Heinemann.

Medical directory: *The London medical directory . . . every physician, surgeon, and general practitioner resident in London*. London: C. Mitchell. 1845. *The London and provincial medical directory*. London: John Churchill. 1848–60. *The London & provincial medical directory, inclusive of the medical directory for Scotland, and the medical directory for Ireland, and general medical register*. London: John Churchill. 1861–9. *The medical directory . . . including the London and provincial medical directory, the medical directory for Scotland, the medical directory for Ireland*. London: J. & A. Churchill. 1870–1905.

Medical who's who: *The medical who's who*. London: London & Counties Press Association [and others]. 1913–27.

Meliss, John Charles. 1875. *St Helena: a physical, historical, and topographical description of the island*. London: L. Reeve & Co.

Men and women of the time: *The men of the time in 1852 or sketches of living notables*. 2d edition, 1853. 3d edition, 1856. 4th edition, 1857. New edition, 1865. 7th edition, 1868. 8th edition, 1872. 9th edition, 1875. 10th edition, 1879. 11th edition, 1884. 12th edition, 1887. 13th edition, 1891. 14th edition, 1895. *Men and women of the time: a dictionary*. 15th edition. By Victor G. Plarr. 1899. London: David Bogue [and others]. 1852–99.

Men-at-the-bar: *Men-at-the-bar: a biographical hand-list of the members of the various inns of court, including Her Majesty's judges, etc.* By Joseph Foster. London: Reeves & Turner. 1885.

Meyen, Franz Julius Ferdinand. 1834. Über das Leuchten des Meeres und Beschreibung einiger Polypen und anderer niederer Thiere. [Read 7 January 1834.] *Nova Acta Physico-Medica Academiae Caesareae Leopoldino-Carolinae Naturae Curiosorum* 16 (suppl. 1): 125–58.

Micklethwait, David. 2000. *Noah Webster and the American dictionary*. Jefferson, N.C., and London: McFarland & Company.

Migotto, Alvaro E., *et al.* 2002. Checklist of the Cnidaria Medusozoa of Brazil. *Biota Neotropica* 2: 1–31.

Mivart, St George Jackson. 1867. On the appendicular skeleton of the primates. [Read 10 January 1867.] *Philosophical Transactions of the Royal Society of London* 157: 299–429.

ML: *More letters of Charles Darwin: a record of his work in a series of hitherto unpublished letters*. Edited by Francis Darwin and Albert Charles Seward. 2 vols. London: John Murray. 1903.

Modern English biography: *Modern English biography, containing many thousand concise memoirs of persons who have died since the year 1850*. By Frederick Boase. 3 vols. and supplement (3 vols.). Truro, Cornwall: printed for the author. 1892–1921.

Moggridge, John Traherne. 1864. Observations on some orchids of the south of France. [Read 3 November 1864.] *Journal of the Linnean Society (Botany)* 8 (1865): 256–8.

Möller, Alfred, ed. 1915–21. *Fritz Müller. Werke, Briefe und Leben*. 3 vols in 5. Jena: Gustav Fischer.

Montagu, Jennifer. 1994. *The expression of the passions: the origin and influence of Charles Le Brun's* Conférence sur l'expression générale et particulière. New Haven, Conn., and London: Yale University Press.

Montgomery, William M. 1988. Germany. In *The comparative reception of Darwinism*, with a new preface, edited by Thomas F. Glick. Chicago and London: University of Chicago Press.

Moore, James Richard. 1977. On the education of Darwin's sons: the correspondence between Charles Darwin and the Reverend G. V. Reed, 1857–1864. *Notes and Records of the Royal Society* 32 (1977–8): 51–70.

———. 1985. Darwin of Down: the evolutionist as squarson-naturalist. In *The Darwinian heritage*, edited by David Kohn. Princeton, N.J.: Princeton University Press in association with Nova Pacifica.

Moore, Raymond Cecil and McCormick, Lavon. 1969. General features of Crustacea. In Pt. R, *Arthropoda 4.*, of *Treatise on invertebrate paleontology*, directed and edited by Raymond C. Moore; revisions and supplements directed and edited by Curt Teichert. Boulder, Colo.: Geological Society of America. Lawrence, Kan.: University of Kansas.

Morelet, Arthur. 1860. *Iles Açores. Notice sur l'histoire naturelle des Açores, suivie d'une description des mollusques terrestres de cet archipel*. Paris: J.-B. Baillière et fils.

Morris, Susan W. 1994. Fleeming Jenkin and 'The origin of species': a reassessment. *British Journal for the History of Science* 27: 313–43.

Morton, Samuel George. 1839. *Crania americana; or a comparative view of the skulls of various aboriginal nations of North and South America: to which is prefixed an essay on the varieties of the human species*. Philadelphia and London: J. Dobson, Simpkin, Marshall & Co.

———. 1846. Some observations on the ethnography and archæology of the American aborigines. *American Journal of Science and Arts* 2d ser. 2: 1–17.

Morton, Vanda. 1987. *Oxford rebels. The life and friends of Nevil Story Maskelyne, 1823–1911, pioneer Oxford scientist, photographer and politician*. Gloucester: Alan Sutton.

Moulinié, Jean Jacques. 1856. *De la reproduction chez les trématodes endo-parasites*. Vol. 3 of *Memoires de l'Institut National Genevois*. Geneva: Kessman.

———, trans. 1865. *Leçons sur l'homme*. By Carl Vogt. Paris: Reinwald.

———, trans. 1868. *De la variation des animaux et des plantes sous l'action de la domestication*.

By Charles Darwin. Preface by Carl Vogt. Paris: C. Reinwald.

Movement in plants: *The power of movement in plants*. By Charles Darwin. London: John Murray. 1880.

Müller, Fritz. 1864. *Für Darwin*. Leipzig: Wilhelm Engelmann.

———. 1865. Notes on some of the climbing-plants near Desterro, in south Brazil. By Herr Fritz Müller, in a letter to C. Darwin. [Read 7 December 1865.] *Journal of the Linnean Society (Botany)* 9 (1867): 344–9.

———. 1868a. Notizen über die Geschlechtsverhältnisse brasilianischer Pflanzen. *Botanische Zeitung* 26: 113–16.

———. 1868b. Ueber Befruchtungserscheinungen bei Orchideen. Aus einem Briefe an Friedrich Hildebrand. *Botanische Zeitung* 26: 629–31.

———. 1869. Ueber einige Befruchtungserscheinungen. Aus einem Briefe an F. Hildebrand. *Botanische Zeitung* 27: 224–6.

———. 1870. Umwandlung von Staubgefässen in Stempel bei Begonia. Uebergang von Switterblüthigkeit in Getrenntblüthigkeit bei Chamissoa. Triandrische Varietät eines monandrischen Epidendrum. Aus einem Briefe an H. Müller. *Botanische Zeitung* 28: 149–53

———. 1881. Farbenwechsel bei Krabben und Garneelen. *Kosmos* 8: 472–3.

Müller, Hermann. 1866a. Nachträge zur Geographie der in Westfalen beobachteten Laubmoose (bis zum 1 November 1865). *Verhandlungen des botanischen Vereins der Provinz Brandenburg* 8: 36–41.

———. 1866b. Thatsachen der Laubmooskunde für Darwin. *Verhandlungen des botanischen Vereins der Provinz Brandenburg* 8: 41–65.

———. 1868. Beobachtungen an westfälischen Orchideeen. *Verhandlungen des naturhistorischen Vereines der preussischen Rheinlande und Westphalens (Botanik)*. 25: 1–62.

———. 1869. Die Anwendung der Darwin'schen Lehre auf Blumen und blumenbesuchende Insekten. *Verhandlungen des naturhistorischen Vereines der preussischen Rheinlande und Westphalens* (Botanik, Correspondenzblatt) 26: 43–66.

———. 1873. *Die Befruchtung der Blumen durch Insekten und die gegenseitigen Anpassungen beider. Ein Beitrag zur Erkenntniss des ursächlichen Zusammenhanges in der organischen Natur*. Leipzig: Wilhelm Engelmann.

Murie, James. 1867. Remarks on an antelope from the White Nile, allied to or identical with the *Kobus sing-sing* of Gray. [Read 10 January 1867.] *Proceedings of the Zoological Society of London* (1867): 3–8.

Murray, Andrew. 1866. *The geographical distribution of mammals*. London: Day and Son.

———. 1867. Dr. Hooker on insular floras. *Gardeners' Chronicle* (1867): 152, 181–2.

———. 1868. Delpino on the apparatus for fecundating phanerogamous plants. *Journal of Travel and Natural History* 1: 181–5.

Murray, John. 1908–9. Darwin and his publisher. *Science Progress in the Twentieth Century* 3: 537–42.

Narrative: *Narrative of the surveying voyages of His Majesty's ships Adventure and Beagle, between the years 1826 and 1836*. [Edited by Robert FitzRoy.] 3 vols. and appendix. London: Henry Colburn. 1839.

Nathusius, Hermann von. 1864. *Vorstudien für Geschichte und Zucht der Hausthiere zunächst am Schweineschädel.* 1 vol. and atlas. Berlin: Wiegandt und Hempel.

Natural selection: Charles Darwin's Natural selection: being the second part of his big species book written from 1856 to 1858. Edited by R. C. Stauffer. Cambridge: Cambridge University Press. 1975.

Navy list: The navy list. London: John Murray; Her Majesty's Stationery Office. 1815–1900.

NBU: Nouvelle biographie universelle depuis les temps les plus reculés jusqu'à nos jours, avec les renseignements bibliographiques et l'indication des sources a consulter. Edited by Jean Chrétien Ferdinand Hoefer. 46 vols. in 23. Paris: Firmin Didot Frères. 1852–66.

NDB: Neue deutsche Biographie. Under the auspices of the Historical Commission of the Bavarian Academy of Sciences. 21 vols. (A–Rohlfs) to date. Berlin: Duncker & Humblot. 1953–.

Neave, S. A., *et al.* 1933. *The history of the Entomological Society of London. 1833–1933.* London: n.p.

Nees von Esenbeck, Theodor Friedrich Ludwig. 1835[–61]. *Genera plantarum florae Germanicae iconibus et descriptionibus illustrata.* 6 vols. Bonn: Henry et Cohen.

Neighbour, Alfred. 1865. *The apiary; or, bees, bee-hives, and bee culture: being a familiar account of the habits of bees, and the most improved methods of management, with full directions, adapted for the cottager, farmer, or scientific apiarian.* London: Kent & Co.; Geo. Neighbour & Sons.

Neumeister, Gottlob. 1837. *Das Ganze der Taubenzucht.* Weimar: B. F. Voigt.

Nevill, Geoffrey. 1868. Notes on some of the species of land Mollusca inhabiting Mauritius and the Seychelles. [Read 23 April 1868.] *Proceedings of the Zoological Society of London* (1868): 257–61.

——. 1869. Additional notes on the land-shells of the Seychelles islands. [Read 28 January 1869.] *Proceedings of the Zoological Society of London* (1869): 61–6.

Newman, William. 1993. Darwin and cirripedology. *History of Carcinology. Crustacean Issues* 8: 349–434.

Newton, Edward. 1867a. Descriptions of some new species of birds from the Seychelles islands. [Read 28 March 1867.] *Proceedings of the Zoological Society of London* (1867): 344–7, 821.

——. 1867b. On the land-birds of the Seychelles archipelago. *Ibis* n.s. 3: 335–60.

Nicols, Arthur. 1882. *The acclimatisation of the Salmonidae at the Antipodes: its history and results.* London: Sampson Low, Marston, Searle, & Rivington.

Notebooks: Charles Darwin's notebooks, 1836–1844. Geology, transmutation of species, metaphysical enquiries. Transcribed and edited by Paul H. Barrett *et al.* Cambridge: Cambridge University Press for the British Museum (Natural History). 1987.

Novara expedition. 1861–75. *Reise der Österreichischen Fregatte Novara um die Erde in den Jahren 1857, 1858, 1859 unter den Befehlen des Commodore B. von Wüllerstorf-Urbair.* 9 vols. in 19. Vienna: Kaiserlich-Königlichen Hof- und Staatsdruckerei.

Nowak, Ronald M. 1999. *Walker's mammals of the world.* 6th edition. 2 vols. Baltimore and London: The Johns Hopkins University Press.

NUC: *The national union catalog. Pre-1956 imprints.* 685 vols. and supplement (69 vols.). London and Chicago: Mansell. 1968–81.

Nunn, John. 1850. *Narrative of the wreck of the 'Favorite' on the island of Desolation: detailing the adventures, sufferings, and privations of John Nunn.* Edited by W. B. Clarke. London: William Edward Painter.

Nyhart, Lynn K. 1995. *Biology takes form. Animal morphology and the German universities, 1800–1900.* Chicago and London: University of Chicago Press.

OBL: *Österreichisches biographisches Lexikon 1815–1950.* Edited by Leo Santifaller *et al.* 11 vols. and 3 fascicles of vol. 12 (A–Slavik Ernst) to date. Vienna: Österreichischen Akademie der Wissenschaften. 1957–.

ODNB: *Oxford dictionary of national biography: from the earliest times to the year 2000.* [Revised edition.] Edited by H. C. G. Matthew and Brian Harrison. 60 vols. and index. Oxford: Oxford University Press. 2004.

OED: *The Oxford English dictionary. Being a corrected re-issue with an introduction, supplement and bibliography of a new English dictionary.* Edited by James A. H. Murray, *et al.* 12 vols. and supplement. Oxford: Clarendon Press. 1970. *A supplement to the Oxford English dictionary.* 4 vols. Edited by R. W. Burchfield. Oxford: Clarendon Press. 1972–86. *The Oxford English dictionary.* 2d edition. 20 vols. Prepared by J. A. Simpson and E. S. C. Weiner. Oxford: Clarendon Press. 1989. *Oxford English dictionary additional series.* 3 vols. Edited by John Simpson *et al.* Oxford: Clarendon Press. 1993–7.

Olby, Robert. 1963. Charles Darwin's manuscript of pangenesis. *British Journal for the History of Science* 1: 251–63.

Oldroyd, David. 1999. Early ideas about glaciation in the English Lake District: the problem of making sense of glaciation in a glaciated region. *Annals of science* 56: 175–203.

O'Leary, Patrick. 1989. *Sir James Mackintosh: the Whig Cicero.* Aberdeen: Aberdeen University Press.

Oliver, William L. R. ed. 1993. *Pigs, peccaries, and hippos: status survey and conservation action plan.* Gland, Switzerland: IUCN.

Oppenheim, Janet. 1985. *The other world: spiritualism and psychic research in England, 1850–1914.* Cambridge: Cambridge University Press.

Orchids: *On the various contrivances by which British and foreign orchids are fertilised by insects, and on the good effects of intercrossing.* By Charles Darwin. London: John Murray. 1862.

Orchids 2d ed.: *The various contrivances by which British and foreign orchids are fertilised by insects, and the good effects of intercrossing.* By Charles Darwin. Revised. London: John Murray. 1877.

Origin: *On the origin of species by means of natural selection, or the preservation of favoured races in the struggle for life.* By Charles Darwin. London: John Murray. 1859.

Origin 2d ed.: *On the origin of species by means of natural selection, or the preservation of favoured races in the struggle for life.* By Charles Darwin. London: John Murray. 1860.

Origin 3d ed.: *On the origin of species by means of natural selection, or the preservation of favoured races in the struggle for life*. With additions and corrections. By Charles Darwin. London: John Murray. 1861.

Origin 4th ed.: *On the origin of species by means of natural selection, or the preservation of favoured races in the struggle for life*. With additions and corrections. By Charles Darwin. London: John Murray. 1866.

Origin 5th ed.: *On the origin of species by means of natural selection, or the preservation of favoured races in the struggle for life*. With additions and corrections. By Charles Darwin. London: John Murray. 1869.

Origin 6th ed.: *The origin of species by means of natural selection, or the preservation of favoured races in the struggle for life*. With additions and corrections. By Charles Darwin. London: John Murray. 1872.

Ornithological notes: Darwin's ornithological notes. Edited by Nora Barlow. *Bulletin of the British Museum (Natural History)*. Historical Series 2 (1959–63): 203–78.

Ovsjannikov, Philipp Vasilyevich and Kovalevsky, Alexander Onufrievich. 1866. Über das Centralnervensystem und das Gehörorgan der Cephalopoden. [Read 22 March 1866.] *Mémoires de l'académie Impériale des Sciences de St.-Pétersbourg* 7th ser. 11.3 (1867): 1–36.

Owen, Alex. 1989. *The darkened room: women, power and spiritualism in late Victorian England*. London: Virago Press.

[Owen, Richard.] 1860. [Review of *Origin* & other works.] *Edinburgh Review* 111: 487–532.

Owen, Richard. 1866. On the osteology of the Dodo (*Didus ineptus* Linn.). [Read 9 January 1866.] *Transactions of the Zoological Society of London* 6 (1869): 49–85.

———. 1866–8. *On the anatomy of vertebrates*. 3 vols. London: Longmans, Green & Co.

Oxford classical dictionary 3d edition, revised. Edited by Simon Hornblower and Anthony Spawforth. Oxford: Oxford University Press. 2003.

Oxford University calendar. Oxford: J. H. Parker *et al.*; Oxford University Press. 1810–.

Pabst, Guido F. J. and Dungs, F. 1975–7. *Orchidaceae Brasilienses*. 2 vols. Hildesheim: Brucke-Verlag Kurt Schmersow.

Paget, John. 1839. *Hungary and Transylvania; with remarks on their condition, social, political, and economical*. 2 vols. London: John Murray.

Paget, Stephen, ed. 1901. *Memoirs and letters of Sir James Paget*. London: Longmans, Green, and Co.

Pancaldi, Giuliano. 1991. *Darwin in Italy. Science across cultural frontiers*. Translated by Ruey Brodine Morelli. Updated and expanded edition. Bloomington and Indianapolis, Ind.: Indiana University Press.

'Parallel roads of Glen Roy': Observations on the parallel roads of Glen Roy, and of other parts of Lochaber in Scotland, with an attempt to prove that they are of marine origin. By Charles Darwin. [Read 7 February 1839.] *Philosophical Transactions of the Royal Society of London* (1839), pt 1: 39–81. [*Collected papers* 1: 89–137.]

[Parker, Henry.] 1862. The *Edinburgh Review* on the supernatural. *Saturday Review*, 15 November 1862, pp. 589–90.

Passerini, Giovanni. 1860. *Gli afidi con un prospetto dei generi ed alcune specie nuove italiane.* Parma: Tipografia Carmigani.

——. 1862–3. Aphididae italicae hucusque observatae. *Archivio per la zoologia, l'anatomia e la fisiologia* 2: 129–212.

Payne, Robert. 1986. *The white rajahs of Sarawak.* Singapore: Oxford University Press.

Peckham, Morse, ed. 1959. *The Origin of Species by Charles Darwin: a variorum text.* Philadelphia: University of Pennsylvania Press.

Penzig, Ottone. 1905. Commemorazione di Federico Delpino. *Malpighia* 19: 294–310.

Peters, James L., *et al.* 1931–87. *Check-list of birds of the world.* 16 vols. Cambridge, Mass.: Harvard University Press and Museum of Comparative Zoology.

Peterson, Rudolph Frederick. 1965. *Wheat: botany, cultivation and utilization.* London: L. Hill Books. New York: Interscience Publishers.

Petteys, Chris. 1985. *Dictionary of women artists. An international dictionary of women artists born before 1900.* Boston, Mass.: G. K. Hall & Co.

Physicians: The roll of the Royal College of Physicians of London. By William Munk. 2d edition, revised and enlarged. 3 vols. London: Royal College of Physicians. 1878. *Lives of the fellows of the Royal College of Physicians of London.* Compiled by G. H. Brown *et al.* 5 vols. London: Royal College of Physicians. Oxford and Washington, D.C.: IRL Press. 1955–89.

'*Planariae*': Brief descriptions of several terrestrial *Planariae*, and of some remarkable marine species, with an account of their habits. By Charles Darwin. *Annals and Magazine of Natural History* 14 (1844): 241–51. [*Collected papers* 1: 182–93.]

Plant, Marjorie. 1965. *The English book trade. An economic history of the making and sale of books.* 2d edition. London: George Allen & Unwin.

Plarr, Victor Gustave. 1930. *Plarr's lives of the fellows of the Royal College of Surgeons of England.* Revised by Sir D'Arcy Power. 2 vols. London: Simpkin Marshall.

Post Office directory of Gloucestershire, with Bath, Bristol, Herefordshire, and Shropshire. London: Kelly & Co. 1856–79.

Post Office directory of the six home counties: Post Office directory of the six home counties, viz., Essex, Herts, Kent, Middlesex, Surrey and Sussex. London: W. Kelly & Co. 1845–78.

Post Office Edinburgh directory: Post-Office annual directory and calendar. Post-Office Edinburgh and Leith directory. Edinburgh: Ballantyne & Hughes [and others]. 1845–1908.

Post Office London directory: Post-Office annual directory. . . . A list of the principal merchants, traders of eminence, &c. in the cities of London and Westminster, the borough of Southwark, and parts adjacent . . . general and special information relating to the Post Office. Post Office London directory. London: His Majesty's Postmaster-General [and others]. 1802–1967.

Post Office London suburban directory: The Post Office London suburban directory. Kelly's London suburban directory. London: Kelly & Co. 1860–1903.

Poulton, Edward Bagnall. 1901. The influence of Darwin upon entomology. *Entomologist's Record* 13: 72–6.

——. 1909. *Charles Darwin and the origin of species. Addresses, etc., in America and England in the year of the two anniversaries*. London: Longmans, Green & Co.

Poulton, Edward Bagnall. 1920. Roland Trimen, 1840–1916. *Proceedings of the Royal Society of London, Series B* 91: xviii–xxvii.

Pridgeon, Alec M., *et al.*, eds. 1999–2003. *Genera orchidacearum*. 3 vols. Oxford: Oxford University Press.

Pritchard, Michael. 1994. *A directory of London photographers 1841–1908*. Revised and expanded edition. Watford, Hertfordshire: PhotoResearch.

Quatrefages de Bréau, Jean Louis Armand de. 1867. [Review of *Mémoire sur les microcéphales ou hommes-singes* by C. Vogt.] *Comptes rendus hebdomadaires des séances de l'académie des sciences* 64: 1226–31.

Qureshi, Sadiah. 2004. Displaying Sara Bartmann, the 'Hottentot Venus'. *History of Science* 42: 233–57.

Raby, Peter. 2001. *Alfred Russel Wallace: a life*. London: Chatto & Windus.

Rachinskii, Sergei A., trans. 1864. *Proiskhozhdenie vidov putem estestvennogo podbora*. [*Origin*.] By Charles Darwin. St Petersburg: A. I. Glazunov.

——, trans. 1865. *Proiskhozhdenie vidov putem estestvennogo podbora*. [*Origin*.] By Charles Darwin. 2d edition, revised. St Petersburg: A. I. Glazunov.

Record of the Royal Society of London: *The record of the Royal Society of London for the promotion of natural knowledge*. 4th edition. London: Royal Society. 1940.

Reed, A. H., ed. 1935. *Early Maoriland adventures of J. W. Stack*. Dunedin and Wellington, New Zealand: A. H. and A. W. Reed.

Reeks, Margaret. 1920. *Register of the associates and old students of the Royal School of Mines and history of the Royal School of Mines*. London: Royal School of Mines (Old Students') Association.

Reichenbach, Heinrich Gustav. 1852. *De pollinis orchidearum genesi ac structura et de orchideis in artem ac systema redigendis*. Leipzig: F. Hofmeister.

Rengger, Johann Rudolph. 1830. *Naturgeschichte der Saeugethiere von Paraguay*. Basel, Switzerland: Schweighausersche Buchhandlung.

Reptiles. Pt 5 of *The zoology of the voyage of HMS Beagle*. By Thomas Bell. Edited and superintended by Charles Darwin. London: Smith, Elder & Co. 1843.

Repton School register: *Repton School register 1557–1910*. Edited for the O. R. Society by the widow of G. S. Messiter. Repton, Derbyshire: A. J. Lawrence. 1910.

'Review of Bates on mimetic butterflies': [Review of 'Contributions to an insect fauna of the Amazon valley', by Henry Walter Bates.] [By Charles Darwin.] *Natural History Review* n.s. 3 (1863): 219–24. [*Collected papers* 2: 87–92.]

Richard, Adolphe. 1852. Essai sur l'anatomie philosophique et l'interprétation de quelques anomalies musculaires du membre thoracique dans l'espèce humaine. [Read 26 January 1852.] *Annales des sciences naturelles (zoologie)* 3d ser. 18: 5–20.

Richardson, Edmund William. 1916. *A veteran naturalist; being the life and work of W. B. Tegetmeier*. London: Witherby & Co.

Rohrbach, Paul. 1866. *Über den Blüthenbau und die Befruchtung von* Epipogium Gmelini. Göttingen: Universitäts-Buchdruckerei von E. A. Huth.

Rolle, Friedrich. 1863. *Chs. Darwin's Lehre von der Entstehung der Arten im Pflanzen- und Thierreich in ihrer Anwendung auf die Schöpfungsgeschichte*. Frankfurt: J. C. Hermann.

Ross, James Clark. 1847. *A voyage of discovery and research in the southern and Antarctic regions, during the years 1839–43*. 2 vols. London: John Murray.

Rossmässler, Emil Adolf. 1863. *Der Wald. Den Freunden und Pflegern des Waldes geschildert*. Leipzig and Heidelberg: C. F. Winter.

Royal Society catalogue of scientific papers: *Catalogue of scientific papers (1800–1900)*. Compiled and published by the Royal Society of London. 19 vols. and index (3 vols.). London: Royal Society of London. Cambridge: Cambridge University Press. 1867–1925.

Royer, Clémence Auguste, trans. 1862. *De l'origine des espèces ou des lois du progrès chez les êtres organisés*. By Charles Darwin. With preface and notes by the translator. [Translated from the 3d English edition.] Paris: Guillaumin & Cie; Victor Masson.

——, trans. 1866. *De l'origine des espèces par sélection naturelle ou des lois de transformation des êtres organisés*. By Charles Darwin. With preface and notes by the translator. 2d edition. Augmented in accordance with notes by the author. Paris: Guillaumin et Cie; Victor Masson et fils.

Rudolphi, Carl Asmund. 1812. *Beyträge zur Anthropologie und allgemeinen Naturgeschichte*. Berlin: Haude & Speuer.

Rudwick, Martin John Spencer. 1974. Darwin and Glen Roy: a 'great failure' in scientific method? *Studies in History and Philosophy of Science* 5: 97–185.

Rupke, Nicolaas A. 1994. *Richard Owen, Victorian naturalist*. New Haven, Conn., and London: Yale University Press.

Rüppell, Eduard. 1835–40. *Neue Wirbelthiere zu der Fauna von Abyssinien gehörig*. 4 vol. in 2. Frankfurt am Main: S. Schmerber.

Rütimeyer, Karl Ludwig. 1861. *Die Fauna der Pfahlbauten in der Schweiz. Untersuchungen über die Geschichte der wilden und der Haus-Säugethiere von Mittel-Europa*. Basel, Switzerland: Bahnmaier's Buchhandlung (C. Detloff).

——. 1862. Eocaene Säugethiere aus dem Gebiet des Schweizerischen Jura. *Neue Denkschriften der allgemeinen Schweizerischen Gesellschaft für die gesammten Naturwissenschaften* 19: 1–98.

——. 1863. *Beiträge zur Kenntniss der fossilen Pferde und zu einer vergleichenden Odontographie der Hufthiere im Allgemeinen*. Basel, Switzerland: Schweighauserische Buchdruckerei.

——. 1865a. Beiträge zu einer palæontologischen Geschichte der Wiederkauer, zunächst an Linné's Genus Bos. *Verhandlungen der Naturforschenden Gesellschaft in Basel* 4 (1867): 299–354.

——. 1865b. *Ueber die Aufgabe der Naturgeschichte. Eine Rectorats-rede*. [Read November 1865, Basel University.] N.p.: n.p.

——. 1866. Über Art und Raçe des zahmen europäischen Rindes. *Archiv für Anthropologie* 1: 219–50.

——. 1867a. Versuch einer natürlichen Geschichte des Rindes in seinen Beziehungen zu den Wiederkauern im Allgemeinen. Erste Abtheilung. *Neue Denkschriften der allgemeinen schweizerischen Gesellschaft für die gesammten Naturwissenschaften* 22: 1–103.

Rütimeyer, Karl Ludwig. 1867b. Versuch einer natürlichen Geschichte des Rindes in seinen Beziehungen zu den Wiederkauern im Allgemeinen. Zweite Abtheilung. *Neue Denkschriften der allgemeinen schweizerischen Gesellschaft für die gesammten Naturwissenschaften* 22: 1–175.

——. 1867c. *Ueber die Herkunft unserer Thierwelt. Eine zoographische Skizze.* Basel and Geneva: H. Georg's Verlagsbuchhandlung.

——. 1867d. Palæontologie. Neue Beiträge zür Kenntnis des Torfschweins. *Verhandlungen der naturforschenden Gesellschaft in Basel* 4: 139–76.

Salter, John William. 1864. On some new fossils from the Lingula-flags of Wales. [Read 23 March 1864.] *Quarterly Journal of the Geological Society* 20: 233–41.

Salter, John William and Henry Hicks. 1865. On some additional fossils from the Lingula-flags. With a note on the genus *Anopolenus*. [Read 7 June 1865.] *Quarterly Journal of the Geological Society* 21: 476–82.

Sarjeant, William A. S. 1980–96. *Geologists and the history of geology: an international bibliography.* 10 vols. including supplements. London: Macmillan. Malabar, Fla.: Robert E. Krieger Publishing.

Scherren, Henry. 1905. *The Zoological Society of London: a sketch of its foundation and development and the story of its farm, museum, gardens, menagerie and library.* London: Cassell.

Scherzer, Karl von. 1861–3. *Narrative of the circumnavigation of the globe by the Austrian frigate Novara (Commodore B. von Wullerstorf-Urbair), undertaken by order of the imperial government, in the years 1857, 1858, & 1859, under the immediate auspices of his I. and R. Highness, the Archduke Ferdinand Maximilian.* 3 vols. London: Saunders, Otley, & Co.

——. 1872. *Fachmännische Berichte über die österreichisch-ungarische Expedition nach Siam, China und Japan (1868–71).* Stuttgart: Verlag Julius Maier.

Schiebinger, Londa L. 1993. *Nature's body: gender in the making of modern science.* Boston: Beacon Press.

S[chlechtenda]l, [D. F. L.]. 1864. Personal-Nachrichten [Hermann Crüger]. *Botanische Zeitung* 22: 119–20.

Schopf, J. William. 2000. Solution to Darwin's dilemma: discovery of the missing Precambrian record of life. *Proceedings of the National Academy of Sciences of the United States of America* 97: 6947–53.

Schultze, Max Johann Sigismund. 1851. *Beiträge zur Naturgeschichte der Turbellarien.* Greifswald: C. A. Koch's Verlagsbuchhandlung.

——. 1856. Beiträge zur Kenntnis der Landplanarien nach Mittheilungen des Dr. Fritz Müller in Brasilien und nach eigenen Untersuchungen von Dr. Max Schultze. *Abhandlungen der Naturforschenden Gesellschaft zu Halle* 4 (1856–7): 19–38.

Schumann, Albert. 1888. *Aargauische Schriftsteller.* Aarau, Switzerland: H. R. Sauerlander.

Sclater, Philip Lutley, ed. 1861–7. *Zoological sketches by Joseph Wolf, from animals in their vivarium, in Regents Park.* 2 vols. London: H. Graves.

———. 1867. Barbets, and their distribution. *Intellectual Observer* 12: 241–6.

Scrope, George Poulett. 1866. The terraces of the chalk downs. *Geological Magazine* 3: 293–6.

Secord, James A. 1985. John W. Salter: the rise and fall of a Victorian palaeontological career. In *From Linnaeus to Darwin: commentaries on the history of biology and geology. Papers from the fifth Easter meeting of the Society for the History of Natural History, 28–31 March, 1983*. London: Society for the History of Natural History.

———. 2000. *Victorian sensation: the extraordinary publication, reception, and secret authorship of* Vestiges of the natural history of creation. Chicago: University of Chicago Press.

Shanafelt, Robert. 2003. How Charles Darwin got emotional expression out of South Africa (and the people who helped him). *Comparative Studies in Society and History* 45: 815–42.

Shattock, Joanne and Wolff, Michael, eds. 1982. *The Victorian press: samplings and soundings*. Leicester: Leicester University Press. Toronto and Buffalo: University of Toronto Press.

Shaw, Alexander. 1844. *An account of Sir Charles Bell's classification of the nervous system*. London: Moyes and Barclay.

Shephard, Sue. 2003. *Seeds of fortune: a gardening dynasty*. London: Bloomsbury.

Shrewsbury School register: Shrewsbury School register. 1734–1908. Edited by J. E. Auden. Oswestry: Woodall, Minshall, Thomas and Co., Caxton Press. 1909.

Shteir, Ann B. 1996. *Cultivating women, cultivating science*. Baltimore and London: Johns Hopkins University Press.

Siebold, Karl Theodor Ernst von. 1857. *On a true parthenogenesis in moths and bees; a contribution to the history of reproduction in animals*. Translated by William S. Dallas. London: John van Voorst.

Smart, John and Wager, Barbara. 1977. George Robert Crotch, 1842–1874: a bibliography with a biographical note. *Journal of the Society for the Bibliography of Natural History* 8 (1976–8): 244–8.

Smith, Andrew. 1849. *Illustrations of the zoology of South Africa . . . collected . . . in the years 1834, 1835, and 1836*. 4 pts. London: Smith, Elder & Co.

Smith, Charles Hamilton. 1839–40. *The natural history of dogs. Canidæ or genus Canis of authors. Including also the genera Hyæna and Proteles*. 2 vols. (Vols. 4 and 5 of Mammalia in *The naturalist's library*, edited by William Jardine.) Edinburgh: W. H. Lizars.

———. 1848. *The natural history of the human species, its typical forms, primæval distribution, filiations, and migrations*. Edinburgh: W. H. Lizars. London: Samuel Highley.

Smith, Frederick. 1861. Observations on the effects of the late unfavourable season on hymenopterous insects; notes on the economy of certain species, on the capture of others of extreme rarity, and on species new to the British fauna. *Entomologist's Annual* (1861): 33–45.

Smith, James Edward. 1824–36. *The English flora*. 5 vols. in 6. Vol. 5, pt 1 (mosses etc.), by William Jackson Hooker; pt 2 (fungi) by Miles Joseph Berkeley. London: Longman, Hurst, Rees, Orme, Brown, and Green.

South America: *Geological observations on South America. Being the third part of the geology of the voyage of the Beagle, under the command of Capt. FitzRoy RN, during the years 1832 to 1836.* By Charles Darwin. London: Smith, Elder & Co. 1846.

Speake, Jennifer, ed. 2003. *The Oxford dictionary of proverbs.* 4th edition. Oxford: Oxford University Press.

'Specific difference in *Primula*': On the specific difference between *Primula veris*, Brit. Fl. (var. *officinalis* of Linn.), *P. vulgaris*, Brit. Fl. (var. *acaulis*, Linn.), and *P elatior*, Jacq.; and on the hybrid nature of the common oxlip. With supplementary remarks on naturally produced hybrids in the genus *Verbascum*. By Charles Darwin. [Read 19 March 1868.] *Journal of the Linnean Society (Botany)* 10 (1869): 437–54.

Spencer, Herbert. 1858–74. *Essays: scientific, political, and speculative ... Reprinted chiefly from the quarterly reviews.* 3 vols. London: Longmans; Williams & Norgate.

———. 1860–2. *First principles.* London: George Manwaring; Williams & Norgate.

———. 1864–7. *The principles of biology.* 2 vols. London: Williams & Norgate.

———. 1867. *First principles.* 2d edition. London: Williams & Norgate.

———. 1904. *An autobiography.* 2 vols. London: Williams and Norgate.

Sprengel, Christian Konrad. 1793. *Das entdeckte Geheimniss der Natur im Bau und in der Befruchtung der Blumen.* Berlin: Friedrich Vieweg.

Stafleu, Frans Antonie. 1971. *Linnaeus and the Linnaeans. The spreading of their ideas in systematic botany, 1735–1789.* Utrecht: A. Oosthoek for the International Association for Plant Taxonomy.

Stanton, William. 1960. *The leopard's spots; scientific attitudes toward race in America 1815–59.* Chicago and London: University of Chicago Press.

Stearn, William T. 1956. Bentham and Hooker's *Genera plantarum*: its history and dates of publication. *Journal of the Society for the Bibliography of Natural History* 3 (1953–60): 127–32.

Stebbins, Robert E. 1988. France. In *The comparative reception of Darwinism*, edited by Thomas F. Glick. Chicago and London: University of Chicago Press.

Stenhouse, John. 1994. The Darwinian enlightenment and New Zealand politics. In *Darwin's laboratory*, edited by Roy Macleod and Philip F. Rehbock. Honolulu, Hawai'i: University of Hawai'i Press.

Stenton, Michael. 1976. *Who's who of British members of Parliament, 1832–1885.* Brighton: Hanover Press.

Sterritt, Neil J., *et al.* 1998. *Tribal boundaries in the Nass watershed.* Vancouver: UBC Press.

Stevens, John Austin, Jr. 1873. A memorial of George Gibbs. [Read 7 October 1873 to the New York Historical Society.] *Annual Report of the Board of Regents of the Smithsonian Institution for 1873*, pp. 219–25.

Stocking, George W., Jr. 1982. *Race, culture, and evolution: essays in the history of anthropology.* Chicago: University of Chicago Press.

———. 1987. *Victorian anthropology.* New York: The Free Press. London: Collier Macmillan.

Stone, S. 1860. Vespidæ in 1860. *Zoologist: a Popular Miscellany of Natural History* 18: 7261–6.

Strick, James. 2000. *Sparks of life: Darwinism and the Victorian debates over spontaneous generation*. Cambridge, Mass.: Harvard University Press.

Strickland, Hugh Edwin and Melville, Alexander Gordon. 1848. *The dodo and its kindred; or, the history, affinities, and osteology of the dodo, solitaire, and other extinct birds of the islands Mauritius, Rodriguez, and Bourbon*. London: Reeve, Benham and Reeve.

Sulivan, Henry Norton, ed. 1896. *Life and letters of the late Admiral Sir Bartholomew James Sulivan, KCB, 1810–1890*. London: John Murray.

Swinhoe, Robert. 1862. On the mammals of the island of Formosa (China). [Read 9 December 1862.] *Proceedings of the Zoological Society of London* (1862): 347–65.

——. 1863. The ornithology of Formosa, or Taiwan. *Ibis* 5: 198–219, 250–311, 377–435.

——. 1870. Catalogue of the mammals of China (south of the river Yangtsze) and of the island of Formosa. [Read 23 June 1870.] *Proceedings of the Zoological Society of London* (1870): 615–53.

Sykes, William Henry. 1834. Catalogue of birds (systematically arranged) of the Rassorial, Grallatorial, and Natatorial orders, observed in the Dukhun. *Journal of the Asiatic Society of Bengal* 3: 597–9, 639–49.

Syme, John T. Boswell, ed. 1863–92. *English botany; or, coloured figures of British plants*. 3d edition. 12 vols. and supplement. London: Robert Hardwicke; George Bell & Sons.

Taxonomic literature: *Taxonomic literature. A selective guide to botanical publications and collections with dates, commentaries and types*. By Frans A. Stafleu and Richard S. Cowan. 2d edition. 7 vols. Utrecht, Netherlands: Bohn, Scheltema & Holkema. The Hague, Netherlands: W. Junk. 1976–88.

Tebbel, John. 1972. *A history of book publishing in the United States*. Vol. 1, *The creation of an industry, 1630–1865*. New York and London: R. R. Bowker.

Tegetmeier, William Bernhard. 1867. *The poultry book: comprising the breeding and management of profitable and ornamental poultry, their qualities and characteristics; to which is added 'The standard of excellence in exhibition birds', authorized by the Poultry Club*. London and New York: George Routledge & Sons.

——. 1868. *Pigeons: their structure, varieties, habits, and management*. London: G. Routledge.

Theile, Friedrich Wilhelm. 1839. Entdeckung von Muskeln, welche die Rückenwirbel drehen (Rotatores dorsi) beim Menschen und den Säugethieren, nebst Bemerkungen über die Processus transversi und obliqui und über die Rückenmuskeln. *Archiv für Anatomie, Physiologie und wissenschaftliche Medizin* (1839): 102–38.

Thompson, John Vaughan. [1828–34.] *Zoological researches, and illustrations; or, natural history of nondescript or imperfectly known animals*. Vol. 1, part 1. Cork: King and Ridings.

Thomson, George Malcolm. 1884–5. Frederick Wollaston Hutton. *New Zealand Journal of Science* 2: 301–6.

Thomson, William. 1862. On the age of the sun's heat. *Macmillan's Magazine* 5: 388–93.

'Three forms of *Lythrum salicaria*': On the sexual relations of the three forms of *Lythrum salicaria*. By Charles Darwin. [Read 16 June 1864.] *Journal of the Linnean Society (Botany)* 8 (1865): 169–96. [*Collected papers* 2: 106–31.]

'Three sexual forms of *Catasetum tridentatum*': On the three remarkable sexual forms of *Catasetum tridentatum*, an orchid in the possession of the Linnean Society. By Charles Darwin. [Read 3 April 1862.] *Journal of the Proceedings of the Linnean Society (Botany)* 6 (1862): 151–7. [*Collected papers* 2: 63–70.]

Times atlas: *'The Times' atlas of the world. Comprehensive edition*. 9th edition. London: Times Books. 1992.

Topham, Jonathan Richard. 1993. 'An infinite variety of arguments': the *Bridgewater treatises* and British natural theology in the 1830s. PhD dissertation: University of Lancaster.

——. 1998. Beyond the 'common context': the production and reading of the Bridgewater Treatises. *Isis* 89: 233–62.

Tort, Patrick. 1996. *Dictionnaire du Darwinisme et de l'evolution*. 3 vols. Paris: Presses Universitaires de France.

Trimen, Roland. 1862–6. *Rhopalocera Africæ Australis; a catalogue of South African butterflies, comprising descriptions of all the known species with notices of their larvæ, pupæ, localities, habits, seasons of appearance, and geographical distribution*. London: Trübner. Cape Town, South Africa: W. F. Mathew.

——. 1863. On the fertilization of *Disa grandiflora*, Linn. . . . drawn up from notes and drawings sent to C. Darwin, Esq., FLS, &c. [Read 4 June 1863.] *Journal of the Proceedings of the Linnean Society (Botany)* 7 (1864): 144–7.

——. 1864. On the structure of *Bonatea speciosa*, Linn. sp., with reference to its fertilisation. [Read 1 December 1864.] *Journal of the Linnean Society (Botany)* 9 (1867): 156–60.

——. 1868. On some remarkable mimetic analogies among African butterflies. [Read 5 March 1868.] *Transactions of the Linnean Society of London* 26 (1868–70): 497–522.

——. 1887–9. *South-African butterflies: a monograph of the extra-tropical species*. With the assistance of James Henry Bowker. 3 vols. London: Trübner.

Trow-Smith, Robert. 1959. *A history of British livestock husbandry 1700–1900*. London: Routledge and Kegan Paul.

Turnbull, C. M. 1972. *The Straits Settlements 1826–67; Indian presidency to crown colony*. London: The Athlone Press.

Turner, William. 1865. *On some malformations of the organs of generation*. Edinburgh: Oliver and Boyd.

——. 1867. On the *musculus sternalis*. [Read 21 January 1867.] *Proceedings of the Royal Society of Edinburgh* 6 (1866–9): 65–6.

'Two forms in species of *Linum*': On the existence of two forms, and on their reciprocal sexual relation, in several species of the genus *Linum*. By Charles Darwin. [Read 5 February 1863.] *Journal of the Proceedings of the Linnean Society (Botany)* 7 (1864): 69–83. [*Collected papers* 2: 93–105.]

Tylor, Edward Burnett. 1865. *Researches into the early history of mankind.* London: John Murray.

Tyndall, John. 1864a. Contributions to molecular physics.— Being the fifth memoir of researches on radiant heat. [Read 17 March 1864.] *Philosophical Transactions of the Royal Society of London* 154: 327–68.

——. 1864b. Contributions to molecular physics. [Read 18 March 1864.] *Proceedings of the Royal Institution of Great Britain* 4 (1862–6): 233–40.

[——.] 1864c. Science and the spirits. *Reader* 4: 725–6.

——. 1865. On calorescence. [Read 23 November 1865.] *Philosophical Transactions of the Royal Society of London* 156 (1866): 1–24.

——. 1867. Miracles and special providences. *Fortnightly Review* n.s. 1: 645–60.

Uschmann, Georg. 1959. *Geschichte der Zoologie und der zoologischen Anstalten in Jena 1779–1919.* Jena: VEB Gustav Fischer.

——. 1984. *Ernst Haeckel. Biographie in Briefen.* Gütersloh: Prisma Verlag.

Uschmann, Georg and Jahn, Ilse. 1959–60. Der Briefwechsel zwischen Thomas Henry Huxley und Ernst Haeckel. *Wissenschaftliche Zeitschrift der Friedrich-Schiller-Universität Jena, Mathematisch–Naturwissenschaftliche Reihe* 9: 7–33.

Usov, Sergei A. trans. 1865. *Uchenie Darvina O proiskhozhdenii vidov obsheponiatno izlozhennoe Fridrikhom Rolle.* [Translation of Rolle 1863.] Moscow: A. I. Glazunov.

Van Lennep, Henry John. 1870. *Travels in little-known parts of Asia Minor: with illustrations of biblical literature and researches in archaeology.* 2 vols. London: John Murray.

Variation: The variation of animals and plants under domestication. By Charles Darwin. 2 vols. London: John Murray. 1868.

Variation 2d ed.: *The variation of animals and plants under domestication.* By Charles Darwin. 2 vols. London: John Murray. 1875.

Vinogradov, Mikhail Evgen'evich, *et al.* 1996. *Hyperiid amphipods (Amphipoda, Hyperiidea) of the world oceans.* Washington, DC: Smithsonian Institution Libraries.

Vladimirskii, M. trans. 1864. *Karla Darvina uchenie o proiskhozhdenie vidov v karstve rastenii i zhivotnykh, primenennoe k istorii mirotvoreniya izlozheno i obyasneno Fridrikhom Rolle.* [Translation of Rolle 1863.] With a supplementary biography of Darwin compiled by S. Seneman. St Petersburg: M. O. Vol'fa.

Vogt, Carl, trans. 1851. *Natürliche Geschichte der Schöpfung des Weltalls, der Erde und der auf ihr befindlichen Organismen, begründet auf die durch die Wissenschaft errungenen Thatsachen.* [Translation of [Chambers] 1844.] Brunswick: F. Vieweg und Sohn.

Vogt, Carl. 1863. *Vorlesungen über den Menschen. Seine Stellung in der Schöpfung und in der Geschichte der Erde.* 2 vols. Giessen: J. Ricker'sche Buchhandlung.

——. 1864. *Lectures on man: his place in creation, and in the history of the earth.* Edited by James Hunt. London: Longman, Green, Longman, and Roberts.

——, trans. 1865. *Über unsere Kenntniss von den Ursachen der Erscheinungen in der organischen*

Natur. Sechs Vorlesungen für Laien, gehalten in dem Museum für praktische Geologie. By Thomas Henry Huxley. Brunswick: Vieweg und Sohn.

Vogt, Carl. 1867. *Mémoire sur les microcéphales ou hommes-singes.* Reprinted from *Mémoires de l'Institute national genévois,* vol. 11. Geneva: Georg, Libraire de l'Institute Genevois.

Vogt, William. 1896. *La vie d'un homme: Carl Vogt.* Paris: Schleicher Frères. Stuttgart: Erwin Nägele.

Volcanic islands: Geological observations on the volcanic islands, visited during the voyage of HMS Beagle, together with some brief notices on the geology of Australia and the Cape of Good Hope. Being the second part of the geology of the voyage of the Beagle, under the command of Capt. FitzRoy RN, during the years 1832 to 1836. By Charles Darwin. London: Smith, Elder & Co. 1844.

Wahlberg, N., *et al.* 2003. Towards a better understanding of the higher systematics of Nymphalidae (Lepidoptera: Papilionoidea). *Molecular Phylogenetics and Evolution* 28: 473–84.

Walker, Francis. 1865. Synopsis of the Diptera of the Eastern Archipelago discovered by Mr. Wallace, and noticed in the 'Journal of the Linnean Society.' [Read 1 June 1865.] *Journal of the Linnean Society (Zoology)* 9 (1868): 1–30.

——. 1867. Catalogue of the homopterous insects collected in the Indian archipelago by Mr. A. R. Wallace, with descriptions of new species. [Read 7 February 1867.] *Journal of the Linnean Society (Zoology)* 10 (1870): 82–193, 276–330.

——. 1868. *Catalogue of the specimens of Blattariae in the collection of the British Museum.* London: Trustees of the British Museum.

——. 1868–70. Notes on Aphides. *Zoologist* 2d ser. 3 (1868): 1048–53, 1118–23, 1296–1301, 1328–33; 5 (1870): 1996–2001.

Wallace, Alfred Russel. 1863. List of birds collected in the island of Bouru (one of the Moluccas), with descriptions of the new species. [Read 13 January 1863.] *Proceedings of the Zoological Society of London* (1863): 18–36.

——. 1864a. On the phenomena of variation and geographical distribution as illustrated by the *Papilionidæ* of the Malayan region. [Read 17 March 1864.] *Transactions of the Linnean Society of London* 25 (1865–6): 1–71.

——. 1864b. The origin of human races and the antiquity of man deduced from the theory of 'natural selection'. *Natural History Review* n.s. 4: 328–36.

——. 1864c. On the varieties of man in the Malay Archipelago. [Read 26 January 1864.] *Transactions of the Ethnological Society of London* 3: 196–215.

——. 1866a. On reversed sexual characters in a butterfly. In *The British Association for the Advancement of Science. Nottingham meeting, August, 1866. Report of the papers, discussions, and general proceedings,* edited by William Tindal Robertson. Nottingham: Thomas Forman. London: Robert Hardwicke.

——. 1866b. On reversed sexual characters in a butterfly, and their interpretation on the theory of modifications and adaptive mimicry (illustrated by specimens). *Report of the thirty-sixth meeting of the British Association for the Advancement of Science, held at Nottingham,* Transactions of the sections, p. 79.

[———.] 1867a. Mimicry and other protective resemblances among animals. *Westminster Review* n.s. 32: 1–43.

———. 1867b. The philosophy of birds' nests. *Intellectual Observer* 11: 413–20.

———. 1867c. Creation by law. *Quarterly Journal of Science* 4: 471–88.

———. 1867d. Ice marks in North Wales (with sketch of glacial theories and controversies). *Quarterly Journal of Science* 4: 33–51.

———. 1867e. The disguises of insects. *Hardwicke's Science Gossip* 3: 193–8.

———. 1867f. Birds' nests and plumage, or the relation between sexual differences of colour and the mode of nidification in birds. *Gardeners' Chronicle and Agricultural Gazette*, 12 October 1867, pp. 1047–8

———. 1868–9. A theory of birds' nests: shewing the relation of certain sexual differences of colour in birds to their method of nidification. *Journal of Travel and Natural History* 1: 73–89.

———. 1869. *The Malay Archipelago: the land of the orang-utan, and the bird of paradise. A narrative of travel, with studies of man and nature.* 2 vols. London: Macmillan and Co.

———. 1870. *Contributions to the theory of natural selection. A series of essays.* London: Macmillan and Co.

———. 1905. *My life: a record of events and opinions.* 2 vols. London: Chapman & Hall.

Wallace, Robert, ed. 1899. *A country schoolmaster: James Shaw.* Edinburgh: Oliver and Boyd.

Waller, Jessie Oriana. 1889. Mental and physical training of children. *Nineteenth Century* 26: 659–67.

Walsh, Benjamin Dann. 1866. On the insects, coleopterous, hymenopterous and dipterous, inhabiting the galls of certain species of willow.— Part 2d and last. *Proceedings of the Entomological Society of Philadelphia* 6: 223–88.

Warington, George. 1867. On the credibility of Darwinism. [Read 4 March 1867.] *Journal of the Transactions of the Victoria Institute* 2: 39–62.

Watson, Hewett Cottrell. 1843–7. Notes of a botanical tour in the western Azores. *London Journal of Botany* 2 (1843): 1–9, 125–31, 394–408; 3 (1844): 582–617; 6 (1847): 380–97.

———. 1845. On the theory of 'progressive development,' applied in explanation of the origin and transmutation of species. *Phytologist* 2: 108–13, 140–7, 161–8, 225–8.

Weale, James Philip Mansel. 1865. Natural history in Natal. *Natural History Review* n.s. 5: 145–6.

———. 1867. Notes on the structure and fertilization of the genus *Bonatea*, with a special description of a species found at Bedford, South Africa. [Read 7 March 1867.] *Journal of the Linnean Society (Botany)* 10 (1869): 470–6.

———. 1870. Observations on the mode in which certain species of Asclepiadeæ are fertilized. [Read 3 November 1870.] *Journal of the Linnean Society (Botany)* 13 (1873): 48–58.

———. 1877. On the variation of Rhopalocerous forms in South Africa. [Read 4 July 1877.] *Transactions of the Royal Entomological Society of London* (1877): 265–75.

Weale, James Philip Mansel. 1878. Notes on South African insects. [Read 3 April 1878.] *Transactions of the Entomological Society of London* (1878): 183–8.

———. 1891. *The truth about the Portuguese in Africa*. London: Swan Sonnenschein.

Webb, Philip Barker, and Berthelot, Sabin. 1836–50. *Histoire naturelle des Iles Canaries*. 3 vols. in 9. and atlas. Paris: Béthune.

Wedgwood, Barbara and Wedgwood, Hensleigh. 1980. *The Wedgwood circle, 1730–1897: four generations of a family and their friends*. London: Studio Vista.

Wedgwood, Hensleigh. 1859–65. *A dictionary of English etymology*. 3 vols. in 4. London: Trübner & Co.

Weir, John Jenner. 1869–70. On insects and insectivorous birds; and especially on the relation between the colour and the edibility of *Lepidoptera* and their larvæ. [Read 1 March 1869 and 4 July 1870.] *Transactions of the Entomological Society of London* (1869): 21–6; (1870): 337–9.

Wellesley index: *The Wellesley index to Victorian periodicals 1824–1900*. Edited by Walter E. Houghton *et al.* 5 vols. Toronto: University of Toronto Press. London: Routledge & Kegan Paul. 1966–89.

Wells, William Charles. 1818. *Two essays: one upon single vision with two eyes; the other on dew. A letter to the Right Hon. Lloyd, Lord Kenyon and an account of a female of the white race of mankind, part of whose skin resembles that of a negro; with some observations on the causes of the differences in colour and form between the white and negro races of men*. London: Archibald Constable and Co. [and others].

West, David A. 2003. *Fritz Müller. A naturalist in Brazil*. Blacksburg, Va.: Pocahontas Press.

Wheeler, George Augustus and Wheeler, Henry Warren. 1878. *History of Brunswick, Topsham, and Harpswell, Maine, including the ancient territory known as Pejepscot*. Boston: Alfred Mudge and Son.

Whitaker, William. 1867. On subaerial denudation, and on cliffs and escarpments of the chalk and lower tertiary beds. *Geological Magazine* 4: 447–54, 483–93.

White, Charles. 1799. *An account of the regular gradation in man, and in different animals and vegetables; and from the former to the latter*. London: C. Dilly.

White, Paul. 2003. *Thomas Huxley. Making the 'man of science'*. Cambridge: Cambridge University Press.

Whitney, Josiah Dwight. 1866. Notice of a human skull recently taken from a shaft near Angels, Calaveras County. [Read 16 July 1866.] *Proceedings of the California Academy of Natural Sciences* 3 (1867): 277–8. [Reprinted in *American Journal of Science and Arts* 43 (1867): 265–7.]

Williams, Trevor Illtyd, ed. 1994. *Collins biographical dictionary of scientists*. 4th edition. London: HarperCollins.

Williamson, M. 1984. Sir Joseph Hooker's lecture on insular floras. *Biological Journal of the Linnean Society* 22: 55–77.

Wingfield, William and Johnson, George William. 1853. *The poultry book; comprising the characteristics, management, breeding, and medical treatment of poultry; being the results of personal observation and the practice of the best breeders, including Captain W. W. Hornby,*

RN; Edward Bond, Esq.; Thomas Sturgeon, Esq.; Charles Punchard, Esq.; Edward Hewitt, Esq.; and others. London: Wm S. Orr and Co.

———. 1856–7. *The poultry book; comprising the characteristics, management, breeding, and medical treatment of poultry; being the results of personal observation and the practice of the best breeders, including Captain W. W. Hornby, RN; Edward Bond, Esq.; Thomas Sturgeon, Esq.; Charles Punchard, Esq.; Edward Hewitt, Esq.; and others.* Another edition. Rearranged and edited by W. B. Tegetmeier. London: Wm S. Orr and Co.

Winner, Christian. 1993. History of the crop. In *The sugar beet crop: science into practice,* edited by David A. Cooke and R. Keith Scott. London: Chapman & Hall.

Wollaston, Alexander Frederick Richmond. 1921. *Life of Alfred Newton, professor of comparative anatomy, Cambridge University, 1866–1907.* With a preface by Sir Archibald Geikie. London: John Murray.

Wollaston, Thomas Vernon. 1854. *Insecta Maderensia; being an account of the insects of the islands of the Madeiran group.* London: John van Voorst.

———. 1856. *On the variation of species with especial reference to the Insecta; followed by an inquiry into the nature of genera.* London: John van Voorst.

Wong, Lin Ken. 1965. *The Malayan tin industry to 1914, with special reference to the states of Perak, Selangor, Negri, Sembilan, and Pahang.* Tucson, Ariz.: Association for Asian Studies, University of Arizona Press.

Wood, Searles Valentine. 1867. On the structure of the postglacial deposits of the south-east of England. [Read 19 June 1867.] *Quarterly Journal of the Geological Society of London* 23: 394–417.

Woolner, Amy. 1917. *Thomas Woolner, R.A., sculptor and poet: his life in letters.* London: Chapman and Hall.

Wrede, Richard and Reinfels, Hans von, eds. 1897. *Das geistige Berlin. Eine Encyklopädie das geistigens Lebens Berlin.* Vols. 1 and 3. Berlin: H. Storm.

Wright, Edward Perceval. 1868. Contributions towards a knowledge of the flora of the Seychelles. [Read 14 December 1868.] *Transactions of the Royal Irish Academy* 24 (1871): 571–8.

Wright, Thomas. 1866. On coral reefs present and past. [Read 13 June 1866.] *Proceedings of the Cotteswold Naturalists' Field Club* 4 (1866–8): 97–173

WWW: Who was who: a companion to Who's who, containing the biographies of those who died during the period [1897–1995]. 9 vols. and cumulated index (1897–1990). London: Adam & Charles Black. 1920–96.

Wydler, H. 1844. Morphologische Mittheilungen. *Botanische Zeitung* 2: 657–60.

———. 1850. Ueber Adoxa moschatellina L. *Flora, oder allgemeine botanische Zeitung* n.s. 8: 433–7.

Yarrell, William. 1843–56. *A history of British birds.* 3 vols. and 2 supplements. London: John van Voorst.

Ziegler, Heinrich Ernst, ed. 1909. *Zoologisches Wörterbuch. Erklärung der zoologischen Fachausdrücke.* Jena: Gustav Fischer.

NOTES ON MANUSCRIPT SOURCES

The majority of the manuscript sources cited in the footnotes to the letters are either in the Darwin Archive, Cambridge University Library, or at Down House, Downe, Kent. Further details about the Darwin Archive are available in the *Handlist of Darwin papers at the University Library Cambridge* (Cambridge: Cambridge University Press, 1960) and the unpublished supplementary handlist available at the library; a new catalogue of the papers is currently being prepared. Further details about the manuscripts at Down House are available in Philip Titheradge, ed. *The Charles Darwin Memorial at Down House, Downe, Kent*, revised ed. ([Downe: Down House Museum], 1981) and from the curator (The Curator, Down House, Downe, Kent, BR6 7JT). In addition, there are a number of named sources that are commonly used in the footnotes: for each of these, the editors have provided brief descriptive notes.

CD's Account books (Down House MS). This series of seventeen account books begins on 12 February 1839, a fortnight after CD and Emma's marriage, and ends with CD's death. The books contain two sets of accounts. From the start, CD recorded his *cash account* according to a system of double-entry book-keeping. On each left-hand page he recorded credits (i.e., withdrawals from the bank, either in the form of cash paid to himself or cheques drawn for others), and on each right-hand page he recorded debits (i.e., cash or cheques paid to others). CD also recorded details of his *banking account* from the start, but only noted them down in a single column at the bottom of the left-hand page of his cash account. In August 1848, however, he began a system of detailing his banking account according to double-entry book-keeping, in a separate chronological section at the back of each account book. On the left, he recorded credits to the account in the form of income (i.e., investments, rent, book sales, etc.). On the right, he recorded debits to the account (i.e., cash or cheque withdrawals).

CD's Classed account books (Down House MS). This series of four account books, covering the years 1839–81, runs parallel to CD's Account books. For each year, September–August (after 1867, January–December), CD divided his expenditure into different classes; in addition, he made a tally for the year of his income, expenditure, cash in hand, and money in the bank. From 1843, CD also compiled at the back of each book a separate account of the total expenditure under the various headings in each year, and from 1844 he added a full account of his income in each year, and of capital invested and 'paid' up.

CD's Experimental notebook (DAR 157a). This notebook contains notes on some of the experiments carried out between 13 November 1855 (with some back references) and 20 May 1868; the majority of the notes date from before 1863. Often only the details of the experiment attempted are given, usually with cross-references to results recorded in CD's portfolios of notes. The notebook also contains a number of letters to CD.

CD's Investment book (Down House MS). This book records for each of CD's investments the income received during the period 1846–81.

CD's 'Journal'. *See* Appendix II.

CD's Library catalogue (DAR 240). This manuscript catalogue of CD's scientific library was compiled by Thomas W. Newton, assistant librarian of the Museum of Practical Geology, London, in August 1875. Additions to the catalogue were later made by Francis Darwin (who inherited most of his father's scientific library) and by H. W. Rutherford, who apparently used this catalogue as a basis for compiling his *Catalogue of the library of Charles Darwin now in the Botany School, Cambridge* (Cambridge: Cambridge University Press, 1908). However, there are items listed in this manuscript catalogue that do not appear in Rutherford's published catalogue, and which must have been dispersed after being listed.

Down Coal and Clothing Club account book (Down House MS). CD was for some years treasurer of this charitable organisation. The account book records subscriptions made by honorary subscribers between 1841 and 1876; between 1848 and 1869 the entries are in CD's handwriting. For the years 1841–8 and 1868–76, there is also a statement of expenditures, though not in CD's handwriting.

Emma Darwin's diary (DAR 242). This is a series of small pocket diaries, in which Emma recorded details of the health of family members, trips made by herself, CD, and their children, school holidays, and visits to Down by others. The collection at CUL comprises diaries for the years 1824, 1833–4, 1839–45, and 1848–96.

H. E. Litchfield's autobiography (DAR 246). This unfinished autobiography, written in 1926 on forty-two loose leaves, and chiefly concerning Henrietta Emma Darwin's childhood, has never been published.

List of pamphlets (DAR 252.4). This is a catalogue of CD's pamphlet collection prepared by CD and Francis Darwin in 1878 (see the letter from Emma Darwin to Henrietta Emma Litchfield, [June 1878] (DAR 219.9: 175)). From about 1878 CD began to arrange the articles, papers, and reprints he received into a numbered collection. CD maintained this reprint collection until his death, when it was taken over by Francis Darwin. Francis continued the collection, adding new items, the numbers running consecutively from those of his father. Evidently, until this catalogue was prepared, CD used a working index similar to that of his 'List of reviews'. The catalogue is in two sections, a list of the quarto collection and one of the general collection. Both sections are alphabetically arranged with the entries pasted on sheets in a loose-leaf folder.

List of reviews (DAR 262.8: 9–18 (English Heritage MS: 88206151–60)). This manuscript, headed 'List Reviews of Origin of Sp & of C. Darwins Books', was CD's working index to his collection of reviews of his own books. It corresponds approximately to the review collection in the Darwin Pamphlet Collection–CUL, but includes some items that were dispersed after being listed.

Reading notebooks. *See Correspondence* vol. 4, Appendix IV. These notebooks are divided into sections entitled 'Books Read' and 'Books to be Read'. CD's entries in 'Books Read' often include a brief opinion of the work.

Scrapbook of reviews (DAR 226.1 and 226.2). Many of the reviews contained in these two volumes bear CD's annotations and thus were evidently collected by CD. However, the scrapbook seems to have been assembled by Francis Darwin: the tables of contents are in the handwriting of H. W. Rutherford, an assistant at Cambridge University Library who acted as a copyist for Francis on several occasions (see *ML*, 1: x, and Francis Darwin, ed. *The foundations of the Origin of Species. Two essays written in 1842 and 1844 by Charles Darwin* (Cambridge: Cambridge University Press, 1909)). In addition, the scrapbook is identified as Francis's in a note (DAR 226.1:132a) made in 1935 by Arthur Keith, whose appeal led to the purchase of Down House as a Darwin memorial (see Arthur Keith, *An autobiography* (London: Watts & Co., 1950)). DAR 226.1 bears the inscription 'Reviews of C. Darwin's works' on the spine, and contains, among others, reviews of *Origin* and *Orchids*; DAR 226.2 is inscribed: 'Reviews. Descent. Expression. Insect. Pl. Eras. D.'

W. E. Darwin's botanical notebook (DAR 117). This notebook contains observational and experimental notes on plants made by William, often in consultation with CD. The first observation bears the date 13 July 1862, and, although the date of the last observation is 26 June 1870, most of the notes were made between 1862 and 1864. The notebook originally contained letters from CD, but these were later removed. William entered notes made from botanical textbooks in a separate notebook (DAR 234).

W. E. Darwin's botanical sketchbook (DAR 186: 43). This sketchbook, which contains entries dated 1862–72, was evidently begun in parallel to William's botanical notebook. It contains ink drawings of various parts of plants, and of sections, together with descriptions, which are sometimes very extensive.

INDEX

The dates of letters to and from Darwin's correspondents are listed in the Biographical register and index to correspondents and are not repeated here. Darwin's works are indexed under the short titles used throughout this volume and listed in the bibliography.

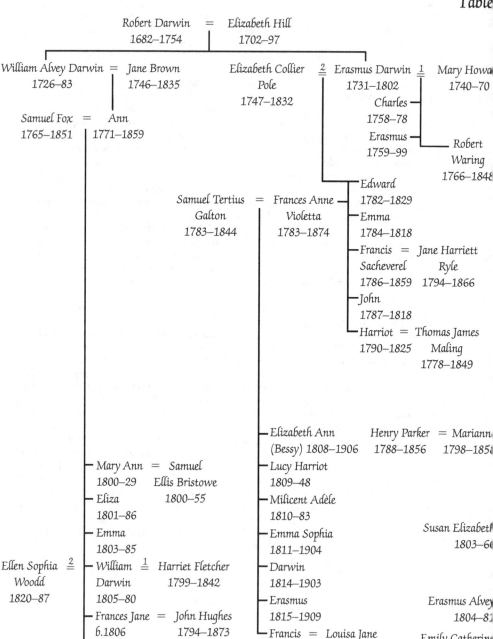

Robert Darwin = Elizabeth Hill
1682–1754   1702–97

William Alvey Darwin = Jane Brown   Elizabeth Collier **2** Erasmus Darwin **1** Mary Howa
1726–83   1746–1835   Pole   1731–1802   1740–70
1747–1832   Charles
Charles
1758–78

Samuel Fox = Ann
1765–1851   1771–1859   Erasmus
1759–99   Robert
Waring
1766–184ε

Edward
1782–1829
Samuel Tertius = Frances Anne   Emma
Galton   Violetta   1784–1818
1783–1844   1783–1874   Francis = Jane Harriett
Sacheverel   Ryle
1786–1859 1794–1866
John
1787–1818
Harriot = Thomas James
1790–1825   Maling
1778–1849

Elizabeth Ann   Henry Parker = Mariann
(Bessy) 1808–1906   1788–1856   1798–185ε
Mary Ann = Samuel   Lucy Harriot
1800–29   Ellis Bristowe   1809–48
Eliza   1800–55   Milicent Adèle
1801–86   1810–83
Emma   Susan Elizabetl
1803–85   Emma Sophia   1803–6ℓ
Ellen Sophia **2** William **1** Harriet Fletcher   1811–1904
Woodd   Darwin   1799–1842   Darwin
1820–87   1805–80   1814–1903
Erasmus   Erasmus Alvey
Frances Jane = John Hughes   1815–1909   1804–8ℓ
b.1806   1794–1873   Francis = Louisa Jane   Emily Catherine
Julia b.1809   1822–1911   Butler   1810–6ℓ
d.1897

ationship

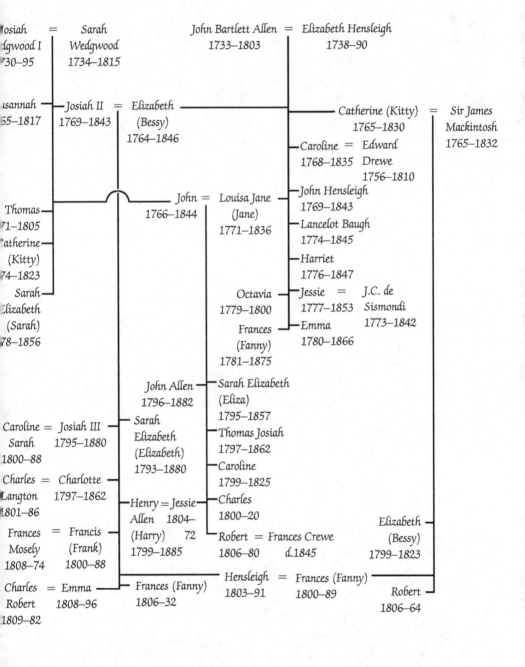

Printed in the United States
By Bookmasters